CHEMISTRY 11

Authors

Christina Clancy
Dufferin-Peel Catholic District
School Board

Ted Doram
Calgary Board of Education

Brian Heimbecker
Dufferin-Peel Catholic District
School Board

Michael Mazza
Science Writer and Consultant

Paul McNulty
Science Writer

Frank Mustoe
Toronto District School Board
and University of Toronto Schools
(retired)

Contributing Authors

Jonathan Bocknek
Science Writer

Trevor Finkle
York Region District School Board

Katherine Hamilton
Science Writer

Michael Jansen
Crescent School
Toronto, Ontario

Bruce McAskill
Science Writer

Consultants

Tigist Amdemichael
Toronto District School Board

Katy Farrow
Thames Valley District School Board

Frank Mustoe
Toronto District School Board
and University of Toronto Schools
(retired)

Trish Thomas
Ottawa-Carleton District School Board

Advisors

Peter Bloch
Toronto District School Board

Trevor Finkle
York Region District School Board

Stewart Grant
Toronto District School Board

Brian Heimbecker
Dufferin-Peel Catholic District
School Board

Barbara Nixon-Ewing
Toronto District School Board

NELSON

NELSON

For more information contact Nelson Education Ltd., 1120 Birchmount Road, Toronto, Ontario M1K 5G4. Or you can visit our website at nelson.com.

Chemistry 11

Every effort has been made to trace ownership of all copyrighted material and to secure permission from copyright holders. In the event of any question arising as to the use of any material, we will be pleased to make the necessary corrections in future printings.

The information and activities in this textbook have been carefully developed and reviewed by professionals to ensure safety and accuracy. However, the publisher shall not be liable for any damages resulting, in whole or in part, from the reader's use of the material. Although appropriate safety procedures are discussed in detail and highlighted throughout the textbook, the safety of students remains the responsibility of the classroom teacher, the principal, and the school board.

ISBN-13: 978-0-07-091575-6
ISBN-10: 0-07-091575-X

2 3 4 5   22 21 20 19

Printed and bound in Canada

PUBLISHER: Diane Wyman
PROJECT MANAGEMENT: Pronk&Associates (Jane McNulty, Sara Goodchild)
SPECIAL FEATURES COORDINATOR: Paula Smith
CONTENT MANAGER: Christine Weber
DEVELOPMENTAL EDITING: Michelle Anderson, Jonathan Bocknek, Lois Edwards, Katherine Hamilton, Tina Hopper, Julie Karner, Charlotte Kelchner, Natasha Marko, David Peebles, Laura L. Prescott, Betty Robinson, Christine Weber
MANAGING EDITOR: Crystal Shortt
SUPERVISING EDITOR: Janie Deneau
COPY EDITORS: May Look, Paula Pettitt-Townsend
PHOTO RESEARCH/PERMISSIONS: Linda Tanaka
ART BUYING: Pronk&Associates
REVIEW COORDINATORS: Jennifer Keay, Alexandra Savage-Ferr
FRONT MATTER/BACK MATTER COORDINATOR: Joanne Chan (Pronk&Associates)
EDITORIAL ASSISTANTS: Michelle Malda, Andrew Wai
MANAGER, PRODUCTION SERVICES: Yolanda Pigden
PRODUCTION COORDINATOR: Sheryl MacAdam
SET-UP PHOTOGRAPHY: David Tanaka
COVER DESIGN: Liz Harasymczuk
INTERIOR DESIGN: Pronk&Associates
ELECTRONIC PAGE MAKE-UP: Pronk&Associates
COVER IMAGES: Large background image: Daryl Benson / Masterfile; Largest circle: J.A. Kraulis / Masterfile; Middle circle: Masterfile Royalty Free; Small circle: PHOTOTAKE Inc. / Alamy

Acknowledgements

Pedagogical Reviewers

Sean Addis
Upper Canada District School Board

Samantha Booth
Niagara Catholic District School Board

Meredith Cammisuli
Halton District School Board

Richard Centritto
Toronto Catholic District School Board

Roman Charabaruk
Toronto District School Board

Laura Chodorowicz
Toronto Catholic District School Board

Michele M. Clayton
Peel District School Board

Connie Fratric
Dufferin-Peel Catholic District School Board

Christine Hazell
Grand Erie District School Board

Kerri Illingworth
Ottawa-Carleton District School Board

Leila A. Knetsch
Toronto District School Board

Diane Lavigne
Niagara Catholic District School Board

Dina Mayr
York Catholic District School Board

Sara McCormick
Peel District School Board

Karrilyn McPhee
Halton District School Board

Andrea C. Miller
Durham District School Board

Robert Miller
Ottawa Catholic School Board

Janine Odreman
District School Board of Niagara

Ranjit Singh-Gill
York Region District School Board

Philip Snider
Lambton-Kent District School Board

Amy Szerminska
Simcoe County District School Board

Frank Villella
Hamilton-Wentworth Catholic District School Board

Sarah Vurma
Simcoe County District School Board

Kimberley Walther
Durham Catholic District School Board

Jennifer Young
Waterloo Region District School Board

Accuracy Reviewers

Prof. Colin Baird
University of Western Ontario

R. Tom Baker
University of Ottawa

Dr. John Carran
Queen's University

John F. Eix
Upper Canada College (retired)

Joe Engemann
Brock University

Dr. Jeff Landry
McMaster University

Safety Reviewer

Jim Agban
Past Chair – S.T.A.O. Safety
Committee

Lab Testers

Mark Ackersviller
Waterloo Region District School
Board

Richard Beddoe
York Catholic District School
Board

Peter Earls
Greater Essex County District
School Board

Tony Jakubczak
Ottawa Catholic School Board

Jennifer Wilson
Toronto District School Board

Bias Reviewer

Bev McMorris
Principal (retired)
Waterloo Region District School
Board

Unit Project Reviewer

Anu Arora
Peel District School Board

Special Features Writers

Jenna Dunlop
Eric Jandciu
Natasha Marko
Adrienne Montgomerie
Margaret McClintock

Project Writer (Unit 5)

Michelle Anderson

Contents

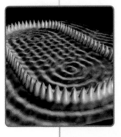

Sample Problems

Activities and Investigations

Activities

Investigations

STSE Special Features

Safety in the Chemistry Lab and Classroom

Scientific investigations are integrated throughout this textbook. Keep in mind at all times that working in a chemistry classroom can involve some risks. *Therefore, become familiar with all facets of laboratory safety, especially for performing investigations safely.* To make the investigations and activities in *Chemistry 11* safe and enjoyable for you and others who share a common working environment,

- become familiar with and use the following safety rules and procedures

- follow any special instructions from your teacher

- *always read* the safety notes before beginning each activity or investigation. Your teacher will tell you about any additional safety rules that are in place at your school.

WHMIS Symbols for Hazardous Materials

Look carefully at the WHMIS (Workplace Hazardous Materials Information System) safety symbols shown here. The WHMIS symbols are used throughout Canada to identify dangerous materials. Make certain you understand what these symbols mean. When you see these symbols on containers, use appropriate safety precautions.

Compressed Gas	Flammable and Combustible Material
Oxidizing Material	Corrosive Material
Poisonous and Infectious Material Causing Immediate and Serious Toxic Effects	Poisonous and Infectious Material Causing Other Toxic Effects
Biohazardous Infectious Material	Dangerously Reactive Material

Safety Symbols

Thes afety symbols in the column on the right are used in *Chemistry 11* to alert you to possible dangers. Be sure you understand each symbol used in an activity or investigation before you begin.

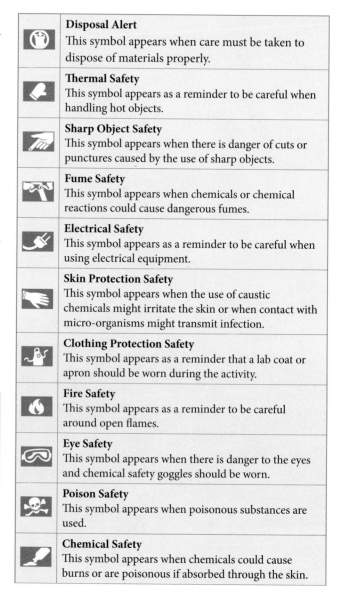

Disposal Alert
This symbol appears when care must be taken to dispose of materials properly.

Thermal Safety
This symbol appears as a reminder to be careful when handling hot objects.

Sharp Object Safety
This symbol appears when there is danger of cuts or punctures caused by the use of sharp objects.

Fume Safety
This symbol appears when chemicals or chemical reactions could cause dangerous fumes.

Electrical Safety
This symbol appears as a reminder to be careful when using electrical equipment.

Skin Protection Safety
This symbol appears when the use of caustic chemicals might irritate the skin or when contact with micro-organisms might transmit infection.

Clothing Protection Safety
This symbol appears as a reminder that a lab coat or apron should be worn during the activity.

Fire Safety
This symbol appears as a reminder to be careful around open flames.

Eye Safety
This symbol appears when there is danger to the eyes and chemical safety goggles should be worn.

Poison Safety
This symbol appears when poisonous substances are used.

Chemical Safety
This symbol appears when chemicals could cause burns or are poisonous if absorbed through the skin.

General Precautions

- Always wear chemical safety goggles and a lab coat or apron in the laboratory. Wear other protective equipment, such as gloves, as directed by your teacher or by the Safety Precautions at the beginning of each investigation.

- If you wear contact lenses, always wear safety goggles or a face shield in the laboratory. Inform your teacher that you wear contact lenses. Generally, contact lenses should not be worn in the laboratory. If possible, wear eyeglasses instead of contact lenses, but remember that eyeglasses are not a substitute for proper eye protection.

- Know the location and proper use of the nearest fire extinguisher, fire blanket, fire alarm, first aid kit, and eyewash station (if available). Read "Fire Safety" on the next page and discuss with your teacher what type of fire-fighting equipment should be used on particular types of fires.

- Do not wear open-toed shoes or sandals in the laboratory. Accessories may get caught on equipment or present a hazard when working with a Bunsen burner. Ties, scarves, long necklaces, and dangling earrings should be removed before starting an investigation.

- Tie back long hair and any loose clothing before starting an investigation.

- Lighters and matches must not be brought into the laboratory.

- Food, drinks, and gum must not be brought into the laboratory.

- Inform your teacher if you have any allergies, medical conditions, or physical problems (including hearing impairment) that could affect your work in the laboratory.

Before Beginning Laboratory Investigations

- Listen carefully to the instructions that your teacher gives you. Do not begin work until your teacher has finished giving instructions.

- Obtain your teacher's approval before beginning any investigation that you have designed yourself.

- Read through all of the steps in the investigation before beginning. If there are any steps that you do not understand, ask your teacher for help.

- Be sure to read and understand the Safety Precautions at the start of each investigation.

- Always wear appropriate protective clothing and equipment, as directed by your teacher and the Safety Precautions.

- Be sure that you understand all safety labels on materials and equipment. Familiarize yourself with the WHMIS symbols in this section.

- Make sure that your work area is clean and dry.

During Laboratory Investigations

- Make sure that you understand and follow the safety procedures for different types of laboratory equipment. Do not hesitate to ask your teacher for clarification if necessary.

- Never work alone in the laboratory.

- Remember that gestures or movements that may seem harmless could have dangerous consequences in the laboratory. For example, tapping people lightly on the shoulder to get their attention could startle them. If they are holding a beaker that contains an acid, for example, the results could be very serious.

- Make an effort to work slowly and steadily in the laboratory. Be sure to make room for other students.

- Organize materials and equipment neatly and logically. For example, place materials where you will not have to reach behind or over a Bunsen burner to get them. Keep your bags and books off your work surface and out of the way.

- Never taste any substances in the laboratory.

- Never touch a chemical with your bare hands.

- Never draw liquids or any other substances into a pipette or a tube with your mouth.

- If you are asked to smell a substance, do not hold it directly under your nose. Keep the object at least 20 cm away from your nose. Take a deep breath and hold it. Waft the fumes towards your nostrils with your hands, then exhale.

- Label all containers holding chemicals. Do not use chemicals from unlabelled containers.

- Hold containers away from your face when pouring liquids or mixing reactants.

- If any part of your body comes in contact with a potentially dangerous substance, report it to your teacher immediately. Wash the affected area immediately and thoroughly with water.

- If you get any material in your eyes, do not rub them. Wash your eyes immediately and continuously for 15 minutes, and make sure that your teacher is informed. A doctor should examine any eye injury. It is recommended not to wear contact lenses when working with chemicals. If you wear contact lenses, take them out immediately. Failing to do so may result in material becoming trapped behind the contact lenses. Flush your eyes with water for 15 minutes, as above.

- Do not touch your face or eyes while in the laboratory unless you have first washed your hands.

- If your clothing catches fire, STOP, DROP, and ROLL. Other students may use the fire blanket to smother the flames. Do not wrap the fire blanket around yourself if on fire. This would result in a chimney effect and choke you.

- Do not look directly into a test tube, flask, or the barrel of a Bunsen burner.
- If you see any of your classmates jeopardizing their safety or the safety of others, let your teacher know.

Heat Source Safety

- When heating any item, wear safety eyewear, heat-resistant safety gloves, and any other safety equipment that your teacher or the Safety Precautions suggest.
- Always use heat-proof, intact containers. Check that there are no large or small cracks in beakers or flasks.
- Never point the open end of a container that is being heated at yourself or others.
- Do not allow a container to boil dry unless specifically instructed to do so.
- Handle hot objects carefully. Be especially careful with a hot plate that may look as though it has cooled down, or glassware that has recently been heated.
- Before using a Bunsen burner, make sure that you understand how to light and operate it safely. Always pick it up by the base. Never leave a Bunsen burner unattended.
- Before lighting a Bunsen burner, make sure that there are no flammable solvents nearby.
- Always have the Bunsen burner secured to a utility stand.

Use EXTREME CAUTION around an open flame.

- If you do receive a burn, run cold water over the burned area immediately. Make sure that your teacher is notified.

- When you are heating a test tube, always slant it. The mouth of the test tube should point away from you and from others.
- Remember that cold objects can also harm you. Wear appropriate gloves when handling an extremely cold object.

Electrical Equipment Safety

- Make sure that the work area and the area of the socket are dry.
- Make sure that your hands are dry when touching electrical cords, plugs, sockets, or equipment.
- When unplugging electrical equipment, do not pull the cord. Grasp the plug firmly at the socket and pull gently.
- Place electrical cords in places where people will not trip over them.
- Use an appropriate length of cord for your needs. Cords that are too short may be stretched in unsafe ways. Cords that are too long may tangle or trip people.
- Never use water to fight an electrical equipment fire. Severe electrical shock may result. Use a carbon dioxide or dry chemical fire extinguisher. (See "Fire Safety" below.)
- Report any damaged equipment or frayed cords to your teacher.

Glassware and Sharp Objects Safety

- Cuts or scratches in the chemistry laboratory should receive immediate medical attention, no matter how minor they seem. Alert your teacher immediately.
- Never use your hands to pick up broken glass. Use a broom and dustpan. Dispose of broken glass in the "Broken Glass Container". Do not put broken glassware into the garbage can.
- Cut away from yourself and others when using a knife or another sharp object.
- Always keep the pointed end of scissors and other sharp objects pointed away from yourself and others when walking.
- Do not use broken or chipped glassware. Report damaged equipment to your teacher.

Fire Safety

- Know the location and proper use of the nearest fire extinguisher, fire blanket, and fire alarm.

- Understand what type of fire extinguisher you have in the laboratory, and what type of fires it can be used on. (See details below.) Most fire extinguishers are the ABC type.
- Notify your teacher immediately about any fires or combustible hazards.
- Water should only be used on Class A fires. Class A fires involve ordinary flammable materials, such as paper and clothing. *Never use water* to fight an electrical fire, a fire that involves flammable liquids (such as gasoline), or a fire that involves burning metals (such as potassium or magnesium).
- Fires that involve a flammable liquid, such as gasoline or alcohol (Class B fires), must be extinguished with a dry chemical or carbon dioxide fire extinguisher.
- Live electrical equipment fires (Class C) must be extinguished with a dry chemical or carbon dioxide fire extinguisher. Fighting electrical equipment fires with water can cause severe electric shock.
- Class D fires involve burning metals, such as potassium and magnesium. A Class D fire should be extinguished by smothering it with sand or salt. Adding water to a metal fire can cause a violent chemical reaction.
- If your hair or clothes catch on fire, STOP, DROP, and ROLL. Other students may use the fire blanket to smother the flames. Do not wrap the fire blanket around yourself if on fire. This would result in a chimney effect and choke you. Do not discharge a fire extinguisher at someone's head.

Clean-Up and Disposal in the Laboratory

- Clean up all spills immediately, as directed by your teacher.
- If you spill acid or base on your skin or clothing, wash the affected area immediately with a lot of cool water.
- You can neutralize spills of acid solutions with sodium hydrogen carbonate (baking soda). You can neutralize spills of basic solutions with sodium hydrogen sulfate or citric acid.
- Clean equipment before putting it away, as directed by your teacher.
- Dispose of materials as directed by your teacher, in accordance with your local school board's policies. Do not dispose of materials in a sink or a drain. Always dispose of them in a special container, as directed by your teacher.

- Wash your hands thoroughly after all laboratory investigations.

Safety in Your On-line Activities

The Internet is like any other resource you use for research—you should confirm the source of the information and the credentials of those supplying it to make sure the information is credible before you use it in your work.

Unlike other resources, however, the Internet has some unique pitfalls you should be aware of, and practices you should follow.

- When you copy or save something from the Internet, you could be saving more than information. Be aware that information you pick up could also include hidden, malicious software code (known as "worms" or "Trojans") that could damage your computer system or destroy data.
- Try to avoid sites that contain material that is disturbing, illegal, harmful, and/or was created by exploiting others.
- *Never* give out personal information on-line. Protect your privacy, even if it means not registering to use a site that looks helpful. Discuss with your teacher ways to use the site while protecting your privacy.
- Report to your teacher any on-line content or activity that you suspect is illegal.

Instant Practice

1. One of the materials you plan to use in a Plan Your Own Investigation bears the following symbols:

Describe the safety precautions you would need to incorporate into your investigation.

2. Online research also comes with safety hazards. Describe the safety practices you would follow when conducting Internet research on WHMIS symbols.

UNIT 1

Matter, Chemical Trends, and Chemical Bonding

BIG IDEAS

- Every element has predictable chemical and physical properties determined by its structure.

- The type of chemical bond in a compound determines the physical and chemical properties of that compound.

- It is important to use chemicals properly to minimize the risks to human health and the environment.

Overall Expectations

In this unit you will…

- **analyze** the properties of commonly used chemical substances and their effects on human health and the environment, and propose ways to lessen their impact

- **investigate** physical and chemical properties of elements and compounds, and use various methods to visually represent them

- **demonstrate** an understanding of periodic trends in the periodic table and how elements combine to form chemical bonds

Unit 1 Contents

Chapter 1
Elements and the Periodic Table

How does the structure of atoms change with an increase in the atomic number and how does the structure of an atom determine its properties?

Chapter 2
Chemical Bonding

How do bonds form between atoms and how do the bonds influence the properties of compounds?

The healing properties of silver particles have long been known, even before bacterial or fungal infections were understood. The ancient Greeks and Romans used silver to treat burns and wounds. Modern scientists have a deeper understanding of infectious agents such as bacteria. They have also learned how silver can interact with bacteria and destroy them. Today, silver is used to treat hard-to-heal wounds. The large photograph shows a cross section of a dressing with a layer of activated charcoal and a layer that is impregnated with tiny silver particles. The inset shows an example of this type of dressing. When the dressing is applied to the wound, silver ions penetrate into the wound and destroy the bacteria. The activated charcoal, consisting mostly of carbon, absorbs the bacterial toxins and also the compounds that cause a wound to have a bad odour. Understanding the properties of elements such as silver and carbon is essential to designing medical treatments. In this unit, you will also learn how understanding the properties of elements and compounds is important in reducing risks to human health and the environment.

As you study this unit, look ahead to the **Unit 1 Project** on pages 94 to 95, which gives you an opportunity to demonstrate and apply your new knowledge and skills. Keep a planning folder so you can complete the project in stages as you progress through the unit.

2

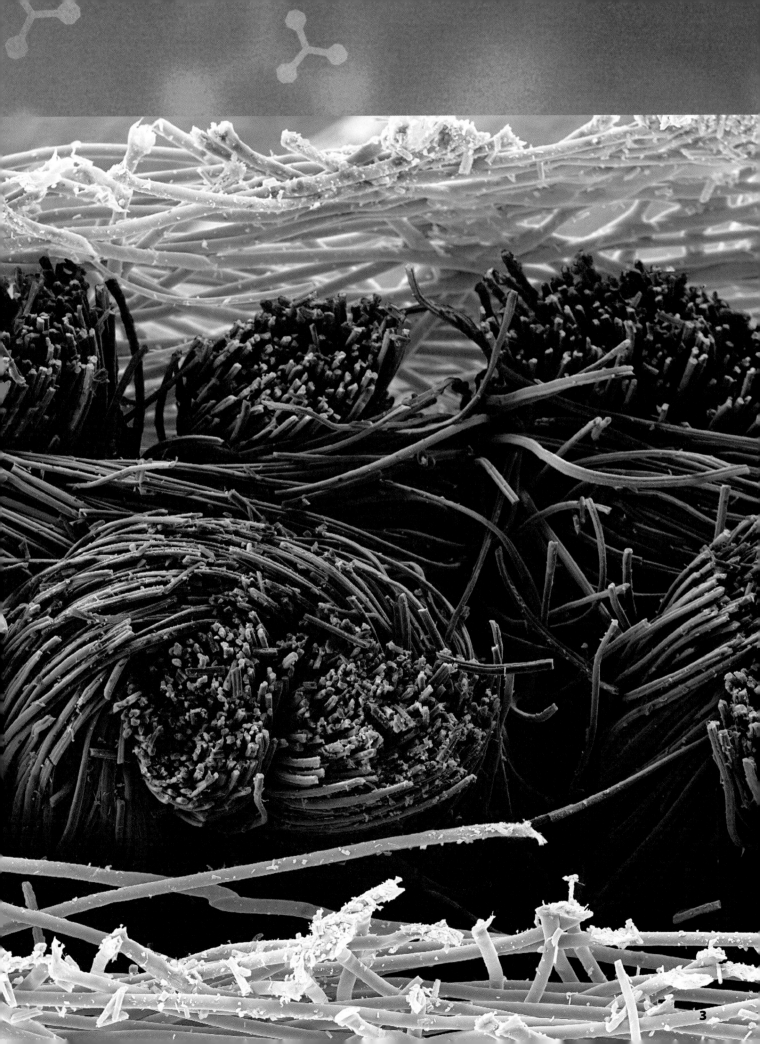

Preparation

Safety in the Laboratory

- Always wear protective clothing, such as safety eyewear and a lab coat or apron, when using materials that could splash or shatter.
- Know which safety equipment, such as a fire blanket, a fire extinguisher, and an eyewash station, are available and where they are located in your classroom.
- Know the proper procedures for using the available safety equipment. For example, if your clothing catches fire, smother the fire with the fire blanket.

- If you get something in your eyes, do not touch them. Use the eyewash station to flush your eyes with water for 15 min, and make sure that someone tells your teacher.
- Follow all instructions for proper disposal of broken glass and chemicals to prevent injury.
- WHMIS (Workplace Hazardous Materials Information System) symbols are used in Canadian schools and workplaces to identify dangerous materials.

1. Which safety equipment should you use if a chemical has splashed into your eyes?
   a. lab apron
   b. protective gloves
   c. fire blanket
   d. safety eyewear
   e. eyewash station

2. Which list of safety equipment includes only equipment that is used after an accident occurs?
   a. a lab apron and protective gloves
   b. protective gloves and a fire blanket
   c. a fire extinguisher and a lab apron
   d. safety eyewear and an eyewash station
   e. an eyewash station and a fire extinguisher

3. Draw a safety map of your classroom. Include a key that identifies the locations of lab aprons, safety eyewear, a fire blanket, a fire extinguisher, and an eyewash station.

4. Examine the fire extinguisher that is available in your classroom. Write a script for a short video that explains the steps needed to use the fire extinguisher.

5. An investigation involves testing how well common kitchen chemicals dissolve in water. Your lab partner thinks that safety eyewear and a lab apron are not necessary. Write an explanation you could use to persuade your lab partner of the necessity of wearing these two pieces of protective clothing.

6. Why is a special container for the disposal of broken glass important?

7. Examine the following WHMIS symbols. Describe the hazard that each symbol represents.

   a.   c.

   b.   d.

8. Describe what you can do to reduce your risk of injury when working with the type of material represented by each WHMIS symbol in question 7.

9. Name the WHMIS symbol that would be used for each of the following chemicals.
   a. carbon monoxide
   b. gasoline
   c. hydrochloric acid
   d. helium

Chemical Symbols

- Every element has a unique chemical symbol that is universally recognized.
- The chemical symbol for an element is composed of one or two letters if the element has been officially named, or three letters if the official name has not yet been selected.

- Only the first letter of a chemical symbol is upper case.
- The names and symbols for the first 20 elements appear frequently and should be memorized to make learning new information in chemistry easier.

10. Which of the following models a correctly written chemical symbol?
- **a.** t
- **b.** Tt
- **c.** tt
- **d.** tT
- **e.** TT

11. Identify the chemical symbol for each element.
- **a.** helium
- **b.** carbon
- **c.** calcium
- **d.** sodium
- **e.** sulfur
- **f.** oxygen
- **g.** argon
- **h.** fluorine

12. State the name of the element that is represented by each symbol.
- **a.** P
- **b.** Al
- **c.** N
- **d.** Be
- **e.** K
- **f.** Li
- **g.** H
- **h.** Ne

13. When learning chemical symbols, it is helpful to identify how the symbols relate to the names of the elements. Examine the chemical symbols and names of the first 20 elements in the periodic table. Identify four categories that describe how the chemical symbols relate to the names of the elements. Provide an example for each category.

14. Why is it important to use correct capitalization when writing chemical symbols?

15. For a quiz on the first 20 elements, a student wrote the incorrect chemical symbols that are listed in the following table. Copy and complete the table. Analyze each symbol to identify the element that the student intended, and write the correct symbol for this element.

Prefixes Used in Naming Molecular Compounds

Incorrect Symbol	Element Intended	Correct Symbol
Bo		
Fl		
Po		
be		
Ch		
Hy		
Ma		
HE		
Ni		
s		
Ox		
Sil		

16. A compound is composed of two or more elements. Examine each chemical formula, and identify the elements that make up the compound.
- **a.** CO
- **b.** H_2O
- **c.** LiF
- **d.** NCl_3
- **e.** $Mg(NO_3)_2$
- **f.** $Al_2(SO_4)_3$
- **g.** CaS

17. Why is it important to have universally recognized symbols for the elements?

18. A student wrote the formula CaO_2 for carbon dioxide. Analyze the formula to determine the error.

- During a chemical reaction, a new substance is produced. The composition and properties of the new substance are different from the composition and properties of the starting materials.
- If the new substance is a gas, bubbling or fizzing of the reaction mixture or a change in odour can provide evidence of its formation.
- If the reaction of two aqueous solutions produces a new substance that is an insoluble solid, or precipitate, evidence of its formation often includes a change in the colour of the reacting solutions or cloudiness of the solution.
- Evidence of a chemical change often includes the release or absorption of energy, such as a change in temperature or the production of light.

19. Which of the following is a chemical change?
    a. An ice cube melts and forms a puddle of water.
    b. The onions in a salad are removed before the salad is eaten.
    c. A wooden log in a fireplace burns to ash.
    d. Red paint and blue paint are mixed, and a new colour forms.
    e. Drops of water form on the outside of a cool glass.

20. Which of the following is a chemical change?
    a. An iron nail rusts.
    b. A model of a molecule is built by a group of students.
    c. Sugar dissolves in hot tea.
    d. A pencil is sharpened.
    e. Water evaporates from a lake.

21. During an investigation, you observe bubbling in a test tube after mixing two substances. Which procedure would you use to check for an odour?
    a. Hold the test tube under your lab partner's nose, and have your lab partner describe the odour.
    b. Hold the test tube under your nose, and waft fumes away from your nose with your hand.
    c. Hold the test tube under your nose, and inhale with a short, shallow breath.
    d. Hold the test tube away from your face, and waft fumes toward your nose with your hand.
    e. Hold the test tube below your chin, and inhale deeply.

22. Identify the evidence of a chemical change in each photograph below.

    a.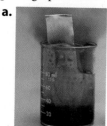
    b.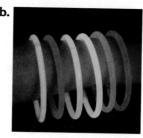

23. Give an example of a chemical change that releases energy in the form of heat and light.

24. How can your sense of touch allow you to determine whether energy is absorbed or released during a chemical change?

25. Boiling is a physical change that might be mistaken for a chemical change. What might cause a person to conclude, incorrectly, that boiling is a chemical change?

26. Is it possible for a chemical change to occur with no visible evidence? Explain your reasoning.

27. Some chemicals are stored in opaque bottles. For example, hydrogen peroxide solution is stored and sold in brown plastic bottles.
    a. Why might it be important for a chemical not to be exposed to light?
    b. Hydrogen peroxide begins to break down into water and oxygen gas when it is exposed to light. What happens to the effectiveness of hydrogen peroxide, as a disinfectant, if it is exposed to light for a long time?

28. Limewater is a clear, colourless solution. The photograph shows a student blowing gently into limewater.

    When doing an experiment like this, never share the straw. Always use a new one.
    a. What evidence of a chemical change does the photograph show?
    b. What chemical in the student's breath might be causing the chemical change?

Writing Formulas and Naming Compounds

- An ionic compound forms when one or more electrons are transferred from a metal atom to a non-metal atom.
- An atom that loses an electron becomes an ion with a positive charge (cation). An atom that gains an electron becomes an ion with a negative charge (anion).
- Ionic compounds can be composed of polyatomic ions, such as nitrate, sulfate, hydroxide, carbonate, and phosphate ions. A polyatomic ion is a group of atoms that act as a single unit and have a charge.
- The name of an ionic compound is made up of the name of the positive ion followed by the name of the negative ion.

- When writing the chemical formula for an ionic compound, you need to use subscripts to make the sum of the charges of the ions equal to zero.
- A molecular compound forms when electrons are shared between non-metal atoms.
- The name of a molecular compound contains prefixes that identify the numbers of atoms of the different elements. You can use these prefixes to help you write the chemical formula for the compound. For example, based on the prefixes in the name diphosphorus tetraiodide, the chemical formula is P_2I_4.

29. Which formula represents an ionic compound?

  a. H_2O

  b. SO_2

  c. CH_4

  d. $NaBr$

  e. NCl_3

30. Which formula represents a molecular compound?

  a. CCl_4

  b. $CaCl_2$

  c. MgF_2

  d. $NaCl$

  e. K_2CO_3

31. Which combination of elements would likely form a molecular compound?

  a. Ca and F

  b. C and Cl

  c. Li and O

  d. Ca and Cl

  e. K and S

32. Describe the charge that is associated with each type of particle.

  a. proton   **c.** anion

  b. electron   **d.** cation

33. Explain why gaining electrons causes an atom to become a negatively charged ion.

34. Name each polyatomic ion.

  a. NO_3^-

  b. SO_4^{2-}

  c. OH^-

  d. CO_3^{2-}

35. What types of elements form an ionic compound?

36. Identify each compound as ionic or molecular.

  a. sodium chloride

  b. carbon monoxide

  c. $Ca(NO_3)_2$

  d. SF_6

37. Identify each compound as ionic or molecular. Then write the chemical formula for the compound.

  a. potassium fluoride

  b. phosphorus tribromide

  c. dinitrogen monoxide

  d. aluminum nitrate

38. What happens to electrons when atoms join to form an ionic compound and when atoms join to form a molecular compound? Compare the two processes.

39. Examine the periodic table shown.

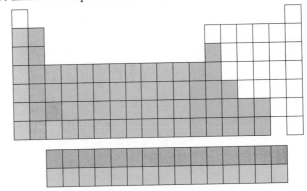

  a. What type of element is shaded?

  b. What type of ion will atoms of the shaded elements form?

40. Explain how you can recognize ionic compounds and molecular compounds using only their names.

Elements and the Periodic Table

Specific Expectations

In this chapter, you will learn how to . . .

- B1.2 **evaluate** the risks and benefits to human health of some commonly used chemical substances (1.1, 1.3)

- B2.1 **use** appropriate terminology related to chemical trends (1.3)

- B2.2 **analyze** data related to the properties of elements within a period to **identify** general trends in the periodic table (1.3)

- B2.3 **use** an inquiry process to **investigate** the chemical reactions of elements with other substances and produce an activity series using the resulting data (1.2)

- B3.1 **explain** the relationship between the atomic number and the mass number of an element, and the difference between isotopes and radioisotopes of an element (1.1)

- B3.2 **explain** the relationship between the isotopic abundance of an element's isotopes and the relative atomic mass of the element (1.1)

- B3.3 **state** the periodic law, and **explain** how patterns in the electron arrangement and forces in atoms result in periodic trends in the periodic table (1.3)

Images such as the one shown here are the best possible "pictures" of atoms that scientists can currently obtain. This image was produced by a scanning tunnelling microscope (STM). The corral shape is created by 48 iron atoms placed on a copper surface.

The STM does not use light. Instead, it scans the surface of a sample with a tiny electric current to detect its shape. The image, including the colour, is then computer generated. Prior to the invention of STMs in 1981, no one had ever seen an image of an atom. Scientists could not have invented the STM, however, without a thorough understanding of the properties of atoms. Now, the STM has not only enabled scientists to learn more about atoms, but it has also made nanotechnology possible. Using nanotechnology, engineers might soon be able to build tiny electronic circuits, one atom at a time, and thus make smaller computers that use less energy than modern computers use.

Organization Reveals Trends

How might you go about organizing different pieces of small hardware, such as nails, bolts, and screws, in a tool chest or cabinet with drawers? You might be surprised to learn that your method of organization can highlight trends (patterns) in the characteristics of the different pieces of hardware. In this activity, you will arrange nails, bolts, and screws and make labels to identify the type of object in each column and row of drawers. You will then look for trends that arise. When your organization is complete, you will compare it with the organization of elements in the periodic table.

Materials

- notebook paper
- pencil
- nails, bolts, and screws of different sizes (4 of each)
- ruler
- balance

Procedure

1. On a piece of notebook paper, draw a grid with three columns and four rows, creating 12 rectangles. Make the rectangles large enough to fit the largest nail, bolt, or screw. These rectangles represent labels for the drawers in a tool chest or cabinet.

2. Arrange the nails, bolts, and screws in the rectangles so they make a pattern that provides a way to easily find what you want. When you are satisfied that the items will be easy to find, write the name of the item (nail, bolt, or screw) in each rectangle.

3. Using the balance, measure the mass of each item and record the mass in its rectangle. Then put the item back in its rectangle.

4. Using the ruler, measure the length of each item and record the length in its rectangle. Then put the item back in its rectangle.

Questions

1. Describe any trends that you observe going down each column.

2. Describe any trends that you observe going across each row.

3. Describe any trends that you observe going diagonally from one corner to the opposite corner.

4. How is your table of labels similar to a periodic table of the elements?

The Nature of Atoms

Key Terms

valence electrons

Lewis diagram

electron pairs

unpaired electrons

isotopes

radioisotopes

atomic mass unit

isotopic abundance

For thousands of years, people have discovered and used matter without understanding the fundamental nature of that matter. For example, elements such as copper, gold, iron, sulfur, and carbon were known to and used by people between 4000 and 11 000 years ago. Before the science of chemistry as we know it today was established, alchemists such as the people shown in **Figure 1.1** added to the known elements. Relying on their keen observation skills, as well as equipment that they often invented, alchemists in the Middle East and Europe discovered arsenic, antimony, bismuth, and phosphorus. And still, the nature of the matter of which these elements are made remained unknown.

Figure 1.1 Between about 500 and 1700 C.E., alchemists such as these combined practical chemistry with an interest in philosophy and mysticism. Some of the great scientists of history, including Robert Boyle and Sir Isaac Newton, either practised alchemy or viewed it favourably.

Imagining the Atom

The first people to ask questions and record their ideas about the nature of matter were philosophers who lived and taught in the region of the ancient Mediterranean and northern Africa. About 2500 years ago, some of these professional thinkers imagined the idea of a fundamental building-block of matter. They reasoned that if you cut an object in half, and kept doing so, you would reach a point at which the object could no longer be cut. This speck of object was called *atomos*, meaning "uncuttable."

The idea of *atomos*—atoms—persisted among some thinkers. However, one of the most influential thinkers of ancient Greece, Aristotle, argued against it. His rejection of atoms was based on philosophical arguments. Neither he nor most other Greek philosophers conducted experiments to demonstrate their ideas. However, Aristotle's ideas were greatly respected by people of authority both in and outside of Greece. As a result, few people thought about atoms or the building-blocks of matter in a way we recognize as scientific for another 2000 years.

Modelling the Atom

The first practical model of the atom was developed in the early 1800s by John Dalton. While studying the properties of atmospheric gases, Dalton inferred that they consisted of tiny particles or atoms. Chemists continued to develop the concept of atoms. They performed numerous experiments and gathered an abundance of data, and they modified the model of the atom to include new discoveries. **Table 1.1** summarizes the most important models of the atom and characteristics of the atom that the models describe. Notice how each model is built on the foundation of the models that came before it.

Table 1.1 Models of the Atom

Model	Development of Model
Dalton's model, or the billiard ball model, of the atom was a solid sphere. Atoms of different elements differed from one another, as implied by the different sizes and colours of these two "atoms."	In 1803, English chemist John Dalton (1766–1844) proposed that all matter consisted of tiny particles called atoms. He proposed that atoms of each element were unique and unlike atoms of any other element. These concepts, illustrated by his model of the atom, are still accepted today. However, Dalton also proposed that atoms were indivisible, or unbreakable. J.J. Thomson was the first person to show that this idea is incorrect.
Thomson's model of the atom is called the "plum pudding" model. It consisted of a positively charged sphere, with negatively charged particles, later called electrons, embedded in it.	In 1897, English physicist Joseph J. Thomson (1856–1940) used cathode ray tubes to demonstrate that atoms could be broken down into smaller particles. He showed that negatively charged particles, later called electrons, could be ejected from atoms, leaving the atom positively charged. His "plum pudding" model of the atom was consistent with all of the characteristics of the atom that were known at the time.
Rutherford's model of the atom is sometimes called the planetary model. Planet-like electrons orbited a positively charged nucleus, which was analogous to the Sun. The nucleus contained most of the mass of the atom.	In 1911, New Zealand physicist and chemist Ernest Rutherford (1871–1937) directed highly energetic, positively charged alpha particles at a very thin gold foil. He traced the paths of the alpha particles after they collided with the gold foil. Using mathematical calculations, he showed that the only configuration of the atom that could explain the paths of the alpha particles was a configuration that had all of the positive charge and nearly all of the mass located in a tiny space at the centre. He inferred that the negatively charged electrons orbited the small nucleus. His experimental data focussed on the nucleus and did not show where the electrons were located, nor how they were moving.
Bohr's model of the atom was similar to Rutherford's model. However, Bohr found that electrons can have only specific amounts of energy. His energy levels are sometimes called electron shells.	In 1913, Danish physicist Niels Bohr (1885–1962) refined Rutherford's model of the atom. Bohr analyzed the pattern of colours of light emitted from heated hydrogen atoms. Using those patterns of light and mathematical calculations, Bohr realized that the light was emitted when electrons dropped from a "higher orbit" to a "lower orbit." He showed that electrons can exist only in specific energy levels. He calculated orbital radii that correspond to these energy levels. Although Bohr's calculations apply to only hydrogen, the general concepts can be applied to all atoms.
Schrödinger showed that electrons move in a region of space, which is often represented visually as a cloud.	In 1926, Austrian physicist Erwin Schrödinger (1887–1961) published a mathematical equation, called the Schrödinger wave equation, that describes the atom in terms of energy. Solutions to this equation show that electrons do not travel in precise orbits but instead exist in defined regions of space. Electrons in each energy level are confined to specific regions, which are represented as electron clouds. This model is usually called the electron cloud model of the atom. Schrödinger's wave equation, along with contributions by Werner Heisenberg (1901–1976), Paul Dirac (1902–1984), Wolfgang Pauli (1900–1958), Friedrich Hund (1896–1997), and others, are explored in advanced chemistry courses.

The Electrons of the Atom

Why was it so important for scientists to develop a detailed model of the atom? An accurate model, or representation, of the atom makes chemistry concepts easier to understand, communicate, and study. In fact, to gain the best understanding of the atom, it is often necessary to rely on more than one model. As shown in **Table 1.1**, the current, most accurate representation of the atom is the electron cloud model, shown in **Figure 1.2**. This model shows the atom as a very small, positively charged nucleus surrounded by clouds of negatively charged electrons. However, this model does not provide any obvious visual information about energy levels. Therefore, chemists often use the model shown in **Figure 1.3** to show energy levels. Because Bohr and Rutherford both contributed to this model, it is often called the Bohr-Rutherford model. Each circle in the Bohr-Rutherford model represents a different energy level, or electron shell. When using this model, it is important to keep in mind that the circles represent energy levels, not actual electron orbits.

Figure 1.2 (A) In the electron cloud model of the atom, electron clouds have different shapes. **(B)** Even when these electron clouds are superimposed to represent a complete atom, their appearance does not give any information about their energy levels.

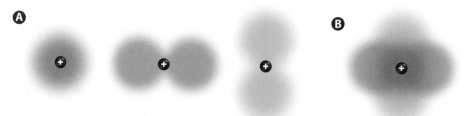

The Bohr-Rutherford model and the electron cloud model look very different. To develop his model, Bohr used the concept of forces of attraction between positive and negative charges. This method allowed him to calculate the radius of the electron shells around the nucleus of a hydrogen atom. Schrödinger, on the other hand, used the energy of electrons to calculate the shape of the electron clouds around the nucleus. While the electron cloud model of the atom does not show any easily visible energy levels, Schrödinger's equation did allow him to find the average distance of electrons from the nucleus. You can think of this as the distance from the nucleus at which electrons spend most of their time. Schrödinger's average distance for an electron in hydrogen is the same as the radius of the electron shell that Bohr calculated, demonstrating how closely their models are related. While the Bohr-Rutherford model provides useful visual information, Schrödinger's equation provides more detailed information about the atom.

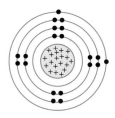

Figure 1.3 The Bohr-Rutherford model of the atom visually depicts energy levels by showing the number of electrons in each electron shell. The shell closest to the nucleus is the lowest energy level.

The Number of Electrons in Each Energy Level

Bohr's work provided information about the energy levels at which electrons can exist. However, further mathematical analysis was required to determine how many electrons can exist in each energy level. Consider the partial periodic table in **Figure 1.4**. The maximum number of electrons that can occupy a shell can be calculated by using $2n^2$, where n is the number of the shell. For example, the number of electrons allowed in the first shell is two ($2 \times 1^2 = 2 \times 1 = 2$), in the second shell is eight ($2 \times 2^2 = 2 \times 4 = 8$), in the third shell is 18 ($2 \times 3^2 = 2 \times 9 = 18$), and so on. When a shell contains the maximum number of electrons, it is said to be filled.

You might expect that the third row of the periodic table would have 18 elements, with 18 electrons in the third shell. Although it is possible for the third shell to contain 18 electrons, elements in the third row do not have more than eight electrons in the outer shell, for reasons that will be explained in advanced chemistry courses. Electrons in the outer shell of an atom are called **valence electrons** and the shell is called the *valence shell*. Note that valence electrons are the only electrons that are involved in the formation of chemical bonds between atoms.

valence electrons
electrons in the outermost shell of an atom

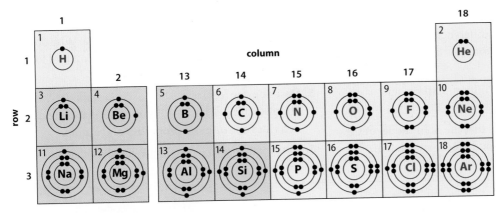

Figure 1.4 This partial periodic table shows Bohr-Rutherford diagrams of the first 18 elements.

Lewis Diagrams and Electrons

To understand the structures of chemical compounds and how electrons are involved in the formation of chemical compounds, it is convenient to use simplified models to represent individual atoms. These models are called **Lewis diagrams**, or sometimes electron dot diagrams. Instead of having plus signs to represent positive charges in the nucleus of an atom, a Lewis diagram has the chemical symbol for the element. As shown in **Figure 1.5**, dots are placed around the chemical symbol to represent the electrons in the valence shell. Only the valence electrons are depicted, since they are the only electrons involved in the formation of bonds between atoms. Lewis diagrams become too complex and are no longer useful, however, if atoms have more than eight electrons in their valence shell. Therefore, Lewis diagrams are used mainly for the elements in columns 1, 2, and 13 through 18 of the periodic table.

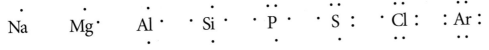

Figure 1.5 Lewis diagrams are quick and easy to use, and they provide important information about the chemical nature of elements.

When drawing the first four dots (electrons) around the symbol for the element in a Lewis diagram, you usually place the first dot at the top of the symbol. You then go clockwise to add the other three dots, spacing the dots equal distances apart. For atoms with five to eight electrons in the valence shell, you place the fifth dot beside the first dot. Additional dots are placed beside the other three dots that are already in place, as shown in **Figure 1.5**.

Pairing dots when drawing a Lewis diagram for an atom with more than four valence electrons is not simply a convenient way to draw the Lewis diagram—it has significance. The first four electrons, being negatively charged, repel one another and remain as far apart as possible in the atom. However, when more than four electrons occupy the same shell, electrons form pairs that interact in a unique way that allows them to be situated closer together. These **electron pairs**, as they are called, are less likely to participate in chemical bond formation than the **unpaired electrons** are (see **Figure 1.6**).

Lewis diagram a model of an atom that has the chemical symbol for the element surrounded by dots to represent the valence electrons of the element

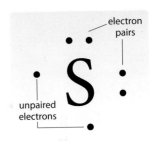

Figure 1.6 This Lewis diagram of sulfur shows that atoms of sulfur have two electron pairs and two unpaired electrons in the valence shell.

electron pairs two electrons that are interacting in a unique way, allowing them to be situated close to each other

unpaired electrons electrons in an unfilled outer shell that are not part of a pair and are, therefore, more likely to participate in bonds with other atoms

1. Describe two significant differences between the Thomson model of the atom and the Rutherford model.

2. Draw Lewis diagrams of lithium, carbon, and fluorine.

3. What information shows that the Bohr model of the atom and the electron cloud model are related?

4. Models are used in many areas of science. Why do you think that models are a very important tool for studying the atom?

5. Describe one feature of Dalton's model, the Thomson model, and the Bohr model that correctly describes one or more properties of the atom.

6. How many electrons could the eighth shell in an atom contain?

The Nucleus of the Atom

Rutherford not only discovered that the nucleus of an atom occupies a very tiny volume at the centre of the atom. He also discovered many important characteristics of the components that make up the nucleus. The results of his research are discussed below.

Protons and Neutrons

After Thomson discovered the electron, chemists and physicists realized that an atom had to contain enough positive charge to balance the number of electrons in the atom. However, they did not know the nature of the matter that carried the positive charge.

When Rutherford was doing experiments with alpha particles, one of the atoms that he bombarded was nitrogen. During one experiment, nitrogen nuclei emitted positively charged particles. Rutherford concluded that the nucleus consisted of individual particles, each with a single positive charge that is equal in strength to the negative charge of an electron. Since hydrogen is the smallest atom, and thus has the smallest nucleus, Rutherford proposed that the positively charged particles in all nuclei are nearly identical to the nucleus of a hydrogen atom. He called these positively charged particles protons. He then calculated the total mass of the number of protons that an atom should have in order to balance the negative charge of the electrons. He discovered that this mass could account for only about half of the actual mass of a typical nucleus.

Rutherford first suggested that a nucleus was composed of enough protons to make up the mass of the nucleus and enough electrons to neutralize the excess protons. As he continued his research, however, he realized that there was no evidence to show that there were electrons in the nucleus. He proposed that the excess mass in the nucleus was due to a new particle, which he called a neutron. He proposed that a neutron had the same mass as a proton but no charge. About 10 years later, Rutherford's assistant, James Chadwick (1891–1974), performed experiments that established the existence of the neutron. The current theory of the atom supports Rutherford's proposal that the nucleus of an atom consists of protons and neutrons. **Figure 1.7** shows how a nucleus is often illustrated.

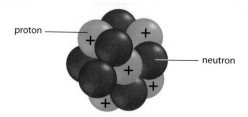

Figure 1.7 Nuclei of atoms are often depicted as a cluster of spheres in different colours, with one colour representing protons and the other representing neutrons.

Neutrons and Isotopes

The chemical nature of an element depends on its atomic number: the number of protons in the nucleus. The atomic number also provides the number of electrons in a neutral atom—an atom in which the positive and negative charges are balanced. Unlike the number of protons, the number of neutrons in the nucleus of atoms of the same element can vary. For example, a hydrogen atom can have zero, one, or two neutrons. Atoms of the same element that have different numbers of neutrons are called **isotopes**.

> **isotopes** atoms that have the same number of protons but different numbers of neutrons

Isotope Notation with Mass Number and Atomic Number

The isotope notation in **Figure 1.8** indicates not only the chemical symbol of the element, but also the isotope of the element. In this notation, the following symbols are used:

- the symbol Z for atomic number, which represents the number of protons in the atom
- the symbol A for mass number, which represents the sum of the number of protons and the number of neutrons

> To find the number of neutrons, N, subtract the atomic number from the mass number: $N = A - Z$.

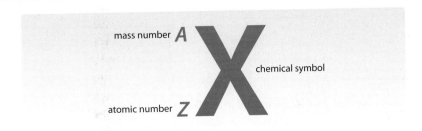

Figure 1.8 The symbol A represents the mass number of the atom, which is the sum of the number of neutrons and the number of protons. The symbol Z is the atomic number, which is the number of protons. In this figure, X represents any chemical symbol.

Using the isotope notation, you can distinguish among the different isotopes of an element. For example, the isotope of hydrogen with no neutrons is represented as ^1_1H. Similarly, the isotope of hydrogen with one neutron is represented as ^2_1H, whereas the isotope with one proton and two neutrons is represented as ^3_1H. The isotope ^2_1H is sometimes called deuterium, or heavy hydrogen, and the isotope ^3_1H is sometimes called tritium. Isotopes of elements other than hydrogen do not have special names, like deuterium and tritium.

The example in **Figure 1.9 (A)** is read as "cobalt-60." Cobalt-60 has 27 protons and $60 - 27 = 33$ neutrons. The example in **Figure 1.9 (B)** is read as "iodine-131." It has 53 protons and $131 - 53 = 78$ neutrons.

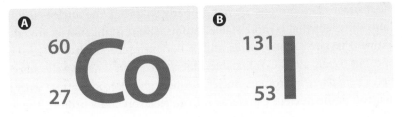

Figure 1.9 (A) Cobalt-60 and **(B)** iodine-131 are radioactive isotopes that are often used in medicine. Cobalt-60 is used in radiation therapy for cancer, and iodine-131 is sometimes used to treat an overactive thyroid.

The Ratio of Neutrons to Protons

Neutrons play an important role in the nucleus. While this role is not immediately obvious, studying examples of known isotopes of different elements, shown in **Table 1.2**, can offer some clues.

Table 1.2 Various Elements and Their Isotopes

Element	A (protons + neutrons)	Z (protons)	N (neutrons)
hydrogen	1	1	0
	2	1	1
	3	1	2
chlorine	35	17	18
	36	17	19
	37	17	20
molybdenum	95	42	53
	96	42	54
	97	42	55
neodymium	142	60	82
	143	60	83
	145	60	85
	146	60	86
mercury	199	80	119
	200	80	120
	201	80	121
	202	80	122

Notice that atoms of the smaller elements in **Table 1.2** have similar numbers of protons and neutrons. As the atomic number of the element gets larger, the number of neutrons exceeds the number of protons by an increasing amount. For instance, molybdenum has roughly 10 more neutrons than protons, neodymium has over 20 more neutrons than protons, and mercury has about 30 more neutrons than protons.

Radioisotopes and Unbalanced Forces in the Nucleus

Many positively charged protons are packed very closely together in the nucleus. The repulsive electric force among the protons is tremendous. As the number of protons increases, however, the number of neutrons increases more rapidly. This has a stabilizing effect, which can be explained by the fact that protons and neutrons also attract one another. The attractive force among protons and neutrons, called the *strong nuclear force*, is approximately 40 times stronger than the repulsive electric force among the positively charged protons. However, it acts over only very short distances. As neutrons are added to the nucleus, they add to the attractive nuclear force but not to the repulsive electric force. A correct balance between protons and neutrons thus stabilizes the nucleus.

radioisotopes isotopes with unstable nuclei that decay into different, often stable, isotopes

When the neutrons and protons are not properly balanced, the nucleus is unstable and decays into a nucleus that is more stable. Atoms with unstable nuclei are said to be radioactive. They are called **radioisotopes**. For example, cobalt-60 ($^{60}_{27}$Co) is a radioisotope, but cobalt-59 ($^{59}_{27}$Co) is stable. Cobalt-60 emits a negatively charged particle, called a *beta particle*, from its nucleus. Although electrons cannot exist in the nucleus, a neutron in the nucleus can decay into an electron and a proton. The newly formed electron is immediately ejected from the nucleus as a beta particle. As a result, the decayed nucleus has one more proton and one less neutron. For example, cobalt-60 becomes nickel-60, as shown below:

$$^{60}_{27}\text{Co} \rightarrow e^- + ^{60}_{28}\text{Ni}$$

Average Atomic Mass

In a periodic table, the atomic mass of an element is usually given in **atomic mass units** (u). The atomic mass unit is based on the mass of an atom of carbon-12 and is defined as one twelfth of the mass of a carbon-12 atom. Because the masses of all other atoms are compared to the mass of carbon-12, these masses are often called *relative atomic masses*. Some time after the atomic mass unit was defined, scientists were able to determine the actual mass of an atom, in grams. One atomic mass unit is now known to be equal to 1.66×10^{-24} g, or $1 \text{ u} = 1.66 \times 10^{-24}$ g.

atomic mass unit one twelfth of the mass of a carbon-12 atom

Many elements have two or more naturally occurring stable isotopes. In order to determine the atomic mass of an element that has more than one isotope, each having a different mass, you must find the *average atomic mass*. Recall how to find the average of several values, by adding the values and dividing the sum by the number of values. If you want to find the average mass of several items, the process becomes slightly more complicated. For example, assume that you want to find the average mass of 10 marbles. Five marbles have a mass of 4.0 g, three marbles have a mass of 3.0 g, and two marbles have a mass of 2.5 g. You would not simply take the average of 4.0, 3.0, and 2.5, because you must account for the number of marbles with each mass. Instead, you would find the average by using one of the four methods below.

Methods for Calculating Averages

Method 1: Add all the masses, and divide the sum by the total number of marbles.

$$\text{average mass} = \frac{\text{sum of the masses}}{\text{number of marbles}}$$

$$= \frac{4.0 \text{ g} + 4.0 \text{ g} + 4.0 \text{ g} + 4.0 \text{ g} + 4.0 \text{ g} + 3.0 \text{ g} + 3.0 \text{ g} + 3.0 \text{ g} + 2.5 \text{ g} + 2.5 \text{ g}}{10}$$

$$= \frac{34 \text{ g}}{10} = 3.4 \text{ g}$$

Method 2: Group the masses by size, and divide the sum by the total number of marbles.

$$\text{average mass} = \frac{(5 \times 4.0 \text{ g}) + (3 \times 3.0 \text{ g}) + (2 \times 2.5 \text{ g})}{10} = \frac{20 \text{ g} + 9 \text{ g} + 5 \text{ g}}{10}$$

$$= \frac{34 \text{ g}}{10} = 3.4 \text{ g}$$

Method 3: Use a weighted average. You know that five tenths, or one half, of the marbles have a mass of 4.0 g. Similarly, three tenths of the marbles have a mass of 3.0 g, and two tenths, or one fifth, have a mass of 2.5 g. You could multiply each mass by the fraction it contributes to the total and then add the products as shown.

$$\text{average mass} = \left(\frac{1}{2} \times 4.0 \text{ g}\right) + \left(\frac{3}{10} \times 3.0 \text{ g}\right) + \left(\frac{1}{5} \times 2.5 \text{ g}\right) = 2.0 \text{ g} + 0.9 \text{ g} + 0.5 \text{ g}$$

$$= 3.4 \text{ g}$$

Method 4: Express the fractions used in method 3 as percentages: $\frac{1}{2}$ is 50%, $\frac{3}{10}$ is 30%, and $\frac{1}{5}$ is 20%. Then convert the percentages into decimal form (for example, $\frac{30\%}{100\%} = 0.30$), and multiply the decimals by their respective masses. Add the resulting values to obtain the average mass.

$$\text{average mass} = (0.50 \times 4.0 \text{ g}) + (0.30 \times 3.0 \text{ g}) + (0.20 \times 2.5 \text{ g})$$

$$= 2.0 \text{ g} + 0.90 \text{ g} + 0.50 \text{ g} = 3.4 \text{ g}$$

You can use method 4 to find the weighted average of any type or number of items, as long as you know the percentage of the whole that is contributed by each subgroup. You do not need to know the total number of items. This means that you can use method 4 to find the average atomic mass of an element with several naturally occurring isotopes.

Isotopic Abundance

Scientists have analyzed the atomic masses of most of the naturally occurring isotopes on Earth. They have determined the percentage contributed by the different isotopes of each element in a large number of samples. These percentages for any given element are so similar to one another that, for most elements, scientists are confident that they represent the percentages for all samples on Earth. Thus, scientists have established a standard percentage for each isotope of an element, which is often called its **isotopic abundance**.

The following Sample Problem demonstrates how to calculate the average mass of the atoms of an element, and the Practice Problems that follow give you an opportunity to practise your skills. The Activity that follows will give you a chance to apply the methods that you have learned.

Sample Problem

Calculating Average Atomic Mass

Problem

The table below provides the atomic mass of each naturally occurring isotope of copper and the percentage of each isotope in a sample of copper. What is the average atomic mass of copper?

Mass and Isotopic Abundance of Each Isotope of Copper

Isotope	Mass (u)	Isotopic Abundance (%)
copper-63	62.93	69.2
copper-65	64.93	30.8

What Is Required?

The problem asks for the average atomic mass of copper.

What Is Given?

You know the mass and isotopic abundance of copper-63:
mass = 62.93 u and isotopic abundance = 69.2%

You know the mass and isotopic abundance of copper-65:
mass = 64.93 u and isotopic abundance = 30.8%

Plan Your Strategy	Act on Your Strategy
Multiply the mass of each isotope by its isotopic abundance, expressed as a decimal, to determine the contribution of each isotope to the average atomic mass.	contribution of isotope copper-63 = 62.93 u × 0.692 = 43.5476 u contribution of isotope copper-65 = 64.93 u × 0.308 = 19.9984 u
Add the contributions of the isotopes to determine the average atomic mass of the element.	average atomic mass of Cu = 43.5476 u + 19.9984 = 63.546 u = 63.5 u

Alternative Solution

The calculation can be combined into one step, as follows:

average atomic mass of Cu = 62.93 u × 0.692 + 64.93 u × 0.308

= 63.5 u

Check Your Solution

The calculated average atomic mass is between the atomic masses of the isotopes, but closer to the atomic mass of the isotope that has the larger isotopic abundance. Three significant digits are appropriate, based on the number of significant digits in the isotopic abundances.

1. Chlorine exists naturally as 75.78% chlorine-35 (mass = 34.97 u) and 24.22% chlorine-37 (mass = 36.97 u). What is the average atomic mass of chlorine?

2. Boron exists naturally as boron-10 (mass = 10.01 u; isotopic abundance = 19.8%) and boron-11 (mass = 11.01 u; isotopic abundance = 80.2%). What is the average atomic mass of boron?

3. Lithium is composed of 7.59% lithium-6 (mass = 6.02 u) and 92.41% lithium-7 (mass = 7.02 u). Calculate the average atomic mass of lithium.

4. Magnesium exists naturally as 78.99% magnesium-24 (mass = 23.99 u), 10.00% magnesium-25 (mass = 24.99 u), and 11.01% magnesium-26 (mass = 25.98 u). What is the average atomic mass of magnesium?

5. Gallium exists naturally as gallium-69 (mass = 68.93 u; isotopic abundance = 60.1%) and gallium-71 (mass = 70.92 u). What is the isotopic abundance of gallium-71? What is the average atomic mass of gallium?

6. Bromine exists naturally as bromine-79 (mass = 78.92 u; isotopic abundance = 50.69%) and bromine-81 (mass = 80.92 u). What is the isotopic abundance of bromine-81? What is the average atomic mass of bromine?

7. A sample of rubidium is 72.17% rubidium-85 (mass = 84.91 u) and 27.83% rubidium-87 (mass = 86.91 u). Calculate the average atomic mass of rubidium.

8. The average atomic mass of nitrogen is 14.01 u. Nitrogen exists naturally as nitrogen-14 (mass = 14.00 u) and nitrogen-15 (mass = 15.00 u). What can you infer about the isotopic abundances for nitrogen?

9. The following isotopes of rhenium are found in nature.
   rhenium-185: m = 184.953 u, abundance = 37.4%
   rhenium-187: m = 186.956 u, abundance = 62.6%
   Analyze the data. Predict the average atomic mass and write down your prediction. Then calculate the average atomic mass and compare it with your prediction.

10. The average atomic mass of iridium is 192.22 u. If iridium-191 has an atomic mass of 190.961 u and an isotopic abundance of 37.3%, and iridium-193 is the only other naturally occurring isotope, what is the atomic mass of iridium-193?

Activity 1.1 Penny Isotopes

All Canadian pennies have a monetary value of one cent. However, not all pennies are alike. Many have different masses, because the Canadian mint has changed the composition of pennies several times. So, you can think of different pennies as "isotopes" of the penny. In this activity, you will determine the isotopic abundances of the penny isotopes in a sample, and the average mass of the penny.

Safety Precaution

Always wash your hands after handling money.

Materials

- sample of at least 25 Canadian pennies
- balance
- calculator

Procedure

1. Sort all your pennies into the following groups of "isotopes:"
   - penny-1: 2000 to the present
   - penny-2: 1997 to 1999
   - penny-3: 1980 to 1996
   - penny-4: 1979 and earlier

2. Count and record the number of pennies in each group.

3. Using the balance, determine the mass of one penny in each group. Record the mass of each penny isotope to the nearest 0.01 g. **Hint:** The final mass will be more accurate if you measure the total mass of all the pennies in each group and then divide the total mass by the number of pennies in the group.

4. Calculate the "isotopic abundance" of each "isotope." You can do this by using the following formula:
$$\text{isotopic abundance} = \frac{\text{number of pennies in category}}{\text{total number of pennies in sample}} \times 100\%$$

5. Calculate the average mass of the penny.

Questions

1. Explain why the average mass of the "element" penny is not necessarily the same as the mass of any one "isotope."

2. Compare the average mass you calculated with the masses obtained by other groups in the class. Explain why the masses are not all the same. Why is the average atomic mass of most elements the same in every sample that has been analyzed?

3. Describe how the penny model is a valid model of isotopes and how it is not a valid model of isotopes.

QUIRKS & QUARKS

with BOB MCDONALD

CBC

THIS WEEK ON QUIRKS & QUARKS

Unearthing an Ancient Andean Element

You may already know that Spain conquered much of South America in the 1500s, plundering its gold and silver mines. You may not know, however, that the Spanish also mined cinnabar—mercury(II) sulfide, $HgS(s)$—, which they valued for its mercury content. Colin Cooke, a PhD student in the Department of Earth and Atmospheric Science at the University of Alberta, studied lake sediment high in the Peruvian Andes of South America, looking for evidence of atmospheric mercury pollution from old Spanish mining operations. Cooke found what he was looking for. However, to his surprise, he also found evidence that people had been mining cinnabar as far back as 1400 B.C.E., long before the Incas settled in the Peruvian Andes. Bob McDonald interviewed Colin Cooke to learn more about the extent of the pollution from ancient cinnabar mines.

Cooke was able to determine when the pollution from mining had occurred by measuring how deeply the pollution was buried below the bottoms of lakes. He found that the pollution from pre-Incan mines was mainly the cinnabar itself. All the pollution was very near the mines. During the Incan Empire, however, cinnabar was mined for its red pigment, which was used for decoration. When processing the ore, the Incas heated it, causing some of the mercury to vaporize. As a vapour, the mercury travelled long distances carried by the wind. The Incas did not know how toxic the vapour was.

When the Spanish arrived, the mining was intensified. The Incan workers mined the ore with picks and then roasted it, causing most of the mercury to vaporize. Although their intent was to condense the vapour in order to collect the mercury, much of the vapour remained in the air. Inhaled mercury vapour is possibly the most toxic form of mercury. The Incan miners inhaled so much mercury that many died within six months after they started mining. Even today, the mines and the miners' skeletons pose a hazard because the mercury that the miners inhaled now sits in pools in their graves.

▶ Related Career

Bob McDonald is a science journalist. He reports on science issues for many radio and television programs. Bob came to science from an arts education and a background in theatre. Science journalists interview experts and share what they learn in ways that non-scientists can understand.

Go to **scienceontario** to find out more

In ancient times, cinnabar was prized for its intense red colour.

QUESTIONS

1. What could the Incan miners have done to protect themselves from inhaling the mercury?

2. Today, mercury is used in many ways, such as in batteries. In nature, mercury waste escapes into waterways and accumulates in fish. Considering the seriousness of mercury poisoning, summarize the risks and benefits of mercury use today.

3. As a science journalist, Bob McDonald, has made a career out of talking to scientists. What credentials do science journalists have? What kinds of jobs do they do, and where do they work?

Section Summary

- As scientists discovered new information about the atom, they modified the model of the atom to reflect each piece of new information.

- The number of electrons that can occupy an electron shell can be determined by using the formula $2n^2$, where n is the shell number.

- The electrons in the outermost shell of an atom are called valence electrons.

- A Lewis diagram is a simplified method for representing an atom.

- The nucleus of an atom contains protons and neutrons and occupies a very small volume at the centre of the atom.

- An appropriate ratio of neutrons to protons stabilizes the nucleus. An atom with an unstable nucleus is called a radioisotope.

- Atomic masses that are reported in data tables are weighted averages, based on isotopic abundances.

Review Questions

1. **K/U** Describe two characteristics of an atom that were determined by Schrödinger and that were not revealed by any previous technique or model.

2. **K/U** In what way does the Bohr-Rutherford model of the atom provide more visual information than the more accurate electron cloud model?

3. **T/I** Examine the Lewis diagram shown here. List at least three details about the atom that you can determine from this diagram.

4. **C** Draw a diagram showing the general isotope notation using the letters Z, A, and X. Include in your diagram the meaning of each symbol.

5. **K/U** Define the term "isotope."

6. **K/U** Explain how an atom of one element can turn into an atom of another element. Under what conditions does this occur?

7. **K/U** Define "isotopic abundance."

8. **K/U** Explain why the atomic mass reported in a periodic table is nearly always an average value.

9. **T/I** Copy and complete the following table.

10. **T/I** The atomic masses and isotopic abundances of the naturally occurring isotopes of silicon are given below. Calculate the average atomic mass of silicon.
$^{28}_{14}Si$: mass = 27.977 u; isotopic abundance = 92.23%
$^{29}_{14}Si$: mass = 28.976 u; isotopic abundance = 4.67%
$^{30}_{14}Si$: mass = 29.97 u; isotopic abundance = 3.10%

11. **T/I** The atomic mass of yttrium-89 is 88.91 u. The average atomic mass of yttrium that is reported in periodic tables is 88.91. Infer why these values are the same.

12. **K/U** Write a paragraph that explains why the ratio of neutrons to protons increases as the atomic number increases.

13. **K/U** You reviewed four methods for calculating average values of a group of items. Which of the methods can you use to calculate average atomic mass? Explain why the other methods cannot be used.

14. **A** Animals consume carbon-containing compounds and incorporate them into their tissues and breathe out carbon dioxide. A very small amount of that carbon is radioactive carbon-14. When the animals die, they no longer exchange carbon with the environment. Based on this information, how do you think that scientists use the percentage of carbon-14 in a fossil to determine its age? (This technique is called carbon dating.)

Isotope Data

	Name of Isotope	Notation for Isotope	Atomic Number	Mass Number	Number of Protons	Number of Electrons	Number of Neutrons
a.			35	81			
b.			22	10			
c.	calcium-44						
d.					47		60

SECTION 1.2

The Periodic Table

Key Terms

periodic law
period
group

The periodic table contains a large amount of useful data. For example, you can use the periodic table to look up chemical symbols, atomic numbers, and average atomic masses of elements. However, the periodic table can reveal much more about the elements when you have a full understanding of its design.

The Development of the Periodic Table

During a timespan of about 200 years, between the 1600s and the 1800s, chemists increased the number of recognized elements to 63 and observed and recorded many of their properties. Chemists began to observe similarities in the chemical nature of groups of elements. English chemist John Newlands (1837–1898) discovered that 56 elements could be classified into 11 groups that have similar chemical properties. Newlands also discovered that the elements in each group differed in atomic mass (then called atomic weight) by factors of eight.

Working independently of each other, German chemist Lothar Meyer (1830–1895) and Russian chemist Dmitri Mendeleev (1834–1907) both developed a table of the known elements according to their atomic masses. Because Mendeleev published his table first (1869), he is considered to be the "father" of the periodic table.

To develop his table, Mendeleev made a card for each of the 63 known elements. He wrote the symbol, atomic mass, and chemical and physical properties of an element on each card. Mendeleev arranged the 63 cards vertically, according to the atomic masses of the elements. When he came to an element that had chemical and physical characteristics that were similar to the first element in the previous column, he started a new column. He then placed the next element beside the element with similar properties. In a few cases, however, Mendeleev had to reverse the order of the elements to make their chemical properties match those of their neighbouring elements. Later, in 1913, British physicist Henry Moseley (1887–1915) developed a method for determining the number of protons in the nucleus of atoms, and thus the atomic number of an element. When the elements were ordered according to atomic number instead of atomic mass, the order agreed with Mendeleev's order.

Mendeleev left several gaps in his periodic table, attributing these gaps to elements that had not yet been discovered. The positions of these gaps enabled him to predict the properties of the missing elements. Several of these elements were discovered within the next few years, and the properties that Mendeleev had predicted were shown to be correct. The discovery of these elements gave validity to Mendeleev's work. The first periodic table that Mendeleev published is shown in **Figure 1.10**.

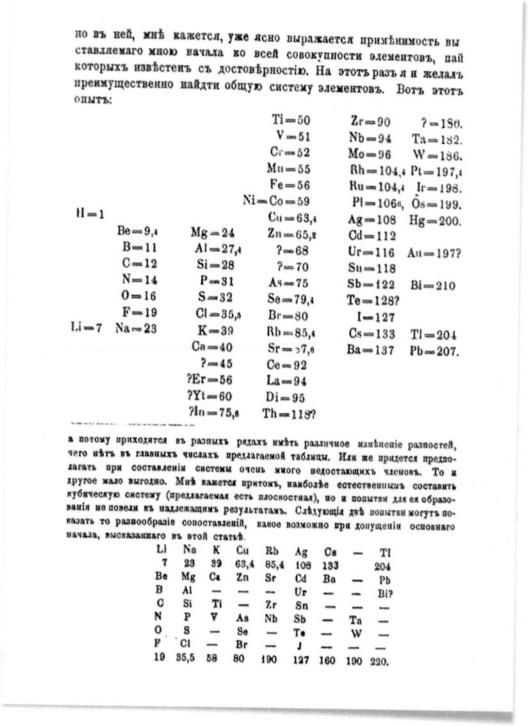
но въ ней, мнѣ кажется, уже ясно выражается примѣнимость выставляемаго мною начала ко всей совокупности элементовъ, пай которыхъ извѣстенъ съ достовѣрностію. На этотъ разъ я и желалъ преимущественно найдти общую систему элементовъ. Вотъ этотъ опытъ:

			Ti=50	Zr=90	?=180.
			V=51	Nb=94	Ta=182.
			Cr=52	Mo=96	W=186.
			Mn=55	Rh=104,4	Pt=197,4
			Fe=56	Ru=104,4	Ir=198.
			Ni=Co=59	Pl=106,6	Os=199.
H=1			Cu=63,4	Ag=108	Hg=200.
	Be=9,4	Mg=24	Zn=65,2	Cd=112	
	B=11	Al=27,4	?=68	Ur=116	Au=197?
	C=12	Si=28	?=70	Sn=118	
	N=14	P=31	As=75	Sb=122	Bi=210
	O=16	S=32	Se=79,4	Te=128?	
	F=19	Cl=35,5	Br=80	I=127	
Li=7	Na=23	K=39	Rb=85,4	Cs=133	Tl=204
		Ca=40	Sr=87,6	Ba=137	Pb=207.
		?=45	Ce=92		
		?Er=56	La=94		
		?Yt=60	Di=95		
		?In=75,6	Th=118?		

а потому приходится въ разныхъ рядахъ имѣть различное измѣненіе разностей, чего нѣтъ въ главныхъ числахъ предлагаемой таблицы. Или же придется предполагать при составленіи системы очень много недостающихъ членовъ. То и другое мало выгодно. Мнѣ кажется притомъ, наиболѣе естественнымъ составить кубическую систему (предлагаемая есть плоскостная), но и попытки для ея образованія не повели къ надлежащимъ результатамъ. Слѣдующія двѣ попытки могутъ показать то разнообразіе сопоставленій, какое возможно при допущеніи основнаго начала, высказаннаго въ этой статьѣ.

Li	Na	K	Cu	Rb	Ag	Cs	—	Tl
7	23	39	63,4	85,4	108	133		204
Be	Mg	Ca	Zn	Sr	Cd	Ba	—	Pb
B	Al	—	—	—	Ur	—	—	Bi?
C	Si	Ti	—	Zr	Sn	—	—	—
N	P	V	As	Nb	Sb	—	Ta	—
O	S	—	Se	—	Te	—	W	—
F	Cl	—	Br	—	J	—	—	—
19	35,5	58	80	190	127	160	190	220.

Figure 1.10 In Mendeleev's first periodic table, he listed the atomic masses vertically and placed elements with similar properties in horizontal rows.

Infer *why some of the columns are longer than others.*

Characteristics of the Modern Periodic Table

The periodic table that you use today looks very different from Mendeleev's table, but it is based on the same periodic, or repeating, relationships. In fact, chemists now call the concept underlying the organization of the periodic table the **periodic law**.

periodic law a statement that describes the repeating nature of the properties of the elements

> **The Periodic Law**
> When elements are arranged by atomic number, their chemical and physical properties recur periodically.

Modern chemists have learned much more about the properties of elements and have filled in all the gaps in Mendeleev's table. As you read in Section 1.1, chemists have also learned much more about the structure of atoms in general. This new information can explain the periodic nature of the table. To begin to understand the periodicity, examine the partial periodic table in **Figure 1.11**. Only the atomic numbers and Lewis diagrams of the elements are shown in the table. The Lewis diagrams clearly show the configuration of the electrons in the valence shell of the atoms of each element.

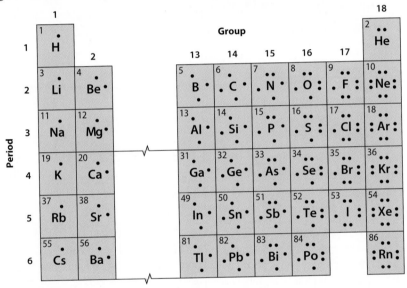

Figure 1.11 Lewis diagrams are shown in this partial periodic table, so you can quickly and easily analyze the electron configuration in each column of the table.

The atomic number increases as you go from left to right in the table. Because the atomic number is the number of protons in the nucleus of an atom, the number of electrons must also increase to make the net charge on the atom zero. The rows are called **periods**. The period number is the number of electron shells that are occupied by one or more electrons. As you go across a period, the outer electron shell is being filled. When the shell contains the maximum number of electrons allowed in that shell, the period ends. Because the first shell can contain only two electrons, the first period has only two elements. The second shell can contain up to eight electrons, so the second period has eight elements. The partial periodic table in **Figure 1.11** omits columns 3 through 12 because, in Period 4, the valence shells of atoms of some elements have more than eight electrons. The analysis of the electron structure of these elements is very complex, so it is left for more advanced chemistry courses.

Figure 1.11 was drawn with only valence electrons to highlight another important feature of the elements as they are arranged in the periodic table. The columns are called **groups**. All elements in a group have the same electron configuration in the valence shell of their atoms. Before chemists knew about valence electrons, they knew that the elements in each group had similar chemical and physical properties. Chemists now know that the configuration of the valence electrons is responsible for these similar properties.

period a row in the periodic table

group a column in the periodic table

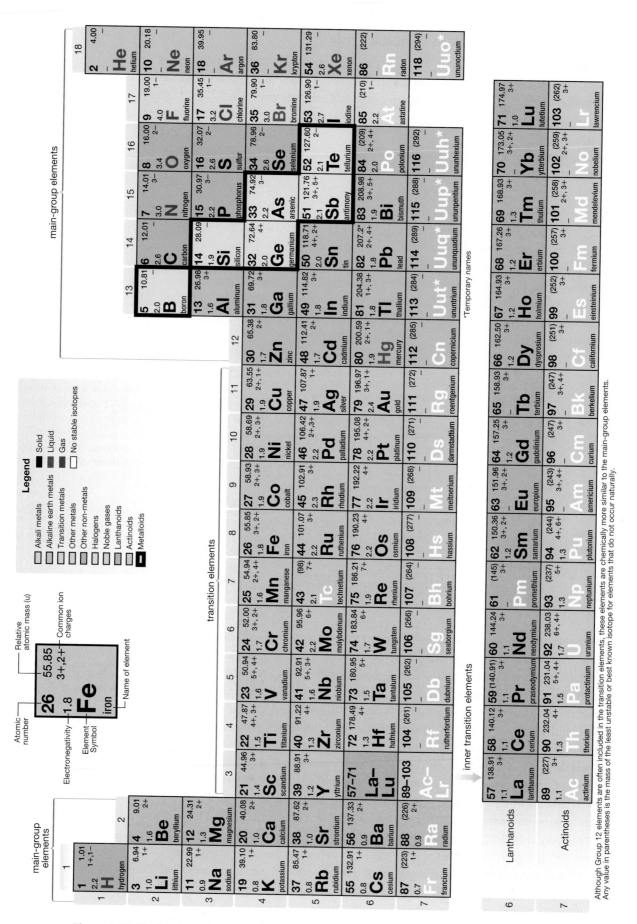

Figure 1.12 The brackets above this table divide the elements into large general categories, while the colours break down the categories according to the chemical properties of the elements.

Categories within the Complete Periodic Table

So far, you have been examining a partial periodic table. The design of the complete, modern, periodic table makes it possible to divide the elements into categories based on their properties. The complete periodic table in **Figure 1.12** has brackets and colour codes that highlight different categories of elements.

Categorizing Elements According to Their Properties

The colours codes in **Figure 1.12** divide the elements into categories that highlight unique chemical and physical properties. Some of the categories have specific names. **Figure 1.13** shows a reduced copy of the periodic table in **Figure 1.12**, with labels that show the names of these categories. You will see references to these categories many times throughout your study of chemistry.

Suggested **Investigation**

Inquiry Investigation 1-A,
Developing an Activity
Series

alkali metals All of the elements in Group 1, except hydrogen, are called alkali metals. They are very reactive. In fact, all alkali metals react vigorously with water. When rubidium and cesium are added to water, the reaction with water is explosive.

other non-metals This category consists of the non-metals that are not halogens or noble gases. These non-metals are neither as reactive as the halogens nor as unreactive as the noble gases. They are the most common elements in the tissues of living organisms.

noble gases The Group 18 elements are called the noble gases. They are extremely unreactive. They do not undergo any naturally occurring reactions. However, chemists have made artificial compounds that contain some of the noble gases.

alkaline earth metals The elements in Group 2 are called the alkaline earth metals. They are reactive, but less so than the alkali metals.

metalloids Metalloids have properties that are between those of the metals and non-metals. Many are shiny solids but they are poor conductors of electric current and are brittle.

halogens The elements in Group 17 are called the halogens. They are reactive non-metals.

transition metals The transition metals in Groups 3 through 11 are very hard metals with very high melting points. Because of their complex electron configuration, they can form a wide variety of compounds with other elements.

other metals This category includes the main-group metals that are not alkali metals and alkaline earth metals. These metals are not as reactive as the Group 1 and 2 metals. They are quite common and useful. For example, you may be familiar with many objects that are made from aluminum or tin.

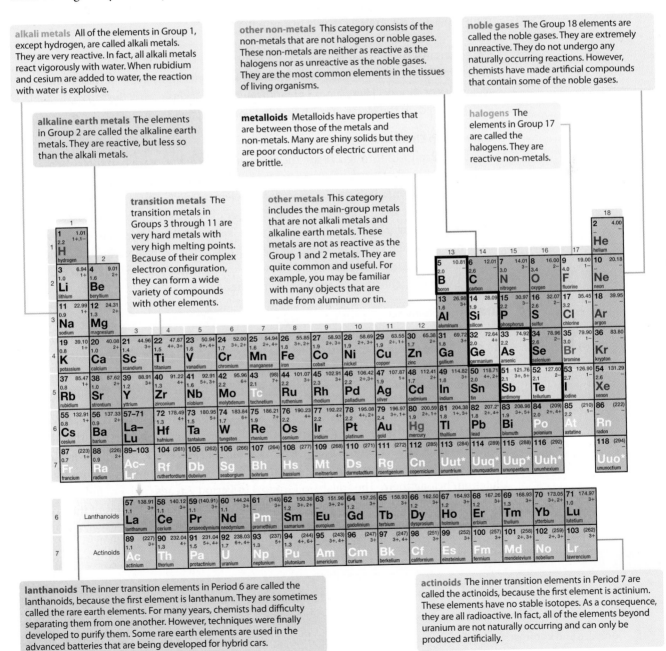

lanthanoids The inner transition elements in Period 6 are called the lanthanoids, because the first element is lanthanum. They are sometimes called the rare earth elements. For many years, chemists had difficulty separating them from one another. However, techniques were finally developed to purify them. Some rare earth elements are used in the advanced batteries that are being developed for hybrid cars.

actinoids The inner transition elements in Period 7 are called the actinoids, because the first element is actinium. These elements have no stable isotopes. As a consequence, they are all radioactive. In fact, all of the elements beyond uranium are not naturally occurring and can only be produced artificially.

Figure 1.13 Each colour in this periodic table represents a group of elements that have similar and unique properties.

Go to **scienceontario** to find out more

Categorizing Elements According to More General Properties

The colours in the periodic table in **Figure 1.14** show another way to categorize the elements. These categories are based on the general properties of the elements.

The elements shaded in blue are metals. Recall from previous science courses that all metals, except mercury, are solid at room temperature. (Mercury is a liquid.) Metals tend to be shiny, are good conductors of electric current, and are malleable and ductile.

The elements shaded in yellow are non-metals. Some non-metals are solid at room temperature, some are gases, and one (bromine) is a liquid. Non-metals are not shiny, do not conduct electric current, and are not malleable or ductile.

The elements shaded in green are metalloids. Their properties are between those of metals and non-metals. In their pure form, many metalloids look like metals, but they are brittle and often poor conductors of electric current. For example, silicon is shiny, but it is brittle and a poor conductor. It is an important semiconductor, used in transistors and integrated circuits in computer chips.

Categorizing According to Blocks of Elements in the Periodic Table

When you look at the periodic table in **Figure 1.12**, you can imagine cutting it into rectangles or blocks. These blocks are indicated by brackets across the top of the table and are based on the electron configurations of the elements.

Main-group Elements

Groups 1, 2, and 12 through 18 are the *main-group elements*. Varied in their chemical and physical properties, they are the most prevalent elements on Earth. The number of valence electrons, with the exception of the elements in Group 12, is completely predictable from the group number. The number of valence electrons increases from one to eight going from left to right (skipping Group 12) across the main-group elements. Although Group 12 does not fit into this pattern, it is often included with the main-group elements because its elements are chemically more similar to the main-group elements than to any other elements.

Transition Elements

The centre of the table contains the *transition elements*, sometimes called the transition metals. These elements make the periodic table 18 columns wide because atoms of some of the elements can have up to 18 electrons in their outer shell. Thus, their electron configurations are more complex than those of the main-group elements.

Inner Transition Elements

At the bottom of the table lie two rows called the *inner transition elements*. These elements fit between columns 3 and 4. Notice that, in **Figure 1.12**, the square in Group 3 of Period 6 reads "La–Lu," meaning that all of the elements from lanthanum through lutetium belong there. The square in Group 3 of Period 7 reads "Ac–Lr," indicating the location of the elements from actinium through lawrencium. Atoms of the inner transition elements can have as many as 32 electrons in some of their valence shells.

Learning Check

7. How did Mendeleev organize his periodic table?

8. State the periodic law and explain its meaning.

9. Compare and contrast groups and periods.

10. Explain why each period ends with a noble gas.

11. Describe the basis for the different categories of the elements in the periodic table.

12. Use a graphic organizer such as a main idea web to distinguish and describe key characteristics of the categories of elements in the periodic table.

Examples of Some Properties and Uses of Metals, Non-metals, and Metalloids

Silver

Silver, a metal, is ductile and can be drawn into a wire and shaped into coils.

Sulfur

Sulfur, a non-metal, is a yellow solid at room temperature. It is used in the processing of rubber and it is an ingredient in gunpowder.

Chlorine

Chlorine, a non-metal, is a greenish-yellow gas. It is very toxic and can damage the respiratory tract.

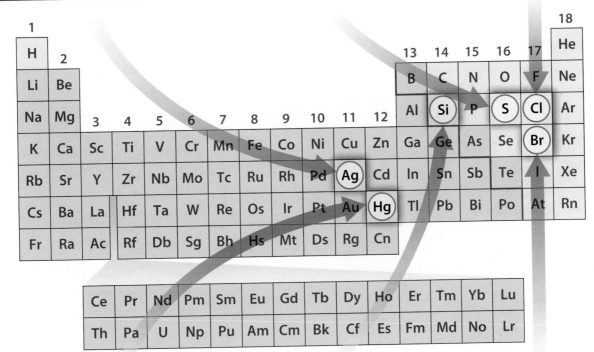

Mercury

Mercury is the only metal that is liquid at room temperature. But, like the other metals, it is shiny and can conduct electric current.

Silicon

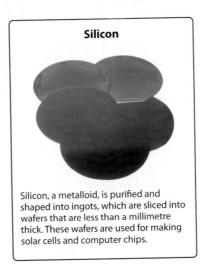

Silicon, a metalloid, is purified and shaped into ingots, which are sliced into wafers that are less than a millimetre thick. These wafers are used for making solar cells and computer chips.

Bromine

Bromine, a non-metal, is a volatile, reddish-brown liquid at room temperature. It is used in water purification and in pesticides.

Figure 1.14 Some properties and uses of metals, metalloids, and non-metals are summarized here.

Alternative Forms of the Periodic Table

The commonly used periodic table, as shown in **Figure 1.12**, is convenient for many purposes. However, it is only one of several possible designs. Some chemists have proposed the designs in **Figure 1.15** for a table of the elements to highlight certain properties of the elements.

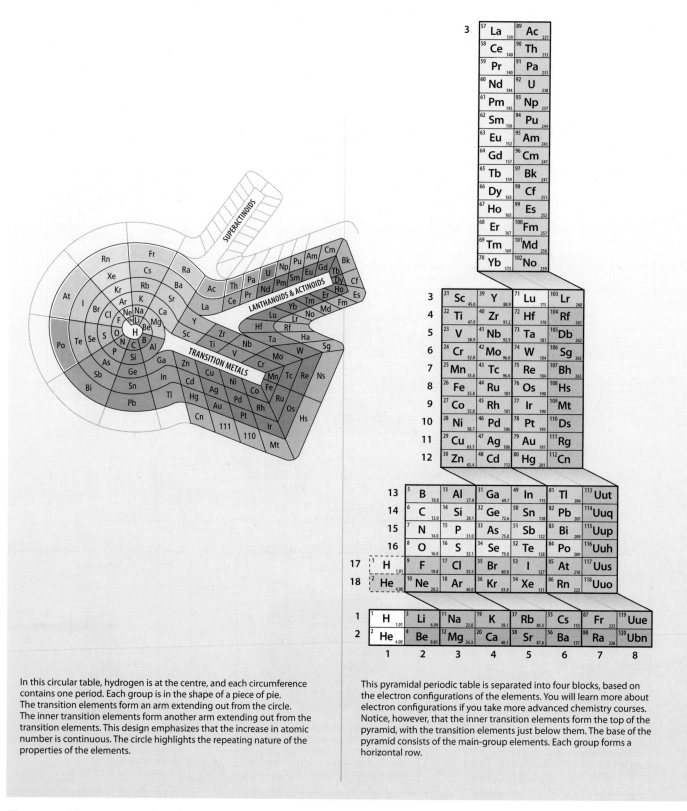

In this circular table, hydrogen is at the centre, and each circumference contains one period. Each group is in the shape of a piece of pie. The transition elements form an arm extending out from the circle. The inner transition elements form another arm extending out from the transition elements. This design emphasizes that the increase in atomic number is continuous. The circle highlights the repeating nature of the properties of the elements.

This pyramidal periodic table is separated into four blocks, based on the electron configurations of the elements. You will learn more about electron configurations if you take more advanced chemistry courses. Notice, however, that the inner transition elements form the top of the pyramid, with the transition elements just below them. The base of the pyramid consists of the main-group elements. Each group forms a horizontal row.

Figure 1.15 These are just a few of the alternative periodic tables that some chemists have suggested.

This student has constructed a unique—and edible—version of the periodic table.

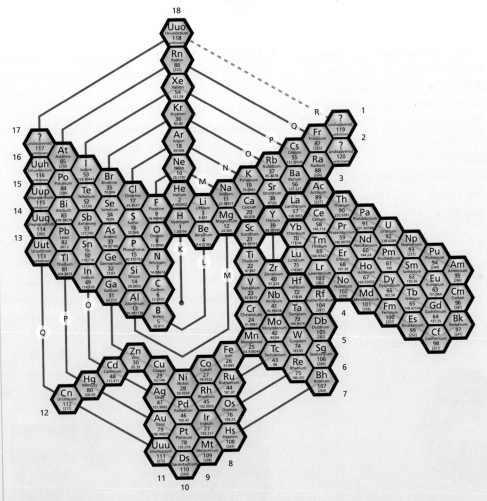

In this "periodic spiral," hydrogen is in the centre. As you go around the spiral clockwise, the atomic number increases. The main-group elements, the alkali and alkaline earth metals, the inner transition metals, and the transition metals all form separate clusters. The noble gases form a line by themselves. Zinc, cadmium, and mercury are also in a line by themselves because they have some characteristics of main-group elements and some characteristics of transition metals. This periodic spiral comes with interactive software that contains a wealth of information about the elements.

Identify *the alkali metals, alkaline earth metals, halogens, noble gases, actinoids, and lanthanoids in this periodic table.*

Section Summary

- Mendeleev developed the periodic table by first listing the elements in order of increasing atomic mass. Then, when the chemical properties of an element resembled those of a previous element in the list, he put this element alongside the previous element. With a little adjusting, Mendeleev developed a table of elements with increasing mass in the columns and similar properties in the rows.

- Elements are categorized as main-group, transition, and inner transition elements, based on their electron configurations.

- Elements are categorized as metals, metalloids, and non-metals, based on their chemical properties.

- Several groups of elements are given specific names due to their uniquely similar properties. Group 1 elements are called the alkali metals, Group 2 elements are called the alkaline earth metals, Group 17 elements are called the halogens, and Group 18 elements are called the noble gases.

- Many alternative forms of the periodic table have been developed to highlight specific properties of the elements.

Review Questions

1. **K/U** When Mendeleev was working on his periodic table, he listed the elements according to their mass. He discovered that he had to reverse the order of a few of the elements to make them fit the pattern of chemical properties. Why were these elements out of order?

2. **K/U** Explain why some elements were missing in Mendeleev's periodic table.

3. **K/U** What characteristic is the same for all the main-group elements in a given period? What characteristic is the same for all the main-group elements in a group?

4. **T/I** Sodium metal is usually stored in oil. Find sodium in the periodic table. Based on its position in the periodic table, why do you think that sodium must be stored in oil?

5. **A** The platter shown here is made from copper. Why could silicon not be used to make objects such as this?

6. **K/U** Name two characteristics that distinguish main-group elements from transition or inner transition elements.

7. **K/U** Which category—metals, metalloids, or non-metals—includes elements that are gases at room temperature?

8. **T/I** List the following elements in order, from best to worst conductor of electric current: silicon, copper, iodine.

9. **T/I** If the inner transition elements were inserted into the periodic table between Groups 3 and 4 in Periods 6 and 7, how many columns wide would the table be? Explain your answer.

10. **K/U** Name three different categories into which the Group 17 elements could be classified.

11. **K/U** Why are the lanthanoids sometimes called rare earth elements?

12. **A** Strontium is sometimes called a "bone seeker" because, when ingested, it is deposited in bones and remains there for long periods of time. Find strontium in the periodic table. Offer a possible explanation for why strontium is called a "bone seeker."

13. **T/I** Examine each of the following pairs of elements. State which element is the most reactive, and explain how you made your choice.
  a. Na or Mg   **b.** Br or Kr   **c.** H or He

14. **K/U** The photograph below shows what happened when a sample of a pure element was dropped in water. Name two elements that could have caused this reaction.

15. **A** Living tissues are made up of just a few elements. In what category are most of these elements found?

16. **C** Create a chart that summarizes the different types of periodic tables presented in this chapter and include the benefits of each model.

Explaining Periodic Trends

In the previous section, you learned that the atomic number increases going from left to right across the periodic table. You also learned that the electron configuration of the valence shell is similar for atoms of the elements within a group. These are only two of the numerous trends in the periodic table. In this section, you will learn about other trends, which vary across a period or down a group in the periodic table, including trends in

- atomic radius
- ionization energy
- electron affinity
- electronegativity

Key Terms

atomic radius
effective nuclear charge
ionization energy
electron affinity
electronegativity

Atomic Radius

Consider the following questions: How would you predict that the size of an atom would change across a period or down a group? On what properties of elements would you base your prediction?

To answer these questions, you need a way to describe the size of an atom. As explained in Section 1.1, electrons move around a nucleus in what is best described as a cloud. In other words, an atom has no clearly defined boundary. To circumvent this problem, chemists define the size of an atom as the space in which the electrons spend 90 percent of their time, as shown in **Figure 1.16**. The size of an atom is usually reported in terms of its **atomic radius**, the distance from the centre of an atom to the boundary within which the electrons spend 90 percent of their time.

The next challenge becomes measuring the size of atoms. There is no way to directly measure the size of the space in which electrons spend 90 percent of their time, so chemists have had to find other methods for measuring atoms. The method that chemists use depends on the type of atom. For elements that can be crystallized in their pure form, chemists use a technique called *X-ray crystallography*. Using X-ray crystallography, chemists can measure the distance between the centres of atoms, as shown in **Figure 1.17 (A)**. The radius of the atom is half the distance between the centres of the atoms. Chemists use similar techniques, called *neutron diffraction* and *electron diffraction*, to measure the distance between the centres of atoms in of the diatomic gas molecules—oxygen, nitrogen, hydrogen, and the halogens. Half of this distance gives the atomic radius of an atom in a diatomic molecule. The method for measuring the radius of a hydrogen atom is illustrated in **Figure 1.17 (B)**.

atomic radius the distance from the centre of an atom to the boundary within which the electrons spend 90 percent of their time

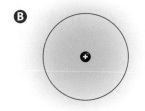

Figure 1.16 (A) This illustration represents the electron cloud of an atom. **(B)** The circle contains the region where the electrons spend 90 percent of their time.

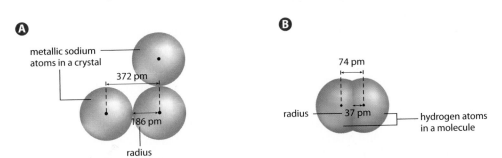

Figure 1.17 (A) These atoms represent sodium atoms in a crystal. **(B)** This molecule represents hydrogen. To obtain the atomic radius of an atom, the centre-to-centre distance is measured and then divided by 2. The unit used in these diagrams is the picometre (pm) which is equal to 1×10^{-12} m.

The development of the scanning tunnelling microscope (STM) has given chemists another method for measuring the size of an atom. In this activity, you will use STM data to estimate the radius of a carbon atom in graphite. Graphite consists of sheets of carbon atoms that are in hexagonal arrays, as shown in the diagram labelled A.

The picture labelled B is a STM image of graphite. The graph below the image shows the electric current through the STM while it was scanning across the black line on the image. The two pointers on the graph are the same as those in the image. They point to the centres of two carbon atoms in one hexagonal unit.

 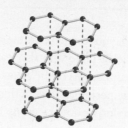

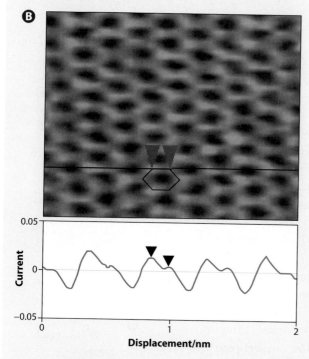

Data obtained from: Chaun-Jian Zhong et al. 2003. Atomic scale imaging: a hands-on scanning probe microscopy laboratory for undergraduates. *Journal of Chemical Education* 80: 194–197.

Materials
- ruler
- calculator

Procedure

1. Use the ruler to measure, in centimetres, the horizontal distance between the pointers on the graph. Estimate the measurement to two decimal places.

2. Measure, in centimetres, the distance that represents 1 nm (1×10^{-9} m) on the scale.

3. Calculate the centre-to-centre distance between the two adjacent carbon atoms by using the following formula:

$$\text{actual distance} = \frac{\text{distance between pointers on graph}}{\text{distance representing 1 nm of scale}}$$

4. From the centre-to-centre distance you calculated in step 3, calculate the radius of a carbon atom.

5. The accepted radius of a carbon atom is 77×10^{-12} m. Calculate the percent error for the radius you calculated in step 4. Go to Measurement in Appendix A for help with calculating percent error.

Questions

1. Compare the structure of graphite to the STM image. What do the bright spots on the STM image represent? What do the dark spots represent?

2. What was your percent error? Give possible reasons for this error.

3. Using the same image and scale, propose another method for calculating the radius of a carbon atom that might be more accurate.

Analyzing Atomic Radii Data

Using a variety of techniques, chemists have been able to measure the atomic radii of atoms of all of the elements with stable isotopes. The results of these measurements are shown in **Figure 1.18** in the form of a periodic table. The values for atomic radii are given in picometres. The spheres in the table show the relative sizes of atoms of the elements.

To analyze the data for size versus atomic number of the elements, recall that all of the positive charge and nearly all of the mass are located in an extremely small volume at the centre of the atom. The electrons, with their negative charge, account for almost the entire volume of the atom. Both the positive charge in the nucleus and the negative charge of the electrons increase as the atomic number increases. The number of occupied shells is the same across any given period. Positive and negative charges attract one another. As the charge becomes greater, the attractive force becomes greater. All of this information helps to explain the pattern shown in **Figure 1.18**.

Trends in Atomic Radius within a Period

As you can see in **Figure 1.18**, the size of an atom decreases going from left to right across a period. In contrast, the number of positive charges in the nucleus increases. The number of electrons also increases, so you might expect the sizes of the atoms to get larger. However, all of the electrons in atoms of elements in the same period are in the same shell. With the increase in the number of positive charges, the attractive force on each electron becomes stronger. As the attractive force becomes stronger, the electrons are drawn closer to the nucleus, and thus the atom is smaller.

Go to **scienceontario** to find out more

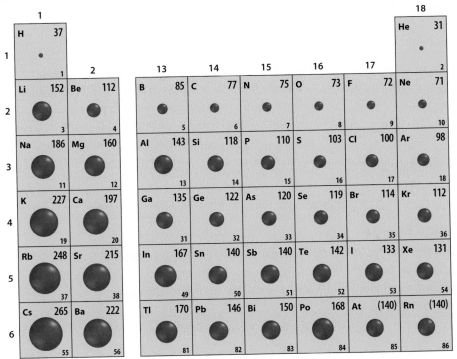

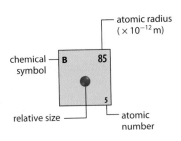

Figure 1.18 This partial periodic table shows the atomic radii of atoms of most of the main-group elements.

Analyze this periodic table with atomic radii and **describe** any trends that you see.

Activity 1.3 Plotting Atomic Radius versus Atomic Number

Data in graphical form are often easier to analyze than data in table form. In this activity, you will plot a graph using data for atomic radius and atomic number.

Materials
- graph paper
- ruler
- pencil

Procedure

1. Plot a graph of atomic radius versus atomic number using the data in the periodic table in **Figure 1.18**. Atomic number will be your independent variable (*x*-axis), and atomic radius will be your dependent variable (*y*-axis). Decide on a scale for your *x*-axis. Notice that elements with atomic numbers 21 through 30, 39 through 48, and 57 through 80 are missing. Decide how you will show that they are missing on your graph.

2. Analyze the atomic radius data in **Figure 1.18** to decide on a suitable scale for your *y*-axis.

3. Plot the data. Connect the data points with straight lines.

4. Label each peak and valley on your graph with the chemical symbol for the element that matches the atomic number on the scale below.

Questions

1. What feature stands out the most on your graph? What does it tell you about the atomic radii of the elements in a particular period or group?

2. How does the atomic radius change going across a period?

3. How does the atomic radius change going down a group?

4. What characteristic of atomic radii is made more obvious by plotting the data in a graph rather than observing the data in the partial periodic table in **Figure 1.18**?

5. What characteristic is more obvious in the periodic table than in your graph?

Trends in Atomic Radius within a Group

effective nuclear charge the apparent nuclear charge, as experienced by the outermost electrons of an atom, as a result of the shielding by the inner-shell electrons

As you go down a group, the atomic number becomes greater, and thus the amount of positive charge also becomes greater. With each step down a group, however, the number of occupied electron shells increases. The inner shells that are filled shield the outer electrons from the positive charge of the nucleus. Because of this shielding, the **effective nuclear charge** appears smaller than the actual charge in the nucleus. As a result, the outer electrons are not attracted to the nucleus as strongly as they would have been without the shielding. Therefore, each additional electron shell, and thus the atomic radius, becomes larger going down a group.

Learning Check

13. How do chemists define the radius of an atom?

14. Why is it not possible to measure the size of an atom directly?

15. How does the amount of charge in the nucleus of an atom affect the size of the atom?

16. Explain the shielding effect of electrons, and describe how it affects the size of an atom.

17. List the following elements in order of increasing atomic number and then in order of increasing size: oxygen, tin, potassium, krypton. Explain why the orders of the elements are different in your two lists.

18. List at least three factors that affect atomic radius.

Ionization Energy

In chemical reactions, atoms can lose, gain, or share electrons with other atoms. When an atom loses an electron, the remaining ion is positively charged. A generalized equation is shown below. The A in the equation represents any atom.

$$A(g) + energy \longrightarrow A^+(g) + e^- \text{ (first ionization)}$$

ionization energy the amount of energy required to remove the outermost electron from an atom or ion in the gaseous state

The amount of energy required to remove an electron from an atom is an important property that plays a role in determining the types of reactions in which atoms of that element might be involved. The amount of energy required to remove the outermost electron from an atom or ion in the gaseous state is called the **ionization energy**. The reason that the definition states that the atom and ion are in the gaseous state is to eliminate any effect of nearby atoms. If ionization energy were measured in the solid or liquid state, the adjacent atoms would affect the measurement.

After one electron has been removed, it is still possible to remove more electrons, leaving the ion with a larger positive charge. A generalized equation for the removal of a second electron is shown below.

$$A^+(g) + energy \longrightarrow A^{2+}(g) + e^- \text{ (second ionization)}$$

The removal of any subsequent electron requires more energy than the removal of the previous electron because there are fewer negative charges repelling each subsequent electron but the same number of positive charges attracting it.

A graph of first ionization energy versus atomic number is shown in **Figure 1.19**. Notice that the elements at all of the peaks in the graph are noble gases. Atoms of the noble gases have filled valence shells. The filled shells make the atoms very stable so you would expect that it would take more energy to remove electrons from atoms of these elements. Notice, also, that the elements at the lowest points on the graph are Group 1 elements. Recall that all of their atoms have only one electron in their valence shell. When that electron is removed, the remaining ion has a filled outer shell, which is a stable configuration.

Suggested Investigation

ThoughtLab Investigation 1-B, Analyzing Ionization Energy Data

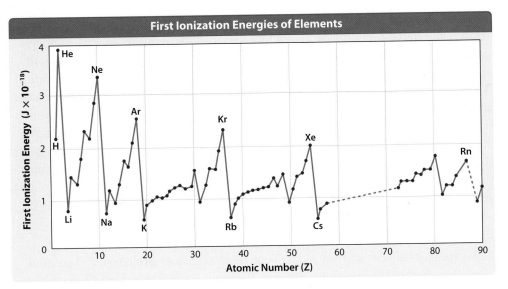

Figure 1.19 The data points on this graph represent the amount of energy required to remove one electron from a neutral atom of the various elements. The numbers on the scale must be multiplied by 10^{-18} to give the correct value in Joules. The dashed lines represent sections in which data points are missing.

Electron Affinity

If one atom loses an electron, another atom must gain this electron. Neutral atoms can gain electrons and become negatively charged. In the process of gaining an electron, a neutral atom can release energy or energy might be needed to add the electron. If energy is needed to add an electron to a neutral atom, the resulting negatively charged ion will be unstable and will soon lose the electron. If energy is released when an electron is added to a neutral atom, the resulting negatively charged ion will be stable. An equation for the generalized reaction in which a stable ion is formed is shown here.

$$A(g) + e^- \rightarrow A^-(g) + energy$$

The energy that is either absorbed or released during the addition of an electron to a neutral atom is called the atom's **electron affinity**. **Figure 1.20** shows a partial periodic table of most of the main-group elements, with electron affinities of elements that become stable ions after gaining an electron. The electron affinities are expressed as negative values because energy is released in the formation of the ion. As the values become more negative, the ions become more stable. The elements with "unstable" in place of a numerical value are the ones that have a positive electron affinity.

electron affinity the energy absorbed or released when an electron is added to a neutral atom

Figure 1.20 The values in this table represent electron affinities. The values must be multiplied by 10^{-19} to get the correct values in Joules.

1			13	14	15	16	17	18
1 H −1.21								**2** He unstable
3 Li −0.989	**4** Be unstable		**5** B −0.448	**6** C −2.02	**7** N unstable	**8** O −2.34	**9** F −5.44	**10** Ne unstable
11 Na −0.877	**12** Mg unstable		**13** Al −0.683	**14** Si −2.22	**15** P −1.19	**16** S −3.32	**17** Cl −5.78	**18** Ar unstable
19 K −0.802	**20** Ca −0.00393		**31** Ga −0.689	**32** Ge −1.97	**33** As −1.30	**34** Se −3.24	**35** Br −5.39	**36** Kr unstable
37 Rb −0.778	**38** Sr −0.00769		**49** In −0.481	**50** Sn −1.78	**51** Sb −1.68	**52** Te −3.16	**53** I −4.90	**54** Xe unstable
55 Cs −0.756	**56** Ba −0.232		**81** Tl −0.320	**82** Pb −0.583	**83** Bi −1.51	**84** Po −3.04	**85** At −4.49	**86** Rn unstable

Trends in Electron Affinity

Analyze the electron affinities in **Figure 1.20**, and look for trends. In general, although not always, the electron affinities become increasingly negative going across a period and up a group. Also, notice that the noble gases do not form a stable ion if an electron is added. This is understandable when you consider that the outer shell of a noble gas is a filled shell. An added electron would be an unpaired electron in a higher shell, which would be very unstable. Next, consider the halogens. This group has the most negative electron affinities. Compared with all the other groups, the largest amount of energy is released when an electron is added, indicating that the ion is very stable. Remember that atoms of the halogens have seven electrons in their outer shell. The addition of one electron fills this shell. As a result, the ion is quite stable.

Electronegativity

Consider what happens when an atom of one element has lost an electron and an atom of another element has gained this electron. The result is oppositely charged ions, which attract each other. **Figure 1.21 (A)** shows a negatively charged bromide ion beside a positively charged potassium ion. The arrows represent the forces acting on the electrons. Notice that the electron lost by the potassium atom and gained by the bromine atom is attracted by both positive nuclei. Therefore, the overall effect is similar to shared electrons, although they are not truly shared. Recall that atoms of some elements do share electrons equally. **Figure 1.21 (B)** represents two chlorine atoms sharing two electrons. The arrows indicate that both electrons are attracted by both positively charged nuclei.

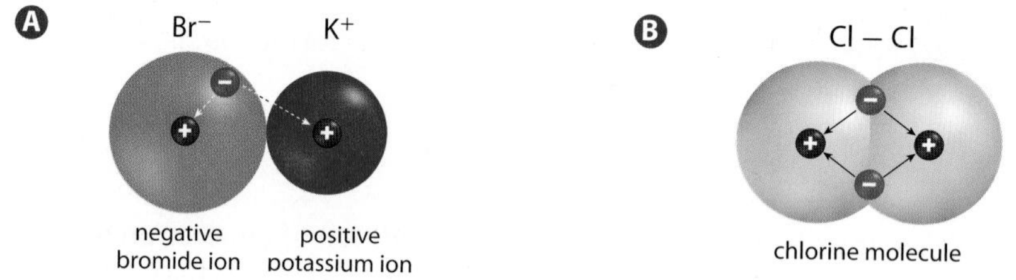

Figure 1.21 (A) When oppositely charged ions attract each other, their positively charged nuclei attract each other's electrons. **(B)** The nuclei of both atoms in the molecule attract the shared electrons.

electronegativity an indicator of the relative ability of an atom to attract shared electrons

The interactions in **Figure 1.21** lead to another important property of atoms, called electronegativity. An element's **electronegativity** is an indicator of the relative ability of an atom of this element to attract shared electrons. Because electronegativity is a relative value, it typically has no units. **Figure 1.22** shows a periodic table with electronegativities. Notice that Group 18 elements have no values. The noble gases do not share electrons with other atoms, so they do not have values for electronegativity.

Electronegativity Values

electronegativity < 1.0
$1.0 \leq$ electronegativity < 2.0
$2.0 \leq$ electronegativity < 3.0
$3.0 \leq$ electronegativity < 4.0

1 H 2.2																	2 He
3 Li 1.0	4 Be 1.6											5 B 2.0	6 C 2.6	7 N 3.0	8 O 3.4	9 F 4.0	10 Ne
11 Na 0.9	12 Mg 1.3											13 Al 1.6	14 Si 1.9	15 P 2.2	16 S 2.6	17 Cl 3.2	18 Ar
19 K 0.8	20 Ca 1.0	21 Sc 1.4	22 Ti 1.5	23 V 1.6	24 Cr 1.7	25 Mn 1.6	26 Fe 1.8	27 Co 1.9	28 Ni 1.9	29 Cu 1.9	30 Zn 1.7	31 Ga 1.8	32 Ge 2.0	33 As 2.2	34 Se 2.6	35 Br 3.0	36 Kr
37 Rb 0.8	38 Sr 1.0	39 Y 1.2	40 Zr 1.3	41 Nb 1.6	42 Mo 2.2	43 Tc 2.1	44 Ru 2.2	45 Rh 2.3	46 Pd 2.2	47 Ag 1.9	48 Cd 1.7	49 In 1.8	50 Sn 2.0	51 Sb 2.1	52 Te 2.1	53 I 2.7	54 Xe
55 Cs 0.8	56 Ba 0.9	57 La 1.1	72 Hf 1.3	73 Ta 1.5	74 W 1.7	75 Re 1.9	76 Os 2.2	77 Ir 2.2	78 Pt 2.2	79 Au 2.4	80 Hg 1.9	81 Tl 1.8	82 Pb 1.8	83 Bi 1.9	84 Po 2.0	85 At 2.2	86 Rn
87 Fr 0.7	88 Ra 0.9	89 Ac 1.1	104 Rf	105 Db	106 Sg	107 Bh	108 Hs	109 Mt	110 Ds	111 Rg	112 Cn	113 Uut	114 Uuq	115 Uup	116 Uuh		118 Uuo

Figure 1.22
Electronegativity is relative, so units are often not used. However, some chemists use the unit Paulings in honour of Linus Pauling, a Nobel prize-winning chemist.

***Identify** the element with the greatest electronegativity and the element with the smallest electronegativity. **Describe** their relationship with each other in the periodic table.*

Analyzing Trends in Electronegativity

Analyzing **Figure 1.22** reveals that the electronegativity of the elements increases going up a group and going from left to right across a period. Thus, fluorine has the largest electronegativity, and francium has the smallest. To discover one of the reasons for these trends, examine **Figure 1.23**. It shows a silicon atom beside a cesium atom, and a silicon atom beside a fluorine atom.

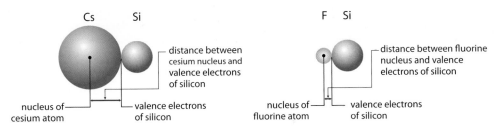

Figure 1.23 As you can see, the nucleus of a small atom can get much closer to the outer-shell electrons of another atom than the nucleus of a larger atom can.

Predict how the distance of the nucleus of an atom from the electrons of an adjacent atom will affect the strength of the attraction between that nucleus and the outer electrons of the adjacent atom.

The elements in **Figure 1.23** were chosen because silicon is a medium-sized atom, cesium is very large, and fluorine is very small. Notice how far the nucleus of the cesium atom is from the outer electrons of the silicon atom and how close the nucleus of the fluorine atom is to the outer electrons of the silicon atom. Because the positively charged nucleus of a small atom can get much closer to the electrons of another atom, this nucleus should be able to exert a stronger attractive force on those electrons. This point raises a new question: Is atomic radius a critical factor in determining the electronegativity of atoms? Examine **Figure 1.24** to find out. The heights of the columns show the electronegativities of the elements, while the spheres on top of the columns show the relative sizes of the atoms. The correlation between size and electronegativity appears to be strong. As the atom becomes smaller, its electronegativity becomes larger.

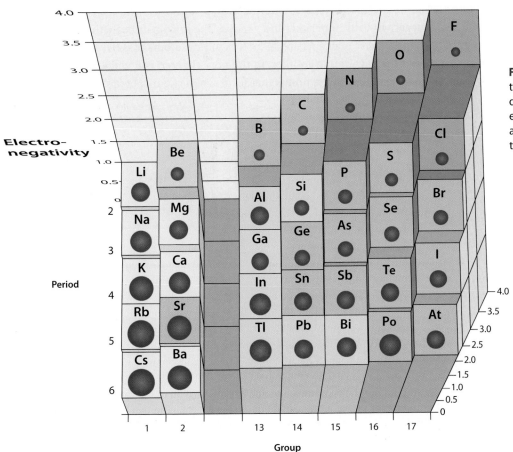

Figure 1.24 This periodic table is shown in three dimensions, so you can easily compare the sizes and electronegativities of the elements.

The halogens are a group of reactive non-metals. Although the atoms all have seven electrons in their valence shell, their properties vary somewhat. The following table lists values for several properties of the halogens.

Properties of Halogens

Halogen	Atomic Radius ($\times 10^{-12}$ m)	Ionization Energy ($\times 10^{-18}$ J)	Electron Affinity ($\times 10^{-19}$ J)	Electro-negativity	Reactivity
fluorine	72	2.79	−5.44	3.98	Highest
chlorine	100	2.08	−5.78	3.16	↓
bromine	114	1.89	−5.39	2.96	↓
iodine	133	1.67	−4.90	2.66	↓
astatine	140	1.53	−4.49	2.2	Lowest

Procedure

1. Before plotting points for graphs, predict what they will look like, by sketching graphs of (A) atomic radius, (B) ionization energy, (C) electron affinity, and (D) electronegativity versus atomic number, for the halogens.

2. On four separate graphs, plot the data for atomic radius, ionization energy, electron affinity, and electronegativity against atomic number.

3. Describe any trends that you see.

Questions

1. Compare your sketches with the graphs plotted from the data. Describe any differences.

2. Suggest a possible explanation for each trend that you listed in step 3, based on what you know about atoms.

3. Suggest the property (or properties) that might be most responsible for the trend in reactivity of the halogens. Explain your reasoning.

Summarizing Trends in the Periodic Table

Figure 1.25 summarizes the trends in atomic radius, ionization energy, electron affinity, and electronegativity within periods and groups. The arrows indicate the direction of increasing values. These concepts are all factors in how atoms of different elements react, or do not react, with one another. In the next chapter, you will learn about bond formation and the nature of the bonds that form when chemical reactions occur.

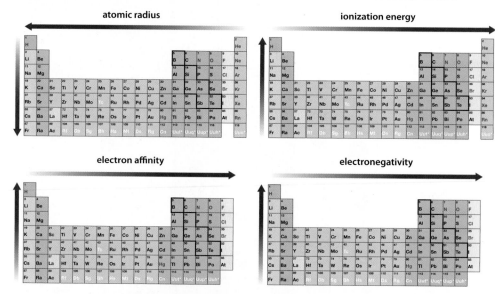

Figure 1.25 You can use these miniature periodic tables as a quick reference when you need to remember the trends in the groups and periods of elements. The two bottom tables do not include Group 18 because the noble gases do not have values for these properties.

*Identify the three properties that follow the same trends. **State** which property follows trends opposite to the other three.*

CHEMISTRY Connections

Elements of the Body

Every time you eat a sandwich or take a breath of air, you are taking in elements that your body needs to function normally. These elements have specific properties, depending on their location in the periodic table. The circle graph below shows the percent by mass of elements in cells in the human body.

OXYGEN In an adult body, there are more than 14 billion billion billion oxygen atoms! Without a constant input of oxygen into the blood, the human body could die in just a few minutes.

CARBON Carbon can form strong bonds with itself and other elements. Carbon forms the long-chained carbon backbones that are an essential part of biological molecules such as carbohydrates, proteins, and lipids. The DNA molecule that determines your characteristics relies on the versatility of carbon and its ability to bond with many different elements.

HYDROGEN Although there are more hydrogen atoms in the human body than there are atoms of all the other elements combined, hydrogen represents only 10 percent of the composition by mass because of its significantly lower mass. Your body requires hydrogen in a variety of essential compounds, like water, rather than in its elemental form. With oxygen and carbon, hydrogen is also a crucial component of carbohydrates and other biological molecules that your body needs for energy.

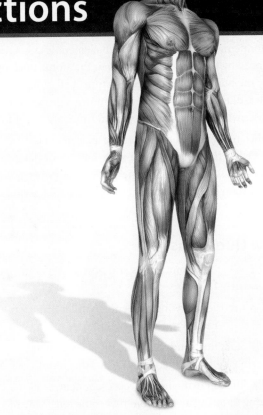

The entire human body is covered with muscle tissue.

NITROGEN As shown in the illustration above, the human body is entirely covered with muscle tissue. Nitrogen atoms are found in the compounds that make up the muscle protein.

OTHER ELEMENTS IN THE BODY Oxygen, carbon, hydrogen, and nitrogen are the most abundant elements in your body and yet are but a few of the elements that your body needs to live and grow. Trace elements, which together make up less than 2 percent of the body's mass, are a critical part of your body. Your bones and teeth could not grow without the constant intake of calcium. Although sulfur comprises less than 1 percent of the human body by mass, it is an essential component in some of the proteins, such as those in your fingernails. Sodium and potassium are crucial for the transmission of electrical signals in your brain.

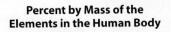

Percent by Mass of the Elements in the Human Body

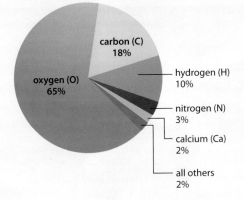

carbon (C) 18%

hydrogen (H) 10%

oxygen (O) 65%

nitrogen (N) 3%

calcium (Ca) 2%

all others 2%

The human body is composed of many different elements.

Connect to Society

Many food manufacturers add sulfites to their products. What are the benefits of adding sulfites to food? How much of the sulfites are needed to provide these benefits? What are the risks of adding sulfites to food? Do sulfites occur naturally in any foods? Carry out research to answer these questions. Then state why you do or do not think that sulfites should be used as a food additive.

Section Summary

- The atomic radius of the atoms of an element is influenced by the amount of charge in the atom's nucleus and by the number of occupied electron shells. The atomic radius increases when going down a group and decreases when going across a period from left to right.

- The ionization energy of the atoms of an element is influenced by the distance between its outermost electron and its nucleus. The ionization energy decreases when going down a group and increases when going across a period.

- The electron affinity of the atoms of an element is influenced by whether the valence shell of the atoms is filled. The electron affinity decreases when going down a group and increases when going across a period.

- The electronegativity of the atoms of an element is influenced by the atomic radius. The electronegativity decreases when going down a group and increases when going across a period.

Review Questions

1. **K/U** What is atomic radius and how is this value obtained?

2. **T/I** List the following elements in order of increasing radius: Ba, Cs, O, Sb, Sn.

3. **K/U** State the trend in atomic radius within a group. Describe the factor that accounts for this trend.

4. **T/I** The following figures represent neutral atoms gaining or losing an electron to become ions. One of these figures represents a chlorine atom, Cl, gaining an electron to become a negatively charged ion, Cl^-. The other figure represents a sodium atom, Na, losing an electron to become a positively charged ion, Na^+. Identify which figure represents each element. Provide possible explanations for the change in size upon ionization. **Note:** The unit pm (picometre) is 10^{-12} m.

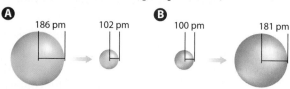

A 186 pm → 102 pm   **B** 100 pm → 181 pm

5. **K/U** Define ionization energy. Use a chemical equation to clarify your definition.

6. **T/I** Review the graph of First Ionization Energies in **Figure 1.19** on page 35. Notice that the differences in ionization energy for the first 20 elements is quite significant. Then, for elements 20 to 30, the values are very similar. Provide a possible explanation for the similarity in these values.

7. **T/I** Write a chemical equation for an atom losing a third electron. Would the ionization energy for this reaction be larger or smaller than the ionization energy for the loss of the first or second electron? Explain.

8. **T/I** Explain why helium does not have a third ionization energy.

9. **T/I** List the following elements in the order of increasing ionization energy for the first ionization: arsenic, cesium, fluorine, helium, phosphorus, strontium.

10. **K/U** What is the significance of an element having a positive electron affinity? Is the resulting ion stable or unstable? Explain why.

11. **K/U** Atoms of what group have the most negative electron affinities? Explain why.

12. **C** Using diagrams, show how electronegativity and electron affinity are different.

13. **K/U** Explain the relationship between electronegativity and the size of an atom.

14. **T/I** Copy the following table in your notebook. Fill in the data for atomic mass, using the periodic table in **Figure 1.12** on page 24. Fill in the data for atomic radius, using **Figure 1.18** on page 33. Compare the data for the two elements. Would a piece of aluminum or a piece of lead be denser? Explain your reasoning.

Data for Aluminum and Lead

Element	Atomic Mass	Atomic Radius
aluminum		
lead		

15. **A** Use the data in **Figure 1.19** and **Figure 1.20** on page 35 to answer the following questions. First ionization data can also be found in the data table in Investigation 1-B on page 42.

   a. What would the total change in energy be to remove an electron from a sodium atom and add the electron to a chlorine atom?

   b. Which of the following combinations would be more stable, a sodium atom and a chlorine atom, or a positively charged sodium ion and a negatively charged chloride ion? Explain your reasoning.

Safety Precautions

- Wear safety eyewear, gloves, and lab coat or apron.
- Hydrochloric acid is corrosive. Avoid contact with skin.
- Use tongs to handle materials.
- Clean up all spills immediately.
- Follow teacher's directions for disposal of chemicals.

Materials

- MSDS for each chemical
- metal strips of copper, $Cu(s)$; iron, $Fe(s)$; magnesium, $Mg(s)$; zinc, $Zn(s)$
- dilute hydrochloric acid, $HCl(aq)$
- grease marker
- 4 small test tubes
- test tube rack
- safety equipment
- tongs
- emery cloth or sandpaper

Developing an Activity Series

Many metals are known to react with dilute acids, resulting in the release of hydrogen gas. Some metals react much more vigorously than others. In this investigation, you will test four metals, supplied by your teacher, for their relative reactivity. The list of metals in the order of most reactive to least reactive is called an activity series.

Pre-Lab Questions

1. When observing a reaction occurring in a test tube, what tells you that a gas is being formed? How would you compare two reaction tubes to determine the one in which the reaction is happening more rapidly?

2. What factors affect the reactivity of metals?

3. What is the WHMIS symbol for acids? Describe the precautions you would need to take when working with acids.

Question

What is the order of reactivity of copper, iron, magnesium, and zinc with acid?

Procedure

1. Gently clean the surfaces of the metal strips with emery cloth or sandpaper.

2. Label the four test tubes by placing the name of one of the metals on each tube.

3. Pour dilute hydrochloric acid to a depth of about 1 cm in each of the four test tubes.

4. One at a time, place the appropriate metal strip in the test tube. Carefully observe any reaction that might be occurring. After 1 min, record your observations. Include information about the rate of the reaction.

5. Dispose of the chemicals as directed by your teacher.

Analyze and Interpret

1. What criteria did you use to determine the rate of a reaction?

2. List the metals in the order of the rate of the reaction, from the fastest reaction to the slowest.

Conclude and Communicate

3. Why do you think your list of metals is called an activity series?

4. Compare your activity series with that of other groups in your class. Offer possible explanations for any differences.

Extend Further

5. **INQUIRY** Design an investigation that you could use to test the same four metals for reactivity to verify your results for this investigation.

6. **RESEARCH** Carry out research to find a use for metals that depends on their reactivity or lack of reactivity.

ThoughtLab
INVESTIGATION

1-B

Skill Check

Initiating and Planning

✓ Performing and Recording

✓ Analyzing and Interpreting

✓ Communicating

Materials

- graph of ionization energies with data points only
- graph paper
- coloured pens or pencils
- data table on this page

Analyzing Ionization Energy Data

Figure 1.19, on page 35, shows data for ionization energies. You can obtain a lot of information from the graph in this figure. In this investigation, you will analyze the graph in **Figure 1.19** in more detail. You will then plot your own graphs from data tables and make more inferences based on the data.

Pre-Lab Questions

1. Review what you have learned about ionization energies in this chapter. Explain the meaning of "ionization energy."

2. Define "second ionization energy."

3. Review the electron configurations in **Figure 1.4** on page 13. Which atoms have a filled valence shell?

4. What is the significance of a filled valence shell?

Question

What characteristics of an atom or ion cause its ionization energy to be very high or very low?

First and Second Ionization Energies (× 10⁻¹⁸ J) for Elements 1 through 20

Z	Element	First Ionization Energy	Second Ionization Energy
1	H	2.18	N/A
2	He	3.94	8.71
3	Li	0.864	12.1
4	Be	1.49	2.92
5	B	1.33	4.04
6	C	1.80	3.90
7	N	2.32	4.75
8	O	2.18	5.63
9	F	2.79	5.60
10	Ne	3.54	6.56
11	Na	0.823	7.57
12	Mg	1.23	2.41
13	Al	0.960	3.02
14	Si	1.31	2.62
15	P	1.68	3.16
16	S	1.66	3.74
17	Cl	2.08	3.81
18	Ar	2.53	4.43
19	K	0.695	5.07
20	Ca	0.979	1.90

Organize the Data

Part 1: Analyzing First Ionization Energies

1. Your teacher will give you a graph of data points for the first ionization energies of the first 60 elements. This graph is similar to the graph on page 35, **Figure 1.19**, but it does not have lines between the data points.

2. Using a coloured pen or pencil, connect the data points for all of the Group 18 elements.

3. Using the same coloured pen or pencil, connect the data points for all of the Group 1 elements.

4. Describe any trends you see in the lines you drew.

5. With a different-coloured pen or pencil, connect the data points for the elements in Periods 1, 2, 3, 4, and 5.

6. Describe any trends that you see in the lines for the elements in each period.

7. On the graph for Periods 4 and 5, there are clusters of data points representing ionization energies that are very close together. Where in the periodic table would you find the elements with the ionization energies represented by these data points?

Part 2: Comparing First and Second Ionization Energies

8. On a clean sheet of graph paper, plot the data in the table of first and second ionization energies on the previous page. Use different-coloured pens or pencils for the two sets of data. Connect the points for each set of data. Note that there is no second ionization energy for hydrogen because hydrogen has only one electron.

9. Label the peaks and valleys with the chemical symbols of the elements, similar to the way the graph in **Figure 1.19** on page 35 is labelled.

10. Draw a partial periodic table, similar to the one in **Figure 1.4** on page 13. Instead of drawing Bohr-Rutherford diagrams for the neutral atoms, draw Bohr-Rutherford diagrams for ions that form when one electron is removed. These diagrams will show the electron configurations of the ions that are formed when one electron has been removed from the atoms. These are the ions from which a second electron will be removed in order to measure the second ionization energy.

Analyze and Interpret

1. Suggest possible reasons for the trends that you described in step 4. These are the trends for the first ionization energies of Group 1 and Group 18 elements.

2. Suggest possible reasons for the trends you described in step 6, for the first ionization energies of elements within a period.

3. What can you infer about the elements with the clustered ionization energies in step 7?

4. Describe any similarities and differences between the graphs for the first ionization energies and the second ionization energies.

5. List the elements that are represented by the peaks in the graph of first ionization energies. Study the electron configurations of these elements in **Figure 1.4**.

   a. What is common to the electron configurations of these elements?

   b. What can you infer about ionization energies and the electron configurations of these elements?

6. Analyze the elements that are represented by the low points or valleys in the graph of first ionization energies in the same way that you analyzed the elements that are represented by the peaks in question 5.

7. Refer to your graph of second ionization energies and to the partial periodic table that you drew to answer question 10. Analyze these data in the same way that you analyzed data for first ionization energies in questions 5 and 6.

Conclude and Communicate

8. Write a paragraph that summarizes the concepts that you learned about electron configurations and the amount of energy needed to remove an electron from an atom or ion.

Extend Further

9. **INQUIRY** Data are available for third, fourth, fifth, and further ionization energies. Develop an inquiry question that you could answer by analyzing these data. Explain how you would use these data to answer your question.

10. **RESEARCH** Use print or Internet resources to research one situation in which data about ionization energies are used in an application.

Chapter 1 | SUMMARY

Section 1.1   The Nature of Atoms

As chemists learned more and more about the nature of atoms, they developed models and symbols to help them understand and communicate about atoms.

KEY TERMS

atomic mass unit
electron pairs
isotopes
isotopic abundance

Lewis diagram
radioisotope
unpaired electrons
valence electrons

KEY CONCEPTS

- As scientists discovered new information about the atom, they modified the model of the atom to reflect each piece of new information.

- The number of electrons that can occupy an electron shell can be determined by using $2n^2$, where n is the shell number.

- The electrons in the outermost shell of an atom are called valence electrons.

- A Lewis diagram is a simplified method for representing an atom.

- The nucleus of an atom contains protons and neutrons and occupies a very small volume at the centre of the atom.

- An appropriate ratio of neutrons to protons stabilizes the nucleus. An atom with an unstable nucleus is called a radioisotope.

- Atomic masses that are reported in data tables are weighted averages, based on isotopic abundances.

Section 1.2   The Periodic Table

Elements in the periodic table can be classified in several different ways, each emphasizing certain properties of the elements.

KEY TERMS

group
period
periodic law

KEY CONCEPTS

- Mendeleev developed the periodic table by first listing the elements in order of increasing mass. Then, when the chemical properties of an element closely resembled those of a previous element in the list, he put this element in line with the previous element. With a little adjusting,

Mendeleev had a table of elements with increasing mass in the columns and similar properties in the rows.

- Elements are categorized as main-group, transition, and inner transition elements, based on their electron configurations.

- Elements are categorized as metals, metalloids, and non-metals, based on their chemical properties.

- Several groups of elements are given specific names due to their uniquely similar properties. Group 1 elements are called the alkali metals, Group 2 elements are called the alkaline earth metals, Group 17 elements are called the halogens, and Group 18 elements are called the noble gases.

- Many alternative forms of the periodic table have been developed to highlight specific properties of the elements.

Section 1.3   Explaining Periodic Trends

By the placement of an element in the periodic table, you can predict the size, ionization energy, electron affinity, and electronegativity of the element, relative to the other elements.

KEY TERMS

atomic radius
effective nuclear charge
electron affinity

electronegativity
ionization energy

KEY CONCEPTS

- The atomic radius of the atoms of an element is influenced by the amount of charge in the atom's nucleus and by the number of occupied electron shells. The atomic radius increases when going down a group and decreases when going across a period from left to right.

- The ionization energy of the atoms of an element is influenced by the distance between its outermost electron and its nucleus. The ionization energy decreases when going down a group and increases when going across a period.

- The electron affinity of the atoms of an element is influenced by whether the valence shell of the atoms is filled. The electron affinity decreases when going down a group and increases when going across a period.

- The electronegativity of the atoms of an element is influenced by the atomic radius. The electronegativity decreases when going down a group and increases when going across a period.

Knowledge and Understanding

Select the letter of the best answer below.

1. Which of the following is the correct Lewis diagram for carbon?

   a.
   b. C
   c. • • C •
   d. $_6^{12}$C
   e. •• •• C ••

2. What is the maximum number of electrons allowed in the third electron shell of an atom?

   a. 18   c. 32   b. 8   d. 2   e. 64

3. The effective nuclear charge is

   a. equal to the sum of the charges of the protons in the nucleus

   b. equal to the sum of the charges of the protons in the nucleus minus the sum of the electrons in the outer shell

   c. less than the sum of the charges of the protons in the nucleus due to shielding by the electrons in the outer shell

   d. less than the sum of the charges of the protons in the nucleus due to shielding by the electrons in the lower, filled shells

   e. greater than the sum of the charges on the protons in the nucleus

4. Isotopes are

   a. atoms with the same number of neutrons but different numbers of protons

   b. atoms with the same number of protons but different numbers of neutrons

   c. atoms with the same sum of neutrons plus protons but different numbers of neutrons and of protons

   d. atoms with the same number of protons and electrons

   e. atoms with the same number of neutrons and electrons

5. Radioisotopes are unstable because

   a. there is an equal number of protons and neutrons in the nucleus

   b. the attractive nuclear forces among the neutrons and protons are too small to balance the repulsive forces among the protons

   c. the number of neutrons is greater than the number of protons

   d. the attractive forces among the protons are less than the repulsive forces among the neutrons

   e. the repulsive forces among the electrons are greater than the repulsive forces among the protons

6. Which statement about the size of the atomic radius is correct?

   a. The atomic radius decreases going down a group.

   b. The atomic radius increases going across a period from left to right.

   c. The atomic radius is unrelated to its position in the periodic table.

   d. The atomic radius increases going up a group.

   e. The atomic radius decreases going across a period from left to right.

7. Which statement about ionization energy is true?

   a. The first ionization energy is greater than the second, third, or fourth ionization energy.

   b. The ionization energy is the same for atoms of all elements in the same group.

   c. For elements in a given period, the ionization energy is greatest for atoms of the element with a filled outer electron shell.

   d. Ionization energy decreases going across a group from left to right.

   e. The ionization energy is greatest for atoms of elements that have only one electron in the valence shell of their atoms.

8. Which statement about electronegativity is true?

   a. Electronegativity is the energy change that occurs when an electron is added to an atom.

   b. Electronegativity is an indicator of the degree to which the nucleus of an atom attracts shared electrons.

   c. The electronegativity of atoms decreases going from left to right across a period.

   d. The electronegativity of atoms increases going down a group.

   e. The electronegativity of a noble gas is greater than the electronegativity of the halogen that is in the same period.

Answer the questions below.

9. Explain what is incorrect about each of the following Lewis diagrams. Draw the correct Lewis diagram for each element.

   a. •• Si ⦙   b. •• B •   c. • Na •

10. Describe how Mendeleev used cards, each having the name and properties of an element, to develop his periodic table.

11. Explain what periodicity is as it applies to the elements. That is, how are they periodic?

12. How do chemists describe the atomic radius of an atom? Why is it not the same as the radius of a circular object such as a coin, or the distance from the centre to the outer edge of the object?

13. Explain the difference between electronegativity and electron affinity.

14. Sketch the following diagram of a periodic table. Outline and label the main-group, transition, and inner transition elements with a coloured pen or pencil. With a different-coloured pen or pencil, outline and label the alkali metals, alkaline earth metals, halogens, noble gases, lanthanoids, and actinoids.

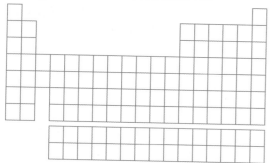

15. Write the general chemical equation that defines the second ionization energy.

16. If the electron affinity of an element is negative, what does this mean about the resulting ion?

17. What metalloids are in Period 5?

Thinking and Investigation

18. Nuclear reactors create highly unstable, or radioactive, waste. This waste is formed when uranium nuclei in the fuel fission, or split, into smaller nuclei. For example, a uranium-235 ($^{235}_{92}U$) nucleus might split into a strontium-95 ($^{95}_{38}Sr$) nucleus and a xenon-137 ($^{137}_{54}Xe$) nucleus. Why are strontium-95 and xenon-137 nuclei unstable? Why do you think that any two smaller nuclei that are produced when a uranium-235 nucleus splits will be unstable? Hint: Study **Table 1.2**.

19. Antimony has two commonly occurring isotopes: antimony-121 and antimony-123. Antimony-121 has a mass of 120.9038 u, and its isotopic abundance is 57.30%. Antimony-123 has a mass of 122.9042 u, and its isotopic abundance is 42.70%. What is the average atomic mass of antimony?

20. Without looking up the electronegativities of the elements shown in the periodic table below, answer the following questions.
    a. Which of the four elements shown has the highest electronegativity?
    b. Which of the four elements shown has the lowest electronegativity?
    c. Explain how you were able to answer parts a. and b.

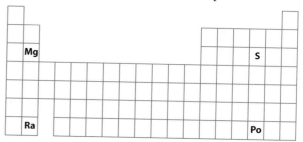

21. The table below is an excerpt from the periodic table, drawn in the modern form. The columns represent groups, and the rows represent periods. The only data in each cell are the chemical symbol and the average atomic weight. The cell with a question mark should contain data for one of the elements that was missing from Mendeleev's periodic table. Three other cells contain data for titanium (Ti), zirconium (Zr), and tantalum (Ta). Titanium is lustrous, conducts electric current, and is ductile. Zirconium is malleable, ductile, and lustrous. Tantalum is strong and very ductile, and conducts electric current. Predict the properties and the approximate atomic mass of the missing element.

Ti 47.88	
Zr 91.22	
?	Ta 180.9

22. When an alkali metal loses an electron and becomes a positively charged ion, its atomic radius decreases dramatically. For example, the radius of a potassium atom, K, is 2.27×10^{-10} m. The radius of a potassium ion, K^+, is 1.38×10^{-10} m. Think about the electron configuration of the alkali metals, and suggest a reason for this significant difference in the atomic radius of a neutral atom and the atomic radius of its positively charged ion.

23. What characteristic of the electron configuration of halogens explains their large negative electron affinities?

24. Draw a graph of electronegativity versus atomic number, using the values in the periodic table in **Figure 1.22** on page 36. Connect the points for the elements of each period with a different-coloured pencil or pen. Examine the graph and answer the following questions.
   a. Describe any forms of periodicity that you observe.
   b. Describe the trends that you observe within any given period.
   c. Describe the trends that you observe within any given group.
   d. Explain the reasons for the trends based on the properties of the elements.

Communication

25. Describe the connection between the radius that Bohr calculated for electrons around the nucleus of a hydrogen atom and the electron cloud that represents the solution to the Schrödinger wave equation.

26. **BIG IDEAS** Every element has predictable chemical and physical properties determined by its electron configuration. Choose an element from Group 1, Group 2, or Group 17. Draw the Bohr-Rutherford diagram for that element and use it as the centre of a spider map with at least four or more legs. For each leg, state a characteristic of that element and relate that characteristic to the electron configuration of the element.

27. Use labelled diagrams to explain how the Bohr model of the atom improved on the model that Rutherford had developed. What new information did Bohr discover that had not been available to Rutherford?

28. Write an e-mail to a classmate who is studying for an exam, explaining how one isotope of magnesium differs from another isotope of magnesium. How are the isotopes the same?

29. Draw a concept map to explain the meaning of "effective nuclear charge." Is the effective nuclear charge of an atom larger or smaller than the actual nuclear charge? How does the effective nuclear charge influence the size of an atom?

30. Use an example to show the difference between a simple average and a weighted average.

31. State the periodic law, and, using a flowchart, describe the observations that led to its development.

32. Prepare an oral presentation for chemistry class in which you discuss some advantages of using a Lewis diagram rather than another type of diagram such as a Bohr-Rutherford diagram or a chemical symbol.

33. Use a diagram to explain how chemists measure the radius of an atom in a solid material, such as a metal.

34. Make a table with the headings Property, Trend Going Down a Group, and Trend Going Across a Period. Under Property, list Atomic Radius, Ionization Energy, Electron Affinity, and Electronegativity. Fill in the table by indicating whether the trend is increasing or decreasing. For each property, write a discussion about the factors that affect that property and why the property follows the trend in your table.

35. Summarize your learning in this chapter using a graphic organizer. To help you, the Chapter 1 Summary lists the Key Terms and Key Concepts. Refer to Using Graphic Organizers in Appendix A to help you decide which graphic organizer to use.

Application

36. The alkali metals are banned from many classrooms. Based on their properties, explain why they are banned.

37. When iodine is taken into the body, it accumulates in the thyroid gland. There, it is used in the synthesis of thyroid hormone. Iodine-131 is a radioactive isotope of iodine that is sometimes used to treat an overactive thyroid. Radioactive substances are usually considered to be dangerous because they damage tissues. Provide a possible reason why this dangerous substance is used for medical purposes.

38. Explain why gold can be used in jewellery, in crowns for teeth, and also as a conductor in electronic devices.

39. Breathing a halogen, such as chlorine or bromine vapour, can seriously harm the nose, throat, and lungs. In contrast, breathing small amounts of a noble gas is not harmful. You might have heard someone talk after breathing helium. Why do you think breathing a halogen is harmful, whereas breathing a noble gas is not?

40. Imagine that you want to pursue a career in computer electronics so you can work with a team that is developing smaller and more efficient microchips. Why would it be important for you to have a strong background in chemistry?

Select the letter of the best answer below.

1. **K/U** The atomic masses for the elements as reported in the periodic table are called relative atomic masses because
   a. the masses of different isotopes are different
   b. they are reported as relative to the mass of carbon-12
   c. the reported mass is an average of the masses of the isotopes
   d. they are reported as relative to the mass of hydrogen-1
   e. they are reported as relative to the atomic number

2. **K/U** In which electron shell is 32 the maximum number of electrons allowed?
   a. first
   b. second
   c. third
   d. fourth
   e. fifth

3. **K/U** How many neutrons does the isotope $^{86}_{36}Kr$ have?
   a. 122
   b. 36
   c. 40
   d. 50
   e. 86

4. **K/U** Elements in Group 2 are called
   a. noble gases
   b. alkaline earth elements
   c. rare earth elements
   d. halogens
   e. alkali elements

5. **K/U** The actinoids and lanthanoids together make up
   a. inner transition elements
   b. alkali elements
   c. transition elements
   d. halogens
   e. main-group elements

6. **K/U** A negative value of electron affinity indicates that when an electron is added to the neutral atom
   a. the resulting ion is unstable
   b. the resulting ion is positively charged
   c. the original atom was unstable
   d. the resulting ion is stable
   e. the original atom was missing an electron

7. **K/U** Which of the following pairs of properties of atoms influences the atomic radius of an atom?
   a. the number of occupied electron shells and the size of the nucleus
   b. the number of electrons in the valence shell and the charge in the nucleus
   c. the number of neutrons in the nucleus and the number of occupied electron shells
   d. the charge in the nucleus and the number of filled electron shells
   e. the charge in the nucleus and the number of occupied electron shells

8. **K/U** Which statement about ionization energy is false?
   a. Ionization energy is the amount of energy required to remove the outermost electron from an atom or ion in the gaseous state.
   b. The ionization energies of Group 1 elements are greater than the ionization energies of the elements in the same period in any other group.
   c. The second ionization energy of a given element is larger than the first ionization energy.
   d. The ionization energies of Group 18 elements are greater than the ionization energies of the elements in the same period in any other group.
   e. Within a period, the ionization energy is largest for atoms or ions that have a filled outer shell.

9. **K/U** Which of the following statements about non-metals is false?
   a. Group 15 elements are all non-metals.
   b. The halogens are non-metals.
   c. The most common elements in living tissues are non-metals.
   d. Non-metals are on the right end of the periodic table.
   e. There are fewer non-metals than there are metals.

10. **K/U** Of the properties listed below, which one is most important in determining the electronegativity of an atom?
    a. the number of neutrons in the nucleus
    b. the size of the atom with which it is sharing an electron
    c. the number of electrons in its valence shell
    d. its atomic number
    e. its atomic radius

Use sentences and diagrams as appropriate to answer the questions below.

11. **K/U** Explain the meaning of isotopic abundance and explain how it affects the value of the relative atomic mass of an element.

12. **K/U** What is the current, most accurate model of the atom and why is it not convenient to use?

13. **T/I** The isotopic abundance of europium-151 is 47.80% and its mass is 150.92 u. The isotopic abundance of europium-153 is 52.20% and its mass is 152.92 u. What is the average atomic mass of europium?

14. **C** Identify the electron pairs and the unpaired electrons in the diagram below. Explain how the role of unpaired electrons differs from the role of electron pairs in the behaviour of an atom.

$$.\ \overset{\displaystyle ..}{\underset{\displaystyle ..}{F}}\ :$$

15. **A** Predict the most reactive element in the periodic table, and explain your choice.

16. **T/I** Silver has two commonly occurring isotopes, silver-107 with an atomic mass of 106.91 u and silver-109 with an atomic mass of 108.90 u. The average atomic mass of silver as reported in the periodic table is 107.87 u. What can you infer about the isotopic abundances of silver-107 and silver-109 from these data?

17. **T/I** Study the graph on the right in which both ionization energy and atomic radius are plotted against atomic number. Based on the graph, answer the following questions.
   a. What trends do you see in ionization energy?
   b. What trends do you see in atomic radius?
   c. Describe the general relationship between the two graphs.
   d. Provide an explanation for the relationship that you described in part c.

18. **K/U** State the periodic law and explain how it is applied in the design of the periodic table.

19. **A** Chlorine is typically used to disinfect water in swimming pools. Bromine is often used to disinfect water in hot tubs. Explain this difference in halogen use.

20. **K/U** Explain why Period 4 is longer than Periods 1, 2, and 3.

21. **K/U** How do chemists define atomic radius?

22. **T/I** Two main-group elements, X and Y, are in the same period. If X has a smaller atomic radius than Y, is X to the left or right of Y in the periodic table? Explain how you made your decision.

23. **T/I** Atoms of Group 1 elements have the lowest first ionization energy of all of the elements but have a larger increase between the first ionization energy and the second ionization energy than other elements have. Explain why this is the case, based on the electronic structure of the neutral atoms and of the ions formed after the first electron is removed.

24. **K/U** Name the group in the periodic table that has the largest negative electron affinities. Describe the electronic structure of the atoms of elements in this group. Explain why these atoms have such large negative electron affinities.

25. **C** Using a sketch, describe the trends in electronegativity going down a group and going from left to right across a period.

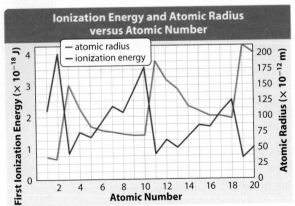

Self-Check

If you missed question …	1	2	3	4	5	6	7	8	9	10	11	12	13	14	15	16	17	18	19	20	21	22	23	24	25
Review section(s)…	1.1	1.1	1.1	1.2	1.2	1.1	1.3	1.3	1.2	1.3	1.1	1.1	1.1	1.1	1.3	1.1	1.3	1.2	1.2	1.2	1.1	1.2	1.3	1.3	1.3

CHAPTER 2

Chemical Bonding

Specific Expectations

In this chapter, you will learn how to . . .

- B1.1 **analyse** on the basis of research, the properties of a commonly used but potentially harmful chemical substance and how that substance affects the environment, and **propose** ways to lessen the harmfulness of the substance or identify alternative substances that could be used for the same purpose (2.3)

- B2.1 **use** appropriate terminology related to chemical bonding (2.1)

- B2.4 **draw** Lewis structures to represent the bonds in ionic and molecular compounds (2.1)

- B2.5 **predict** the nature of a bond, using electronegativity values of atoms (2.1)

- B2.6 **build** molecular models, and **write** structural formulas, for molecular compounds containing single and multiple bonds and for ionic crystalline structures (2.2)

- B2.7 **write** chemical formulas for binary and polyatomic compounds, including those with multiple valences, and name the compounds using the IUPAC nomenclature system (2.2)

- B3.4 **explain** the differences between the formation of ionic bonds and the formation of covalent bonds (2.1)

- B3.5 **compare** and **contrast** the physical properties of ionic and molecular compounds (2.3)

Salt is used in the preparation of many types of food. It is possibly the most common food additive. The technical name for common table salt is sodium chloride. As you learned in Chapter 1, sodium is an alkali metal which is very reactive, and thus dangerous to handle. It reacts vigorously with water. Chlorine gas is toxic. Exposure to even small amounts of this gas can cause irritation to the eyes, nose, and throat. So, why is sodium chloride safe to consume? Sodium and chlorine, in their elemental form, are chemically very different from the compound they form when they are bonded together as sodium chloride. In this chapter, you will learn about the properties of ionic and molecular compounds, including the nature and formation of chemical bonds.

Searching for Clues

In this activity, you will observe four different substances that are normally found in the kitchen. Then you will examine your observations for clues to help you determine the ionic or molecular nature of each compound.

Safety Precautions

- Wear safety eyewear throughout this activity.
- Wear a lab coat or apron throughout this activity.
- Do not taste any materials in a laboratory.

Materials

- table salt
- table sugar
- baking soda
- cornstarch
- distilled water
- watch glass
- magnifying lens
- 5 beakers (100 mL)
- marker and labels
- scoopula
- stirring rod
- conductivity tester

Procedure

1. Place a small amount of table salt in the watch glass, and observe it with the magnifying lens. Note whether the particles have a characteristic shape.

2. Repeat step 1 with the other three solid substances.

3. Label one beaker "control." Label each of the other four beakers with the name of one of the substances. Pour 50 mL of distilled water into each beaker.

4. Add a scoopula of table salt to the appropriate beaker. Stir with the stirring rod, and observe what happens. Observe whether the salt does not dissolve, dissolves slowly, or dissolves quickly.

5. Repeat step 3 with the other three substances. Make sure that you rinse the stirring rod between substances.

6. Test each solution, including the control, for conductivity. Record your results.

7. Based on each test—shape, solubility, and conductivity—predict whether you think it indicates that the substance is an ionic or molecular compound.

Questions

1. Why did you measure the conductivity of the control?

2. Compare all three predictions you made about the nature of each individual compound. Were your three predictions the same or not?

3. Write down your final conclusion about whether each substance is an ionic compound or a molecular compound. Compare your conclusions with the conclusions of the other groups in your class.

4. Which property—shape, solubility, or conductivity—do you think is the best one for predicting whether the compound is ionic or molecular?

The Formation of Ionic and Covalent Bonds

Key Terms

octet rule

ionic bond

ionic compound

covalent bond

molecular compound

single bond

double bond

triple bond

bonding pair

lone pair

Lewis structure

polyatomic ion

polar covalent bond

electronegativity difference

Ninety-two naturally occurring elements combine to form the millions of different compounds that are found in nature. Very few of these elements, however, are found in their elemental form in nature. Some of the elements that are found in their elemental form are the noble gases, as illustrated in **Figure 2.1**. What property of atoms causes them to combine with atoms of other elements? Why are some combinations of elements much more common than others? Answers to these questions are based on the types of bonds that form between atoms of elements. Over the next few pages, you will examine some naturally occurring compounds and look for patterns in these compounds to find clues about the nature of chemical bonds.

Clues in Naturally Occurring Compounds

Scientists often study patterns in nature to better understand scientific concepts. Chemists learn a great deal about the nature of chemical bonds by observing trends in naturally forming compounds. For example, ores are metal compounds that are mined, as shown in **Figure 2.2**, to extract the pure metals. Ores are solid and consist of a metal combined with a non-metal, such as oxygen, sulfur, or a halogen, or with polyatomic ions such as carbonate ions, CO_3^{2-}. Very few metals are found in their elemental form in nature. The few metals that are found in their elemental form, such as gold and silver, are called precious metals. For a compound, such as an ore, to be in solid form, some type of strong attractive force must be holding the individual particles together.

Figure 2.1 The helium that was used to inflate these balloons is a noble gas. Noble gases are some of the very few elements that are found in nature in their elemental form.

Figure 2.2 Ores, consisting of metals combined with non-metals, are sometimes obtained from open pit mines, such as the copper mine shown here.

Clues in the Atmosphere

You can gain more insight into the nature of chemical bonds by examining the atmosphere. It contains the oxygen that you inhale and the carbon dioxide that you exhale. The atmosphere also contains water vapour that condenses to form clouds, which can then become snow or rain. The major component of the atmosphere is nitrogen. As well, there are traces of argon, methane, ozone, and hydrogen. Of these gases, only argon, a noble gas, is found as individual atoms, not bonded to any other atoms. The non-metal elements, oxygen, nitrogen, and hydrogen, are found in the atmosphere as diatomic molecules. This means that they are made up of two identical atoms bonded together. Carbon dioxide, water, and methane are examples of atoms of non-metal elements bonded together.

The following patterns can be discerned from these observations:

- Metals usually form bonds with non-metals. The compounds they form are solid.
- Non-metals can bond with one another to form gases, liquids, or solids.
- The only elements that are *never* found in a combined form in nature are the noble gases.

Stability of Atoms and the Octet Rule

Because atoms of the noble gases are always found as monatomic gases, and because atoms of all other elements are usually found chemically bonded to other atoms, you can infer that there is something very unique about the chemistry of noble gases, which prevents the atoms from forming bonds. Recall, from Chapter 1, that the noble gases are the only elements whose atoms have a filled valence shell, as shown in **Figure 2.3**. This leads to the conclusion that atoms that have filled valence shells do not tend to form chemical bonds with other atoms. Such atoms are referred to as stable.

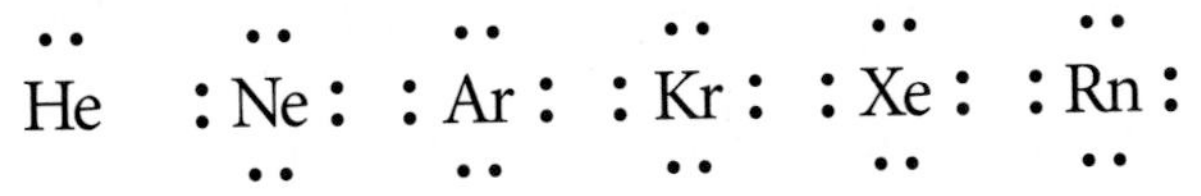

Figure 2.3 Atoms of each of the noble gases except helium have eight electrons in their outer shell, giving them filled valence shells. Because helium is in Period 1, only its first shell, which holds a maximum of two electrons, is occupied. Thus, for helium, two electrons constitute a filled valence shell.

The observation that a filled valence shell makes atoms stable led early chemists to propose that when bonds form between atoms, they do so in a way that gives each atom a filled valence shell. Because, for most main-group elements, a filled valence shell contains eight electrons, this configuration is often called an *octet*. These observations led to the **octet rule** for bond formation, which is stated below.

octet rule a "rule of thumb" that allows you to predict the way in which bonds will form between atoms

> **The Octet Rule**
> When bonds form between atoms, the atoms gain, lose, or share electrons in such a way that they create a filled outer shell containing eight electrons.

As you read in Chapter 1, atoms of the transition elements and inner transition elements can have complex electron configurations. They can have more than eight electrons in their valence shells and, therefore, they do not follow the octet rule. Because main-group elements are much more common on Earth, however, a very large number of compounds that you study will follow the octet rule. Thus, the octet rule provides an important basis on which to predict how bonds will form.

The Formation of Ionic Bonds

ionic bond the attractive electrostatic force between a negative ion and a positive ion

ionic compound a chemical compound composed of ions that are held together by ionic bonds

An **ionic bond** is the attractive electrostatic force between oppositely charged ions. Thus, before an ionic bond can form, atoms must be ionized. According to the octet rule, atoms gain or lose electrons to attain a filled valence shell. In Chapter 1, Section 1.3, you learned that an atom of an element with fewer than four electrons in its valence shell, especially an alkali metal atom, can lose electrons relatively easily. You also learned that an atom with more than four electrons in its valence shell can gain electrons and form a stable ion. Thus, in general, a metal loses all of its valence electrons and becomes an ion with an octet of electrons in its outer shell. A non-metal gains enough electrons to fill its valence shell. These oppositely charged ions exert attractive electrostatic forces on each other, resulting in the formation of an ionic bond. A compound that is held together by ionic bonds is called an **ionic compound**.

Because ionic compounds must have an overall charge of zero, the number of electrons that are lost by the metal atoms must be equal to the number of electrons gained by the non-metal atoms. Two such examples are shown in **Figure 2.4**, in the form of Lewis diagrams. Notice that the electrons of the metals are depicted as open circles and the electrons of the non-metals are depicted as dots, so you can follow them throughout the process.

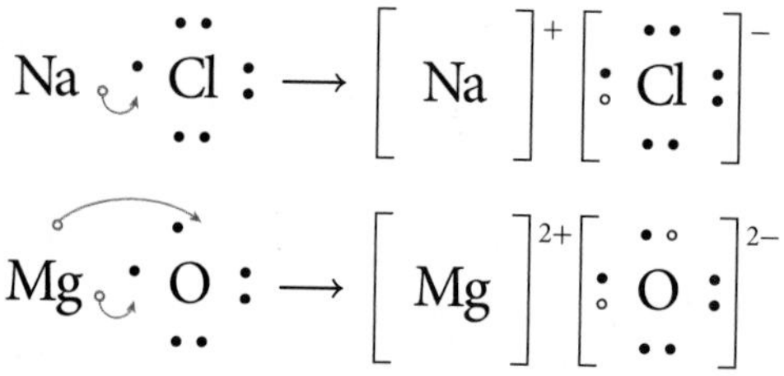

Figure 2.4 When metal atoms, such as sodium and magnesium, lose electrons, they have no valence electrons remaining. Therefore, there are no dots around the symbols for the metal ions.

In each example in **Figure 2.4**, the number of electrons gained by the non-metal atom is exactly the same as the number of electrons lost by the metal atom. It is also possible for the number of electrons gained by a non-metal atom to be different from the number of electrons lost by a metal atom. However, the total number of electrons gained by non-metal atoms must be the same as the total number of electrons lost by metal atoms. Examples of three such situations are shown in **Figure 2.5**.

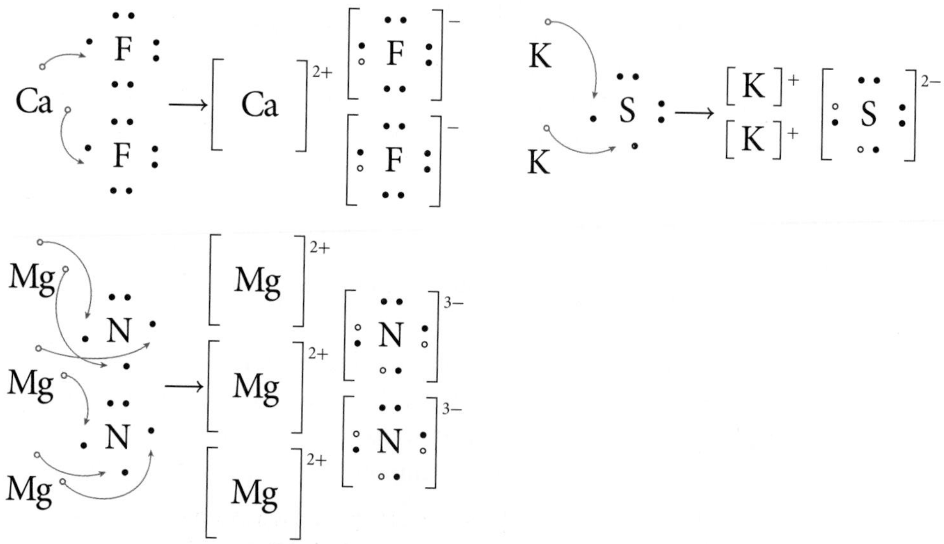

Figure 2.5 In each example, you can see that the total number of positive charges (electrons lost) on the metal ions is equal to the total number of negative charges (electrons gained) on the non-metal ions.

Ionic Compounds Containing Transition Metals

All of the examples in **Figures 2.4** and **2.5** include only main-group elements. You can determine the number of valence electrons of an atom of a main-group element by its group number. Occasionally, however, you will be working with transition metals. In Chapter 1, you read that the electron configuration of transition metals is quite complex. Therefore, it is not possible to predict the number of electrons that a transition metal atom can lose from its group number. In fact, the number of electrons that a transition metal can lose can vary. For example, an iron atom can lose either two electrons or three electrons. You can find the number of electrons that atoms of a transition element can lose by checking the periodic table. **Figure 2.6** shows you how to find the possible charges on the resulting ions after the metal atoms have become ionized. Notice that the possible charges on the ions are highlighted. The figure shows a few common transition metals that can form more than one possible ion. As stated above, iron atoms can lose two or three electrons. Thus iron atoms can form ions with charges of 2+ or 3+.

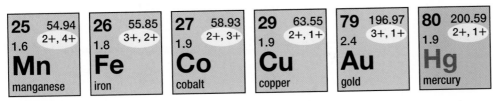

Figure 2.6 These cells are taken directly from the periodic table on page 24. The common ion charges are highlighted.

When you are working with transition metals, you will be given the charge or enough information to determine the charge on the ions. For example, you might be told that two iron atoms have combined with three oxygen atoms and asked to draw a Lewis diagram of the compound. If you do not know that the oxygen ion has a charge of 2−, you can find it in the periodic table. Since there are three oxygen ions, the total negative charge in the compound will be 6−. Thus, the total positive charge on the two iron ions must be 6+. Therefore, there must be a charge of 3+ on each iron ion. The Lewis diagram for this compound is shown in **Figure 2.7 (A)**.

You might also be asked to draw the Lewis diagram of a compound that contains one iron ion and two chloride ions. Since a chloride ion has a charge of 1−, the single iron ion must have a charge of 2+. The Lewis diagram of this compound is shown in **Figure 2.7 (B)**. It is important to remember that the electron configuration for iron atoms is complex. Iron atoms do not actually have two valence electrons or three valence electrons as shown in **Figure 2.7**. The iron atoms are drawn as though they have either two or three valence electrons only because these are the numbers of electrons that they can lose when they become ionized.

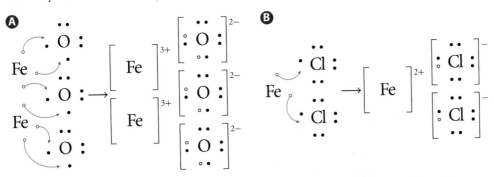

Figure 2.7 (A) Each of the two iron atoms loses three electrons to oxygen atoms. Thus, the resulting ions have a charge of 3+. **(B)** An iron atom loses one electron to one chlorine atom and a second electron to another chlorine atom. The resulting iron ion has a charge of 2+.

The Formation of Covalent Bonds

covalent bond the attraction between atoms that results from the sharing of electrons

molecular compound a chemical compound that is held together by covalent bonds

The octet rule states that atoms can also acquire a filled outer shell by sharing electrons. When the nuclei of two atoms are both attracted to one or more pairs of shared electrons, the attraction is called a **covalent bond**. A compound that is held together by covalent bonds is called a **molecular compound**. Molecular compounds consist of non-metal elements only. Examples of molecular compounds are water and carbon dioxide. Molecular compounds can be solid, liquid, or gas at room temperature.

Only unpaired electrons are likely to participate in chemical bonds. **Figure 2.8** shows how covalent bonds form when **(A)** hydrogen atoms share their only electrons and when **(B)** chlorine atoms share their only unpaired electrons.

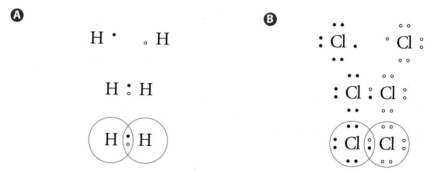

Figure 2.8 In both **(A)** and **(B)**, electrons of one atom are shown as open circles and electrons of the other atom are shown as dots, to help you follow the electrons. In the third row, circles surrounding each atom in the molecule show the filled shell of electrons for each atom.

Multiple Bonds

In some molecules, there are not enough valence electrons for two atoms to share one pair of electrons and form filled valence shells. For example, in carbon dioxide, the carbon atom has four valence electrons and each of the two oxygen atoms has six valence electrons, as shown in **Figure 2.9 (A)**. If the carbon atom shared one pair of electrons with each oxygen atom, each of the oxygen atoms would have only seven electrons and the carbon atom would have only six electrons in the valence shell, as shown in **Figure 2.9 (B)**. This configuration would not provide all atoms with filled outer shells. Instead, to complete an octet for each atom, the unpaired electrons on all atoms are rearranged, as shown in **Figure 2.9 (C)**, to become shared. The atoms now share four electrons as shown in **Figure 2.9 (D)**. It is important to remember that Lewis diagrams are just that—diagrams. It would be more correct to show electron clouds overlapping and forming new electron clouds with different shapes. However, it is more difficult to visualize the number of electrons in valence shells when using the electron cloud model.

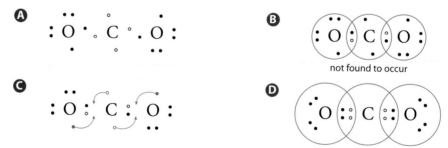

Figure 2.9 **(A)** Each oxygen atom has six valence electrons and the carbon atom has four. **(B)** If each oxygen atom shared two electrons with the carbon atom, neither would have a filled outer shell. **(C)** Instead, the remaining unpaired electrons in both oxygen atoms and the carbon atom are rearranged so that they can also be shared by the atoms. **(D)** When each oxygen atom shares four electrons (two pair) with the carbon atom, all of the atoms acquire an octet of electrons.

Explain how you could predict the number of bonds that an atom could form with other atoms.

Types of Covalent Bonds and Electron Pairs

While one pair of shared electrons constitutes a **single bond**, two pairs of shared electrons make up a **double bond**. Compounds can also have triple bonds, which consist of three pairs of shared electrons. Nitrogen, the gas that makes up most of the atmosphere, is an example of a molecule that has a **triple bond**, as shown in **Figure 2.10**.

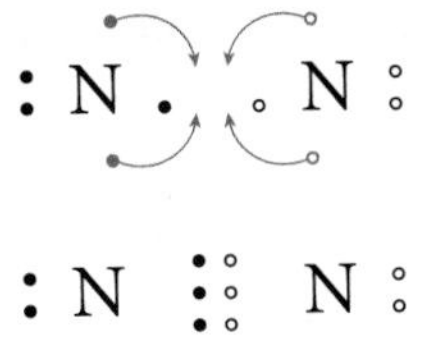

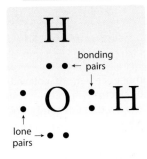

Figure 2.10 Two nitrogen atoms must share three pairs of electrons to complete an octet of electrons around each nitrogen atom.

Although there are rarely any unpaired electrons in molecular compounds, some electron pairs are shared while others are not. A pair of shared electrons is called a **bonding pair**. A pair of electrons that is not involved in a covalent bond is called a **lone pair**. These types of electron pairs are labelled in the water molecule in **Figure 2.11**. When a Lewis diagram is used to portray a complete molecular compound, as done in this figure, the diagram is called a **Lewis structure**.

H

bonding ← pairs

: O : H

lone → pairs

Figure 2.11 The oxygen atom in water has two bonding pairs and two lone pairs. The hydrogen atoms each have one bonding pair.

Drawing Lewis Structures

Although there are no specific steps that you can always follow to draw a Lewis structure, there are some guidelines that will help you. First, draw a Lewis diagram for each atom in the structure. Start with the atom that has the most unpaired electrons. Determine the number of bonds that each atom can form with other atoms. That number is the same as the number of unpaired electrons. Finally, try to fit the atoms together in a way that will create a filled outer shell for each atom. The example used in **Figure 2.12 (A)** has one carbon atom, two oxygen atoms, and two hydrogen atoms. Begin with the carbon atom. If both oxygen atoms are bonded to the carbon atom with double bonds, there will be no way to add the hydrogen atoms. If both hydrogen atoms are bonded to the carbon atom, there will be bonds for only one oxygen atom. The final result is shown in **Figure 2.12 (B)**.

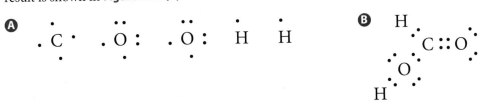

Figure 2.12 If you take the atoms in **(A)** and test different ways of connecting them, you will find that the Lewis structure in **(B)** creates filled outer shells for all of the atoms. This compound is commonly called formic acid, HCOOH.

single bond a covalent bond that results from atoms sharing one pair of electrons

double bond a covalent bond that results from atoms sharing two pairs of electrons

triple bond a covalent bond that results from atoms sharing three pairs of electrons

bonding pair a pair of electrons that is shared by two atoms, thus forming a covalent bond

lone pair a pair of electrons that is not part of a covalent bond

Lewis structure a Lewis diagram that portrays a complete molecular compound

Go to **scienceontario** to find out more

Chapter 2 Chemical Bonding · **NEL** 57

Polyatomic Ions and Bond Formation

polyatomic ion a molecular compound that has an excess or a deficit of electrons, and thus has a charge

When you first look at the structure in **Figure 2.13 (A)**, (ignoring the colour), it appears to be a typical Lewis structure. However, when you count the electrons, you will find a new feature in this compound. Count the number of electrons that are the same colour as the symbol, and you will find that they represent the number of valence electrons that an atom of that element has. The carbon atom has four black electrons, and each oxygen atom has six red electrons. The colour-coded electrons account for all of the valence electrons that are available. If there were no additional electrons, the atoms would not all have filled valence shells. To fill the shells, two electrons were added, as shown in green in **Figure 2.13 (A)**. These two electrons give the compound a negative charge of 2−. Nevertheless, it is a valid Lewis structure. Some molecular compounds, like non-metal atoms, can gain electrons to complete octets on all of their atoms. Such compounds are called **polyatomic ions** because they consist of two or more atoms. The correct diagram for polyatomic ions includes brackets and a number and sign, as shown in **Figure 2.13 (B)**.

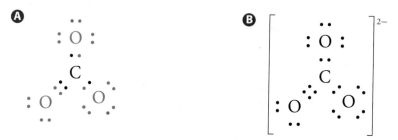

Figure 2.13 **(A)** If you count the number of electrons, you will get 24. Because this is two more electrons than the sum of the valence electrons in three oxygen atoms and one carbon atom, the compound has a charge of 2−. **(B)** To show that this compound is a polyatomic ion, it is bracketed and a 2− is placed outside the brackets.

Typically, electrons in Lewis structures are not colour coded so you cannot easily see whether there are any extra electrons. Nevertheless, you can quickly determine whether a Lewis structure represents a neutral molecular compound or a polyatomic ion by first counting the electrons and comparing that number with the total number of valence electrons that each atom would have. For example, the structure in **Figure 2.13 (A)** has 24 electrons. Add up the number of valence electrons by reasoning that each oxygen atom has six valence electrons and the carbon atom has four valence electrons, giving a total of 22 electrons. You immediately know that you must have two extra electrons, giving you a negatively charged polyatomic ion.

A negatively charged polyatomic ion can bond to a positively charged ion to form an ionic compound in the same way that a metal ion and a non-metal ion can bond to form an ionic compound. **Figure 2.14** shows two examples of ionic compounds that contain the carbonate ion.

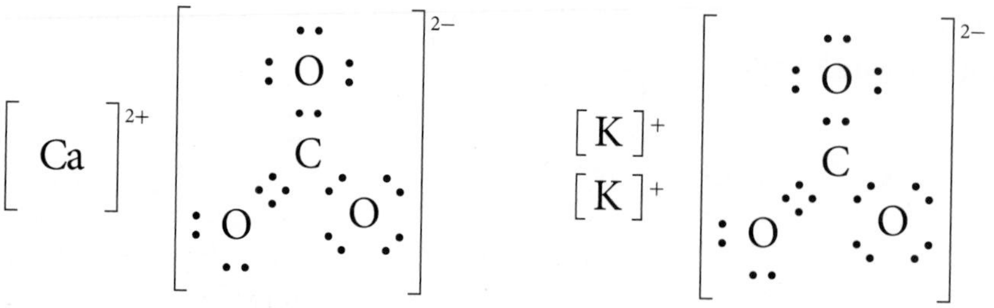

Figure 2.14 Polyatomic ions, like simple ions, must combine with oppositely charged ions that will give the final compound a neutral charge.

Positively Charged Polyatomic Ions

Positively charged polyatomic ions also exist, but the only one that is common is the ammonium ion. The ammonium ion forms when the molecular compound ammonia combines with a hydrogen ion, as shown in **Figure 2.15**. Ammonia has three bonded pairs and one lone pair. The hydrogen ion bonds with the lone pair to form an ammonium ion.

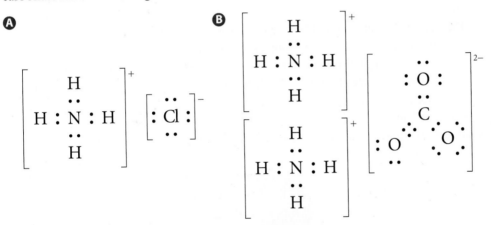

Figure 2.15 The ammonium ion is the only common positively charged polyatomic ion. Try not to confuse it with ammonia, which is a neutral molecular compound.

The ammonium ion forms ionic compounds by bonding with negatively charged ions. It can bond with a simple negatively charged ion, such as the chloride ion, as shown in **Figure 2.16 (A)**. It can also bond with a negatively charged polyatomic ion, such as the carbonate ion shown in **Figure 2.16 (B)**.

A **B**

Figure 2.16 (A) The ammonium ion behaves like any other positively charged ion and forms ionic compounds, such as ammonium chloride, by bonding with negatively charged simple ions. **(B)** The ammonium ion can also form ionic compounds, such as ammonium carbonate, by bonding with negatively charged polyatomic ions.

Learning Check

1. State the octet rule, and give one example of how it can be applied.

2. When a calcium atom becomes ionized, it has a charge of 2+. When a bromine atom becomes ionized, it has a charge of 1−. Explain how ionic bonds can form between calcium and bromine to produce a compound that has a zero net charge.

3. Given a Lewis structure with four non-metal atoms, how would you determine whether it is a molecular compound with no charge or a polyatomic ion?

4. Draw a Lewis structure of two oxygen atoms that are covalently bonded together to form an oxygen molecule. Identify the bonding pairs and the lone pairs.

5. How do double bonds and triple bonds form? Why do they form?

6. Describe a situation in which two atoms that are covalently bonded together can be part of an ionic compound.

The Importance of Electronegativity in Bond Formation

Based on what you just learned about ionic bonds and covalent bonds, you might assume that they are two separate and distinct types of connections between atoms. However, like ionic bonds, covalent bonds also involve electrostatic attractions between positively charged nuclei and negatively charged electrons. To understand how electrostatic attraction influences the nature of bonds, recall the concept of electronegativity, which you learned about in Chapter 1.

Electronegativity is an indicator of the relative ability of an atom of a given element to attract shared electrons. Shared electrons constitute a covalent bond. Thus, the relative electronegativities of the elements of the two atoms that are bonded together should provide information about the nature of the bond. Although Lewis diagrams are drawn as though no electrons are shared between two nuclei in ionic compounds, the positively charged nucleus of each ion is attracting the negatively charged electrons of the other ions. Thus, the concept of electronegativity also applies to ionic compounds.

Electronegativity Difference and Bond Type

What do the relative electronegativities of elements tell you about the nature of bonds? If the electronegativity of one of the two atoms that are bonded together is greater than the electronegativity of the other atom, the electrons will be attracted more strongly to the first atom. In general, electrons spend more time around the atoms with the greater electronegativity.

Figure 2.17 illustrates a bond between a carbon atom and a chlorine atom. Of course, the carbon atom is bonded to other atoms as well as the chlorine atom. The electronegativity of the chlorine atom (3.2) is higher than the electronegativity of the carbon atom (2.6). The arrow indicates that the shared electrons are more strongly attracted to the chlorine atom, and thus spend more time there. The Greek letter delta, δ, is often used to represent "partial." Therefore, the symbols $\delta+$ and $\delta-$ indicate that the carbon atom is partially positively charged and the chorine atom is partially negatively charged.

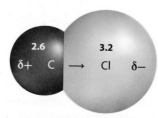

Figure 2.17 Because the shared electrons in this bond spend more time near the chlorine nucleus, the chlorine atom is slightly negatively charged. This leaves the carbon atom slightly positively charged.

Describe *How do you know that the electrons will spend more time near the chlorine atom than the carbon atom? Describe the data that tell you this.*

polar covalent bond
a covalent bond around which there is an uneven distribution of electrons, making one end slightly positively charged and the other end slightly negatively charged

electronegativity difference the difference between the electronegativities of two atoms

Covalent bonds, in which the electron distribution is unequal, are called **polar covalent bonds**. These bonds are often referred to simply as polar bonds. Because these bonds have a positive "pole" and a negative "pole," they are sometimes also called *bond dipoles*. Depending on the difference in the electronegativities of the bonded atoms, some covalent bonds are only slightly polar while others are extremely polar. Chemists have devised a system for classifying the extent of the polarity of the bonds by calculating the **electronegativity difference** (ΔEN) for the two elements involved in the bond. You can calculate the electronegativity difference for any two elements by finding the electronegativity of each element in a table, such as the one in **Figure 1.22** on page 36, and then subtracting the smaller electronegativity from the larger electronegativity.

Applying Electronegativity Difference

As shown in **Figure 2.18**, bonds in which the electronegativity difference of the atoms is greater than 1.7 are classified as *mostly ionic*. The term "mostly" is used because there is always some attraction between the nucleus of one atom and the electrons of the other atom involved in the bond. If the electronegativity difference of two atoms that are bonded together is between 0.4 and 1.7, the bond is classified as *polar covalent*. If the electronegativity difference is less than 0.4, the bond is classified as *slightly polar covalent*. It is only when the electronegativity difference is zero that the bond can be classified as a *non-polar covalent* bond. The images on the right of **Figure 2.18** are electron cloud models of atoms bonded together. The image at the top shows a positive ion and a negative ion beside each other, indicating that the bond is mostly ionic. The next image shows atoms joined by a polar covalent bond. Chemists often use an arrow, like the one above this image, to show a polar bond. The tail of the arrow above this image looks like a plus sign, to signify the slightly positively charged end of the bond. The arrow points in the direction in which the electrons spend more time. The bottom image shows two atoms equally sharing electrons in a non-polar covalent bond.

The following examples show you how to use the information in **Figure 2.18** to calculate the electronegativity difference for two atoms. The first example involves the bond between a potassium atom and a fluorine atom. The electronegativity of fluorine is 4.0, and the electronegativity of potassium is 0.8. The electronegativity difference is calculated by subtracting the smaller number from the larger number, as shown below. Because 3.2 is much larger than 1.7, the bond between potassium and fluorine is mostly ionic.

$$\Delta EN = EN_\text{F} - EN_\text{K}$$
$$\Delta EN = 4.0 - 0.8$$
$$\Delta EN = 3.2$$

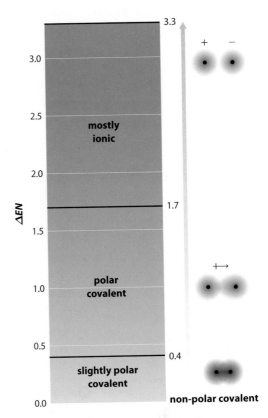

Figure 2.18 The shading in the diagram indicates that the character of bonds changes gradually from mostly ionic at the top to non-polar covalent at the bottom. The electronegativity difference values on the right are the transition points that separate the types of bonds. The images on the far right are models of compounds with the bond character in the different ranges of electronegativity difference.

Next, consider the bond between two oxygen atoms in an oxygen molecule. The electronegativity of oxygen is 3.4. The electronegativity difference, as shown below, is zero. Therefore, the bond between two oxygen atoms is non-polar covalent.

$$\Delta EN = EN_\text{O} - EN_\text{O}$$
$$\Delta EN = 3.4 - 3.4$$
$$\Delta EN = 0.0$$

Finally, consider the bond between a carbon atom and a chlorine atom, discussed on the previous page. The electronegativity of carbon is 2.6, and the electronegativity of chlorine is 3.2. The electronegativity difference is 0.6, as shown below. This value is between 1.7 and 0.4, indicating that the bond is a polar covalent bond.

$$\Delta EN = EN_\text{Cl} - EN_\text{C}$$
$$\Delta EN = 3.2 - 2.6$$
$$\Delta EN = 0.6$$

Percent Ionic and Covalent Character

Chemists have devised another approach for describing the bond character, using percentages of either ionic or covalent character. **Table 2.1** relates electronegativity differences to *percent ionic character* and *percent covalent character*. In the following activity, you will analyze the relationship between electronegativity differences and percent ionic character.

Table 2.1 Character of Bonds

Electronegativity Difference	0.00	0.65	0.94	1.19	1.43	1.67	1.91	2.19	2.54	3.03
Percent Ionic Character	0	10	20	30	40	50	60	70	80	90
Percent Covalent Character	100	90	80	70	60	50	40	30	20	10

Classifying bond type is not always simple. The bond between a hydrogen atom and a chlorine atom provides a good example of overlap in ionic character and covalent character. The electronegativity difference for hydrogen and chlorine is 1.0, placing it in the polar covalent category. As a gas, the compound behaves as a polar molecule. When the compound is dissolved in water, however, the atoms become separate ions, both surrounded by water molecules. Thus, the bond type of this compound varies, depending on whether it is a gas or dissolved in water. **Figure 2.19** shows Lewis diagrams for the two states.

Figure 2.19 The electronegativity difference for hydrogen and chlorine indicates that a bond between these atoms results in a polar covalent molecule when HCl is in a gaseous state (**A**). Its interaction with water molecules causes HCl to behave as an ionic compound (**B**).

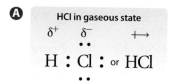
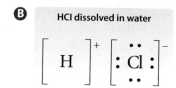

Activity 2.1 — Electronegativity Difference versus Percent Ionic Character

Why did chemists choose the electronegativity differences of 0.4 and 1.7 for the transition points for slightly polar covalent, polar covalent, and mostly ionic bonds? Analyzing the relationship between electronegativity difference and percent ionic character in this activity will help you understand the reasons behind the choice of these values.

Materials
• graph paper   • ruler   • pencil

Procedure

1. Construct a graph using the data in the first two rows of **Table 2.1**. Put electronegativity difference on the *x*-axis and percent ionic character on the *y*-axis. Choose scales for the axes that will make the graph take up more than half of a sheet of graph paper.

2. After you plot all the points, draw a smooth curved line of best fit through the points.

3. Draw a straight, vertical line on the graph through the point where the electronegativity difference is 1.7. At the point at which the vertical line crosses the curve, draw a horizontal line across the graph. Record the value of the percent ionic character at the point where your horizontal line touches the axis.

4. Repeat step 3 for the point where the electronegativity difference is 0.4.

Questions

1. What is the percent ionic character when the electronegativity difference is 1.7? Do you think this is a reasonable value for the transition point between polar covalent and mostly ionic bonds? Explain your reasoning.

2. What is the percent ionic character when the electronegativity difference is 0.4? Do you think this is a reasonable value for the transition point between polar covalent and slightly polar covalent bonds? Explain your reasoning.

3. Why do you think it was important to make your graph spread out to more than half of the sheet of graph paper?

4. Imagine that you were to draw a graph of percent covalent character versus electronegativity difference. Predict the values of percent covalent character that you would find when the electronegativity differences are 0.4 and 1.7. Explain why you think you would find these results.

Section Summary

- The octet rule can be used to predict how bonds will form.
- An ionic bond forms when a negatively charged ion and a positively charged ion are attracted to each other.
- A covalent bond forms when two atoms share one or more pairs of electrons.
- A polyatomic ion consists of two or more atoms that are covalently bonded together and carry a charge. A polyatomic ion can form an ionic compound with a simple ion or another polyatomic ion of the opposite charge.
- A chemical bond can be non-polar covalent, slightly polar covalent, polar covalent, or mostly ionic, depending on the electronegativity difference between the two atoms that are bonded together.

Review Questions

1. **K/U** What property of the noble gases led to the octet rule? Explain.

2. **K/U** Explain why metal atoms tend to lose electrons to form ions and why non-metal atoms tend to gain electrons to form ions.

3. **C** Draw Lewis diagrams of calcium and bromine. Use these diagrams to show how ionic bonds form between these atoms. Explain how these structures satisfy the octet rule.

4. **T/I** For each of the following, use Lewis diagrams to predict the number of atoms of each element that will be present in an ionic compound formed by the two elements.
   a. calcium and fluorine   c. magnesium and nitrogen
   b. sodium and oxygen

5. **T/I** Draw Lewis diagrams of two oxygen atoms. Use your diagrams to show how an oxygen molecule forms from two oxygen atoms. Explain why there must be a double bond between the two oxygen atoms.

6. **K/U** How many electrons make up a triple bond?

7. **K/U** Draw a Lewis structure of a hydrogen atom covalently bonded to a fluorine atom. Identify all the bonding pairs and all the lone pairs.

8. **K/U** Assume that you are shown a Lewis structure with one nitrogen atom and three oxygen atoms. How would you determine whether the structure represented a neutral molecule or a polyatomic ion?

9. **T/I** Explain why the following compound can be considered an ionic compound, even though it does not contain any metal ions.

$$\left[\begin{array}{c} H \\ \overset{\displaystyle ..}{\underset{\displaystyle ..}{H : N : H}} \\ H \end{array} \right]^{+} \qquad \left[\begin{array}{c} .. \\ : I : \\ .. \end{array} \right]^{-}$$

10. **T/I** Predict whether the bond between each pair of atoms will be non-polar covalent, slightly polar covalent, polar covalent, or mostly ionic.
    a. carbon and fluorine   e. silicon and hydrogen
    b. oxygen and nitrogen   f. sodium and fluorine
    c. chlorine and chlorine   g. iron and oxygen
    d. copper and oxygen   h. manganese and oxygen

11. **T/I** For each polar and slightly polar covalent bond in question 10, indicate the locations of the partial positive and partial negative charges. Explain how you made each decision.

12. **T/I** Arrange the bonds in each group below in order of increasing polarity.
    a. hydrogen bonded to chlorine, oxygen bonded to nitrogen, carbon bonded to sulfur, sodium bonded to chlorine
    b. carbon bonded to chlorine, magnesium bonded to chlorine, phosphorus bonded to oxygen, nitrogen bonded to nitrogen

13. **C** Make a sketch that shows the relationship between electronegativity difference and percent ionic character of a chemical bond. Why do you think that the transition points between types of chemical bonds are reported in electronegativity difference rather than percent ionic character?

14. **K/U** Explain the meaning of the symbol above these chemical symbols. $\overset{\longleftrightarrow}{NO}$

15. **A** Toward the beginning of this section, you read that metals are usually found in combination with non-metals in nature, and that these compounds are solid. From what you now know, how would you classify these compounds? Give an example.

16. **A** The atmosphere consists mostly of nitrogen and oxygen, along with small amounts of carbon dioxide and trace amounts of hydrogen. Does the atmosphere consist almost entirely of polar compounds or non-polar compounds? Explain your reasoning.

Writing Names and Formulas for Ionic and Molecular Compounds

"Please pass the sodium chloride."

"Do you have enough sucrose for your tea?"

"May I please have some more dihydrogen monoxide?"

"These biscuits are so hard and flat! I must have forgotten to put the sodium hydrogen carbonate in the dough."

You might have heard statements like these while having a meal with family or friends, as in **Figure 2.20**, but the terminology was probably quite different. Four of the terms in these statements are chemical names for common substances. Do you know what the substances are? In this section, you will learn how to name and write the formulas for ionic and molecular compounds. As you read this section, try to figure out the common names for the chemicals identified in the earlier statements.

Figure 2.20 Would other friends or family members know what you meant if you asked, "May I please have some more dihydrogen monoxide?" during a meal?

Standardized Naming

Imagine sitting at a table with six people, all of whom speak a different language. If you said, "Please pass the potatoes," no one would know what you wanted. Someone else might say "Kartoffel?" but you would probably not know what he or she meant.

Chemistry is a language that has millions of "words." Chemists had begun to recognize the need to "speak the same language" as early as the late 1700s. Several chemistry organizations began to develop rules for naming compounds. The current standards are set by the International Union for Pure and Applied Chemistry (IUPAC). The organization was founded in 1919 and still holds meetings to maintain and improve on the rules that allow chemists throughout the world to communicate clearly and concisely.

Naming Binary Ionic Compounds

Binary compounds are among the simplest compounds to name. A *binary ionic compound* is an ionic compound that consists of atoms of only two (bi-) different elements. Because ionic compounds nearly always consist of metals and non-metals, one of these two elements must be a metal and the other must be a non-metal. Study the steps below to review the rules for naming binary ionic compounds.

Go to **scienceontario** to find out more

Rules for Naming Binary Ionic Compounds

1. The name of the metal ion is first, followed by the name of the non-metal ion.

2. The name of the metal ion is the same as the name of the metal atom.

3. If the metal is a transition metal, it might have more than one possible charge. In these cases, a roman numeral is written in brackets after the name of the metal to indicate the magnitude of the charge.

4. The name of the non-metal ion has the same root as the name of the atom, but the suffix is changed to -*ide*.

The names of several common non-metal ions are listed in **Table 2.2**.

As you learned in Section 2.1, when forming an ionic compound, the positive and negative ions must combine in numbers that result in a zero net charge. There is no need to indicate these numbers in the name, however, because they are determined by the charges on the ions. The examples in **Table 2.3** will help you review the rules for naming binary ionic compounds, starting with Lewis diagrams.

Table 2.2 Names of Some Common Non-metal Ions

Formula for Ion	Name of Ion
F^-	fluoride
Cl^-	chloride
Br^-	bromide
I^-	iodide
O^{2-}	oxide
S^{2-}	sulfide
N^{3-}	nitride

Table 2.3 Examples of Naming Binary Ionic Compounds

Steps			
1. Name the metal ion first.	The metal is sodium.	The metal is calcium.	The metal is iron.
2. The name of the ion is the same as the name of the metal.	The name of the sodium ion is *sodium*.	The name of the calcium ion is *calcium*.	The name of the iron ion is *iron*.
3. If the metal ion can have more than one charge, indicate the charge with a roman numeral in brackets.	Sodium ions always have a charge of 1+ so no roman numeral is needed.	Calcium ions always have a charge of 2+ so no roman numeral is needed.	Iron ions can have a charge of 2+ or 3+. The Lewis diagram shows that the iron ions have a charge of 3+, so the roman numeral III must be added. The name of the metal ion becomes *iron(III)*.
4. Name the non-metal ion second. Use the root name of the atom with the suffix -*ide*.	The non-metal is chlorine. Change *chlorine* to *chloride*. Add the name chloride to sodium. The name of the compound is **sodium chloride**.	The non-metal is fluorine. Change *fluorine* to *fluoride*. Add the name fluoride to calcium. The name of the compound is **calcium fluoride**.	The non-metal is oxygen. Change *oxygen* to *oxide*. Add the name oxide to iron(III). The name of the compound is **iron(III) oxide**.

Naming Ionic Compounds with Polyatomic Ions

The rules for naming ionic compounds that have polyatomic ions are fundamentally the same as the rules for naming binary ionic compounds. Each polyatomic ion that you encounter has its own name and is treated as a single unit in a compound. The names and structures of all of the polyatomic ions that you are likely to encounter are listed in **Table 2.4**. Recall that the only common positively charged polyatomic ion is the ammonium ion, NH_4^+, which you saw in **Figure 2.15**.

Table 2.4 Some Common Polyatomic Ions

Name	Formula	Name	Formula
ammonium	NH_4^+	nitrate	NO_3^-
acetate or ethanoate	CH_3COO^-	nitrite	NO_2^-
benzoate	$C_6H_5COO^-$	oxalate	$OOCCOO^{2-}$
borate	BO_3^{3-}	hydrogen oxalate	$HOOCCOO^-$
carbonate	CO_3^{2-}	permanganate	MnO_4^-
hydrogen carbonate	HCO_3^-	phosphate	PO_4^{3-}
perchlorate	ClO_4^-	hydrogen phosphate	HPO_4^{2-}
chlorate	ClO_3^-	dihydrogen phosphate	$H_2PO_4^-$
chlorite	ClO_2^-	sulfate	SO_4^{2-}
hypochlorite	ClO^-	hydrogen sulfate	HSO_4^-
chromate	CrO_4^{2-}	sulfite	SO_3^{2-}
dichromate	$Cr_2O_7^{2-}$	hydrogen sulfite	HSO_3^-
cyanide	CN^-	cyanate	CNO^-
hydroxide	OH^-	thiocyanate	SCN^-
iodate	IO_3^-	thiosulfate	$S_2O_3^{2-}$

There are no comprehensive rules for naming polyatomic ions, so it is best just to learn the names. There are some generalizations, however, that will help you remember some of the names. If you read through the names and structures in **Table 2.4**, you will notice that several groups, or families, of polyatomic ions have names with similar roots and have compositions that vary only in the number of oxygen atoms. **Table 2.5** lists the prefixes and suffixes and shows how they are assigned to each family of ions.

Table 2.5 Prefixes and Suffixes for Families of Polyatomic Ions

Relative Number of Oxygen Atoms	Prefix	Suffix	Example	
Family of Four				
most	per-	-ate	ClO_4^-	perchlorate
second most	(none)	-ate	ClO_3^-	chlorate
second fewest	(none)	-ite	ClO_2^-	chlorite
fewest	hypo-	-ite	ClO^-	hypochlorite
Family of Two				
most	(none)	-ate	NO_3^-	nitrate
fewest	(none)	-ite	NO_2^-	nitrite

It is important to notice that the suffix, *-ate* or *-ite*, does not specify a certain number of oxygen atoms. Instead it indicates the relative number of oxygen atoms. For example, nitrate has three oxygen atoms and nitrite has two oxygen atoms, whereas sulfate has four oxygen atoms and sulfite has three oxygen atoms.

In **Table 2.4**, you will also notice that some polyatomic ions with a charge of 2— or 1— have *hydrogen* or *dihydrogen* at the beginning of their name. This term describes the number of hydrogen ions added to the original polyatomic ion. For example, the phosphate ion, PO_4^{3-}, has no hydrogen ions. Hydrogen phosphate, HPO_4^{2-}, has one hydrogen ion and one less negative charge than the phosphate ion. Dihydrogen phosphate, $H_2PO_4^{-}$, has two hydrogen ions and two fewer negative charges than the phosphate ion.

As well, you will notice the prefix *thio-* in front of two of the polyatomic ions. This prefix indicates that a sulfur atom has taken the place of an oxygen atom. For example, the sulfate ion, SO_4^{2-}, has one sulfur atom and four oxygen atoms. The thiosulfate ion, $S_2O_3^{2-}$, has two sulfur atoms and three oxygen atoms.

Writing Chemical Formulas for Ionic Compounds

A chemical name provides all the information that you need to write the chemical formula for a compound. The following steps summarize the rules for writing the chemical formulas for ionic compounds.

Rules for Writing Chemical Formulas for Ionic Compounds

1. Identify the positive ion and the negative ion.

2. Find the chemical symbols for the ions, either in the periodic table or in the table of polyatomic ions. Write the symbol for the positive ion first and the symbol for the negative ion second.

3. Determine the charges of the ions. If you do not know the charges, you can find them in the periodic table on page 24.

4. Check to see if the charges differ. If the magnitudes of the charges are the same, the formula is complete. If they differ, determine the number of each ion that is needed to create a zero net charge. Write the numbers of ions needed as subscripts beside the chemical symbols, with one exception. When only one ion is needed, leave the subscript blank. A blank means one. If a polyatomic ion needs a subscript, the formula for the ion must be in brackets and the subscript must be outside the brackets.

When the charges of the ions are not the same, you have to determine the number of each ion that is needed to create a zero net charge. To do this, you could simply "guess and check." However, the cross-over method shown in **Figure 2.21** is a more direct way to determine the number of each ion that is needed. As shown in **Figure 2.21**, use the magnitude of the charge of each ion as the subscript for the opposite ion. Below each diagram is a calculation that demonstrates why the subscripts always give you the numbers of ions that result in a zero net charge for the compound. **Table 2.6**, on the next page, shows examples of applying the rules for writing formulas for ionic compounds.

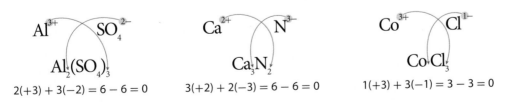

$$2(+3) + 3(-2) = 6 - 6 = 0 \qquad 3(+2) + 2(-3) = 6 - 6 = 0 \qquad 1(+3) + 3(-1) = 3 - 3 = 0$$

Figure 2.21 When you make the number of ions of each element (the subscript) equal in magnitude to the charge of the opposite ion, you will create a compound with a zero net charge.

Notice, in the first example in **Figure 2.21**, that you use only the net charge and ignore the subscript on the polyatomic ion. Also note that the new subscript that indicates the number of polyatomic ions in the compound goes outside the brackets.

Examples of Using Rules for Formulas for Ionic Compounds

The following examples will show you how the rules are applied to writing formulas for ionic compounds.

Table 2.6 Examples of Writing Formulas for Ionic Compounds

Steps / Name	aluminum chloride	calcium iodide	potassium permanganate
1. Identify the positive ion and the negative ion.	Aluminum is first, so it is the positive ion. Chloride is second and has the suffix *-ide*, so it is the negative ion.	Calcium is first, so it is the positive ion. Iodide is second and has the suffix *-ide*, so it is the negative ion.	Potassium is first, so it is the positive ion. Permanganate is second, so it is the negative ion. It does not end with *-ide*, so it is a polyatomic negative ion.
2. Find the chemical symbols for the ions. Write the symbol for the positive ion first and the symbol for the negative ion second.	The symbol for aluminum is Al, and the symbol for chloride is Cl. The formula without subscripts is $Al\_Cl\_$.	The symbol for calcium is Ca, and the symbol for iodide is I. The formula without subscripts is $Ca\_I\_$.	The symbol for potassium is K, and the symbol for permanganate is MnO_4. The formula without subscripts is $K\_MnO_4\_$.
3. Determine the charges of the ions.	The aluminum ion has a charge of 3+, and the chloride ion has a charge of 1−.	The calcium ion has a charge of 2+, and the iodide ion has a charge of 1−.	The potassium ion has a charge of 1+, and the permanganate ion has a charge of 1−.
4. Check to see if the charges differ. If the charges are the same, the formula is complete. If they differ, determine the number of each ion that is needed to create a zero net charge. Write the numbers of ions needed as subscripts beside the chemical symbols.	The charges differ, so use the method in **Figure 2.21** to find the number of ions needed. Al^{3+} Cl^{1-} $AlCl_3$ You need three chloride ions for one aluminum ion. The formula is **$AlCl_3$**.	The charges differ, so use the method in **Figure 2.21** to find the number of ions needed. Ca^{2+} I^{1-} CaI_2 You need two iodide ions for one calcium ion. The formula is **CaI_2**.	The charges are the same, so the formula is **$KMnO_4$**.

Writing Names and Formulas for Acids and Bases

Go to **scienceontario** to find out more

Acids are compounds that ionize, or come apart, in water and release a hydrogen ion, H^+. Thus, the positive ion in acids is the hydrogen ion. For example, on page 62, you read that when HCl dissolves in water, it separates into a hydrogen ion, H^+, and a chloride ion, Cl^-.

Bases are compounds that produce a *hydroxide ion*, OH^-, when they dissolve in water. Thus, the negative ion in bases is usually the hydroxide ion. Notice that the hydroxide ion was listed in **Table 2.4** among the polyatomic ions. Sodium hydroxide, NaOH(s), is a common example of a base. When NaOH(s) dissolves in water, it separates into a sodium ion, Na^+, and a hydroxide ion, OH^-.

Naming and Writing Formulas for Bases

The rules for naming bases and for writing their formulas are the same as the rules for naming and writing formulas for all other ionic compounds. For example, NaOH is sodium hydroxide because Na^+ is the metal ion and the name of the ion is the same as the name of the metal. Hydroxide is the name of the polyatomic ion, OH^-. Thus, any compound with a metal ion or a positively charged polyatomic ion combined with the hydroxide ion is a base. Pure bases are often solids. You can distinguish between the pure base and the basic solution simply by noting its state, which is often indicated by a symbol in brackets after the formula. For example, NaOH(s), where (s) represents solid, is the pure compound. NaOH(aq), where (aq) means "aqueous solution," is the solution of sodium hydroxide in water.

Terminology Involving Bases

The term **alkali** is often used to refer to a base that is soluble in water. This term comes from soap making in the Middle East thousands of years ago. Warm water was poured on the ashes from burnt wood or dried plants, dissolving the bases that were in the ashes. The resulting solution was then boiled with animal fats to make soap. The Arabic word *al-qali* means "the ashes." The earliest records of soap making date from around 2800 BCE in ancient Babylonia, which is now part of Iraq.

You probably recognize the term alkali from the Group 1 metals in the periodic table, which are called alkali metals. As you know, the alkali metals react violently with water. One of the products of that reaction is the hydroxide of the alkali metal, which is a base that is dissolved in water.

alkali a base that is soluble in water

Naming and Writing Formulas for Acids

Acids, in their pure form, are molecular compounds. However, they are named according to the rules for ionic compounds. For example, pure HCl is hydrogen chloride. Hydrogen is named as though it was the positively charged ion and its name is not changed. Chloride is named as though it was the negatively charged ion. The rules for writing formulas for acids are the same as the rules for writing formulas for other ionic compounds.

When an acid is dissolved in water, the name is changed. The current naming system recommended by IUPAC is relatively new and the older, classical naming system is used so frequently that it is helpful to learn both systems. In the IUPAC naming system, the name of the pure acid is simply preceded by the term "aqueous." For example, when hydrogen chloride is dissolved in water, it becomes aqueous hydrogen chloride. The classical names, however, are not quite as simple. To learn the classical names, it is convenient to separate acids into two categories, those that contain oxygen and those that do not.

Acids That Do Not Contain Oxygen

The classical name for acids that do not contain oxygen is formed by omitting the word hydrogen, adding the prefix *hydro-* and the suffix *-ic* and *acid* to the root name. For example, hydrogen chloride becomes hydrochloric acid. **Table 2.7** lists some examples.

Table 2.7 Names of Some Common Acids without Oxygen

Pure Substance (name)	Formula H(negative ion)(aq)	Classical Name hydro (root)ic acid	IUPAC Name aqueous hydrogen (negative ion)
hydrogen fluoride	HF(aq)	hydrofluoric acid	aqueous hydrogen fluoride
hydrogen cyanide	HCN(aq)	hydrocyanic acid	aqueous hydrogen cyanide
hydrogen sulfide	H_2S(aq)	hydrosulfuric acid	aqueous hydrogen sulfide

Notice, in **Table 2.7**, that all of the examples except HCN are binary acids. That is, they contain only hydrogen and a non-metal.

Acids That Contain Oxygen

Acids that contain oxygen are called **oxoacids**. They are composed of hydrogen, oxygen, and atoms of at least one other element, which is usually, but not always, a non-metal. The combination of oxygen and an atom of another element is essentially, a negatively charged polyatomic ion. In fact, almost any of the negatively charged polyatomic ions in **Table 2.4** can be found in acids. However, notice what would happen if you combined a hydrogen ion with a hydroxide ion, which is a polyatomic ion. You would have HOH, which is water.

oxoacid an acid composed of hydrogen, oxygen, and atoms of at least one other element

Naming Oxoacids

The rules for determining IUPAC names for oxoacids are the same as the rules for naming acids with no oxygen atoms. To learn the classical naming system, you need to refer to the system for naming polyatomic ions with varying numbers of oxygen atoms in **Table 2.5**. Just as there are families of ions with varying numbers of oxygen atoms, there are families of acids with varying numbers of oxygen atoms. **Table 2.8** relates the prefixes and suffixes of the polyatomic ions to those of the corresponding acids. Notice that the prefixes remain the same while the suffix *-ite* changes to *-ous acid* and the suffix *-ate* changes to *-ic acid*.

Table 2.8 Classical Naming System for Families of Oxoacids

Name of Ion	Name of Acid (dissolved in water)	Examples	
		Name of Ion	Name of Acid (dissolved in water)
hypo(root)ite	hypo(root)ous acid	hypochlorite, ClO^-	hypochlorous acid, $HClO$
(root)ite	(root)ous acid	chlorite, ClO_2^-	chlorous acid, $HClO_2$
(root)ate	(root)ic acid	chlorate, ClO_3^-	chloric acid, $HClO_3$
per(root)ate	per(root)ic acid	perchlorate, ClO_4^-	perchloric acid, $HClO_4$

To name other oxoacids, look for the prefix (if any) and the suffix and match them to the prefix (if any) and suffix of the acid in **Table 2.8**. For example, the ion nitrate has no prefix and the suffix is *-ate*. The acid would then have no prefix and would have the suffix *-ic acid*. The name of the pure substance, hydrogen nitrate, when dissolved in water would be nitric acid. When the ion already includes one hydrogen atom, such as hydrogen carbonate, HCO_3^-, simply add another hydrogen, $H_2CO_3(aq)$. The name of the ion would be carbonate, and the acid would be carbonic acid.

Learning Check

7. What is a binary ionic compound?

8. Write the names and chemical formulas for the compounds containing the following.
   a. potassium and sulfur
   b. oxygen and magnesium
   c. chlorine and iron
   d. magnesium and nitrogen
   e. hydrogen and iodine
   f. calcium and hydroxide ion

9. Write the name of each compound.
   a. $CrBr_2$ **c.** $HgCl$ **e.** $HNO_3(aq)$
   b. Na_2S **d.** PbI_2 **f.** KOH

10. Write the chemical formula for each compound.
   a. zinc bromide   **d.** magnesium chloride
   b. aluminum sulfide   **e.** hydrogen nitride
   c. copper(II) nitride   **f.** copper(II) hydroxide

11. The root of the names of the following ions is *fluor*. Name each ion, and explain how you decided on the name.
   a. FO^- **b.** FO_2^- **c.** FO_3^- **d.** FO_4^-

12. Write the chemical formula for each compound.
   a. iron(II) sulfate   **d.** magnesium phosphate
   b. sodium nitrate   **e.** hydrogen carbonate
   c. copper(II) chromate   **f.** aluminum hydroxide

Writing Names and Formulas for Binary Molecular Compounds

The names of molecular compounds include more details than the names of ionic compounds, because non-metals can combine in a variety of ratios. For example, nitrogen and oxygen can combine to form six different molecular compounds: NO, NO_2, N_2O, N_2O_3, N_2O_4, and N_2O_5. Clearly, the name *nitrogen oxide* could mean any of these compounds. The rules for naming binary molecular compounds make it possible for each compound to have its own name, which clearly describes the numbers of atoms in the compound. These rules are listed on the following page.

Naming Binary Molecular Compounds

The rules listed below explain how to name binary molecular compounds. The prefixes that are used for naming these compounds are listed in **Table 2.9**. Three examples follow, in **Table 2.10**.

Table 2.9 Prefixes for Binary Molecular Compounds

Number	Prefix
1	mono-
2	di-
3	tri-
4	tetra-
5	penta-
6	hexa-
7	hepta-
8	octa-
9	nona-
10	deca-

Rules for Naming Binary Molecular Compounds

1. Name the element with the lower group number first. Name the element with the higher group number second.

2. The one exception to the first rule occurs when oxygen is combined with a halogen. In this situation, the halogen is named first.

3. If both elements are in the same group, name the element with the higher period number first.

4. The name of the first element is unchanged.

5. To name the second element, use the root name of the element and add the suffix -*ide*.

6. If there are two or more atoms of the first element, add a prefix to indicate the number of atoms.

7. Always add a prefix to the name of the second element to indicate the number of atoms of this element in the compound. (If the second element is oxygen, an "o" or "a" at the end of the prefix is usually omitted.)

Table 2.10 Examples of Naming Molecular Compounds

Steps \ Atoms in Compound	two nitrogen atoms and one oxygen atom	five iodine atoms and one phosphorus atom	two chlorine atoms and seven oxygen atoms
1. Name the element with the lower group number (to the left in the periodic table) first. Name the element with the higher group number (to the right in the periodic table) second. **2.** The one exception to the first rule occurs when oxygen is combined with a halogen. In this situation, the halogen is named first. **3.** If both elements are in the same group, name the element with the higher period number first.	Nitrogen is in Group 15 and oxygen is in Group 16, so nitrogen comes first. **_nitrogen _oxygen**	Iodine is in Group 17 and phosphorus is in Group 15, so phosphorus comes first. **_phosphorus _iodine**	Chlorine is in Group 17 and oxygen is in Group 16, so oxygen should be first and chlorine should be second. However, when oxygen is combined with a halogen, the halogen is named first. **_chlorine _oxygen**
4. The name of the first element is unchanged. **5.** To name the second element, use the root name of the element and add the suffix -*ide*.	The name *nitrogen* is unchanged, but *oxygen* is changed to *oxide*. **_nitrogen _oxide**	The name *phosphorus* is unchanged, but *iodine* is changed to *iodide*. **_phosphorus _iodide**	The name *chlorine* is unchanged, but oxygen is changed to *oxide*. **_chlorine _oxide**
6. If there are two or more atoms of the first element, add a prefix to indicate the number of atoms. **7.** Always add a prefix to the name of the second element to indicate the number of atoms of this element in the compound. (If the second element is oxygen, an "o" or "a" at the end of the prefix is usually omitted.)	There are two nitrogen atoms, so the prefix is *di-*. There is one oxygen atom, so the prefix is *mono-*. Because the second element is oxygen, use *mon-*. The name of the compound is **dinitrogen monoxide**.	There is only one phosphorus atom, so no prefix is added. There are five iodine atoms, so the prefix is *penta-*. The name of the compound is **phosphorus pentaiodide**.	There are two chlorine atoms, so the prefix *di-* is added. There are seven oxygen atoms so the prefix should be *hepta-*. However, the second element is oxygen so the "a" on *hepta-* is omitted. The prefix is *hept-*. The name of the compound is **dichlorine heptoxide**.

Writing Chemical Formulas for Binary Molecular Compounds

An important exception to all of the rules for naming and writing formulas for binary molecular compounds occurs when the two elements in a compound are carbon and hydrogen. Combinations of carbon and hydrogen constitute a large group of compounds called hydrocarbons, which are a subgroup of a larger group of compounds called organic compounds. Organic compounds consist of all compounds that contain carbon atoms, other than carbon monoxide (CO), carbon dioxide (CO_2), carbonates (CO_3^{2-}), cyanides (CN^-), and carbides (several forms). Organic compounds have a unique naming system. You will study organic chemistry in more advanced chemistry courses. The following rules apply to inorganic compounds, which are all compounds other than organic compounds. **Table 2.11** provides three examples of naming binary molecular compounds.

Rules for Writing Chemical Formulas for Binary Molecular Compounds

1. Write the symbol for the element with the lowest group number first.

2. Write the symbol for the element with the highest group number second.

3. The one exception to the first two rules occurs when oxygen is combined with a halogen. In this case, the symbol for the halogen is written first.

4. If both elements are in the same group, write the symbol for the one with the higher period number first.

5. If the number of atoms of either or both elements is greater than one, write the number as a subscript beside the symbol. The absence of a subscript is understood to mean one.

Table 2.11 Examples of Writing Chemical Formulas for Binary Molecular Compounds

Steps \ Atoms in Compound	two nitrogen atoms and one oxygen atom	two chlorine atoms and one oxygen atom	four bromine atoms and one silicon atom
1. Write the symbol for the element with the lowest group number first. 2. Write the symbol for the element with the highest group number second.	Nitrogen is in Group 15 and oxygen is in Group 16, so the symbol for nitrogen is written first. **N_O_**	Chlorine is in Group 17 and oxygen is in Group 16, so the symbol for oxygen should be written first. **O_Cl_**	Bromine is in Group 17 and silicon is in Group 14, so the symbol for silicon is written first. **Si_Br_**
3. The one exception to the first two rules occurs when oxygen is combined with a halogen. In this case, the symbol for the halogen is written first. 4. If both elements are in the same group, write the symbol for the one with the higher period number first.	Oxygen is not combined with a halogen. Oxygen and nitrogen are not in the same group. No changes are needed. **N_O_**	Oxygen is combined with the halogen chlorine, so the symbol for chlorine comes first. **Cl_O_**	Oxygen is not in the compound. Silicon and bromine are not in the same group. No changes are needed. **Si_Br_**
5. If the number of atoms of either or both elements is greater than one, write the number as a subscript beside the symbol. The absence of a subscript is understood to mean one.	There are two nitrogen atoms, so the subscript 2 is written beside N. There is only one oxygen atom, so there is no subscript beside O. The formula is **N_2O**.	There are two chlorine atoms, so the subscript 2 is written beside Cl. There is only one atom of oxygen, so there is no subscript beside O. The formula is **Cl_2O**.	There is one silicon atom, so there is no subscript beside Si. There are four bromine atoms, so the subscript 4 is written beside Br. The formula is **$SiBr_4$**.

Names from Formulas and Formulas from Names

Problem

Write the names of the compounds for parts *a* and *b*. Write the formulas for parts *c* and *d*.

a. SF_6 **b.** $Cu(NO_3)_2$ **c.** aluminum sulfide   **d.** sulfur trioxide

What Is Required?

You need to determine the names of SF_6 and $Cu(NO_3)_2$

You also need to determine the formulas for aluminum sulfide and sulfur trioxide.

What Is Given?

You are given the formulas for two compounds: SF_6 and $Cu(NO_3)_2$

You are given the names of two compounds: aluminum sulfide and sulfur trioxide

Plan Your Strategy	Act on Your Strategy
a. SF_6 S is sulfur and F is fluorine. They are both non-metals so the compound is molecular and you need prefixes. Fluorine is the second element so you change the ending to *–ide*.	_sulfur _fluoride
There is one sulfur atom so the prefix would be *mono*. However, it is the first element so no prefix is needed. There are six fluorine atoms so the prefix is *hexa-*.	**sulfur hexafluoride**
b. $Cu(NO_3)_2$ Cu is copper and it is a metal. NO_3 is a polyatomic ion and the name is nitrate. The compound is ionic so you do not need prefixes.	copper_ nitrate
A copper ion can have a charge of $1+$ or $2+$. Nitrate has a charge of $1-$ and there are two of the ions. Therefore the copper ion must have a charge of $2+$ to make the compound neutral. Add (II) to the name of copper.	**copper(II) nitrate**
c. aluminum sulfide Aluminum is a metal. Its symbol is Al. Its charge is $3+$. Sulfur is a non-metal. Its symbol is S. Its charge is $2-$.	$Al^{3+}S^{2-}$
The charges are not the same so you need subscripts. The subscript (number of atoms) of each element is the same as the magnitude of the charge of the other ion.	**Al_2S_3**
d. sulfur trioxide Sulfur and oxygen are both non-metals, so the compound is molecular. You need subscripts. The symbol for sulfur is S. The symbol for oxygen is O.	S_O_
Sulfur has no prefix, so the number of atoms is assumed to be one and therefore no subscript is needed. Oxygen has the prefix *tri-*, meaning there are three oxygen atoms. Its subscript is 3.	**SO_3**

Check Your Solution

When you add up the charges on the ionic compounds, they add up to zero. The names and symbols for the molecular compounds describe the same number of atoms of the same elements.

1. Write the name of P_4S_7.

2. Write the name of $Pb(NO_3)_2$.

3. Write the formula for manganese(IV) chloride.

4. Write the formula for nitrogen triiodide.

5. Write the name of CuBr.

6. Write the formula for iron(III) oxide.

7. Write the formula for silicon dioxide.

8. Write the name of SeF_6.

9. Write the name of CaO.

10. Write the formula for cobalt(III) nitrate.

Drawing Structural Formulas for Molecular Compounds

A chemical name or formula tells you how many atoms of each element are in a molecule. However, it does not provide information about how the atoms are bonded to one another. A Lewis structure shows you how the atoms are connected to each other, but it is cumbersome to draw. Because chemists need an easier way to show the connections between atoms, they developed structural formulas. You can draw a **structural formula** from a Lewis structure by drawing a single line to represent a pair of bonding electrons and omitting the lone pairs. Thus, one straight line represents one bond. **Figure 2.22** shows some Lewis structures you have seen before, as well as some new Lewis structures, along with their structural formulas.

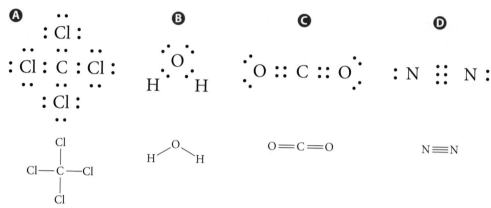

Figure 2.22 Compound (**A**) is carbon tetrachloride. Instead of drawing 32 dots, as you would for a Lewis structure, you need to draw only four lines for a structural formula. Compound (**B**) is dihydrogen monoxide. Compound (**C**) is carbon dioxide. Notice that the double bonds are drawn as two lines. Compound (**D**) is nitrogen, which has a triple bond.

State the common name for dihydrogen monoxide.

Suggested **Investigation**

Inquiry Investigation 2-B, Building Molecular Models

Although structural formulas provide more information than the simpler chemical formulas, they are still only two-dimensional, while real molecules are three-dimensional. Thus, to visualize a molecule completely, you need to build a model. You can use something as simple as toothpicks and Styrofoam® balls to build three-dimensional models. You can also use kits that have different sizes and colours of balls to represent atoms of different elements. Regardless of the materials, these models can help you visualize and analyze molecules.

Building your own models gives you an understanding of the three-dimensional structure of molecules that you cannot attain any other way. However, modern computers can now generate molecular models of very small to very large molecules. The image in **Figure 2.23** is a space-filling model. The relative sizes of the spheres and the way they fit together is an excellent representation of the shape of the actual molecule.

Figure 2.23 This is a computer-generated model of ascorbic acid (vitamin C), $C_6H_8O_6$.

Section Summary

- The name of a binary ionic compound starts with the name of the metal element and, if necessary, a roman numeral indicating the charge on the ion. This is followed by the name of the non-metal element with the ending changed to -*ide*.

- The formula for a binary ionic compound starts with the symbol for the metal element followed by the symbol for the non-metal element. Subscripts indicate the numbers of atoms of the two elements.

- Bases are named according to the rules for ionic compounds.

- When acids are dissolved in water, they are named according to different rules than when they are in their pure form.

- The name of a binary molecular compound starts with the name of the element that has the lower group number. The name of the element that has the higher group number is last, and the ending is changed to -*ide*. Prefixes are used to indicate the numbers of atoms of the two elements. However, a prefix is not used for the first element if there is only one atom of this element.

- The formula for a binary molecular compound starts with the symbol for the element with the lower group number, followed by the symbol for the element with the higher group number. Subscripts indicate the numbers of atoms of the two elements in the compound.

- A structural formula shows how the atoms in a compound are attached to each other.

Review Questions

1. **T/I** Turn to page 64 and read the "dinner table" statements at the top of the page. Then answer the following questions.
   a. What do you think are the common names for sodium chloride, dihydrogen monoxide, and sodium hydrogen carbonate? Write the chemical formulas for these compounds.
   b. What do you think is the common name for sucrose?
   c. Identify each compound as an ionic compound or a molecular compound. Explain your reasoning.

2. **K/U** Explain why prefixes that indicate the numbers of atoms of the different elements are not needed in the names of ionic compounds.

3. **K/U** What is a polyatomic ion?

4. **K/U** What is the difference between a sulfate ion and a sulfite ion? How would you be able to determine the difference without looking up the names in a table?

5. **T/I** Write the name of each compound.
   a. Al_2O_3 c. Na_3P e. NH_4Cl g. $HNO_3(aq)$
   b. HgI_2 d. K_3PO_4 f. $LiClO_4$ h. $LiOH(aq)$

6. **T/I** Write the formula for each compound.
   a. zinc oxide d. magnesium iodide
   b. iron(II) sulfide e. cobalt(III) chloride
   c. potassium hypochlorite f. sodium cyanide

7. **K/U** Why must the name of a molecular compound include prefixes to indicate the numbers of atoms of the elements in the compound?

8. **T/I** The following six compounds contain nitrogen and oxygen: NO, NO_2, N_2O, N_2O_3, N_2O_4, and N_2O_5. Write the names of these compounds.

9. **T/I** Write the formula for each compound.
   a. phosphorus pentachloride
   b. difluorine monoxide
   c. sulfur trioxide
   d. silicon tetrabromide
   e. cobalt(II) hydroxide
   f. sulfur hexafluoride

10. **T/I** Write the name of each compound.
    a. CO c. CS_2 e. SiO_2 g. $Ba(OH)_2$
    b. BCl_3 d. CCl_4 f. PI_3 h. $H_3BO_3(s)$

11. **K/U** Explain why the name of C_3H_8 is not tricarbon octahydride.

12. **T/I** Draw Lewis structures for these compounds. From your Lewis structures draw structural formulas.
    a. NF_3 b. HCN c. $ClNO$

13. **C** In a group, discuss the advantages and disadvantages of using structural formulas.

14. **T/I** First draw a Lewis structure for each compound. Then, using your diagram, draw a structural formula.
    a. two carbon atoms bonded to each other, and two hydrogen atoms bonded to each carbon atom
    b. two carbon atoms bonded to each other, with three hydrogen atoms bonded to one carbon atom, and one hydrogen atom and one oxygen atom bonded to the second carbon atom

Comparing the Properties of Ionic and Molecular Compounds

> **melting point** the temperature at which a compound changes from a solid to a liquid

It is not a coincidence that water melts at 0°C and boils at 100°C. The Celsius temperature scale is based on the melting point and boiling point of water. All compounds have melting points and boiling points, but these temperatures vary widely with the type of substance. What factors determine the melting point and boiling point of a compound? Do the same factors affect the other properties of a compound? This section will answer these and other questions concerning the properties of ionic and molecular compounds.

Melting Points and Boiling Points of Compounds

As you read above, boiling points and melting points are unique to each pure compound. Thus, they can provide important information about the characteristics of the compound. For example, the melting point and boiling point of a compound reveal information about the strength of the attractions that are holding the particles (ions or molecules) of the compound together. Consider what is happening to a compound when it melts or boils.

Melting Point

The **melting point** of a compound is the temperature at which it changes from a solid to a liquid at standard atmospheric pressure (the pressure exerted on the ground by dry air at sea level, or 101.325 kPa). In a solid, the particles—ions or molecules—are so strongly attracted to one another that they cannot pull apart. You can imagine a solid as particles held together by springs, as shown in **Figure 2.24 (A)**.

You might recall from previous science courses that, no matter how low the temperature, all particles have some kinetic energy. So, although the particles in a solid cannot pull away from the surrounding particles, they are always vibrating. You have probably learned that temperature is directly related to the kinetic energy of the particles in a substance. As energy in the form of heat enters a substance, the kinetic energy, and thus the temperature of the substance, increases. When the kinetic energy of the particles is great enough for the particles to pull away from one another, as shown in **Figure 2.24 (B)**, the temperature stops increasing and the compound melts. If the melting point of a compound is very high, you know that a large amount of energy is needed for the particles to pull away from one another. Therefore, the forces holding them together must be very strong. A low melting point tells you that the particles are easily pulled apart, and thus the forces attracting them to one another are relatively weak.

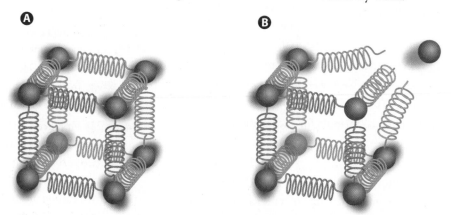

Figure 2.24 **(A)** Even in a solid, all particles are moving. **(B)** As the temperature of a substance increases, more and more particles have enough energy to break away from their nearest neighbour. Note that, in the case of molecular compounds, the spheres represent the entire molecule and the springs represent attractive interactions between individual molecules.

Boiling Point

The **boiling point** of a compound is the temperature at which it changes from a liquid to a gas at standard atmospheric pressure. In a liquid, particles have enough kinetic energy to pull away from one neighbouring particle, only to be attracted to another neighbouring particle. The particles slide past one another. At the boiling point, the particles have enough kinetic energy to completely break away from all the other particles and the compound becomes a gas. Gas particles have enough energy to bounce off one another when they collide rather than sticking together. Thus, the boiling point of a compound, like the melting point, provides information about the strength of the forces between the particles. A high boiling point indicates that the attractive forces between the particles in a liquid are very strong. A low boiling point tells you that these forces are relatively weak.

boiling point the temperature at which a compound changes from a liquid to a gas

Forces between Particles in a Compound

As you read, a comparison of the melting and boiling points of a variety of substances can provide information about the strength of the forces between ions in ionic compounds and between molecules in molecular compounds. Note that when a molecular compound melts or boils, the covalent bonds remain intact.

- A low melting point or boiling point means that particles with small amounts of kinetic energy can break away from the adjacent particles. Thus the forces between particles are weak.

- A very high melting point or boiling point means that the particles must have a very large amount of kinetic energy to break away, and thus the forces between particles are strong.

Keeping these relationships in mind, consider the data in **Table 2.12**.

Table 2.12 Melting Points and Boiling Points of Some Common Compounds

Compound	Melting Point (°C)	Boiling Point (°C)
ethanol (grain alcohol), C_2H_5OH	−114	+78.3
ammonia, NH_3	−77.7	−33.3
cesium bromide, CsBr	+636	+1300
hydrogen, H_2	−259	−253
hydrogen chloride, HCl	−114	−85
magnesium oxide, MgO	+2825	+3600
methane (natural gas), CH_4	−182	−161
nitrogen, N_2	−210	−196
sodium chloride, NaCl	+801	+1465
water, H_2O	0	+100

If you analyze the data in **Table 2.12** and classify the compounds into three categories, you will get the results in **Table 2.13**. An analysis of the melting points would give the same categories.

Table 2.13 Categories of Compounds Based on Boiling Point

High Boiling Point	Intermediate Boiling Point	Low Boiling Point
cesium bromide, CsBr	ethanol, C_2H_5OH	hydrogen, H_2
magnesium oxide, MgO	ammonia, NH_3	nitrogen, N_2
sodium chloride, NaCl	hydrogen chloride, HCl	methane, CH_4
	water, H_2O	

Compounds with High Melting Points and Boiling Points

Consider the compounds with high boiling points in **Table 2.13**. These compounds are all ionic. Their high boiling points are explained by the fact that the attractive electrostatic forces between oppositely charged particles create very strong bonds. An examination of the structure of ionic compounds will reveal why so much energy is needed to break these bonds. **Figure 2.25** shows the arrangement of sodium and chloride ions in a crystal of sodium chloride. The same structure continues throughout an entire crystal. Notice that each chloride ion is attracted to six adjacent sodium ions, and each sodium ion is attracted to six adjacent chloride ions. Because the attractive forces are all the same, there are no specific pairs of sodium and chloride ions that you could identify as "molecules." Each ion is strongly attracted to all the adjacent ions of the opposite charge. There are continuous chains of ions that are attracted to each other throughout the entire crystal, making the structure very stable. The formula, NaCl, simply means that there is a 1:1 ratio of sodium to chloride ions in the entire crystal. One sodium ion and one chloride ion are referred to as a formula unit of sodium chloride, never as a molecule of sodium chloride.

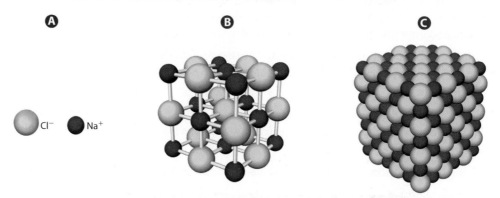

Figure 2.25 **(A)** The yellow spheres represent chloride ions and the blue spheres represent sodium ions. **(B)** This model is called a "ball and stick" model. The balls represent the ions, and the sticks represent the bonds. **(C)** This model is called a "space-filling model." It shows that, in the actual crystal, the ions are packed tightly together.

Compounds with Intermediate Melting Points and Boiling Points

Now consider the compounds with intermediate boiling points in the second column of **Table 2.13**. If you look for a similarity among these compounds, you will find that they are all molecular compounds. As well, they all have one or more polar bonds. Depending on the overall structure of a molecule that has polar bonds, the entire molecule can be polar. **Figure 2.26** shows models of water and ammonia to illustrate why they are polar. One end of each molecule is slightly negative, while the other end is slightly positive.

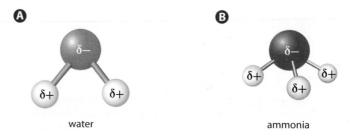

water   ammonia

Figure 2.26 The white spheres represent hydrogen atoms, the red sphere represents an oxygen atom, and the blue sphere represents a nitrogen atom. You can think of the polarity as being caused by the electrons spending more time around the oxygen atom in water **(A)** and the nitrogen atom in ammonia **(B)** and less time around the hydrogen atoms.

Representing Polar Molecules

A polar molecule is often represented as an oval shape with a slightly positively charged end (positive pole) and a slightly negatively charged end (negative pole). Because a polar molecule has one slightly positive end and one slightly negative end, it is often called a **dipole**. **Figure 2.27** shows how the positive ends of polar molecules attract the negative ends of other polar molecules. This attractive force, called a **dipole-dipole force**, is much smaller than the forces between ions. The dipole-dipole force is the main attractive force that acts between polar molecules. The intermediate strength of this force results in the intermediate boiling points of compounds that are composed of polar molecules.

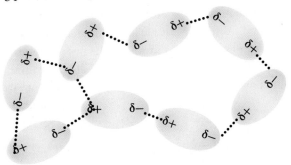

Figure 2.27 Each oval represents a polar molecule. As the positively charged end of one molecule is attracted to the negatively charged end of another, the molecules form a continuous network.

Compounds with Low Melting Points and Boiling Points

Finally, consider the compounds in **Table 2.13** that have low boiling points. Notice that their molecules are all non-polar. The bonds between the carbon and hydrogen atoms in methane are slightly polar, but the molecule, as shown in **Figure 2.28**, is symmetrical. Therefore, the polarities of the bonds cancel one another in the whole molecule. Nevertheless, some attractive forces exist between the molecules. Although non-polar molecules have no distinct separation of charge, it is still possible for the positive nuclei of atoms in one molecule to attract the electrons in a neighbouring molecule. These attractions are very weak. As a consequence of these weak forces, compounds that are composed of non-polar molecules have much lower boiling points than compounds that are composed of polar molecules of a similar size.

In summary, these three interactions (strong attractive forces between ions, weaker dipole-dipole attractive forces between polar molecules, and very weak attractive forces between non-polar molecules), determine the boiling points and melting points of pure substances. Because the dipole-dipole forces and the weak attractive forces act between molecules, they are called **intermolecular forces**. This distinguishes them from the covalent bonds that act within molecules. The intermolecular forces determine the melting points and boiling points of molecular compounds.

dipole a molecule with a slightly positively charged end (positive pole) and a slightly negatively charged end (negative pole)

dipole-dipole force the attractive force between the positive end of one molecule and the negative end of another molecule

intermolecular forces attractive forces that act between molecules

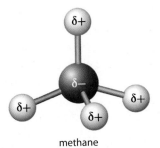

methane

Figure 2.28 The black sphere represents a carbon atom and the white spheres represent hydrogen atoms in this methane molecule. Each bond is slightly polar, but the symmetry of the molecule makes it non-polar.

Learning Check

13. Explain what is happening, on the level of ions and molecules, when a substance is melting.

14. One compound has a melting point of 714°C. Another compound, which is similar in size and appearance, has a melting point of 146°C. How would you classify these compounds based on their melting points?

15. Why is it incorrect to refer to a "molecule" of a compound such as potassium iodide?

16. What is a dipole-dipole force?

17. Why do non-polar molecules have very low melting and boiling points?

18. What forces are included within the category of intermolecular forces?

Other Properties of Ionic and Molecular Compounds

The strength and the type of bonds and intermolecular forces that exist among ions and molecules affect several properties, in addition to melting points and boiling points. Among these properties are solubility in water and electrical conductivity.

Solubility in Water

Whether or not a substance dissolves in water is an important property. For example, many vitamins and nutrients in food (**Figure 2.29**) move through your bloodstream from your digestive system to all of the tissues in your body because they are soluble in water. Similarly, waste materials that are water soluble are carried to your kidneys where they are eliminated from your body. Many chemical processes can take place only when the compounds are dissolved in water. It is not always possible to predict whether a compound will dissolve in water. However, differing trends in solubility can be clearly seen when considering the polarities of substances.

Figure 2.29 All of the nutrients in these foods are critical to good health. The nutrients that are soluble in water reach your bloodstream and are carried to your tissues quickly.

For a substance to dissolve in water, the water molecules must be more strongly attracted to particles of that substance than to other water molecules. As you know, water molecules are polar, having a slightly positive end and a slightly negative end. The positive end will attract a negative ion or the negatively charged end of another polar molecule. Likewise, the negative end of a water molecule will attract a positive ion or the positively charged end of another polar molecule. Consequently, water will dissolve many ionic compounds and polar compounds. For example, table sugar (sucrose) is a polar molecular compound, and table salt (sodium chloride) is an ionic compound. Both are soluble in water.

Water molecules are much more strongly attracted to each other than to non-polar molecules. Therefore, most non-polar compounds do not dissolve in water. For example, fats and oils are mixtures of non-polar compounds and they do not dissolve in water.

Note that not all ionic compounds and polar molecular compounds are soluble in water. In Unit 4, you will read more about solubility and learn how to determine whether a particular compound is soluble in water.

Electrical Conductivity

Electrical conductivity is the ability of a substance to allow an electric current to exist within it. A substance can conduct an electric current only if charges (electrons or ions) can move independently of one another. In a pure metal, electrons can move somewhat freely because they are not tightly bound to the metal atoms. When a pure metal is conducting an electric current, electrons are moving with ease from one metal atom to the next.

electrical conductivity the ability of a substance or an object to allow an electric current to exist within it

In any type of compound, electrons are held tightly by the atoms. In an ionic compound, electrons have moved from a metal atom to a non-metal atom. Once they are bound to the non-metal atom, however, they are held tightly. A pure ionic compound can only conduct an electric current under conditions in which entire ions can move independently of one another.

As you know, ionic compounds are solid at room temperature. In a solid, the oppositely charged ions are held rigidly together. Therefore, in their solid form, ionic compounds cannot conduct an electric current. When an ionic compound is in the liquid state, however, the ions are free to move independently of one another. This occurs only at very high temperatures, but, at these temperatures, ionic compounds can conduct an electric current.

Ionic compounds can also conduct an electric current when in an aqueous state, as shown in **Figure 2.30**. When an ionic compound is dissolved in water, the ions are free of other ions because they are surrounded by water molecules. Thus, ionic solutions can also conduct an electric current.

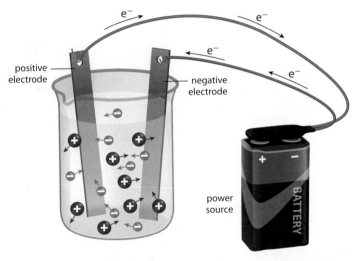

Figure 2.30 Positive ions are attracted to a negative electrode, and negative ions are attracted to a positive electrode. The ions create an electric current as they move around each other in opposite directions.

When atoms are bound together in a molecular compound, they are sharing electrons. The electrons never leave one atom completely. Therefore, there are no positive and negative charges that are independent of one another. This means that molecular compounds cannot conduct an electric current regardless of whether they are non-polar or polar. If a polar compound is dissolved in water and electrodes are placed in the water, the molecules will orient themselves so that their positively charged end is directed toward the negative electrode and their negatively charged end is directed toward the positive electrode. However, the charges never leave the molecules. Thus, even in a water solution, molecular compounds cannot conduct an electric current. You might recall that acids are molecular compounds when in a pure form but come apart and become ionic when dissolved in water. Therefore, aqueous solutions of acids do conduct an electric current.

Section Summary

- The strength of the attractive forces acting between ions or molecules determines the melting point and boiling point of a compound.
- Ionic compounds usually have the highest melting points and boiling points. Polar molecules have intermediate melting points and boiling points, and non-polar molecules have the lowest melting points and boiling points for molecules of similar sizes.
- Ionic and polar compounds are likely to be soluble in water. Non-polar compounds are insoluble in water.
- For a substance to conduct an electric current, oppositely charged particles must be free to move independently of one another.

Review Questions

1. **K/U** Explain the basis of the Celsius temperature scale.

2. **K/U** Describe, on the level of individual particles, what happens to a substance when it is heated.

3. **K/U** What property of particles determines whether they will pull away from adjacent particles?

4. **K/U** How would you classify a compound that has a boiling point of −182°C? Explain your answer.

5. **K/U** Explain why compounds consisting of polar molecules are likely to have a higher melting point than compounds consisting of non-polar molecules.

6. **T/I** What would you predict about the melting point of a compound that will not dissolve in water? Explain your thinking.

7. **K/U** Explain how an attractive force can exist between non-polar molecules.

8. **T/I** If a compound has very high melting and boiling points, is the compound likely to be soluble in water? Explain the relationship between these two properties of a compound.

9. **C** Use sketches to show how a non-polar molecule can have polar bonds.

10. **K/U** Describe what must happen, on a particle level, for a substance to dissolve in water.

11. **K/U** Glycerol is a compound that dissolves readily in water. The water solution of glycerol, however, will not conduct an electric current. What would you predict about the properties of glycerol?

12. **K/U** Under what two conditions can an ionic compound conduct an electric current?

13. **K/U** Can polar molecular compounds conduct electric current under either of the conditions that you described in question 12? Explain why or why not.

14. **A** To be transported throughout the body in the bloodstream, fat molecules must be bound to protein molecules, as shown in the following figure. Explain why you think this is necessary.

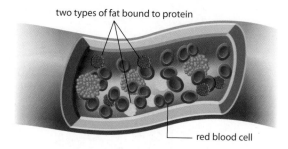

15. **A** You might have heard the saying, "Like dissolves like." From what you have learned about solubility, comment on the validity of this statement.

16. **T/I** Two molecular compounds, X and Y, have similar masses. Compound X is solid at room temperature, has a melting point of 146°C, and is soluble in water. Compound Y is liquid at room temperature, has a melting point of −10°C, and is not soluble in water.

    a. What would you predict about the polarities of compound X and compound Y?

    b Based on your predictions, explain the differences in their melting points and solubilities.

Skill Check

Initiating and Planning

✓ **Performing and Recording**

✓ **Analyzing and Interpreting**

✓ **Communicating**

Materials

- crystal structure model kit

  or

- polystyrene balls of two different sizes

- toothpicks

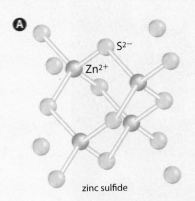

Ⓐ

S^{2-}

Zn^{2+}

zinc sulfide

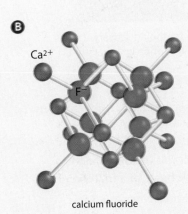

Ⓑ

Ca^{2+}

F^-

calcium fluoride

These diagrams represent "repeating units" for each of the ionic compounds. Note that the ions on the outer edges (sulfur in the zinc sulfide and calcium in the calcium fluoride) are bonded to more oppositely charged ions in the adjacent "repeating units." This is the reason that some sulfur ions in **(A)** do not appear connected to other ions.

Modelling Ionic Compounds

In this investigation, you will build models of crystals of three different ionic compounds to better visualize their structures.

Pre-Lab Questions

1. What information is provided by a formula for an ionic compound?

2. Describe the bonding that occurs between ions in an ionic compound.

3. What do the "balls" and "sticks" represent in a ball-and-stick model of a compound?

Question

What can you predict about the structures of crystals by building models?

Procedure

1. Your teacher will give you a crystal structure model kit, or polystyrene balls and toothpicks.

2. Carefully study the arrangement of the ions in the sodium chloride model in **Figure 2.25**.

3. Choose which size or colour of balls you will use to represent the sodium and chloride ions. Discuss, with your partner, how the ions are arranged and how you will connect the "ions." Build a model of a sodium chloride crystal.

4. Study the illustrations of the zinc sulfide and calcium fluoride crystals shown here. Each illustration shows one "repeating unit" for a crystal of each of the two compounds.

5. Repeat step 3 for zinc sulfide and for calcium fluoride. Build models of at least two "repeating units" for each compound.

6. Compare your models with another group's models. If your models are not the same, discuss the differences and decide which, if any, are the correct models.

Analyze and Interpret

1. What is the ratio of metal ion to non-metal ion in each of your models?

2. Provide a possible explanation as to why the ratio of metal to non-metal ions can be the same and the structures of the crystals can be different.

Conclude and Communicate

3. How well do you think your models represent real crystals? Describe ways in which your models are similar to real crystals and ways in which they are different.

Extend Further

4. **RESEARCH** Using print and Internet resources, research the technique first used by chemists to determine the crystal structure of ionic compounds.

Inquiry
INVESTIGATION

2-B

Skill Check

Initiating and Planning

✓ Performing and Recording

✓ Analyzing and Interpreting

✓ Communicating

Materials

- molecular model kit
- pen
- paper

Building Molecular Models

Models are very important for chemists. You cannot see detailed features of a molecule, even with a microscope. However, you can build a model that shows some of the properties that chemists have determined through experimentation. In this investigation, you will use a molecular model kit to assemble models of a few molecules.

James Watson and Francis Crick won a Nobel prize for their discovery of the structure of DNA (deoxyribonucleic acid). In this photograph, they are discussing their model.

Pre-Lab Questions

1. When drawing a Lewis structure, what basic rule tells you where the electrons must be?

2. What characteristics of a Lewis structure tell you whether a bond between two atoms is a single bond, a double bond, or a triple bond?

3. What is the difference between a two-dimensional image and a three-dimensional image?

Question

What can you predict about the structures of molecules by building models?

Procedure

1. Your teacher will provide a molecular model kit for you to use.

2. Copy the table shown below into your notebook. Use your table for drawings in Procedure steps 3, 4, and 5. Each cell in the last two columns must be large enough for you to draw Lewis structures and sketches of your molecular models.

Data Table for Model Building

Name	Formula	Lewis Structure	Sketch of Shape

3. In your table, write the name and formula, and draw a Lewis structure of each molecule below.
   a. hydrogen bonded to hydrogen
   b. chlorine bonded to chlorine
   c. oxygen bonded to two hydrogen atoms
   d. carbon bonded to two oxygen atoms
   e. nitrogen bonded to three hydrogen atoms
   f. carbon bonded to four chlorine atoms
   g. nitrogen bonded to three fluorine atoms

4. Look through your textbook, and choose three molecules that are not in the list above. Record the names and formulas for these compounds in your data table. Draw Lewis structures of these molecules in your table.

5. Based on your Lewis structures, build models of all the molecules. Make a sketch of each of your models. Consult the directions that came with the kit for information about assembling the models.

Analyze and Interpret

1. Compare your models and sketches with those of your classmates. Discuss any differences.

Conclude and Communicate

2. What can you learn from models that you cannot learn from Lewis structures?

3. Summarize the strengths and limitations of creating molecular models using kits. What can you infer from the models? What features of the molecules cannot be inferred from the models?

Extend Further

4. **INQUIRY** Describe the difference between ball-and-stick models and space-filling models. Discuss the advantages and disadvantages of using each type of model.

5. **RESEARCH** Using print and Internet resources, research a discovery of a structure in chemistry or biochemistry that depended heavily on model building.

Case Study

Feminization of Male Fish
Monitoring the Effects of Environmental Estrogens

Scenario

To meet your community involvement requirement for your Grade 12 diploma, you have begun volunteering at a conservation area in your region. The Conservation Authority has been actively developing a new watershed protection plan. This plan is especially important because a new wastewater treatment plant was built along a river in the middle of the watershed several years ago. Local citizens are concerned about how this wastewater treatment plant might impact the quality of drinking water in the area. The Conservation Authority has been collecting data related to the health and status of organisms in local stream and river ecosystems. You have recently been helping to collect data by sampling fish populations in the watershed.

The sampling data that you have collected in this year show some startling changes in the ecosystem. In particular, approximately 50% of the male fish sampled also have some female anatomy. As well, there are about five times as many females compared to males in the population. These results suggest that some type of environmental estrogen has been entering the water from some, as yet unknown, source.

Natural and Synthetic Estrogens

There are many natural estrogens. These include estrogens produced in living plants as well as those that are produced in animals. However, there are also many environmental estrogens. These environmental estrogens are synthetic compounds that mimic estrogen activity and that are released into the environment as by-products of industry. More than 60 chemical substances, including dioxin and DDT, have been identified as environmental estrogens. These compounds are produced for use in many different sectors, including the pharmaceutical, plastics, and detergent-manufacturing industries.

Nonylphenol ethoxylates (NPEs)

Data previously collected by the Conservation Authority in your region indicate that one particular class of environmental estrogens, called nonylphenol ethoxylates (NPEs), is the likely cause of the changes in the fish you have observed. These chemicals have been used for more than 40 years as detergents, emulsifiers (which keep oils dispersed in a liquid to prevent clumping) wetting agents (that lower the surface tension of a liquid so that liquids mix together more easily) and dispersing agents (that are used to keep particles in a suspension from clumping and coming out of suspension). NPEs are not produced naturally–their presence in the environment is entirely the result of human activity. The two most likely sources of environmental NPEs are wastewater from industrial operations and the water output from municipal wastewater treatment plants.

Waterways and marshlands are the breeding grounds for many species of fish. Environmental estrogens that mimic the action of the female hormone, estrogen, can get into these waters from waste treatment plants or industrial waste. When developing male fish are exposed to these estrogens, they can become feminized.

Conservation Ontario

Status of NPEs in Canada, the European Union, and the United States, in 2010.

Canada NPEs are considered to be "toxic" as defined in the *Canadian Environmental Protection Act, (CEPA), 1999.* Nevertheless, their use is not legally banned. In the event that they are banned in the future, research is underway to develop and test alternative chemicals that can serve the purposes of the compounds that are currently responsible for the production of NPEs.

Canadian studies have shown that, where adequate sewage treatment practices are employed, normal exposure to NPEs does not pose a significant risk to human health. As well, the current use of NPEs does not generally pose a risk to the aquatic environment. However, discharge directly to the aquatic environment of untreated or partially treated waste is likely to harm aquatic organisms.

European Union NPEs are legally banned in the European Union. They have been replaced with more expensive but safer products. The new products are alcohol-based compounds.

Officials in the European Union have taken the position that NPEs and their by-products can cause serious illness or even death in humans with pre-existing conditions that make them susceptible, even at very low exposures. They have established that NPEs are a contributing factor to breast cancer, among other diseases.

United States NPEs are not banned in the United States. In fact, they are widely used. NPEs are favoured as cheap and readily available compounds for industrial processes. Officials in the United States believe that NPEs and their by-products that remain in water after treatment at a sewage treatment plant are within established scientific limits for toxicity.

The position of officials in the United States is that they are aware of the impact that established limits of NPEs have on the environment, and believe these impacts to be acceptable. On the other hand, they feel that other chemicals that might be considered as replacements for NPEs are not well understood, and thus, it would not be possible to set safe limits for them.

Research and Analyze

1. As you read, more than 60 chemical substances, including NPEs, dioxin, and DDT have been identified as environmental estrogens. Research sources of environmental estrogens other than NPEs, including products manufactured by the pharmaceutical industry. What properties of some common pharmaceutical products allow them to persist in water systems and influence the growth and development of organisms?

2. Research some effects of estrogen-mimicking agents (environmental estrogens) on human health. What are the accepted limits and concentrations of some of the common environmental estrogens in ecosystems? How consistently and effectively are ecosystems monitored in Canada for the presence of environmental estrogens?

3. Environmental estrogens appear to have a significant impact on biological organisms and ecosystems. Proven environmental estrogens, for example, pesticides such as DDT, have been banned from use in many parts of the world. However, these synthetic estrogens persist in the environment and organisms, including humans, continue to be exposed to them due to natural cycling of air and water. Analyze the impact of banning a substance such as an environmental estrogen in certain parts of the world but not in others.

Take Action

1. **Plan** In a group, discuss the controversy surrounding the use of environmental estrogens. What are some key issues to consider when analyzing how to reduce or eliminate the impacts of these compounds? Share the results of the research and analysis you conducted in questions 1 to 3 above.

2. **Act** Prepare an informational brochure that could be handed out by the Conservation Authority at their public information session on the watershed protection plan. Ensure that you create a brochure that will enable your audience to understand the properties of environmental estrogens, the potential risks associated with their presence in the environment, and possible ways to reduce the impact of environmental estrogens. Support your position with information from credible sources.

Chapter 2 | SUMMARY

Section 2.1 | The Formation of Ionic and Covalent Bonds

An ionic bond forms when oppositely charged ions attract each other. A covalent bond forms when two atoms share one or more pairs of electrons.

KEY TERMS

bonding pair
covalent bond
double bond
electronegativity difference
ionic bond
ionic compound
Lewis structure

lone pair
molecular compound
octet rule
polar covalent bond
polyatomic ion
single bond
triple bond

KEY CONCEPTS

- The octet rule can be used to predict how bonds will form.
- An ionic bond forms when a negatively charged ion and a positively charged ion are attracted to each other.
- A covalent bond forms when two atoms share one or more pairs of electrons.
- A polyatomic ion consists of two or more atoms that are covalently bonded together and carry a charge. A polyatomic ion can form an ionic compound with a simple ion or another polyatomic ion of the opposite charge.
- A chemical bond can be non-polar covalent, slightly polar covalent, polar covalent, or mostly ionic, depending on the electronegativity difference between the two atoms that are bonded together.

Section 2.2 | Writing Names and Formulas for Ionic and Molecular Compounds

The name and chemical formula for a compound describe exactly how many atoms of each element are present in one particle of the compound.

KEY TERM

alkali
oxoacid
structural formula

KEY CONCEPTS

- The name of a binary ionic compound starts with the name of the metal element and, if necessary, a roman numeral indicating the charge on the ion. This is followed by the name of the non-metal element with the ending changed to -ide.
- The formula for a binary ionic compound starts with the symbol for the metal element followed by the symbol for the non-metal element. Subscripts indicate the numbers of atoms of the two elements.

- Bases are named according to the rules for ionic compounds.
- When acids are dissolved in water, they are named according to different rules than when they are in their pure form.
- The name of a binary molecular compound starts with the name of the element that has the lower group number. The name of the element that has the higher group number is last, and the ending is changed to -ide. Prefixes are used to indicate the numbers of atoms of the two elements. However, a prefix is not used for the first element if there is only one atom of this element.
- The formula for a binary molecular compound starts with the symbol for the element with the lower group number, followed by the symbol for the element with the higher group number. Subscripts indicate the numbers of atoms of the two elements in the compound.
- A structural formula shows how the atoms in a compound are attached to each other.

Section 2.3 | Comparing the Properties of Ionic and Molecular Compounds

The type of bonds and the shape of the particles influence the properties of compounds, such as their melting and boiling points, solubility in water, and electrical conductivity.

KEY TERMS

boiling point
dipole
dipole-dipole force

electrical conductivity
intermolecular forces
melting point

KEY CONCEPTS

- The strength of the attractive forces acting between ions or molecules determines the melting point and boiling point of a compound.
- Ionic compounds usually have the highest melting points and boiling points. Polar molecules have intermediate melting points and boiling points, and non-polar molecules have the lowest melting points and boiling points.
- Many ionic and polar compounds are soluble in water. Non-polar compounds are insoluble in water.
- For a substance to conduct an electric current, oppositely charged particles must be free to move independently of one another.

Knowledge and Understanding

Select the letter of the best answer below.

1. Which statement about ionic compounds is false?
   a. An ionic compound is comprised of ions held together by an electrostatic force.
   b. An ionic compound typically consists of a metal ion and a non-metal ion.
   c. An ionic compound contains the same number of oppositely charged ions.
   d. An ionic compound has a zero net charge.
   e. The composition of an ionic compound can often be predicted by the octet rule.

2. The circled electrons in this Lewis diagram are called
   a. unpaired electrons
   b. free electrons
   c. an electron pair
   d. a bonding pair
   e. an unbound pair

3. The electronegativity of magnesium is 1.3, and the electronegativity of oxygen is 3.4. The bond that forms between them is
   a. mostly ionic
   b. polar covalent
   c. slightly polar covalent
   d. non-polar covalent
   e. none of the above

4. The chemical name of $Mg(ClO_3)_2$ is
   a. magnesium chloride
   b. magnesium dichlorite
   c. magnesium chlorite
   d. magnesium chlorate
   e. magnesium hypochlorite

5. The element that comes second in the name of a binary molecular compound is the element that
   a. has the lower group number
   b. has the higher group number
   c. has the higher period number
   d. is the non-metal
   e. has the greater mass

6. The chemical name of $SiBr_4$ is
   a. monosilicon tetrabromide
   b. silicon hexabromide
   c. monosilicon pentabromide
   d. silicon octabromide
   e. silicon tetrabromide

7. Which statement about the properties of compounds is true?
   a. A compound that has a very high melting point is a liquid at room temperature.
   b. Ionic bonds are stronger than intermolecular forces.
   c. Non-polar molecules experience no intermolecular forces.
   d. A compound that has a very low boiling point is a liquid at room temperature.
   e. Dipole-dipole forces are stronger than the force between oppositely charged ions.

8. Which compound is *most* likely to be soluble in water?
   a. a non-polar compound
   b. a slightly polar compound
   c. a polar compound
   d. an ionic compound
   e. all of the above

Answer the questions below.

9. In this chapter, you read that ores are metals combined with non-metals. How would you classify the compounds that are found in ores? Why?

10. Several different gaseous compounds that consist of non-metals are found in the atmosphere. How would you classify these gaseous compounds? Why?

11. Aluminum ions have a charge of 3+ and oxide ions have a charge of 2−. How can aluminum ions and oxide ions combine to form a compound with a net charge of zero?

12. Copy the following diagram and complete a Lewis structure for the compound. Draw a circle around each atom and its electrons and describe how each atom satisfies the octet rule.

    H   H

    H   C   C   F

    Cl   H

13. Explain the meaning of the term "bond dipole."

14. The boiling point of a compound depends on a balance between two conditions. What are these conditions? Explain.

15. Describe the two forces that make up intermolecular forces.

16. State which type of compound, ionic or molecular, can conduct electric current. What conditions are necessary for this type of compound to conduct electric current?

Thinking and Investigation

17. Use Lewis diagrams to predict the ratio of metal to non-metal ions in a compound formed by each pair of elements.
   a. magnesium and fluorine
   b. potassium and bromine
   c. rubidium and chlorine
   d. calcium and oxygen

18. Each of the following Lewis structures has an error in it. State what the error is, and draw the correct Lewis structure.

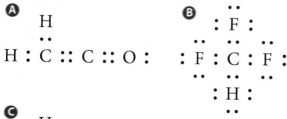

19. Name each compound.
   a. $MgCl_2$
   b. Na_2O
   c. $FeCl_3$
   d. CuO
   e. $Ba(ClO)_2$
   f. NH_4NO_3
   g. $H_2CrO_4(aq)$
   h. $H_3PO_4(s)$
   i. KOH
   j. $Cd(OH)_2$

20. Write the formula for each compound.
   a. gold(III) chloride
   b. magnesium oxide
   c. lithium nitrite
   d. calcium phosphide
   e. manganese(II) sulfide
   f. calcium hypochlorite
   g. aqueous hydrogen chloride
   h. sulfuric acid
   i. cobalt(II) hydroxide
   j. lithium hydroxide

21. Draw a Lewis structure of each molecule consisting of the following combinations of atoms.
   a. one carbon atom bonded to three hydrogen atoms and one chlorine atom
   b. one carbon atom bonded to two sulfur atoms
   c. two iodine atoms bonded together
   d. three carbon atoms bonded together in a chain; three hydrogen atoms bonded to each of the carbon atoms on the ends; an oxygen atom bonded to the central carbon atom

22. Name each compound.
   a. SO_2
   b. N_2O_4
   c. CO
   d. Cl_2O

23. Write the formula for each compound.
   a. dihydrogen monoxide
   b. sulfur trioxide
   c. silicon tetrachloride

24. Identify the errors in each phrase or statement, and rewrite it correctly.
   a. four molecules of potassium bromide
   b. The compound $NaHSO_4$ is sodium sulfate.
   c. The compound KNO_2 is potassium nitrate.

Communication

25. BIG IDEAS The type of chemical bond in a compound determines the physical and chemical properties of that compound. Name and sketch two different types of chemical bonds. For each bond type, describe two ways in which it influences the properties of the compound.

26. BIG IDEAS It is important to use chemicals properly to minimize the risks to human health and the environment. You read that when sodium, a highly reactive metal, is combined with chlorine, a toxic gas, the product, sodium chloride, is very safe. Using print and Internet resources, research another element or compound that can be made safe by reacting it with another element or compound. Share your findings in the format of your choosing.

27. In some Lewis diagrams, one of the chemical symbols might have no dots. Draw an example of this, and explain why one of the symbols has no dots.

28. Identify the chemical bonds in the following compounds as mostly ionic, polar covalent, slightly polar covalent, or non-polar covalent. Show and explain the calculations you used to identify the bonds.
   a. calcium chloride
   b. carbon dioxide
   c. nitrogen
   d. silicon tetrachloride

29. "If there were no intermolecular forces, all molecular compounds would be gases." Do you agree or disagree with this statement? Explain your reasoning, as if you were explaining it to a classmate who did not understand intermolecular forces.

30. The boiling points of argon ($-186°C$) and fluorine ($-188°C$) are quite similar. Write a paragraph that you could read to help a Grade 10 student understand why these boiling points are similar, based on intermolecular forces.

31. Molecules of methane, CH_4, and water, H_2O, have similar masses. However, their boiling points are very different. The boiling point of methane is $-161°C$, and the boiling point of water is $+100°C$. Draw sketches of these molecules, and use your sketches to explain why their boiling points are so different.

32. Write the names of the following ions: I^-, IO^-, IO_2^-, IO_3^-, IO_4^-. The last four ions are polyatomic ions. Design a different naming system that you think would be descriptive of the ions and easy to remember.

33. Draw a structural formula based on the Lewis structure shown here. Explain, in detail, the relationship between the two diagrams.

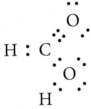

34. Summarize your learning in this chapter using a graphic organizer. To help you, the Chapter 2 Summary lists the Key Terms and Key Concepts. Refer to Using Graphic Organizers in Appendix A to help you decide which graphic organizer to use.

Application

35. You have two white crystalline solids. One is an ionic compound, and the other is a molecular compound. Design an investigation to determine which is which. Assume that your investigation cannot involve dissolving them in water.

36. Water and methanol, CH_3OH (a type of alcohol), mix together in any proportions. Find their boiling points. Then, based on the boiling points you find, design a method you could use to separate water and methanol that are mixed together.

37. Suppose that you have two colourless solutions. One is a solution of an ionic compound in water, and the other is a solution of a molecular compound in water. Design an investigation to determine which solution is which. Describe the tests you would perform and the results you would expect for each solution.

38. Pure sodium can be extracted from sodium chloride using a process called electrolysis. Sodium ions can pick up electrons from one electrode and form sodium atoms. Chloride ions can give up electrons to the other electrode and form chlorine atoms, which then combine to form molecules of chlorine gas. The diagram shown here is a simplified sketch of the apparatus. Imagine that you were asked to design the containers and other equipment for this process. Review what you have learned about compounds that carry an electric current and about the properties of sodium metal and chlorine gas. Describe the challenges you would have to overcome when designing the equipment. Present some possible solutions to these challenges.

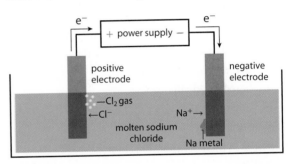

39. In 1906, the Nobel Prize in Chemistry was awarded to French chemist, Henri Moissan, for isolating fluorine in its pure elemental form. Why would this achievement be deserving of such a prestigious honour? Use your understanding of the properties of the elements, as well as chemical bonds, to explain your answer.

40. You might have heard advertisements about detergents that "break up grease." Oil and grease consist of large non-polar molecules, which are very insoluble in water. Nevertheless, detergents, which seem to dissolve in water, can remove oil and grease from clothing in water. A space-filling model of a typical detergent molecule is shown below. Study the model, and provide a possible explanation for how detergents can remove grease from clothing.

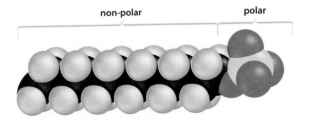

Select the letter of the best answer below.

1. **K/U** Ionic bonds form between which two types of elements?
   a. metals and metalloids
   b. metals and non-metals
   c. metalloids and non-metals
   d. two non-metals
   e. two metals

2. **K/U** What is the correct name of the compound formed from Fe^{3+} and Cl^-?
   a. iron chloride
   b. iron(I) chloride
   c. iron chloride(I)
   d. iron(II) chloride
   e. iron(III) chloride

3. **K/U** The electrons in a non-polar bond are
   a. gained by one atom and lost by the other
   b. shared equally
   c. shared unequally
   d. gained
   e. lost

4. **K/U** Which compound is ionic?
   a. KBr
   b. H_2O
   c. HCl(g)
   d. NH_3
   e. CH_4

5. **K/U** Which compound is molecular?
   a. NaOH
   b. $PbCl_2$
   c. MnO_2
   d. $CaCrO_4$
   e. SiO_2

6. **K/U** The name of the compound N_2O_4 is
   a. nitrogen oxide
   b. dinitrogen dioxide
   c. nitrogen tetraoxide
   d. dinitrogen tetroxide
   e. tetranitrogen dioxide

7. **K/U** What type of bonding occurs within a water molecule and between water molecules?
   a. non-polar covalent within a water molecule and ionic between water molecules
   b. polar covalent within a water molecule and dipole-dipole between water molecules
   c. polar covalent within a water molecule and ionic between water molecules
   d. non-polar covalent within a water molecule and dipole-dipole between water molecules
   e. ionic within a water molecule and dipole-dipole between water molecules

8. **K/U** Which statement about ionic compounds is false?
   a. They may be soluble in water.
   b. They have very high melting points.
   c. They cannot conduct electric current when melted.
   d. They are held together with ionic bonds.
   e. They are solid at room temperature.

9. **K/U** Which statement about molecular compounds is false?
   a. They can be solid, liquid, or gas at room temperature.
   b. They conduct electric current when dissolved in water.
   c. They are held together with covalent bonds.
   d. They have low to moderate boiling points.
   e. They can be polar.

10. **K/U** Intermolecular forces consist of
    a. ionic and covalent bonds
    b. dipole-dipole forces and ionic bonds
    c. dipole-dipole forces and weak attractive forces
    d. covalent bonds and dipole-dipole forces
    e. weak attractive forces and ionic bonds

Use sentences and diagrams, as appropriate, to answer the questions below.

11. **A** State the octet rule, and give an example of how you would apply it to determine the number of sodium ions that would be needed to form an ionic compound with sulfur.

12. **A** Draw Lewis diagrams of the following elements in the order given: hydrogen, carbon, carbon, hydrogen. Connect the atoms to form a Lewis structure of a molecule. (Make sure that you keep the atoms in the order given.) Is the bond between the two carbon atoms a single, double, or triple bond? Explain how you know what type of bond it is.

13. **K/U** Draw the Lewis structure of a nitrogen molecule. Circle and label the bonding pairs and the lone pairs.

14. **C** Draw a diagram to show how you would use the charges on ions to determine the subscripts for the chemical formula for an ionic compound. Explain your diagram. Under what circumstances would you use this method?

15. **T/I** For each pair of atoms below, predict whether the bond between the atoms will be non-polar covalent, slightly polar covalent, polar covalent, or mostly ionic. For each slightly polar bond or polar covalent bond, indicate which atom will be slightly positive and which atom will be slightly negative.
  a. carbon and fluorine
  b. oxygen and nitrogen
  c. chlorine and chlorine
  d. manganese and oxygen

16. **T/I** Name each compound.
  a. $Mg_3(PO_4)_2$
  b. $NaIO_3$
  c. $AlPO_4$
  d. $NaHCO_3$

17. **T/I** Write the chemical formula for each compound.
  a. potassium thiocyanate
  b. yttrium chloride
  c. iron(III) sulfide
  d. tin(II) fluoride

18. **T/I** Name each compound.
  a. Si_3N_4
  b. PCl_5
  c. SF_6
  d. ClF_3

19. **T/I** Write the formula for each compound.
  a. sulfur trioxide
  b. carbon monoxide
  c. diselenium dibromide
  d. nitrogen triiodide

20. **C** Use a Venn diagram to compare Lewis structures and structural formulas. Give an example of each. Under what circumstances would you use a structural formula instead of a Lewis structure?

21. **T/I** The melting points of three compounds are listed below. Predict the type of attractive forces between the particles of each compound when the compound is in its solid form.

Melting Points of Three Compounds

Compound	Melting Point (°C)
scandium oxide	2489
nitrogen trichloride	−40
ethane	−182.79

22. **C** A classmate asks, "How could there possibly be any intermolecular forces between non-polar compounds?" Answer your classmate's question, using a diagram to support your explanation.

23. **T/I** List the following compounds in the order of their boiling points, from lowest to highest without knowing any exact boiling points. Explain your reasoning for your order, based on the structures given.

methanol   chlorine   potassium oxide

24. **C** Make a sketch to represent dipole-dipole forces. Explain what is happening in your sketch. What effect do dipole-dipole forces have on the properties of a polar compound?

25. **K/U** Describe the circumstances that are necessary for a compound to conduct an electric current.

Self-Check

If you missed question …	1	2	3	4	5	6	7	8	9	10	11	12	13	14	15	16	17	18	19	20	21	22	23	24	25
Review section(s)…	2.1	2.2	2.1	2.1	2.1	2.2	2.1	2.3	2.1 1.3	2.3	2.1	2.1	2.1	2.2	2.1	2.2	2.2	2.2	2.2	2.2	2.3	2.3	2.3	2.3	2.3

Home Clean Home

Canadians are generally aware of the importance of cleanliness to prevent the spread of disease. Most families use a variety of cleaning products to kill bacteria and other infectious agents. One of the most widely used products for this purpose is household bleach. However, some consumers, who are concerned about the health and environmental impacts of household bleach, choose to clean and disinfect their homes with a "green" substitute for bleach. One of the common green choices for this purpose is household vinegar. As a science journalist, you have been asked to write a report for a consumer group comparing these two products. You will be focusing on the products' toxicity (harmful effects to the environment and on human health), as well as their effectiveness as disinfectants. The report may take the form of an article, a web page, a blog, a podcast, a short video, or any other medium approved by your teacher.

How do household bleach and vinegar compare in terms of environmental and human toxicity? How do they compare in terms of effectiveness as a disinfectant?

Initiate and Plan

1. Establish a format you will use to reference the print and electronic resources used to write your report. Determine how you will assess your resources for accuracy, reliability, and bias.

chemical household products

potential substitutes

Perform and Record

2. Using electronic and print resources, and your own knowledge of chemistry, answer the following questions about household bleach and vinegar.
   a. What is the name of the active ingredient in each product?
   b. Write the chemical formula for this ingredient.
   c. Is the active ingredient an ionic or molecular compound?
   d. What is the average concentration of this ingredient in the product when used as a household disinfectant?

3. Use electronic and print resources to research how use of household bleach and vinegar may harm the environment and/or human health.

   The following terms can help guide your research:
   - toxicity
   - environmental impact
   - pollutants
   - MSDS
   - chemical interaction

4. Use electronic and print resources to research the effectiveness of each product as a disinfecting agent.

   The following terms can help guide your research:
   - disinfectant
   - microbial reduction
   - antimicrobial efficacy
   - pathogens

5. Select an appropriate graphic organizer to summarize your research. Go to Using Graphic Organizers in Appendix A for help with choosing a graphic organizer.

Analyze and Interpret

1. Prepare a risk-benefit table outlining, from the perspectives you just researched, the risks and benefits associated with the use of each product as a household disinfectant.

2. Based on the risks and benefits, make a recommendation to consumers as to the suitability of each product as a household disinfectant.

Communicate Your Findings

3. Prepare your report in the form of an article, a web page, a blog, a podcast, or a short video in the style in which a science journalist would present information to the general public. In the report, provide information about the appropriate government ministry to which consumers could write to communicate any concerns they have regarding the use of specific household cleaning products.

Assessment Criteria

Once you complete your project, ask yourself these questions. Did you . . .

☑ **T/I** analyze your resources for accuracy, reliability, and bias?

☑ **K/U** describe the active ingredient in each product in terms of its chemical make-up?

☑ **A** identify any health concerns associated with each product?

☑ **A** identify the environmental impact of each product?

☑ **A** determine the effectiveness of each product as a disinfectant?

☑ **A** make a recommendation, based on your risk-benefit analysis, as to the suitability of each product for household use?

☑ **C** organize your research using an appropriate format and appropriate academic documentation?

☑ **C** select a format to present your findings in a way that is appropriate for both purpose and audience?

☑ **C** use scientific terminology accurately?

BIG IDEAS

- Every element has predictable chemical and physical properties determined by its structure.
- The type of chemical bond in a compound determines the physical and chemical properties of that compound.
- It is important to use chemicals properly to minimize the risks to human health and the environment.

Overall Expectations

In this unit you learned how to…

- analyze the properties of commonly used chemical substances and their effects on human health and the environment, and propose ways to lessen their impact
- investigate physical and chemical properties of elements and compounds, and use various methods to visually represent them
- demonstrate an understanding of periodic trends in the periodic table and how elements combine to form chemical bonds

Chapter 1 | Elements and the Periodic Table

KEY IDEAS

- The atomic number of an atom is the number of protons in the atom. The mass number of an atom is the sum of the protons and neutrons that make up the nucleus of the atom.
- An appropriate ratio of neutrons to protons stabilizes the nucleus. An atom with an unstable nucleus is called a radioisotope.
- Atomic masses that are reported in data tables are weighted averages, based on isotopic abundances.
- The periodic law states that when elements are arranged by atomic number, their chemical and physical properties recur periodically.

- The atomic radius of an element is influenced by the amount of charge in its nucleus and by the number of occupied electron shells.
- The ionization energy of an element is influenced by the distance between its outermost electron and its nucleus.
- The electron affinity of an element is influenced by whether the valence shells of the atoms are filled. It is also influenced by whether the added electron is paired or unpaired.
- The electronegativity of an element is influenced by the atomic radius of the atoms of the element.

Chapter 2 | Chemical Bonding

KEY IDEAS

- The octet rule can be used to predict the way that bonds will form. An ionic bond is the attraction between oppositely charged ions. A covalent bond forms when two atoms share one or more pairs of electrons.
- A polyatomic ion consists of two or more atoms that are covalently bonded together and carry a charge.
- Chemical bonds can be non-polar covalent, slightly polar covalent, polar covalent, or mostly ionic, depending on the electronegativity difference between the two atoms that are bonded together.
- The name of an ionic compound is composed of the name of the positive ion followed by the name of the negative ion. In the formula for an ionic compound, the symbol for the positive ion is followed by the symbol for the negative ion. Subscripts are used to indicate the numbers of the different types of ions that are needed to make the net charge equal to zero.
- Bases are named according to the rules for naming ionic compounds. The rules for naming acids dissolved in water are

different from the rules for naming acids that are in their pure form.
- The name of a binary molecular compound includes prefixes for the number of atoms of each element in a molecule of the compound. The chemical formula for a molecular compound represents the numbers and kinds of atoms that make up each individual molecule.

- The strengths of the attractive forces acting among ions and among molecules determine the melting point and boiling point of a compound.
- Ionic compounds tend to be crystalline solids (at room temperature) that are soluble in water, have high melting points and boiling points, and can conduct an electric current in the liquid state or when dissolved in water.
- Molecular compounds can be gases, liquids, or solids at room temperature. Many are not soluble in water, whereas most have low melting points and boiling points, and cannot conduct an electric current in any form.

Knowledge and Understanding

Select the letter of the best answer below.

1. As scientists gathered new information about the atom, they revised the model of the atom to incorporate the new information. Examine the atomic model shown. Which statement best describes the discovery that resulted in this model?

   a. When alpha particles are directed at a thin gold foil, some of them bounce back.

   b. Atoms are indivisible and join together in whole-number ratios.

   c. When atoms are heated, they give off specific patterns of coloured light.

   d. Atoms can be broken down into smaller particles using a cathode-ray tube.

   e. Mathematical analyses show that electrons do not travel in precise orbits.

2. Which atom has six valence electrons?

   a. lithium   d. neon

   b. carbon   e. boron

   c. sulfur

3. Which atom has a nucleus that contains 20 protons?

   a. lithium   d. calcium

   b. krypton   e. potassium

   c. carbon

4. Which element in the periodic table below is a brittle solid at room temperature, does not conduct an electric current, and is shiny?

   a. 1   d. 4

   b. 2   e. 5

   c. 3

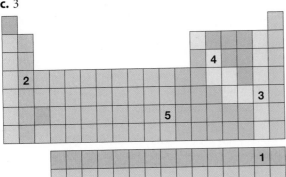

5. Which of the following atoms has the largest radius?

   a. lithium

   b. lead

   c. cesium

   d. fluorine

   e. carbon

6. Which of the following atoms has the highest first ionization energy?

   a. helium

   b. xenon

   c. potassium

   d. hydrogen

   e. carbon

7. Which of the following atoms has the highest electronegativity?

   a. iodine

   b. tellurium

   c. sulfur

   d. strontium

   e. chlorine

8. Which statement correctly describes the formation of an ion from an atom?

   a. A sodium atom gives away one electron to form an ion that has a charge of 1−.

   b. An oxygen atom gives away two electrons to form an ion that has a charge of 2+.

   c. A helium atom gains one electron to form an ion that has a charge of 1−.

   d. A magnesium atom gives away two electrons to form an ion that has a charge of 2+.

   e. A fluorine atom gives away seven electrons to form an ion that has a charge of 7+.

9. Which pair of atoms would form an ionic bond?

   a. bromine and oxygen

   b. chlorine and lithium

   c. carbon and nitrogen

   d. sodium and neon

   e. copper and chromium

10. Which compound likely has the lowest boiling point?

   a. calcium oxide, CaO

   b. magnesium chloride, $MgCl_2$

   c. methane, CH_4

   d. aluminum chloride, $AlCl_3$

   e. sodium bromide, NaBr

Answer the questions below.

11. Describe the contributions of Thomson and of Rutherford in understanding the structure and composition of the atom.

12. What are valence electrons? How do valence electrons affect the properties of atoms?

13. Compare isotopes in terms of their subatomic particles and their bonding patterns.

14. Compare the electrical conductivities of metals and non-metals.

15. Using only the periodic table, place the following atoms in order of increasing atomic radius: C, Ba, O, Ca, and Ge.

16. Using only the periodic table, place the following atoms in order of decreasing electronegativity: Mg, P, Ca, K, and Cl.

17. How does electronegativity change as effective nuclear charge increases? Explain your reasoning.

18. Describe three ways that an oxygen atom can bond to satisfy the octet rule.

19. Describe the formation of single, double, and triple covalent bonds. Use diagrams to support your description.

20. Calculate ΔEN for each of the following bonds and predict the nature of the bond.
   a. sodium and nitrogen
   b. sulfur and oxygen
   c. fluorine and fluorine

21. What types of elements are bonded together to make an ionic compound? Give three examples of ionic compounds. Identify the types of elements that make up the compound in each example.

22. The data in the table describes the properties of three substances. Based on the data, identify whether each substance is ionic, polar covalent, or non-polar covalent.

Properties of Three Substances

Substance	Melting Point (°C)	Solubility in Water	Electrical Conductivity
A	1800	Very soluble	In molten state and in aqueous solution
B	−200	Insoluble	None
C	10	Insoluble	None

23. Describe the location of metalloids in the periodic table. Explain how this location reflects the properties of metalloids, compared with the properties of metals and non-metals.

24. What is meant by the term "stable octet"?

25. To better understand the steps that occur in a series of reactions, a scientist plans an experiment to identify which products contain the oxygen atoms that were part of the water molecules that reacted. How might the scientist use radioisotopes in the experiment?

26. Describe how the numbers of electrons in metal atoms and in non-metal atoms change when the elements form ionic compounds.

27. Describe how the physical state of a substance at room temperature depends on the strength of the intermolecular forces between the particles of the substance.

28. What difference between the solid state and liquid state of an ionic compound accounts for the difference in electrical conductivity for the two states.

Thinking and Investigation

29. Copy and complete the following table.

Particles Contained in Isotopes

Isotope	Atomic Number	Mass Number	Number of Protons	Number of Neutrons
$^{44}_{20}Ca$				
			10	10
		14	6	
	17			20
$^{28}_{12}Mg$				
	30	66		
		138		82

30. Compare the trends in electronegativity with the trends in atomic radius within a period in the periodic table as the atomic number increases.

31. Write the chemical formula for each of the following compounds.
   a. calcium nitride
   b. sulfur diiodide
   c. lead(II) bromide
   d. aluminum phosphite

32. Write the name of each compound.

    a. NCl_3

    b. K_2CO_3

    c. FeO

    d. N_2O_4

33. Identify which element in Period 3 is represented by X in each formula.

    a. XF_4

    b. CaX

    c. XH_3

34. Write two names for compounds a, b, and c and name compound d.

    a. $HNO_3(aq)$

    b. $HI(aq)$

    c. $HOOCCOOH(aq)$

    d. $Co(OH)_3(s)$

35. Write the chemical formula for each compound.

    a. aqueous hydrogen hypochlorite

    b. ammonium hydroxide

    c. nitrous acid

    d. magnesium hydroxide

36. The following Bohr-Rutherford diagram shows the formation of a compound:

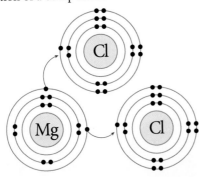

    a. What type of compound is formed? Explain your reasoning.

    b. How many valence electrons does each atom have?

    c. After the electrons move as shown, how many valence electrons does each particle have?

37. The forces between ions in an ionic compound are stronger than the forces between molecules in a non-polar molecular compound, yet ionic compounds are more soluble in water than non-polar molecular compounds are. Explain this observation.

38. Which compound, $NaCl$ or $BrCl$, would you expect to have a higher melting point? Explain your reasoning.

39. A metal atom has an equal number of positively charged protons and negatively charged electrons. It can form an ionic compound by giving up two electrons.

    a. What charge will the metal ion have when it forms a compound?

    b. If a single non-metal atom gains both electrons, what charge will the non-metal ion have? What will the overall charge of the compound be?

    c. If the metal atom gives its electrons to two non-metal atoms, what charge will each non-metal ion have? What will the overall charge of the compound be?

40. The following diagrams show the use of a conductivity tester during an investigation.

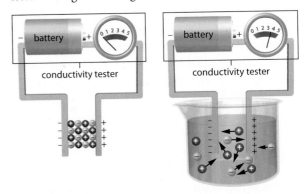

    a. What type of compound is being tested? How can you tell?

    b. Describe the results that are shown in the diagrams. Include the state of each substance.

    c. Explain the difference in conductivity for the two conditions that are shown in the diagrams.

Communication

41. Draw and label a Bohr-Rutherford model to show and explain how electrons fill the first three energy levels of an atom.

42. **BIG IDEAS** Every element has predictable chemical and physical properties determined by its structure. Draw Lewis diagrams of an atom of oxygen, an atom of fluorine, an atom of neon, and an atom of sodium. Identify the charge of the ion that is formed from each atom, and use your diagrams to support your answers.

43. Describe the steps you would use to determine the numbers of protons, neutrons, and electrons in an atom of carbon-14.

44. Sketch the shape of the periodic table. Indicate the trends for atomic number, atomic radius, electronegativity, electron affinity, and ionization energy within a period and within a group.

45. **BIG IDEAS** The type of chemical bond in a compound determines the physical and chemical properties of that compound. Chlorine and sodium form a compound that is a crystalline solid at room temperature, but chlorine and bromine form a compound that is a gas under the same conditions. Explain the differences in the properties of the compounds, in terms of their chemical bonds.

46. **BIG IDEAS** It is important to use chemicals properly to minimize the risks to human health and the environment. Examine the recommended use and safety precautions for a household product, such as a cleaning product or a lawn-care product. Write a short script for a video that could be used to inform users about the types of compounds that make up the product, the properties of these compounds, the chemical formulas of these compounds, and the proper use of the product. Use Internet or print resources to gather the information you need.

47. Write two or three sentences to explain how the term "covalent" describes what happens to electrons in a covalent bond. Use sketches to support your explanation.

48. You have learned that elements join in a definite ratio to form a compound. What does this mean? Use words, sketches, or both to explain how the ratio of elements in a compound compares with the ratio of components in a mixture.

49. Create a sequence of sketches that clearly convey how the sharing of bonding electrons differs in a non-polar bond, a slightly polar bond, and a polar covalent bond.

50. Draw a Venn diagram to compare ionic compounds with molecular compounds. Include information about the structure and properties of these compounds.

51. Prefixes are used in the names of molecular compounds to identify how many atoms of each element make up each molecule. Write down at least three words that use prefixes to represent numbers, and identify these numbers.

52. Write a blog about the exceptions to the rules for naming and writing formulas for binary ionic and molecular compounds.

53. While studying for a test on chemical bonding, a student writes the following statements on index cards: Statement A: All polar molecules must have polar bonds. Statement B: All non-polar molecules must have non-polar bonds. Analyze these statements. Decide whether each statement is accurate, and write a brief statement to explain your reasoning. Use examples to support your reasoning.

54. Write the procedure for an investigation to classify several solids as ionic compounds or molecular compounds by testing their solubility in water, their electrical conductivity in the solid state and when dissolved in water. What evidence would you expect to observe for each type of compound?

Application

55. Models are helpful for representing natural phenomena and objects that are too large or too small to observe directly. As new technology and discoveries lead to new information, scientists must revise their models to reflect this new information. Examine the models of atoms that are shown below.

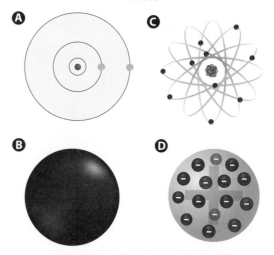

a. In which order were the models developed?

b. Identify the scientist who is associated with each model.

c. Briefly describe each model, and explain how it differs from the previous model.

d. Sketch one other model of the atom, identify the scientist associated with it, and explain how it differs from the other models.

56. Describe a model that helped you understand a difficult concept in this or another science class that you have taken.

57. Calculate the atomic mass of vanadium, which is composed of 0.2504% vanadium-50 (mass 49.95 u) and 99.75% vanadium-51 (mass 50.94 u).

58. Draw a Lewis diagram for each atom. Then predict the change that must occur in each atom to form an ion. Determine the charge of the resulting ion.
- **a.** phosphorus
- **b.** oxygen
- **c.** strontium
- **d.** lithium

59. An atom of aluminum gives away three electrons when it joins with one or more non-metal atoms to form an ionic compound. What ratio of aluminum and each non-metal atom below would be required to have a net charge of zero?
- **a.** chlorine, which can gain one electron
- **b.** nitrogen, which can gain three electrons
- **c.** oxygen, which can gain two electrons

60. Predict whether an ionic compound or a molecular compound will form in the reaction between each pair of elements.
- **a.** carbon and chlorine
- **b.** oxygen and fluorine
- **c.** calcium and sulfur

61. Which is more polar: NH_3 or PH_3? Explain your reasoning.

PH₃   NH₃

62. Locate three cleaning products in your home, and read the list of ingredients. Use Internet or print resources to identify each chemical as either an ionic compound or a molecular compound (if possible) and write the chemical formula for the compound.

63. Your liver contains enzymes (compounds that speed up chemical reactions) that convert insoluble compounds into soluble compounds. Why do you think it is beneficial to convert insoluble compounds into soluble compounds in your body? What do you think happens to the compounds after they become soluble?

64. This ball-and-stick model represents a molecule that is composed of atoms of two different elements.
- **a.** Describe the shape of the molecule.
- **b.** Could the molecule be non-polar if the bonds between the atoms have an electronegativity difference of 0.85? Explain your reasoning.
- **c.** What would you expect the state of this compound to be at room temperature and normal atmospheric pressure? Explain your reasoning, based on the intermolecular forces you would predict for this compound to exhibit.

65. Oven cleaners and drain cleaners, like the one shown here, usually contain sodium hydroxide or another strong base. These strong bases are very corrosive. Why do you think that a corrosive chemical would be used to clean ovens and drains? Suggest a possible alternative.

66. Examine the following model of a compound.

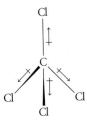

- **a.** Describe the polarity of the bonds, and identify the locations of the partial charges.
- **b.** Describe the polarity of the molecule.
- **c.** Would you expect this compound to be a solid at room temperature? Explain your reasoning.

Select the letter of the best answer below.

1. **K/U** As scientists gathered new information about the atom, they revised the model of the atom to incorporate the new knowledge. Examine the atomic model shown. Which statement best describes the discovery that resulted in this model?

   a. When alpha particles are directed at a thin gold foil, some of them bounce back.
   b. Atoms are indivisible and join together in whole-number ratios.
   c. When atoms are heated, they give off specific patterns of coloured light.
   d. Atoms can be broken down into smaller particles using a cathode-ray tube.
   e. Mathematical analyses show that electrons do not travel in precise orbits.

2. **K/U** Which statement correctly describes the relationship between the atomic number and the mass number of an element?
   a. The mass number equals the atomic number plus the number of protons.
   b. The mass number equals the atomic number plus the number of neutrons.
   c. The atomic number equals the mass number plus the number of neutrons.
   d. The mass number plus the atomic number equals the number of neutrons.
   e. The mass number plus the number of neutrons equals the atomic number.

3. **K/U** Based on the locations of the elements on the periodic table, which of the following atoms has the largest radius?
   a. rubidium   **d.** helium
   b. tin   **e.** boron
   c. lithium

4. **K/U** Based on the locations of the elements on the periodic table, which of the following atoms has the highest ionization energy?
   a. strontium   **d.** hydrogen
   b. calcium   **e.** beryllium
   c. fluorine

5. **K/U** Based on the locations of the elements on the periodic table, which of the following atoms has the lowest electronegativity?
   a. iodine   **d.** strontium
   b. tellurium   **e.** chlorine
   c. sulfur

6. **K/U** Which group of elements has atoms that satisfy the octet rule without forming chemical bonds?
   a. transition metals   **d.** lanthanoids
   b. halogens   **e.** noble gases
   c. alkali metals

7. **K/U** Which pair of atoms would form a polar covalent bond?
   a. sodium and nitrogen   **d.** sulfur and chlorine
   b. hydrogen and hydrogen   **e.** calcium and bromine
   c. carbon and sulfur

8. **K/U** Which statement about ionic bonding is correct?
   a. A non-metal atom shares an electron with another non-metal atom to form an ionic bond.
   b. A non-metal atom transfers an electron to a metal atom to form an ionic bond.
   c. A non-metal atom transfers an electron to another non-metal atom to form an ionic bond.
   d. A metal atom transfers an electron to a non-metal atom to form an ionic bond.
   e. A metal atom shares an electron with a non-metal atom to form a covalent bond.

9. **K/U** What is the correct name for $AlPO_4$?
   a. aluminum phosphite
   b. aluminum phosphate
   c. aluminum phosphorus tetroxide
   d. aluminum phophorous
   e. aluminum(III) phosphoroxide

10. **K/U** What is the correct formula for sulfur dioxide?
    a. SO_2 **d.** S_2O_2
    b. S_2O **e.** SO_2^{2-}
    c. $(SO)_2$

Use sentences and diagrams, as appropriate, to answer the questions below.

11. **C** Draw a Bohr-Rutherford model of an atom of phosphorus. Explain how you can examine the electron arrangement to determine where phosphorus is located in the periodic table. Identify the period and group number.

12. **K/U** Compare isotopes with radioisotopes.

13. **T/I** Silver exists naturally as 51.84% silver-107 (mass = 106.91 u) and 48.16% silver-109 (mass = 108.90 u). What is the average atomic mass of silver?

14. **C** Hydrogen and helium have similar physical properties but different chemical properties.
    a. How are hydrogen and helium alike?
    b. Draw a Lewis diagram for an atom of each element. How do the electron arrangements in the atoms explain the differences between the chemical properties of the two elements?

15. **K/U** Explain the connection between the trend in atomic radius and the trend in electronegativity across a period. Use diagrams to help support your explanation.

16. **A** Some groups in the periodic table contain elements that have very similar properties. Group 14, however, includes elements that differ in certain properties.
    a. Categorize the elements in Group 14 based on their properties.
    b. Use the trend in atomic radius to explain the differences in the properties of the elements in this group.

17. **A** The alkali metals become more reactive as you move down their group in the periodic table. Explain this observation in terms of ionization energy, atomic radius, and the shielding effect.

18. **T/I** Draw a Lewis diagram for each atom. Then predict the change that must occur in each atom to form an ion.
    a. barium
    b. sulfur
    c. potassium
    d. nitrogen

19. **T/I** Identify each compound as ionic or covalent. Then use the correct method to write the chemical formula for each compound.
    a. magnesium nitride
    b. oxygen difluoride
    c. tin(II) bromide
    d. aluminum phosphate
    e. cobalt(III) sulfite

20. **T/I** Identify each compound as ionic or covalent. Then use the correct method to write the name of each compound.
    a. PCl_5
    b. Li_2CO_3
    c. CuO
    d. N_2O_3
    e. NH_4NO_2

21. **C** Create a flowchart that outlines the steps you would follow to write the name of a binary compound if you were given its formula. Keep in mind that the compound could be ionic or molecular, and it could contain a multivalent metal.

22. **C** Explain how ambiguity is avoided in the names of binary molecular compounds and in the names of ionic compounds.

23. **K/U** Compare substances containing ionic compounds and molecular compounds with respect to the properties of melting point, electrical conductivity, state at room temperature, and solubility in water.

24. **T/I** Table salt and sugar are white solids that dissolve in water. However, one is a molecular compound and one is an ionic compound. Describe two laboratory procedures that you could use to identify unmarked samples of these substances. Describe the results you would expect to observe.

25. **A** The differences in the properties of compounds often reflect the type of chemical bonding in the compounds. This is especially true for molecular compounds that are composed of similar molecules. Examine the following data for the compounds chloromethane, CH_3Cl, and methane, CH_4.

Comparing Properties of Two Compounds

Compound	Melting Point (°C)	Boiling Point (°C)
CH_3Cl	−97.7	−24.2
CH_4	−182.5	−161.5

a. Draw a Lewis structure for each molecule.
b. Compare the polarity of the C-H bond with the polarity of the C-Cl bond, and explain any differences.
c. Explain the differences in the properties listed in the table, based on your understanding of polarity and intermolecular forces.

Self-Check

If you missed question …	1	2	3	4	5	6	7	8	9	10	11	12	13	14	15	16	17	18	19	20	21	22	23	24	25
Review section(s)…	1.1	1.1	1.3	1.3	1.3	2.1	2.1	2.1	2.2	2.2	1.1, 1.2	1.1	1.1	1.2	1.3	1.3	1.2, 1.3	1.1; 2.1	2.2	2.2	2.2	2.2	2.3	2.3	2.1; 2.3

Chemical Reactions

BIG IDEAS

- Chemicals react in predictable ways.

- Chemical reactions and their applications have significant implications for society and the environment.

Overall Expectations

In this unit, you will…

- **analyze** chemical reactions used in a variety of applications, and assess their impact on society and the environment

- **investigate** different types of chemical reactions

- **demonstrate** an understanding of the different types of chemical reactions

Unit Contents

Chapter 3
Synthesis, Decomposition, and Combustion Reactions

What are important examples of synthesis, decomposition, and combustion reactions and their applications?

Chapter 4
Displacement Reactions

What roles do single displacement reactions and double displacement reactions play in your daily life and in the environment?

Our society relies on many different types of chemical reactions for essential products and technologies. The applications of chemical reactions are numerous and range from life-saving medicines to the formation of our transport systems, such as the railroad tracks shown here. Each year, Canada's railway system transports more than four million passengers and 300 million tonnes of cargo to destinations in Canada and the United States. In fact, over 48 000 km of steel railroad tracks crisscross North America. In the past, the only way to connect sections of tracks was to bolt them together. However, this method meant that intense vibrations and stress could weaken the joints. Today, railroad tracks are connected by welding them together. This produces a continuous track that requires less maintenance, provides a smoother ride, and allows trains to travel at higher speeds. A common method of track welding uses a chemical reaction called a thermite reaction. In a thermite reaction, powdered iron oxide reacts with powdered aluminum. The reaction releases a tremendous amount of heat, which melts steel that is held in a heat-resistant container. The molten steel then flows from the container into a mould encasing the ends of two sections of track. After the steel has hardened and the mould is removed, a grinding tool is used to smooth the rail surface.

As you study this unit, look ahead to the **Unit 2 Project** on pages 206 to 207, which gives you an opportunity to demonstrate and apply your new knowledge and skills. Keep a planning folder so you can complete the project in stages as you progress through the unit.

Preparation

Safety in the Laboratory

- Always wear a lab apron and safety eyewear when working in the laboratory with materials that could pose a safety problem. These include harmful chemicals such as acids and bases, unidentified substances, and substances that are being heated.

- Know any safety precautions before beginning an investigation. Also, make sure you understand the associated safety and WHMIS symbols that have been provided.

- Light the flame of a Bunsen burner quickly once the flow of gas begins. The air intake can be adjusted to produce a blue flame.

- When heating a substance over a flame, point the open end of the container that is being heated away from yourself and others. Never allow a container to boil dry.

- Immediately wash any part of your body that comes in contact with a laboratory chemical.

- If you are asked to smell a substance, hold the container slightly in front of and beneath your nose, and then waft the fumes toward your nostrils.

1. Describe what each of the following symbols means and any precautions that must be taken when you see the symbol.

   a.

   b.

   c.

   d.

   e.

2. Why should you point the open end of a container heated by a Bunsen burner flame away from yourself and others?

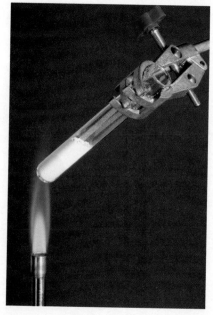

3. You have finished your investigation procedure and are washing out the test tubes that you used. Should you continue wearing your safety eyewear? Explain your reasoning.

4. Make a table in your notebook that
   a. identifies the parts of the Bunsen burner shown below.
   b. describes the function of each component.

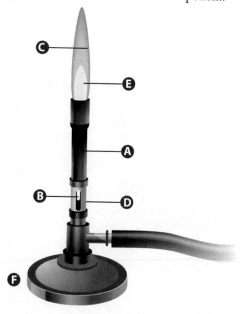

5. During an investigation, you turn on the gas to your Bunsen burner but then have trouble lighting it. Describe the steps you should take.

6. An investigation requires you to add two solutions to a test tube and heat the mixture gently until it changes colour. Describe a safe method for performing this procedure.

Naming Compounds and Writing Their Formulas

- Ionic compounds form as a result of the transfer of electrons from metal atoms to non-metal atoms.

- The name of an ionic compound identifies the positive and negative ions that make up the compound. In the chemical formula of an ionic compound, subscripts are used to indicate the number of atoms of each element in the compound. To write the chemical formula, place the subscripts after the corresponding symbol for each element in the compound.

- Molecular compounds form as a result of the sharing of electrons between non-metal atoms.

- The name of a molecular compound identifies the elements that make up the compound. Prefixes are used to identify the number of atoms of each element in the compound. If a prefix is not included in the name, then only one atom of the element is present in each molecule of the compound.

- In the chemical formula of a molecular compound, subscripts are used to indicate the number of atoms of each element and are based on the prefixes in the compound's name. To write the chemical formula, place the subscripts after the corresponding symbol for each element in the compound.

7. Which of the following elements is a non-metal?
   a. Li
   b. Cd
   c. Ti
   d. Mn
   e. Cl

8. Which of the following is a molecular compound?
   a. NaCl
   b. $CaSO_4$
   c. AlF_3
   d. CBr_4
   e. LiH

9. Ionic compounds and molecular compounds form different types of chemical bonds.
   a. What types of elements tend to make up ionic compounds?
   b. What type of elements tend to make up molecular compounds?
   c. Describe the difference between how electrons are involved in the formation of ionic compounds and how they are involved in the formation of molecular compounds.

10. Identify each compound as ionic or molecular, and write its chemical formula.
    a. lithium fluoride
    b. phosphorus trichloride
    c. dinitrogen trioxide
    d. aluminum sulfate
    e. carbon tetrabromide
    f. copper(II) chloride

11. The mineral cuprite, shown in the photograph below, is dark red in colour. Cuprite is made of the compound $Cu_2O(s)$. What is the correct name for $Cu_2O(s)$?
    a. copper oxide
    b. copper(I) oxide
    c. cobalt oxide
    d. copper(II) oxide
    e. dicobalt monoxide

12. Copy the table below in your notebook. Complete the table by providing the number of atoms and prefixes that are used when naming molecular compounds.

Prefixes Used in Naming Molecular Compounds

Number of Atoms	Prefix	Number of Atoms	Prefix
1	mono-	2	
	hepta-	5	
3			octa-
6		4	
	nona-		deca-

The Law of Conservation of Mass

- Chemical equations represent a chemical reaction using symbols and chemical formulas.
- The law of conservation of mass states that in a chemical reaction, the total mass of the products is always the same as the total mass of the reactants.
- Coefficients are used to balance an equation to reflect the law of conservation of mass. The same number of each kind of atom must appear in both the reactants and the products.

13. How many atoms of oxygen are represented in the formula $Ca(ClO_2)_3$?

    a. 2

    b. 4

    c. 6

    d. 8

    e. 10

14. In which of the following equations does the same number of each kind of atom appear in both the reactants and the products?

    a. $H_2(g) + O_2(g) \rightarrow H_2O(g)$

    b. $H_2(g) + O_2(g) \rightarrow 2H_2O(g)$

    c. $2H_2(g) + O_2(g) \rightarrow H_2O(g)$

    d. $2H_2(g) + O_2(g) \rightarrow 2H_2O(g)$

    e. $H_2(g) + 2O_2(g) \rightarrow 2H_2O(g)$

15. In your own words, describe the law of conservation of mass.

16. The starting materials of a reaction are placed in a sealed container, and the container is placed on a balance. The mass reading on the balance is 50.0 g. The chemicals are mixed, and a chemical reaction begins. What do you predict will happen to the mass reading as the reaction continues? Explain your reasoning.

17. Why is changing a subscript in a chemical equation an incorrect method of representing the law of conservation of mass?

18. A student says that one formula unit of $Mg_3Al_2(SiO_4)_3(s)$ contains more silicon atoms than one formula unit of $Mg_2Si_2O_6(s)$ does. Do you agree? Explain your answer.

Identifying Evidence of a Chemical Change

- During a chemical reaction, a new substance is produced that has a different composition and different properties compared to the starting materials.
- Evidence of a chemical change may include one or more of the following: formation of a gas, formation of a precipitate, change in odour, change in colour, change in temperature, and production of light.

19. What is the result of every chemical reaction?

    a. A solid forms.

    b. A new substance forms.

    c. Atoms are created.

    d. Energy is released.

    e. Compounds are created.

20. What evidence of a chemical reaction can you identify in each of the following images?

    a.

b.

21. What evidence of a chemical reaction happens when you ignite a Bunsen burner?

22. Explain why the following events suggest, but do not prove, that a chemical change has occurred.

    a. the production of a gas

    b. the release of energy

Chemical Properties

- A chemical property describes the ability (or inability) of a substance to react with another substance to form new substances.

- A chemical property can be determined only by trying to change the chemical make-up of the substance.

- Predicting the results of chemical reactions requires a good understanding of chemical properties, such as reactivity with water, with oxygen, and with acids.

23. You wish to determine the reactivity of a metal with acid. Describe a procedure you might follow.

24. The instructions to build a model volcano say to mix baking soda with something to model the eruption. Unfortunately, you cannot read what the other substance is. When you mix baking soda with water, nothing happens. When you use vinegar, bubbles of gas form. What can you conclude about the chemical properties of baking soda?

25. Oxygen gas makes up about 20 percent of the atmosphere. Many elements are reactive with oxygen and form compounds when exposed to air.

a. The elements gold and platinum are found in their uncombined forms in nature. What can you conclude about the reactivity of these elements with oxygen?

b. Based on the photograph below, what evidence do you see that different parts of the object have different reactivities with oxygen? What conclusions can you draw from this evidence?

Chemical Reactions and Laboratory Tests

- Four types of chemical reactions are synthesis, decomposition, single displacement, and double displacement.

- When studying a chemical reaction, identifying the products that are formed is important. A burning or glowing wooden splint can be used to identify hydrogen, oxygen, and carbon dioxide gases that are produced from chemical reactions.

- A burning splint inserted into hydrogen gas in a test tube causes the gas to ignite, resulting in a popping sound.

- A burning splint inserted into carbon dioxide gas in a test tube ceases to burn.

- A glowing splint inserted into oxygen gas in a test tube will start burning again.

- Acids and bases can react to produce a neutral solution, which has a pH of 7. An acidic solution has a pH of less than 7, and a basic solution has a pH of more than 7.

- A pH indicator changes colour to show the acidity or basicity of a solution. For example, litmus is red in acidic solutions and blue in basic solutions. Bromothymol blue is yellow in acidic solutions and blue in basic solutions.

26. Identify each of the following as a synthesis reaction, a decomposition reaction, a single displacement reaction, or a double displacement reaction.

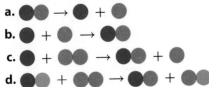

27. What role can pH indicators play in testing the products of a chemical reaction?

28. Litmus paper is often used to test the products of a chemical reaction. What can you conclude about the products if you observe the following changes?
a. Red litmus paper turns blue.
b. Blue litmus paper turns red.

29. Gases are sometimes produced during a chemical reaction. Explain how a wooden splint can be used to test for each of the following gases.
a. hydrogen
b. oxygen
c. carbon dioxide

Synthesis, Decomposition, and Combustion Reactions

Specific Expectations

In this chapter, you will learn how to . . .

- C1.1 **analyze**, on the basis of research, chemical reactions used in various industrial processes that can have an impact on the health and safety of local populations (3.1, 3.2, 3.3)

- C2.1 **use** appropriate terminology related to chemical reactions (3.1, 3.2, 3.3)

- C 2.2 **write** balanced chemical equations to represent synthesis, decomposition, and combustion reactions, using the IUPAC nomenclature system (3.1, 3.2, 3.3)

- C2.3 **investigate** synthesis and decomposition reactions by **testing** the products that are formed in each reaction (3.2)

- C2.4 **predict** the products that are formed in different types of synthesis and decomposition reactions (3.2)

- C2.7 **design** an inquiry to **demonstrate** the difference between a complete and incomplete combustion reaction (3.3)

- C2.8 **plan** and **conduct** an inquiry to **compare** the properties of non-metal oxide solutions and metal oxide solutions (3.2)

- C3.1 **identify** various types of chemical reactions, including synthesis, decomposition, and combustion (3.2, 3.3)

- C3.2 **explain** the difference between a complete combustion reaction and an incomplete combustion reaction (3.3)

- C3.3 **explain** the chemical reactions that result in the formation of acids and bases from metal oxides and non-metal oxides (3.2)

In 2008, the Phoenix Mars Lander travelled from Earth to Mars. Among the devices on Phoenix was a Canadian meteorological station, designed to measure Martian weather conditions. Phoenix made its final descent on pulses of gases fired from thruster rockets, as shown in this artist's depiction. These gases—ammonia, nitrogen, and hydrogen—were the products of the chemical decomposition of the rocket fuel hydrazine, $N_2H_4(\ell)$. This application is just one example of the many technologies that rely on an understanding of chemical reactions.

Causing Chemical Reactions

The Phoenix Mars Lander relied on chemical reactions to settle gently on the surface of Mars. Here on Earth, chemical reactions are everywhere in your daily life. In this activity, you will observe various changes that take place during chemical reactions.

Safety Precautions

- Wear safety eyewear throughout this activity.
- Tie back loose hair and clothing.
- Use EXTREME CAUTION when you are near an open flame.

Materials

- 10 mL of blue food colouring solution
- 5 mL of household bleach (sodium hypochlorite solution, $NaOCl(aq)$) in a dropper bottle
- 4 mL of 3% hydrogen peroxide solution, $H_2O_2(aq)$
- 4 mL of household bleach, $NaOCl(aq)$
- wooden splint
- 10 mL graduated cylinder
- 2 beakers (50 mL)
- test tube (36 mL) with stopper
- test tube rack
- Bunsen burner secured to a utility stand
- igniter for Bunsen burner

Procedure

1. Read the Procedure, and then create a table to record your observations. Give your table an appropriate title.

2. Use a graduated cylinder to pour 5 mL of the blue food colouring solution into each of two 50 mL beakers. Set aside the first beaker as a control.

3. Use the dropper bottle to add 20 drops of household bleach to the second beaker. After a few minutes, record your observations.

4. Put the test tube in the test tube rack. Pour 4 mL of hydrogen peroxide into the test tube. Add 4 mL of household bleach, and then quickly put the stopper in the test tube. Record your observations.

5. Light the Bunsen burner. Ignite the splint, and then blow out its flame. Remove the test tube stopper and put the glowing end of the splint into the top of the test tube. Record your observations.

Questions

1. What evidence do you have that a chemical reaction occurred between the food colouring and the bleach?

2. Based on your observations of the glowing splint, what was the gas produced in the reaction between hydrogen peroxide and bleach?

Writing Chemical Equations

Key Terms

chemical reaction

reactant

product

chemical equation

skeleton equation

balanced chemical
equation

coefficient

chemical reaction
a process in which
substances interact,
causing different
substances with different
properties to form

reactant a starting
substance in a chemical
reaction

product a substance
that is formed in a
chemical reaction

chemical equation
a condensed statement
that expresses chemical
change using symbols
and chemical names or
formulas

You probably have no trouble understanding the message in **Figure 3.1**. Over time, you have learned the shortened words and abbreviations that are used to keep text messages brief, yet still express the intended information. In Unit 1, you learned how chemists use chemical formulas to convey a large amount of information about chemical compounds using a few simple characters—much like text messaging.

Chemists also need to convey information about what happens to chemicals and how they change when they interact in a chemical reaction. A **chemical reaction** is a process in which substances interact to form one or more different substances. In this section, you will learn two ways to represent chemical reactions.

Figure 3.1 Text messages use widely recognized abbreviations to convey information in a small amount of space. Similarly, chemists use symbols to communicate information about chemical processes.

Describing Chemical Reactions

When you see a rusty car or bridge, such as the bridge in **Figure 3.2**, you know that the rust is not a material that was originally used to build the structure. The iron in the steel has reacted with oxygen, causing rust to form. Rust is a substance with chemical and physical properties that are very different from those of iron. In every chemical reaction, the **reactants**, or starting substances, interact and form different substances, called the **products**. In this example, iron and oxygen are the reactants, and rust is the product of the reaction.

Suppose that you wanted to describe a chemical reaction, such as the burning of charcoal, C(s), or the bubbling that occurs when acetic acid, $CH_3COOH(aq)$, in vinegar reacts with sodium hydrogen carbonate, $NaHCO_3(s)$, in baking soda. You could write a sentence that described the reactants and products. You could also give some details about the reaction process. However, this is a long and tedious way to share the information. A **chemical equation** is a condensed statement that expresses chemical change using symbols and chemical names or formulas. A chemical equation can be a word equation, a skeleton equation, or a balanced chemical equation. In this section, you will learn how to write three forms of a chemical equation: a word equation, a skeleton equation, and a balanced chemical equation.

Figure 3.2 The rust that has formed on this bridge is the product of a chemical reaction in which iron and oxygen are the reactants.

Writing Word Equations

In a word equation, the reactants and products are identified by their names only. For example, a word equation for one of the reactions involved in the formation of rust is shown below.

$$\text{iron} + \text{oxygen} \rightarrow \text{iron(III) oxide}$$

reactants   **product**

This equation can be read as follows: iron and oxygen react to form iron(III) oxide. The reactants are on the left side of an arrow, and the products are on the right side. The arrow shows the direction of the chemical change that is taking place. The arrow stands for *yields* or *reacts to produce*. If there are two or more substances on either side of the arrow, a plus sign is placed between their names. **Table 3.1** lists symbols commonly used in word equations and other chemical equations.

Note that some reactions are reversible, which means that the products can be changed back into the reactants. To illustrate this, a special type of arrow symbol is used, as shown in **Table 3.1**. For example, the chlorine-bleaching process that removes colour during the manufacturing of white paper involves a reversible chemical reaction. In this reaction, aqueous solutions of hypochlorous acid, HOCl(aq), and hydrochloric acid react to form chlorine gas and liquid water. These products, chlorine gas and liquid water can, in turn, react to form the original products. Here is the word equation for this reaction:

$$\text{hypochlorous acid} + \text{hydrochloric acid} \rightleftharpoons \text{chlorine} + \text{water}$$

Table 3.1 Symbols Used in Chemical Equations

Symbol	Purpose
+	Indicates that two or more reactants or products are involved
→	Shows the direction of the chemical change that is taking place
⇌	Indicates a reversible reaction

Writing Skeleton Equations

A **skeleton equation** is a chemical equation in which chemical formulas are used to represent the substances that are involved in a chemical reaction. The relative quantities of the substances, however, are not included.

Because skeleton equations are written using chemical formulas, they convey more information than word equations do. Skeleton equations clearly show the number of atoms of each element in the reactants and products, while word equations do not. In addition, skeleton equations generally show the physical state of each substance by including one of the symbols shown in **Table 3.2** after each chemical formula. The word equation and skeleton equation for the formation of iron (III) oxide, commonly referred to as rust, are shown below. **Figure 3.3** shows powdered iron (III) oxide.

Word equation: iron $+$ oxygen $\rightarrow$ iron(III) oxide

Skeleton equation: $Fe(s) + O_2(g) \rightarrow Fe_2O_3(s)$

Figure 3.3 Iron(III) oxide is an inorganic solid with the formula Fe_2O_3. In nature, this substance is the mineral hematite, which is a principal source of iron for the mining industry. Since prehistoric times, people around the world have used iron(III) oxide as a colouring agent (pigment) for painting. In this context, the substance is often called red ochre.

Table 3.2 Symbols Used in Chemical Equations for Physical State

Symbol	Purpose
(s)	Identifies a solid state
(ℓ)	Identifies a liquid state
(g)	Identifies a gaseous state
(aq)	Identifies an aqueous solution

In the case of the reversible reaction involving the chlorine-bleaching process, the reactants are in aqueous solution, and the products are a gas and liquid water. The word equation and skeleton equation for this reaction are given below.

Word equation:

hypochlorous acid $+$ hydrochloric acid $\rightleftharpoons$ chlorine $+$ water

Skeleton equation:

$HClO(aq) + HCl(aq) \rightleftharpoons Cl_2(g) + H_2O(\ell)$

You can practise writing skeleton equations that represent chemical reactions by working through the Sample Problem and Practice Problems on the next page.

Writing Skeleton Equations

Problem

When solid sodium carbonate is added to an aqueous solution of hydrochloric acid, liquid water, carbon dioxide gas, and an aqueous solution of sodium chloride are formed. What is the skeleton equation for this chemical reaction?

What Is Required?

A skeleton equation that represents the chemical reaction is required.

What Is Given?

You are given the reactants: solid sodium carbonate and an aqueous solution of hydrochloric acid.

You are given the products: liquid water, carbon dioxide gas, and an aqueous solution of sodium chloride.

Plan Your Strategy	Act on Your Strategy
Determine the chemical formula for each substance. Include the state.	Solid sodium carbonate: $Na_2CO_3(s)$ Aqueous solution of hydrochloric acid: $HCl(aq)$ Liquid water: $H_2O(\ell)$ Carbon dioxide gas: $CO_2(g)$ Aqueous solution of sodium chloride: $NaCl(aq)$
Write the skeleton equation. Use an arrow to show the direction of the chemical change that is taking place. Use a plus sign to separate two or more reactants or products.	$Na_2CO_3(s) + HCl(aq) \rightarrow H_2O(\ell) + CO_2(g) + NaCl(aq)$

Check Your Solution

The reactants are written on the left side of the arrow, and the products are written on the right side. The chemical formula for each substance is written correctly. The physical state of each reactant and product is shown.

Write a skeleton equation for each chemical reaction. Indicate the state of each reactant and product in the skeleton equation. Remember that the following seven elements are diatomic: hydrogen, $H_2(g)$; nitrogen, $N_2(g)$; oxygen, $O_2(g)$; fluorine, $F_2(g)$; chlorine, $Cl_2(g)$; bromine, $Br_2(\ell)$; and iodine, $I_2(s)$.

1. Gaseous hydrogen and oxygen react to form gaseous water.

2. Solid sodium metal reacts with liquid water to form an aqueous solution of sodium hydroxide and hydrogen gas.

3. Solid potassium chlorate breaks down to form solid potassium chloride and oxygen gas.

4. Solid copper reacts with oxygen gas to form solid copper(II) oxide.

5. When aqueous solutions of silver nitrate and sodium chloride are combined, the reaction produces an aqueous solution of sodium nitrate and a precipitate of silver chloride.

6. The complete combustion of propane gas, $C_3H_8(g)$, in the presence of oxygen gas forms gaseous water and carbon dioxide.

7. Sulfur trioxide gas reacts with liquid water to form an aqueous solution of sulfuric acid.

8. Solid ammonium chloride is formed when hydrogen chloride gas reacts with gaseous ammonia.

9. Solid aluminum and gaseous fluorine form when solid aluminum fluoride breaks down.

10. Liquid mercury reacts with oxygen gas to form solid mercury(II) oxide.

In a chemical reaction, matter is conserved. All the atoms in the reactant(s) are incorporated into the product(s). A different quantity of molecules may be formed, but the amount of each type of atom is conserved. In this activity, you will model the reaction of molecular hydrogen with molecular oxygen to form water.

Materials

- models of 8 hydrogen atoms
- models of 4 oxygen atoms
- 8 connectors

Procedure

1. Divide the models of atoms into two equal sets of four hydrogen atoms and two oxygen atoms. One set of atoms will represent the reactants. The other set will represent the products.

2. Using the set for the reactants, assemble two hydrogen molecules and one oxygen molecule to represent the reactants.

3. Using the set for the products, assemble as many water molecules as possible to represent the products.

Questions

1. a. How many models of atoms were left over after you assembled the reactants and the products?

   b. What do your results tell you about the conservation of atoms in the reaction of molecular hydrogen with molecular oxygen to form water?

2. a. How many models did you assemble to represent reactant molecules?

   b. How many models did you assemble to represent product molecules?

   c. When hydrogen and oxygen react, are molecules conserved? Explain.

Beyond Skeleton Equations

A skeleton equation is more useful than a word equation because it shows the formulas and states of the substances in a chemical reaction. However, it is still not a complete description of what happens in the reaction. Compare the reactants and products in **Figure 3.4**:

Figure 3.4 As this model shows, a skeleton equation is an incomplete representation of a reaction because it does not have an equal number of atoms of each element in both the reactants and the products.

Compare *the number of atoms of each element shown on the reactant side and the product side of the chemical equation.*

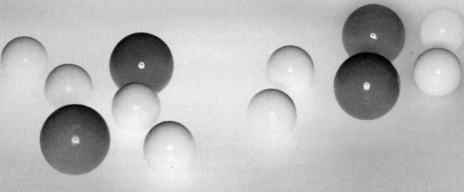

| aluminum | + | bromine | → | aluminum bromide |
| Al(s) | + | $Br_2(g)$ | → | $AlBr_3(s)$ |

one aluminum atom   two bromine atoms   one aluminum atom and three bromine atoms

From the models in **Figure 3.4**, you can see that the number of atoms of bromine on the reactant side is not equal to the number of atoms of bromine on the product side. While skeleton equations clearly show the substances that are involved in a reaction, they do not show the relative numbers of atoms, molecules, and ions.

1. Describe the connection between a chemical reaction and a skeleton equation.

2. According to **Table 3.1**, what symbol is used to represent the physical state of a substance that is dissolved in water?

3. In photosynthesis, plants produce glucose, $C_6H_{12}O_6(s)$, and oxygen from carbon dioxide and water.
   a. What information would be included in a skeleton equation representing this chemical reaction?
   b. What information would not be included?

4. Explain the meaning of the following symbol in a chemical equation:
$$\rightleftharpoons$$

5. The following skeleton equation was written to represent solid sodium metal reacting with liquid water to form an aqueous solution of sodium hydroxide and hydrogen gas.
$$Na(s) + H_2O(g) + Na(OH)_2(\ell) + H(g)$$
   a. Analyze the skeleton equation, and identify the errors in it.
   b. Write the skeleton equation in correct form.

6. Identify the reactants and products in each reaction, and write a skeleton equation for each reaction.
   a. Bubbles of hydrogen gas and oxygen gas form as water is broken down using an electric current.
   b. Liquid bromine and zinc chloride form as chlorine gas is bubbled through an aqueous solution of zinc bromide.

Writing Balanced Chemical Equations

To reflect the ratios of substances involved in a chemical reaction accurately, an equation must show an equal number of atoms of each element in both the reactants and the products. A **balanced chemical equation** is a statement that uses chemical formulas and coefficients to show the identities and ratios of the reactants and products in a chemical reaction. A **coefficient** is a positive number that is placed in front of a chemical formula to show the relative number of particles (atoms, ions, molecules, or formula units) of the substance that are involved in the reaction.

Figure 3.5 shows the balanced chemical equation for the reaction of aluminum and bromine. There are two aluminum atoms and six bromine atoms on each side of the equation. The whole numbers 2, 3, and 2 in front of the formulas are coefficients.

> **balanced chemical equation** a statement that uses chemical formulas and coefficients to show the identities and ratios of the reactants and products in a chemical reaction
>
> **coefficient** in a balanced chemical equation, a positive number that is placed in front of a formula to show the relative number of particles of the substance that are involved in the reaction

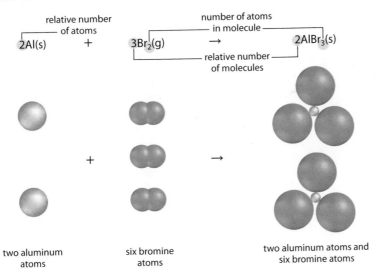

two aluminum atoms   six bromine atoms   two aluminum atoms and six bromine atoms

Figure 3.5 In a balanced chemical equation, each atom on the reactant side must have a corresponding atom of the same element on the product side.

Explain why six bromine atoms are shown on each side of the equation.

How to Balance a Chemical Equation

Most chemical equations can be balanced by following the steps given in **Table 3.3**. As you gain more experience balancing chemical equations, you may discover patterns when balancing certain types of equations. These patterns will help you become more proficient and faster at balancing the equations. As you read each step in the table, closely examine how it applies to balancing the equation for the reaction of hydrogen and chlorine to form hydrogen chloride.

Table 3.3 Steps for Balancing Chemical Equations

Step	Process	Example
1	**Write the skeleton equation for the reaction.** Make sure that the chemical formulas correctly represent the substances. Use an arrow to separate the reactants from the products. Use a plus sign to separate multiple reactants and products. Show the physical state of each reactant and product.	$H_2(g)$ + $Cl_2(g)$ → $HCl(g)$ two hydrogen atoms · two chlorine atoms · one hydrogen atom and one chlorine atom
2	**Count the atoms of each element in the reactants. Identify any polyatomic ions that are present.** The reaction at the right does not involve any polyatomic ions. If the reactants and products in another reaction contain the same polyatomic ion, count it as a single item rather than counting each atom within it.	$H_2(g) + Cl_2(g) \rightarrow$ Two atoms of hydrogen and two atoms of chlorine appear in the reactants. There are no polyatomic ions present.
3	**Count the atoms of each element in the products. Identify any polyatomic ions that are present.**	$HCl(g)$ One atom of hydrogen and one atom of chlorine appear in the products. There are no polyatomic ions present.
4	**Add coefficients so that you have an equal number of atoms of each element on both sides of the equation. If different numbers of the same polyatomic ion are present on either side of the equation, add a coefficient to the compound that will cause the number of the ion to be equal on both sides.** Do *not* change a subscript in a chemical formula to balance an equation. This would change the formula to the formula for a different substance—a substance that is not involved in the reaction.	$H_2(g)$ + $Cl_2(g)$ → $2HCl(g)$ 2 atoms H · 2 atoms Cl · 2 atoms H + 2 atoms Cl two hydrogen atoms · two chlorine atoms · two hydrogen atoms and two chlorine atoms
5	**Write the coefficients in their lowest possible ratio.**	$H_2(g) + Cl_2(g) \rightarrow 2HCl(g)$ The ratio 1 hydrogen to 1 chlorine to 2 hydrogen chloride (1:1:2) is the lowest possible ratio because the coefficients cannot be reduced further and still be whole numbers.
6	**Check your work.** Make sure that the chemical formulas are written correctly. Then check that the number of atoms of each element is equal on both sides of the equation. You may want to examine each element in a polyatomic ion separately when checking your solution.	$H_2(g) + Cl_2(g) \rightarrow 2HCl(g)$ There are two hydrogen atoms and two chlorine atoms on both sides of the equation.

Tips for Balancing Equations

Balancing equations is an essential skill to master in your study of chemistry. The steps that are summarized in **Figure 3.6** will help you balance many equations throughout this course. If you are having trouble balancing a particularly difficult or complex equation, the following tips may be helpful:

Go to **scienceontario** to find out more

- Try balancing the substance with the largest number of atoms, on either side of the equation, first. Balance hydrogen, oxygen, and any element that appears in more than two substances later.
- Balance polyatomic ions as single items, as mentioned in **Table 3.3**.
- If the addition of coefficients to balance one type of atom or polyatomic ion causes another atom or polyatomic ion to be unbalanced, then repeat steps 2 through 4 in **Table 3.3**.
- If you find yourself repeatedly adjusting the same two coefficients, double-check that the formulas are correct.
- An incorrect formula can prevent an equation from balancing. Keep in mind the seven diatomic elements: hydrogen, $H_2(g)$; nitrogen, $N_2(g)$; oxygen, $O_2(g)$; fluorine, $F_2(g)$; chlorine, $Cl_2(g)$; bromine, $Br_2(\ell)$; and iodine, $I_2(s)$.

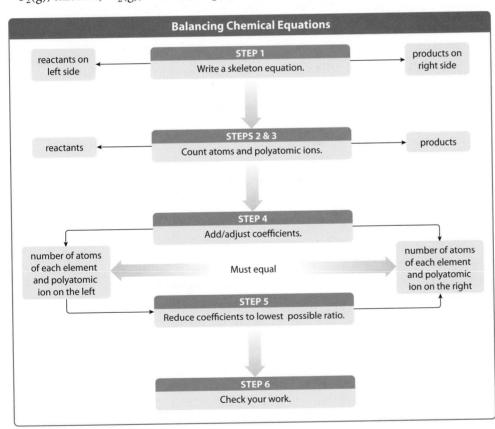

Figure 3.6 The numbered steps in this flowchart show how to balance chemical equations. These steps correspond to the steps in **Table 3.3**.

Uses of Balanced Chemical Equations in Industry

Balanced chemical equations are used in the production of a vast array of products, including pharmaceuticals, fragrances, food additives and flavours, cleaning products, fuels, ceramics, and fertilizers. The chemical engineers who oversee industrial chemical processes make sure that the correct amounts of reactants are used. In Unit 3, you will learn how to use balanced chemical equations to calculate the quantities of reactants and products.

You can practise balancing equations that represent chemical reactions by working through the Sample Problem and Practice Problems on the next page.

Balancing Chemical Equations

Problem

Write the balanced chemical equation for the reaction that occurs between aqueous solutions of silver nitrate and calcium chloride. Solid silver chloride and an aqueous solution of calcium nitrate are formed.

What Is Required?

A balanced chemical equation that represents the chemical reaction is required.

What Is Given?

You are given the reactants: aqueous silver nitrate and aqueous calcium chloride.
You are given the products: solid silver chloride and aqueous calcium nitrate.

Plan Your Strategy	Act on Your Strategy
Write a skeleton equation. Show the physical state of each reactant and product.	$AgNO_3(aq) + CaCl_2(aq) \rightarrow AgCl(s) + Ca(NO_3)_2(aq)$
Count the atoms or ions of each element in the reactants. Consider a polyatomic ion as a single unit if it appears in the reactants and the products.	Reactants: 1 Ag^+, 1 NO_3^-, 1 Ca^{2+}, 2 Cl^-
Count the atoms or ions of each element in the products.	Products: 1 Ag^+, 2 NO_3^-, 1 Ca^{2+}, 1 Cl^-
Insert the coefficient 2 in front of $AgNO_3$ to balance the nitrate ions.	$2AgNO_3(aq) + CaCl_2(aq) \rightarrow AgCl(s) + Ca(NO_3)_2(aq)$
Insert the coefficient 2 in front of AgCl to balance the silver and chloride ions.	$2AgNO_3(aq) + CaCl_2(aq) \rightarrow 2AgCl(s) + Ca(NO_3)_2(aq)$
Adjust the coefficients if necessary, so they are in the lowest possible ratio.	The ratio 2:1:2:1 cannot be reduced. It is the lowest possible ratio.
Check to make sure that the number of atoms of each element and the number of the same polyatomic ion are equal on both sides of the equation.	Reactants: 2 Ag^+, 2 NO_3^-, 1 Ca^{2+}, 2 Cl^- Products: 2 Ag^+, 2 NO_3^-, 1 Ca^{2+}, 2 Cl^-

Check Your Solution

The chemical formula for each substance is written correctly. The number of atoms of each element and the number of the same polyatomic ion are equal on both sides of the equation. The coefficients are written in the lowest possible ratio. The balanced chemical equation for this reaction is

$$2AgNO_3(aq) + CaCl_2(aq) \rightarrow 2AgCl(s) + Ca(NO_3)_2(aq)$$

Write a balanced chemical equation for each reaction.

11. $NO(g) + O_2(g) \rightarrow NO_2(g)$

12. $Mg(s) + AlCl_3(aq) \rightarrow Al(s) + MgCl_2(aq)$

13. $NaOH(aq) + CuCl_2(aq) \rightarrow NaCl(aq) + Cu(OH)_2(s)$

14. $C_2H_4(g) + O_2(g) \rightarrow CO_2(g) + H_2O(g)$

15. $Cu(s) + AgNO_3(aq) \rightarrow Cu(NO_3)_2(aq) + Ag(s)$

16. $Al(s) + MnO_2(s) \rightarrow Al_2O_3(s) + Mn(s)$

17. Propane, $C_3H_8(g)$, burns in the presence of oxygen gas to form carbon dioxide gas and water vapour.

18. Gaseous ammonia and oxygen react to form nitrogen dioxide gas and liquid water.

19. Aqueous solutions of potassium sulfide and cobalt(II) chloride react to form a solution of potassium chloride and a precipitate of cobalt(II) sulfide.

20. Carbon dioxide gas, liquid water, and an aqueous solution of sodium chloride form when hydrogen chloride gas is bubbled through a solution of sodium carbonate.

Section Summary

- A chemical equation is a representation of a chemical reaction. An arrow separates the reactants from the products. Plus signs separate multiple reactants or products.

- A skeleton equation represents a chemical reaction by using chemical formulas for the reactants and products, and symbols for the physical states of these substances. A skeleton equation is an incomplete representation because it does not show that all the atoms in the reactants must appear in the same quantities in the products.

- A balanced chemical equation uses chemical formulas and coefficients to show the relative number of particles of each substance that is involved in a chemical reaction.

- A balanced chemical equation reflects the law of conservation of mass.

- A balanced equation shows the same number of atoms of each element in the reactants and the products.

Review Questions

1. **K/U** List three forms of a chemical equation that can be used to represent a chemical reaction.

2. **T/I** The only reactant in a chemical reaction is potassium chlorate, $KClO_3(s)$. Can this reaction have phosphoryl chloride, $POCl_3(\ell)$, as a product? Explain.

3. **C** Create a Venn diagram to compare and contrast a word equation and a balanced chemical equation.

4. **C** What symbols are used to differentiate liquid water from an aqueous solution when writing a chemical equation?

5. **A** A fume hood provides ventilation to protect people from toxic gases or dust released during a chemical reaction. Consider the reaction
$NaOCl(aq) + 2HCl(aq) \rightarrow Cl_2(g) + NaCl(aq) + H_2O(\ell)$
Identify the reactant or product that requires the use of a fume hood.

6. **K/U** Why is a skeleton equation considered to be an incomplete description of a reaction?

7. **A** Adding an effervescent (fizzy) tablet to a glass of water produces the result shown below. Based on this photograph, what symbol should be used in the skeleton equation for at least one of the products?

8. **K/U** Where are the coefficients placed when balancing a chemical equation?

9. **C** Make a flowchart to show the sequence of steps that are required to write a balanced chemical equation.

10. **K/U** Why can you not change a subscript to balance a chemical equation?

11. **T/I** Write a balanced chemical equation for each chemical reaction.
   a. Solid potassium reacts with gaseous chlorine to form solid potassium chloride.
   b. Aluminum foil is placed in an aqueous solution of copper(II) sulfate, producing solid copper metal and a solution of aluminum sulfate.
   c. Nitrogen gas and hydrogen gas react to form gaseous ammonia.
   d. An aqueous solution of calcium chloride reacts with fluorine gas to form an aqueous solution of calcium fluoride and chlorine gas.

12. **T/I** Barium hydroxide can react with phosphoric acid to form barium hydrogen phosphate and water:
$Ba(OH)_2(s) + H_3PO_4(aq) \rightarrow BaHPO_4(s) + H_2O(\ell)$
   a. Identify the polyatomic ions in the reactants.
   b. Can each polyatomic ion be treated as a unit when balancing the equation? Explain why or why not.
   c. Write the balanced chemical equation.

13. **T/I** Sodium hydroxide and sulfuric acid react to form sodium sulfate and water. Does the following chemical equation correctly describe the reaction? If not, explain the error and write the correct chemical equation.
$4NaOH(aq) + 2H_2SO_4(aq) \rightarrow 2Na_2SO_4(aq) + 4H_2O(\ell)$

14. **T/I** Your friend has written the following balanced equation for the reaction in which aluminum and chlorine form aluminum(III) chloride:
$$Al(s) + Cl(g) \rightarrow AlCl(s)$$
Is this correct? If not, how would you suggest that your friend correct it?

Synthesis Reactions and Decomposition Reactions

Key Terms

synthesis reaction
decomposition reaction
electrolysis

The bullsnake (*Pituophis melanoleucus*), shown in **Figure 3.7**, is a non-venomous snake that is found in much of North America, including southern Canada. People often mistake it for a venomous rattlesnake because it may have similar colours and markings. In addition, when a bullsnake is threatened, it imitates a rattlesnake by hissing and rapidly beating its tail against plants or leaves on the ground to make noise. Another animal mimic, the hoverfly (subfamily Syrphidae), is also found in Ontario. Many species of hoverfly have yellow and black markings. Therefore, they strongly resemble bees or wasps. Hoverflies, however, do not sting or bite. People who understand how to classify these animals understand their different characteristics and can predict where they are likely to be encountered.

Like these animals, chemical reactions can be classified. By understanding how to classify chemical reactions, you can better understand them, recognize patterns in them, and make predictions about their products.

Figure 3.7 When disturbed, this bullsnake may imitate the aggressive posture of a rattlesnake and shake its tail. Unlike a rattlesnake, however, a bullsnake produces no poisonous venom. Understanding how to classify these two animals is very important to biologists. Likewise, chemists need to understand how to classify chemical reactions.

Classifying Chemical Reactions

Chemical reactions can be classified by type, including

- synthesis reactions

- decomposition reactions

- combustion reactions

- single displacement reactions

- double displacement reactions

In this chapter, you will learn how to recognize and predict the products of synthesis reactions, decomposition reactions, and combustion reactions. In Chapter 4, you will study single displacement reactions and double displacement reactions.

Characteristics of Synthesis Reactions

In a **synthesis reaction**, two or more substances react to produce a single compound. A synthesis reaction can be represented by the following general form:

synthesis reaction
a chemical reaction in which two or more reactants combine to produce a single compound

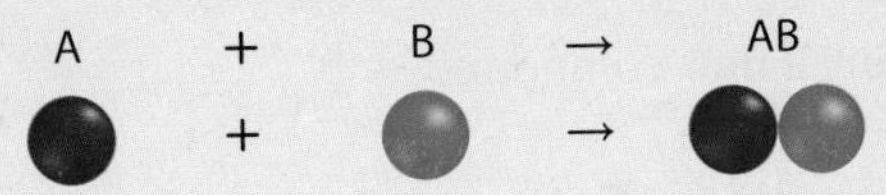

$$A \quad + \quad B \quad \rightarrow \quad AB$$

In a synthesis reaction, the reactants can be any combination of elements and compounds. The product, however, is always a single compound. Therefore, the determining factor in classifying a chemical reaction as a synthesis reaction is the formation of a single compound. An example of a synthesis reaction that occurs in the environment is the formation of acid precipitation. **Figure 3.8** models one step in this process.

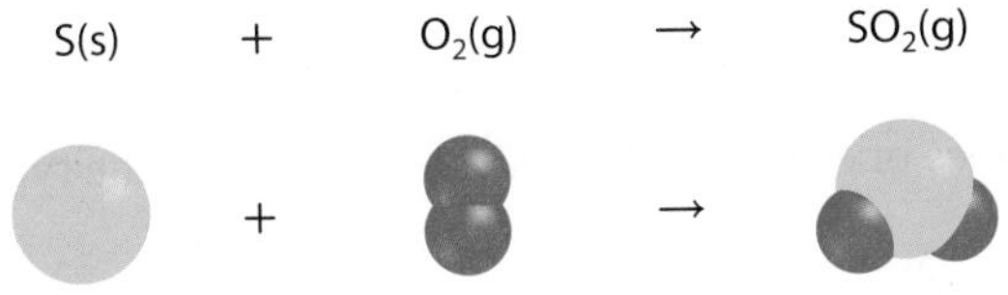

$$S(s) \quad + \quad O_2(g) \quad \rightarrow \quad SO_2(g)$$

Figure 3.8 In a synthesis reaction, a single compound forms from two or more reactants. This model shows sulfur dioxide forming from the elements sulfur and oxygen.

Explain the relationship between the general form of a synthesis reaction and the reaction shown here.

Synthesis Reactions and Acid Precipitation

Normally, precipitation such as, snow, and fog is slightly acidic because carbon dioxide dissolves in water in the atmosphere and reacts with the water according to the following synthesis equation:

$$\text{carbon dioxide} \; + \; \text{water} \; \rightarrow \; \text{carbonic acid}$$
$$CO_2(g) \quad + \quad H_2O(\ell) \rightarrow \quad H_2CO_3(aq)$$

Other natural processes contribute to the acidity of atmospheric precipitation. For example, lightning strikes produce nitrogen compounds that form nitric acid, $HNO_3(aq)$. Volcanic gases add sulfur compounds that form sulfuric acid, $H_2SO_4(aq)$. But the major sources of acid-forming compounds are coal-burning and oil-burning power plants. These power plants produce sulfur compounds, which can be carried many hundreds of kilometres by winds. In the atmosphere, sulfur compounds combine with atmospheric moisture to cause rain, snow, and fog with increased acidity.

There are three key reactions in the formation of acid precipitation involving sulfur. These reactions are all synthesis reactions. In the first reaction in the sequence, sulfur and oxygen (two elements) combine to form sulfur dioxide, as shown in **Figure 3.8**. Note that diatomic molecules, such as oxygen, are molecular elements and not compounds, because they consist of two atoms of one element. In the second reaction, sulfur dioxide (a compound) combines with additional oxygen (an element) to form sulfur trioxide. In the third reaction, sulfur trioxide and water (two compounds) combine to form sulfuric acid (another compound).

$$S(s) + O_2(g) \rightarrow SO_2(g)$$
$$2SO_2(g) + O_2(g) \rightarrow 2SO_3(g)$$
$$SO_3(g) + H_2O(\ell) \rightarrow H_2SO_4(aq)$$

Notice how, in each equation, a single product is formed from the reactants.

Types of Synthesis Reactions

On the next few pages, you will learn about the following types of synthesis reactions:

- two elements forming a binary compound

- an element and a compound forming a new compound

- two compounds forming a new compound

Two Elements Forming a Binary Compound

A common type of synthesis reaction occurs when two elements react. These reactions include those listed below. Note that a *univalent* metal is a metal that can form ions with only one charge, while a *multivalent* metal is a metal that can form ions with more than one charge.

- a univalent metal reacting with a non-metal to form an ionic compound

- a multivalent metal reacting with a non-metal to form various compounds

- two non-metals combining to form a molecular compound

The product of the reaction of two elements is a binary compound composed of the elements. The types of elements that react are related to the type of compound that forms.

A Univalent Metal Reacting with a Non-metal to Form an Ionic Compound

When a metal reacts with a non-metal, an ionic compound forms. Recall that the electronegativity difference between a metal and a non-metal tends to be large, so electrons are transferred from the metal atoms to the non-metal atoms. The charges of the resulting ions determine the chemical formula for the compound. **Figure 3.9** shows the reaction between sodium and chlorine. Through the transfer of electrons, sodium ions and chloride ions are formed. Sodium is a member of the alkali metals and forms only one ion, Na^+. Chlorine is a halogen and forms the chloride ion, Cl^-. You can predict the formula for the product that forms, $NaCl$. When predicting the formula for the product that forms from the synthesis reaction of a metal and a non-metal, remember to balance the charges so that the net charge of the compound is zero.

Go to **scienceontario** to find out more

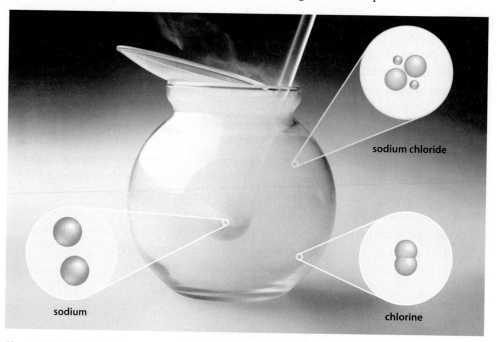

sodium chloride

sodium

chlorine

Figure 3.9 When sodium reacts with chlorine, the two elements undergo a synthesis reaction in which a binary ionic compound, sodium chloride, is produced.

7. What is the general form of a synthesis reaction?

8. When a metal reacts with a non-metal during a synthesis reaction, what type of compound forms?

9. Figure 3.9 shows a synthesis reaction between solid sodium and chlorine gas.

    a. What main characteristic of this chemical reaction causes it to be classified as a synthesis reaction?

    b. Write a balanced chemical equation for this reaction.

10. Solid calcium and chlorine gas can produce a solid product in a synthesis reaction.

    a. Predict the product of this reaction.

    b. Write a balanced chemical equation for this reaction.

11. As you have learned, three chemical reactions involving sulfur are associated with the formation of acid precipitation. Create a graphic organizer to compare the types of reactants in these reactions.

12. A product of the chemical reaction of ethane, $C_2H_6(g)$, with oxygen is carbon dioxide. Could this reaction be a synthesis reaction? Explain.

A Multivalent Metal Reacting with a Non-metal to Form Various Compounds

The periodic table provides the charges that metal ions can have. Thus, the periodic table is an important tool for predicting the formulas of the compounds that are formed in synthesis reactions.

If a metal has more than one possible ion charge, you cannot accurately predict the product of a synthesis reaction that involves this metal. For example, copper can have a charge of 1+ or 2+. For the reaction of copper with a non-metal, you need to consider each charge when predicting the possible product. Thus, for the reaction of copper with chlorine, you need to consider the following reactions:

Formation of Copper(I) Chloride

$$\text{copper} + \text{chlorine} \rightarrow \text{copper(I) chloride}$$
$$2Cu(s) + Cl_2(g) \rightarrow 2CuCl(s)$$

Formation of Copper(II) Chloride

$$\text{copper} + \text{chlorine} \rightarrow \text{copper(II) chloride}$$
$$Cu(s) + Cl_2(g) \rightarrow CuCl_2(s)$$

As these chemical equations indicate, the reaction might produce copper(I) chloride or copper(II) chloride. Both compounds are shown in **Figure 3.10**.

copper(I) chloride

copper(II) chloride

Figure 3.10 Because copper is a multivalent metal, it can form two different binary compounds with chlorine.

Analyze *Manganese can form the oxides MnO(s), MnO₂(s), Mn₂O₃(s), and Mn₂O₇(s). How are these manganese oxides similar to the copper compounds shown here?*

Two Non-metals Combining to Form a Molecular Compound

When two non-metals react, the electronegativity difference between the elements tends to be too low for the transfer of electrons to occur. Instead, the non-metal atoms share electrons, and the elements combine to form a molecular compound. Because electrons are shared, the non-metal atoms do not form ions, so there are no charges to help you determine the chemical formula for the product. In addition, many non-metals can combine in several different ratios. For example, carbon and oxygen can react to form either carbon dioxide or carbon monoxide, as shown below.

Formation of Carbon Dioxide

carbon + oxygen → carbon dioxide

$$C(s) + O_2(g) \rightarrow CO_2(g)$$

Formation of Carbon Monoxide

carbon + oxygen → carbon monoxide

$$2C(s) + O_2(g) \rightarrow 2CO(g)$$

If you only know that the reactants are carbon and oxygen, you cannot predict with certainty which compound will form without additional information. This information can be determined by collecting and analyzing the products formed during the reaction.

Sample Problem

Predicting Synthesis Products

Problem

Write balanced chemical equations that show the likely products of the synthesis reactions of chromium with sulfur.

What Is Required?

The chemical formulas of the likely products of the synthesis reactions are required. Balanced chemical equations for the reactions are also required.

What Is Given?

You are given the reactants: chromium and sulfur.
You know the type of reaction: synthesis.

Plan Your Strategy	Act on Your Strategy
Identify the types of elements involved. Determine the type of compound they will form.	Chromium, Cr: metal Sulfur, S: non-metal They will react to form an ionic compound.
If there is a metal, determine whether it is multivalent. If so, determine its possible charges. If there is no metal, examine the information about the products given in the problem.	Chromium is a multivalent metal. Cr^{3+}, Cr^{2+}
Use the appropriate method to write the formulas of the products.	Cr_2S_3 CrS
Write a balanced chemical equation for each reaction.	$2Cr(s) + 3S(s) \rightarrow Cr_2S_3(s)$ $Cr(s) + S(s) \rightarrow CrS(s)$

Check Your Solution

The overall charge of each formula is zero. The chemical equations are balanced.

Predict the product that is likely to form in each reaction, and write a balanced chemical equation for the reaction.

21. lithium and oxygen

22. strontium and fluorine

23. iron and bromine

24. phosphorus and hydrogen, forming gaseous phosphorus trihydride

25. calcium and iodine

26. tin and oxygen

27. bismuth and sulfur

28. aluminum and iodine

29. silver and oxygen

30. nitrogen and oxygen, forming nitrogen dioxide

An Element and a Compound Forming a New Compound

In some synthesis reactions, an element and a compound are the reactants. Earlier in this section, you saw one such reaction, which is involved in the formation of acid precipitation:

$$\text{sulfur dioxide} + \text{oxygen} \rightarrow \text{sulfur trioxide}$$
$$2SO_2(g) \quad + \quad O_2(g) \quad \rightarrow \quad 2SO_3(g)$$

Another reaction that occurs between an element and a compound is the formation of phosphorus pentachloride from phosphorus trichloride and chlorine. This is a reversible reaction, as shown below.

$$\text{phosphorus trichloride} + \text{chlorine} \rightleftharpoons \text{phosphorus pentachloride}$$
$$PCl_3(\ell) \quad + \quad Cl_2(g) \quad \rightleftharpoons \quad PCl_5(s)$$

Phosphorus pentachloride is used as a source of chlorine in various industrial chemical reactions, including the production of flame-retardant products, such as the fabric shown in **Figure 3.11**. Another use of phosphorus pentachloride is in lithium-based batteries.

Figure 3.11 This hot-air balloon is made from a flame-resistant fabric. The same type of fabric is also used for such applications as making protective clothing for firefighters.

Two Compounds Forming a New Compound

The final combination of reactants that can undergo a synthesis reaction is two compounds. Because a synthesis reaction has a single compound as the product, the reacting compounds tend to be small, simple compounds, such as oxides and water. An oxide is a compound that is composed of oxygen and another element.

A Non-metal Oxide Reacting with Water

Earlier in this section, you saw how carbonic acid forms in the reaction between carbon dioxide and water, and how sulfuric acid forms in the reaction between sulfur trioxide and water. In both of these reactions, a non-metal oxide combines with water to form an acid:

$$CO_2(g) + H_2O(\ell) \rightarrow H_2CO_3(aq)$$
$$SO_3(g) + H_2O(\ell) \rightarrow H_2SO_4(aq)$$

Suggested Investigation

Plan Your Own Investigation 3-A, Testing the Acidity of Oxides

Reactions involving carbon dioxide and sulfur trioxide contribute to acid precipitation. Acid precipitation formed in the reaction between non-metal oxides and water can damage structures, such as buildings and bridges, and plants, as shown in **Figure 3.12**. Acid precipitation can also cause bodies of water to become too acidic to support fish and other aquatic life. To help reduce the harm done by acid precipitation, governments have passed laws that limit emissions of oxides from industrial plants.

Figure 3.12 These trees have died due to acid precipitation.

A Metal Oxide Reacting with Water

As you have learned, a non-metal oxide can react with water to form an acid. Similarly, a metal oxide can react with water to form a metal hydroxide (or base). For example, the formation of sodium hydroxide can be represented by the following equation:

sodium oxide + water → sodium hydroxide
$$Na_2O(s) + H_2O(\ell) \rightarrow 2NaOH(aq)$$

Sodium hydroxide has a wide variety of industrial uses, including the production of paper, soaps, and detergents.

Metal hydroxides are ionic compounds. You can use the charge of the metal ion in the oxide to predict the formula for the metal hydroxide that is formed during the synthesis reaction between a metal oxide and water.

Characteristics of Decomposition Reactions

A **decomposition reaction** is a reaction in which a compound breaks down into two or more elements or simpler compounds. It is often the reverse of a synthesis reaction. A decomposition reaction can be represented by the following general form:

$$AB \rightarrow A + B$$

In a decomposition reaction, a single reactant, which is always a compound, is changed into more than one product. These products can be elements, elements and compounds, or multiple compounds.

Decomposition Reactions and Rocket Thrusters

The chapter opener showed how the Phoenix Mars Lander used decomposition reactions within its thruster rockets to fire pulses of gases, which controlled its final descent to the Martian surface. One of the thruster rockets is shown in **Figure 3.13**.

> **decomposition reaction** a chemical reaction in which a compound breaks down into elements or simpler compounds

Figure 3.13 The Phoenix Mars Lander has twelve thruster rockets **(A)**. The thruster-rocket system in the partial model of the Phoenix Mars Lander **(B)** is being tested for its ability to withstand the stresses related to the decomposition of rocket fuel.

The reactions in the thruster rockets of the Phoenix Mars Lander broke down liquid hydrazine, $N_2H_4(\ell)$, according to the following decomposition reactions:

$$3N_2H_4(\ell) \rightarrow 4NH_3(g) + N_2(g)$$
$$N_2H_4(\ell) \rightarrow N_2(g) + 2H_2(g)$$

Hydrazine is an effective fuel for thruster rockets because its decomposition very quickly produces a large volume of hot gases from a small amount of fuel. The amount of hydrazine being released controls the volume of hot gases being produced, and therefore the degree of thrust being provided. The final descent of the Phoenix Mars Lander could be controlled very precisely by varying the amount of hydrazine being burned in each of the lander's twelve thruster rockets. Hydrazine is also effective for use in thruster rockets because it does not need to be combined with another chemical for a decomposition reaction to occur. Therefore, it does not require a complex storage or mixing system.

Decomposition Reactions and Automobile Safety

When a vehicle with air bags is in a collision, a decomposition reaction causes the air bags to inflate within thousandths of a second. The air bags cushion the people inside the vehicle, preventing impact with hard surfaces.

How does an air bag work? During a crash, the abrupt slowing of the vehicle triggers a sensor in the air bag. The sensor generates an electrical impulse that ignites a pellet of sodium azide, $NaN_3(s)$. The sodium azide undergoes a very rapid decomposition reaction into sodium and nitrogen as represented in **Figure 3.14**. The nitrogen gas inflates the air bag. The sodium is a very reactive metal and is hazardous. Other chemical reactions take place within the air bag to convert the sodium into compounds that are harmless and stable.

Sodium azide is a toxic chemical. It is changed into harmless substances by the reactions that occur when an air bag inflates. However, most vehicles are never involved in crashes that cause their air bags to inflate. When the vehicles are no longer useful, they are often recycled by being crushed flat and shredded into small pieces. Various metals and plastics can then be separated and used again. Before the recycling begins, the air bags are removed or caused to inflate to prevent the release of sodium azide.

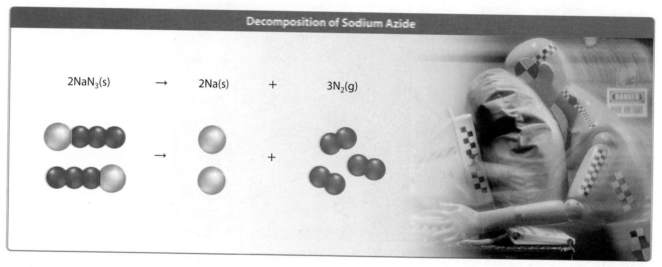

Decomposition of Sodium Azide

$2NaN_3(s) \rightarrow 2Na(s) + 3N_2(g)$

Figure 3.14 In a decomposition reaction, a single compound breaks down. The decomposition of sodium azide is used to inflate air bags.

Analyze *Which product of the decomposition of sodium azide causes an air bag to inflate?*

Types of Decomposition Reactions

Just as there are different types of synthesis reactions, there are different types of decomposition reactions. On the following pages, you will learn more about these decomposition reactions:

- a binary compound decomposing into its elements
- a metal nitrate decomposing into a metal nitrite and oxygen gas
- a metal carbonate decomposing into a metal oxide and carbon dioxide
- a metal hydroxide decomposing into a metal oxide and water

Depending on the type of compound that breaks down, you might be able to predict the products that are likely to form because of patterns in the ways that the substances react. The general rules presented in this section reflect a small number of decomposition reactions but include examples of some common reactions you might encounter.

A Binary Compound Decomposing into Its Elements

A compound that is composed of only two elements will usually decompose into those elements. High temperature or an electric current is often used to break a compound into its elements. **Electrolysis** is a process that uses electrical energy to cause a chemical reaction.

electrolysis a process that uses electrical energy to cause a chemical reaction

Electrolysis was used by early chemists to isolate and identify elements, and it is used by modern industries to produce elements for commercial use. For example, one method of generating hydrogen gas for hydrogen-fuelled vehicles is through the decomposition of water, as shown in **Figure 3.15**. An electric current is used to break down the water molecules into hydrogen molecules and oxygen molecules.

Decomposition of Water

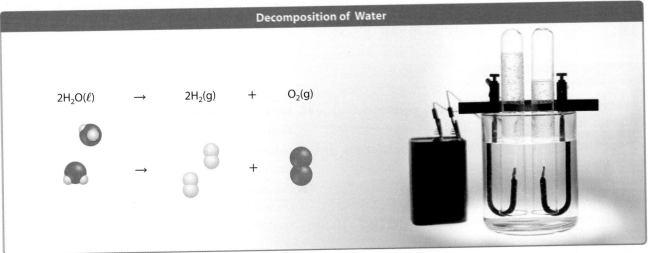

$$2H_2O(\ell) \longrightarrow 2H_2(g) + O_2(g)$$

Figure 3.15 The electrolysis of water uses electrical energy to break down water into the elements hydrogen and oxygen. On an industrial scale, hydrogen-generating plants can use the electrolysis of water to produce hydrogen for use as a fuel.

Another important application of electrolysis is the decomposition of sodium chloride, according to the following chemical equation:

$$\text{sodium chloride} \rightarrow \text{sodium} + \text{chlorine}$$
$$2NaCl(\ell) \rightarrow 2Na(\ell) + Cl_2(g)$$

Notice that the reactant must be in the liquid state. This is because an ionic compound, such as sodium chloride, is a good electrical conductor only when its ions are free to move.

The products of the decomposition of sodium chloride are widely used. Sodium is used as a coolant in nuclear reactors. It is also used in sodium vapour lamps, as shown in **Figure 3.16**. Chlorine is used in water purification, and it is an important substance in the manufacture of an assortment of products, such as bleach, paper, paints, and plastics.

Heat can also be used to decompose a compound into its elements. In this process, which is called thermal decomposition, heat breaks chemical bonds in a compound. The *thermal decomposition* of mercury(II) oxide was important in the identification of oxygen as an element. Mercury(II) oxide decomposes into its elements according to the following chemical equation:

$$\text{mercury(II) oxide} \rightarrow \text{mercury} + \text{oxygen}$$
$$2HgO(s) \rightarrow 2Hg(\ell) + O_2(g)$$

Uses of mercury include the manufacture of thermometers and barometers. Mercury is mixed with other metals to make a material used to fill cavities in teeth. In nature, mercury is commonly found in combination with sulfur in the mineral cinnabar, $HgS(s)$. Powdered cinnabar is used as a pigment to make bright red paint. Mercury is also used in certain types of electrochemical batteries.

Figure 3.16 The sodium used in this street lamp was generated through the decomposition of sodium chloride. The yellow glow is characteristic of a sodium vapour lamp.

A Metal Nitrate Decomposing into a Metal Nitrite and Oxygen Gas

Compounds that are composed of more than two elements generally do not decompose into their individual elements. For example, sodium nitrate, $NaNO_3(s)$, will not decompose into sodium, nitrogen, and oxygen. You can often predict the products that are likely to form, however, based on the polyatomic ion in the compound. For example, compounds that are composed of the nitrate ion often decompose into a nitrite-containing compound and oxygen gas. Sodium nitrate, which can be used in explosives, decomposes as follows:

$$\text{sodium nitrate} \rightarrow \text{sodium nitrite} + \text{oxygen}$$
$$2NaNO_3(s) \rightarrow 2NaNO_2(s) + O_2(g)$$

Other decomposition reactions are also possible for nitrates, depending on the conditions in which the reactions occur.

Learning Check

13. What type of product forms when a metal oxide reacts with water?

14. Make a graphic organizer to compare the solutions formed when water reacts with a metal oxide and with a non-metal oxide.

15. Give the general form of a decomposition reaction. Describe the main characteristic of this type of reaction.

16. Is it possible for a decomposition reaction to have an element as a reactant? Explain.

17. In what state, other than the liquid state, is electrolysis possible? Explain your reasoning.

18. Describe the role of thermal decomposition in the isolation of elemental mercury, and give two examples of how mercury is used.

A Metal Carbonate Decomposing into a Metal Oxide and Carbon Dioxide

Quicklime, or calcium oxide, is a main component of cement. Cement is used to make concrete, such as the sidewalk shown in **Figure 3.17**. The decomposition of calcium carbonate in limestone is a major step in the production of cement. A metal carbonate generally decomposes to form a metal oxide compound, while giving off carbon dioxide gas. The following chemical equation shows the decomposition of calcium carbonate:

$$\text{calcium carbonate} \rightarrow \text{calcium oxide} + \text{carbon dioxide}$$
$$CaCO_3(s) \rightarrow CaO(s) + CO_2(g)$$

Cement is mixed with water, sand, and gravel to make concrete. As the concrete dries, quicklime reacts to form compounds that bind the mixture together.

Figure 3.17 Quicklime, the main component of the cement in the concrete on which this mural was drawn, was produced in the decomposition of calcium carbonate. In addition, the main component of the chalk used to draw the mural is calcium carbonate.

Calcium carbonate is the main component of limestone, marble, and seashells. In this activity, you will thermally decompose calcium carbonate. You will also cause a reaction between an aqueous solution of a product of the decomposition reaction and carbon dioxide that causes particles of calcium carbonate to form.

Safety Precautions

- Wear safety eyewear throughout this activity.
- Tie back loose hair and clothing.
- Use EXTREME CAUTION when you are near an open flame.
- Use clean a drinking straw in this activity. Get the straw from your teacher when it is needed. Do not let it touch the lab bench.

Materials

- small piece of calcium carbonate, $CaCO_3(s)$
- 15 mL of distilled water
- phenolphthalein indicator solution in a dropper bottle
- metal gauze
- tripod
- Bunsen burner secured to a utility stand
- igniter for Bunsen burner
- tongs
- 2 test tubes
- test tube stopper
- 10 mL graduated cylinder
- test-tube rack
- filter funnel
- filter
- clean drinking straw

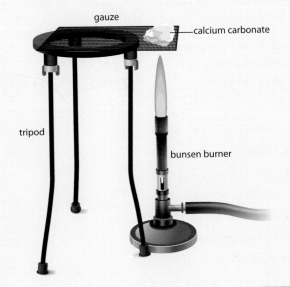

gauze

calcium carbonate

tripod

bunsen burner

Procedure

1. Place a piece of calcium carbonate on metal gauze supported on a tripod.

2. Ignite the Bunsen burner. Adjust the air vents to obtain a blue flame. Use a clamp to secure the burner to a utility stand.

3. Heat the calcium carbonate with the Bunsen burner flame for 8 minutes. Allow the solid material to cool for 2 minutes. Record your observations.

4. Use the tongs to transfer the solid material to a test tube.

5. Use the graduated cylinder to add 10 mL of water to the test tube. Put the stopper in the test tube and then shake the test tube for about 15 seconds. Place the test tube in the test-tube rack. Record your observations.

6. Place the filter funnel into the opening of the second test tube. Put the filter paper into the funnel. Pour about half of the contents of the first test tube into the funnel and allow the liquid to pass through the filter.

7. Use a clean drinking straw to gently blow a stream of bubbles through the filtered liquid collected in the second test tube. Carefully observe the colour and cloudiness of the solution. Record your observations.

8. Add 3 drops of phenolphthalein indicator solution to the first test tube. Record your observations.

Questions

1. Write a balanced chemical equation for the thermal decomposition of calcium carbonate.

2. Explain why this is a decomposition reaction.

3. What is the material that remained after the calcium carbonate was heated? How did the solidity of this material compare to that of calcium carbonate? Explain your answer.

4. In step 5, calcium oxide and water reacted to produce an aqueous solution of calcium hydroxide. Write a balanced chemical equation for this reaction.

5. What type of reaction is the one for which you wrote a balanced equation in question 4? Explain your answer.

6. An aqueous solution of calcium hydroxide is commonly called limewater. Limewater reacts with carbon dioxide gas to form solid calcium carbonate, according to the following equation:

   $Ca(OH)_2(aq) + CO_2(g) \rightarrow CaCO_3(s) + H_2O(\ell)$

   What was the source of carbon dioxide gas in step 7? How did the appearance of the solution change in step 7 to indicate the formation of calcium carbonate?

7. Is an aqueous solution of calcium hydroxide acidic or basic? Explain your answer.

A Metal Hydroxide Decomposing into a Metal Oxide and Water

A metal oxide is also produced in the decomposition of a metal hydroxide. When heated, a metal hydroxide will generally break down to form a metal oxide and water. For example, calcium oxide can be formed by the decomposition of calcium hydroxide according to the following chemical equation:

$$\text{calcium hydroxide} \rightarrow \text{calcium oxide} + \text{water}$$
$$\text{Ca(OH)}_2(s) \rightarrow \text{CaO}(s) + \text{H}_2\text{O}(g)$$

Based on the few general patterns of the decomposition reactions presented in this section, you will now be able to predict the likely products of the decomposition of many compounds.

Sample Problem

Predicting Decomposition Products

Problem

Write a balanced chemical equation that shows the likely products of the decomposition of rubidium carbonate.

What Is Required?

Determine the products that are likely to form when rubidium carbonate decomposes, and write a balanced chemical equation for the reaction.

What Is Given?

You are given the type of reaction: decomposition.

You are given the reactant: rubidium carbonate.

Plan Your Strategy	Act on Your Strategy
Identify the type of compound that is decomposing.	A metal carbonate is decomposing.
Determine the types of products that usually form in this type of reaction.	A metal oxide and carbon dioxide are the usual products of the decomposition of a metal carbonate.
Write a word equation for the reaction.	rubidium carbonate → rubidium oxide + carbon dioxide
Write and balance a chemical equation for the reaction.	$\text{Rb}_2\text{CO}_3(s) \rightarrow \text{Rb}_2\text{O}(s) + \text{CO}_2(g)$

Check Your Solution

Each chemical formula is correct, and the chemical equation is balanced. The products are those that you would expect to be produced from the decomposition of a metal carbonate.

Practice Problems

Determine the products that are likely to form in the decomposition of each compound, and write a balanced chemical equation for the reaction.

31. potassium bromide

32. aluminum oxide

33. magnesium hydroxide

34. calcium nitrate

35. copper(II) carbonate

36. chromium(III) chloride

37. barium carbonate

38. rubidium nitrate

39. lithium hydroxide

40. magnesium chloride

Other Decomposition Reactions

Some decomposition reactions produce more than two products. One of these reactions is the decomposition of the explosive trinitrotoluene (TNT), shown in **Figure 3.18**. The chemical formula for TNT is $C_7H_5N_3O_6(s)$. Its decomposition produces four products:

$$\text{trinitrotoluene} \rightarrow \text{nitrogen} + \text{water} + \text{carbon monoxide} + \text{carbon}$$
$$2C_7H_5N_3O_6(s) \rightarrow 3N_2(g) + 5H_2O(g) + 7CO(g) + 7C(s)$$

The explosive force of TNT comes from the gases that are produced. These gases are heated by the energy that is released during the reaction, and they rapidly expand outward. TNT has applications for military, demolition, and industrial uses. It is also used for underwater blasting because it is insoluble in water.

Figure 3.18 In 1917, the ship *SS Mont-Blanc* collided with another ship and caught on fire in the harbour of Halifax, Nova Scotia. *SS Mont-Blanc* had been loaded with explosives, including hundreds of rounds of ammunition and about 180 000 kg of TNT. The fire caused the explosives to detonate, sending a huge cloud into the sky.

Another decomposition reaction that produces multiple products is the decomposition of ammonium dichromate, $(NH_4)_2Cr_2O_7(s)$. This reaction is often used in chemistry demonstrations, as shown in **Figure 3.19**, because of the dramatic and energetic change in the compound. Three products form in the decomposition of ammonium dichromate:

$$\text{ammonium dichromate} \rightarrow \text{nitrogen} + \text{chromium(III) oxide} + \text{water}$$
$$(NH_4)_2Cr_2O_7(s) \rightarrow N_2(g) + Cr_2O_3(s) + 4H_2O(g)$$

Go to **scienceontario** to find out more

Figure 3.19 Ammonium dichromate is a bright orange powder **(A)**. As it decomposes **(B)**, it releases heat and light. The transformation of ammonium dichromate into a dark green powder, chromium(III) oxide **(C)**, makes an ammonium dichromate "volcano," a popular chemistry demonstration.

Section Summary

- In a synthesis reaction, two or more reactants combine to form a single product. This reaction can be represented by the general form A + B → AB.

- The product of a synthesis reaction between a metal and a non-metal is usually a binary ionic compound. The product of a synthesis reaction between a multivalent metal and a non-metal depends on the charge of the metal ion formed.

- In a synthesis reaction, a non-metal can combine with another non-metal in several different ratios.

- A non-metal oxide reacts with water to form an acid in a synthesis reaction. A metal oxide reacts with water to form a metal hydroxide, which is a base.

- In a decomposition reaction, a single reactant breaks apart to form two or more products. This reaction can be represented by the general form AB → A + B. A binary compound generally decomposes into its elements.

- A metal nitrate generally decomposes into a metal nitrite and oxygen gas.

- A metal carbonate generally decomposes into a metal oxide and carbon dioxide gas. A metal hydroxide generally decomposes into a metal oxide and water.

Review Questions

1. **K/U** Describe the two main types of chemical reactions presented in this section.

2. **C** Using a graphic organizer, compare and contrast synthesis and decomposition reactions.

3. **A** Some electrical power plants use the energy that is released by burning coal to generate electrical energy. Why would using coal that has a low sulfur content help to protect the environment?

4. **K/U** What types of reactants form a binary ionic compound in a synthesis reaction?

5. **T/I** Predict the product(s) and write a balanced chemical equation for the reaction of aluminum and sulfur.

6. **C** Describe the steps you would take to predict the products of the reaction between a metal and a non-metal.

7. **T/I** Predict whether each compound would form an acid or a base in a reaction with water.
   a. dinitrogen trioxide
   b. lithium oxide
   c. sulfur trioxide
   d. calcium oxide

8. **A** Explain why techniques that are used to combat acid precipitation focus on removing the non-metal oxide and not the other reactant in the reaction.

9. **K/U** In a decomposition reaction, what type of reactant generally forms two elements?

10. **K/U** If you wanted to generate carbon dioxide gas by using a decomposition reaction, what type of reactant would you use?

11. **T/I** The photograph below shows the decomposition of mercury(II) oxide, HgO(s), a red powder.
    a. Which product is visible in the photograph?
    b. Which product is not visible?
    c. How could you test for the presence of the product that is not visible?

12. **C** Using a graphic organizer, compare the types of reactants in each of the synthesis reactions associated with the formation of acid precipitation.

13. **K/U** What products would you expect to form when a metal nitrate decomposes?

14. **C** A student thinks that sodium carbonate will decompose into sodium, carbon, and oxygen. Create a labelled diagram to show why the student is incorrect.

15. **A** If you wanted to generate lithium oxide, what are two decomposition reactions you could try?

16. **A** The term *pyrotechnics* refers to the use of chemicals to produce explosive displays for special effects and similar purposes. Infer why ammonium dichromate, rather than trinitrotoluene, would be used in a theatrical pyrotechnics display.

Combustion Reactions

In some areas of the Arctic, large amounts of methane, $CH_4(g)$, are entering the atmosphere. As the climate becomes warmer and the ground thaws, bacteria produce methane from the remains of dead plants and animals. The release of large amounts of methane into the atmosphere is a concern because methane is an important greenhouse gas that contributes to global warming. Molecule for molecule, methane has 25 times the warming effect that carbon dioxide has. **Figure 3.20** shows how researchers ignite gas bubbles that are released from the ground and from lakes to test whether they contain methane. As the methane burns, it undergoes chemical changes and new substances form. The reaction that is taking place is known as a combustion reaction.

Key Terms

combustion reaction

hydrocarbon

soot

Figure 3.20 University of Alaska Fairbanks researcher Katey Walter lights a pocket of methane on a thermokarst lake in Siberia in March of 2007. Igniting the gas is a way to demonstrate, in the field, that it contains methane.

Characteristics of Combustion Reactions with Hydrocarbons

In a **combustion reaction**, like the one shown in **Figure 3.20**, oxygen combines with another substance and releases energy in the form of heat and light. One or more oxides, compounds that are composed of oxygen, are produced. For example, the combustion of methane, modelled in **Figure 3.21**, produces the oxides carbon dioxide and water (dihydrogen monoxide). Methane is a **hydrocarbon**, a compound that is composed only of the elements carbon and hydrogen. The combustion of hydrocarbons can be either complete or incomplete, as you will learn in this section.

combustion reaction the reaction of a substance with oxygen, producing one or more oxides, heat, and light

hydrocarbon a compound that is composed only of the elements carbon and hydrogen

$$CH_4(g) \quad + \quad 2O_2(g) \quad \rightarrow \quad CO_2(g) \quad + \quad 2H_2O(g)$$

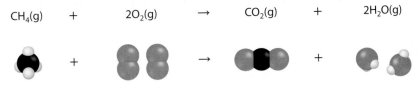

Figure 3.21 Oxygen is always a reactant in a combustion reaction, such as the complete combustion of methane, which is modelled here.

Reactions similar to combustion reactions can occur with other substances, such as fluorine, taking the place of oxygen. You will learn more about these reactions, called oxidation reactions, in Grade 12.

Complete Combustion of Hydrocarbons

Like many of the other hydrocarbons shown in **Table 3.4**, methane is a fuel.

Table 3.4 Hydrocarbons and Their Uses

Name	Formula	Use
methane	$CH_4(g)$	Fuel used for heating and cooking
ethane	$C_2H_6(g)$	Raw material used for making plastics
propane	$C_3H_8(g)$	Fuel used for heating and cooking
butane	$C_4H_{10}(g)$	Fuel used in lighters; propellant
acetylene (ethyne)	$C_2H_2(g)$	Fuel used for welding
benzene	$C_6H_6(\ell)$	Substance used in the manufacture of paints, plastics, and detergents

These hydrocarbons, as well as many others, can undergo combustion reactions to produce carbon dioxide and water. When a hydrocarbon burns, the products of the reaction depend on the amount of oxygen that is present. When oxygen is available in sufficient amounts, complete combustion occurs. All the carbon atoms and hydrogen atoms from the hydrocarbon molecules combine with oxygen atoms to form molecules of the oxides carbon dioxide and water vapour. The general form of the complete combustion of a hydrocarbon is

$$\text{hydrocarbon} + \text{oxygen} \rightarrow \text{carbon dioxide} + \text{water}$$
$$C_xH_y + O_2(g) \rightarrow CO_2(g) + H_2O(g)$$

Welding, shown in **Figure 3.22**, is an example of a process that depends on hydrocarbon combustion.

Figure 3.23 A blue flame indicates that complete combustion is occurring.

Figure 3.22 This welding torch uses acetylene, a hydrocarbon, as fuel.

The colour of the flame in a combustion reaction can indicate whether complete combustion is occurring. Natural gas, which is mainly composed of methane, is used in many homes for heating and cooking. It is also used in laboratory burners, such as the one in **Figure 3.23**. When using a gas laboratory burner, it is important to adjust the air vents to allow the proper amount of oxygen to enter the burner. If the flame is yellow, more oxygen is needed for complete combustion to occur. The blue flame that forms during complete combustion generates mostly heat and very little light. In other words, complete combustion is a more efficient process for generating heat.

Incomplete Combustion of Hydrocarbons

As you have learned, having the correct amount of oxygen results in complete combustion and having too little oxygen results in incomplete combustion. Incomplete combustion is sometimes useful because it produces a bright yellow flame, such as the candle flame in **Figure 3.24**. However, it can also be a hazard. In addition to producing carbon dioxide and water vapour, an incomplete combustion reaction can produce elemental carbon, or **soot**, and toxic carbon monoxide, $CO(g)$. For example, the incomplete combustion of propane, $C_3H_8(g)$, can occur according to this chemical equation:

propane $+$ oxygen $\rightarrow$ carbon $+$ carbon monoxide $+$ carbon dioxide $+$ water

$$2C_3H_8(g) + 7O_2(g) \rightarrow 2C(s) + 2CO(g) + 2CO_2(g) + 8H_2O(g)$$

In addition, the incomplete combustion of propane can result in many other ratios of carbon products.

soot fine particles consisting mostly of carbon, formed during the incomplete combustion of a hydrocarbon

Figure 3.24 The yellow flame of this candle indicates that incomplete combustion is occurring. The colour of the candle flame is produced by carbon particles within it that are hot enough to glow.

Identify the product in this combustion reaction that can cause deposits of a dark substance to form on nearby surfaces.

The production of carbon monoxide is a serious concern when incomplete combustion reactions are taking place. Carbon monoxide is a toxic gas with no colour or odour. Its molecules are similar in size and shape to diatomic oxygen molecules. This similarity allows carbon monoxide to bind to oxygen binding sites in the blood more tightly than oxygen itself. The longer a person inhales carbon monoxide, the fewer binding sites that are available to carry oxygen. Initial symptoms of carbon monoxide poisoning include headache, dizziness, or nausea. Prolonged or high exposure can result in vomiting, collapse, loss of consciousness, and eventually suffocation.

Carbon monoxide can build up to harmful levels in some industries, such as pulp and paper production, petroleum refineries, and steel production. Taxi drivers, welders, forklift operators, and other people who work in enclosed spaces where combustion reactions occur are at risk. Proper ventilation and a procedure for reporting signs of possible exposure are key to avoiding carbon monoxide poisoning. In addition, carbon monoxide detectors are important for monitoring the level of this deadly gas.

The Sample Problems on the next two pages show chemical equations for complete combustion reactions and incomplete combustion reactions. You can practise balancing these types of equations in the Practice Problems that follow.

Suggested Investigation

Plan Your Own Investigation 3-B, Comparing Complete and Incomplete Combustion

19. What types of energy do combustion reactions release?

20. What is a hydrocarbon?

    a. In what kinds of combustion reactions can hydrocarbons take part?

    b. What are the products of a complete combustion reaction involving a hydrocarbon and oxygen?

21. During the reaction of nitrogen gas and oxygen gas to form nitrous oxide gas, $N_2O(g)$, energy in the form of heat is absorbed.

    a. Is this reaction a combustion reaction? Explain your reasoning.

    b. If you concluded that this reaction is not a combustion reaction, identify which type of reaction it is.

22. Compare **Figure 3.23** and **Figure 3.24**.

    a. In terms of the type of combustion reaction occurring, explain why the flames in the two figures are different colours.

    b. Explain whether the flame shown in **Figure 3.23** is generating more heat or more light.

    c. List the products that are being formed by the combustion reaction shown in each figure.

23. Do you expect a gas stove to be designed so that complete combustion occurs? Explain your reasoning.

24. A science fiction movie shows a scene on the outside of a spaceship in outer space. In the scene, a stray shot from a blaster ignites an antenna array, which begins to burn. Analyze the scientific accuracy of this scene based on your understanding of combustion reactions.

Writing Balanced Chemical Equations for Complete Combustion Reactions

Problem

What is the balanced chemical equation for the complete combustion of butane, $C_4H_{10}(g)$?

What Is Required?

A balanced chemical equation for the complete combustion of butane is required.

What Is Given?

You are given the reactant: butane.
You are given the type of reaction: complete combustion.

Plan Your Strategy	Act on Your Strategy
Identify the reactants. Butane is one of the reactants. Because the reaction is a combustion reaction, the other reactant must be oxygen.	Reactants: $C_4H_{10}(g)$, $O_2(g)$
Identify the products. Because the reaction is a complete combustion reaction, the products must be carbon dioxide and water.	Products: $CO_2(g)$, $H_2O(g)$
Write a balanced chemical equation for the reaction.	$2C_4H_{10}(g) + 13O_2(g) \rightarrow 8CO_2(g) + 10H_2O(g)$

Check Your Solution

The chemical formula for each substance is written correctly. The number of atoms of each element is equal on both sides of the equation. The coefficients are written in the lowest possible ratio.

Writing Balanced Chemical Equations
for Incomplete Combustion Reactions

Problem

Write a balanced chemical equation for the incomplete combustion of hexane, $C_6H_{14}(\ell)$, given the following partial chemical equation:

$$C_6H_{14}(\ell) + 5O_2(g) \rightarrow CO_2(g) + CO(g) + \underline{\quad} + \underline{\quad} H_2O(g)$$

What Is Required?

A balanced chemical equation for the incomplete combustion of hexane is required, based on the given partial chemical equation.

What Is Given?

You are given a partial chemical equation:

$$C_6H_{14}(\ell) + 5O_2(g) \rightarrow CO_2(g) + CO(g) + \underline{\quad} + \underline{\quad} H_2O(g)$$

You know the type of reaction: incomplete combustion.

Plan Your Strategy	Act on Your Strategy
Identify the reactants. Because there is insufficient oxygen present for complete combustion to occur, the coefficient for oxygen is provided in the question.	Reactants: $C_6H_{14}(\ell)$, $5O_2(g)$
Identify the possible products. The products of an incomplete combustion reaction may include either elemental carbon or carbon monoxide gas, or both.	Possible products: $CO_2(g)$, $H_2O(g)$, $C(s)$, $CO(g)$
Examine the partial chemical equation, and identify the missing information. One product is not given, and the coefficient for $H_2O(g)$ is not given. Begin balancing the equation with the information you have. Identify the missing product.	$C_6H_{14}(\ell) + 5O_2(g) \rightarrow CO_2(g) + CO(g) + \underline{\quad} + 7H_2O(g)$ The numbers of hydrogen atoms and oxygen atoms are equal on both sides of the equation, but there are more carbon atoms in the reactants than in the products. The missing product must be elemental carbon.
Write the balanced chemical equation for the incomplete combustion reaction.	$C_6H_{14}(\ell) + 5O_2(g) \rightarrow CO_2(g) + CO(g) + 4C(s) + 7H_2O(g)$

Check Your Solution

The chemical formula for each substance is written correctly. The number of atoms of each element is equal on both sides of the equation. The coefficients are written in the lowest possible ratio.

Write a balanced chemical equation for each chemical reaction.

41. complete combustion of heptane, $C_7H_{16}(\ell)$

42. complete combustion of nonane, $C_9H_{20}(\ell)$

43. complete combustion of acetylene, $C_2H_2(g)$

44. complete combustion of benzene, $C_6H_6(\ell)$

45. complete combustion of octane, $C_8H_{18}(\ell)$

46. incomplete combustion of octane:
$2C_8H_{18}(\ell) + 17O_2(g) \rightarrow \underline{\ }CO(g) + \underline{\ }H_2O(g)$

47. incomplete combustion of pentane:
$2C_5H_{12}(\ell) + 11O_2(g) \rightarrow \underline{\quad} + 12H_2O(g)$

48. incomplete combustion of propane:
$C_3H_8(g) + 2O_2(g) \rightarrow \underline{\quad} + \underline{\ }H_2O(g)$

49. incomplete combustion of heptane: $4C_7H_{16}(\ell) + 37O_2(g) \rightarrow \underline{\ }CO_2(g) + \underline{\ }CO(g) + \underline{\ }H_2O(g)$

50. incomplete combustion of cyclohexane:
a. $C_6H_{12}(\ell) + 6O_2(g) \rightarrow \underline{\quad} + 6H_2O(g)$
b. $C_6H_{12}(\ell) + 3O_2(g) \rightarrow \underline{\quad} + 6H_2O(g)$

Characteristics of Combustion Reactions with Other Chemicals

In addition to hydrocarbons, many other substances can combine with oxygen in combustion reactions. For instance, when hydrogen is burned as a fuel for a spacecraft, such as the space shuttle, it undergoes a combustion reaction in which water is produced. This reaction is modelled in **Figure 3.25 (A)**. **Figure 3.25 (B)** shows a demonstration of the combustion of hydrogen. The combustion reaction between hydrogen and oxygen is also a synthesis reaction because multiple reactants combine to form a single product.

Figure 3.25 Hydrogen and oxygen react to form water during a combustion reaction **(A)**. A hydrogen-filled balloon is ignited in **(B)** to demonstrate the combustion of hydrogen.

Analyze *Why is this reaction classified as a combustion reaction?*

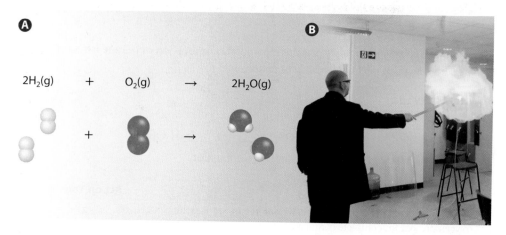

Suggested Investigation

Inquiry Investigation 3-C, Conducting Synthesis, Decomposition, and Combustion Reactions

You have also encountered combustion reactions involving other chemicals in your study of synthesis reactions. For example, the reaction of sulfur with oxygen, which is involved in the formation of acid precipitation, is both a combustion reaction and a synthesis reaction. Similarly, the reaction of magnesium with oxygen can be classified as both a synthesis reaction and a combustion reaction. In these reactions, energy in the form of heat and light is released, as shown in **Figure 3.26**.

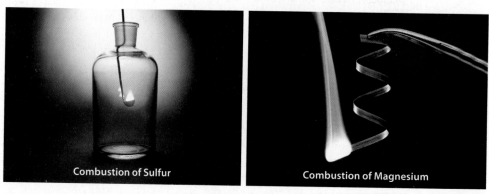

Combustion of Sulfur   Combustion of Magnesium

Figure 3.26 Each of these elements undergoes a combustion reaction with oxygen.

Compare *How are the products of these reactions similar?*

Combustion Reactions and Life Processes

So far, you have learned about combustion reactions mainly in industrial and atmospheric applications—applications that are non-living. For example, when glucose, a common form of sugar, is burned, it reacts with oxygen gas to produce carbon dioxide, water, and a large amount of energy. The combustion of glucose can be represented by the chemical equation below.

$$C_6H_{12}O_6(s) + 6O_2(g) \rightarrow 6CO_2(g) + 6H_2O(g)$$

However, this chemical equation also applies to reactions taking place in living cells that are necessary to sustain life.

Cellular Respiration

The equation for the combustion of glucose also summarizes a series of reactions that occur during a process called cellular respiration, which is the metabolism of glucose in living cells. Some of the energy that is released during cellular respiration is captured to fuel metabolic processes. The rest is released as heat. **Table 3.5** highlights some of the differences between combustion and cellular respiration.

Table 3.5 Comparison of Combustion and Cellular Respiration

Characteristics	Combustion	Cellular Respiration
Speed	Is a faster process	Is a slower process
Temperature	Generally occurs at high temperatures	Occurs at body temperatures
Complete or incomplete reaction	May be incomplete, producing soot and carbon monoxide	Always complete, producing carbon dioxide, water, and energy
Duration	Is not a continuous process	Occurs at all times within living cells

Learning Check

25. Describe incomplete combustion.

26. Why is incomplete combustion potentially hazardous?

27. List three industries in which carbon monoxide exposure can occur.

28. What determines whether complete or incomplete combustion will occur?

29. When can a synthesis reaction also be classified as a combustion reaction?

30. Bacteria that are involved in the decomposition of once-living material carry out the process represented by the following overall chemical equation:

glucose + oxygen → carbon dioxide + water

Would you describe this process as combustion or cellular respiration? Explain your answer.

Summarizing Synthesis, Decomposition, and Combustion Reactions

In this chapter, you have learned about synthesis, decomposition, and combustion reactions. By identifying the type of reaction, you can better predict how the reactants will interact and what products will likely form. **Table 3.6** summarizes the key characteristics of each type of reaction discussed in this chapter. You can use **Table 3.6** to help you classify chemical reactions. In Chapter 4, you will learn about two more types of reactions. Keep in mind that all the types of reactions you will learn about in this course represent only a very small portion of known reactions.

Table 3.6 Characteristics of Synthesis, Decomposition, and Combustion Reactions

Type of Reaction	General Form	Key Characteristics
Synthesis reaction	$A + B \rightarrow AB$	Two elements or small compounds combine to form a single product.
Decomposition reaction	$AB \rightarrow A + B$	A single reactant breaks apart to form two or more products.
Combustion reaction	element or compound + $O_2(g) \rightarrow$ oxides	Oxygen reacts with an element or compound to form one or more oxides. Energy, in the form of heat and light, is produced.

QUIRKS & QUARKS

with BOB MCDONALD

CBC

Power, Sweet Power

You likely use batteries every day in devices such as watches, cellphones, and laptops.

Most batteries are made from metals and acids. They convert the chemical energy in an electrolyte into electrical energy that can do work. The electrolyte is usually an acid paste, and the electrodes commonly contain toxic metals, such as lead, cadmium, and mercury. Unfortunately, old or discarded batteries can break down and release these toxic substances into the environment.

Because of the potential hazards of batteries, some scientists are trying to develop other types of batteries, made from less toxic substances. Dr. Shelley Minteer from St. Louis University has developed a prototype of an electricity-producing fuel cell that runs on sugar. Bob McDonald interviewed Dr. Minteer to learn more.

A Natural Solution

Minteer saw that batteries are inefficient, and that the heavy metals they use pose environmental risks. By contrast, living cells use a very efficient system, in which a carbohydrate, glucose, goes through a series of reactions, collectively known as cellular respiration, to produce all the energy that is needed by the cells. Minteer wanted to mimic cellular respiration at an electrode surface.

When a fuel cell runs on sugar, its fuel is always close at hand. In the future, cellphones could possibly be powered by maple syrup or soft drinks!

The sugar fuel cell Minteer developed uses a few millilitres of sugar water and two electrodes. Each electrode is about the size of a postage stamp and coated with enzymes like those found in living cells. The enzymes break down sucrose into simpler sugars, such as glucose, which react at the electrodes to produce carbon dioxide, water, and electrical energy.

Three or four times longer-lasting than a typical cellphone battery, the internal components of the sugar fuel cell are completely naturally sourced (the enzymes come from potatoes), and biodegradable. To recharge the fuel cell, sugar is added. The fuel cell uses no toxic metals or potentially dangerous chemicals, unlike most regular batteries and fuel cells. This technology may be well suited to small portable electronic devices that currently use rechargeable batteries.

▶ Related Career

Electrochemists study the relationship between electricity and chemical changes, such as electrolysis, corrosion, and the chemical changes that occur in batteries. Electrochemists also study lightning, luminescence, and neurons.

Go to **scienceontario**
to find out more

Dr. Shelley Minteer

QUESTIONS

1. The overall reaction that takes place in Dr. Minteer's sugar fuel cell is the same as the overall reaction for cellular respiration. In both cases, glucose, $C_6H_{12}O_6(aq)$, reacts with oxygen to form water and carbon dioxide. What is the chemical equation for cellular respiration, and how would you classify it?

2. How could Dr. Minteer's invention help to protect the environment?

3. Electrochemistry has a wide range of applications. Research and describe one job, outside the field of chemistry, that uses electrochemical knowledge.

Section Summary

- A combustion reaction of a substance with oxygen produces one or more oxides. Energy, in the form of heat and light, is released.

- A hydrocarbon is a compound that is composed of only the elements hydrogen and carbon. The combustion of a hydrocarbon can be either complete or incomplete.

- The products of complete combustion reactions are carbon dioxide and water vapour.

- The products of incomplete combustion reactions include carbon, carbon monoxide, carbon dioxide, and water vapour.

- Some reactions are both a combustion reaction and a synthesis reaction.

Review Questions

1. **K/U** In a laboratory investigation, what evidence might indicate that a combustion reaction is occurring?

2. **K/U** Identify three hydrocarbons that are used as fuels and undergo combustion reactions.

3. **C** Create a flowchart to describe how you would correctly adjust the air vents on a laboratory burner.

4. **T/I** A student says that $C_{12}H_{22}O_{11}(s)$ is a type of hydrocarbon. Do you agree? Explain your answer.

5. **T/I** Identify the missing product in the following balanced chemical equation for the complete combustion of pentane:

   $C_5H_{12}(\ell) + 8O_2(g) \rightarrow$ _____ $+ 6H_2O(g)$

6. **T/I** Write a balanced chemical equation for the complete combustion of each hydrocarbon.
   - **a.** ethene, $C_2H_4(g)$
   - **b.** decane, $C_{10}H_{22}(\ell)$
   - **c.** butene, $C_4H_8(g)$
   - **d.** hexane, $C_6H_{14}(\ell)$

7. **T/I** Would the presence of excess oxygen cause hazardous products to be formed during combustion? Explain your reasoning.

8. **K/U** Is the combustion reaction shown in **Figure 3.23** likely to be producing soot? Explain your answer.

9. **T/I** Refer to **Figure 3.26**.
   - **a.** In the reactions shown in the photographs, what types of energy are being released?
   - **b.** What types of chemical products are being formed?

10. **A** Carbon monoxide poisoning can be lethal.
    - **a.** What type of chemical reaction produces carbon monoxide?
    - **b.** What are the early signs of carbon monoxide poisoning?
    - **c.** Why can prolonged exposure lead to death?

11. **A** The photograph below shows a device that is designed to protect people from a toxic gas.
    - **a.** What gas is the device made to detect?
    - **b.** In what type of reaction is this gas likely to be formed?

12. **K/U** Why is it not possible to balance a chemical equation representing incomplete combustion without having additional information about the reaction?

13. **C** Create a Venn diagram to compare complete combustion of hydrocarbons with incomplete combustion of hydrocarbons.

14. **K/U** What causes a candle flame to appear yellow?

15. **T/I** Classify each reaction as a synthesis or combustion reaction, or both. Give reasons for your classification.
    - **a.** $2Ca(s) + O_2(g) \rightarrow 2CaO(s)$
    - **b.** $2C_4H_{10}(g) + 13O_2(g) \rightarrow 8CO_2(g) + 10H_2O(g)$
    - **c.** $6K(s) + N_2(g) \rightarrow 2K_3N(s)$

16. **A** What would happen to a living organism if its cells were no longer able to carry out the series of reactions that make up cellular respiration?

Safety Precautions

- Wear safety eyewear throughout this investigation.
- Wear a lab coat or apron throughout this investigation.

Suggested Materials

- can or bottle of soda water, $CO_2(aq)$, unopened
- acid-base indicator
- magnesium oxide, $MgO(s)$
- distilled water
- 2 beakers (50 mL)
- stirring rod
- any other equipment suggested by your teacher

Testing the Acidity of Oxides

Oxides are chemical compounds that contain oxygen bonded to at least one other element. Earth's crust contains many oxides, including quartz, $SiO_2(s)$, a common component of sand; corundum, $Al_2O_3(s)$, an aluminum ore; and hematite, or rust, $Fe_2O_3(s)$. Oxides of metals and non-metals exhibit different behaviour when they react with water. In this investigation, you will test the acidity of metal and non-metal aqueous oxides.

Pre-Lab Questions

1. How would you test the acidity of an aqueous solution?
2. What type of compound is magnesium oxide, $MgO(s)$?
3. What type of compound is carbon dioxide, $CO_2(g)$?
4. Why is it important to wear safety eyewear throughout this investigation?

Question

What patterns exist in the acidity of products formed in reactions of oxides with water?

Plan and Conduct

1. With your group, design an investigation to test the acidity of non-metal oxides and metal oxides in aqueous solutions.
2. As a group, decide how you will test the acidity of each aqueous solution.
3. Design a table like the one below to record your data. Give your table an appropriate title.

Sample Data Table

Non-metal Oxide	Acidity of Aqueous Solution	Metal Oxide	Acidity of Aqueous Solution

4. Show your experimental plan to your teacher before beginning your investigation. Conduct your investigation, and record your results.
5. When you have completed your investigation, dispose of the reaction products as instructed by your teacher.

Use these suggested materials in your experimental plan.

Analyze and Interpret

1. What techniques did you use to determine the acidity of the solutions?

2. What can you conclude about aqueous solutions of non-metal oxides?

3. Write a balanced chemical equation to represent the reaction between one of the non-metal oxides you used and water.

4. Based on your observations, what can you conclude about aqueous solutions of metal oxides? How could you test your conclusions?

5. Write a balanced chemical equation to represent the reaction between one of the metal oxides you used and water.

6. Classify, by type, the reactions represented by your balanced chemical equations in questions 3 and 5.

Conclude and Communicate

7. Use your observations to write a summary describing how the reaction of water with a metal oxide differs from the reaction of water with a non-metal oxide.

8. **INQUIRY** Design an investigation to test the acidity of aqueous solutions of oxides containing phosphorus, iron, sulfur, and sodium. Include safety precautions in your design. With your teacher's approval, conduct your investigation.

9. **RESEARCH** Reactions that involve non-metal oxides and water in the atmosphere can cause precipitation (such as rain, snow, and even fog) to be quite acidic. Use Internet or print resources to answer the following questions.

   a. Which oxides that cause acid precipitation are most prevalent in the atmosphere?

   b. Describe the sources of the oxides that cause acid precipitation. Estimate the percentages contributed to the atmosphere from natural and human sources.

   c. Describe some techniques that are used to mitigate the effects of acid precipitation.

Safety Precautions

- Wear safety eyewear throughout this investigation.
- Tie back loose hair and clothing.
- Use EXTREME CAUTION when you are near an open flame.

Suggested Materials

- candle
- wooden splint
- Bunsen burner secured to a utility stand
- igniter for Bunsen burner
- retort clamp
- 2 Erlenmeyer flasks (250 mL)
- watch glass
- 5 cm × 5 cm cardboard
- tongs
- wire mesh
- any other equipment suggested by your teacher

Comparing Complete and Incomplete Combustion

The combustion of hydrocarbons can be either complete or incomplete, depending on the amount of oxygen that is available and on the conditions during the reaction. In this investigation, you will design a procedure to combust natural gas, which is mainly methane, $CH_4(g)$, and candle wax, $C_{25}H_{52}(s)$, under different conditions to compare the products of complete and incomplete combustion. Note that candle wax is a mixture of hydrocarbons, and the chemical formula $C_{25}H_{52}(s)$ is used as a representation of the mixture.

Pre-Lab Questions

1. Compare and contrast complete and incomplete combustion.
2. Brainstorm how you can show the difference between complete and incomplete combustion, using methane and wax.
3. Write a list of safety issues concerning combustion reactions and the compounds you will be working with in this investigation.

Questions

How can complete and incomplete combustion of a hydrocarbon be recognized? How can conditions during a reaction affect the combustion of a hydrocarbon?

Plan and Conduct

Part 1: Combustion of Methane

1. With your group, design a procedure to collect the products of the complete combustion of methane. Decide what observations and tests you will use to identify the combustion products you collect.
2. Show your experimental plan to your teacher before beginning your investigation.
3. Conduct your investigation, and record your results.

The products of combustion can be collected by inverting an Erlenmeyer flask over the flame of a Bunsen burner.

Part 2: Combustion of Candle Wax

4. As a group, determine a method to keep a lit candle upright without holding it in your hand.

5. Design a procedure to collect the products of the incomplete combustion of candle wax. Decide what observations and tests you will use to identify the combustion products you collect.

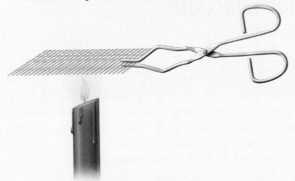

Holding a cool, nonflammable item in or very near the flame will help you capture all the products of incomplete combustion.

6. Show your experimental plan to your teacher before beginning your investigation.

7. Conduct your investigation, and record your results.

Analyze and Interpret

1. What products of combustion did you collect in Part 1? How did you identify these products?

2. How did your results indicate that the methane in Part 1 was undergoing complete combustion?

3. What products of combustion did you collect in Part 2? How did you identify these products?

4. What evidence did you use to conclude that the candle wax in Part 2 was undergoing incomplete combustion?

5. What do your observations in Part 2 suggest about the role of heat in the combustion process?

Conclude and Communicate

6. Write a balanced chemical equation for the complete combustion of methane, $CH_4(g)$.

7. **a.** Write a skeleton equation for the incomplete combustion of candle wax, which is represented by the chemical formula $C_{25}H_{52}(s)$.

   b. Why is it impossible to balance the equation for the incomplete combustion of a hydrocarbon without additional information?

8. **a.** Explain how your procedure distinguished between complete and incomplete combustion.

   b. Based on your experimental results and your knowledge of complete and incomplete combustion, compare the products of these two reactions.

Extend Further

9. **INQUIRY** Design an investigation to capture the products of combustion from burning, $C_4H_{10}(s)$, butane in a disposable butane lighter. In addition to using direct observation to determine the products of combustion, you might use cobalt chloride paper to test for the presence of water, and limewater to test for the presence of carbon dioxide. Describe all safety precautions required for your proposed investigation. With your teacher's approval, conduct your investigation. Determine whether complete or incomplete combustion occurred.

10. **RESEARCH** The Ontario Fire Marshall's office stresses the importance of using carbon monoxide detectors in homes. Use Internet or print resources to answer the following questions.

    a. Where in the home should carbon monoxide detectors be placed?

    b. How do carbon monoxide detectors work?

Safety Precautions

- Wear safety eyewear throughout this investigation.

- Tie back loose hair and clothing.

- Use EXTREME CAUTION when you are near an open flame.

- Wear a lab coat or apron throughout this investigation.

Materials

- 3 cm of copper wire, $Cu(s)$

- 10 mL of a 3% solution of hydrogen peroxide, $H_2O_2(aq)$

- 3 g of powdered manganese(IV) oxide, $MnO_2(s)$

- 5 wooden splints

- water

- small piece of calcium carbide, $CaC_2(s)$

- 20 mL of freshly prepared limewater, $Ca(OH)_2(aq)$

- fine sandpaper

- crucible tongs

- Bunsen burner secured to a utility stand

- igniter for Bunsen burner

- 5 test tubes

- grease marker

- test-tube clamp

- scoopula

- test-tube rack

- 500 mL beaker

- 4 test-tube stoppers

Conducting Synthesis, Decomposition, and Combustion Reactions

In this investigation, you will observe synthesis, decomposition, and combustion reactions. You will perform a synthesis reaction between copper and oxygen, and you will perform a decomposition reaction of hydrogen peroxide, $H_2O_2(aq)$. Then you will prepare acetylene, $C_2H_2(g)$, and use it to observe complete and incomplete combustion.

Pre-Lab Questions

1. Write the general form for the following types of reactions:
   a. synthesis reaction
   b. decomposition reaction
   c. complete combustion of a hydrocarbon

2. Describe why it is important to wear a lab coat or apron when working with a solution of hydrogen peroxide.

Question

What happens to the elements and compounds in the synthesis, decomposition, and combustion reactions that are carried out in this investigation?

Procedure

Part 1: Synthesis

1. Use fine sandpaper to clean the copper wire until the wire is shiny.

2. Light the Bunsen burner. Use crucible tongs to hold the copper wire in the hottest part of the flame for 1 to 2 min.

3. Examine the wire as it cools. Observe any changes caused by the synthesis reaction between copper and oxygen.

Part 2: Decomposition

4. Hold a test tube with a clamp. Add 10 mL of hydrogen peroxide and 3 g of powdered manganese(IV) oxide to the test tube.

5. Observe the decomposition reaction of hydrogen peroxide. Note that manganese(IV) oxide acts as a catalyst—it remains unchanged during the reaction, but it speeds up the decomposition of hydrogen peroxide.

6. Insert a glowing splint into the test tube to identify the gas being produced.

7. Place the test tube in a rack to cool.

8. Dispose of the contents of the test tube as directed by your teacher.

Part 3: Complete and Incomplete Combustion

9. Label four test tubes as follows: $1, \frac{1}{2}, \frac{1}{4}, \frac{1}{10}$. Mark a line on the test tube labelled "$\frac{1}{2}$" to show where the test tube would be $\frac{1}{2}$ full. Similarly, mark lines on the test tubes labelled "$\frac{1}{4}$" and "$\frac{1}{10}$" to show where the test tubes would be $\frac{1}{4}$ full and $\frac{1}{10}$ full.

10. Half fill a 500 mL beaker with water. Fill the test tubes completely with water. Place stoppers in the test tubes, and put the test tubes in the rack.

11. Drop a small piece of calcium carbide into the beaker of water to produce acetylene gas:
$$CaC_2(s) + 2H_2O(\ell) \rightarrow C_2H_2(g) + Ca(OH)_2(aq)$$

12. Invert the test tube labelled "1" in the beaker so that its mouth is underwater. Remove the stopper. Move the test tube directly over the bubbling calcium carbide to collect acetylene gas. The gas will displace the water in the test tube. When the test tube is filled with acetylene gas, remove it from the beaker and replace the stopper. Keeping the test tube inverted, place it in the rack.

The rising gas displaces the water in the test tube.

13. Repeat step 12 to collect acetylene gas in the remaining test tubes. When the appropriate amount of gas has been collected in each test tube, the water level will reach the mark you made on the test tube in step 9. Lift the test tube to allow the remaining water to drain, and then replace the stopper. Place the test tube in the rack.

14. Use a clamp to hold the test tube labelled "1" in a horizontal position. Point the top of the test tube away from you and other nearby people. Remove the stopper, and immediately bring a burning splint to the mouth of the test tube to ignite the acetylene gas. As soon as the combustion reaction is complete, add about 5 mL of limewater to the test tube. Replace the stopper, and shake the test tube vigorously. Observe the combustion reaction and the appearance of the limewater. Record your observations.

15. Repeat step 14 for each test tube.

Analyze and Interpret

1. Refer to Part 1.
   a. What evidence of a chemical change did you observe?
   b. Write a balanced chemical equation for this reaction.

2. Refer to Part 2.
   a. What gas was produced during the reaction? Explain your answer.
   b. Write a balanced chemical equation for the reaction.

3. Refer to Part 3.
   a. In which test tube(s) did you observe complete combustion? Explain your answer.
   b. Write a balanced chemical equation for the complete combustion of acetylene gas.
   c. In which test tube(s) did you observe incomplete combustion? Explain your answer.
   d. What products formed during the incomplete combustion of acetylene gas?

Conclude and Communicate

4. In Part 1, what evidence of a chemical change did you observe?

5. The decomposition reaction in Part 2 produces oxygen. Why is a decomposition reaction useful when an element needs to be isolated from a compound?

6. In Part 3, why did the amount of soot that was produced in the test tubes vary?

Extend Further

7. **INQUIRY** Design an investigation to observe a reaction that is both a synthesis reaction and a combustion reaction. With your teacher's approval, conduct your investigation. Write a balanced chemical equation for the chemical reaction. Identify the type(s) of reaction that occurred, and explain your reasoning.

8. **RESEARCH** Torches with acetylene gas are commonly used to weld and cut metals. Use Internet or print resources to answer the following questions.
   a. Why is an oxyacetylene torch generally attached to two cylinders? What is the gas in each cylinder?
   b. Why is an oxyacetylene torch more effective than an acetylene torch for welding metals?
   c. Describe some specific applications in which an oxyacetylene torch is used.

Case Study

Lithium Recycling

Concerns Associated with Living Near a Recycling Facility

Scenario

Suppose that your community is located near a plant that recycles lithium batteries. The plant has had several fires. One fire was so large that 52 firefighters were needed to extinguish it. From your home, you could see dark clouds of smoke rising from the plant, and you heard several loud explosions. You later learned that the explosions sent flaming debris into the air. The debris spread the fire to a nearby building by igniting materials, mainly cardboard, that had been stacked outside. While the fire was burning, you and the other members of your community were told to stay inside to avoid inhaling the contaminants in the outside air. Fortunately, the explosions and the fire did not cause any reported injuries.

Lithium + Water = Flammable Gas

The fire at the battery recycling plant involved lithium, which is highly reactive with water. When lithium is exposed to water, it reacts according to the following chemical equation:

$$2Li(s) + 2H_2O(\ell) \rightarrow 2LiOH(aq) + H_2(g)$$

The products of this single displacement reaction are lithium hydroxide and hydrogen gas. Hydrogen gas is highly flammable. Initially, the firefighters could not use water to douse the flames because water would have made the fire more intense. Their only choice was to contain the fire until the lithium burned away. It was not until several hours after the fire began that firefighters could use water on it.

The fire also released poisonous sulfur dioxide gas into the air. This gas can be deadly when inhaled. Because it dissipated quickly, however, nearby residents were not evacuated. Since the fire, authorities have been monitoring the air quality around your community. They have reported that any health and environmental effects of the fire were short term and confined to the area around the plant. Local authorities are working with the community and the company that runs the plant to address the concerns of local residents, like you.

A statement issued by the company says that the company has strict safety measures in place to protect the people and environment near the plant. For example, the company uses earth-covered concrete bunkers to store lithium products before they are recycled. The company has promised to review its safety practices in order to prevent similar fires in the future. It has reported that its employees are cleaning up after the fire and investigating the cause.

Even after firefighters worked through the night to extinguish the fire, it still smouldered the next day.

The statement issued by the company includes the following information:

A B R Anycity Battery Recycling

Lithium Batteries and Recycling

- *Lithium batteries are used in many electronic devices, such as mobile phones, other hand-held devices, and laptop computers.*
- *Lithium batteries are also used in hybrid electric vehicles and for oil and gas exploration.*
- *Unsafe use and disposal of lithium batteries can result in fires, the release of toxic gases, and the contamination of soil, water, and air.*
- *Proper recycling of lithium batteries can reduce hazardous waste by preventing the batteries and their contents from ending up in landfills. Recycling also conserves natural resources and can result in the recovery of marketable products.*

You and many members of your community are concerned that the recycling of lithium batteries poses too great a risk, considering the complications involved in putting out a fire involving lithium. Many of your neighbours believe that battery recycling plants should not be located near people's homes. They want the plant to close or relocate to another area. With some of your neighbours, you have formed a community group in response to the fire. The goal of your group is to find solutions to minimize the risk from chemical fires for your community and the surrounding environment.

Research and Analyze

1. Conduct research to find out more about the properties of the element lithium. What properties make lithium useful in batteries? What properties make lithium dangerous?

2. Conduct research to find out what rules and regulations companies that process or manufacture chemicals are required to follow, to ensure the safety of people in the surrounding community and the environment.

3. Analyze the risks and benefits of locating plants that process or manufacture chemicals, such as lithium, near residential communities.

Take Action

1. **Plan** In a group, discuss the concerns associated with the location of chemical plants, such as the lithium battery recycling plant, close to residential areas. What are the different points of view about plant locations? What might be some benefits of locating chemical plants and other industries near established neighbourhoods? In addition, what are some points of view about developing residential areas that encroach on the locations of established plants? Share the results of the research and analysis you conducted in questions 1 to 3.

2. **Act** Prepare a multimedia presentation to get more members of your community involved in the issue. Your presentation should represent your opinion about the location of the battery recycling plant in your community, and it should outline the reasons why members of your community are concerned. In your presentation, propose possible solutions that may help to ensure the safety of your community.

Lithium is widely used in batteries that power electronic devices, such as this laptop computer battery.

Section 3.1 | Writing Chemical Equations

A chemical equation describes what occurs in a chemical reaction in shortened form. A balanced chemical equation provides the most accurate representation of a reaction.

KEY TERMS

balanced chemical equation   product
chemical equation   reactant
chemical reaction   skeleton equation
coefficient

KEY CONCEPTS

- A chemical equation is a representation of a chemical reaction. An arrow separates the reactants from the products. Plus signs separate multiple reactants or products.

- A skeleton equation represents a chemical reaction by using chemical formulas for the reactants and products, and symbols for the physical states of these substances. A skeleton equation is an incomplete representation because it does not show that all the atoms in the reactants must appear in the same quantities in the products.

- A balanced chemical equation uses chemical formulas and coefficients to show the relative number of particles of each substance that is involved in a chemical reaction.

- A balanced chemical equation reflects the law of conservation of mass.

- A balanced equation shows the same number of atoms of each element in the reactants and the products.

Section 3.2 | Synthesis Reactions and Decomposition Reactions

In a synthesis reaction, two reactants combine into a single product. In a decomposition reaction, a single compound breaks down into two or more products.

KEY TERMS

decomposition reaction   synthesis reaction
electrolysis

KEY CONCEPTS

- In a synthesis reaction, two or more reactants combine to form a single product. This reaction can be represented by the general form $A + B \rightarrow AB$.

- The product of a synthesis reaction between a metal and a non-metal is usually a binary ionic compound. The product of a synthesis reaction between a multivalent metal and a non-metal depends on the charge of the metal ion formed.

- In a synthesis reaction, a non-metal can combine with another non-metal in several different ratios.

- A non-metal oxide reacts with water to form an acid in a synthesis reaction. A metal oxide reacts with water to form a metal hydroxide, which is a base.

- In a decomposition reaction, a single reactant breaks apart to form two or more products. This reaction can be represented by the general form $AB \rightarrow A + B$. A binary compound generally decomposes into its elements.

- A metal nitrate generally decomposes into a metal nitrite and oxygen gas.

- A metal carbonate generally decomposes into a metal oxide and carbon dioxide gas. A metal hydroxide generally decomposes into a metal oxide and water.

Section 3.3 | Combustion Reactions

When a substance reacts with oxygen during a combustion reaction, one or more oxides are produced, and energy is released as heat and light.

KEY TERMS

combustion reaction
hydrocarbon
soot

KEY CONCEPTS

- A combustion reaction of a substance with oxygen produces one or more oxides. Energy, in the form of heat and light, is released.

- A hydrocarbon is a compound that is composed of only the elements hydrogen and carbon. The combustion of a hydrocarbon can be either complete or incomplete.

- The products of complete combustion reactions are carbon dioxide and water vapour.

- The products of incomplete combustion reactions include carbon, carbon monoxide, carbon dioxide, and water vapour.

- Some reactions are both a combustion reaction and a synthesis reaction.

Knowledge and Understanding

Select the letter of the best answer below.

1. Which of the following represents both the chemical composition and the relative amount of each substance in a chemical change?
   a. word equation
   b. skeleton equation
   c. balanced chemical equation
   d. chemical property
   e. chemical reaction

2. Consider the chemical reaction $CaCO_3(s) + HCl(aq) \rightarrow CaCl_2(aq) + CO_2(g) + H_2O(\ell)$. Which compound is in a solid state?
   a. $CaCO_3$
   b. HCl
   c. $CaCl_2$
   d. CO_2
   e. H_2O

3. Which formula correctly represents sodium chloride that has been dissolved in water?
   a. $NaCl$
   b. $NaCl(\ell)$
   c. $NaCl(s)$
   d. $NaCl(H_2O)$
   e. $NaCl(aq)$

4. Ammonia reacts with oxygen to form nitrogen monoxide and water according to the chemical equation $4NH_3(g) + 5O_2(g) \rightarrow 4NO(g) + 6H_2O(g)$. In this chemical equation, which numbers are examples of coefficients?
   a. 2 and 3
   b. 2 and 6
   c. 4 and 6
   d. 3 and 5
   e. 2 and 4

5. Which reactant(s) would likely form a metal hydroxide?
   a. a metal oxide and carbon dioxide
   b. a hydrocarbon and oxygen
   c. a metal oxide and water
   d. a metal nitrate
   e. a metal carbonate

6. What are the reactants in a complete combustion reaction?
   a. oxygen and a hydrocarbon
   b. carbon dioxide and a metal oxide
   c. water vapour and carbon dioxide
   d. a metal nitrite and oxygen
   e. water vapour and a metal oxide

7. Which reactant(s) would undergo a reaction that is both a synthesis reaction and a combustion reaction?
   a. water vapour and a metal oxide
   b. a binary compound
   c. a hydrocarbon and oxygen
   d. a metal and oxygen
   e. water vapour and carbon dioxide

8. Which reactant(s) would form an acidic solution?
   a. a metal oxide and water
   b. a metal and a non-metal
   c. a hydrocarbon and oxygen
   d. a metal carbonate
   e. a non-metal oxide and water

Answer the questions below.

9. Which molecules always have the formula X_2 in a balanced chemical equation?

10. In the chemical equation $2H_2(g) + O_2(g) \rightarrow 2H_2O(g)$, what is the value of the coefficient for $O_2(g)$?

11. Compare the solution that is formed from a metal oxide with the solution that is formed from a non-metal oxide.

12. Describe the process of electrolysis.

13. What products are formed during incomplete combustion of a hydrocarbon?

14. What is the danger of having an incomplete combustion reaction occur in an enclosed space?

15. What type of reaction is often used in explosives?

16. Why might a synthesis reaction involving a metal such as copper have more than one possible product?

17. Identify three industrial products that are produced using electrolysis.

Thinking and Investigation

18. Copy and balance each chemical equation.
    a. $Mg_3N_2(s) \rightarrow Mg(s) + N_2(g)$
    b. $Mn(s) + O_2(g) \rightarrow Mn_2O_3(s)$
    c. $CO_2(g) + H_2(g) \rightarrow CH_4(g) + H_2O(g)$
    d. $PbO(s) \rightarrow Pb(s) + O_2(g)$
    e. $C_2H_6(g) + O_2(g) \rightarrow CO_2(g) + H_2O(g)$
    f. $Cu(s) + AgNO_3(aq) \rightarrow Ag(s) + Cu(NO_3)_2(aq)$
    g. $C_3H_8(g) + O_2(g) \rightarrow CO_2(g) + H_2O(g)$
    h. $PbCl_4(aq) + K_3PO_4(aq) \rightarrow KCl(aq) + Pb_3(PO_4)_4(s)$

19. A chemical equation is balanced, but it does not accurately represent a chemical reaction that occurred. How is this possible?

20. Predict the likely product(s) of the synthesis reaction between the given reactants. Then write a balanced chemical equation for each reaction.
  a. potassium and sulfur
  b. chromium and chlorine
  c. silver and oxygen
  d. sulfur and chlorine, forming a gaseous hexachloride

21. Determine the products that are likely to form when each compound decomposes. Write a balanced chemical equation for the decomposition.
  a. magnesium iodide
  b. copper(II) nitrate
  c. barium carbonate

22. Is there more than one possible product that might form in the synthesis reaction between an alkaline-earth metal and a halogen? Explain your reasoning.

23. Write a balanced chemical equation for the complete combustion of each hydrocarbon.
  a. $C_2H_6(g)$
  b. $C_5H_{12}(\ell)$
  c. $C_8H_{18}(\ell)$

24. Most elements are combined with other elements in nature. A chemist wants to isolate an element from a compound to obtain a sample of the element. Which type of chemical reaction is the chemist likely to use? Explain.

Communication

25. Research the proper maintenance of a gas or wood-burning heating system, as well as the hazards associated with an improperly maintained heating system. Present your findings in a pamphlet.

26. Draw and label diagrams that model each type of reaction.
  a. synthesis
  b. decomposition
  c. combustion

27. **BIG IDEAS** Chemicals react in predictable ways. Develop a diagram showing how all alkaline-earth metals react in a similar way. Include the balanced chemical equations for the synthesis reactions that produce calcium bromide, magnesium bromide, and strontium bromide.

28. **BIG IDEAS** Chemical reactions and their applications have significant implications for society and the environment. Write an e-mail to a local politician, explaining why scrubbers need to be installed on smokestacks at a local or regional coal-fired power plant to reduce sulfur dioxide emissions.

29. Create a table to summarize the types of synthesis reactions you have studied in this chapter.

30. Create a table to summarize the types of decomposition reactions you have studied in this chapter.

31. At high pressure and temperature in a closed container, nitrogen gas and hydrogen gas react to form ammonia gas in a reversible chemical reaction. A student wrote the following chemical equation to describe the reaction: $N(g) + H_3(g) \rightarrow NH_3(g)$. State why this chemical equation is incorrect, and write it correctly.

32. Summarize your learning in this chapter using a graphic organizer. To help you, the Chapter 3 Summary lists the Key Terms and Key Concepts. Refer to Using Graphic Organizers in Appendix A to help you decide which graphic organizer to use.

Application

33. Following the development of electric batteries (voltaic piles at that time), a rapid succession of elements were isolated and identified. Why do you think these advances in technology and science were likely connected?

34. In the engine of an automobile, nitrogen gas and oxygen gas combine to form various nitrogen oxides, including nitrogen dioxide, $NO_2(g)$. Nitrogen oxides are a component of smog.
  a. What kind of reaction is this?
  b. Write a balanced chemical equation for this reaction.
  c. A catalytic converter helps to break apart nitrogen oxides before they leave the exhaust pipe of an automobile. What kind of reaction takes place in the catalytic converter? Explain your reasoning.

Catalytic Converter

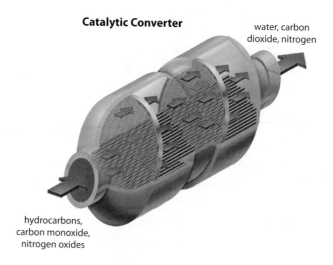

water, carbon dioxide, nitrogen

hydrocarbons, carbon monoxide, nitrogen oxides

35. Hydrogen peroxide, $H_2O_2(\ell)$, decomposes to form water and oxygen gas. Write a balanced chemical equation for this reaction.

36. The inside of a fireplace is coated with a fine, black powder, as shown in the photograph below.
  a. What is the black powder?
  b. What does the presence of the black powder indicate about the chemical reaction in the fireplace?
  c. Why is the proper maintenance of a fireplace important for safety?

37. Many power plants burn coal or oil as their energy source to generate electricity. These fuels can contain sulfur that can enter the atmosphere and lead to the formation of acid precipitation. The diagram below shows scrubber technology that power plants can use to remove sulfur from their exhaust gases.

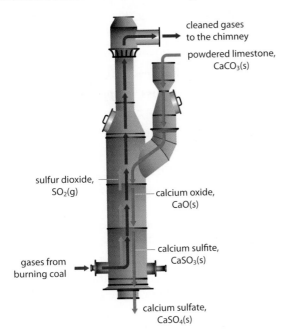

cleaned gases to the chimney
powdered limestone, $CaCO_3(s)$
sulfur dioxide, $SO_2(g)$
calcium oxide, $CaO(s)$
calcium sulfite, $CaSO_3(s)$
gases from burning coal
calcium sulfate, $CaSO_4(s)$

a. When powdered limestone, $CaCO_3(s)$, is added to the hot exhaust gases, calcium oxide, $CaO(s)$, and carbon dioxide, $CO_2(g)$, form. Identify the type of reaction this is, and write a balanced chemical equation for it.

b. The calcium oxide reacts with sulfur dioxide, $SO_2(g)$, to form calcium sulfite, $CaSO_3(s)$. Identify the type of reaction this is, and write a balanced chemical equation for it.

c. The calcium sulfite reacts with oxygen in air to convert it into a more stable form, calcium sulfate, $CaSO_4(s)$. Identify the type of reaction this is, and write a balanced chemical equation for it.

d. Create a flowchart showing the sequence of reactions this process uses to prevent the escape of sulfur to the atmosphere.

e. Research the effectiveness of this technology in preventing the escape of sulfur to the atmosphere.

38. Balanced chemical equations are used in a wide array of applications. Research and describe one job, outside the field of chemistry, that requires the frequent use of balanced chemical equations.

39. How does the state of a compound affect electrolysis?

40. Carbon monoxide is a toxic chemical that forms during a combustion reaction. Many other toxic materials can also form during a fire. The table shows some of the harmful chemicals that formed during a fire in Hamilton, Ontario, in July 1997.
  a. Research the effects of several of these chemicals on human health and the environment.
  b. In what ways do toxic chemical fires affect local communities?

Some Atmospheric Contaminants Released by the 1997 Hamilton, Ontario, Fire

Type of Contaminant	Compounds or Elements	
Volatile organic compounds	benzene, $C_6H_6(\ell)$ 1,3-butadiene, $C_4H_6(g)$ chlorobenzene, $C_6H_5Cl(\ell)$ naphthalene, $C_{10}H_8(s)$ styrene, $C_8H_8(\ell)$ toluene, $C_7H_8(\ell)$ vinyl chloride, $C_2H_3Cl(g)$	
Halogen compounds	hydrogen chloride, $HCl(g)$	
Metals	chromium copper iron lead	manganese nickel vanadium zinc

Select the letter of the best answer below.

1. **T/I** Which coefficient completes the following balanced chemical equation?

$$3Ca(OH)_2(aq) + 2H_3PO_4(aq) \rightarrow$$
$$Ca_3(PO_4)_2(s) + \underline{\quad}H_2O(\ell)$$

   a. 2
   b. 3
   c. 4
   d. 5
   e. 6

2. **K/U** Consider the chemical reaction $CaCO_3(s) + HCl(aq) \rightarrow CaCl_2(aq) + CO_2(g) + H_2O(\ell)$. Which compound(s) involved in this reaction would be dissolved in water?
   a. $CaCO_3$ and HCl
   b. HCl and $CaCl_2$
   c. H_2O
   d. $CaCl_2$, CO_2, and H_2O
   e. $CaCl_2$, HCl, and H_2O

3. **K/U** Which of the following is a correct description of a chemical reaction?
   a. A hydrocarbon produces carbon dioxide and water in a decomposition reaction.
   b. A metal and a non-metal form a binary compound in a decomposition reaction.
   c. A metal oxide reacts with water to form a metal hydroxide in a synthesis reaction.
   d. A metal carbonate forms a metal oxide and carbon dioxide in a combustion reaction.
   e. A hydrocarbon reacts with oxygen to form carbon dioxide and water in a synthesis reaction.

4. **A** Which chemical is least likely to be found in the exhaust of a poorly maintained gas furnace?
   a. carbon dioxide
   b. carbon monoxide
   c. carbon (soot)
   d. water
   e. calcium oxide

5. **T/I** What are the likely products of the decomposition of silver nitrate, $AgNO_3(aq)$?
   a. a metal and nitrate ions
   b. nitrogen and a metal oxide
   c. a metal, nitrogen, and oxygen
   d. a metal nitrite and oxygen
   e. water vapour and a metal oxide

6. **K/U** When the following substances react, which forms a basic solution?
   a. a metal oxide and water
   b. a metal and a non-metal
   c. a hydrocarbon and oxygen
   d. a metal carbonate
   e. a non-metal oxide and water

7. **K/U** Which oxide, when it reacts with water in air, causes rain to be naturally acidic?
   a. $CO_2(g)$
   b. $CaO(s)$
   c. $O_2(g)$
   d. $BaO(s)$
   e. $SiO_2(s)$

8. **K/U** Which formula has a coefficient of 3 and represents six atoms of oxygen?
   a. $2O_3(g)$
   b. $3KNO_3(aq)$
   c. $3H_2O(g)$
   d. $3CO_2(g)$
   e. $C_6H_{12}O_6(s)$

9. **T/I** What are the likely products of the decomposition of a metal carbonate?
   a. oxygen and a hydrocarbon
   b. carbon dioxide and a metal oxide
   c. water vapour and carbon dioxide
   d. a metal nitrite and oxygen
   e. water vapour and a metal oxide

10. **K/U** Which equation represents both a combustion reaction and a synthesis reaction?
   a. $H_2CO_3(aq) \rightarrow H_2O(\ell) + CO_2(g)$
   b. $2Cd(s) + O_2(g) \rightarrow 2CdO(s)$
   c. $2K(s) + Cl_2(g) \rightarrow 2KCl(s)$
   d. $C_3H_8(g) + 5O_2(g) \rightarrow 4H_2O(g) + 3CO_2(g)$
   e. $Zn(s) + 2HCl(aq) \rightarrow ZnCl_2(aq) + H_2(g)$

Use sentences and diagrams, as appropriate, to answer the questions below.

11. **K/U** What gas is generated in the decomposition of a metal carbonate?

12. **T/I** When sodium carbonate is heated, it breaks down to form solid sodium oxide and carbon dioxide gas.
   a. Write a balanced chemical equation for this reaction.
   b. Describe a procedure that could be used to determine the mass of the carbon dioxide formed when 10 g of sodium carbonate breaks down.

13. **A** Describe a situation in which incomplete combustion would be preferred and a situation in which complete combustion would be preferred.

14. **C** Write a procedure for the proper adjustment of a Bunsen burner, including the type of combustion reaction that could occur.

15. **T/I** Balance each skeleton equation.
  a. $Cr(ClO_3)_2(s) \rightarrow CrCl_2(s) + O_2(g)$
  b. $Rb(s) + O_2(g) \rightarrow Rb_2O(s)$
  c. $C_2H_4(g) + O_2(g) \rightarrow CO_2(g) + H_2O(g)$
  d. $KOH(s) \rightarrow K_2O(s) + H_2O(g)$

16. **T/I** Classify each skeleton equation as a synthesis, decomposition, or combustion reaction. Then balance the equation.
  a. $C_3H_8(g) + O_2(g) \rightarrow CO_2(g) + H_2O(g)$
  b. $KBrO_3(s) \rightarrow KBr(s) + O_2(g)$
  c. $CaO(s) + SO_2(g) \rightarrow CaSO_3(s)$
  d. $Ca(NO_3)_2(s) \rightarrow Ca(NO_2)_2(s) + O_2(g)$
  e. $C_{12}H_{22}O_{11}(s) \rightarrow C(s) + H_2O(\ell)$
  f. $C_2H_6(g) + O_2(g) \rightarrow CO_2(g) + H_2O(g)$

17. **T/I** Predict the product(s) that would likely form in each synthesis reaction. Then write a balanced chemical equation for the reaction.
  a. $Al(s) + Cl_2(g) \rightarrow$ **b.** $BaO(s) + H_2O(\ell) \rightarrow$

18. **T/I** Predict the product(s) that would likely form in each decomposition reaction. Then write a balanced chemical equation for the reaction.
  a. $Ca_3N_2(s) \rightarrow$ **b.** $H_2SO_3(aq) \rightarrow$

19. **T/I** Write and balance a chemical equation for the complete combustion of each compound.
  a. pentane, $C_5H_{12}(\ell)$
  b. propene, $C_3H_6(g)$

20. **K/U** Explain why more than one binary compound might result from the reaction of iron with chlorine, but not from the reaction of zinc with chlorine.

21. **K/U** List two sources of gases that contribute to acid precipitation.

22. **K/U** From what type of reactant do carbon dioxide and a metal oxide form?

23. **A** Gasoline is a complex mixture of hydrocarbons. The main components of gasoline are usually heptane, octane, pentane, and benzene. Hydrazine, $N_2H_4(\ell)$, is being explored as an alternative fuel for vehicles. However, because hydrazine is toxic and highly unstable, many technological advances would need to be made before it could be widely used for this application.
  a. The incomplete combustion of gasoline can lead to the production of nitrogen monoxide and nitrogen dioxide. How do these compounds contribute to air pollution?
  b. As a fuel source, what advantage does the decomposition of hydrazine have, compared with the combustion of hydrocarbons in gasoline, in relation to emissions of oxides?

24. **C** Use the balanced chemical equation below to develop a diagram that shows how to identify the reactants, products, and coefficients, as well as the state of each substance in a chemical equation.
  $2HCl(g) + Na_2CO_3(aq) \rightarrow CO_2(g) + H_2O(\ell) + 2NaCl(aq)$

25. **A** A company that is developing a chemistry kit wants to include materials that can generate water.
  a. Identify the type of reaction that each material would undergo to generate water.
    i. $LiOH(s)$ **iii.** $H_2(g) + O_2(g)$
    ii. $H_2SO_4(aq)$ **iv.** $CH_4(g)$
  b. Which material would you recommend for the kit? Explain your answer.

Self-Check

If you missed question …	1	2	3	4	5	6	7	8	9	10	11	12	13	14	15	16	17	18	19	20	21	22	23	24	25
Review section(s)…	3.1	3.1	3.2, 3.3	3.3	3.2	3.2	3.2	3.1	3.2	3.2, 3.3	3.2	3.2	3.3	3.3	3.1	3.2. 3.3	3.2	3.2	3.3	3.2	3.2	3.2	3.3	3.1	3.2

Displacement Reactions

Specific Expectations

In this chapter, you will learn how to . . .

- C1.1 **analyze**, on the basis of research, chemical reactions used in various industrial processes that can have an impact on the health and safety of local populations (4.3)

- C1.2 **assess** the effectiveness of some applications that are used to address social and environmental needs and problems (4.3)

- C2.2 **write** balanced chemical equations to represent single displacement and double displacement reactions using the IUPAC nomenclature system (4.1, 4.2)

- C2.3 **investigate** single displacement and double displacement reactions by **testing** the products of each reaction (4.1, 4.2)

- C2.5 **predict** the products of single displacement reactions, using the metal activity series and the halogen series (4.1)

- C2.6 **predict** the products of double displacement reactions (4.2)

- C2.9 **investigate** neutralization reactions (4.2)

- C2.10 **plan** and **conduct** an inquiry to demonstrate a single displacement reaction, using elements from the metal activity series (4.1)

- C3.1 **identify** various types of chemical reactions, including single displacement and double displacement (4.1, 4.2, 4.3)

Displacement reactions play an important role in the mining industry. An ore may be mined as a source of a metal, such as copper, nickel, or gold. Often, different types of displacement reactions are used first to change the metal to a soluble form and then to remove the metal from solution in its solid form. Similar extractions can occur naturally when waste materials from a mining operation are exposed to air and rain, as shown in this photograph of Sudbury, Ontario.

Displacement reactions are also used to address environmental issues. For example, these reactions are used to remove toxic metal ions from water or contaminated soils.

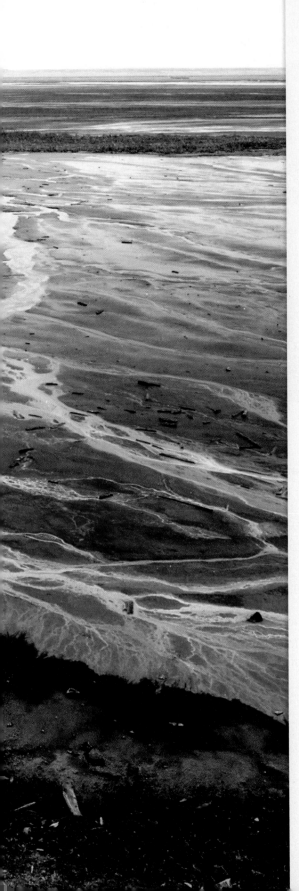

Observing Displacement Reactions

In this activity, you will see two examples of displacement reactions, which you will learn about in the chapter.

Safety Precautions

- Wear safety eyewear throughout this activity.
- Tie back loose hair and clothing.
- Use EXTREME CAUTION when you are near an open flame.
- Wear a lab coat or apron throughout this activity.

Materials

- 110 mL of 1 mol/L hydrochloric acid, HCl(aq)
- 3 drops of phenolphthalein indicator solution
- 15 cm of clean magnesium, Mg(s), ribbon
- test tube
- 120 mL of 1 mol/L sodium hydroxide solution, NaOH(aq)
- wooden splint
- 600 mL beaker
- Bunsen burner secured to a utility stand
- igniter for Bunsen burner

Procedure

1. Pour 100 mL of 1 mol/L hydrochloric acid, into a 600 mL beaker. Add three drops of phenolphthalein indicator solution. Record the appearance of the solution.

2. Add 10 cm of magnesium ribbon to the solution. Record your observations.

3. In a small test tube, place about 10 mL of hydrochloric acid and 5 cm of magnesium ribbon. Light the Bunsen burner. Ignite the splint, and then blow out its flame. Put the glowing end of the splint into the top of the test tube. Record your observations.

4. When the magnesium in the beaker has stopped reacting, add 120 mL of 1 mol/L sodium hydroxide solution. Record your observations.

5. Dispose of the solutions as directed by your teacher.

Questions

1. What colour is phenolphthalein indicator in an acidic solution and in a basic solution?

2. What happens when magnesium is added to hydrochloric acid?

3. Magnesium hydroxide, $Mg(OH)_2$(s), is not soluble in water. Use this fact, as well as your answer to question 1, to explain your observations in step 4.

4. In this lab activity, you observed two reactions involving magnesium. Why do you think these reactions are called displacement reactions?

Single Displacement Reactions

Key Terms

single displacement reaction

activity series

single displacement reaction a chemical reaction in which one element in a compound is replaced (displaced) by another element

In a game of basketball, shown in **Figure 4.1**, a team can have five players on the court at one time. However, a substitute player can take the place of another player. Substitution also occurs in chemical reactions. A **single displacement reaction** is a reaction in which one element takes the place of another element in a compound.

Figure 4.1 On a basketball court, one player can be replaced with a different player. Similarly, in a single displacement reaction, one element in a compound is replaced by another element.

Characteristics of Single Displacement Reactions

In Chapter 3, you learned that the decomposition reaction used to inflate air bags produces nitrogen gas and sodium metal. A single displacement reaction then converts sodium, which is a very reactive element, into the stable compound sodium oxide:

$$\text{sodium} \ + \ \text{iron(III) oxide} \ \rightarrow \ \text{sodium oxide} \ + \ \text{iron}$$
$$6\text{Na(s)} \ + \ \text{Fe}_2\text{O}_3\text{(s)} \ \rightarrow \ 3\text{Na}_2\text{O(s)} \ + 2\text{Fe(s)}$$

In this reaction, the element sodium takes the place of, or displaces, iron from the compound iron(III) oxide.

The products of a single displacement reaction are an element and a compound that are different from the reactants. Two general forms for single displacement reactions are shown below. A and B represent metals, and X and Y represent non-metals.

Reactions in which a metal displaces another metal:

$$\text{A} \quad + \quad \text{BX} \quad \rightarrow \quad \text{AX} \quad + \quad \text{B}$$

Reactions in which a non-metal displaces another non-metal:

$$\text{AX} \quad + \quad \text{Y} \quad \rightarrow \quad \text{AY} \quad + \quad \text{X}$$

Types of Single Displacement Reactions

The general forms you just read show two types of single displacement reactions:

- a metal displacing another metal from an ionic compound
- a non-metal displacing another non-metal from an ionic compound

A single displacement reaction does not always occur between two metals or two non-metals. The following single displacement reaction also occurs:

- a metal displacing hydrogen from an acid or water

A Metal Displacing Another Metal from an Ionic Compound

Most single displacement reactions involve a metal displacing another metal from an ionic compound. **Figure 4.2** shows the reaction that occurs when a piece of copper metal is placed in a solution of silver nitrate. The balanced chemical equation for this reaction is

$$\text{copper} + \text{silver nitrate} \rightarrow \text{copper(II) nitrate} + \text{silver}$$
$$Cu(s) + 2AgNO_3(aq) \rightarrow Cu(NO_3)_2(aq) + 2Ag(s)$$

Copper begins as a metallic element and becomes metallic ions dissolved in a solution. Silver begins as metallic ions dissolved in a solution and becomes a metallic element. This pattern of change is regularly seen in single displacement reactions.

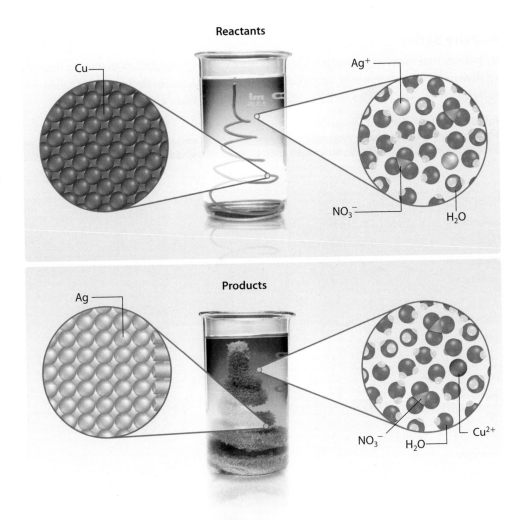

Figure 4.2 Copper displaces silver from silver nitrate dissolved in water. Solid silver forms, and the solution turns blue due to the formation of copper(II) ions.

Infer What happens to the nitrate ions during the reaction of copper and silver nitrate?

When a Reaction Does Not Occur

What happens if the metals are reversed? Look at **Figure 4.3**, which shows silver metal in a copper(II) nitrate solution. As you can see, nothing happens. No reaction takes place. The silver does not displace the copper from the solution. So, a single displacement reaction occurs only for certain combinations of metals and ionic compounds.

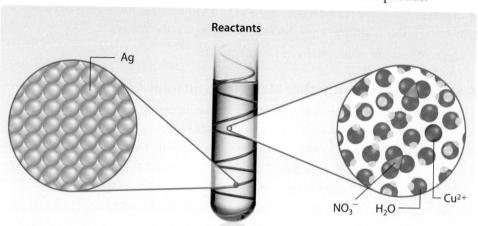

Figure 4.3 When silver metal is placed in a copper(II) nitrate solution, no reaction happens. The silver cannot displace the copper from the copper(II) nitrate dissolved in water.

Activity Series

By performing many experiments, chemists were able to develop lists that show the relative reactivity of elements, specifically metals and halogens. These lists are called **activity series**. The activity series of metals is shown in **Table 4.1**.

Suggested Investigation

Plan Your Own Investigation 4-A, Making an Activity Series of Metals

Table 4.1 Activity Series of Metals

Metal	Displaces Hydrogen ...	Reactivity
lithium		most reactive
potassium		
barium		
calcium		
sodium	from cold water	
magnesium		
aluminum		
zinc		
chromium		
iron		
cadmium		
cobalt		
nickel		
tin		
lead	from acids	
hydrogen		
copper		
mercury		
silver		
platinum		
gold		least reactive

Using the Activity Series of Metals

The activity series allows chemists to predict whether a single displacement reaction between a metal and an ionic compound will occur. As you can see in **Table 4.1**, the elements in the activity series are placed in order from the most reactive to the least reactive.

Compare the locations of copper and silver in **Table 4.1**. Copper is more reactive than silver and is higher than silver in the activity series. As a result, a single displacement reaction will occur when copper metal is placed in an aqueous solution of a silver compound. A reaction will not occur when silver metal is placed in an aqueous solution of a copper compound because silver is less reactive than copper.

Predicting Products of a Single Displacement Reaction

To predict the products of a single displacement reaction, look at the activity series of metals. If the single element is higher in the activity series, and therefore more reactive than the element it might replace in the compound, a reaction will occur. The products will be the less active metal (as an element) and an ionic compound composed of the more active metal ion and the anion (negatively charged ion) of the original compound.

Go to **scienceontario** to find out more

Applications and the Activity Series of Metals

The reactivity of metals affects how appropriate they are for various applications. For example, many older homes have water pipes made of steel with an inner coating of zinc. Over time, however, the zinc flakes off, allowing the water to directly contact the steel. As the steel rusts, the pipe can become clogged and even break. Modern homes may have water pipes made of copper, shown in **Figure 4.4**, or plastic, both of which are resistant to corrosion.

Titanium is a metal that has many applications because it is strong, light in weight, and resistant to corrosion. It is commonly used to make replacement hip and knee joints. It is also used in dentistry for tooth replacement. A titanium implant is placed into the jawbone, and the top of the implant is covered with an artificial tooth. Titanium is also used for screws and pins inserted to stabilize broken bones.

Figure 4.4 Copper pipes, which are often used in home plumbing systems, are resistant to corrosion. The activity series of metals shows that copper is much less reactive than iron, which is a chief component of steel pipes.

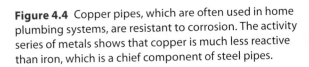

Learning Check

1. What is the general form of a single displacement reaction in which a metal displaces another metal?

2. What is the main characteristic of a single displacement reaction?

3. How is an activity series developed for a group of elements?

4. Refer to **Table 4.1**. Why are some metal objects coated with a thin layer of platinum or gold to prevent corrosion?

5. In an investigation, a piece of copper wire is added to a solution that contains lead ions.
   a. Predict whether a reaction will occur.
   b. Explain your prediction.

6. Titanium is not listed in **Table 4.1**.
   a. Would you expect titanium to appear closer to the top or the bottom of the activity series?
   b. Explain your reasoning.

A Metal Displacing Hydrogen from Acid or Water

The element hydrogen is just below lead in the activity series of metals. Although hydrogen is not a metal, its single valence electron allows it to form hydrogen ions that have a 1+ charge. Because hydrogen forms cations (positively charged ions) like metals do, it is often involved in single displacement reactions with metals. When hydrogen is replaced by a metal, hydrogen gas, $H_2(g)$, is produced. Compounds from which hydrogen can be displaced include acids and water.

Displacement from an Acid

The reaction of magnesium with hydrochloric acid, $HCl(aq)$, shown in **Figure 4.5**, is an example of a metal displacing hydrogen from an acid. The balanced chemical equation for this reaction is

$$\text{magnesium} \; + \; \text{hydrochloric acid} \; \rightarrow \; \text{magnesium chloride} \; + \; \text{hydrogen}$$
$$Mg(s) \quad + \quad 2HCl(aq) \quad \rightarrow \quad MgCl_2(aq) \quad + \quad H_2(g)$$

When an acid is a reactant in a single displacement reaction, you can think of the acid in terms of the ions it forms when it is dissolved in water. The ions in hydrochloric acid can be treated as H^+ and Cl^-. The metal displaces the hydrogen and forms an ionic compound with the chloride ion.

The location of hydrogen in the activity series of metals shows the relative reactivity of hydrogen in acids. Every metal above hydrogen in the activity series can displace hydrogen from an acid, as noted in **Table 4.1**, but the metals below hydrogen cannot.

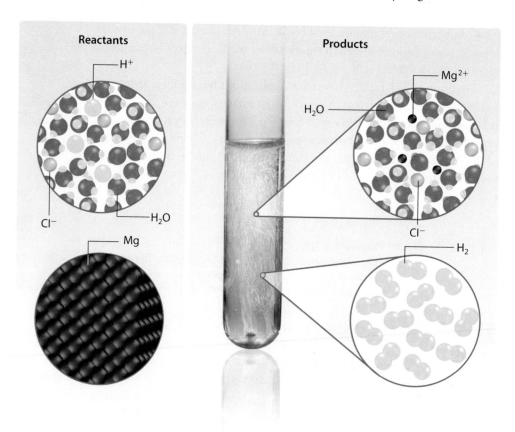

Figure 4.5 When magnesium displaces hydrogen from hydrochloric acid, bubbles of hydrogen gas, $H_2(g)$, form.

Identify Give an example of another metal that can displace hydrogen from acids, and give an example of a metal that cannot displace hydrogen from acids.

Displacement from Water

Hydrogen can also be displaced from water. However, the hydrogen atoms in water are harder to displace than the hydrogen atoms in an acid. As a result, only very active metals can displace hydrogen from water, as shown in **Table 4.1**. In **Figure 4.6**, you can see sodium reacting with water in a single displacement reaction. The balanced chemical equation for this reaction is

$$\text{sodium} + \text{water} \rightarrow \text{sodium hydroxide} + \text{hydrogen}$$
$$2Na(s) + 2H_2O(\ell) \rightarrow 2NaOH(aq) + H_2(g)$$

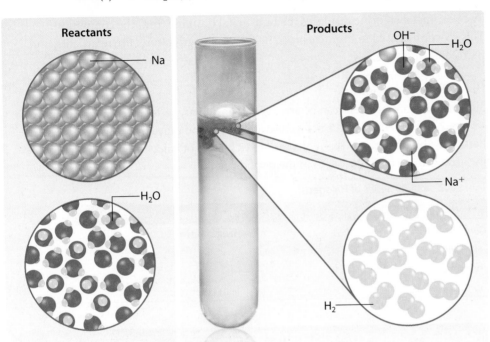

Figure 4.6 Sodium is reactive enough to displace hydrogen from water. The change in colour of the phenolphthalein that was added indicates the presence of hydroxide ions from the formation of sodium hydroxide during the reaction.

The products of this reaction include diatomic hydrogen gas and sodium hydroxide. When predicting the products of a single displacement reaction that involves water, think of water as being composed of hydrogen and hydroxide ions, H^+ and OH^-. The metal displaces the hydrogen ions, which form hydrogen gas, and it bonds with the hydroxide ions to form an ionic compound.

Making Alkali Metals Safer

The reactivity of sodium and other alkali metals makes these metals useful in a wide range of chemical reactions, including the generation of hydrogen for fuel cells through single displacement reactions. However, the reactivity of alkali metals also makes them dangerous. Alkali metals react easily with water and oxygen in the air, which means that handling them can be hazardous. In several industrial processes, alkali metals are used in liquid ammonia to reduce contact with water and oxygen. Liquid ammonia must be kept well below its boiling point of $-33°C$ to prevent the release of toxic ammonia gas.

A new technique provides the reactivity of alkali metals without the danger of using very cold ammonia. An alkali metal is absorbed into a porous silica gel to form a powder. The powder can still undergo reactions, but because the particles of the alkali metal are so small and are surrounded by silica, the reactions can take place safely without the need for cold ammonia. As well as making the reaction conditions safer, the powder forms more stable products than those formed using an alkali metal in ammonia.

A Non-metal Displacing Another Non-metal from an Ionic Compound

Non-metals also undergo single displacement reactions. When one of the reactants is a diatomic halogen molecule and the other reactant is an ionic compound that contains a halogen, the halogen in the compound may be replaced. For example, when chlorine gas is bubbled through an aqueous solution of sodium bromide, the following reaction occurs:

$$\text{chlorine} \quad + \quad \text{sodium bromide} \quad \rightarrow \quad \text{sodium chloride} \quad + \quad \text{bromine}$$
$$Cl_2(g) \quad + \quad 2NaBr(aq) \quad \rightarrow \quad 2NaCl(aq) \quad + \quad Br_2(\ell)$$

The non-metal chlorine displaces the non-metal bromine from the compound. This reaction is used to produce bromine for use in agriculture, fire retardants, and petroleum additives.

Halogen Activity Series

Similar to metals, halogens can be arranged in an activity series, as shown in Table 4.2. The reactivity of the halogens decreases as you move from top to bottom within the group. Therefore, the activity series of halogens mirrors the arrangement of the halogens in the periodic table.

Table 4.2 Activity Series of Halogens

Halogen	Reactivity
fluorine	most reactive
chlorine	
bromine	↑
iodine	least reactive

Activity 4.1 Predicting Trends in the Reactivity of Halogens

The halogens, found in Group 17 in the periodic table, share some common characteristics. For example, the atoms of all the halogens have seven electrons in their outermost energy level. As with most groups in the periodic table, the halogens exhibit trends, such as their ability to react with other substances.

Properties of Halogens

Halogen	Atomic Radius (picometres)	Ionization Energy (kJ/mol)	Electronegativity
fluorine	72	1681	3.98
chlorine	100	1251	3.16
bromine	114	1140	2.96
iodine	133	1008	2.66
astatine	140	920	2.20

Procedure

Construct a line graph for each property. Plot the halogens along the x-axis. Label the axes and units as necessary.

Questions

1. Identify any periodic trends in your graphs. Relate these trends to the activity series of halogens shown in Table 4.2.

2. Astatine does not appear in the activity series of halogens, shown in Table 4.2, because it is a very rare element. Predict the location of astatine in the activity series. Provide a brief explanation of your reasoning.

Predicting Products of Single Displacement Reactions

When predicting the products of a single displacement reaction involving halogens, you can use the activity series of halogens in the same way that you use the activity series of metals. If the uncombined halogen is higher in the activity series than the halogen in the compound that it might replace, then a reaction will occur. From the activity series, you can see that fluorine can replace any other halogen, but iodine can replace none.

Predicting Products in Single Displacement Reactions

Problem

Using the activity series of metals and halogens, write a balanced chemical equation for each single displacement reaction. If you predict that no reaction will occur, write "NR."

a. $Ca(s) + H_2O(\ell) \rightarrow$ **b.** $Fe(s) + CrSO_4(aq) \rightarrow$ **c.** $Br_2(\ell) + NaI(aq) \rightarrow$

What Is Required?

If a single displacement reaction will occur, the chemical formulas of the products are required.

What Is Given?

Reactants: **a.** calcium and water **b.** iron and chromium(II) sulfate **c.** bromine and sodium iodide

Type of reaction: single displacement

Plan Your Strategy	Act on Your Strategy
Identify the elements that are involved in the displacement.	**a.** calcium and hydrogen **b.** iron and chromium **c.** bromine and iodine
Locate the elements in the activity series. Determine whether a reaction will occur.	**a.** Calcium is above hydrogen in the activity series of metals, and it is reactive enough to displace hydrogen from water. Therefore, a reaction will occur. **b.** Iron is below chromium in the activity series of metals. Iron is not reactive enough to displace chromium, so no reaction will occur. **c.** Bromine is above iodine in the activity series of halogens, so it is reactive enough to displace iodine. Therefore, a reaction will occur.
Predict the products that will form.	**a.** Calcium hydroxide and hydrogen gas will form. **b.** NR **c.** Sodium bromide and iodine will form.
Write the formulas of the products. Balance each chemical equation.	**a.** $Ca(s) + 2H_2O(\ell) \rightarrow Ca(OH)_2(aq) + H_2(g)$ **b.** $Fe(s) + CrSO_4(aq) \rightarrow NR$ **c.** $Br_2(\ell) + 2NaI(aq) \rightarrow 2NaBr(aq) + I_2(aq)$

Check Your Solution

Based on the activity series of metals, calcium can displace hydrogen from water, but iron cannot displace chromium from chromium(II) sulfate. Based on the activity series of halogens, bromine can displace iodine from sodium iodide.

Using the appropriate activity series, write a balanced chemical equation for each single displacement reaction. If you predict that no reaction will occur, write "NR."

1. $Mg(s) + CrSO_4(aq) \rightarrow$

2. $Br_2(\ell) + KF(aq) \rightarrow$

3. $Zn(s) + H_2SO_4(aq) \rightarrow$

4. $F_2(g) + MgI_2(aq) \rightarrow$

5. $Cl_2(g) + NaI(aq) \rightarrow$

6. $Ni(s) + H_2O(\ell) \rightarrow$

7. $Pb(s) + Sn(ClO_3)_4(aq) \rightarrow$

8. $K(s) + H_2O(\ell) \rightarrow$

9. $HCl(aq) + Cd(s) \rightarrow$

10. $Pb(ClO_3)_4(aq) + Al(s) \rightarrow$

Section Summary

- In a single displacement reaction, one element replaces another element in a compound to produce a new element and a new compound.

- The general form for single displacement reactions in which a metal displaces another metal, where A and B are metals, is $A + BX \rightarrow AX + B$. The general form for single displacement reactions in which a non-metal displaces another non-metal, where X and Y are non-metals, is $AX + Y \rightarrow X + AY$.

- An activity series lists elements in order, from most reactive to least reactive.

- The activity series of metals and the activity series of halogens are used to predict whether a single displacement reaction will occur.

- A single displacement reaction can only occur when an uncombined element is higher in an activity series than the element it would replace in a compound.

Review Questions

1. **K/U** Describe the displacement reaction that occurs in an air bag to change sodium into a less harmful chemical.

2. **K/U** Explain why "single displacement" is a suitable term to describe the reactions discussed in this section.

3. **T/I** Look again at **Figures 4.2** and **4.3**. Describe how the evidence in these figures can be used to determine the relative placement of silver and copper in an activity series.

4. **A** Based on the activity series of metals, explain why large amounts of gold jewellery and coins have survived from ancient civilizations.

5. **T/I** For each pair of reactants, write a balanced chemical equation if a single displacement reaction will occur. If you predict that no reaction will occur, write "NR."
   a. copper and magnesium sulfate
   b. zinc and iron(II) chloride
   c. magnesium and aluminum sulfate
   d. zinc and hydrochloric acid
   e. copper and zinc nitrate
   f. magnesium and sulfuric acid

6. **A** Aluminum is much more abundant than iron in Earth's crust. Use **Table 4.1** to infer why aluminum was very expensive and not widely used until the late 1800s.

7. **K/U** Why is hydrogen included in the activity series of metals?

8. **A** Describe two methods you could use to produce hydrogen gas by a single displacement reaction.

9. **C** Draw a model of a single displacement reaction in which hydrogen is displaced.

10. **K/U** When a metal displaces hydrogen from water, what types of substances form?

11. **C** Describe how to use the periodic table to compare the relative reactivities of the halogens.

12. **T/I** If the liquid in the photograph is water, could the metal be zinc? If the liquid is an acid, could the metal be zinc? Explain your reasoning.

13. **K/U** Which halogens can be replaced by chlorine?

14. **C** Design a concept map that shows the relationships among the following terms:
    - single displacement reaction
    - activity series
    - metals
    - non-metals
    - hydrogen
    - acids
    - halogens

15. **T/I** For each pair of reactants, write a balanced chemical equation if a single displacement reaction will occur. If you predict that no reaction will occur, write "NR."
    a. iron and hydrobromic acid
    b. bromine and magnesium iodide
    c. potassium and aluminum sulfate
    d. lithium and water
    e. cobalt and water
    f. bromine and iron(II) chloride

16. **T/I** Which halogen is capable of displacing only one halogen and is itself replaced by two halogens?

Double Displacement Reactions

X rays are highly useful for creating images of bones in the human body, but they generally do not show soft tissues clearly. To help doctors diagnose conditions involving the digestive tract, a patient may be asked to drink a liquid that contains tiny particles of barium sulfate, $BaSO_4(s)$. These particles block X rays, allowing organs such as the stomach to appear in high contrast on X-ray images, as shown in **Figure 4.7**.

Particles of barium sulfate can be produced by the reaction of barium chloride with sodium sulfate:

barium chloride + sodium sulfate → sodium chloride + barium sulfate

$$BaCl_2(aq) \quad + \quad Na_2SO_4(aq) \quad \rightarrow \quad 2NaCl(aq) \quad + \quad BaSO_4(s)$$

Notice that the positive ions, Ba^{2+} and Na^+, change partners. This reaction is called a **double displacement reaction**. A double displacement reaction involves the exchange of positive ions between two ionic compounds to form two new ionic compounds.

Key Terms

double displacement
   reaction
precipitate
neutralization

> **double displacement reaction** a chemical reaction in which the positive ions of two ionic compounds exchange places, resulting in the formation of two new ionic compounds

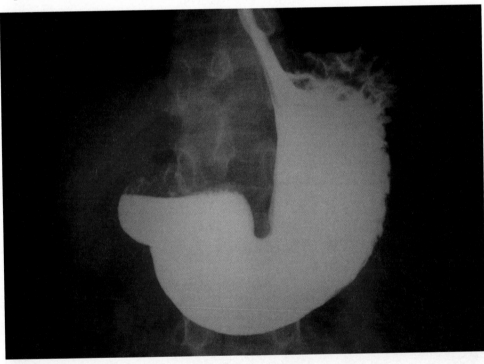

Figure 4.7 Drinking a liquid that contains barium sulfate allows organs to show up well on X-ray images. Barium sulfate, which is not soluble in water, can be produced by a double displacement reaction involving two soluble compounds, barium chloride and sodium sulfate.

Characteristics of Double Displacement Reactions

A double displacement reaction generally occurs between compounds that are in aqueous solution. The general form of a double displacement reaction is

AX + BY → AY + BX

In this equation, A and B are positively charged ions, or cations, and X and Y are negatively charged ions, or anions.

Types of Double Displacement Reactions

Just as there are different types of single displacement reactions, there are different types of double displacement reactions. On the following pages, you will learn about these double displacement reactions:

- a reaction that forms a solid

- a reaction that forms a gas

- a reaction that forms water

You will also learn about guidelines you can use to predict whether the products that are likely to form are solids, gases, or water.

Determining the Products of a Double Displacement Reaction

To determine the products of a double displacement reaction, you must first determine the ions that make up the reactants. For example, consider the reaction between lithium chloride and lead(II) nitrate:

$$LiCl(aq) + Pb(NO_3)_2(aq) \rightarrow$$

The reactants are composed of four ions. Lithium chloride is composed of lithium ions, Li^+, and chloride ions, Cl^-. Lead(II) nitrate is composed of lead(II) ions, Pb^{2+}, and nitrate ions, NO_3^-.

To determine the products of the reaction, simply change the pairs of ions. In this example, one of the products is lithium nitrate, formed from lithium ions pairing with nitrate ions. The other product is lead(II) chloride, formed from lead(II) ions pairing with chloride ions. When you write the formula of each ionic compound, remember to balance the charges of the ions. The final balanced equation for the reaction between lithium chloride and lead(II) nitrate is

$$\begin{array}{ccccccc} \text{lithium chloride} & + & \text{lead(II) nitrate} & \rightarrow & \text{lithium nitrate} & + & \text{lead(II) chloride} \\ 2LiCl(aq) & + & Pb(NO_3)_2(aq) & \rightarrow & 2LiNO_3(aq) & + & PbCl_2(s) \end{array}$$

A common use of lithium nitrate is the manufacture of fireworks and flares that produce a red colour. Lead(II) chloride is used to make glass that transmits infrared radiation for use as lenses in night-vision goggles, as shown in **Figure 4.8**.

Figure 4.8 Night-vision equipment detects infrared radiation. Such equipment can be used to observe the activities of animals, such as the lion and rhinoceros shown here, when there is insufficient visible light to see clearly.

7. What is the general form of a double displacement reaction?

8. In what state are the reactants of most double displacement reactions?

9. Explain whether describing a double displacement reaction as the exchange of cations between two compounds is correct.

10. When predicting the products of a double displacement reaction, why do you first need to determine the ions in each reactant?

11. Would you expect an element to form during a double displacement reaction? Explain your reasoning.

12. Potassium bromide can react with silver nitrate in a double displacement reaction to form an aqueous potassium compound and a solid silver compound.

a. What are the names and formulas of the products that form during this reaction?

b. Write a balanced chemical equation for this reaction.

A Reaction That Forms a Solid

A common observation during many double displacement reactions is the formation of a solid **precipitate**. In the previous example, the product, lead(II) chloride, is a precipitate. A precipitate is also formed during the reaction between silver nitrate and sodium chloride, shown in **Figure 4.9**. The balanced chemical equation for this reaction is

silver nitrate + sodium chloride → silver chloride + sodium nitrate

$$AgNO_3(aq) + NaCl(aq) \rightarrow AgCl(s) + NaNO_3(aq)$$

The solid product, silver chloride, is the precipitate.

precipitate an insoluble solid that is formed by a chemical reaction between two soluble compounds

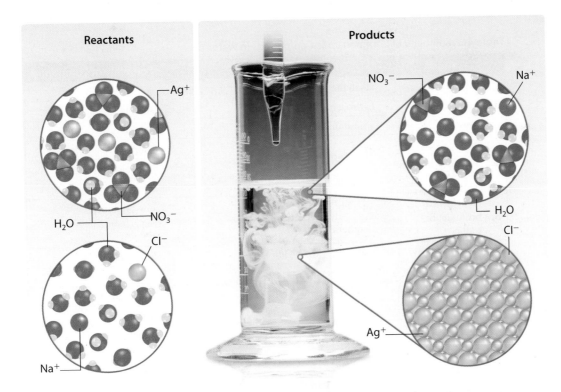

Figure 4.9 When aqueous solutions of silver nitrate and sodium chloride are mixed, a double displacement reaction occurs. A precipitate, silver chloride, is formed. The second product, sodium nitrate, remains in aqueous solution.

Identify the positive ions in this reaction, and describe what happens to them.

Solubility Guidelines

How do chemists know whether a double displacement reaction is likely to produce a precipitate? A set of solubility guidelines has been assembled, based on the experimental results. These guidelines are shown in **Table 4.3**. Compounds that are insoluble or have low solubility in water form precipitates in double displacement reactions.

Solubility Guidelines

1. The hydrogen ion, ammonium ion, and all Group 1 (alkali metal) ions form soluble compounds with nearly all anions.

2. Nitrate and acetate ions form soluble compounds with nearly all cations.

3. Chloride, bromide, and iodide ions form compounds that have low solubility with silver, lead(II), mercury(I), copper(I), and thallium cations only.

4. Fluoride forms compounds that have low solubility with magnesium, calcium, barium, and lead(II) cations only.

5. The sulfate ion forms compounds that have low solubility with calcium, strontium, barium, and lead(II) cations only.

6. The sulfide ion forms soluble compounds only with the ions listed in guideline 1 and with Group 2 cations.

7. The hydroxide ion forms compounds that are soluble only with the cations listed in guideline 1, and with strontium, barium, and thallium cations.

8. Phosphate, carbonate, and sulfite ions form compounds that have low solubility with all cations except for those listed in guideline 1.

Table 4.3 Solubility of Common Ionic Compounds in Water

	Anion	+	Cation	→	Solubility of Compound
1.	most		alkali metal ions (Li^+, K^+, Rb^+, Cs^+, Fr^+)		soluble
	most		hydrogen ion, H^+		soluble
	most		ammonium ion, NH_4^+		soluble
2.	nitrate, NO_3^-		most		soluble
	acetate (ethanoate),		Ag^+		low solubility
	CH_3COO^-		most others		soluble
3.	chloride, Cl^- bromide, Br^-		Ag^+, Pb^{2+}, Hg_2^{2+}, Cu^+, Tl^+		low solubility
	iodide, I^-		all others		soluble
4.	fluoride, F^-		Mg^{2+}, Ca^{2+}, Ba^{2+}, Pb^{2+}		low solubility
			most others		soluble
5.	sulfate, SO_4^{2-}		Ca^{2+}, Sr^{2+}, Ba^{2+}, Pb^{2+}		low solubility
			all others		soluble
6.	sulfide, S^{2-}		alkali ions and H^+, NH_4^+, Be^{2+}, Mg^{2+}, Ca^{2+}, Sr^{2+}, Ba^{2+}		soluble
			all others		low solubility
7.	hydroxide, OH^-		alkali ions and H^+, NH_4^+, Sr^{2+}, Ba^{2+}, Tl^+		soluble
			all others		low solubility
8.	phosphate, PO_4^{3-} carbonate, CO_3^{2-}		alkali ions and H^+, NH_4^+		soluble
	sulfite, SO_3^{2-}		all others		low solubility

"Soluble" here means that more than 1 g of a substance will dissolve in 100 mL of water at 25°C.

Predicting the Precipitate in a Double Displacement Reaction

Problem

The double displacement reaction between magnesium chloride and lead(II) acetate forms a precipitate. Predict the products, and write a balanced chemical equation that identifies the precipitate.

What Is Required?

Determine the products that form when magnesium chloride and lead(II) acetate react, and write a balanced chemical equation that shows which product is a precipitate.

What Is Given?

You know the reactants: magnesium chloride and lead(II) acetate

You know the type of reaction: double displacement

Plan Your Strategy	Act on Your Strategy
Identify the ions that make up each reactant.	Magnesium chloride: • magnesium ions, Mg^{2+} • chloride ions, Cl^- Lead(II) acetate: • lead(II) ions, Pb^{2+} • acetate ions, CH_3COO^-
Switch the pairs of ions to determine the products.	One product is magnesium acetate (magnesium ions paired with acetate ions). The second product is lead(II) chloride (lead(II) ions paired with chloride ions).
Write a word equation for the reaction.	magnesium chloride + lead(II) acetate → magnesium acetate + lead(II) chloride
Use the solubility guidelines to determine the precipitate.	Magnesium acetate is not the precipitate because acetate ions form a soluble compound with magnesium ions. Lead(II) chloride is the precipitate because the compound formed from chloride ions and lead(II) ions has low solubility.
Write a balanced chemical equation for the reaction.	$MgCl_2(aq) + Pb(CH_3COO)_2(aq) \rightarrow Mg(CH_3COO)_2(aq) + PbCl_2(s)$

Check Your Solution

All the chemical formulas are correct, and the chemical equation is balanced. The products are correctly formed by switching the ions in the reactants. The precipitate is correctly identified, based on the solubility guidelines.

Practice Problems

Determine the products that form in the double displacement reaction between each pair of reactants, and identify the precipitate. Then write a balanced chemical equation.

11. potassium sulfate and calcium chloride

12. barium nitrate and sodium carbonate

13. iron(III) chloride and sodium hydroxide

14. rubidium sulfide and copper(II) iodide

15. zinc bromide and copper(I) acetate

16. lithium hydroxide and magnesium chloride

17. aluminum sulfate and lead(II) nitrate

18. lithium phosphate and magnesium chloride

19. calcium nitrate and magnesium sulfate

20. silver nitrate and magnesium chloride

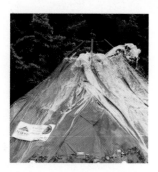

Figure 4.10 The bubbling of this model volcano is caused by the formation of carbon dioxide gas, which is produced in the reaction between sodium hydrogen carbonate (commonly known as baking soda) and the acetic acid in vinegar.

A Reaction That Forms a Gas

Sometimes, the production of a gas, rather than a precipitate, indicates that a double displacement reaction has occurred. Many of these double displacement reactions are, in fact, two reactions that occur in rapid succession. A double displacement occurs, but then one of the products quickly decomposes into water and a gas. To predict the products in one of these reactions, you need to recognize the product that decomposes.

Formation of Carbon Dioxide

Have you ever made a vinegar and baking soda volcano like the one in **Figure 4.10**? Vinegar contains acetic acid, $CH_3COOH(aq)$, and baking soda is sodium hydrogen carbonate, $NaHCO_3(s)$. To determine the products, identify the ions and change the pairs of ions. The balanced chemical equation for the reaction is

acetic acid + sodium hydrogen $\rightarrow$ sodium acetate + carbonic acid
carbonate (hydrogen carbonate)

$$CH_3COOH(aq) + NaHCO_3(s) \rightarrow NaCH_3COO(aq) + H_2CO_3(aq)$$

This cannot be the overall reaction because there is no carbon dioxide.

Decomposition of Carbonic Acid

In these reactions, the product to watch for is carbonic acid, $H_2CO_3(aq)$. Carbonic acid decomposes into liquid water and carbon dioxide gas according to the following equation:

carbonic acid $\rightarrow$ water + carbon dioxide
$$H_2CO_3(aq) \rightarrow H_2O(\ell) + CO_2(g)$$

Overall Chemical Reaction

Combining the double displacement reaction with the decomposition reaction gives the overall balanced equation:

acetic acid + sodium hydrogen $\rightarrow$ sodium acetate + water + carbon
carbonate dioxide

$$CH_3COOH(aq) + NaHCO_3(s) \rightarrow NaCH_3COO(aq) + H_2O(\ell) + CO_2(g)$$

Formation of Ammonia

Gaseous ammonia, $NH_3(g)$, forms in a similar way when a double displacement reaction is followed by a decomposition reaction. In this type of double displacement reaction, the product to look for is ammonium hydroxide, $NH_4OH(aq)$. For example, the equation for the double displacement reaction between ammonium chloride and sodium hydroxide is

ammonium chloride + sodium hydroxide $\rightarrow$ sodium chloride + ammonium hydroxide
$$NH_4Cl(aq) + NaOH(aq) \rightarrow NaCl(aq) + NH_4OH(aq)$$

Decomposition of Ammonium Hydroxide

Ammonium hydroxide decomposes to form water and ammonia according to the following equation:

ammonium hydroxide $\rightarrow$ water + ammonia
$$NH_4OH(aq) \rightarrow H_2O(\ell) + NH_3(g)$$

Overall Chemical Reaction

So, the overall equation for the chemical reaction that occurs is

ammonium + sodium $\rightarrow$ sodium chloride + water + ammonia
chloride hydroxide

$$NH_4Cl(aq) + NaOH(aq) \rightarrow NaCl(aq) + H_2O(\ell) + NH_3(g)$$

Suggested **Investigation**

Inquiry Investigation 4-B, Observing Double Displacement Reactions

Predicting Products of Double Displacement Reactions That Form Gases

These two types of double displacement reactions, which form the gases carbon dioxide and ammonia, are important to remember when you need to predict products. They are summarized in **Table 4.4**.

Table 4.4 Double Displacement Reactions That Form Gases

Reactants	Products
acid + compound containing carbonate ion	ionic compound + water + carbon dioxide
compound containing ammonium ions + compound containing hydroxide ions	ionic compound + water + ammonia

Learning Check

13. What are the characteristics of a precipitate?

14. In **Figure 4.9**, why is silver chloride shown in a different form than the other compounds?

15. Do the solubility guidelines shown in **Table 4.3** apply to the solubility of compounds in all solvents, including alcohol and oil? Explain.

16. State the two types of reactions that occur when a double displacement reaction produces a gas. Give the general form of each reaction.

17. According to **Table 4.3**, which substance in each of the following pairs is more soluble?
  a. $FeS(s)$ or $Ba(OH)_2(s)$
  b. $TlCl(s)$ or $MgS(s)$
  c. $H_3PO_4(s)$ or $SrCO_3(s)$
  d. $PbSO_4(s)$ or $Na_2SO_3(s)$

18. Write a balanced chemical equation for the double displacement reaction between calcium carbonate, $CaCO_3(s)$, and hydrochloric acid, $HCl(aq)$.

A Reaction That Forms Water

When two solutions are mixed, the formation of a solid precipitate or a gas is evidence that a reaction has occurred. However, there is one type of double displacement reaction that occurs with no outward evidence of a reaction. This type of reaction forms water as a product, but, because the reactants are in aqueous solution, there is no visible sign that water molecules have formed.

Neutralization Reactions

Water can form when an acid and a base are combined in a process called **neutralization**. Water forms when hydrogen ions from the acid join with hydroxide ions from the base according to the following reaction:

$$H^+(aq) + OH^-(aq) \rightarrow H_2O(\ell)$$

Because the hydrogen ions and hydroxide ions combine to form water, the amounts of these potentially harmful ions decrease. As a result, the solution that forms from a neutralization reaction may be neutral—neither acidic nor basic.

neutralization the process of making a solution neutral (pH = 7) by adding a base to an acidic solution or by adding an acid to an alkaline (basic) solution

Uses of Neutralization Reactions

Neutralization reactions between an acid and a base are important for treating acid or base spills. For example, when sulfuric acid, $H_2SO_4(aq)$, spilled from derailed tanker cars near Englehart, Ontario, a solution of the base calcium hydroxide, $Ca(OH)_2(aq)$, was used to help neutralize the acid during clean-up. The balanced chemical equation for this reaction is

$$\text{sulfuric acid} + \text{calcium hydroxide} \rightarrow \text{calcium sulfate} + \text{water}$$
$$H_2SO_4(aq) + Ca(OH)_2(aq) \rightarrow CaSO_4(s) + 2H_2O(\ell)$$

Suggested Investigation

Inquiry Investigation 4-C, Observing a Neutralization Reaction

Inquiry Investigation 4-D, Modelling Neutralization Reactions Used in Scrubber Technology

Other Uses of Neutralization Reactions

Reactions between acids and bases are also important for optimizing soil conditions. For example, lettuce and celery grow better in neutral to basic soil, but strawberries and tomatoes grow better in acidic soil. **Figure 4.11** shows soil being tested to determine whether it is acidic, neutral, or basic. Depending on the types of plants that will be grown, substances can be added to the soil to neutralize acids or bases. On a smaller scale, when someone takes antacid tablets to treat acid indigestion, basic substances in the tablets help to neutralize the excess acid in the stomach. You will learn more about acids, bases, and neutralization in Unit 4.

Figure 4.11 Depending on the results of this soil test, lime, CaO(s), might be added to neutralize excess acids in the soil, or a substance such as ammonium sulfate might be added to neutralize bases.

Summarizing Double Displacement Reactions

Most double displacement reactions form a precipitate. Therefore, when you are asked to determine whether a reaction occurs, you will usually need to examine the solubility guidelines to determine whether one of the products is insoluble. Nevertheless, the double displacement reactions that form gases or water are important to remember. The flowchart in **Figure 4.12** summarizes the double displacement reactions you have studied in this section.

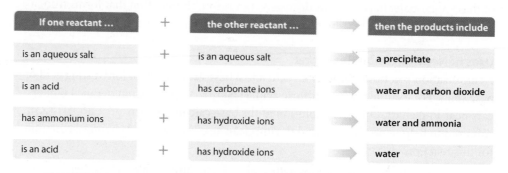

If one reactant ...	+	the other reactant ...	→	then the products include
is an aqueous salt	+	is an aqueous salt	→	a precipitate
is an acid	+	has carbonate ions	→	water and carbon dioxide
has ammonium ions	+	has hydroxide ions	→	water and ammonia
is an acid	+	has hydroxide ions	→	water

For any other reactant pairs, check the solubility guidelines. The product might include a precipitate.

Figure 4.12 Use this flowchart to help you predict the products of a double displacement reaction.

Predicting Gases and Water in Double Displacement Reactions

Problem

Predict the products in each double displacement reaction, and write a balanced chemical equation.

a. sodium carbonate reacting with hydrobromic acid

b. sodium hydroxide reacting with hydrobromic acid

What Is Required?

Determine the products that form from the given reactants, and write a balanced chemical equation for each reaction.

What Is Given?

You know the reactants: a. sodium carbonate and hydrobromic acid

b. sodium hydroxide and hydrobromic acid

You know the type of reaction: double displacement

Plan Your Strategy	Act on Your Strategy	
Identify the ions that make up each reactant.	a. Sodium carbonate: • sodium ions, Na^+ • carbonate ions, CO_3^{2-}	Hydrobromic acid: • hydrogen ions, H^+ • bromide ions, Br^-
	b. Sodium hydroxide: • sodium ions, Na^+ • hydroxide ions, OH^-	Hydrobromic acid: • hydrogen ions, H^+ • bromide ions, Br^-
Switch the pairs of ions to determine the products.	a. sodium bromide and hydrogen carbonate (carbonic acid) b. sodium bromide and water	
Look for carbonic acid or ammonium hydroxide, which break down into water and a gas.	a. Carbonic acid will break down into water and carbon dioxide. b. Neither carbonic acid nor ammonium hydroxide is formed.	
Write a word equation for each reaction.	a. sodium carbonate + hydrobromic acid → sodium bromide + water + carbon dioxide b. sodium hydroxide + hydrobromic acid → sodium bromide + water	
Write and balance a chemical equation for each reaction.	a. $Na_2CO_3(aq) + 2HBr(aq) \rightarrow 2NaBr(aq) + H_2O(\ell) + CO_2(g)$ b. $NaOH(aq) + HBr(aq) \rightarrow NaBr(aq) + H_2O(\ell)$	

Check Your Solution

All the chemical formulas are correct, and each chemical equation is balanced.
The products are correctly formed by switching the ions of the reactants.

Practice Problems

Determine the products that form in the double displacement reaction between each pair of reactants. Then write a balanced chemical equation for the reaction.

21. potassium carbonate and hydrochloric acid

22. sulfuric acid and sodium carbonate

23. ammonium chloride and sodium hydroxide

24. rubidium hydroxide and hydrochloric acid

25. calcium carbonate and acetic acid

26. lithium hydroxide and ammonium bromide

27. sulfuric acid and lithium hydroxide

28. lithium hydrogen carbonate and acetic acid

29. calcium hydroxide and nitric acid

30. ammonium chloride and magnesium hydroxide

Section Summary

- In a double displacement reaction, the positive ions in two compounds trade places to form two new compounds.
- The general form of a double displacement reaction is $AX + BY \rightarrow AY + BX$.
- A double displacement reaction generally produces a precipitate, a gas, or water.

- Solubility guidelines can be used to predict whether a product of a double displacement reaction will be a precipitate.
- Neutralization occurs when there is a double displacement reaction between an acid and a base. Water is a product of neutralization.

Review Questions

1. **T/I** The following general form of a double displacement reaction is incorrect:

$$CW + DZ \rightarrow CD + WZ$$

   a. What is wrong with this general equation?

   b. Write a correct equation using the reactants shown.

2. **C** Create a graphic organizer that you could use to predict the products of a double displacement reaction.

3. **K/U** What is a precipitate?

4. **K/U** What evidence would help you decide if a double displacement reaction has occurred?

5. **T/I** What two anions are almost never found in a precipitate formed during a reaction?

6. **T/I** What would be most likely to happen if you combined the solutions of two compounds in which the positive ions are alkali metals?

7. **T/I** What would be most likely to happen if you combined the solutions of two compounds that have the same anion?

8. **C** Suppose that you see an online video in which a person adds sodium nitrate to a sample of tap water and a precipitate forms. The person in the video concludes that there is lead in the water. Write a comment to address the error in this conclusion.

9. **A** To determine whether a rock sample is limestone, a geologist places several drops of hydrochloric acid on it. Limestone is calcium carbonate, $CaCO_3(s)$.

   a. What evidence is the geologist expecting to see if the rock sample is limestone?

   b. Write a balanced chemical equation for the reaction between hydrochloric acid and calcium carbonate.

10. **A** During a neutralization reaction, there is no visible evidence that a reaction is occurring. Describe a safe way by means of which you could determine whether a neutralization reaction is occurring.

11. **T/I** The photograph below shows the reaction between sodium hydroxide and copper(II) chloride.

   a. What evidence of a double displacement reaction do you see?

   b. What are the names of the products that formed?

   c. Which product is the precipitate? Explain your reasoning.

   d. Write a balanced chemical equation for the reaction.

12. **T/I** What gas forms in the reaction between ammonium bromide and sodium hydroxide? Write a balanced chemical equation for this reaction.

13. **K/U** What reactants are involved in a neutralization reaction?

14. **K/U** Which ions combine to form water during the reaction between an acid and a base?

15. **T/I** What products should form during the reaction between hydrochloric acid, $HCl(aq)$, and a solution of calcium hydroxide, $Ca(OH)_2(aq)$? Write a balanced chemical equation for the reaction.

16. **C** Analyze the following neutralization reaction:

$$3NaOH(aq) + H_3PO_4(aq) \rightarrow H_2O(aq) + Na_3PO_4(s)$$

   In a brief paragraph, describe any errors you find and explain how you would correct them.

Reactions in Industry

At the beginning of Unit 2, you read about a chemical reaction that is used to weld railroad tracks together. This is just one of the many reactions that are involved in the processes required to make the products you use in your daily life. Chemical reactions are also involved in cleaning up the problems that result from these processes.

Thermite Reactions

Now that you have learned more about chemical reactions, you can re-examine thermite reactions. When this type of reaction is used to weld railroad tracks, the reactants, aluminum and iron oxide are powdered and thoroughly mixed. The reaction is a single displacement reaction in which aluminum displaces iron according to the following balanced chemical equation:

$$\text{aluminum} + \text{iron oxide} \rightarrow \text{iron} + \text{aluminum oxide}$$
$$8Al(s) + 3Fe_3O_4(s) \rightarrow 9Fe(\ell) + 4Al_2O_3(s)$$

Once the reaction begins, it releases enough energy to melt the iron.

In **Figure 4.13**, an aluminum wrench is hitting rusty iron, causing a small-scale thermite reaction that produces sparks. In industrial settings or workshops, the grinding or cutting of iron or steel can produce powdered iron oxides. Care must be taken in such areas to avoid an unexpected thermite reaction when using aluminum objects.

In addition to iron oxide, other metals can be used in thermite reactions. For example, a thermite reaction involving copper(II) oxide can be used to produce pure copper according to the following chemical equation:

$$\text{aluminum} + \text{copper(II) oxide} \rightarrow \text{copper} + \text{aluminum oxide}$$
$$2Al(s) + 3CuO(s) \rightarrow 3Cu(\ell) + Al_2O_3(s)$$

This reaction is often used to produce pure copper for welding electrical conductors. The copper(II) oxide is held in a heat-resistant reaction chamber. The thermite reaction heats the copper enough to melt it. The liquid copper is then allowed to flow into a mold that surrounds the ends of the conductors. When the copper cools into a solid, it forms a weld that allows an electric current to flow between the conductors.

Key Terms

matte

leaching

Go to **scienceontario**
to find out more

Figure 4.13 The friction from the impact of an aluminum wrench with rusty iron generated enough heat to initiate a small thermite reaction, as shown by the bright sparks flying away from the point of impact.

Analyze The thermite reaction shown here occurs in the solid state. How is this different from the other single displacement reactions you have studied?

Magnesium Mining from Seawater

Many metals, such as copper, zinc, and gold, are extracted from solid ores. Although magnesium is abundant in Earth's crust, it is extracted from seawater, not rocks. As a result, plants that produce magnesium, such as the one shown in **Figure 4.14**, are usually located on the coast. Magnesium ions are the second most abundant cations found in seawater, with only sodium ions in greater abundance. The process of producing metallic magnesium requires several chemical reactions.

Steps in the Process of Magnesium Mining

Figure 4.15 shows the main steps in extracting magnesium from seawater. The steps in the flowchart match the reactions described below.

Figure 4.14 This plant processes seawater to extract magnesium.

Calcium carbonate is decomposed to produce calcium oxide.

Calcium hydroxide is produced by a synthesis reaction between calcium oxide and water.

Magnesium hydroxide precipitates when calcium hydroxide is mixed with seawater.

A neutralization reaction between magnesium hydroxide and hydrochloric acid produces magnesium chloride.

Magnesium and chlorine are produced by the decomposition of magnesium chloride.

Figure 4.15 Use this flowchart to help you understand the sequence of chemical reactions used in the extraction of magnesium from seawater.

Decomposition of Calcium Carbonate

First, calcium carbonate from seashells is decomposed to produce calcium oxide. Recall, from Chapter 3, that carbon dioxide is a product when a metal carbonate decomposes. The chemical equation is

$$\text{calcium carbonate} \rightarrow \text{calcium oxide} + \text{carbon dioxide}$$
$$CaCO_3(s) \rightarrow CaO(s) + CO_2(g)$$

Synthesis of Calcium Hydroxide

The calcium oxide undergoes a synthesis reaction to form calcium hydroxide:

$$\text{calcium oxide} + \text{water} \rightarrow \text{calcium hydroxide}$$
$$CaO(s) + H_2O(\ell) \rightarrow Ca(OH)_2(aq)$$

Double Displacement between Calcium Hydroxide and Magnesium Compounds

The calcium hydroxide reacts with the magnesium ions in seawater in a double displacement reaction. The magnesium ions are separated from the other ions in seawater, such as sodium, chloride, and bromide ions, as the precipitate magnesium hydroxide:

$$\text{calcium hydroxide} + \text{magnesium ions} \rightarrow \text{magnesium hydroxide} + \text{calcium ions}$$
$$Ca(OH)_2(aq) + Mg^{2+}(aq) \rightarrow Mg(OH)_2(s) + Ca^{2+}(aq)$$

Neutralization of Magnesium Hydroxide

The solid magnesium hydroxide is filtered out and undergoes neutralization with hydrochloric acid:

$$\text{magnesium hydroxide} + \text{hydrochloric acid} \rightarrow \text{magnesium chloride} + \text{water}$$
$$Mg(OH)_2(s) + 2HCl(aq) \rightarrow MgCl_2(aq) + 2H_2O(\ell)$$

Decomposition of Magnesium Chloride

The magnesium chloride is dried, melted, and then decomposed through electrolysis to form magnesium metal:

magnesium chloride $\longrightarrow$ magnesium + chlorine

$$MgCl_2(\ell) \longrightarrow Mg(\ell) + Cl_2(g)$$

Industrial Uses of Magnesium

The main industrial use for magnesium is in the manufacturing of aluminum-magnesium alloys. An *alloy* is a mixture of two or more metals. Because magnesium is a less dense metal than aluminum, their alloys are lighter in weight than pure aluminum. In addition, the alloys are stronger and more resistant to corrosion than pure aluminum. Beverage cans are often made of aluminum to which a small amount of magnesium has been added, making the metal stronger and easier to shape. Magnesium is also commonly used in electronic devices because of its light weight and its electrical properties. The kayak shown in **Figure 4.16** is made, in part, of an alloy of aluminum and magnesium.

Figure 4.16 The frame of this kayak is constructed using tubing that is made from an alloy of aluminum and magnesium.

Infer *What properties of aluminum-magnesium tubing make it suitable for constructing a kayak?*

Learning Check

19. Describe the displacement that occurs in a thermite reaction involving iron oxide.

20. What product makes a thermite reaction useful for welding?

21. Explain how a thermite reaction can be used to produce pure copper.

22. How are seashells used to help extract magnesium from seawater?

23. During the extraction of magnesium from seawater, why is a precipitate of magnesium formed, if it is converted into soluble magnesium chloride in the next step?

24. Refer to **Figure 4.15**. What is an advantage of the production of chlorine in the final step of extracting magnesium from seawater?

Extracting Metals from Ores

Gold and copper are important metals for Canada's economy. Several different methods are used for extracting these metals. Each method combines physical processes, such as grinding and filtering, and chemical processes to obtain the desired product.

Copper Smelting

A smelter is a facility that uses heat to extract metal from ore. **Figure 4.17** shows a smelter that is used to refine copper at the Kidd Creek copper and zinc mine in Timmins, Ontario. An important part of the design of this type of smelter is the different heights of the furnaces. Molten products flow from one furnace continuously into the next furnace because of the height difference, reducing the need for large buckets and transfer equipment.

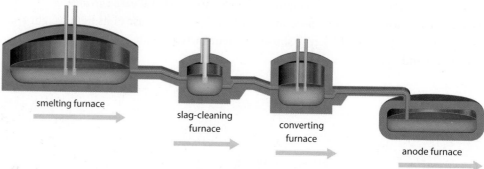

smelting furnace

slag-cleaning furnace

converting furnace

anode furnace

Figure 4.17 These furnaces produce increasingly pure copper through a series of reactions.
Identify how the construction of the smelter allows gravity to move the material from one furnace to the next.

Smelting Furnace

The smelting furnace is the first stage of copper production. An impure copper sulfide **matte**, $Cu_2S(\ell)$, forms. Oxygen is used to separate the iron from the copper in the mineral chalcopyrite, $CuFeS_2(s)$, ore according to this overall reaction:

$$2CuFeS_2(s) + 4O_2(g) \rightarrow Cu_2S(\ell) + 2FeO(\ell) + 3SO_2(g)$$

The iron(II) oxide is reacted with sand and limestone to convert it into a low density compound called *slag* which floats to the top of the molten mixture. The copper matte is more dense and sinks to the bottom.

The sulfur dioxide gas that is produced in these reactions and in later steps of the process is collected and sent to a plant that produces sulfuric acid. This prevents the release of large amounts of pollutants that contribute to the formation of acid precipitation.

Slag-Cleaning Furnace

The products of the smelting furnace move to the slag-cleaning furnace, where they separate due to differences in density. The matte flows into the converting furnace, while the undesired material, called *slag*, is sent to storage.

Converting Furnace and Anode Furnace

In the next step, matte is purified further, to about 99 percent, in the converting furnace through further displacement reactions. Air is blown through the molten mixture and oxygen in the air reacts with the copper matte in a two step reaction. First, some copper(I) sulfide is converted to copper(I) oxide. The copper(I) oxide then reacts with more copper(I) sulfide to form metallic copper and sulfur dioxide.

$$2Cu_2S(\ell) + 3O_2(g) \rightarrow 2Cu_2O(\ell) + 2SO_2(g)$$
$$Cu_2S(\ell) + 2Cu_2O(\ell) \rightarrow 6Cu(\ell) + SO_2(g)$$

matte an impure copper(I) sulfide mixture that is formed by smelting the sulfide ore

Suggested **Investigation**

Inquiry Investigation 4-E, From Copper to Copper

As in the earlier furnaces, the sulfur dioxide gas is collected, so it can be sent to a plant that produces sulfuric acid. The small furnaces make it easier to recover the gases and provide a steady stream of sulfur dioxide to the acid plant. Meanwhile, the copper undergoes a final purification step in the anode furnace.

Because copper is an excellent conductor of electricity, a major use of copper is in electrical wiring. Copper is also melted with other metals to make alloys. For example, bronze is mainly an alloy of copper and tin, and brass is mainly an alloy of copper and zinc.

Gold and Cyanide Leaching

Gold is a relatively non-reactive metal. As a result, it can be found in nature in its uncombined form. As shown in **Figure 4.18**, however, most gold is mixed into the rock that surrounds it and must be separated. The most cost-effective method for removing the gold involves treating the crushed rock with a sodium cyanide solution to dissolve the gold.

Go to **scienceontario** to find out more

Figure 4.18 Much of the rock mined as gold ore contains particles of gold that are too tiny to be seen without a microscope. Some rocks, however, such as the one shown here, contain visible particles of gold.

Leaching is the process of converting a metal to a soluble form to extract the metal. A commonly used process for the extraction of gold from ore is to react it with a solution of sodium cyanide, $NaCN(aq)$, to form sodium dicyanoaurate(I), $Na[Au(CN)_2](aq)$, and sodium hydroxide. The overall reaction for leaching gold is

$$4Au(s) + 8NaCN(aq) + O_2(g) + 2H_2O(\ell) \rightarrow 4Na[Au(CN)_2](aq) + 4NaOH(aq)$$

The gold is recovered from the solution through displacement by zinc, according to the following equation:

$$2Na[Au(CN)_2](aq) + Zn(s) \rightarrow 2NaCN(aq) + Zn(CN)_2(aq) + 2Au(s)$$

The cyanide solution can be recycled and used again to convert gold to a soluble form in an aqueous solution.

In industry, the most important use of gold is in the manufacture of electronics components. Gold is an excellent conductor of electricity and is resistant to corrosion, so it is used in small amounts in electronic devices, including cell phones and computers. Gold is also an ingredient in some medications. In addition, it is used to fill cavities in teeth and to make crowns to cover and protect teeth. Gold is highly suited to such uses because it does not corrode, it does not trigger allergic reactions, and it is easy to shape into the desired form. Gold coins have been used for many centuries as currency. Other uses of gold include the production of jewellery, watches, and art objects.

leaching a process that is used to extract a metal by dissolving the metal in an aqueous solution

Waste and Spill Treatment

Because of the potentially harmful effects of the chemicals that are used and formed during metal production, steps are taken to reduce emissions and to respond to spills.

Sulfur Dioxide Waste

Some industries, such as coal-burning power plants, use scrubbers to remove sulfur dioxide, $SO_2(g)$, from exhaust gases to prevent its release into the atmosphere. If sulfur dioxide is released into the atmosphere, it can eventually become sulfuric acid in rain and snow.

Because many metals, such as copper, are found as sulfide ores, large amounts of sulfur dioxide are commonly formed during metal extraction and purification. Sulfur dioxide can be converted into sulfuric acid, which is either used in some of the purification steps or sold. You may recognize the synthesis reactions that are involved in forming sulfuric acid from Chapter 3:

$$2SO_2(g) + O_2(g) \rightarrow 2SO_3(g)$$
$$SO_3(g) + H_2O(\ell) \rightarrow H_2SO_4(aq)$$

Cyanide Spills

Cyanide leaching allows gold to be extracted from ores that would have too low a gold content to be profitable. Unfortunately, cyanide is deadly in very small amounts if ingested. A cyanide spill may occur if the wall of a holding pond breaks or if a storm produces a large amount of rain and causes the holding pond to overflow. Two methods are commonly used to treat a cyanide spill.

Use of Sodium Hypochlorite

The first method is a two-reaction process. In the first reaction, sodium hypochlorite, the active ingredient in many chlorine bleaches, is added to the cyanide solution:

$$NaCN(aq) + NaOCl(aq) \rightarrow NaCNO(aq) + NaCl(aq)$$

Sodium cyanate, $NaCNO(aq)$, much less toxic than sodium cyanide. However, the second reaction entirely eliminates any toxicity. Additional sodium hypochlorite is used to convert the sodium cyanate into non-toxic compounds:

$$2NaCNO(aq) + 3NaOCl(aq) + H_2O(\ell) \rightarrow 3NaCl(aq) + N_2(g) + 2NaHCO_3(aq)$$

Use of Iron(II) Sulfate

The second method involves adding iron(II) sulfate, which binds the toxic and soluble cyanide ions into non-hazardous, complex iron(II) cyanide ions. The iron(II) cyanide ions form precipitates with many metal ions, such as zinc and iron, as shown in **Figure 4.19**.

First, a double displacement reaction occurs:

$$FeSO_4(aq) + 2NaCN(aq) \rightarrow Fe(CN)_2(aq) + Na_2SO_4(aq)$$

Next, a synthesis reaction forms complex ion, iron(II) cyanide ions:

$$Fe(CN)_2(aq) + 4NaCN(aq) \rightarrow Na_4Fe(CN)_6(aq)$$

Finally, double displacement reactions, such as the reaction below, cause the iron(II) cyanide ions to form precipitates with other metal ions that are present.

$$3Na_4Fe(CN)_6(aq) + 4FeCl_3(aq) \rightarrow Fe_4[Fe(CN)_6]_3(s) + 12NaCl(aq)$$

Figure 4.19 The precipitate that is formed in the reaction between sodium iron(II) cyanide and iron(III) chloride is a pigment known as Prussian blue. This pigment is used in blueprints and paints.

Infer *Why is forming a precipitate helpful when cleaning up a cyanide spill?*

Section Summary

- A thermite reaction, used to weld railroad tracks, is a single displacement reaction between aluminum and iron oxide.

- Extracting magnesium from seawater involves several chemical reactions, including a double displacement reaction that forms a magnesium hydroxide precipitate and a neutralization reaction that forms soluble magnesium chloride.

- One method for refining copper involves several reactions in which oxygen displaces sulfur in an ore.

- After gold is leached from crushed rock using cyanide, it is recovered from the solution through displacement by zinc.

- The sulfur dioxide gas that is produced during metal refining can be collected and converted into sulfuric acid, through synthesis reactions, to prevent its release into the atmosphere.

- Sodium hypochlorite and iron(II) sulfate can be used to treat a toxic cyanide spill, making the spill less toxic.

Review Questions

1. **T/I** In a thermite reaction, why are the aluminum and metal oxide in powdered form rather than large pieces?

2. **A** What are some benefits of using reactants in the solid state in a thermite reaction, rather than the states of the reactants in most other single displacement reactions you have studied?

3. **C** Create a Venn diagram to compare the extraction of magnesium from seawater with the extraction of gold and copper from their sources.

4. **K/U** What type of reaction provides the calcium oxide that is used to extract magnesium from seawater?

5. **C** Create a graphic organizer to summarize the reactions involved in the extraction of magnesium from seawater.

6. **K/U** In the extraction of magnesium from seawater, what is the key process that is used to separate magnesium ions from all the other ions that are dissolved in seawater?

7. **K/U** What is the purpose of the neutralization step in the extraction of magnesium from seawater?

8. **A** A power plant built on the coast of an ocean can be designed to use the motion of water between high and low tides to generate electricity.
   a. What step of magnesium extraction might benefit from construction of such a power plant? Explain your reasoning.
   b. Give the balanced chemical equation that is associated with the step you identified.

9. **K/U** What chemical compounds are produced in the smelting furnace during copper production?

10. **C** Compare matte with slag in copper refining.

11. **T/I** Why is zinc able to displace gold that was leached from an ore?

12. **A** What process in the extraction of gold is modelled in this photograph?

13. **A** Instead of releasing the sulfur dioxide gas produced during metal refining into the environment, a company may convert it into sulfuric acid.
   a. What is an environmental benefit of this process?
   b. What is an economic benefit of this process for the company?
   c. Give the balanced chemical equations that describe the production of sulfur dioxide gas during the smelting of copper.

14. **K/U** Name two chemicals that are used to clean up a cyanide spill.

15. **A** During a clean-up of a contaminated site, a barrel containing a cyanide solution is discovered. In terms of the final products formed, which treatment of the solution would be preferred? Explain your reasoning.

16. **K/U** Why are iron(II) cyanide ions important in the treatment of a cyanide spill with iron(II) sulfate?

Safety Precautions

- Wear safety eyewear throughout this investigation.
- Wear a lab coat or apron throughout this investigation.

Suggested Materials

- 7 small pieces of each of the following metals: copper, $Cu(s)$, magnesium, $Mg(s)$, and zinc, $Zn(s)$
- sandpaper or emery paper
- dropper bottles containing dilute solutions (0.1 mol/L) of copper(II) sulfate, $CuSO_4(aq)$; zinc sulfate, $ZnSO_4(aq)$; magnesium sulfate, $MgSO_4(aq)$, tin(II) sulfate, $SnSO_4(aq)$; iron(II) sulfate, $FeSO_4(aq)$; and hydrochloric acid, $HCl(aq)$
- 24-well reaction plate
- wash bottle with distilled water
- 6 test tubes
- test-tube rack

Making an Activity Series of Metals

While metals share common properties, they do not all behave the same way in single displacement reactions. For example, zinc displaces many metal cations from their aqueous ionic compounds. Gold, on the other hand, is extremely unreactive. Potassium reacts violently with water. Zinc does not react with water, but it reacts with acids. These observations raise a question: How can metals be ranked in order of their reactivity?

Pre-Lab Questions

1. Why do you need to clean magnesium metal with sandpaper before using it in this investigation?

2. How could you set up this investigation to make the chemical reactions easier to see?

3. Explain why you may need to repeat one or more of the reactions using a test tube, after first using the well plate.

4. Identify the WHMIS symbols that apply to hydrochloric acid. What safety precautions are required when handling hydrochloric acid?

5. Describe how you will properly dispose of the chemicals after this investigation.

Question

What is the order of reactivity of the metals copper, iron, magnesium, tin, and zinc in single displacement reactions?

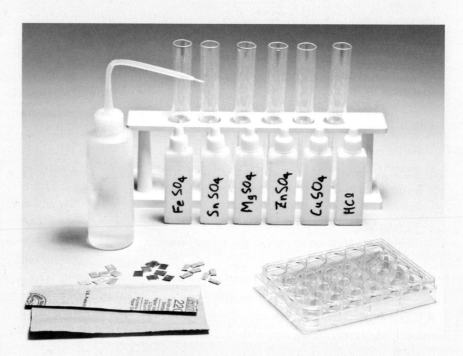

Plan and Conduct

1. Design a procedure to react each metal with each of the given solutions, as well as with water. In your procedure, identify the identities and amounts of the materials that you will need to combine and observe. Prepare a data table to record your observations.

2. Have your teacher approve your procedure before you begin.

3. Carry out your procedure. If any of the metals you are given appear dull (rather than shiny), remove this oxide coating using sandpaper or emery paper.

4. Record any changes in the appearance of each metal or solution. Look for colour changes on the surface of the metal or in the solution. If you have difficulty determining whether a reaction has occurred, repeat the test using a test tube to better observe the reaction. Remember that some chemical reactions are slow, so plan to re-examine any combinations that do not react immediately.

5. At the end of the investigation, dispose of the solutions in the waste beaker provided. Do not pour anything down the drain.

Analyze and Interpret

1. For each single displacement reaction that occurred, write the corresponding balanced chemical equation. Keep in mind that hydrogen is a diatomic element.

2. Lithium metal reacts with water.
   a. Is lithium more or less reactive than magnesium?
   b. Write a balanced chemical equation to represent the reaction of lithium with water.

Conclude and Communicate

3. Rank the metals you tested in order of reactivity, beginning with the most reactive metal. Include lithium in your ranking. (You have just created an *activity series of metals*.)

4. Explain how an activity series of metals can be used to predict single displacement reactions.

5. Given that lithium reacts with water and that magnesium does not, do you expect lithium to react with hydrochloric acid? If so, write a balanced chemical equation to represent the reaction.

6. What evidence shows that hydrogen reacts differently in water than it does in hydrochloric acid?

Extend Further

7. **INQUIRY** Imagine that you are given an unknown metal, which could be magnesium, zinc, or aluminum. Design an investigation involving single displacement reactions that would allow you to identify the unknown metal.

8. **RESEARCH** Use print or Internet resources to research the use of sacrificial anodes. What are they? Where are they used? How do they work? Include balanced chemical equations in your answer.

Safety Precautions

- Wear safety eyewear throughout this investigation.

- Tie back loose hair and clothing.

- Use EXTREME CAUTION when you are near an open flame.

- Wear a lab coat or apron throughout this investigation.

Materials

- dropper bottles containing dilute aqueous solutions (0.1 mol/L) of magnesium chloride, $MgCl_2(aq)$; sodium hydroxide, $NaOH(aq)$; iron(III) chloride, $FeCl_3(aq)$; potassium chloride, $KCl(aq)$; sodium sulfate, $Na_2SO_4(aq)$; calcium chloride, $CaCl_2(aq)$; silver nitrate, $AgNO_3(aq)$; copper(II) sulfate, $CuSO_4(aq)$; sodium phosphate, $Na_3PO_4(aq)$; and hydrochloric acid, $HCl(aq)$

- sheet of white paper

- a few particles of sodium carbonate, $Na_2CO_3(s)$

- wooden splint

- a few particles of ammonium chloride, $NH_4Cl(s)$

- red litmus paper

- 24-well reaction plate

- small scoop

- 2 beakers (50 mL)

- tongs

- 10 mL graduated cylinder

- Bunsen burner secured to a utility stand

- igniter for Bunsen burner

Observing Double Displacement Reactions

A double displacement reaction involves the exchange of cations between ionic compounds, usually in an aqueous solution.

$$AX + BY \rightarrow AY + BX$$

Most double displacement reactions result in the formation of a precipitate. Some double displacement reactions result in the formation of an unstable compound that decomposes to water and a gas. The reaction between an acid and a base—neutralization—is a double displacement reaction in which a salt and water are formed.

Pre-Lab Questions

1. Write a balanced chemical equation for each possible reaction in **Table A**. If you do not think a precipitate will form, write "NR."

2. Carbonic acid, $H_2CO_3(aq)$, decomposes into carbon dioxide, and water. Ammonium hydroxide, $NH_4OH(aq)$, decomposes into ammonia and water. Given this information, repeat question 1 for each potential reaction in **Table B**.

3. Look up the MSDS information for the chemicals used in this investigation. Make a table to record the safety measures you should take when working with these chemicals.

Questions

How can you tell if a double displacement reaction has occurred? How can you predict the products of a double displacement reaction?

Procedure

1. Copy **Table A** and **Table B** into your notebook.

Table A Double Displacement Reactions That May Form a Precipitate

Reactants in Aqueous Solutions	Observations
$MgCl_2(aq) + NaOH(aq)$	
$FeCl_3(aq) + NaOH(aq)$	
$KCl(aq) + Na_2SO_4(aq)$	
$CaCl_2(aq) + AgNO_3(aq)$	
$CuSO_4(aq) + Na_3PO_4(aq)$	

Table B Double Displacement Reactions That May Form a Gas

Reactants	Observations
$Na_2CO_3(s) + 2HCl(aq)$	
$NH_4Cl(s) + NaOH(aq)$	

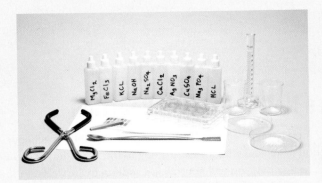

Conclude and Communicate

4. How did you know that a double displacement reaction occurred?

   a. What type of double displacement reaction may not produce a visible product?

   b. Illustrate your answer with an example of a reaction that was not part of this investigation.

Extend Further

5. INQUIRY When aqueous aluminum bromide is added to aqueous potassium dichromate, a precipitate forms. (The formula for the dichromate ion is $Cr_2O_7^{2-}$.)

   a. Write a balanced chemical equation to represent this double displacement reaction.

   b. The dichromate ion is not listed in your solubility guidelines. How can you determine the identity of the precipitate?

6. RESEARCH In the 1820s, Friedrich Wöhler, a German physician, carried out a simple double displacement reaction between aqueous silver isocyanate, AgOCN(aq), and aqueous ammonium chloride. Use the Internet to research Wöhler's experiment.

   a. Write a balanced chemical equation to represent this double displacement reaction. Use your solubility guidelines to predict the precipitate.

   b. After the reaction, Wöhler used filtration to remove the precipitate, which he discarded. He then evaporated the water from the resulting solution by gentle heating. Wöhler found, to his surprise, that the crystals he obtained were not what he expected. What compound did he obtain?

   c. Why was Wöhler's experiment so important?

Procedure

2. Place the reaction plate on the sheet of white paper.

3. Carry out each reaction in **Table A**. If you are unsure about your observations for a reaction, repeat the reaction using a test tube for improved visibility.

4. Use the scoop to add a few particles of sodium carbonate to a 50 mL beaker. Use the graduated cylinder to add 10 mL of hydrochloric acid. When the reaction subsides, lower a burning splint into the beaker to test the gas produced.

5. Use the scoop to add a few particles of ammonium chloride to a 50 mL beaker. Add about 2 mL of sodium hydroxide solution. Waft your hand over the mouth of the beaker, toward your nose, to detect any odour. Holding the beaker with tongs, gently warm the solution over the lit Bunsen burner. Hold a moistened piece of red litmus paper in the fumes given off. Record your observations in **Table B**.

Analyze and Interpret

1. How accurate were your predictions in Pre-Lab question 1?

2. Explain how solubility guidelines can be used to predict the formation of a precipitate.

3. What gases were formed in steps 3 and 4?

   a. How were you able to identify each gas?

   b. How well do your balanced chemical equations from Pre-Lab question 2 support your observations?

Inquiry
INVESTIGATION

Skill Check

Initiating and Planning

✓ Performing and Recording

✓ Analyzing and Interpreting

✓ Communicating

4-C

Safety Precautions

- Wear safety eyewear throughout this investigation.
- Tie back loose hair and clothing.
- Use EXTREME CAUTION when you are near an open flame.
- Wear a lab coat or apron throughout this investigation.

Materials

- 25 mL of 0.10 mol/L sodium hydroxide solution, NaOH(aq)
- phenolphthalein indicator solution in a dropper bottle
- 5 mL of 1.0 mol/L hydrochloric acid, HCl(aq), in a dropper bottle
- about 0.5 g of sodium chloride, NaCl(s)
- distilled water
- silver nitrate solution, $AgNO_3$(aq), in a dropper bottle
- 25 mL graduated cylinder
- 250 mL beaker
- glass dropper or plastic pipette
- stirring rod
- hot plate, or retort stand, ring clamp, and metal gauze (to use with Bunsen burner)
- Bunsen burner secured to a utility stand
- igniter for Bunsen burner
- test tube
- test-tube holder
- nichrome loop or water-soaked wooden splint
- 50 mL beaker

Observing a Neutralization Reaction

Acid-base neutralization is a type of double displacement reaction. In general, it involves the reaction of an acid with a base, leading to the formation of a salt and water. In this investigation, you will neutralize a sodium hydroxide solution with hydrochloric acid. Then you will isolate and test the product.

Pre-Lab Questions

1. In step 4 of the Procedure, you will evaporate the solution to dryness.
   a. Why do you think you will need to swirl the solution as you do this?
   b. Why will you need to use beaker tongs to hold the beaker when you are swirling the solution?
2. What is the expected flame test colour for sodium chloride?
3. Why will you need to carry out a flame test and a silver nitrate test on both a sample of the test-tube residue and a sample of sodium chloride?

Question

What are the products of the neutralization of a sodium hydroxide solution with hydrochloric acid?

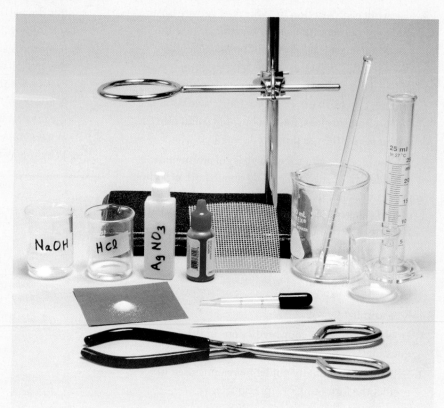

Procedure

1. Using the graduated cylinder, add exactly 25 mL of 0.10 mol/L sodium hydroxide solution to a 250 mL beaker. Add two drops of phenolphthalein indicator solution. Note the colour.

2. Using the medicine dropper or plastic pipette, add 1.0 mol/L hydrochloric acid, one drop at a time, to the sodium hydroxide solution. Stir the solution after every few drops.

3. As the colour of the solution begins to fade, stir after you add each drop of hydrochloric acid, until the colour *just* disappears. This is the neutralization point.

4. Transfer the solution to a test tube. With gentle heat, either from a hot plate or a Bunsen burner, evaporate the solution to dryness. Keep moving the test tube gently through the flame. Make sure that you hold the test tube with a test-tube holder.

5. Record the colour of the residue in the test tube. Using either a nichrome loop or a water-soaked wooden splint, pick up a small crystal of the residue. Hold the sample near the top of a Bunsen burner flame. Record the colour of the flame.

6. Obtain about 0.5 g of sodium chloride in a 50 mL beaker. Repeat the flame test using the sodium chloride crystals. Record the colour of the flame.

7. Add a few millilitres of water to dissolve the solid residue from the neutralization. Add a drop of silver nitrate solution. Record your observations.

8. Repeat step 7 with the sodium chloride crystals. Record your observations.

Analyze and Interpret

1. **a.** Compare the results of the flame tests you performed.

   b. Compare the results of the silver nitrate tests you performed.

   c. When the solution containing the neutralization products was heated, what evaporated?

2. Use your answers for question 1 to write a balanced chemical equation for the reaction of sodium hydroxide with hydrochloric acid.

Conclude and Communicate

3. Explain how a neutralization reaction is a type of double displacement reaction.

4. To a chemist, what does the term "salt" mean?

5. Explain why it is unsafe to taste the product of the neutralization of sodium hydroxide with hydrochloric acid.

6. Write a balanced chemical equation to represent each neutralization reaction.

   a. $H_2SO_4(aq) + Al(OH)_3(aq) \rightarrow$

   b. $H_3PO_4(aq) + Ca(OH)_2(aq) \rightarrow$

Extend Further

7. **INQUIRY** You performed this investigation qualitatively, but you could gain additional information by performing it quantitatively. You would need to make very careful measurements of the volumes of the solutions used, however. Design a procedure to determine the average volume of one drop from a medicine dropper or plastic pipette.

8. **RESEARCH** Use print or Internet resources to research how antacids neutralize stomach acid. Include at least one balanced chemical equation in your answer.

Safety Precautions

- Wear safety eyewear throughout this investigation.
- Wear a lab coat or apron throughout this investigation.
- Use clean drinking straws in this investigation. Get the straws from your teacher when they are needed. Do not let them touch the lab bench.

Materials

- 200 mL of distilled water
- bromothymol blue indicator in a dropper bottle
- 200 mL of freshly prepared limewater, $Ca(OH)_2(aq)$
- 2 Erlenmeyer flasks (250 mL)
- 2 clean drinking straws

Modelling Neutralization Reactions Used in Scrubber Technology

In a submarine or spacecraft, exhaled carbon dioxide, $CO_2(g)$, needs to be removed, or "scrubbed," from the air. A simple way to do this is to allow the carbon dioxide to react with a base to form a solid. One method used in submarines and the International Space Station is to use lithium hydroxide monohydrate, $LiOH \cdot H_2O(s)$, to remove carbon dioxide according to the following chemical equation:

$$2LiOH \cdot H_2O(s) + CO_2(g) \rightarrow Li_2CO_3(s) + 3H_2O(g)$$

In this investigation, you will exhale into limewater, $Ca(OH)_2(aq)$, a basic solution, to convert the carbon dioxide in your breath into a precipitate of calcium carbonate, $CaCO_3(s)$, through a neutralization reaction.

When a submarine is submerged in the ocean, scrubber technology is used to remove excess carbon dioxide from the air that the crew breathes.

Pre-Lab Questions

1. What is the purpose of the bromothymol blue indicator in this investigation?

2. Write a balanced chemical equation for the reaction of carbon dioxide gas with water to produce carbonic acid, $H_2CO_3(aq)$.

3. What safety precautions should you take when handling a caustic substance, such as limewater?

Questions

How can a solution of limewater remove carbon dioxide from the air?
How does this reaction model modern gas scrubbers?

Procedure

1. Add about 200 mL of distilled water to a 250 mL Erlenmeyer flask. Add three drops of bromothymol blue indicator.

2. To another 250 mL Erlenmeyer flask, add 200 mL of freshly prepared limewater. Add three drops of bromothymol blue indicator.

3. Using a clean drinking straw, exhale into the flask containing the distilled water. Record the number of seconds required to turn the solution yellow.

4. Repeat step 3 for the limewater solution. Continue to exhale into the solution for up to 15 s. Carefully observe the solution in terms of colour and cloudiness. Record your observations.

Analyze and Interpret

1. **a.** How long, in seconds, did it take for your exhaled breath to make pure water acidic?

   b. What can you conclude about the product of the reaction between carbon dioxide and water?

2. Did 15 s of exhaling into a limewater solution have a noticeable effect on the acidity of the limewater solution?

3. Write a balanced chemical equation to support your answer to question 2.

Conclude and Communicate

4. **a.** What was the role of the calcium hydroxide in the limewater, in relation to carbon dioxide?

   b. Explain how this reaction models modern scrubber technology.

Extend Further

5. **INQUIRY** Introducing carbon dioxide into a limewater solution produces cloudiness, which you observed. What would occur if carbon dioxide was allowed to bubble through a limewater solution for a considerable length of time? If you have access to dry ice (frozen carbon dioxide), investigate this. Record your observations and, with the help of external resources, interpret your observations using a chemical equation.

6. **RESEARCH** Various devices are being proposed and prototyped to capture and remove carbon dioxide from the air.

   a. Research one of these devices, and explain how it works.

   b. Describe one major advantage and one major disadvantage of the device.

   c. Find out what happens to the carbon dioxide that is captured by the device.

Inquiry
INVESTIGATION

Skill Check

Initiating and Planning

✓ Performing and Recording

✓ Analyzing and Interpreting

✓ Communicating

4-E

From Copper to Copper

You have learned about several different types of chemical reactions, including synthesis, decomposition, single displacement, and double displacement reactions. In this investigation, you will follow copper through a sequence of reactions, which will eventually produce metallic copper again.

Pre-Lab Questions

The following chemical equations represent the reactions that you will observe in this investigation:

A. $Cu(s) + 4HNO_3(aq) \rightarrow Cu(NO_3)_2(aq) + 2NO_2(g) + 2H_2O(\ell)$

B. $Cu(NO_3)_2 + NaOH \rightarrow Cu(OH)_2 + NaNO_3$

C. $Cu(OH)_2 \rightarrow CuO + H_2O$

D. $CuO + H_2SO_4 \rightarrow CuSO_4 + H_2O$

E. $CuSO_4 + Zn \rightarrow Cu + ZnSO_4$

1. a. Copy equations A to E into your notebook. Balance equations B through E.
   b. Include the physical state (s, ℓ, aq, g) of each reactant and product in equations B through E. Refer to the solubility guidelines as needed.

2. What type of chemical reaction is represented by each equation (A through E) above?

3. When testing the acidity or alkalinity of a solution with litmus paper, why do you need to use a glass rod to place a drop of the solution on a piece of litmus paper, rather than simply putting the litmus paper into the solution?

4. Read the safety precautions listed on the left side of this page. Identify the reasoning behind each precaution.

Question

What type of chemical reaction is involved in each step of the conversion of copper to copper?

Procedure

Reaction A: Reaction of Copper with Nitric Acid to Form Copper(II) Nitrate

1. Examine the reaction in the photograph. Record your observations. Identify the gas and solution formed.

Safety Precautions

- Wear safety eyewear throughout this investigation.

- Wear a lab coat or apron throughout this investigation.

- Make sure that the electrical cord of the hot plate is not dangling over the edge of the work area. Unplug the hot plate when it is not in use.

- Sodium hydroxide can cause blindness if it gets in the eyes.

- NaOH and H_2SO_4 solutions are corrosive. If you spill one of these solutions on your skin, immediately wash the area with plenty of cold water. Inform your teacher.

- Dispose of solutions according to your teacher's directions.

Materials

- about 4 mL of 0.4 mol/L copper(II) nitrate solution, $Cu(NO_3)_2(aq)$

- 6 mol/L sodium hydroxide solution, NaOH(aq), in a dropper bottle

- red litmus paper

- 3 mol/L sulfuric acid, $H_2SO_4(aq)$, in dropping bottle

- about 0.8 g of powdered zinc, Zn(s)

- 50 mL beaker

- glass rod

- hot plate

- wash bottle with distilled water

- 250 mL beaker with water and one or two ice cubes (ice-water bath)

The reaction of copper with concentrated nitric acid gives off heat. A poisonous gas is released.

Reaction B: Preparation of Copper(II) Hydroxide

2. Obtain about 4 mL of 0.4 mol/L copper(II) nitrate solution in a 50 mL beaker. Add 6 mol/L sodium hydroxide solution one drop at a time, to the beaker until the solution is basic. To test the alkalinity of the solution, use the glass rod to transfer a drop of the solution to the red litmus paper. (Do not put the litmus paper in the solution.) Red litmus paper turns blue in a basic solution. Record your observations.

Reaction C: Preparation of Copper(II) Oxide

3. Heat the solution formed in reaction B on a hot plate, constantly stirring with a glass rod, until all the pale blue precipitate reacts to form a black precipitate. Use a wash bottle to rinse any pale blue precipitate from the walls of the beaker, using as little water as possible. Constantly stir or swirl any precipitate-containing solution that is being heated to avoid a sudden boiling over, or *bumping*.

4. Cool the beaker containing the black precipitate in an ice-water bath for several minutes, until the outside of the beaker feels cold.

Reaction D: Preparation of Copper(II) Sulfate Solution

5. Carefully add 3 mol/L sulfuric acid, one drop at a time, to the cold mixture until the black precipitate just dissolves. (**CAUTION:** Sulfuric acid is corrosive.) Avoid adding more sulfuric acid than necessary. Record your observations.

Reaction E: Regeneration of the Copper Metal

6. Carefully add about 0.80 g of powdered zinc to the solution of copper(II) sulfate. Stir or swirl the solution until the blue colour disappears. Add more zinc if necessary. Record your observations.

The copper produced in reaction E is called "spongy" copper, because of its appearance.

Analyze and Interpret

1. Think about step 6 in the Procedure.
   a. Explain why zinc reacts with sulfuric acid, but copper does not.
   b. Explain why magnesium could be used instead of zinc, but sodium could not.
   c. Explain why powdered zinc was used instead of a lump of zinc.

2. If this investigation started with 0.10 g of copper metal in reaction A, what mass of copper should be recovered at the end of reaction E?

3. Explain why reaction A is not a double displacement reaction.

4. Suggest a procedure you could use to recover the copper(II) oxide.

5. At the end of step 6, what dissolved ions are present in significant quantities?

Conclude and Communicate

6. Summarize this investigation using a flowchart. Include balanced chemical equations in your flowchart. If possible, identify the type of reaction beside each equation.

Extend Further

7. **INQUIRY** Do you think you could perform this type of reaction sequence with zinc, so that you began and ended with zinc? Design a procedure, based on the procedure you used for copper. Include a balanced chemical equation for each step. If your teacher approves your procedure, carry out the reactions.

8. **RESEARCH** Use print or Internet resources to research the compound(s) responsible for the formation of the green coating, or *patina*, on old copper roofs. Does this patina lead to the type of rusting exhibited by iron?

Case Study

Smelting Emissions
Determining Acceptable Levels of Risk from Exposure

Scenario

Suppose that you live in a community that is home to one of Canada's 15 base metal smelters. Base metal smelting produces common useful metals, such as zinc, lead, copper, and nickel, from metal ores. It is an important industry in Canada. However, harmful chemicals, such as nickel sulfides and oxides, cadmium, arsenic, sulfur dioxide, and mercury, can be released into the environment during smelting. These chemicals can endanger the health and safety of local populations.

The smelter in your community has become controversial. Some of the people in your community are concerned about their exposure to toxic chemical emissions. You have joined an organization whose goal is to reduce air, soil, and water pollution in Canada. The organization is interested in finding out more about base metal smelting and the industry's effects on the environment and human health. As a member of the organization, you have volunteered to research this issue.

What Is Base Metal Smelting?

Base metal smelting involves industrial processes that use chemical reactions at high temperatures and pressures to recover base metals from their ores. For example, heat and chemicals are used to extract nickel from crushed ore. Once the nickel is separated from other metals in the ore, it is present as nickel(II) sulfate, NiSO4(aq). The nickel(II) sulfate is then reacted with hydrogen gas under high temperatures and pressures to produce nickel metal, according to the following single displacement reaction:

$$NiSO_4(\ell) + H_2(g) \rightarrow Ni(s) + H_2SO_4(\ell)$$

Base metal smelting uses chemical reactions to separate common metals from their ores. The facility shown in this photograph is located in Sault Ste. Marie, Ontario.

Health Effects

Some nickel oxides, sulfides, and other nickel compounds can be released during the processing of nickel. Breathing in nickel and its compounds has been associated with chronic bronchitis and certain cancers.

Many of the substances that are emitted by base metal smelters are listed as toxic by the Canadian Environmental Protection Act. The goal of the act is to protect the environment and human health by reducing or almost eliminating certain toxic substances in the environment. However, some people think that the standards in the act are not strict enough. For example, several chemicals that are released by base metal smelters are thought to increase the risk of cancer at any level of exposure, no matter how small. Many people agree that there should be no risk of exposure to these chemicals.

Economics

Some people think that pressure from the base metal smelting industry has prevented the government from providing strict standards emissions. The smelting industry employs thousands of Canadians and contributes billions of dollars to the economy. Many smelters are located in remote areas. Finding other types of employment in these areas can be challenging. Smelting companies argue that setting stricter limits on emissions would cause them to lose business to competitors in other countries, which would result in job losses in local communities. They also argue that improved technologies and processes have already reduced emissions significantly.

Fact Sheet

Metals and Their Uses

- Cobalt is used in superalloys, to provide high strength and resistance to corrosion and abrasion. These properties are required for products such as jet aircraft engines, stationary gas turbines for pipeline compressors, magnet steel, and stainless steel. Cobalt is also used in electroplating and, as a blue dye additive, in paints, glass, and ceramics.

- Copper is used in the production of cables, wires, and electrical products; pipes for plumbing, heating, and ventilation; and building wire and sheet metal facings.

- Lead is mainly used in the manufacture of automobile batteries. It is also used in radiation shielding.

- Nickel is an integral component of stainless steel. It provides durability, strength, and resistance to corrosion. As well, nickel is used in alloy production for specialized applications.

- Zinc is most commonly used in galvanizing, to protect steel structures from corrosion. Zinc is also used in a wide range of chemicals for consumer products, agricultural feed supplements and fertilizers, and tires.

Percentage of Canadian Emissions Released from the Base Metal Smelting Sector in 2002

Substance	Units	Base Metal Smelting*	Canadian Total**	Percent Emissions (%)
Arsenic*	tonnes	153	201	76
Cadmium	tonnes	31	33	94
Lead	tonnes	196	223	88
Mercury	kg	1 700	2 949	58
Nickel	tonnes	258	475	54
Total particulate matter	tonnes	10 757	523 319	2
Sulfur dioxide	tonnes	669 967	1 419 520	47

* Rubinoff Environmental, Pollution Prevention and Pollution Control Initiatives in the Base Metals Smelting and Refining Sector, February 2004.

** Environment Canada, 2005.

Research and Analyze

1. Research the health effects of some of the chemicals that are released to the environment from base metal smelters, including cadmium, arsenic, sulfur dioxide, and mercury. What factors influence the effects of exposure to a harmful chemical? What segments of a local population might be most vulnerable to the negative effects of exposure to toxic smelter emissions?

2. Research base metal smelting in Ontario. What metals do Ontario smelters process? How do the smelters contribute to Ontario's economy? Where are they located? What specific information can you find about the positive and negative effects of some of these smelters?

3. What level of risk from exposure to harmful smelter emissions do you think is acceptable? Should people who live near smelters have to accept even a low level of risk? What is the responsibility of provincial and federal governments for protecting people from exposure to toxic substances? What effects might closing a smelter have on a community?

Take Action

1. **Plan** In a group, discuss the concerns associated with base metal smelting. Identify different points of view regarding the reduction or elimination of emissions. Share the results of the research and analysis you conducted for questions 1 to 3 above.

2. **Act** Prepare a report based on your findings. Provide recommendations that might help to reduce or eliminate the exposure of local populations to toxic emissions from base metal smelters. Support your recommendations with information from credible sources.

Section 4.1 | Single Displacement Reactions

In a single displacement reaction, an element displaces a less active element in a compound.

KEY TERMS
- activity series
- single displacement reaction

KEY CONCEPTS
- In a single displacement reaction, one element replaces another element in a compound to produce a new element and a new compound.

- The general form for single displacement reactions in which a metal displaces another metal, where A and B are metals, is $A + BX \rightarrow AX + B$. The general form for single displacement reactions in which a non-metal displaces another non-metal, where X and Y are non-metals, is $AX + Y \rightarrow X + AY$.

- An activity series lists elements in order, from most reactive to least reactive.

- The activity series of metals and the activity series of halogens are used to predict whether a single displacement reaction will occur.

- A single displacement reaction can only occur when an uncombined element is higher in an activity series than the element it would replace in a compound.

Section 4.2 | Double Displacement Reactions

In a double displacement reaction, ions of two ionic compounds trade places to form new compounds.

KEY TERMS
- double displacement reaction
- neutralization
- precipitate

KEY CONCEPTS
- In a double displacement reaction, the positive ions in two compounds trade places to form two new compounds.

- The general form of a double displacement reaction is $AX + BY \rightarrow AY + BX$.

- A double displacement reaction generally produces a precipitate, a gas, or water.

- Solubility guidelines can be used to predict whether a product of a double displacement reaction will be a precipitate.

- Neutralization occurs when there is a double displacement reaction between an acid and a base. Water is a product of neutralization.

Section 4.3 | Reactions in Industry

Many chemical reactions are used in industrial processes and for cleaning up hazardous waste.

KEY TERMS
- leaching
- matte

KEY CONCEPTS
- A thermite reaction, used to weld railroad tracks, is a single displacement reaction between aluminum and iron oxide.

- Extracting magnesium from seawater involves several chemical reactions, including a double displacement reaction that forms a magnesium hydroxide precipitate and a neutralization reaction that forms soluble magnesium chloride.

- One method for refining copper involves several reactions in which oxygen displaces sulfur in an ore.

- After gold is leached from crushed rock using cyanide, it is recovered from the solution through displacement by zinc.

- The sulfur dioxide gas that is produced during metal refining can be collected and converted into sulfuric acid, through synthesis reactions, to prevent its release into the atmosphere.

- Sodium hypochlorite and iron(II) sulfate can be used to treat a toxic cyanide spill, making the spill less toxic.

Knowledge and Understanding

Circle the letter of the best answer below.

1. Which metal is the most reactive?
   a. aluminum
   b. copper
   c. sodium
   d. gold
   e. iron

2. What is produced during a neutralization reaction?
   a. water
   b. a precipitate
   c. oxygen
   d. a metal
   e. a non-metal

3. Which metal does not react with hydrochloric acid?
   a. chromium
   b. copper
   c. sodium
   d. magnesium
   e. zinc

4. Which compound could be a precipitate in a double displacement reaction?
   a. sodium bromide
   b. copper(II) nitrate
   c. calcium acetate
   d. lead(II) sulfate
   e. potassium chloride

5. Which reactants would likely form a solid product?
   a. $NH_4Br(aq) + NaOH(aq)$
   b. $HI(aq) + K_2CO_3(aq)$
   c. $ZnCl_2(aq) + AgNO_3(aq)$
   d. $NaOH(aq) + HClO_3(aq)$
   e. $Mg(OH)_2(aq) + HNO_3(aq)$

6. Which compound is used to leach gold from its ore?
   a. iron(II) sulfate
   b. zinc metal
   c. sodium hypochlorite
   d. sodium cyanide
   e. oxygen

7. When the following pairs of substances react, which pair is likely to have water as a product?
   a. $AgBr(aq) + NaCl(aq)$
   b. $NaI(aq) + K_2CO_3(aq)$
   c. $Ca(OH)_2(aq) + AgClO_3(aq)$
   d. $NaOH(aq) + H_2SO_4(aq)$
   e. $Mg(ClO_3)_2(aq) + HNO_3(aq)$

8. Which process occurs during the mining of magnesium from seawater, but is not a chemical reaction?
   a. The formation of liquid magnesium metal from liquid magnesium chloride.
   b. The formation of liquid magnesium chloride from solid magnesium chloride.
   c. The formation of a solution of magnesium chloride from solid magnesium hydroxide.
   d. The formation of solid calcium oxide from solid calcium carbonate.
   e. The formation of aqueous calcium hydroxide from solid calcium oxide.

Answer the questions below.

9. Which family of metals tends to form soluble ionic compounds?

10. What is the least reactive metal that can still displace hydrogen from cold water?

11. How does a decomposition reaction result in the formation of a gas in a double displacement reaction?

12. Compare the reactivity of bromine with the reactivities of the other halogens.

13. What type of chemical spill could be cleaned up using a neutralization reaction?

14. Oxygen is an important reactant in many industrial processes.
    a. Which metal is most dependent on oxygen in its production?
    b. Explain your reasoning.

15. What is the general form of a single displacement reaction that involves halogens?

16. What type of solid material is likely to be a reactant in a single displacement reaction?

Thinking and Investigation

17. Describe what you would expect to see if you placed a piece of potassium in an aqueous solution of lithium chloride. Explain your reasoning.

18. Suppose that you observed the reaction that is shown in the photograph during an investigation.

   a. What is the solid material called?

   b. What type of reaction is most likely occurring?

19. Classify each displacement reaction.

   a. $RbCl(aq) + AgNO_3(aq) \rightarrow RbNO_3(aq) + AgCl(s)$

   b. $Zn(s) + 2AgNO_3(aq) \rightarrow Zn(NO_3)_2(aq) + 2Ag(s)$

   c. $Br_2(\ell) + 2KI(aq) \rightarrow I_2(s) + 2KBr(aq)$

   d. $Pb(CH_3COO)_2(aq) + 2NaCl(aq) \rightarrow$
                             $2Na(CH_3COO)_2(aq) + PbCl_2(s)$

20. Using the appropriate activity series, write a balanced chemical equation for each single displacement reaction. If you predict that no reaction will occur, write "NR."

   a. $Mg(s) + Co(NO_3)_3(aq) \rightarrow$

   b. $Cl_2(g) + LiBr(aq) \rightarrow$

   c. $Zn(s) + HClO_4(aq) \rightarrow$

   d. $Ni(s) + Cd(NO_3)_3(aq) \rightarrow$

   e. $Al(s) + NiCl_2(aq) \rightarrow$

   f. $K(s) + H_2O(\ell) \rightarrow$

   g. $Cl_2(g) + CaF_2(aq) \rightarrow$

21. Describe how you would conduct an investigation to determine an activity series for a group of metals.

22. The following reactions occur during the smelting of copper:

$$2CuFeS_2(s) + 4O_2(g) \rightarrow Cu_2S(\ell) + 2FeO(\ell) + 3SO_2(g)$$

$$2FeS(\ell) + 3O_2(g) \rightarrow 2FeO(\ell) + 2SO_2(g)$$

$$2FeS_2(\ell) + 9O_2(g) \rightarrow 2FeO(\ell) + 4SO_2(g)$$

   a. Based on these reactions, compare the reactivities of oxygen and sulfur.

   b. Explain your reasoning.

23. Determine the products that are formed in the double displacement reaction between each pair of reactants, and write a balanced chemical equation. Identify the states of the products. One of the products in each reaction will be a precipitate, a gas, or liquid water.

   a. potassium sulfate and barium bromide

   b. nitric acid (aqueous hydrogen nitrate) and lithium carbonate

   c. copper(II) bromide and sodium hydroxide

   d. rubidium sulfide and lead(II) nitrate

   e. ammonium sulfate and potassium hydroxide

   f. iron(II) bromide and silver nitrate

   g. lithium hydroxide and sulfuric acid (hydrogen sulfate)

24. Samples of an unknown metal, X, are placed in several solutions, and the results are recorded in the table below. Based on the results, determine the relative reactivity of the unknown metal compared with the metals in the solutions used.

Observations

Solution	Result
aluminum nitrate	no reaction
nickel(II) nitrate	nickel metal formed
calcium nitrate	no reaction

25. Aluminum sulfate is used in water purification. It is added to water to cause small particles to form larger clumps. The clumps then settle to the bottom of large tanks, such as the ones shown in the photograph. Write balanced chemical equations to show how aluminum sulfate can be produced from a single displacement reaction and from a double displacement reaction.

Communication

26. **BIG IDEAS** Chemicals react in predictable ways. Describe the process in which ammonia gas forms from a double displacement reaction. Provide balanced chemical equations to illustrate your answer.

27. Using a Venn diagram, compare single displacement reactions with double displacement reactions.

28. Explain why no reaction occurs if a double displacement reaction does not produce a precipitate, a gas, or water.

29. **BIG IDEAS** Chemical reactions and their applications have significant implications for society and the environment. Imagine that you are writing a newspaper article about the use of cyanide to recover gold at a local mine. What are some points for and against the use of cyanide?

30. Imagine that you are a producer for a radio science show. Write a script for a podcast to describe what occurs when zinc metal is placed into a solution of hydrochloric acid.

31. The photograph below shows a Roman aqueduct made from a series of arches with a channel on top to carry water. The aqueduct had a slight downward slope to allow the water to run downhill. Once the water reached a city, pipes carried it throughout the city. Lead was often used to make the water pipes. In a brief paragraph, describe the health concern related to displacement reactions that is associated with the use of lead water pipes.

32. One of your friends does not understand how to predict whether a single displacement reaction will occur. Write an explanation to help your friend.

33. Summarize your learning in this chapter using a graphic organizer. To help you, the Chapter 4 Summary lists the Key Terms and Key Concepts. Refer to Using Graphic Organizers in Appendix A to help you decide which graphic organizer to use.

Application

34. Describe a procedure for using a displacement reaction to produce a sample of the element copper. Explain your reasoning.

35. During the final stage of magnesium production, the electrolysis of magnesium chloride, the toxic gas chlorine is produced. Use print or Internet resources to determine how the chlorine gas can be put to use within the magnesium production process so that it does not require disposal.

36. Iron(II) sulfate is used to treat cyanide spills. Based on the reactions that lead to the precipitation of the cyanide as a complex iron(II) cyanide ion, explain why the correct amount of iron(II) sulfate must be used for the cyanide ions to be successfully removed.

37. Using print or Internet resources, research the causes and hazards of acid mine drainage. Imagine that a retaining wall around a tailings pile breaks after a week of heavy rains, and the rainwater that had collected is released.

    a. Explain why there might be a concern that ions of metals, such as lead, nickel, and cadmium, have also been released.

    b. Describe the training an engineer would need to have to determine how to mitigate hazards associated with acid drainage at a particular mine.

38. Magnesium phosphate provides the human body with the minerals magnesium and phosphorus. How could solid magnesium phosphate be prepared and collected, using hydrochloric acid, sodium phosphate solution, water, and magnesium metal?

    a. Write a clear step-by-step procedure.

    b. Include a balanced chemical equation for each reaction you suggest.

39. Sodium hydrogen carbonate, commonly known as baking soda, is used in recipes that include an acidic ingredient.

    a. Describe the reaction that occurs.

    b. Explain why sodium hydrogen carbonate is used.

40. Acid indigestion can result from the overproduction of hydrochloric acid. A traditional remedy for treating acid indigestion is drinking a glass of baking soda (sodium hydrogen carbonate) dissolved in water. Write a balanced chemical equation for the reaction that occurs.

Select the letter of the best answer below.

1. **K/U** Magnesium is involved in several reactions as it is mined from seawater. How would you classify these reactions?
   a. synthesis
   b. decomposition and single displacement
   c. combustion, double displacement, and decomposition
   d. combustion and synthesis
   e. double displacement and decomposition

2. **K/U** Which reactants would likely form carbon dioxide gas?
   a. $NH_4Cl(aq) + KOH(aq)$
   b. $HCl(aq) + Na_2CO_3(aq)$
   c. $ZnCl_2(aq) + K_2CO_3(aq)$
   d. $NaOH(aq) + HClO_3(aq)$
   e. $Mg(s) + H_2CO_3(aq)$

3. **K/U** Which metal reacts with hydrochloric acid but not with water?
   a. barium
   b. copper
   c. sodium
   d. silver
   e. zinc

4. **K/U** Which reactants would likely form a solid product?
   a. $KOH(aq) + NH_4Cl(aq)$
   b. $HCl(aq) + Zn(s)$
   c. $Na_2CO_3(aq) + HClO_3(aq)$
   d. $ZnCl_2(aq) + Pb(NO_3)_2(aq)$
   e. $Mg(s) + HNO_3(aq)$

5. **T/I** Which materials do not react with each other?
   a. $Cr(s) + CdCl_2(aq)$
   b. $Fe(s) + AlCl_3(aq)$
   c. $NaCl(aq) + F_2(g)$
   d. $AgNO_3(aq) + Cu(s)$
   e. $Cl_2(g) + NiBr_2(aq)$

6. **T/I** Which chemical does not form a precipitate when it is added to a sodium phosphate solution?
   a. lead(II) nitrate
   b. calcium acetate
   c. silver chlorate
   d. potassium nitrate
   e. magnesium chloride

7. **T/I** Which observation would you not expect to make during the reaction of sodium with water?
   a. bubbling
   b. a precipitate forming
   c. a popping sound when testing with a burning splint
   d. the sodium disappearing
   e. red litmus turning blue

8. **K/U** Which of the following groups of metals contains only metals that cannot react to produce hydrogen gas during a displacement reaction?
   a. silver, gold, mercury
   b. copper, tin, zinc
   c. lithium, barium, calcium
   d. potassium, aluminum, platinum
   e. gold, silver, iron

9. **C** Which statement describes what happens when a precipitate forms?
   a. A metal displaces a less reactive metal from a solution.
   b. Positive ions switch places, and a compound breaks down.
   c. Negative ions switch places, and an insoluble compound forms.
   d. A halogen displaces iodine from a solution.
   e. An acid and a base react to form liquid water.

10. **K/U** Which ion is least likely to be found in a precipitate?
    a. iodide
    b. silver
    c. phosphate
    d. nitrate
    e. calcium

Use sentences and diagrams, as appropriate, to answer the questions below.

11. **K/U** Classify each displacement reaction based on the reactants shown.
    a. $LiCl(aq) + AgNO_3(aq) \rightarrow$
    b. $Ag(s) + CuNO_3(aq) \rightarrow$
    c. $Cl_2(g) + KI(aq) \rightarrow$
    d. $Pb(NO_3)_2(aq) + Na_2SO_4(aq) \rightarrow$

12. **T/I** Predict the products that are formed in each reaction in question 11, and write a balanced chemical equation for the reaction. If you predict that no reaction will occur, write "NR."

13. **C** Write instructions for using the periodic table to determine whether a single displacement reaction that involves halogens will occur.

14. **A** Several drops of sodium sulfate solution are added to a sample of water to test for contaminants. The possible contaminants in the water are sodium nitrate, calcium chloride, and lead(II) acetate.
   a. Which of these contaminants might be present if a precipitate forms?
   b. Write the balanced chemical equation for the reaction that involves this contaminant.

15. **T/I** Identify the errors in each equation, and then write the correct balanced chemical equation.
   a. $2K(s) + H_2O(aq) \rightarrow K_2O(aq) + H_2(g)$
   b. $2LiCl(aq) + Pb(NO_3)_2(aq) \rightarrow LiNO_3(s) + PbCl_2(aq)$

16. **C** While researching chemical reactions online, a student encounters a site that makes the following statement: "Double displacement reactions occur when the anions of two ionic compounds switch partners." Do you agree with this statement? Why or why not?

17. **K/U** Why is sulfur dioxide often a by-product of metal refining?

18. **K/U** Why can a single displacement reaction not occur in the reverse direction?

19. **T/I** The results of two laboratory tests are given below:
$$CX + B \rightarrow BX + C$$
$$A + CX \rightarrow \text{no reaction}$$
Use these results to write an activity series for metals A, B, and C.

20. **A** Hard water contains large amounts of dissolved calcium and magnesium ions. When soap is used with hard water, a solid soap scum forms on surfaces, such as sinks and bathtubs.
   a. What term describes the formation of soap scum?
   b. How would you classify the reaction that results in the formation of soap scum?

21. **C** Describe the process of neutralization in terms of the types of reactants used, the type of reaction that occurs, and the products that form.

22. **C** Describe what happens during leaching, and explain why leaching can be both helpful and harmful.

23. **A** Underwater photographers often use a rebreather to prevent bubbles from interfering with their work. A rebreather removes carbon dioxide from their exhaled air through a series of chemical reactions. The carbon dioxide is converted into carbonic acid, $H_2CO_3(aq)$, which then reacts with sodium hydroxide to form a product that reacts with calcium hydroxide.
   a. Write a balanced chemical equation for the reaction between carbonic acid and sodium hydroxide.
   b. What two terms can be used to describe this reaction?
   c. Write a balanced chemical equation for the final reaction in the rebreather.

24. **C** Give two meanings of the term "precipitation," and explain how each meaning is associated with chemical reactions.

25. **K/U** Which halogen could not be isolated during a displacement reaction?

Self-Check

If you missed question …	1	2	3	4	5	6	7	8	9	10	11	12	13	14	15	16	17	18	19	20	21	22	23	24	25
Review section(s)…	4.3	4.2	4.1	4.2	4.1	4.2	4.1	4.1	4.2	4.2	4.1, 4.2	4.1, 4.2	4.1	4.2	4.1, 4.2	4.2	4.3	4.1	4.1	4.2	4.2	4.3	4.2	4.2	4.1

Demonstrating Chemical Reactions at a Science Museum

Suppose you work as an assistant science educator at a science museum that hosts visiting school groups. You have been asked to assist in a chemistry workshop by demonstrating five types of chemical reactions—synthesis, decomposition, combustion, single displacement, and double displacement. With the help of your supervisor, you will develop and conduct the demonstrations.

First, you must figure out how to demonstrate each type of chemical reaction. Once you have identified the chemical reactions, you will then develop a step-by-step procedure with safety precautions and a detailed materials list for each demonstration. Finally, you will prepare a worksheet and answer key for the visiting students to complete as they observe your demonstrations.

How can you safely and effectively demonstrate the five major types of chemical reactions?

Initiate and Plan

1. Identify five simple, effective, and safe chemical reactions that you could perform to demonstrate the five reaction types. Assume that any materials and equipment in your chemistry classroom are available.

2. For each chemical reaction that you will demonstrate, list the chemicals and equipment that are required.

3. Write a detailed, step-by-step procedure that outlines how to perform each demonstration. Describe all safety precautions that must be followed and refer to all relevant Material Safety Data Sheet (MSDS) information. You should also include any suggestions or hints that will help the audience to better understand the material.

4. To add interest to each demonstration, research at least one example of an application for each of the five types of chemical reactions.

5. Your teacher will act as the supervisor in this scenario. Have your teacher check and approve your list of materials, procedures, and safety precautions.

6. Prepare the worksheet for students to fill out during each demonstration. Include question prompts and space for students to write notes about each chemical reaction. Students could

   • identify reactants

   • predict the reaction type and products before the reaction begins

   • record their observations

   • write and classify a balanced chemical equation for each reaction

   • demonstrate an understanding of the real-world application of each reaction

7. Prepare an answer key to accompany the student worksheet you prepared. Have your teacher check it for accuracy.

A common double displacement reaction occurs when commercially available antacids, which are basic, neutralize stomach acids that cause a burning sensation called heartburn.

Many medicines, such as acetylsalicylic acid (ASA), are commercially produced by synthesis reactions.

Perform and Record

8. With your teacher's approval, carry out a demonstration of each reaction for your own class as a practice run. Discuss observations and results with your classmates.

9. Have your classmates complete the student worksheet as you demonstrate the reactions.

Analyze and Interpret

1. Based on feedback from your classmates, were your demonstrations successful? How could you modify each demonstration to ensure better understanding of the five types of chemical reactions?

2. Indicate which reactions, if any, can be classified under more than one heading. For example, is there a combustion reaction that is also a synthesis reaction?

Communicate Your Findings

3. Make any necessary modifications based on your practice run. Then, compile the procedures, materials, safety precautions, and worksheet into a legible, detailed summary document that you could submit to your supervisor at the science museum. Include diagrams or digital photographs to make it as easy to follow as possible. Provide an assessment of how well your demonstration taught the topics based on your classmates' feedback from the practice run.

Assessment Criteria

Once you complete your project, ask yourself these questions. Did you …

☑ **K/U** identify and safely demonstrate one example of each type of chemical reaction?

☑ **T/I** write an appropriate procedure for each chemical reaction that included safety precautions and MSDS information?

☑ **A** identify a real-world application of each chemical reaction type?

☑ **C** prepare an appropriate worksheet for students to complete during the demonstration?

☑ **C** prepare a complete and accurate answer key for the student worksheet?

☑ **T/I** make any needed modifications based on feedback from classmates who viewed the practice demonstration?

☑ **C** prepare a detailed summary outlining the materials, procedures, and overall assessment of whether the demonstrations meet the teaching goals for students visiting a science museum?

☑ **C** prepare a document that is suitable for submission in a workplace?

☑ **K/U** use appropriate scientific terminology for the given audience and purpose of the demonstrations?

Over time, silver becomes tarnished. Tarnish is silver sulfide that forms on the surface. Cleaning solutions contain metals, such as aluminum, which undergo a single displacement reaction with the silver sulfide to remove the tarnish.

Hydrogen peroxide that consumers buy undergoes a decomposition reaction to produce oxygen gas and water. This is why hydrogen peroxide bottles have a "best before" date.

BIG IDEAS

- Chemicals react in predictable ways.
- Chemical reactions and their applications have significant implications for society and the environment.

Overall Expectations

In this unit you learned how to…

- **analyze** chemical reactions used in a variety of applications, and **assess** their impact on society and the environment
- **investigate** different types of chemical reactions
- **demonstrate** an understanding of the different types of chemical reactions

Chapter 3 — Synthesis, Decomposition, and Combustion Reactions

KEY IDEAS

- A chemical equation is a representation of a chemical reaction. A balanced chemical equation uses chemical formulas and coefficients, and reflects the law of conservation of mass, which states that matter is neither created nor destroyed in a chemical reaction. A balanced equation shows the same number of atoms of each element in the reactants and the products.
- In a synthesis reaction, two or more reactants combine to form a single product. This reaction can be represented by the general form $A + B \rightarrow AB$.
- A non-metal oxide reacts with water to form an acid in a synthesis reaction. A metal oxide reacts with water to form a metal hydroxide, which is a base.
- In a decomposition reaction, a single reactant breaks apart to form two or more products. This reaction can be represented by the general form $AB \rightarrow A + B$. A binary compound generally decomposes into its elements.

- A metal nitrate generally decomposes into a metal nitrite and oxygen gas.
- A metal carbonate generally decomposes into a metal oxide and carbon dioxide gas. A metal hydroxide generally decomposes into a metal oxide and water.
- A combustion reaction of a substance with oxygen produces one or more oxides. Energy, in the form of heat and light, is released.
- The products of complete combustion reactions are carbon dioxide and water vapour. The products of incomplete combustion reactions include carbon, carbon monoxide, carbon dioxide, and water vapour.

Chapter 4 — Displacement Reactions

KEY IDEAS

- In a single displacement reaction, one element replaces another element in a compound to produce a new element and a new compound.
- The activity series of metals and the activity series of halogens are used to predict whether a single displacement reaction will occur. A single displacement reaction can occur only when the uncombined element is higher in an activity series than the element it would replace in a compound.
- In a double displacement reaction, the positive ions in two compounds trade places to form two new compounds, including a precipitate, a gas, or water.
- Solubility guidelines can be used to predict whether a product of a double displacement reaction will be a precipitate.

- Neutralization occurs when a double displacement reaction happens between an acid and a base. Water is a product of neutralization.
- Chemical reactions of many different types are used in industry to extract and purify metals, including gold, copper, and magnesium; to produce other useful materials; and to prevent or clean up potentially harmful conditions caused by chemicals.
- Sulfur dioxide gas that is produced during industrial processes can be collected and converted into sulfuric acid through synthesis reactions to prevent its release into the atmosphere.

Knowledge and Understanding

Select the letter of the best answer below.

1. What type of reaction is likely to occur between two elements?
   a. double displacement
   b. decomposition
   c. single displacement
   d. synthesis
   e. neutralization

2. Which statement correctly describes a neutralization reaction?
   a. a synthesis reaction between an acid and a base
   b. a single displacement reaction between an acid and a base
   c. a single displacement reaction between an acid and a non-metal oxide
   d. a double displacement reaction between an acid and a base
   e. a double displacement reaction between a metal oxide and a base

3. Propane is a common portable fuel source for many lanterns and camping stoves, as in the photograph below. What is the balanced equation for the complete combustion of propane, $C_3H_8(g)$?
   a. $C_3H_8(g) + O_2(g) \rightarrow 3CO_2(g) + 4H_2O(g)$
   b. $C_3H_8(g) + O_2(g) \rightarrow C_3H_8O_2(g)$
   c. $C_3H_8(g) + 5O_2(g) \rightarrow 3CO_2(g) + 4H_2O(g)$
   d. $C_3H_8(g) + 2O_2(g) \rightarrow 3C(s) + 4H_2O(g)$
   e. $C_3H_8(g) + 3O_2(g) \rightarrow 3CO_2(g) + 4H_2(g)$

4. Which of the following metals is the most chemically active?
   a. copper
   b. gold
   c. lithium
   d. aluminum
   e. zinc

5. Which of the following reactants would likely undergo a single displacement reaction?
   a. $Na(s) + Cl_2(g) \rightarrow$
   b. $NaBr(aq) + Cl_2(g) \rightarrow$
   c. $NaBr(aq) + AgNO_3(aq) \rightarrow$
   d. $AgNO_3(s) \rightarrow$
   e. $C(s) + O_2(g) \rightarrow$

6. Which of the following metals will displace hydrogen from acids but not from cold water?
   a. copper
   b. lithium
   c. calcium
   d. gold
   e. aluminum

7. Which of the following statements about combustion reactions is true?
   a. Oxygen is a reactant in complete combustion reactions but not in incomplete combustion reactions.
   b. Carbon monoxide is a product in incomplete combustion reactions but not in complete combustion reactions.
   c. Water is a product in incomplete combustion reactions but not in complete combustion reactions.
   d. A hydrocarbon is a product in both complete combustion reactions and incomplete combustion reactions.
   e. Carbon dioxide is a product in complete combustion reactions but not in incomplete combustion reactions.

8. Which substance could be a precipitate in a double displacement reaction?
   a. $Cu(CH_3COO)_2(s)$
   b. $H_2O(\ell)$
   c. $BaSO_4(s)$
   d. $NaCl(s)$
   e. $CO_2(g)$

9. Which of the following materials could be used to neutralize a solution of dissolved carbon dioxide?
   a. $NaCl(aq)$
   b. $CaCl_2(s)$
   c. $HCl(aq)$
   d. $SO_3(g)$
   e. $CaO(s)$

10. Which substances are used in treating cyanide spills?
    a. $Zn(s)$ and $H_2SO_4(aq)$
    b. $NaOCl(aq)$ and $FeSO_4(aq)$
    c. $CaCO_3(s)$ and $HCl(aq)$
    d. $H_2O(\ell)$ and $O_2(g)$
    e. $Mg(OH)_2(s)$ and $NaOCl(\ell)$

Answer the questions below.

11. Complete the missing information in the following skeleton equation and balance the chemical equation:

$$NaOH(aq) + \underline{\quad} \rightarrow 3NaCl(aq) + Al(OH)_3(s)$$

12. A metal and a non-metal react in a synthesis reaction.
   a. Describe how electrons are involved in the reaction process.
   b. Identify the type of bond that forms.

13. Copy the following table and complete it by providing the type of compounds that decompose to form the indicated products.

Decomposition Products

Type of Compound	Products
	Metal oxide and carbon dioxide
	Two elements
	Metal nitrite and oxygen

14. What type of reaction do you expect to happen if electrolysis is used?

15. What type of reaction might be able to reverse a decomposition reaction?

16. Copy and balance each chemical equation.
   a. $Ca_3N_2(s) \rightarrow Ca(s) + N_2(g)$
   b. $Cr(s) + O_2(g) \rightarrow Cr_2O_3(s)$
   c. $CH_4(g) + O_2(g) \rightarrow CO_2(g) + H_2O(g)$
   d. $BaO(s) \rightarrow Ba(s) + O_2(g)$

17. Classify each of the reactions in question 16.

18. Classify each of the following reactions as a synthesis reaction, a decomposition reaction, a single displacement reaction, a double displacement reaction, or a combustion reaction. Explain your answers.
   a. $2Al(s) + 3CuCl_2(aq) \rightarrow 2AlCl_3(aq) + 3Cu(s)$
   b. $2H_2(g) + O_2(g) \rightarrow 2H_2O(g)$
   c. $2C_7H_5N_3O_6(s) \rightarrow$
      $3N_2(g) + 5H_2O(g) + 7CO(g) + 7C(s)$
   d. $3Mg(s) + N_2(g) \rightarrow Mg_3N_2(s)$
   e. $NaCl(aq) + AgNO_3(aq) \rightarrow AgCl(s) + NaNO_3(aq)$
   f. $2K_2O(\ell) \rightarrow 4K(\ell) + O_2(g)$

19. What types of reactions could *not* have oxygen as a reactant?

20. For each oxide listed below, state whether it would form an acid or a base upon reaction with water.
   a. $BaO(s)$
   b. $P_2O_3(s)$
   c. $CO_2(s)$
   d. $Li_2O(s)$

21. What type of compound can combine with water to form an acid?

22. What type of reaction is used in the final step to prepare each of the following metals?
   a. gold
   b. magnesium

Thinking and Investigation

23. The following reaction occurs in a closed vessel at high temperature:
   $$H_2(g) + I_2(g) \rightleftharpoons 2HI(g)$$
   Is this a synthesis reaction or a decomposition reaction? Explain your answer.

24. The images below show two flames that can result when using a burner in the laboratory. Describe each flame and identify which represents complete combustion and which represents incomplete combustion. Explain your reasoning.

25. An unknown solid is heated over a burner. During the reaction, a glowing splint is placed into the container and reignites.
   a. What gas is produced during the reaction? Explain your reasoning.
   b. What type of reaction is occurring in the container?
   c. What happens to the mass of the container and its contents as the container is heated?

26. A solid compound is decomposed by heating. The gas given off extinguishes a burning splint.
   a. What is the identity of the gas? Explain your reasoning.
   b. The gas is bubbled through water, and the resulting solution is tested with litmus paper. Describe the effect of the solution on litmus paper, and write a balanced chemical equation to support your answer.

27. A student completed the table below by identifying the reactions that are likely to happen, given the reactants listed. Which reaction type(s) did the student identify correctly?

	Reactants	Likely Reaction
I	$Be(s) + Cl_2(g)$	Synthesis
II	$C_4H_{10}(g) + O_2(g)$	Single displacement
III	$Br_2(\ell) + KCl(aq)$	Double displacement
IV	$NaNO_3(s)$	Decomposition

a. I only
b. II and IV only
c. I and IV only
d. III only
e. II and III only

28. Write a balanced chemical equation for each of the following reactions.
  a. Liquid benzene, $C_6H_6(\ell)$, reacts with oxygen gas to form carbon dioxide gas and liquid water.
  b. Carbon monoxide gas reacts with oxygen to form carbon dioxide gas.
  c. Solid sodium chloride and gaseous bromine form when chlorine gas reacts with solid sodium bromide.
  d. Solid calcium carbonate reacts to form solid calcium oxide and carbon dioxide gas.

29. Classify each of the reactions in question 28.

30. A single displacement reaction occurs between copper and silver nitrate. When 63.5 g of copper reacts with 339.8 g of silver nitrate, 215.8 g of silver is produced.
  a. What other product is formed?
  b. Write a balanced chemical equation for this reaction.
  c. What is the mass of the second product?

31. For an investigation, a student adds hydrochloric acid to two beakers. One beaker contains an aqueous solution of sodium nitrate and the other beaker contains an aqueous solution of silver nitrate. The results of these additions are shown below.

a. Based on the results, identify which beaker had the aqueous solution of sodium nitrate and which beaker had an aqueous solution of silver nitrate. Explain your answer.
b. Provide the name and chemical formula for the precipitate in beaker B.
c. Classify the reaction that occurred in beaker B.
d. Write a chemical equation showing the reaction that occurred in beaker B.
e. Explain why there is no apparent reaction in beaker A.

32. Predict whether each of the following reactions will occur in aqueous solutions. If you predict that a reaction will not occur, explain your reasoning. If a reaction does occur, write a balanced chemical equation for the reaction.
  a. sodium hydroxide + ammonium sulfate →
  b. niobium(V) sulfate + barium nitrate →
  c. strontium bromide + silver nitrate →

33. You must plan and conduct an investigation of a neutralization reaction.
  a. What types of chemicals will you use?
  b. What do you predict will be the end result of the reaction?
  c. What could you do to help demonstrate that you predicted the products correctly?

34. A student has a solution that contains one or more metal ions. The possible ions that might be present are silver ions and lead(II) ions. Solutions of sodium chloride and of sodium sulfate can be used to test for the presence of the metal ions. Write a procedure that would correctly identify which ions are present, and identify the observations that would indicate the presence of each ion.

Communication

35. Draw diagrams or build models to illustrate the process of balancing a chemical equation.

36. Create a Venn diagram to compare and contrast subscripts and coefficients in balanced chemical equations.

37. While balancing the chemical equation for the complete combustion of propene, $C_3H_6(g)$, a student writes the following: $C_3H_6(g) + O_2(g) \rightarrow 3CO_2(g) + 3H_2O(g)$. At this point, the student realizes that there is an odd number of oxygen atoms on the products side but there must be an even number of oxygen atoms on the reactants side. Write a procedure that describes how to finish balancing this equation.

38. The statement "Matter is neither created nor destroyed" is often used in discussions of the law of conservation of mass. Describe how this statement applies to atoms during the process of balancing chemical equations.

39. Draw a graphic organizer that illustrates how to classify a chemical reaction based solely on the reactants.

40. **BIG IDEAS** Chemicals react in predictable ways. Write a procedure for testing the acidity of a solution. Include all materials, safety cautions, and expected results.

41. **BIG IDEAS** Chemical reactions and their applications have significant implications for society and the environment. Draw a diagram to illustrate the sources of compounds that result in the formation of acid precipitation and the reactions that are involved.

42. The general forms of various types of chemical reactions are shown below.

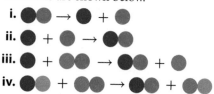

a. Identify each type of chemical reaction.
b. Describe each chemical reaction in terms of the reactants and products involved.

43. How would you explain to a classmate why you are more likely to consult a solubility table for double displacement reactions than for single displacement reactions?

44. Write a set of instructions for using solubility guidelines to determine whether either product of a double displacement reaction is a precipitate.

45. While researching chemical reactions on-line, you see a posting in a forum by someone looking for tips on using sodium metal to recover silver from aqueous solutions of silver nitrate. Write a response to this person's post that demonstrates your understanding of sodium and its reactivity. Include information about the reaction the person is considering, and whether you think the proposed process is feasible. Include balanced chemical equations to illustrate your points.

Application

46. Effervescent tablets contain two chemicals that produce bubbles when placed in water, as shown below. Is it correct to identify these chemicals as solids in the balanced chemical equation? Explain your answer.

47. Both incomplete combustion and complete combustion pose risks. Identify the potential problem caused by each and the chemical that poses the risk.

48. Suppose you want to generate magnesium oxide.
a. What are two decomposition reactions you could use?
b. Write a balanced chemical equation for each decomposition reaction.

49. Acid precipitation can leach metals, such as zinc, from the soil. This can result in elevated levels of metal ions in lakes and rivers.
a. What type of reaction could be used to remove these ions?
b. What hazards could the treatment itself pose?

50. Liming is one way to help renew an acidified lake. Liming involves adding crushed limestone (calcium carbonate) to the water, as shown in the photograph below.

a. Write a balanced chemical equation for the reaction that happens between the limestone and sulfuric acid, which is a major component of acid precipitation.

b. What harm could result from liming?

51. Catalytic converters installed in cars, trucks, and newer-model motorcycles play an important role in reducing the emission of harmful gases into the atmosphere. Research and classify the reactions that take place in a catalytic converter.

52. The process of iron rusting involves the release of heat. Normally, this occurs slowly and the heat that is released serves no useful purpose. However, some single-use instant heat packs, such as the one shown in the photograph below, contain a mixture of materials that speed up the rusting process and, therefore, speed up the generation of heat. Although several steps are involved, the overall reaction in these packs is the formation of iron(III) oxide from iron powder and oxygen.

a. Write a balanced chemical equation for the overall reaction in the heat pack.

b. Explain how you would classify this reaction, and give reasons for your answer.

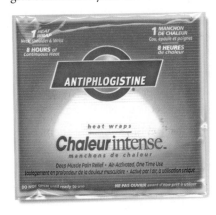

53. Two Canadian scientists, H.O. Pritchard and P.Q.E. Clothier, published a paper in which they suggested that the compound di-tert-butyl peroxide (DTBP) be used as a fuel in internal combustion engines. If oxygen is plentiful, DTBP can undergo complete combustion. If there is insufficient oxygen for combustion, DTBP can undergo a decomposition reaction and power the same engine. Research the chemical formula for DTBP and write balanced equations for the complete combustion of DTBP and for the decomposition of DTBP. Why is a fuel that would power an engine in the absence of oxygen useful?

54. During the isolation and purification of nickel, the nickel can be separated from other metals present. This is done through a synthesis reaction between nickel and carbon monoxide gas, which forms $Ni(CO)_4(g)$.

a. Write the balanced chemical equation for this reaction.

b. What type of reaction could be a source of the carbon monoxide reactant?

c. The purified nickel is recovered by decomposing the $Ni(CO)_4(g)$. Why would recycling the carbon monoxide gas help to make the refinement process more efficient and safe?

55. Although copper can be found in its elemental form in nature, as shown in the photograph below, it is often extracted from carbonate or oxide ores.

a. Why is copper found in its elemental form but iron is not?

b. One method of extracting copper from its oxide carbonate ore involves the reaction of copper(II) carbonate with dilute sulfuric acid. The resulting copper compound is then reacted with iron, which forms an iron(II) compound. Write a balanced chemical equation for each of these reactions.

56. On December 6, 1917, an explosion rocked the port city of Halifax, Nova Scotia. This explosion claimed the lives of 2000 people and injured about 9000 others. Do research to investigate the causes and consequences of this chapter in Canadian history.

a. What chemical reactions were involved in the explosion, and how did the reactants come to be chemically combined?

b. What were the economic and social costs of the explosion?

c. What was done to address the immediate and the long-term effects?

Select the letter of the best answer below.

1. **K/U** Which of the following equations represents a synthesis reaction?
   a. $AC + BD \rightarrow AD + BC$
   b. $A + BC \rightarrow AC + B$
   c. $AB \rightarrow A + B$
   d. $A + B \rightarrow AB$
   e. $AD + BC \rightarrow ABC + D$

2. **A** Which statement best describes the use of scrubber technology to reduce harmful emissions that could lead to acid precipitation?
   a. decomposition reactions break down a non-metal oxide
   b. synthesis reactions trap a non-metal oxide
   c. synthesis reactions trap a metal oxide
   d. single displacement reactions replace the metal from a metal oxide
   e. combustion reactions produce a non-metal oxide

3. **K/U** Two elements are produced during a reaction. What type of reaction has occurred?
   a. complete combustion
   b. single displacement
   c. synthesis
   d. incomplete combustion
   e. decomposition

4. **K/U** Which of the following single displacement reactions will occur?
   a. $NaCl(aq) + Al(s) \rightarrow$
   b. $KBr(aq) + Cl_2(g) \rightarrow$
   c. $HCl(aq) + Ag(s) \rightarrow$
   d. $CaCl_2(aq) + I_2(g) \rightarrow$
   e. $AgNO_3(aq) + Au(s) \rightarrow$

5. **T/I** How does a synthesis reaction differ from a decomposition reaction?

6. **K/U** Which of the following is a correct description of a chemical reaction?
   a. A metal oxide reacts with water to form an acid in a synthesis reaction.
   b. A metal displaces a more active metal from a compound in a single displacement reaction.
   c. A metal carbonate forms a metal oxide and carbon dioxide in a decomposition reaction.
   d. A hydrocarbon breaks down to form carbon dioxide and water in a decomposition reaction.
   e. Two metals form a binary compound in a synthesis reaction.

7. **T/I** Complete the following analogy: solubility guidelines are to double displacement reactions as activity series are to _____.
   a. complete combustion reactions

   b. single displacement reactions
   c. synthesis reactions
   d. incomplete combustion reactions
   e. decomposition reactions

8. **K/U** Which substance remains in aqueous solution in a double displacement reaction?
   a. PbI_2
   b. $BaSO_4$
   c. $(NH_4)_3PO_4$
   d. CO_2
   e. FeS

9. **K/U** Solubility guidelines are used to determine whether which type of reaction will occur?
   a. double displacement
   b. decomposition
   c. single displacement
   d. synthesis
   e. combustion

10. **T/I** Which of the following is the balanced chemical equation for the neutralization reaction between hydrochloric acid and calcium hydroxide?
    a. $2HCl(aq) + Ca(OH)_2(aq) \rightarrow CaCl_2(aq) + 2H_2O(\ell)$
    b. $2HClO_3(aq) + Ca(OH)_2(aq) \rightarrow$
    $Ca(ClO_3)_2(aq) + 2H_2O(\ell)$
    c. $H_2Cl(aq) + 2CaOH(aq) \rightarrow 2CaCl(aq) + 2H_2O(\ell)$
    d. $HCl(aq) + CaOH(aq) \rightarrow CaCl(aq) + H_2O(\ell)$
    e. $2HClO_4(aq) + Ca(OH)_2(aq) \rightarrow$
    $Ca(ClO_4)_2(aq) + 2H_2O(\ell)$

Use sentences and diagrams as appropriate to answer the questions below.

11. **C** Describe the difference between coefficients and subscripts and whether each can be manipulated when balancing chemical equations.

12. **T/I** Balance each of the following chemical equations.
    a. $Br_2(\ell) + NaI(aq) \rightarrow NaBr(aq) + I_2(s)$
    b. $Al(s) + Cu(NO_3)_2(aq) \rightarrow Cu(s) + Al(NO_3)_3(aq)$
    c. $Fe_2O_3(s) \rightarrow Fe(s) + O_2(g)$
    d. $Cl_2(g) + NaBr(aq) \rightarrow NaCl(aq) + Br_2(\ell)$
    e. $Li(s) + N_2(g) \rightarrow Li_3N(s)$
    f. $AgNO_3(aq) + CaCl_2(aq) \rightarrow AgCl(s) + Ca(NO_3)_2(aq)$

13. **A** When developing a chemical manufacturing process, scientists and engineers must consider all possible methods for producing a desired substance. A single compound can be the product of many different types of reactions. Each type of reaction requires different reactants and produces different by-products to form the same substance. Why might using a synthesis reaction to produce the desired substance be the best choice when considering the impact of industrial processes on the environment?

14. **K/U** Describe the process of neutralization.

15. **T/I** Predict the products and write a balanced chemical equation for each reaction.
  a. the complete combustion of $C_4H_8(g)$
  b. $Al(s) + Br_2(\ell) \rightarrow$
  c. $RbNO_3(s) \rightarrow$

16. **A** The effort to isolate alkali metals using electrolysis was often hazardous.
  a. Write the balanced chemical equation for the electrolysis of molten lithium chloride.
  b. Describe the hazardous condition that would result if water contaminated the electrolysis apparatus. Write a balanced chemical equation to support your answer.

17. **C** Describe each of the following types of reactions in terms of the reactants and products involved.
  a. decomposition reaction
  b. double displacement reaction

18. **A** Silicon is an important element in the manufacture of semiconductors for computers, electronic devices, and photovoltaic cells such as the ones shown below. Liquid silicon can be produced from the reaction of silicon dioxide, the main component of sand, with solid carbon. Carbon dioxide is the other product.

  a. Write a balanced chemical equation for this reaction.

b. How would you classify this reaction? Explain your reasoning.

c. Why would a process that produces silicon from silicon dioxide through decomposition be considered more environmentally friendly than the reaction with carbon?

19. **K/U** What types of compound can decompose to produce an oxide that forms a basic solution when added to water?

20. **A** Lakes that have been acidified through acid precipitation are sometimes limed, or treated with crushed limestone. Limestone is mostly calcium carbonate.
  a. What is the purpose of liming lakes?
  b. Write a balanced chemical equation to show the effect of adding limestone to a lake that contains sulfuric acid.

21. **T/I** Describe a method you could use to identify an unknown solid as a metal carbonate or a metal nitrate.

22. **A** Why is reducing non-metal oxide emissions an effective way to reduce acid precipitation?

23. **T/I** Predict the products of and write a balanced chemical equation for each of these displacement reactions. If no reaction will occur, write "No reaction."
  a. $MgS(aq) + Cu(NO_3)_2(aq) \rightarrow$
  b. $Cu(s) + Mg(NO_3)_2(aq) \rightarrow$
  c. $Br_2(g) + KI(aq) \rightarrow$
  d. $KNO_3(aq) + Na_3PO_4(aq) \rightarrow$

24. **K/U** Compare the ions that are produced by acids and bases in aqueous solutions.

25. **T/I** If both reactants in a double displacement reaction are composed of alkali metals, what can you conclude about the products?

Self-Check

If you missed question …	1	2	3	4	5	6	7	8	9	10	11	12	13	14	15	16	17	18	19	20	21	22	23	24	25
Review section(s)…	3.2	3.2	3.2	4.1	3.2	3.2, 4.1, 4.2	3.2, 3.3, 4.1, 4.2	4.2	4.2	4.2	3.1	3.1	3.1, 3.2	4.2, 4.3	3.2, 3.3	4.1	3.2, 4.2	4.1	3.2	3.2, 4.2	3.2	3.2, 4.3	4.1, 4.2	4.2	4.2

Quantities in Chemical Reactions

BIG IDEAS

- Relationships in chemical reactions can be described quantitatively.

- The efficiency of chemical reactions can be determined and optimized by applying an understanding of quantitative relationships in such reactions.

Overall Expectations

In this unit, you will ...

- **analyze** processes in the home, the workplace, and the environmental sector that use chemical quantities and calculations, and **assess** the importance of quantitative accuracy in industrial chemical processes

- **investigate** quantitative relationships in chemical reactions, and solve related problems

- **demonstrate** an understanding of the mole concept and its significance to the quantitative analysis of chemical reactions

Unit 3 Contents

What do koalas and pharmaceutical companies have in common? They both like eucalyptus trees. Koalas eat eucalyptus leaves almost exclusively. Pharmaceutical companies have developed products that contain the oil of eucalyptus leaves to help relieve the symptoms of colds and influenzas. Quantitative accuracy is crucial in the development and use of consumer products. For example, it is safe for people to ingest oil of eucalyptus at the low dosages recommended on cold and flu products. However, oil of eucalyptus can be deadly at higher dosages and in pure form.

Understanding chemical quantities is important because many processes that are used in the home, workplace, and environmental sector involve chemical quantities and calculations. In this unit, you will learn more about chemical quantities and the importance of quantitative accuracy in chemical reactions and processes. As well, you will have an opportunity to carry out investigations and solve problems related to chemical quantities.

As you study this unit, look ahead to the **Unit 3 Project** on pages 336 to 337. The Unit 3 Project will give you an opportunity to demonstrate and apply your new knowledge and skills. Keep a planning folder so that you can complete the project in stages as you progress through this unit.

Preparation

Ionic and Molecular Compounds

- Ionic compounds form as a result of the transfer of electrons from metal atoms to non-metal atoms.
- The name of an ionic compound identifies the positive and negative ions that make up the compound. To write the chemical formula of an ionic compound, you must use subscripts so that the sum of the ions' charges is zero.

- Molecular compounds form when electrons are shared by two or more non-metal atoms.
- Each prefix in the name of a molecular compound identifies the number of atoms of an element that must be shown when writing the chemical formula of the compound. For example, based on the prefixes in the name dinitrogen tetrahydride, the chemical formula is $N_2H_4(\ell)$.

1. Which of the following is a molecular compound? Choose the letter of the correct answer.
   a. $Mg(NO_3)_2(s)$ **d.** $P_2S_3(s)$
   b. $BaS(s)$ **e.** $Li_2SO_4(aq)$
   c. $CaO(s)$

2. What is the name of the compound $SiO_2(s)$? Choose the letter of the correct answer.
   a. silicon oxide   **d.** silicon dioxide
   b. monosilicon oxide   **e.** silicon dioxygen
   c. monosilicon dioxide

3. Explain how you could distinguish an ionic compound from a molecular compound based on its name or chemical formula.

4. State the number of atoms of each element present in a compound with the chemical formula $Ca_3(PO_4)_2(s)$.

5. Identify each compound as ionic or molecular, and then write its chemical formula.
   a. lithium iodide   **d.** dinitrogen pentoxide
   b. carbon tetrafluoride   **e.** aluminum trihydride
   c. tin(IV) sulfide   **f.** chlorine trifluoride

Chemical Reactions

- During a chemical reaction, a new substance is produced. The new substance has a different composition and different properties than the starting materials.
- Evidence of a chemical change may include one or more of the following: formation of a gas, formation of a precipitate, change in odour, change in colour, or change in energy.

- A reaction is often classified as a synthesis, decomposition, single displacement, or double displacement reaction based on how the reactant(s) are changed into the product(s).
- A reaction can also be classified based on the specific reactants and products involved. A combustion reaction can be identified in this way.

6. Which of the following observations is the most reliable indication that a chemical reaction has occurred? Choose the letter of the correct answer.
   a. A colourless solid changes into a colourless liquid with the addition of heat.
   b. Two clear, colourless solutions are mixed, resulting in a milky-white, cloudy solution.
   c. A metallic solid is shaped into a thin, flat sheet when it is pounded with a hammer.
   d. A brown, opaque liquid solidifies into a hard, brown, opaque solid as the temperature decreases.
   e. A powdery white solid is mixed with a clear liquid, resulting in a clear, colourless liquid.

7. Match each type of reaction to the correct format below.
   Types of Reactions
   I. synthesis
   II. decomposition
   III. single displacement
   IV. double displacement
   V. combustion

   Formats
   a. $A + BC \rightarrow B + AC$
   b. fuel + oxygen $\rightarrow$ oxides of the fuel
   c. $DEF \rightarrow D + E + F$
   d. $WX + YZ \rightarrow WZ + YX$
   e. $F + G \rightarrow FG$

8. What evidence of a chemical reaction can you identify in each image?

a.

b.

9. Describe each type of reaction in terms of the reactants and products involved.
  a. synthesis reaction
  b. single displacement reaction
  c. decomposition reaction
  d. double displacement reaction
  e. complete combustion reaction
  f. incomplete combustion reaction

10. Classify each reaction as a synthesis reaction, a decomposition reaction, a single displacement reaction, or a double displacement reaction.
  a. $2Al(s) + 3CuCl_2(aq) \rightarrow 2AlCl_3(aq) + 3Cu(s)$
  b. $2H_2(g) + O_2(g) \rightarrow 2H_2O(g)$
  c. $2C_7H_5N_3O_6(s) \rightarrow$
  $\qquad\qquad 3N_2(g) + 5H_2O(g) + 7CO(g) + 7C(s)$
  d. $3Mg(s) + N_2(g) \rightarrow Mg_3N_2(s)$
  e. $NaCl(aq) + AgNO_3(aq) \rightarrow AgCl(s) + NaNO_3(aq)$
  f. $2K_2O(\ell) \rightarrow 4K(\ell) + O_2(g)$

Balanced Chemical Equations

- A balanced chemical equation represents a chemical reaction using symbols and chemical formulas.
- The law of conservation of mass states that in a chemical reaction, the total mass of the products is equal to the total mass of the reactants.

- Coefficients are used to balance an equation to reflect the law of conservation of mass. The same number of each kind of atom must appear in both the reactants and the products.

---

11. Which of the following is a balanced chemical equation? Choose the letter of the correct answer.
  a. $C_7H_{16}(\ell) + O_2(g) \rightarrow 7CO_2(g) + H_2O(g)$
  b. $C_7H_{16}(\ell) + 11O_2(g) \rightarrow 7CO_2(g) + 8H_2O(g)$
  c. $2C_7H_{16}(\ell) + O_2(g) \rightarrow 14CO_2(g) + 16H_2O(g)$
  d. $C_7H_{16}(\ell) + 7O_2(g) \rightarrow 7CO_2(g) + H_2O(g)$
  e. $C_7H_{16}(\ell) + O_2(g) \rightarrow 7CO_2(g) + 8H_2O(g)$

12. How many atoms of nitrogen are represented in the expression, $5Ca(NO_3)_2(s)$, which is a term from a chemical equation? Choose the letter of the correct answer.
  a. 1   **d.** 10
  b. 2   **e.** 30
  c. 5

13. In your own words, explain why it is necessary to balance chemical equations.

14. Your class is given the following equation to balance:
  $H_2SO_4(aq) + Al(s) \rightarrow Al_2(SO_4)_3(aq) + H_2(g)$
  A fellow student has attempted to balance the equation as follows:
  $H_2(SO_4)_3(aq) + 2Al(s) \rightarrow Al_2(SO_4)_3(aq) + H_2(g)$
  What error has the student made in balancing the equation?

15. What is the correct balanced equation for the equation given in question 14?

16. Balance each chemical equation.
  a. $Br_2(\ell) + NaI(aq) \rightarrow NaBr(aq) + I_2(s)$
  b. $Al(s) + Cu(NO_3)_2(aq) \rightarrow Cu(s) + Al(NO_3)_3(aq)$
  c. $Fe_2O_3(s) \rightarrow Fe(s) + O_2(g)$
  d. $C_5H_{12}(\ell) + O_2(g) \rightarrow H_2O(g) + CO_2(g)$
  e. $Li(s) + N_2(g) \rightarrow Li_3N(s)$
  f. $AgNO_3(aq) + CaCl_2(aq) \rightarrow AgCl(s) + Ca(NO_3)_2(aq)$

Predicting Chemical Equations

- To predict the products in a chemical reaction, it is necessary to know the types of chemical equations.
- The activity series of metals and the activity series of halogens are useful in predicting whether a single displacement reaction will occur.
- A complete chemical equation includes the chemical formulas of the elements and compounds, the state of each substance, and the correct balancing coefficients.
- A word equation is sometimes used to describe a chemical reaction.

17. Complete and balance each chemical equation. Indicate the type of reaction.

  a. $Fe(s) + O_2(g) \rightarrow$

  b. $CaCl_2(aq) + NH_4OH(aq) \rightarrow$

  c. $C_3H_8(g) + O_2(g) \rightarrow$

18. Write a complete balanced equation, including states, for each reaction described below. Then describe the evidence that indicates a reaction has occurred.

  a. Zinc metal reacts with solid yellow sulfur when the two elements are mixed.

  b. Carbonic acid in a soft drink spontaneously decomposes to form carbon dioxide and water when the bottle cap is removed.

  c. A precipitate is formed when solutions of barium chloride and potassium chromate are mixed.

  d. Hydrogen gas is burned as a fuel in many types of spacecraft.

19. Balance each chemical equation, write a word equation for the reaction, and indicate the type of reaction.

  a. $KClO_3(s) \rightarrow KCl(s) + O_2(g)$

  b. $Al(s) + H_2SO_4(aq) \rightarrow H_2(g) + Al_2(SO_4)_3(aq)$

  c. $Ba(OH)_2(aq) + H_3PO_4(aq) \rightarrow$
  $$H_2O(\ell) + Ba_3(PO_4)_2(s)$$

20. Which of the following equations represents a single displacement reaction that will occur?

  a. $2NaOH(aq) + Mg(s) \rightarrow 2Na(s) + Mg(OH)_2(aq)$

  b. $2KF(s) + Cl_2(g) \rightarrow 2KCl(s) + F_2(g)$

  c. $HgCl_2(s) + Fe(s) \rightarrow FeCl_2(s) + Hg(\ell)$

  d. $3AgNO_3(s) + Au(s) \rightarrow Au(NO_3)_3(s) + 3Ag(s)$

  e. $Pt(s) + H_2SO_4(aq) \rightarrow PtSO_4(aq) + H_2(g)$

Tests for Identifying Products

- A wooden splint is used to identify hydrogen, oxygen, and carbon dioxide gases. A burning splint causes a popping sound with hydrogen and is extinguished in carbon dioxide. A glowing splint reignites in oxygen.
- Indicators change colour in the presence of acids and bases. For example, litmus paper is red in an acidic solution and blue in a basic solution.

21. You perform a burning splint test on the products of a chemical reaction, and find that the flame disappears and the splint begins to smoke. Which of the following reactions most likely occurred?

  a. $Mg(s) + 2HCl(aq) \rightarrow MgCl_2(aq) + H_2(g)$

  b. $H_2CO_3(aq) \rightarrow CO_2(g) + H_2O(\ell)$

  c. $2Na(s) + 2H_2O(\ell) \rightarrow 2NaOH(aq) + H_2(g)$

  d. $2HgO(s) \rightarrow Hg(\ell) + O_2(g)$

  e. $CO_2(g) + H_2O(\ell) \rightarrow H_2CO_3(aq)$

22. Look at all of the products of the reactions in question 21. Would any of these products cause litmus paper to change colour? Describe any colour change(s) that you think would occur.

23. Define the term "neutralization" with respect to the reaction between an acid and a base.

24. Compare the ions that are characteristic of acids and bases.

25. How could you use litmus paper to determine when a neutralization reaction is complete?

In the following section, you will

- distinguish between accuracy and precision
- convert formal notation into scientific notation, and vice versa
- solve for a variable in a mathematical equation
- solve a problem that involves proportions
- round off a measurement to the proper number of significant digits

- solve multi-step problems that include addition, subtraction, multiplication, and division
- use the rules for significant digits correctly in mathematical operations
- solve percent calculations
- read measurements and assess the uncertainty of each measurement

26. A student was practising target shooting and made three sets of shots, as shown below in the targets. For each target, describe whether the shots demonstrated low or high precision, and low or high accuracy.

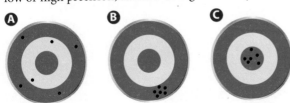

27. How many significant digits does each measurement contain?
    a. 3.64 cm
    b. 11.0 g
    c. 0.0016 L
    d. 0.000 504 0 km
    e. 70 mL

28. Round each measurement to the number of significant digits given in brackets.
    a. 1.864 93 kg (3)
    b. 0.449 99 L (1)
    c. 0.645 87 m (2)
    d. 32 692.7 kg (4)
    e. 15.250 J (3)
    f. 0.007 958 km (2)

29. Which of the following descriptions refer to exact numbers?
    a. The elevation of Brampton, Ontario, is 221 m.
    b. There are 12 eggs in one dozen.
    c. One foot is equal to 0.3048 m.
    d. One kilometre is equal to 1000 m.
    e. The attendance at a hockey game was 16 487.

30. Express each number in scientific notation and to two significant digits.
    a. 238.6
    b. 0.005 28
    c. 4.150
    d. 1 604 387.2
    e. 0.000 000 000 200

31. A package of lean ground beef weighs 4.82 lb. If there are 2.21 lb in a kilogram, what is the mass of the package in kilograms?

32. Perform the indicated mathematical operation in each expression, and round your answer to the correct number of significant digits.
    a. 3.894 m × 2.16 m
    b. $\dfrac{4.8 \times 10^2 \text{ g}}{9.231 \text{ mL}}$
    c. 2.96 L + 8.1 L + 5.0214 L
    d. 3 880.95 cm + 0.24 km + 2.86 m
    e. 2.46 kg/pup × 7 pups
    f. 9.146 Mg − 91.37 kg

33. Perform each calculation, and round your answer to the correct number of significant digits. Assume that all the numbers were obtained by measurement.
    a. 0.022 mg/kg × 315 kg
    b. $\dfrac{(33.4 \text{ m} + 112.74 \text{ m} + 0.008 \text{ m})}{(6.488 \text{ s}^2)}$
    f. $\dfrac{(101.3 \text{ kPa} \times 22.4 \text{ L})}{(1.00 \text{ mol} \times 273.15 \text{ K})}$
    d. $\dfrac{(315.44 \text{ kg} - 208.1 \text{ kg}) \times 8.8175 \text{ m}}{(3.16 \text{ s} \times 2.1 \text{ s})}$
    e. $\dfrac{(7.6 \times 10^3 \text{ m} + 8.16 \times 10^3 \text{ m})}{(3.85 \times 10^4 \text{ s} - 2.1 \times 10^4 \text{ s})}$

34. a. If you received 35 out of 43 on a test, what percentage of answers did you get correct?
    b. Suppose that the regular price of an item you want to buy is $15.99 and the item is on sale for 30% off. What is the sale price of the item? If you had to add 13% H.S.T., how much would you pay for the item?

35. Solve each equation for x.
    a. $x = 3 + 4 \times 8$
    b. $3x = \dfrac{18}{3}$
    c. $5 = \dfrac{2}{x}$
    d. $6 = \dfrac{2x}{4}$
    e. $16 = \dfrac{x - 5}{10}$
    f. $26 = 3(4) + \dfrac{(x)2^2}{2}$

Specific Expectations

In this chapter, you will learn how to...

- D1.1 **analyze** processes in the home, workplace, and environmental sector that involve the use of chemical quantities and calculations (5.1)

- D2.1 **use** appropriate terminology related to quantities in chemical reactions (5.1, 5.2)

- D2.3 **solve** problems related to quantities in chemical reactions by performing calculations that involve quantities in moles, number of particles, and atomic mass (5.2)

- D3.2 **describe** the relationships between the Avogadro constant, the mole concept, and the molar mass of any given substance (5.1, 5.2)

In everyday life, you often measure a mass or count the number of items you need. For example, you might buy a dozen eggs to make an egg salad or you might buy 0.5 kg of rice to cook for dinner. The method you use to buy a quantity is usually based on convenience. For example, it is not practical to count a specific number of rice grains, but it is practical to measure a mass of rice grains.

Chemists also count items, but the items they count are very small. Recall that chemical reactions occur among individual particles, such as atoms, ions, molecules, and formula units. So, chemists need to know how many reactant particles are present to determine how many new particles are formed during a chemical reaction. Because these numbers are usually large, scientists use a unit that contains many particles when specifying the amounts of substances in a chemical reaction. Because it is impractical to count individual particles in a laboratory, chemists have developed a unit they can use to count particles by taking their mass. In this chapter, you will learn about this new unit and how to use it.

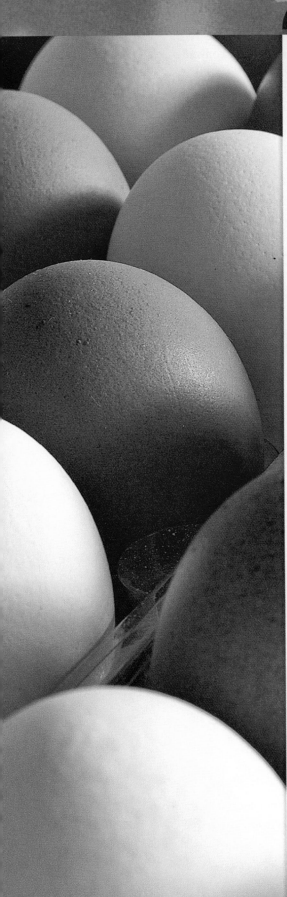

Quantitative Accuracy Matters

In this activity, you will combine baking soda, NaHCO$_3$(s), and road salt, CaCl$_2$(s), two seemingly safe chemicals. How will the quantities you use affect the results of the reaction?

Safety Precautions

- Use caution when performing this activity. The bag could burst after the chemicals have been mixed. Do not stand over the bag while the reaction is occurring.
- Calcium chloride is an irritant. Wash your hands immediately if they touch the calcium chloride.
- Wear safety eyewear and a lab coat or apron throughout this activity.

Materials

- clear, resealable plastic bag
- 1 scoop of sodium hydrogen carbonate (baking soda), NaHCO$_3$(s)
- 1 scoop of calcium chloride (road salt), CaCl$_2$(s)
- 10 drops of phenol red or bromothymol blue indicator solution
- scoopula
- 10 mL graduated cylinder
- small plastic vial

Procedure

1. Create a table to record your observations.
2. Place one scoopula (5 mL) of baking soda in one corner of the plastic bag. Place one scoopula (5 mL) of road salt in the other corner of the bag.
3. Put 10 drops of indicator solution in the small vial.
4. Carefully stand the vial in the bag, between the two substances.
5. Seal the bag. Holding the sealed bag, pour some of the indicator solution in the vial onto the substance in each corner of the bag. Feel and observe each corner of the bag. Record your observations.
6. Mix the contents of the bag completely. Record your observations.
7. Repeat steps 1 to 6 using different small quantities of road salt and baking soda. For example, try using half a scoop of baking soda and one scoop of road salt, or one scoop of baking soda and three scoops of road salt.

Questions

1. What did you observe when you poured the indicator solution onto the substance in each corner of the bag?
2. What happened once the contents of the bag were mixed completely?
3. What happened when you changed the quantities of the chemicals?
4. What does this activity demonstrate about the importance of the quantities of chemicals used in a reaction and how they relate to safety?

The Mole and the Avogadro Constant

Key Terms

mole

Avogadro constant

People have learned to group things in different ways for convenience and efficiency, both at home and in the workplace. Eggs are sold by the dozen. Recordable DVDs are often sold in packs of 50 or 100.

Chemists work with large quantities of atoms and molecules on a daily basis. In this section, you will learn how to count and measure these large quantities in order to conduct research.

Grouping Items for Convenience

Table 5.1 shows how various consumer items are organized into convenient units that represent groups of items. Each unit can be used for different kinds of items, but the unit is always constant. For example, a dozen can be used to describe items such as doughnuts, flowers, and eggs, but a dozen always means 12. Paper, like the kind shown in **Figure 5.1**, is often sold in reams. Unlike some of the other quantities, such as a pair or a dozen, a ream is generally used only to describe paper products.

Table 5.1 Common Units

Item	Unit	Number
socks	pair	2
eggs	dozen	12
pencils	gross (12 dozen)	144
paper	ream (one package of sheets of paper)	500

Figure 5.1 Paper is counted by the ream. Each ream contains 500 sheets of paper.

Describe *What are other counting units for groups of items?*

The Mole

Items such as those listed in **Table 5.1** are easy to count because they are macroscopic (large enough to see). Chemists also use a convenient unit when counting individual particles (atoms, molecules, ions, and formula units) of a substance. These entities, however, are obviously too small and numerous to count directly. For convenience, chemists use a special counting unit, called the mole, to count and measure individual particles of substances.

Using the Mole to Count Particles

The **mole** is the SI base unit that is used to measure the amount of a substance. One mole (mol) of a substance is the amount of the substance that contains as many particles (atoms, molecules, ions, or formula units) as the number of atoms in exactly 12 g of the isotope carbon-12, or 6.022 141 79 × 10²³ particles of the substance. This value is called the **Avogadro constant**, named in honour of Italian scientist Amedeo Avogadro (1776–1856). Its symbol is N_A.

The value of the Avogadro constant is usually rounded to 6.02×10^{23} for convenience. Scientists continue to develop more accurate methods for determining the experimental value of the Avogadro constant. The value has changed somewhat since it was first proposed more than 100 years ago. Even though the value is experimental, chemists use the Avogadro constant to determine the number of representative particles, such as atoms, molecules, ions and formula units, in a substance. For example, if you have one mole of iron atoms, you have 6.02×10^{23} atoms of Fe(s). One mole of water represents 6.02×10^{23} molecules of $H_2O(\ell)$. Also, one mole of sodium chloride contains 6.02×10^{23} formula units of NaCl(s).

The Avogadro constant can describe any substance, just as a dozen can describe any item. Chemists work with moles in a way that is similar to how a grocer might work with dozens of oranges. For example, three dozen oranges is the same as 36 oranges.

$$3 \text{ dozen} \times \frac{12}{1 \text{ dozen}} = 36$$

In comparison, chemists might work with the fructose (fruit sugar), $C_6H_{12}O_6(s)$, molecules in the oranges. They could calculate the number of fructose molecules in three moles (3.0 mol) of fructose in a similar way.

$$3.0 \text{ mol} \times \frac{6.02 \times 10^{23} \text{ molecules}}{1 \text{ mol}} = 1.8 \times 10^{24} \text{ molecules}$$

What Is a "Particle"?

Chemists need to "count" different types of particles. It is important to know what kind of particle is being discussed. For example, when a problem refers to oxygen, $O_2(g)$, you usually want to find out the number of molecules. However, there might be a reason you would want to count the number of oxygen atoms, O. The representative particles for different elements and compounds are listed below.

Representative Particles for Elements and Compounds
- The representative particles for pure, monatomic elements, such as iron, Fe(s), are *atoms*.
- The representative particles for diatomic molecules, such as oxygen, $O_2(g)$, and compounds, such as water, $H_2O(\ell)$, are molecules.
- The representative particles for pure ionic compounds, such as sodium chloride, NaCl(s), are *formula units*.

mole the SI base unit that is used to measure the amount of a substance; it contains as many particles (atoms, molecules, ions, or formula units) as exactly 12 g of the isotope carbon-12

Avogadro constant the number of particles in one mole of a substance; a value that is equal to 6.02×10^{23} particles

Go to **scienceontario** to find out more

1. Why has the value of the Avogadro constant changed over the years, since it was first proposed?

2. Explain what a mole is, in your own words.

3. If you had two moles of hydrogen atoms, how many individual hydrogen atoms would you have?

4. Suppose that you were in a space station and looked down at Earth. Do you think you would be able to see one person standing in a field? What about one mole of people standing in a field? Explain.

5. Suppose that you need 2000 sheets of paper. How many reams of paper do you need?

6. **Table 5.1** shows the different units that are often used to count different kinds of items in a group. Why might a dozen not be an appropriate unit to use to count paper?

7. Use print and Internet sources to research why the isotope carbon-12 was used to define the mole. Write a brief summary of your findings.

Visualizing the Mole

The Avogadro constant is a huge quantity. Because atoms, molecules, ions, and formula units are so small, scientists need to use quantities nearly as large as the Avogadro constant for practical applications. When it is written out, the Avogadro constant looks like this:

$$602\ 214\ 179\ 000\ 000\ 000\ 000\ 000$$

What does a mole look like? **Figure 5.2** shows samples of several different substances. Each sample has 6.02×10^{23} atoms, molecules, or formula units. In other words, each sample has one mole of particles.

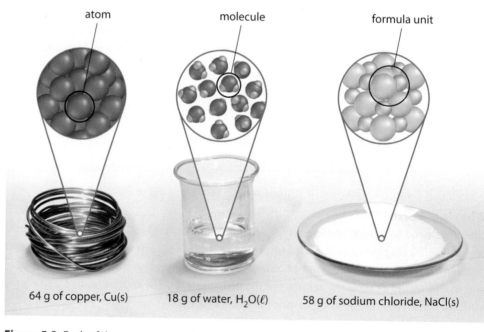

atom   molecule   formula unit

64 g of copper, Cu(s)   18 g of water, H₂O(ℓ)   58 g of sodium chloride, NaCl(s)

Figure 5.2 Each of these samples contains one mole (1.0 mol) of a different substance.

Figure 5.3 Imagine how many individual grains of sand this beach contains. If a beach contained one mole of sand grains , it would be 1000 m wide, 10 m deep, and 5.5 million km long!

Even though you have seen what a mole of several different substances looks like, it is still sometimes difficult to visualize the size of a mole. For example, try to imagine the size of a beach, like the one in **Figure 5.3**, that contains one mole of grains of sand. Activity 5.1 provides a hands-on example that demonstrates the size of a mole.

Practice working with moles and the Avogadro constant helps to reinforce your understanding of quantities in chemical reactions. Examine the Sample Problem on the next page, and then complete the Practice Problems that follow.

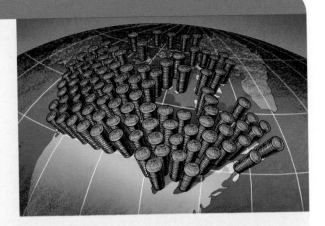

Activity 5.1 One Big Pile of Pennies

Natural Resources Canada states that the area of Canada (land and water) is 9 984 670 km². If this entire area were covered with one mole of pennies, how deep do you think the pile would be?

Materials
- 20 pennies
- centimetre ruler

Procedure

1. Stack 20 pennies.

2. Measure the height of the stack. Then measure the diameter of one penny.

Questions

1. Calculate the radius of a penny.

2. Use the formula for the volume of a cylinder, $V = \pi r^2 h$, to find the volume of your stack of pennies in cubic centimetres.

3. Use a proportion to find the volume of one mole of pennies in cubic centimetres, using the volume for your stack of pennies.

4. Since the area of Canada is in square kilometres, convert the volume you calculated in step 3 to cubic kilometres.

5. To determine the depth of one mole of pennies spread over Canada, divide the volume you calculated in step 4 by the area of Canada.

6. The land area of Ontario is 1 076 395 km². If only Ontario were covered with one mole of pennies, how deep would the pile be?

7. Why do you think chemists work with moles of chemicals when they perform chemical reactions?

Sample Problem

Using the Avogadro Constant

Problem
Earth has an equatorial circumference of 40 076 km. If you lined up 22 grains of salt, end to end, the line would be 1.0 cm long. If you could line up one mole of grains of salt, end to end, around Earth, how many times would the line encircle the planet?

What Is Required?
You need to determine how long a line of 6.02×10^{23} grains of salt is. Then you can figure out how many times the line would encircle Earth.

What Is Given?
You know the length of a line of 22 grains of salt: 1.0 cm
You know the circumference of Earth: 40 076 km

Plan Your Strategy	Act on Your Strategy
Calculate the length of a line of 6.02×10^{23} grains of salt in centimetres.	$6.02 \times 10^{23} \text{ grains} \times \dfrac{1.0 \text{ cm}}{22 \text{ grains}}$ $= 2.736\ 36 \times 10^{22} \text{ cm}$
Convert the length of the line to kilometres. There are 100 cm in 1 m, and 1000 m in 1 km.	$2.736\ 36 \text{ cm} \times 10^{22} \text{ cm} \times \dfrac{1 \text{ m}}{100 \text{ cm}} \times \dfrac{1 \text{ km}}{1000 \text{ m}}$ $= 2.736\ 36 \times 10^{17} \text{ km}$
Calculate the number of times the line would go around Earth.	$2.736\ 36 \times 10^{17} \text{ km} \times \dfrac{1 \text{ Earth circumference}}{40\ 076 \text{ km}}$ $= 6.8 \times 10^{12} \text{ Earth circumferences}$

Check Your Solution
The units cancelled properly. The size of the final answer seems appropriate.

1. An average refrigerator has a volume of 0.6 m³. If a grain of salt has a volume of 9.39×10^{-11} m³, how many refrigerators would one mole of salt grains fill?

2. If you drove for 6.02×10^{23} days at a speed of 110 km/h, how far would you travel?

3. How long would it take to count 6.02×10^{23} raisins, if you counted at a rate of one raisin per second?

4. A ream of paper (500 sheets) is 4.8 cm high. If you stacked 6.02×10^{23} sheets of paper on top of each other, how high would the stack be, in kilometres?

5. The total volume of the Rogers Centre is 1.6×10^6 m³. If the volume of 100 peas is about 55 cm³, how many Rogers Centres would 6.02×10^{23} peas fill?

6. Canada's coastline is 243 042 km long. If you laid 6.02×10^{23} metre sticks end to end along the coast of Canada, how many rows of sticks would there be?

7. Earth's oceans contain about 1.31×10^9 km³ of water. One tablespoon is equal to 15 cm³. If you could remove 6.02×10^{23} tablespoons of water from Earth's oceans, would you completely drain them?

8. Suppose that you were given 6.02×10^{23} pennies when you were born and you lived for 100 years. How much money, in dollars, would you have to spend each second if you wanted to spend all this money in your lifetime?

9. How would the mass of pennies you were given in question 8 compare with the mass of Earth? The mass of 10 pennies is about 24 g. The mass of Earth is 5.98×10^{24} kg.

10. If a row of approximately 5.0×10^7 atoms measured 1.0 cm, how long would a row of 6.02×10^{23} atoms be?

From Dozens to Moles

You have learned how to convert between individual items and dozens of items. Now you are going to use the same steps to work with individual particles and moles of particles. A comparison of dozens and moles is shown in **Figure 5.4**.

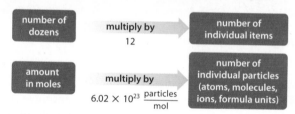

Figure 5.4 This graphic organizer shows how to convert from amount in moles to number of particles.

Converting Moles to Number of Particles

Again, because atoms, molecules, ions, and formula units are so small, chemists work with moles of these particles instead of dozens of them. For example, chemists know that 1 mol of carbon contains 6.02×10^{23} atoms of carbon, and 2 mol of carbon dioxide contains $2 \times 6.02 \times 10^{23} = 1.20 \times 10^{24}$ molecules of carbon dioxide. The relationship between moles, individual particles, and the Avogadro constant can be expressed as

$$n = \frac{N}{N_A}, \text{ where } n \text{ is the amount in moles}$$

N is the number of individual particles
N_A is the Avogadro constant

If you know the amount of particles (in mol) and you want to find the number of individual particles, you can rearrange the above relationship as follows:

$$N = n \times N_A$$

The following Sample Problem demonstrates how to work with this relationship.

Converting Amount in Moles to Number of Particles

Problem

Hydrazine, $N_2H_4(\ell)$, is a versatile compound that is used in pharmaceuticals, rocket fuels, and airbags. Suppose that a chemical sample in an airbag contains 3.65 mol of hydrazine.

a. How many molecules are in the sample?

b. How many atoms are in the sample?

What Is Required?

You need to find the number of molecules in the hydrazine sample.

You need to find the number of atoms in the hydrazine sample.

What Is Given?

You know the amount of hydrazine: 3.65 mol

You know the Avogadro constant, N_A: 6.02×10^{23}

You know the relationship between moles, individual particles, and the Avogadro constant: $N = n \times N_A$

You know the number of atoms in each hydrazine molecule: two nitrogen atoms + four hydrogen atoms = six atoms total

Plan Your Strategy	Act on Your Strategy
a. Use the relationship between moles, individual particles, and the Avogadro constant to find the number of molecules of hydrazine.	$N = n \times N_A$ $N = 3.65 \; \cancel{mol} \times \dfrac{6.02 \times 10^{23} \text{ molecules}}{1 \; \cancel{mol}}$ $= 2.197\,300 \times 10^{24}$ molecules $= 2.20 \times 10^{24}$ molecules
b. Multiply the number of hydrazine molecules by the number of atoms in each molecule.	$2.197\,300 \times 10^{24} \; \cancel{molecules} \times \dfrac{6 \text{ atoms}}{1 \; \cancel{molecule}}$ $= 1.318\,38 \times 10^{25}$ atoms $= 1.32 \times 10^{25}$ atoms

Check Your Solution

In part a, since 22 is between 3 and 4 times 6.02 (based on comparing 22.0×10^{23} to 6.02×10^{23}), the answer is reasonable.

In part b, since there are 6 times as many atoms as molecules (based on comparing 2.2 to 13.2), the answer is reasonable.

11. A Canadian penny contains 0.106 mol of copper. How many atoms of copper are in a Canadian penny?

12. The head of a small pin contains about 8×10^{-3} mol of iron. How many iron atoms are in the head of the pin?

13. How many molecules of oxygen gas are in a room that contains 8.5×10^3 mol of oxygen gas?

14. If a marble countertop contains 849 mol of calcium carbonate, $CaCO_3(s)$, how many formula units of calcium carbonate are in the countertop?

15. A recipe calls for half a teaspoon of salt, which contains 5.23×10^{-2} mol of sodium chloride. How many formula units of sodium chloride are needed?

16. A window-cleaning solution contains 3.86 mol of acetic acid, $CH_3COOH(\ell)$. How many molecules of acetic acid are in the solution?

17. A fuel tank used in a barbecue contains 2.0×10^2 mol of propane, $C_3H_8(g)$. What is the total number of atoms in the tank?

18. Freon™, $CCl_2F_2(g)$, is a refrigerant that is no longer used in car air conditioners because it damages the ozone layer. A sample contains 4.82 mol of Freon™.
   a. How many molecules of Freon™ are in the sample?
   b. How many atoms, in total, are in the sample?

19. Glauber's salt is a common name for sodium sulfate decahydrate, $Na_2SO_4 \cdot 10H_2O(s)$. It is used in the manufacture of detergents. Suppose that a sample of 36.2 mol of sodium sulfate decahydrate is required.
   a. What number of sodium atoms would be in the sample?
   b. What number of water molecules would be in the sample?

20. A sample of sucrose, $C_{12}H_{22}O_{11}(s)$, contains 0.16 mol of oxygen, atoms (O).
   a. What amount in moles of sucrose is in the sample?
   b. How many atoms of carbon are in the sample?

Converting Number of Particles to Amount in Moles

As mentioned earlier, working with individual particles of different chemicals is not practical for chemists in a lab. Instead, chemists work with amounts in moles of chemicals. So, rather than describing a particular sample of carbon dioxide, as 2.4×10^{24} molecules of carbon dioxide, chemists describe it as 4.0 mol of carbon dioxide. Similarly, in the equation for a chemical reaction, the coefficients can represent the amount in moles of chemicals that react. Moles can also be used to describe the amount of carbon dioxide in a soft drink, such as the one in **Figure 5.5**.

To convert from individual particles to moles of particles, you can use the relationship involving the Avogadro constant, which you learned earlier:

$$n = \frac{N}{N_A}$$

This conversion is illustrated in **Figure 5.6**. The following Sample Problem demonstrates how to use it.

Figure 5.5 Even though there seem to be a lot of bubbles in a soft drink, there is only about 0.1 mol of carbon dioxide in 1 L.

Determine the number of carbon dioxide molecules in 1 L of a soft drink.

Figure 5.6 The Avogadro constant is used to convert individual particles to moles of particles, and vice versa.

Converting Number of Particles to Amount in Moles

Problem

A commercial product that people sometimes take to settle an upset stomach contains magnesium hydroxide, $Mg(OH)_2(aq)$, a base. If a teaspoon of stomach medicine contains 4.1×10^{21} formula units of magnesium hydroxide, what amount in moles of magnesium hydroxide is in the teaspoon?

What Is Required?

You are asked to find the amount in moles in a sample of magnesium hydroxide.

What Is Given?

You know the number of formula units: 4.1×10^{21}.

You know the Avogadro constant: 6.02×10^{23}.

You know that you need to use the following relationship: $n = \dfrac{N}{N_A}$.

Plan Your Strategy	Act on Your Strategy
Write the relationship between moles, individual particles, and the Avogadro constant. Substitute the known values into the formula.	$n = \dfrac{N}{N_A}$ $= \dfrac{4.1 \times 10^{21} \text{ formula units}}{6.02 \times 10^{23} \text{ formula units/mol}}$ $= 6.8 \times 10^{-3} \text{ mol}$

Check Your Solution

If you substitute your answer back into the relationship $N = n \times N_A$, the answer is the same as the given number of formula units.

Practice Problems

21. A gold coin contains 9.51×10^{22} atoms of gold. What amount in moles of gold is in the coin?

22. A patient in a dentist's office is given 1.67×10^{23} molecules of dinitrogen monoxide (laughing gas), $N_2O(g)$, during a procedure. What amount in moles of dinitrogen monoxide is the patient given?

23. A sheet of drywall contains 1.2×10^{26} formula units of gypsum (calcium sulfate dihydrate), $CaSO_4 \cdot 2H_2O(s)$. What amount in moles of gypsum is in the sheet of drywall?

24. Limewater, a weak solution of calcium hydroxide, $Ca(OH)_2(s)$, is used to detect the presence of carbon dioxide gas. Suppose that you are given a solution that contains 8.7×10^{19} formula units of calcium hydroxide. What amount in moles of calcium hydroxide is in the solution?

25. If there are a total of 7.3×10^{29} atoms in a sample of glucose, $C_6H_{12}O_6(s)$, what amount in moles of glucose is in the sample?

26. A sample of aluminum oxide, $Al_2O_3(s)$, contains 8.29×10^{25} total atoms. Calculate the amount in moles of aluminum oxide in the sample. **Hint:** This is a two-step problem. Calculate the number of formula units first.

27. Trinitrotoluene, or TNT for short, has the chemical formula $C_7H_5N_3O_6(s)$. If a stick of dynamite is pure TNT and it contains 2.5×10^{25} atoms in total, what amount in moles of TNT does it contain?

28. A sample of rubbing alcohol solution contains ethanol, $C_2H_5OH(\ell)$. If the sample contains 1.25×10^{23} atoms of hydrogen in the ethanol, what amount in moles of ethanol is in the sample?

29. A cleaning solution contains 7.9×10^{26} molecules of ammonia, $NH_3(aq)$. What amount in moles of ammonia is in the solution?

30. A muffin recipe calls for cream of tartar, or potassium hydrogen tartrate, $KHC_4H_4O_6(s)$. The amount of cream of tartar that is required contains 2.56×10^{23} atoms of carbon. What amount in moles of potassium hydrogen tartrate is required?

Section Summary

- People group common items into units of quantity for convenience. Chemists also group atoms, molecules, ions, and formula units into units of quantity for convenience.

- A mole is the amount of substance that contains as many particles (atoms, molecules, ions, or formula units) as the number of atoms in exactly 12 g of the isotope carbon-12.

- The mole is the SI base unit that is used to measure the amount of a substance.

- Chemists use the mole to count very small particles, such as atoms, molecules, ions, and formula units.

- The Avogadro constant allows chemists to convert back and forth between individual particles and moles of particles.

Review Questions

1. **K/U** Define "mole" in your own words. Use an example to illustrate your definition.

2. **K/U** Compare a mole and a dozen. Provide one similarity and one difference.

3. **K/U** Is the value that is presently accepted for the mole, $6.022\ 141\ 79 \times 10^{23}$ particles, an exact number? Explain.

4. **T/I** If the volume of one mole of water at room temperature is 18.02 mL, what is the volume of one molecule of water, in litres?

5. **C** Use a flowchart to show the process of converting moles of phosphorus, $P_4(s)$, to atoms of phosphorus.

6. **T/I** Calculate the number of formula units of sodium chloride in 0.0578 mol.

7. **T/I** Calculate the number of particles in each sample. Indicate the correct type of particle (atom, molecule, ion, or formula unit).
   a. 0.156 mol Au(s)
   b. 7.8 mol $MgCl_2(s)$
   c. 15.2 mol $H_2O_2(\ell)$

8. **T/I** Calculate the amount in moles of molecules for each substance listed.
   a. a sample of ammonia, $NH_3(g)$, containing 8.1×10^{20} atoms of hydrogen
   b. a sample of diphosphorus pentoxide, $P_2O_5(s)$, containing a total of 4.91×10^{22} atoms of phosphorus and oxygen

9. **T/I** A sample of ethanol, $C_2H_5OH(\ell)$, contains 2.49×10^{23} molecules of ethanol. A sample of carbon contains 1.65 mol of carbon atoms. Which sample, the ethanol or the carbon, contains the greater amount in moles of carbon?

10. **T/I** If there are 4.28×10^{21} atoms of hydrogen and oxygen in a sample of water, what amount in moles of water is in the sample?

11. **A** Aluminum oxide, $Al_2O_3(s)$, forms a thin coating on an aluminum surface, such as the body of an airplane, when aluminum is exposed to oxygen in air. This coating helps to reduce further corrosion of the aluminum. Suppose that a sample contains 2.6 mol of aluminum oxide.
   a. How many formula units are in the sample?
   b. How many atoms, in total, are in the sample?
   c. How many aluminum atoms are in the sample?

12. **A** Zinc chloride hexahydrate, $ZnCl_2 \cdot 6H_2O(s)$, is used in the textile industry, often to prepare fireproofing agents. In a particular fireproofing solution, 0.46 mol of zinc chloride hexahydrate is used.
   a. How many formula units of zinc chloride hexahydrate are in the solution?
   b. How many atoms of chlorine are in the solution?

13. **C** Draw a graphic organizer that shows the conversions between number of particles and amount in moles.

14. **A** Arrange the following three samples from largest to smallest in terms of their numbers of representative particles:
   - 3.92 mol of octane, $C_8H_{18}(\ell)$
   - 6.52×10^{23} atoms of copper, $Cu(s)$
   - 1.25×10^{24} formula units of sodium hydrogen carbonate, $NaHCO_3(s)$.

15. **T/I** Calculate the number of atoms in 6.0 mol of fluorine, $F_2(g)$, molecules.

Mass and the Mole

In the previous section, you learned that working with moles is more convenient than working with individual particles. But is it convenient to measure substances in moles when doing an experiment? To do this, you would have to count all the particles each time you wanted to perform a reaction. That would be inconvenient because there is no device for counting particles.

How do chemists measure the quantities of substances they need when performing a chemical reaction? Scientists have developed sophisticated instruments, such as the one shown in **Figure 5.7**, to measure the relative masses of compounds accurately. They also have devised a way to determine the mass of a given number of particles, instead of counting individual particles.

Molar Mass: The Mass of One Mole

Recall that one atom of carbon-12 has a mass of exactly 12 u, and one mole (6.02×10^{23} atoms) of carbon-12 has a mass of exactly 12 g. Also recall that the atomic mass unit is defined using the carbon-12 isotope and that the atomic masses of all the other elements are defined using carbon-12 as the standard. These relationships are important, because it means that one mole of any element has a mass that is numerically equal to the element's atomic mass expressed in grams. For example, an atom of iron has an atomic mass of 55.85 u, and one mole of iron atoms has a mass of 55.85 g. Additional examples are shown in **Table 5.2**.

The relationship between mass and moles gives chemists a practical way to count atoms—by measuring their mass on a balance. Because of these relationships, you can use the periodic table to find the mass of one mole of any element.

Key Term

molar mass

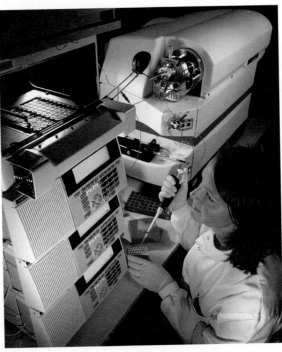

Figure 5.7 The mass spectrometer is a device used to make precise mass measurements and to determine molecular structure.

Table 5.2 Average Atomic Mass and Mass of One Mole of Atoms of Four Elements

Element	Average Atomic Mass (u)	Mass of One Mole of Atoms (g)
boron	10.81	10.81
potassium	39.10	39.10
bromine	79.90	79.90
gold	196.97	196.97

Go to **scienceontario** to find out more

Molar Mass: The Mass of One Mole

The term for the mass of one mole of a substance is **molar mass**. The symbol for molar mass is *M*, and the unit is g/mol. The mass of one mole of atoms of any element in the periodic table is called its *atomic molar mass*. The units for this mass are g/mol. Similarly, you would call the mass of one mole of molecules of a substance its *molecular molar mass* and the mass of one mole of formula units of a substance its *formula unit molar mass*. The unit for all of these terms is g/mol.

molar mass the mass of one mole of a substance; the symbol for molar mass is *M*, and the unit is g/mol

8. Why is it not convenient to measure substances in moles?

9. What is the difference between atomic molar mass and molar mass?

10. What is the relationship between the atomic mass of an element in atomic mass units and the atomic molar mass of the same element in grams?

11. What is the mass of 6.02×10^{23} copper atoms?

12. In what other way can the mass of 6.02×10^{23} atoms be described? What is the unit for this value?

13. Why is the molar mass of most elements—for example, gold as shown in **Table 5.2**—not a whole number?

Linking Moles and Mass: The Molar Mass of a Compound

You have already seen that the value of the atomic mass of an element in the periodic table is the same as the value of its molar mass in grams. How do you find the molar mass of a compound?

To find the molar mass of a compound, such as water, you find the molar mass of each element in the periodic table, multiply by the number of atoms of each element, and add the values.

$$M_{H_2O} = 2M_H + 1M_O$$
$$= 2(1.01 \text{ g/mol}) + 1(16.00 \text{ g/mol})$$
$$= 18.02 \text{ g/mol}$$

Therefore, the molar mass of water is 18.02 g/mol.

The following Sample Problem shows how to determine the molar mass of a more complex compound. Study this problem, and then try the Practice Problems.

Determining the Molar Mass of a Compound

Problem

What is the molar mass of aluminum sulfate, $Al_2(SO_4)_3(s)$?

What Is Required?

You need to find the mass of one mole of aluminum sulfate.

What Is Given?

You know the chemical formula: $Al_2(SO_4)_3(s)$

Plan Your Strategy	Act on Your Strategy
Use the periodic table to find the atomic molar mass of each element in aluminum sulfate. Multiply the atomic molar masses by the number of atoms of the element in the compound.	$2M_{Al} = 2(26.98 \text{ g/mol}) = 53.96 \text{ g/mol}$ $3M_S = 3(32.07 \text{ g/mol}) = 96.21 \text{ g/mol}$ $12M_O = 12(16.00 \text{ g/mol}) = 192.0 \text{ g/mol}$
Add the molar masses of the atoms of each element to get the molar mass of the compound.	$M_{Al_2(SO_4)_3} = 53.96 \text{ g/mol} + 96.21 \text{ g/mol} + 192.0 \text{ g/mol}$ $\qquad = 342.2 \text{ g/mol}$ The molar mass of aluminum sulfate is 342.2 g/mol.

Check Your Solution

Use rounded values for the atomic molar masses to get an estimate of the answer.

$(2 \times 27) + (3 \times 32) + (12 \times 16) = 342$ This estimate is close to the answer, 342.2 g/mol.

31. State the molar mass of each element.
   a. sodium   **c.** xenon
   b. tungsten   **d.** nickel

32. Calculate the molar mass of phosphorus, $P_4(s)$.

33. Determine the molar mass of calcium phosphate, $Ca_3(PO_4)_2(s)$.

34. Calculate the molar mass of lead(II) nitrate, $Pb(NO_3)_2(s)$.

35. Determine the molar mass of the iron(III) thiocyanate ion, $FeSCN^{2+}(aq)$.

36. Calculate the molar mass of sodium stearate, $NaC_{17}H_{35}COO(s)$.

37. Calculate the molar mass of barium hydroxide octahydrate, $Ba(OH)_2 \cdot 8H_2O(s)$. (**Hint:** The molar mass of a hydrate must include the water component.)

38. Determine the molar mass of tetraphosphorus decoxide, $P_4O_{10}(s)$.

39. Calculate the molar mass of iron(II) ammonium sulfate hexahydrate, $(NH_4)_2Fe(SO_4)_2 \cdot 6H_2O(s)$.

40. The formula for a compound that contains an unknown element, A, is A_2SO_4. If the molar mass of the compound is 361.89 g/mol, what is the atomic molar mass of A?

Relationships among Moles, Mass, and Particles

One mole of a substance contains 6.02×10^{23} particles (atoms, molecules, ions, or formula units). The molar mass (the mass of one mole of a substance), then, represents the mass of 6.02×10^{23} particles. This means that connections exist among the number of particles, the molar mass, and the mass of a substance. It also means that you can count particles using a balance (with the help of some calculations), as shown in **Figure 5.8**.

Figure 5.8 By using a balance and mathematics, you can convert between mass and number of particles.

Notice that the amount in moles of a substance plays a central role in the relationships shown in **Figure 5.9**. You will find that these amounts will also play a central role in many of the calculations you do in this course.

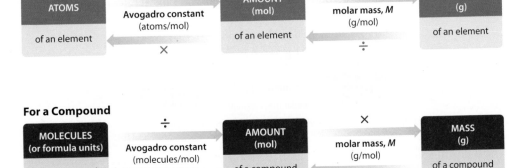

For an Element

ATOMS of an element   ÷ Avogadro constant (atoms/mol)   AMOUNT (mol) of an element   × molar mass, M (g/mol)   MASS (g) of an element

For a Compound

MOLECULES (or formula units) of a compound   ÷ Avogadro constant (molecules/mol)   AMOUNT (mol) of a compound   × molar mass, M (g/mol)   MASS (g) of a compound

Figure 5.9 The Avogadro constant relates the number of particles of a substance to the amount in moles of the substance. Similarly, the molar mass relates the mass of a substance to the amount in moles.

Applying the Relationships among Moles, Mass, and Particles

Suppose that you have one mole of the element neon as shown in **Figure 5.10**. One mole of neon contains 6.02×10^{23} atoms and has a mass of 20.18 g. Its molar mass is therefore 20.18 g/mol. So, 6.02×10^{23} atoms of neon is 1 mol of neon or 20.18 g of neon.

The same holds true for compounds. Suppose that you have one mole of water. One mole of water contains 6.02×10^{23} molecules of water, and it has a mass of 18.02 g. Its molar mass is therefore 18.02 g/mol. So, 6.02×10^{23} molecules of water is 1 mol of water or 18.02 g of water. Could you use this relationship to calculate the amount in moles and the number of particles in 36.04 g of water?

Figure 5.10 One mole of the element neon is 20.18 g.

Determine *If this sign contains 10.0 g of neon, what amount in moles of neon is in the sign?*

Converting Amount in Moles to Mass

As mentioned earlier, if you wanted to perform a reaction, you would not count the particles required. This would not be practical. Instead, the relationships you have learned allow you to calculate the mass of each required reactant, using the amount required and the molar mass.

Suppose that you were asked to perform the following reaction:

$$Zn(s) + 2HCl(aq) \rightarrow ZnCl_2(aq) + H_2(g)$$

You have been told to use 1.50 mol of zinc. How would you measure the zinc? The first step is to use molar mass to go from amount to mass using the equation below:

$$\text{mass} = \text{amount in moles} \times \text{molar mass}$$

The equation can also be written in symbols, as follows:

$$m = n \times M$$

This conversion is represented in **Figure 5.11**. Using the equation above, you would find the molar mass of zinc in the periodic table (65.38 g/mol) and then multiply it by the 1.50 mol of zinc required. You would get 98.07 g as your answer, which you could then measure on a balance.

$$m = 1.50 \text{ mol} \times 65.38 \ \frac{\text{g}}{\text{mol}}$$
$$= 98.1 \text{ g}$$

| amount (mol) | multiply by molar mass (g/mol) → | mass (g) |

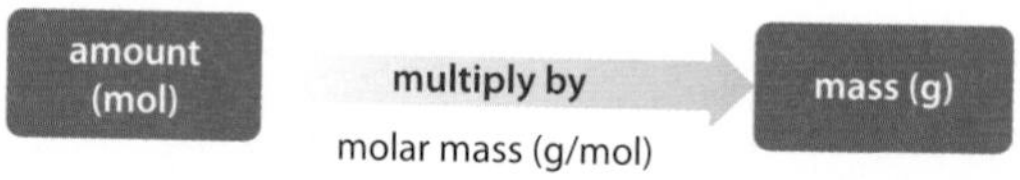

Figure 5.11 This graphic organizer shows how to convert from amount in moles to mass in grams.

The following Sample Problem and Practice Problems will help you practise converting amount to mass.

Converting Amount to Mass

Problem

A chemist performs a reaction that produces 0.258 mol of silver chloride precipitate, AgCl(s). What mass of precipitate is produced?

What Is Required?

You need to calculate the mass of silver chloride that is produced.

What Is Given?

You know the amount of silver chloride produced: 0.258 mol

You know the formula for silver chloride: AgCl(s).

Plan Your Strategy	Act on Your Strategy
Use the periodic table to find the atomic molar masses of silver and chlorine. Multiply the atomic molar masses by the number of atoms of each element in the compound. Add the values calculated above to find the molar mass of the compound.	$1M_{Ag} = 1(107.87 \text{ g/mol}) = 107.87 \text{ g/mole}$ $2M_{Cl} = 2(35.45 \text{ g/mol}) = 70.90 \text{ g/mole}$ $M_{AgCl} = 1M_{Ag(s)} + 2M_{Cl(s)}$ $\qquad = 107.87 \text{ g/mole} + 70.90 \text{ g/mole}$ $\qquad = 178.77 \text{ g/mole}$ The molar mass of silver chloride is 178.77 g/mol.
Write the formula that relates mass to amount and molar mass. Substitute in the known values to calculate the mass.	$m = n \times M$ $\quad = 0.258 \text{ mol} \times 178.77 \dfrac{\text{g}}{\text{mol}}$ $\quad = 46.1 \text{ g}$

Check Your Solution

0.258 mol is about $\frac{1}{4}$ of a mole, and 46 is about $\frac{1}{4}$ of 178, so the answer seems reasonable.

Practice Problems

41. Calculate the mass of 3.57 mol of vanadium.

42. Calculate the mass of 0.24 mol of carbon dioxide.

43. Calculate the mass of 1.28×10^{-3} mol of glucose, $C_6H_{12}O_6$(s).

44. Calculate the mass of 0.0029 mol of magnesium bromide, $MgBr_2$(s), in milligrams.

45. Name each compound, and then calculate its mass. Express this value in scientific notation.
  a. 4.5×10^{-3} mol of $Co(NO_3)_2$(s)
  b. 29.6 mol of $Pb(S_2O_3)_2$(s)

46. Determine the chemical formula for each compound, and then calculate its mass.
  a. 4.9 mol of ammonium nitrate
  b. 16.2 mol of iron(III) oxide

47. What is the mass of 1.6×10^{-3} mol of calcium chloride dihydrate, $CaCl_2 \cdot 2H_2O$(s), in milligrams?

48. A litre of water contains 55.56 mol of water molecules. What is the mass of a litre of water, in kilograms?

49. For each group of three samples, determine the sample with the largest mass.
  a. 2.34 mol of bromine, Br_2(ℓ); 9.80 mol of hydrogen sulfide, H_2S(g); 0.568 mol of potassium permanganate, $KMnO_4$(s)
  b. 13.7 mol of strontium iodate, $Sr(IO_3)_2$(s); 15.9 mol of gold(III) chloride, $AuCl_3$(s); 8.61 mol of bismuth silicate, $Bi_2(SiO_3)_3$(s)

50. Which has the smallest mass: 0.215 mol of potassium hydrogen sulfite, $KHSO_3$(s); 1.62 mol of sodium hydrogen sulfite, $NaHSO_3$(s); or 0.0182 mol of aluminum iodate, $Al(IO_3)_3$(s)?

Converting Mass to Amount in Moles

Often, the mass of a substance is known, but the amount in moles of the substance is not known. For example, sulfur dioxide, $SO_2(g)$, which is emitted in the exhaust gases from many power plants, is well known for its contribution to acid rain. As a result, many power plants, such as the one shown in **Figure 5.12**, remove the sulfur dioxide in their exhaust stacks using a reaction with substances such as calcium hydroxide, $Ca(OH)_2(s)$. By converting the mass of sulfur dioxide emitted to the amount emitted, technicians at the power plant can calculate and measure the amount of calcium hydroxide required to react with the sulfur dioxide.

Determining the amount of a substance from the mass of the substance involves rearranging the equation $m = n \times M$, as follows:

Figure 5.12 Power plants have developed the technology to "scrub" emissions, such as sulfur dioxide, out of the exhaust, using chemicals such as calcium hydroxide.

$$n = \frac{m}{M}$$

In other words,

$$\text{amount in moles} = \frac{\text{mass}}{\text{molar mass}}$$

Again, the molar mass of the substance is used as the conversion factor. This conversion is represented in **Figure 5.13**. A sample problem follows.

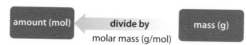

Figure 5.13 To convert from mass to amount, use the molar mass as a conversion factor.

Sample Problem

Converting Mass to Amount

Problem

What amount in moles is 15.3 g of sulfur dioxide, $SO_2(g)$, taken from a power plant exhaust stack?

What Is Required?

You need to convert 15.3 g of sulfur dioxide to the amount in moles.

What Is Given?

You know the mass of sulfur dioxide: 15.3 g

Plan Your Strategy	Act on Your Strategy
Use the periodic table to find the atomic molar masses of sulfur and oxygen. Multiply the atomic molar masses by the number of atoms of each element in the compound. Add these values to calculate the molar mass of the compound.	$M_{SO_2} = 1(M_S) + 2(M_O)$ $M_{SO_2} = 1(32.07 \text{ g/mol}) + 2(16.00 \text{ g/mol})$ $\quad = 64.07 \text{ g/mol}$ The molar mass of sulfur dioxide is 64.07 g/mol.
Divide the given mass by the molar mass to find the amount in moles.	$n = \dfrac{m}{M}$ $\quad = \dfrac{15.3 \text{ g}}{64.07 \text{ g/mol}}$ $\quad = 0.239 \text{ mol}$ The amount of sulfur dioxide is 0.239 mol.

Check Your Solution

If you work backward, your answer in moles (0.239 mol) multiplied by the molar mass (64.07 g/mol) equals the mass given (15.3 g).

51. Convert 29.5 g of ammonia to the amount in moles.

52. Determine the amount in moles of potassium thiocyanate, KSCN(s), in 13.5 kg.

53. Determine the amount in moles of sodium dihydrogen phosphate, NaH_2PO_4(s), in 105 mg.

54. Determine the amount in moles of xenon tetrafluoride, XeF_4(s), in 22 mg.

55. Write the chemical formula for each compound, and then calculate the amount in moles in each sample.

   a. 3.7×10^{-3} g of silicon dioxide

   b. 25.38 g of titanium(IV) nitrate

   c. 19.2 mg of indium carbonate

   d. 78.1 kg of copper(II) sulfate pentahydrate

56. The characteristic odour of garlic comes from allyl sulfide, $(C_3H_5)_2S(\ell)$. Determine the amount in moles of allyl sulfide in 168 g.

57. Road salt, $CaCl_2$(s), is often used on roads in the winter to prevent the build-up of ice. What amount in moles of calcium chloride is in a 20.0 kg bag of road salt?

58. Calculate the amount in moles of trinitrotoluene, $C_7H_5(NO_2)_3$(s), an explosive, in 3.45×10^{-3} g.

59. Arrange the following substances in order from largest to smallest amount in moles:

- 865 mg of $Ni(NO_3)_2$(s)
- 9.82 g of $Al(OH)_3$(s)
- 10.4 g of AgCl(s)

60. Place the following substances in order from smallest to largest amount in moles, given 20.0 g of each:

- glucose, $C_6H_{12}O_6$(s)
- barium perchlorate, $Ba(ClO_4)_2$(s)
- tin(IV) oxide, SnO_2(s)

Conversions between Number of Particles and Mass

You have already learned how to convert number of particles to amount in moles, and amount in moles to number of particles. You have also converted between amount and mass. What about converting number of particles to mass, and vice versa? These conversions are shown in **Figure 5.14**.

Suppose that a chemist had a piece of copper wire with a mass of 2.34 g. How could the chemist calculate the number of copper atoms in the wire? Or suppose that you were given the challenge of calculating the mass of a sample of nitrogen gas which contained 5.34×10^{24} molecules of nitrogen. How would you figure out the steps in the calculation? Once again, the concept of the mole plays a central role in these types of calculations. Whether you are given the number of particles or the mass, your first step is to find the amount in moles of the substance.

Suggested Investigation

Inquiry Investigation 5-A, Exploring Conversions between Mass and Particles

Figure 5.14 This graphic organizer shows how to convert among number of particles, amount in moles, and mass in grams.

Study the following Sample Problems to see examples of conversions between mass and number of particles. As the Alternative Solutions in these Sample Problems will show, you do not necessarily need to use mathematical formulas for the calculations if you are able to visualize the relationships.

Converting Number of Particles to Mass

Problem

What is the mass of 4.72×10^{23} formula units of chromium(III) iodide, $CrI_3(s)$?

What Is Required?

You are asked to find the mass of chromium(III) iodide.

What Is Given?

You know the number of formula units of chromium(III) iodide: 4.72×10^{23}

You know the Avogadro constant: $N_A = 6.02 \times 10^{23}$

You know the conversion factors you will need to use: the Avogadro constant and the molar mass of chromium(III) iodide

Plan Your Strategy	Act on Your Strategy
Calculate the amount of chromium(III) iodide from the given number of formula units, using the Avogadro constant.	$n = \dfrac{N}{N_A}$ $= \dfrac{4.72 \times 10^{23}\ \text{formula units}}{6.02 \times 10^{23}\ \text{formula units/mol}}$ $= 0.784\ 053\ \text{mol}$
Calculate the mass from the amount in moles using the molar mass of chromium(III) iodide.	$m = n \times M$ $= 0.784\ 053\ \text{mol} \times 432.70\ \dfrac{g}{mol}$ $= 339\ g$

Alternative Solution 1

Work with the units to set up the steps in your calculation. You are still finding the amount in moles first and then mass, but you are not using the mathematical formulas $n = \dfrac{N}{N_A}$ and $m = n \times M$.

Plan Your Strategy	Act on Your Strategy
Calculate the amount of chromium(III) iodide from the given number of formula units, using the Avogadro constant.	$n = 4.72 \times 10^{23}\ \text{formula units} \times \dfrac{1\ \text{mol}}{6.02 \times 10^{23}\ \text{formula units}}$ $= 0.784\ 053\ \text{mol}$
Calculate the mass from the amount in moles, using the molar mass of chromium(III) iodide.	$m = 0.784\ 053\ \text{mol} \times \dfrac{432.70\ g}{1\ \text{mol}}$ $= 339\ g$

Alternative Solution 2

The calculation can be arranged on one line, as follows:

$$m = 4.72 \times 10^{23}\ \text{formula units} \times \frac{1\ \text{mol}}{6.02 \times 10^{23}\ \text{formula units}} \times \frac{432.70\ g}{1\ \text{mol}}$$

$$= 339\ g$$

This arrangement can be used in most multi-step problems, if you are using the units to calculate the answer.

Check Your Solution

The number 4.72 is about $\frac{3}{4}$ of 6.02 (with the same power of 10), and 339 g is about $\frac{3}{4}$ of the numerical value of the molar mass 432.70 g/mol, so the answer is reasonable.

Converting Mass to Number of Particles

Problem
Phosphoryl chloride, $POCl_3(\ell)$, is an important compound in the production of flame retardants. How many molecules of phosphoryl chloride are in a 25.2 g sample?

What Is Required?
You are asked to find the number of molecules of phosphoryl chloride.

What Is Given?
You are given the mass of a phosphoryl chloride sample: 25.2 g
You know the Avogadro constant: $N_A = 6.02 \times 10^{23}$

Plan Your Strategy	Act on Your Strategy
Calculate the amount of phosphoryl chloride from the given mass, using the molar mass.	$n = \dfrac{m}{M}$ $= \dfrac{25.2 \text{ g}}{153.32 \text{ g/mol}}$ $= 0.164\,362 \text{ mol}$
Calculate the number of molecules of phosphoryl chloride from the amount in moles, using the Avogadro constant.	$N = n \times N_A$ $= 0.164\,362 \text{ mol} \times \dfrac{6.02 \times 10^{23} \text{ molecules}}{1 \text{ mol}}$ $= 9.89 \times 10^{22} \text{ molecules}$

Alternative Solution
Work with the units to set up the steps in your calculation. You are still finding amount in moles first and then molecules, but you are not using the mathematical formulas $n = \dfrac{m}{M}$ and $N = n \times N_A$.

Plan Your Strategy	Act on Your Strategy
Calculate the amount of phosphoryl chloride from the given mass, using the molar mass.	$n = 25.2 \text{ g} \times \dfrac{1 \text{ mol}}{153.32 \text{ g}}$ $= 0.164\,362 \text{ mol}$
Calculate the number of molecules of phosphoryl chloride from the amount, using the Avogadro constant.	$N = 0.164\,362 \text{ mol} \times \dfrac{6.02 \times 10^{23} \text{ molecules}}{1 \text{ mol}}$ $= 9.89 \times 10^{22} \text{ molecules}$

Again, this calculation can be arranged on one line, as follows:

$$N = 25.2 \text{ g} \times \frac{1 \text{ mol}}{153.32 \text{ g}} \times \frac{6.02 \times 10^{23} \text{ molecules}}{1 \text{ mol}}$$

$$= 9.89 \times 10^{22} \text{ molecules}$$

Check Your Solution
The number 25.2 g is about $\frac{1}{6}$ of the molar mass, so the number of molecules should be about $\frac{1}{6}$ of the Avogadro constant. Since 9.89×10^{22} is about $\frac{1}{6}$ of 6.02×10^{23}, the answer is reasonable.

61. Calculate the mass of each sample.

   a. 1.05×10^{26} atoms of neon, $Ne(g)$

   b. 2.7×10^{24} molecules of phosphorus trichloride, $PCl_3(\ell)$

   c. 8.72×10^{21} molecules of karakin, $C_{15}H_{21}N_3O_{15}(s)$

   d. 6.7×10^{27} formula units of sodium thiosulfate, $Na_2S_2O_3(s)$

62. Determine the number of molecules or formula units in each sample.

   a. 32.4 g of lead(II) phosphate, $Pb_3(PO_4)_2(s)$

   b. 8.62×10^{-3} g of dinitrogen pentoxide, $N_2O_5(s)$

   c. 48 kg of molybdenum(VI) oxide, $MoO_3(s)$

   d. 567 g of tin(IV) fluoride, $SnF_4(s)$

63. Sodium hydrogen carbonate, $NaHCO_3(s)$, is the principal ingredient in many stomach-relief medicines.

   a. A teaspoon of a particular brand of stomach-relief medicine contains 6.82×10^{22} formula units of sodium hydrogen carbonate. What mass of sodium hydrogen carbonate is in the teaspoon?

   b. The bottle of this stomach-relief medicine contains 350 g of sodium hydrogen carbonate. How many formula units of sodium hydrogen carbonate are in the bottle?

64. Riboflavin, $C_{17}H_{20}N_4O_6(s)$, is an important vitamin in the metabolism of fats, carbohydrates, and proteins in your body.

   a. The current recommended dietary allowance (RDA) of riboflavin for adult men is 1.3 mg/day. How many riboflavin molecules are in this RDA?

   b. The RDA of riboflavin for adult women contains 1.8×10^{18} molecules of riboflavin. What is the RDA for adult women, in milligrams?

65. What is the mass, in grams, of a single atom of platinum?

66. Rubbing alcohol often contains propanol, $C_3H_7OH(\ell)$. Suppose that you have an 85.9 g sample of propanol.

   a. How many carbon atoms are in the sample?

   b. How many hydrogen atoms are in the sample?

   c. How many oxygen atoms are in the sample?

67. a. How many formula units are in a 3.14 g sample of aluminum sulfide, $Al_2S_3(s)$?

   b. How many ions (aluminum and sulfide), in total, are in this sample?

68. Which of the following two substances contains the greater mass?

   • 6.91×10^{22} molecules of nitrogen dioxide, $NO_2(g)$

   • 6.91×10^{22} formula units of gallium arsenide, $GaAs(s)$

69. Many common dry-chemical fire extinguishers contain ammonium phosphate, $(NH_4)_3PO_4(s)$, as their principal ingredient. If a sample of ammonium phosphate contains 4.5×10^{21} atoms of nitrogen, what is the mass of the sample?

70. Place the following three substances in order, from greatest to smallest number of hydrogen atoms:

   • 268 mg of sucrose, $C_{12}H_{22}O_{11}(s)$

   • 15.2 g of hydrogen cyanide, $HCN(\ell)$

   • 0.0889 mol of acetic acid, $CH_3COOH(\ell)$

Suggested Investigation

Inquiry Investigation 5-B, Using the Mole for Measuring and Counting Particles

Using the Mole

In this chapter, you learned about an important concept in chemistry: the mole. You also learned about the relationships among the number of particles in a substance, the amount of a substance, and the mass of a substance. Knowing these important relationships will help you explore the mole concept and its uses. In the next chapter, you will learn how the mass proportions of the elements in a compound relate to the formula for the compound and how chemists use their understanding of molar mass to find out important information about chemical compounds.

Section Summary

- The mass of one atom of an element in atomic mass units has the same numerical value as the mass of one mole of atoms of the same element in grams.

- Molar mass is an important tool for converting back and forth between the mass and amount of an element or a compound.

- Conversions can be made between the number of individual particles of a substance and the mass of the substance, using both the Avogadro constant and the molar mass.

- All of these conversions are important when calculating and measuring the amounts of substances in a chemical reaction.

Review Questions

1. **K/U** Compare and constrast the following pairs of terms. Give examples of each.
   a. atomic molar mass, molar mass
   b. molecular molar mass, formula unit molar mass
   c. formula unit molar mass, atomic molar mass
   d. molar mass, molecular molar mass

2. **C** Use a flowchart to show the relationships among number of particles, amount in moles, and mass in grams.

3. **C** Use a flowchart to show how the atomic mass and atomic molar mass of an element are related.

4. **T/I** Determine the amount in moles of gallium oxide, $Ga_2O_3(s)$, in a 45.2 g sample.

5. **T/I** What is the mass of 3.2×10^2 mol of cerium nitrate, $Ce(NO_3)_3(s)$?

6. **T/I** Calculate the amount in moles of strontium chloride, $SrCl_2(s)$, in a 28.6 kg sample.

7. **T/I** What is the mass of 0.68 mol of iron(III) sulfate, $Fe_2(SO_4)_3(s)$?

8. **T/I** What is the mass of 2.9×10^{26} molecules of dinitrogen pentoxide, $N_2O_5(g)$?

9. **T/I** Calculate the number of oxygen atoms in 15.2 g of trinitrotoluene, $C_7H_5(NO_2)_3(s)$.

10. **T/I** Which has more sulfur atoms: 13.4 g of potassium thiocyanate, $KSCN(s)$, or 0.067 mol of aluminum sulfate, $Al_2(SO_4)_3$?

11. **T/I** Which has a greater amount in moles: a sample of sulfur trioxide containing 4.9×10^{22} atoms of oxygen or a 4.9 g sample of carbon dioxide?

12. **A** Imagine that you are an environmental chemist who is testing drinking water for water hardness. A sample of water you test has 3.5×10^{-2} mg of dissolved calcium carbonate, $CaCO_3(aq)$.
    a. How many formula units of calcium carbonate are in the sample?
    b. How many oxygen atoms are in the calcium carbonate in the sample?

13. **A** A chemist is testing for lead content in water samples taken downstream from a battery manufacturing plant. Health Canada suggests that drinking water should have a maximum lead content of 0.010 mg/L of water. If a test reveals that a 1 L water sample contains 3.1×10^{17} atoms of lead, is the water safe to drink? Explain.

14. **T/I** Copy and complete the following table.

Number of Particles, Amount, and Mass of Selected Elements and Compounds

Substance	Number of Particles	Amount (mol)	Mass (g)
$P_4(s)$		13.2	
$Ba(MnO_4)_2(s)$	6.7×10^{20}		
$C_5H_9NO_4(s)$			19.62

15. **A** Methyl salicylate, $C_6H_4(OH)COOCH_3(s)$, is used in many consumer products, such as mouthwash, as a flavouring. A mouthwash sample contains 1.38×10^{18} molecules of methyl salicylate. What is the mass of the methyl salicylate in the sample?

16. **T/I** Determine the order of the following three substances, from smallest to greatest number of carbon atoms: 5.6×10^{23} molecules of benzoic acid, $C_6H_5COOH(s)$; 1.3 mol of acetic acid, $CH_3COOH(\ell)$; 0.17 kg of oxalic acid, $HOOCCOOH(s)$.

Materials

- 1 dozen dried kidney beans (or another type of large bean)
- 1 dozen dried chick peas (or another type of bean of similar size)
- 1 dozen dried lentils (or another type of bean of similar size)
- small piece of copper wire, Cu(s)
- 10 mL of sodium chloride (table salt), NaCl(s)
- 10 mL distilled water, $H_2O(\ell)$
- 50 mL beaker
- balance
- graduated cylinder

Exploring Conversions between Mass and Particles

In this investigation, you will work with two units of quantity: the dozen and the mole. You will practise converting a dozen into individual items, and vice versa. You might ask, "How is working with a dozen useful in chemistry?" In the second part of this investigation, you will apply the skills you learned while working with a dozen to determine quantities of different sample substances.

Pre-Lab Questions

1. How would you convert 10 dozen into a number of items? Describe the steps you would follow, and perform the calculation.

2. How would you convert 10 moles into a number of particles? Describe the steps you would follow, and perform the calculation.

3. Read the Procedure and then explain why it is important that you use the same 50 mL beaker in Part 1 and Part 2 of the investigation.

4. How would you determine the amount in moles of a sample, if you are given only its mass?

Questions

How can you use units of quantity, such as a dozen and a mole, to count small objects and particles and to determine mass? How can you use mass to determine quantities of a substance and number of particles?

Procedure

Part 1: Counting and Measuring the Mass of Beans

1. Read steps 2 to 5 in this Procedure, and create a table to record your results.

2. Count out one dozen of one type of bean.

3. Measure the mass of the 50 mL beaker. Record your result.

4. Fill the beaker with the dozen beans. Measure the mass of the beans and beaker. Record your result.

5. Repeat steps 2 to 5 for each type of bean you have been given.

Part 2: Measuring the Mass of Chemicals

6. Read steps 7 to 11, and create a table to record your results.

7. Measure the mass of the sample of copper given to you by your teacher. Record your result.

8. Determine the molar mass of copper from the periodic table. Record this value in your table.

9. You have already measured the mass of the 50 mL beaker in Part 1. Add 10 mL of sodium chloride to the beaker. Measure the mass of the sodium chloride and the beaker. Record your result.

10. Determine the molar mass of sodium chloride from the periodic table. Record this value in your table.

11. Repeat steps 9 and 10 using distilled water instead of sodium chloride.

Analyze and Interpret

Part 1

1. For each type of bean, calculate the mass of the dozen beans in the beaker.

2. From the net mass found in step 1, calculate the average mass of a single bean for each type of bean.

3. For each type of bean, calculate the number of beans in 3.5 dozen. Then calculate the mass of 3.5 dozen beans.

4. For each type of bean, calculate how many dozen beans you would have if you had 25.0 g of beans. How many individual beans would you have?

Part 2

5. Use the mass and the molar mass of each substance you measured to calculate the amount in moles of each sample. Remember to calculate the net mass for the sodium chloride and water samples.

6. Determine the number of atoms of copper, molecules of water, and formula units of sodium chloride in the samples you were given.

7. Calculate the mass of 1 mol of each substance.

8. Calculate the mass of one particle (atom of copper, molecule of water, and formula unit of sodium chloride) of each substance.

9. Calculate the number of particles in 3.0 mol of each substance.

10. Calculate the mass of 5.0 mol of each substance.

11. Calculate the amount in moles of each substance in 100 g of the substance.

Conclude and Communicate

12. What general strategy or formula could you use to convert a dozen items to individual items, and vice versa?

13. What general strategy or formula could you use to convert a number of dozen items to the mass of the items, and vice versa?

14. Given the mass of a substance, what general strategy or formula could you use to determine the mass of one particle of the substance?

15. What general strategy or formula could you use to convert the mass of a substance to the amount in moles of the substance?

16. What general strategy or formula could you use to convert the amount of a substance in moles to the number of particles of the substance?

17. Compare the strategies or formulas you used in Part 1 with the strategies or formulas you used in Part 2. How are they similar?

Extend Further

18. **INQUIRY** Design an investigation that will enable you to answer the following question: What amount in moles of calcium carbonate, $CaCO_3(s)$, is in one tablet of a given calcium supplement? Assume that the supplement contains only calcium carbonate.

19. **RESEARCH** Research how knowledge of chemical quantities is important in the education, training, and work of environmental chemists.

Safety Precautions

- Tie back loose hair and clothing.
- Do not touch the lit candle.
- Be careful not to spill any molten wax on your hands.

Materials

- paraffin wax candle (tea light)
- wooden splints
- water
- sodium chloride (table salt), $NaCl(s)$
- sucrose (table sugar), $C_{12}H_{22}O_{11}(s)$
- weighing paper
- paper
- wax crayon
- balance
- clock or timer
- small paper clip
- beaker
- 10 mL graduated cylinder
- medicine dropper
- 50 mL graduated cylinder
- ruler

Using the Mole for Measuring and Counting Particles

The mole is used to group large numbers of particles (atoms, molecules, ions, and formula units) of a substance. Chemists can use the mole to determine how many particles of a substance they are working with. In this investigation, you will use the mole concept to practise measuring and counting atoms, molecules, and other particles of different substances.

Pre-Lab Questions

1. List safety precautions you should take when working with a candle in a laboratory.

2. Write a plan to figure out the amount of candle wax that burns over a period of time and the rate at which it burns.

3. Describe the correct method for measuring a crystalline solid, such as table salt or table sugar, on a balance.

Question

What are some methods for measuring and counting the number of atoms, molecules, formula units, and ions in samples of different chemicals?

Procedure

1. Read through this Procedure, and create a table you can use to record your results for each step.

2. Measure the mass of a candle. Light the candle, and burn the paraffin wax for 10 min. Put out the candle, and let it sit for 4 min to allow the melted wax to solidify. Measure the mass of the candle again.

3. Measure the mass of a paper clip. Try floating the paper clip on water in a beaker. Hold the paper clip horizontally (flat). Very carefully, lower the paper clip onto the surface of the water. The paper clip should float.

4. Measure the mass of a 10 mL graduated cylinder. Using a dropper full of water, count the number of drops of water you need to get 1 mL of water in the cylinder. Measure the mass of the water and the cylinder.

5. a. Calculate the mass of 0.10 mol of sodium chloride. Using weighing paper, measure the mass on a balance, and place it in a 50 mL graduated cylinder. Measure and record the volume of the salt. Return the salt to the weighing paper.

   b. Calculate the mass of 0.10 mol of sucrose. Using weighing paper, measure the mass, and place it in the 50 mL graduated cylinder. Measure and record the volume of the sugar. Return the sugar to the weighing paper.

6. a. Draw a 10 cm by 10 cm square on a piece of paper. Measure the mass of your favourite colour of wax crayon.

   b. Colour in the entire square as completely as possible with the wax crayon. Measure the mass of the crayon again.

   c. Repeat the process with a 20 cm by 20 cm square on a piece of paper.

Analyze and Interpret

1. Refer to your results from Procedure step 2 to answer this question.

   a. Calculate the number of wax molecules that burned. Assume that the formula for paraffin wax is $C_{25}H_{52}(s)$.

   b. Calculate the number of carbon atoms that burned.

2. Refer to your results from Procedure step 3 to answer this question: How many atoms of iron is the water holding up? Assume that the paper clip is pure iron.

3. Refer to your results from Procedure step 4 to answer this question.

   a. What is the mass of 1 mL of water?

   b. How many water molecules are in 1 mL of water?

   c. How many hydrogen atoms are in one drop of water?

   d. Calculate the volume of one mole of water.

4. Refer to your results from Procedure step 5 to answer this question.

   a. Compare the volumes of the 0.10 mol samples of table salt and table sugar, and account for any differences.

   b. How many sodium and chloride ions are in the salt sample?

   c. How many oxygen atoms are in the sugar sample?

5. Refer to your results from Procedure step 6 to answer this question. Assume that the wax crayon, like the candle, is paraffin wax. Based on the mass of wax crayon used, calculate the number of molecules of wax in the 10 cm by 10 cm square and the 20 cm by 20 cm square. Compare the difference in the areas of the two squares with the difference in the masses of the wax. How do they compare?

Conclude and Communicate

6. When burning a candle like the one in this investigation, how long would it take to burn 1 mol of paraffin wax?

7. In terms of the wax crayon, what would the dimensions of a square coloured on a page have to be to use an entire mole of wax?

8. Calculate the mass of one mole of water. Then, using your calculations for the volume of one mole of water, calculate the density of water at room temperature in grams per millilitre (g/mL). Check a reference source to see if this value is reasonable.

Extend Further

9. **INQUIRY** Design an investigation to determine the number of particles (molecules or formula units) of chalk it would take to write your name. First, research the main chemical that makes up chalk. Perform your investigation after your teacher approves it.

10. **RESEARCH** In this investigation, you assumed that the formula for the paraffin wax you used was $C_{25}H_{52}(s)$. Research the different formulas for paraffin wax. How did the formula you used affect your results? Why was it important to make an assumption about the formula in order to calculate your results?

Case Study

Prescription for Safety

Ensuring Unit and Dosage Precision

Scenario

Every day, tens of thousands of people in Canada drop off prescriptions to be filled at their local pharmacies. Then they wait... and wait. What accounts for the frequently long wait times?

Dosages of prescription medications must be calculated accurately and safely before they are dispensed to a customer. This can require precision in applying fractions, decimals, ratios, percentages, and metric conversions. Unfortunately, however, dosing errors do occur on rare occasions, and these errors can lead to adverse effects.

For example, the Food and Drug Administration (FDA) in the United States released an advisory concerning possible dosing errors associated with the liquid form of TAMIFLU®. TAMIFLU® is an antiviral agent, which is a type of medication that helps to prevent some flu viruses from replicating in the human body. When a virus enters a human body, it uses body cells to create more copies of itself. TAMIFLU® may be recommended to help stop the replication process. If TAMIFLU® is prescribed within two days of the appearance of flu symptoms, it can reduce the recovery time by about one day. TAMIFLU® can also help to prevent the spread of some types of flu.

Unfortunately, the FDA discovered a problem. The FDA advisory stated that there had been reports of cases in which the instructions on the label of the liquid form of TAMIFLU® gave dosage information in millilitres (mL) or teaspoons (tsp),

while the measuring device provided with the medication was marked in milligrams (mg). This led to dosing errors and confusion among patients, doctors, nurses, and pharmacists.

The Canadian federal government released its own advisory to health-care professionals on October 13, 2009, about the risk of TAMIFLU® dosing errors when administered as a 12 mg/mL oral suspension. According to the advisory, no known dosing errors had occurred in Canada, but the government was nonetheless launching an investigation. The advisory emphasized the following information:

- Pharmacists should ensure that the units used in the prescription instructions match the units on the dosing dispenser provided. For example, a dispenser marked in milligrams should be used for a prescription given in milligrams.

- In Canada, 30 mg, 45 mg, and 60 mg graduations appear on the dispenser provided with TAMIFLU®, not graduations in millilitres (mL) or teaspoons (tsp). Therefore, the recommended doses for both adults and children should be given in milligrams.

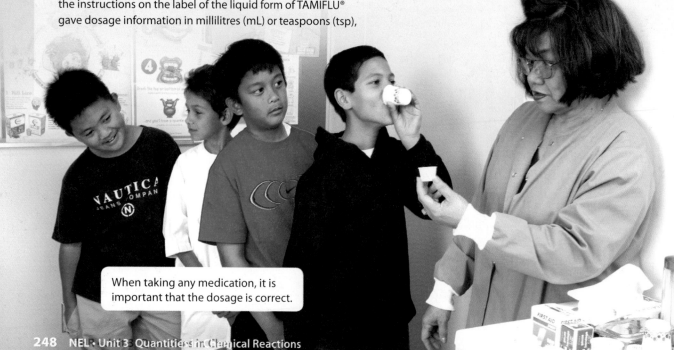

When taking any medication, it is important that the dosage is correct.

Risk of TAMIFLU® dosing errors

The Canadian federal government advisory highlights the importance of accurate ratio calculations and unit conversions for administering the correct dose. For example, in one case in the United States, a six-year-old child was prescribed three quarters of a teaspoon of TAMIFLU® twice daily. However, the child's parents were given an oral dosing dispenser with 30 mg, 45 mg, and 60 mg markings, instead of a dispenser with teaspoon markings. The units given in the instructions on the bottle were also different from the units on the dispenser.

Fortunately, the parents knew that one teaspoon is 5 mL and therefore three quarters of a teaspoon is 3.75 mL. They had read that the concentration of TAMIFLU® in liquid suspension is 12 mg/mL. They used this information to convert the dosing instructions they received to the correct quantity of medication, in milligrams, for each dose:

dose = concentration of TAMIFLU® × dose of TAMIFLU®
        in liquid suspension

= 12 mg/mL × 3.75 mL

= 45 mg

These parents were able to determine that the correct dose for their child was 45 mg. However, many people would have been unable to figure out this conversion and might have given a child an incorrect dose.

The recommended dose of TAMIFLU®, like all medications, depends on a person's age and body mass. The recommended doses for adults and children are given in milligrams in the table below.

Dose of TAMIFLU® Powder for Oral Suspension (12 mg/mL)

Body Mass	Recommended Dose for Five Days	Volume of Liquid TAMIFLU® per Dose
≤ 15 kg	30 mg twice daily	2.5 mL
> 15 kg to 23 kg	45 mg twice daily	3.8 mL
> 23 kg to 40 kg	60 mg twice daily	5.0 mL
> 40 kg	75 mg twice daily	6.2 mL

Source: Health Canada

Research and Analyze

1. In December 2009, the Institute for Safe Medication Practices Canada reported a case in which a school-aged child mistakenly received an adult prescription of 150 mg of TAMIFLU® twice daily, instead of 60 mg twice daily, the amount appropriate for the child's body mass. Conduct research to find out why dosage guidelines, based on body mass, age, and other factors, exist for medications such as TAMIFLU®.

2. Conduct research to find out what steps doctors, nurses, and pharmacists can take to avoid dosing errors.

3. Using a graphic organizer, analyze the possible consequences of not following dosage guidelines.

Take Action

1. **Plan** In a group, discuss the issue of dosing errors. Share the results of the research and analysis you conducted for questions 1 to 3. What suggestions do you have for minimizing the risk of dosing errors?

2. **Act** Create a pamphlet, website, or public service announcement about how and why patients and health-care providers should work together to minimize the risk of dosing errors.

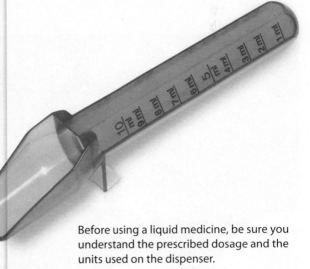

Before using a liquid medicine, be sure you understand the prescribed dosage and the units used on the dispenser.

Section 5.1   The Avogadro Constant and the Mole

Chemists use the mole as a convenient way to count atoms, molecules, ions, and formula units.

KEY TERMS
Avogadro constant
mole

KEY CONCEPTS

- People group common items into units of quantity for convenience. Chemists also group atoms, molecules, ions, and formula units into units of quantity for convenience.
- A mole is the amount of substance that contains as many particles (atoms, molecules, ions, or formula units) as the number of atoms in exactly 12 g of the isotope carbon-12.
- The mole is the SI base unit that is used to measure the amount of a substance.
- Chemists use the mole to count very small particles, such as atoms, molecules, ions, and formula units.
- The Avogadro constant allows chemists to convert back and forth between individual particles and moles of particles.

Section 5.2   Mass and the Mole

Conversion factors, such as molar mass and the Avogadro constant, allow chemists to convert among number of individual particles, amount, and mass.

KEY TERM
molar mass

KEY CONCEPTS

- The mass of one atom of an element in atomic mass units has the same numerical value as the mass of one mole of atoms of the same element in grams.
- Molar mass is an important tool for converting back and forth between the mass and amount of an element or a compound.
- Conversions can be made between the number of individual particles of a substance and the mass of the substance, using both the Avogadro constant and the molar mass.
- All of these conversions are important when calculating and measuring the amounts of substances in a chemical reaction.

Knowledge and Understanding

Select the letter of the best answer below.

1. The Avogadro constant allows you to convert between
   a. mass and amount in moles
   b. mass and number of particles
   c. amount in moles and number of particles
   d. mass and atomic number
   e. number of particles and atomic number

2. Why do chemists use moles when working with substances instead of individual atoms or molecules?
   a. The mole is the more practical unit to use.
   b. The mole is a unique unit for chemists to use.
   c. Individual atoms or molecules can be different in different samples of a compound.
   d. The mole determines which products form.
   e. The mole starts a chemical reaction.

3. How is the mass of one atom of an element related to the mass of one mole of atoms of the element?
   a. They both have the same number of particles.
   b. The mass of one atom of an element in atomic mass units has the same numerical value as the mass of one mole of atoms of the same element in grams.
   c. The mass of one mole of atoms of an element in atomic mass units has the same numerical value as the mass of one atom of an element in grams.
   d. The mass of one atom is 12 times the mass of one mole of the same element.
   e. There is no relationship because the masses vary from element to element.

4. What is the symbol for molar mass?
   a. *M*
   b. *m*
   c. *N*
   d. *n*
   e. Mol

5. What are the units for molar mass?
   a. mol
   b. g
   c. mol/g
   d. g/mol
   e. g/u

6. The substance with the largest atomic molar mass is
   a. cadmium
   b. vanadium
   c. thallium
   d. selenium
   e. lanthanum

Use the following information to answer questions 7 and 8.

   i. divide by the molar mass
   ii. divide by the Avogadro constant
   iii. multiply by the molar mass
   iv. multiply by the Avogadro constant

7. When converting from the number of individual particles of a substance to the mass of the substance, you would first _____ and then _____.
   a. i, iii
   b. i, iv
   c. ii, iii
   d. ii, iv
   e. iv, i

8. When converting from the mass of a substance to the number of individual particles of the substance, you would first _____ and then _____.
   a. i, iii
   b. i, iv
   c. ii, iii
   d. ii, iv
   e. iv, i

Answer the questions below.

9. Is the amount in moles of atoms the same as the amount in moles of molecules in water? Explain.

10. Fill in the blanks: _____ g of zinc is 1.00 mol of zinc and is _____ atoms of zinc.

11. a. Compare the number of particles in one mole of chromium, with one mole of aluminum.
    b. Compare the mass of one mole of chromium with one mole of aluminum.

12. Describe the steps you would take to find the molar mass of a compound.

13. Describe the method you would use to calculate the mass of a compound from a given amount in moles.

14. What is the name and value of the number used to convert moles of carbon dioxide to molecules of carbon dioxide?

15. Suppose that you are given the value of 65.38 for the element zinc. Name two units that can describe this value. Describe what each unit means.

16. Why would you work in moles when dealing with substances in a chemical equation, but work in grams when measuring those substances?

Thinking and Investigation

17. Calculate the amount in moles of each substance.
  a. 0.045 kg of lithium thiosulfate, $Li_2S_2O_3(g)$
  b. 9.52×10^{24} molecules of benzene, $C_6H_6(\ell)$
  c. 356 formula units of tin(II) borate, $Sn_3(BO_3)_2(s)$

18. Calculate the mass of each substance.
  a. 1.3×10^{22} molecules of difluorine monoxide, $F_2O(g)$
  b. 0.659 mol of glucose, $C_6H_{12}O_6(s)$
  c. 3.5×10^{26} formula units of cobalt(III) nitrate, $Co(NO_3)_3(s)$

19. Ibuprofen, $C_{13}H_{18}O_2(s)$, is used to treat fever, pain, and inflammation. A common adult dose is 400 mg of ibuprofen. What amount in moles of ibuprofen is in a 4.00×10^2 mg dose?

20. What mass of chromium metal is in a sample that contains 6.9×10^{25} formula units of ammonium dichromate, $(NH_4)_2Cr_2O_7(s)$?

21. Determine the amount in moles in 20.0 g of glucose, $C_6H_{12}O_6(s)$, and 20.0 g of propane, $C_3H_8(g)$.

22. Which substance has the larger molar mass: nitrogen trichloride, $NCl_3(s)$, or sodium chloride, $NaCl(s)$?

23. What amount in moles is present in 10.00 g of each of the following compounds?
  a. calcium chloride dihydrate, $CaCl_2 \cdot 2H_2O(s)$, used as road salt for icy roads
  b. octane, $C_8H_{18}(\ell)$, a component of gasoline
  c. cysteine, $C_3H_7O_2S(\ell)$, an amino acid

24. Using your own words, explain the difference between the atomic mass unit and the atomic molar mass. What is the advantage of using atomic molar mass?

25. Copy and complete the table below in your notebook.

26. The circle graph below shows the mass of each element that makes up one mole of tryptophan, an amino acid. What is the chemical formula for tryptophan?

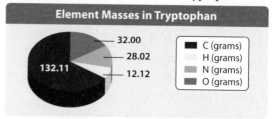

Element Masses in Tryptophan

- 32.00
- 28.02
- 132.11
- 12.12
- C (grams)
- H (grams)
- N (grams)
- O (grams)

Communication

27. Use a flowchart to show how you would find the mass of a sample from the number of individual molecules it contains.

28. Use a flowchart to show how you would find the number of individual molecules from the mass of a compound.

29. Draw a graphic organizer to illustrate the relationships among number of molecules, amount in moles, and mass in grams of a sample of carbon dioxide.

30. The structure of a trinitrotoluene (TNT) molecule is shown below. A web site states that there are 21 atoms in each mole of TNT. Write a correction to this statement that could be posted on the web site. Explain why the statement is incorrect in a manner that could be understood by the general public.

Particles, Amount, and Mass of Various Pure Substances

Substance	Total Number of Atoms	Number of Molecules or Formula Units	Molar Mass (g/mol)	Amount of Substance (mol)	Mass (g)
$C_3H_6O_2(\ell)$				0.0675	
$NaC_6H_5COO(s)$	3.56×10^{19}				
$Al(H_2PO_4)_3(s)$					32.71
CCl_2F_2 (g)		4.75×10^{25}			
$C_4H_{10}O_2(\ell)$				0.0574	
$NaHCO_3(s)$					41.34

31. A YAG, or yttrium aluminum garnet, $Y_3Al_5O_{12}(s)$, is a synthetic gemstone. Create a bar graph to show the amount in moles of each element in a 5.67 carat yttrium aluminum garnet. (1 carat = 0.20 g)

32. Using a Venn diagram, show the similarities and differences between working with dozens and working with moles.

33. Summarize your learning in this chapter using a graphic organizer. To help you, the Chapter 5 Summary lists the Key Terms and Key Concepts. Refer to Using Graphic Organizers in Appendix A to help you decide which graphic organizer to use.

Application

34. Ibuprofen is a common household medication.
   a. Why is it important for the manufacturer to put the unit mass of the medication on the container?
   b. Why is it important to follow the directions for taking the medication, as found on the container?
   c. Research why it is recommended that youth take ibuprofen, instead of acetylsalicylic acid (ASA) during a viral illness.

35. A compound is composed of two elements and has the formula X_2Y_3. The atomic molar mass of X is 47.87 g/mol, and 0.25 mol of the compound has a mass of 47.99 g. What is the most likely identity of compound X_2Y_3, by formula and name?

36. Lycopene, $C_{40}H_{56}(s)$, is a complex molecule that is a highly potent antioxidant. Antioxidants help to reduce the breakdown of cells. Some studies indicate that these effects occur with a daily intake of 6.5 mg of lycopene.
   a. Calculate the amount in moles of lycopene necessary to provide its protective effects.
   b. Lycopene is responsible for the red colour in tomatoes. Conduct research to find out how much lycopene is in an average tomato.

37. Vitamin A plays an important role in several body functions, including vision and skin health. The most common form of vitamin A is retinol, $C_{20}H_{30}O(s)$.
   a. The recommended daily intake of vitamin A is 600 μg. What amount in moles is in this mass of vitamin A?
   b. Unlike vitamins B and C, vitamin A is fat-soluble. This means that excess vitamin A is harder to remove from the body than excess vitamins B and C. You must therefore be careful how much vitamin A you take in vitamin supplements. Research the effects of taking too much vitamin A. How much vitamin A can you take before it becomes toxic?

38. In ancient Egypt, a material called kohl, a black cosmetic, was applied around the eyes to repel flies and to reduce glare from the Sun. Kohl contained a mineral called galena, which is a natural source of lead(II) sulfide, PbS(s).
   a. Quantities as high as 84 g of lead(II) sulfide/100 g of kohl have been found in some preparations of this cosmetic. What mass of lead(II) sulfide would a 0.50 g sample of such a kohl preparation contain?
   b. What amount in moles of lead(II) sulfide is this?
   c. Research Health Canada's position on lead in cosmetics.

39. Toothpaste has many additives whose quantities have to be carefully controlled for safety reasons. Brands of toothpaste made for sensitive teeth usually contain potassium nitrate, $KNO_3(s)$, at a concentration of 5% w/w (which means 5 g of potassium nitrate/100 g of toothpaste). Toothpaste may also contain fluoride, in the form of sodium fluoride, NaF(s), at a concentration of 0.11% w/w.
   a. If the mass of a tube of toothpaste for sensitive teeth is 250 g, what masses of potassium nitrate and sodium fluoride are in the tube?
   b. What amounts in moles of potassium nitrate and sodium fluoride are in the tube?
   c. Research some of the benefits and risks of using potassium nitrate and sodium fluoride in toothpaste.

40. **BIG IDEAS** The efficiency of chemical reactions can be determined and optimized by applying an understanding of quantitative relationships in such reactions. An important aspect of industrial chemical processes is the need for quantitative accuracy. There can be many potential environmental impacts if accuracy is not observed. Outline the results you obtained in the Launch Lab. How did the results you obtained illustrate some safety precautions that industries might take when working with similar chemical processes?

Select the letter of the best answer below.

1. **K/U** The Avogadro constant is
   a. a number that is equivalent to one mole of a substance
   b. a value that is equal to 6.02×10^{23} particles
   c. a value in atomic mass units that describes the atomic mass of an element
   d. a and c above
   e. a and b above

2. **K/U** Which of the following describes a mole?
   a. a unit that is used to measure mass
   b. a unit that is equivalent to the atomic mass
   c. a term that is used to describe a type of particle
   d. the SI base unit that is used to measure the amount of a substance
   e. a term that is used to describe the physical state of a substance

3. **T/I** The number of particles and amount, respectively, in 45.2 g of potassium nitrate is
   a. 2.69×10^{22} particles, 0.0447 mol
   b. 2.69×10^{23} particles, 0.447 mol
   c. 3.20×10^{23} particles, 0.532 mol
   d. 3.94×10^{23} particles, 0.654 mol
   e. 3.20×10^{23} particles, 0.531 mol

4. **T/I** What is the mass of 5.8 mol of nitrogen dioxide, $NO_2(g)$?
   a. 2.7 g
   b. 27 g
   c. 2.7×10^2 g
   d. 17 g
   e. 1.7×10^2 g

5. **K/U** To get the number of particles of a compound from the mass of a compound, you ____ the molar mass and then ____ the Avogadro constant.
   a. multiply by ; divide by
   b. divide by; divide by
   c. divide by; multiply by
   d. multiply by; multiply by
   e. none of the above; none of the above

6. **T/I** How many hydrogen atoms are in 3.25 g of acetic acid, $CH_3COOH(\ell)$?
   a. 3.26×10^{22}
   b. 6.74×10^{22}
   c. 9.77×10^{22}
   d. 1.30×10^{23}
   e. 3.26×10^{23}

7. **K/U** To calculate the mass of a compound from the number of particles of the compound, you _____ the Avogadro constant and then _____ the molar mass.
   a. divide by; multiply by
   b. divide by; divide by
   c. multiply by; multiply by
   d. multiply by; divide by
   e. add; subtract

8. **T/I** Determine the mass of one formula unit of barium hexafluorosilicate, $BaSiF_6(s)$.
   a. 1.68×10^{26} g
   b. 2.16×10^{21} g
   c. 4.64×10^{-22} g
   d. 6.02×10^{-23} g
   e. 1.68×10^{-26} g

9. **T/I** Examine the pie graph below. What is the molar mass of compound X?

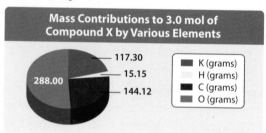

Mass Contributions to 3.0 mol of Compound X by Various Elements

117.30
15.15
288.00
144.12

K (grams)
H (grams)
C (grams)
O (grams)

   a. 1693.71 g
   b. 564.57 g
   c. 367.38 g
   d. 188.19 g
   e. 110.25 g

10. **T/I** Determine the number of atoms in 0.625 mol of germanium, Ge(s).
    a. 2.73×10^{25}
    b. 3.76×10^{23}
    c. 6.99×10^{25}
    d. 9.63×10^{23}
    e. 2.73×10^{23}

Use sentences and diagrams, as appropriate, to answer the questions below.

11. **K/U** Describe one similarity and one difference between the atomic molar mass of an element and the molecular molar mass of a compound.

12. **T/I** Determine the mass of 5.2×10^{21} molecules of propanol, $C_3H_7OH(\ell)$, in grams.

13. **T/I** Calculate the amount in moles of zinc ions in 0.0679 g of zinc phosphate, $Zn_3(PO_4)_2$.

14. **T/I** Convert 0.834 mol of lead(IV) nitrate, $Pb(NO_3)_4(s)$, into mass.

15. **C** Citric acid, $C_6H_8O_7(aq)$, is often used as an additive to make candies "sour."
   a. Draw a flowchart that shows how to calculate the molar mass of citric acid.
   b. Verify the steps in your flowchart by calculating the molar mass of citric acid.

16. **T/I** Acetylsalicylic acid (ASA), $C_9H_8O_4(s)$, is used as a pain reliever by millions of people. Calculate the number of molecules of ASA in a 325 mg tablet.

17. **T/I** Triclosan, $C_{12}H_7Cl_3O_2(s)$, is an agent used to give some soaps their antibacterial properties. Determine the number of atoms of each element in 0.678 g of Triclosan.

18. **T/I** The following graph shows the atoms in a compound used to make instant glue.

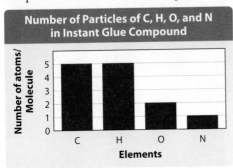

Number of Particles of C, H, O, and N in Instant Glue Compound

   a. What is the molar mass of the compound?
   b. What would be the number of each type of atom in 4.00 mol of this compound?

19. **A** Aspartame, $C_{14}H_{18}N_2O_5(s)$, is used as an artificial sweetener in low calorie foods. Arrange the elements in aspartame in order from largest to smallest contribution to the molar mass of the compound. Is this order the same as the order from largest to smallest number of atoms in the formula for the compound? Explain.

20. **K/U** Explain how the Avogadro constant is used as a conversion factor.

21. **T/I** What is the mass of each compound, in kilograms?
   a. 34.8 mol of sodium metasilicate, $Na_2SiO_3(s)$
   b. 4.96×10^{25} molecules of ascorbic acid, $C_6H_8O_6(s)$

22. **T/I** If you had a 2.00 mol sample of each of the following substances, which sample would have more molecules?
   • phosphorus pentoxide, $P_2O_5(s)$
   • phosphorus trichloride, $PCl_3(\ell)$
   • dinitrogen tetraoxide, $N_2O_4(g)$

23. **A** Ethanol, $C_2H_5OH(\ell)$, and water can be mixed in any proportions. This is useful for making a variety of substances, from fuel to antiseptic. If you were asked to make an 80.00 g mixture of ethanol and water, containing equal molar amounts of both, what mass of each substance would you need?

24. **C** Suppose that you are given the following materials to do a small investigation:
   • a dozen marbles sealed in a thin plastic bag
   • a balance
   • a sealed jar containing 365 of the same marbles

Write a step-by-step procedure outlining how you could do the following, without opening the bag or the jar:
   a. Estimate the average mass of one marble.
   b. Estimate the mass of the marbles in the jar (two ways).
   c. Estimate the mass of one mole of marbles (two ways).

Explain how the principles of this small investigation apply to working with mass, amount in moles, and molar mass.

25. **A** Dioxins, such as 2, 3, 6, 7-tetrachlorodibenzo-4-dioxin (TCDD), $C_{12}H_4Cl_4O_2(s)$, are some of the most toxic chemicals on the planet. If the average level of exposure to TCDD by humans on Earth is 3.70×10^{-13} mol/day, what is the value in picograms per day? $(1 \text{ g} = 1 \times 10^{12} \text{ pg})$

Self-Check

If you missed question …	1	2	3	4	5	6	7	8	9	10	11	12	13	14	15	16	17	18	19	20	21	22	23	24	25
Review section(s)…	5.1	5.1	5.1	5.2	5.2	5.2	5.2	5.2	5.2	5.1	5.2	5.2	5.2	5.2	5.2	5.2	5.2	5.2	5.2	5.1, 5.2	5.2	5.1	5.2	5.1, 5.2	5.2

CHAPTER 6

Proportions in Chemical Compounds

Specific Expectations

In this chapter, you will learn how to . . .

- D2.2 **conduct** an inquiry to **calculate** the percentage composition of a compound (6.1)

- D2.4 **determine** the empirical formulas and molecular formulas of various chemical compounds, given molar masses and percentage composition or mass data (6.2)

- D3.1 **explain** the law of definite proportions (6.1)

- D3.3 **explain** the relationship between the empirical formula and the molecular formula of a chemical compound (6.2)

Willow tree bark has been used to reduce pain and fever for thousands of years. It is still used today as a traditional medicine by Aboriginal peoples of North America. In the 1800s, scientists performed chemical analyses to identify the compound in willow tree bark that was responsible for its healing effects. Chemists were then able to determine the composition of and chemical formula for the compound, known as salicin. This knowledge enabled them to change its composition through various chemical reactions. Through their efforts, salicin's composition was altered to form salicylic acid, a more potent medication than salicin. However, salicylic acid is very corrosive to the stomach lining. By further changing the compound, chemists were able to reduce this harmful effect. The new compound they created was called acetylsalicylic acid (ASA). ASA is the main ingredient in Aspirin™, a medication that has been used to treat pain and inflammation since 1899.

Comparing Compounds

Salicylic acid, $C_7H_6O_3(s)$, is often found in lotions that are used to relieve muscle and joint pain. In higher concentrations, it can be used to remove warts. Acetylsalicylic acid (ASA), $C_9H_8O_4(s)$, is found in medications that are used to relieve pain, fever, and inflammation. These compounds are made of the same elements, and the proportions of these elements in the compounds are similar. Do the compounds have similar properties or do they differ? In this lab, you will observe the effects of each compound on different surfaces to answer this question in terms of each compound's corrosive properties.

Materials

- 1 scoop of acetylsalicylic acid powder, $C_9H_8O_4(s)$
- 1 scoop of salicylic acid powder, $C_7H_6O_3(s)$
- water
- various surfaces: aluminum foil, wax, Styrofoam®, gelatin spread on a glass slide
- two small beakers
- scoop
- medicine droppers or disposable pipettes
- clock or timer

Safety Precautions

- The chemicals you will be using can be toxic and caustic. Wear protective clothing, gloves, and safety eyewear.

Procedure

1. Read the procedure and create a data table to record your findings.

2. Add one scoop (5 mL) of acetylsalicylic acid to one beaker and one scoop (5 mL) of salicylic acid to another beaker. Add an equal number of drops of water to the two beakers using a medicine dropper or pipette. Stir after adding each drop. Continue adding drops and stirring until the liquid can just barely flow on its own.

3. Using a medicine dropper or pipette, transfer equal-sized amounts (about the size of a dime) of the two liquids onto two separate pieces of aluminum foil. Begin timing.

4. Observe the appearance of the surface every minute for 5 min. Record your observations.

5. Repeat steps 2 to 4 for the other surfaces.

Questions

1. Describe the effects of each chemical on each surface.

2. What can you conclude about the corrosiveness of acetylsalicylic acid and salicylic acid?

Chemical Proportions and Percentage Composition

Key Terms

law of definite proportions

mass percent

percentage composition

Figure 6.1 NASA discovered water on the Moon in 2009 by intentionally crashing a rocket into the lunar crater Cabeus A and then analyzing the debris and dust from the impact.

law of definite proportions a law stating that a chemical compound always contains the same proportions of elements by mass

On October 9, 2009, an extraordinary mission to find a life-sustaining substance on the Moon occurred. Hoping to find traces of water on the Moon, National Aeronautics and Space Administration's (NASA) Lunar Crater Observation and Sensing Satellite (LCROSS) mission sent a rocket hurtling into the crater Cabeus A, shown in **Figure 6.1**. Cabeus A, which is 100 km wide, is found near the south pole of the Moon. The impact tossed debris and dust over the lunar surface. Using special instruments, scientists found evidence that this lunar material contained significant amounts of water.

Compounds That Contain the Same Elements in the Same Proportions

Does lunar water have the same formula as water on Earth? It turns out that the answer to this question is yes. More than 200 years ago, scientists discovered that compounds contain elements with fixed mass proportions. It does not matter where water is found. Whether it comes from the moon, a lake, snow, or ice, as shown in **Figure 6.2**, a molecule of water always contains two hydrogen atoms and one oxygen atom. It also does not matter how large the sample is. The discovery that elements in the same compound are always found in fixed mass proportions was accepted as a law approximately 100 years after the initial discovery.

The Law of Definite Proportions

Around the turn of the 19th century, French chemist Joseph Louis Proust analyzed samples of copper(II) carbonate, $CuCO_3(s)$, from both natural and synthetic sources. He discovered that all the samples contained the same proportions of copper, carbon, and oxygen by mass, no matter what their source. Based on his discovery, Proust developed the **law of definite proportions** (also known as the law of definite composition or the law of constant composition).

Law of Definite Proportions

The elements in a chemical compound are always present in the same proportions by mass.

Figure 6.2 Humans use water in its many forms for recreation.

Identify What is the chemical composition of water in all of these photographs?

Mass Percent

According to the law of definite proportions, all water molecules have the same chemical composition. The average atomic mass of water molecules is 18.02 u (2.02 u of hydrogen plus 16.00 u of oxygen). Translated into moles, 1 mol of water has a mass of 18.02 g, which consists of 2.02 g of hydrogen and 16.00 g of oxygen. Note, however, that the law of definite proportions deals with the proportion of each element in a compound by mass, rather than the mass of each element. As a result, the proportion of each element is usually expressed as a percentage of the total mass of the compound. This is known as a **mass percent**. To determine the mass percent of each element in the compound, divide the mass of each element by the mass of the compound and multiply by 100 percent. As shown in the calculations below, the mass percent of hydrogen in water is 11.2 percent.

mass percent the mass of an element in a compound, expressed as a percentage of the total mass of the compound

$$\text{mass percent of H} = \frac{\text{mass of H}}{\text{mass of H}_2\text{O}(\ell)} \times 100\%$$

$$= \frac{2.02 \text{ g}}{18.02 \text{ g}} \times 100\%$$

$$= 11.2\%$$

This example for the hydrogen in water shows how the mass percent of an element can be calculated using the chemical formula for the compound. Mass percent can also be calculated using the results from an experiment. The following Sample Problem shows how mass percent can be calculated based on mass data obtained from an experiment. The Practice Problems and activity that follow will give you the opportunity to practise determining mass percent using the chemical formula for a compound and using experimental mass data.

Sample Problem

Determining the Mass Percent of an Element in a Compound

Problem

A 13.8 g sample of a compound contains 8.80 g of iron. Determine the mass percent of iron in the sample.

What Is Required?

You need to find the mass percent of iron in the sample.

What Is Given?

You know the mass of the sample: 13.8 g
You also know the mass of iron in the sample: 8.80 g

Plan Your Strategy	Act on Your Strategy
Divide the mass of iron by the mass of the sample, and then multiply by 100%.	$\text{mass percent of Fe} = \dfrac{\text{mass of Fe}}{\text{mass of sample}} \times 100\%$ $= \dfrac{8.80 \text{ g}}{13.8 \text{ g}} \times 100\%$ $= 63.8\%$ The mass percent of iron in the sample is 63.8%.

Check Your Solution

The mass of iron is about 9 g of the approximately 14 g sample, or about 64%, so the answer is reasonable.

Practice Problems

1. Calculate the mass percent of oxygen in iron(II) oxide, FeO(s).

2. Calculate the mass percent of oxygen in dinitrogen tetroxide, $N_2O_4(g)$.

3. A 650 mg sample is analyzed and found to contain 51.0 mg of hydrogen. What is the mass percent of hydrogen in the sample?

4. Calculate the mass percent of oxygen in acetic acid, $CH_3COOH(\ell)$.

5. Potassium dichromate, $K_2Cr_2O_7(s)$, is used in the production of safety matches. A 50.0 g sample of potassium dichromate contains 12.8 g of potassium. What is the mass percent of potassium in the potassium dichromate sample?

6. A 30.0 g sample of a compound contains 8.2 g of carbon. What is the mass percent of carbon in the sample?

7. Which substance has the greater mass percent of chromium: chromic acid, $H_2CrO_4(aq)$, or dichromic acid, $H_2Cr_2O_7(aq)$?

8. Which substance has the greater mass percent of sulfur: sulfurous acid, $H_2SO_3(aq)$, or peroxodisulfuric acid, $H_2S_2O_8(aq)$?

9. Calcium chloride, $CaCl_2(s)$, is used as a de-icer. Calculate the mass percent of chlorine in calcium chloride.

10. Many metals are refined from sulfide mineral deposits that were laid down by volcanoes billions of years ago. List the following sulfide compounds in order, from greatest to least mass percent of sulfur: lead(II) sulfide, PbS(s): zinc sulfide, ZnS(s): copper(I) sulfide, $Cu_2S(s)$.

Activity 6.1 Mass Percent of Oxygen

Imagine that you are an early chemist, studying compounds that contain one atom of potassium, one atom of chlorine, and varying numbers of atoms of oxygen. You are testing one compound to experimentally determine the mass percent of oxygen so you can calculate the relative number of oxygen atoms in the compound. You know that your compound will decompose when heated, losing the oxygen as a gas and leaving potassium chloride as a solid. You first measure the mass of your empty test tube. You then place some of the compound in the test tube and again measure the mass. You heat the test tube and then measure the mass of the test tube with the potassium chloride residue. Use the observations in the table below to determine the mass percent of oxygen in potassium chlorate.

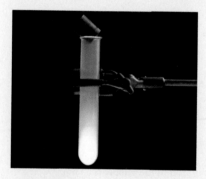

Data Table

Mass of test tube	19.85 g
Mass of test tube + original compound	24.62 g
Mass of test tube + potassium chloride residue	22.83 g

Procedure

1. Calculate the mass of your original compound.

2. Calculate the mass of potassium chloride residue that was produced.

3. Calculate the mass of oxygen that was lost as a gas.

4. Calculate the mass percent of oxygen in your original compound, based on the experimental data.

5. Determine whether your compound contains one, two, three or four oxygen atoms by calculating the mass percent of oxygen for each of the possible compounds.

6. Calculate the percentage error in the experiment, using the difference between the theoretical mass percent of oxygen (expected value for the compound that you determined in step 5) and the mass percent of oxygen measured by experiment. Go to Measurement in Appendix A for help with percentage error.

Questions

1. How does the law of conservation of mass allow you to calculate the mass of oxygen that was produced?

2. Why do you think chemists use mass percent rather than a percent of the number of atoms in a compound?

3. Write the chemical formula for your original compound. State the name of the compound.

4. Write the balanced chemical equation for the decomposition reaction that occurred when you heated your compound.

Compounds That Contain the Same Elements in Different Proportions

Whether you take a sample of water from a glacier, the Moon, or your own tap, the mass percent of hydrogen and oxygen atoms always are the same. Each unique compound contains the same percent by mass of each of its elements. However, there are numerous compounds that have the same elements in different proportions.

Different Proportions, Different Properties

If you completed the Launch Lab at the beginning of this chapter, you saw how different compounds with the same elements in different proportions—salicylic acid and acetylsalicylic acid—have different properties. Two other compounds with the same elements but very different properties are nitrogen dioxide, $NO_2(g)$, and dinitrogen monoxide, $N_2O(g)$, shown in **Figure 6.3**. Nitrogen dioxide, which is emitted from combustion engines, is highly toxic. In comparison, dinitrogen monoxide, a chemical with a similar formula, is used to calm patients during dental procedures.

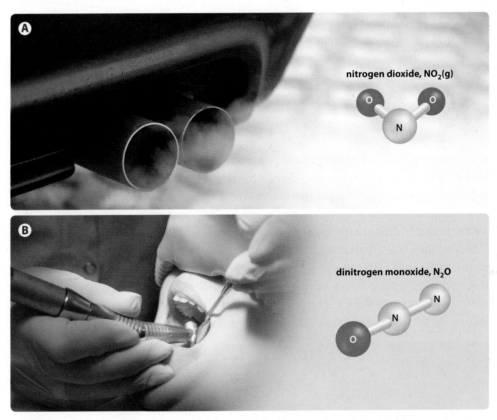

nitrogen dioxide, $NO_2(g)$

dinitrogen monoxide, N_2O

Figure 6.3 The car exhaust **(A)** contains nitrogen dioxide, a toxic brown gas that contributes to pollution. The dental patient **(B)** is receiving dinitrogen monoxide, commonly known as "laughing gas." Dinitrogen monoxide is a helpful compound, when used properly, with completely different properties than nitrogen dioxide. Both of these compounds contain nitrogen and oxygen, but in different proportions.

Similarly, water, $H_2O(\ell)$, and hydrogen peroxide, $H_2O_2(\ell)$, contain the same elements but in different proportions. Their properties differ greatly. Water is generally stable, whereas hydrogen peroxide is quite reactive. Carbon dioxide, $CO_2(g)$, and carbon monoxide, $CO(g)$, are another example. Both contain carbon and oxygen. But whereas carbon dioxide is a gas you breathe out, carbon monoxide can be a deadly poison.

1. The launch of the rocket that NASA intentionally crashed into the Moon burned hydrogen and formed water as the main product. Was this water the same as the water that is found naturally on Earth and the Moon? Explain.

2. Explain how nitrogen and oxygen can combine to form two different compounds, as shown in **Figure 6.3**.

3. Why do you think the law of definite proportions is also known as the law of definite composition or the law of constant composition?

4. Can the mass percent of carbon in carbon dioxide, $CO_2(g)$, change? Explain.

5. Can the mass percent of carbon in different compounds be different? Explain.

6. The mass percent of carbon in carbon monoxide, $CO(g)$, is 42.9%.
   a. What is the mass percent of oxygen?
   b. Why is the mass percent of carbon atoms different from the mass percent of oxygen atoms, even though they are in the same ratio in the formula CO?

Percentage Composition

percentage composition the percent by mass of each element in a compound

When you combine the mass percent for all of the elements in a compound, you have the **percentage composition** of the compound. Two examples are shown below.

Percentage composition of $NO_2(g)$
mass percent of N = 30.45%
mass percent of O = 69.55%
Total   100%

Percentage composition of $N_2O(g)$
mass percent of N = 63.65%
mass percent of O = 36.25%
Total   100%

Determining Percentage Composition from Mass Data

Knowing the percentage composition of a compound can help scientists determine the identity and composition of newly discovered chemical compounds. For example, vitamin C, an important molecule for human health, was first discovered in the early 1900s. Humans cannot synthesize their own vitamin C. Most people get the vitamin C they require from their diet, mainly from fruits and vegetables, such as the oranges shown in **Figure 6.4**. Because some people do not get enough vitamin C from their diet, synthetic vitamin C is produced and sold as a dietary supplement. In order for manufacturers to produce a synthetic copy of a molecule, however, scientists must first determine the composition and structure of the original molecule.

vitamin C
$C_6H_8O_6$

Figure 6.4 Vitamin C is a complex organic molecule that comes from plant sources. Some foods, such as citrus fruits, strawberries, and rosehips, are especially rich in this molecule.

Explain   How might knowledge of the percentage composition of vitamin C help scientists produce it synthetically in a laboratory?

An Application of Percentage Composition

Vitamin C, also known as ascorbic acid, $C_6H_8O_6(s)$, was first produced synthetically in a laboratory in 1933. This was such an important accomplishment that one of the scientists involved, Sir Walter Norman Haworth, received the Nobel Prize in Chemistry for his part in this achievement. Determining the proportions of the different elements in ascorbic acid was a crucial part of the process. The Sample Problem below shows how to calculate the percentage composition of elements in a simpler compound. Study this problem, and then complete the Practice Problems that follow.

Go to **scienceontario** to find out more

Sample Problem

Determining Percentage Composition from Mass Data

Problem
A sample of a compound that is found in gasoline has a mass of 35.8 g. The sample contains 30.10 g of carbon and 5.70 g of hydrogen. What is the percentage composition of the compound?

What Is Required?
You need to calculate the mass percent of carbon and hydrogen in the compound.

What Is Given?
You know the mass of the compound sample: 35.8 g
You know the mass of carbon in the sample: 30.10 g
You know the mass of hydrogen in the sample: 5.70 g

Plan Your Strategy	Act on Your Strategy
Find the mass percent of carbon. To do this, divide the mass of carbon by the mass of the sample and then multiply by 100 to get the percent.	mass percent of C $= \dfrac{\text{mass of C}}{\text{mass of sample}} \times 100\%$ $= \dfrac{30.10 \text{ g}}{35.8 \text{ g}} \times 100\%$ $= 0.8407 \times 100\%$ $= 84.1\%$
Find the mass percent of hydrogen. Divide the mass of hydrogen by the mass of the sample, and then multiply by 100 to get the percent.	mass percent of H $= \dfrac{\text{mass of H}}{\text{mass of sample}} \times 100\%$ $= \dfrac{5.70 \text{ g}}{35.8 \text{ g}} \times 100\%$ $= 0.1592 \times 100\%$ $= 15.9\%$ Therefore, the percentage composition of the compound is 84.1% carbon and 15.9% hydrogen.

Alternative Solution
Because there are only two elements in the compound, the mass percent of hydrogen can also be found by subtracting the mass percent of carbon from 100%.

Mass percent of H $= 100\% - 84.1\%$
$= 15.9\%$

Check Your Solution
The sum of the calculated percentages is 100%, so the answers are reasonable.

11. A 19.6 g sample of compound A contains 16.1 g of nitrogen and 3.5 g of hydrogen. Determine the percentage composition of the compound.

12. A 304 g sample of compound B contains 207.9 g of chromium and 96.1 g of oxygen. Determine the percentage composition.

13. A 60.0 mg sample of compound C contains 24.0 mg of carbon, 4.0 mg of hydrogen, and 32.0 mg of oxygen. Determine the percentage composition.

14. A 15 g sample of compound D contains 7.22 g of nickel, 2.53 g of phosphorus, and 5.25 g of oxygen. Determine the percentage composition.

15. A sample of an unknown compound contains 84.05 g of carbon, 5.00 g of hydrogen, 42.02 g of nitrogen, and 96.08 g of oxygen. Determine the percentage composition of the compound.

16. What is the percentage composition of a compound with a mass of 48.72 g, if it contains 32.69 g of zinc and 16.03 g of sulfur?

17. The percentage composition of a compound is 79.9% copper and 20.1% sulfur. The mass of a sample of the compound is 160.0 g. What are the masses of the copper and sulfur in the sample?

18. A chemist analyzed a sample of a compound that was known to contain potassium, manganese, and oxygen. She obtained the following results:
Total mass of sample = 316.08 g
Mass of potassium = 78.20 g
Mass of manganese = 109.88 g
Determine the percentage composition of the compound.

19. A volatile chemical was removed from a crime scene. Analysis of a 35.20 mg sample of the chemical showed that it contained 3.56 mg of carbon and 0.28 mg of hydrogen. Later, investigators learned that the chemical was chloroform, $CHCl_3(\ell)$. What is the percentage composition of chloroform?

20. A 68.2 g sample of an unknown alcohol was tested by a student and found to contain 44.2 g of carbon, 9.3 g of hydrogen, and 14.7 g of oxygen. The student concluded that the unknown alcohol was ethanol, $C_2H_5OH(\ell)$, which contains 52.1% carbon, 13.2% hydrogen, and 34.7% oxygen. Was the student correct? Explain, using percentage composition calculations.

Percentage Composition in Industry

It is important to be able to calculate percentage composition from mass values. However, it is more common to calculate percentage composition from a chemical formula and molar masses. There are many applications of the latter type of calculation in industry, such as in metal refining.

For example, copper is a valuable metal with a wide range of uses, such as in coins (pennies), computer circuit boards, and electrical wiring. In nature, copper is found mainly in the form of ores, which are ionic compounds. One example of a copper ore is chalcocite, or copper(I) sulfide, $Cu_2S(s)$, which is shown in **Figure 6.5**.

Figure 6.5 Chalcocite is a copper ore. Most metals are found in impure rocks and mineral ores, which must be refined to obtain the pure metals. Knowledge of percentage composition is necessary for metallurgists who want to assess the values of ores and mineral deposits.

Determining Percentage Composition from a Chemical Formula

Metallurgists need to know the percentage composition of different metal ores so they can determine the amount and value of the metal that can be extracted from an ore. To determine the percentage composition of copper(I) sulfide from its chemical formula, $Cu_2S(s)$, a sample mass must be chosen. Recall that the law of definite proportions states that the proportions of the elements in a compound remain the same, no matter what their size or source. Therefore, any mass could be picked for the sample size. The most convenient sample size, however, is equivalent to the molar mass of the compound or, in other words, 1 mol of the compound. This allows the masses of the elements in the sample to be determined directly from the periodic table.

Thus, the mass percent of copper in copper(I) sulfide can be calculated as follows:

$$\text{mass percent of Cu} = \frac{\text{mass of Cu in 1 mol of } Cu_2S}{\text{mass of 1 mol of } Cu_2S} \times 100\%$$

$$= \frac{(2 \times 63.55 \text{ g})}{159.17 \text{ g}} \times 100\%$$

$$= 0.798517 \times 100\%$$

$$= 79.9\%$$

Because there are only two elements in the compound, the mass percent of sulfur can be found by subtracting the mass percent of Cu(s) from 100.

$$\text{mass percent of S} = 100\% - 79.9\%$$

$$= 20.1\%$$

Therefore, the percentage composition of copper(I) sulfide is 79.9 percent copper and 20.1 percent sulfur.

The following flowcharts illustrate the steps needed to calculate percentage composition from the chemical formula and the steps needed to calculate percentage composition from mass data. Note that most chemical compounds are composed of more than two elements. The Sample Problem on the following page shows you how the percentage composition of a larger compound is determined.

Suggested **Investigation**

Inquiry Investigation 6-A, Determining the Percentage Composition of Magnesium Oxide

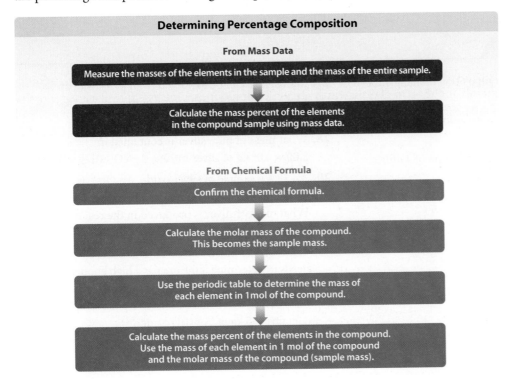

Determining Percentage Composition

From Mass Data

Measure the masses of the elements in the sample and the mass of the entire sample.

Calculate the mass percent of the elements in the compound sample using mass data.

From Chemical Formula

Confirm the chemical formula.

Calculate the molar mass of the compound. This becomes the sample mass.

Use the periodic table to determine the mass of each element in 1 mol of the compound.

Calculate the mass percent of the elements in the compound. Use the mass of each element in 1 mol of the compound and the molar mass of the compound (sample mass).

Determining Percentage Composition from a Chemical Formula

Problem

The most common copper ore comes from the mineral chalcopyrite, $CuFeS_2(s)$.
Determine the percentage composition of chalcopyrite.

What Is Required?

To determine the percentage composition of chalcopyrite, you need to calculate the mass
percents of copper, iron, and sulfur in the compound.

What Is Given?

You know the formula for chalcopyrite: $CuFeS_2(s)$

Plan Your Strategy	Act on Your Strategy
Calculate the molar mass of chalcopyrite. This becomes your sample mass.	$M_{CuFeS_2} = 1M_{Cu} + 1M_{Fe} + 2M_S$ $\qquad = 1(63.55 \text{ g/mol}) + 1(55.85 \text{ g/mol}) + 2(32.07 \text{ g/mol})$ $\qquad = 183.54 \text{ g/mol}$ The molar mass of $CuFeS_2(s)$ is 183.54 g/mol.
Calculate the mass percents of copper, iron, and sulfur. For each calculation, use the mass of the element in 1 mol of the compound and the mass of 1 mol of the compound (the sample mass).	$\text{mass percent of Cu} = \dfrac{\text{mass of Cu}}{\text{mass of } CuFeS_2} \times 100\% = \dfrac{63.55 \text{ g}}{183.54 \text{ g}} \times 100\% = 34.62\%$ $\text{mass percent of Fe} = \dfrac{\text{mass of Fe}}{\text{mass of } CuFeS_2} \times 100\% = \dfrac{55.85 \text{ g}}{183.54 \text{ g}} \times 100\% = 30.43\%$ $\text{mass percent of S} = \dfrac{\text{mass of S}}{\text{mass of } CuFeS_2} \times 100\% = \dfrac{2(32.07 \text{ g})}{183.54 \text{ g}} \times 100\% = 34.95\%$ Therefore, the percentage composition of chalcopyrite is 34.62% copper, 30.43% iron, and 34.95% sulfur by mass.

Check Your Solution

The masses of the three elements in $CuFeS_2(s)$ are very similar. Therefore, each mass is
about $\frac{1}{3}$ of the total mass of the sample and the three mass percents add to 100%.

21. What is the percentage composition of manganese(II) sulfide, $MnS(s)$?

22. What is the percentage composition of silver oxide, $Ag_2O(s)$?

23. Calculate the percentage composition of sulfuric acid, $H_2SO_4(\ell)$.

24. What is the percentage composition of aluminum hydroxide, $Al(OH)_3(s)$?

25. Calculate the percentage composition of strontium nitrate, $Sr(NO_3)_2(s)$.

26. Indigo, $C_{16}H_{10}N_2O_2(s)$, is a blue dye. Determine the percentage composition of indigo.

27. Manganese can be extracted from pyrolusite ore, $MnO_2(s)$. Calculate the mass of manganese that can be extracted from 325 kg of pyrolusite ore.

28. What mass of pure silver is contained in 2.00×10^2 kg of silver nitrate, $AgNO_3(s)$?

29. An 18.4 g sample of silver oxide, $Ag_2O(s)$, is decomposed into silver and oxygen by heating. What mass of silver is produced in the reaction?

30. An industrial chemist is testing some iron ore samples that contain iron(III) oxide, $Fe_2O_3(s)$. The chemist needs to determine if the ore has a high enough mass percent of iron to be suitable for mining. Calculate the mass of iron that could be refined from a sample that contains 355 kg of iron(III) oxide.

Section Summary

- The law of definite proportions states that the elements in a chemical compound are always present in the same proportions by mass.

- The same elements can form different compounds when combined in different whole-number ratios.

- The mass percent of an element in a compound is the mass of the element expressed as a percentage of the total mass of the compound.

- The percentage composition of a compound is the percent by mass of each element in a compound.

- Percentage composition can be calculated using either mass data for a substance or the chemical formula for the substance.

Review Questions

1. **K/U** Samples of carbon dioxide are taken from the atmosphere and from a block of dry ice. Are the chemical formulas for the carbon dioxide in the two samples different? Explain.

2. **K/U** When determining the percentage composition of a compound from its formula, the calculations are usually based on 1 mol of a sample. Explain why this is the most convenient amount to use.

3. **K/U** Acetylene, $C_2H_2(g)$, is made up of two carbon atoms and two hydrogen atoms. Explain why acetylene does not contain 50% of each element by mass.

4. **T/I** What is the mass percent of each element in sucrose, $C_{12}H_{22}O_{11}(s)$?

5. **C** Draw a flowchart that shows the steps needed to determine the percentage composition of a compound, using the masses of the elements in the compound.

6. **T/I** A sample of an unknown compound is analyzed and found to contain 0.90 g of calcium and 1.60 g of chlorine. The sample has a mass of 2.50 g. Determine the percentage composition of the compound.

7. **T/I** A 19.00 kg sample of an unknown compound contains 0.06 kg of hydrogen, 11.01 kg of gold, and 7.93 kg of chlorine. Determine the percentage composition of the compound.

8. **K/U** In your own words, explain the difference between mass percent and percentage composition.

9. **T/I** Phosphoric acid, $H_3PO_4(\ell)$, is used in some carbonated beverages to give them a tangy flavour. Determine the percentage composition of phosphoric acid.

10. **T/I** Determine the percentage composition of magnesium phosphate, $Mg_3(PO_4)_2(s)$.

11. **T/I** What mass of boron would be in a 35.0 g sample of sodium tetraborate, $Na_2B_4O_7(s)$?

12. **A** Calcium carbonate, $CaCO_3(s)$, can have many different forms. Coral, shown below, as well as marble and chalk, are substances that have $CaCO_3(s)$ as a principal component.

a. Determine the percentage composition of this compound.

b. When acid rain reacts with the marble in a statue, carbon dioxide gas is formed. What does this do to the mass percent of carbon and oxygen in the statue?

13. **A** Washing soda is the common name for a compound that is used to make soap and glass. Washing soda contains 43.4% sodium by mass and 45.3% oxygen by mass, as well as one other element. If the molar mass of washing soda is 105.99 g/mol, identify the other element.

14. **A** A typical soap molecule is made up of a polyatomic anion associated with a cation. The polyatomic anion contains hydrogen, carbon, and oxygen. One soap molecule has 18 carbon atoms and contains 70.5% carbon, 11.5% hydrogen, and 10.4% oxygen by mass. It also contains one alkali metal ion. Identify this alkali metal ion.

Empirical and Molecular Formulas

molecular formula
the formula for a compound that shows the number of atoms of each element that make up a molecule of that compound

empirical formula
a formula that shows the smallest whole-number ratio of the elements in a compound

ratio relative amount; proportional relationship

In Section 6.1, you read that chemists can determine the percentage composition of a compound from mass data alone. However, the percentage composition only tells chemists which elements the compound contains and the relative masses of these elements. For more specific information about the composition of a compound, you need to know the relative number of atoms of each element.

Comparing Molecular and Empirical Formulas of Compounds

Chemical formulas communicate which elements are found in a compound, as well as their proportions. For example, the formula for hydrogen peroxide, H_2O_2, indicates that a molecule of hydrogen peroxide contains two atoms of hydrogen and two atoms of oxygen. Thus, this formula shows the actual number of atoms of each element in one molecule of the compound. Such a formula is known as the **molecular formula**. Molecular formulas can often be quite complex, as some molecules, such as glucose, $C_6H_{12}O_6$, contain many atoms.

A simpler method of writing a chemical formula involves showing only the proportional relationship or relative number of the atoms of each element in a compound. Such a formula, known as an **empirical formula**, shows the smallest whole-number **ratio**, or proportional relationship, of the elements in a compound. For instance, the molecular formula for hydrogen peroxide is H_2O_2, and the ratio of hydrogen to oxygen atoms is 2:2. The ratio 2:2 is not the smallest ratio because you can reduce it to 1:1. Thus, the empirical formula for this compound is HO.

Note that, for some compounds, the empirical formula may be the same as the molecular formula. For example, water has two hydrogen atoms and one oxygen atom, as shown in **Figure 6.6**. Thus, the molecular formula for water is H_2O. As this formula shows the smallest whole-number ratio of the elements in this molecule, it is also the molecule's empirical formula. Additionally, it is important to recognize, as you learned in Chapter 2, that ionic compounds only have one possible atomic configuration. As a result, they are always represented by empirical formulas. They never have a molecular formula. As the name indicates, molecular formulas apply only to molecules.

Figure 6.6 The empirical and molecular formulas for water are the same, as the molecular formula for water already shows the smallest whole-number molar ratio of the elements in this molecule.

Chemical Formulas and Specific Composition

The empirical formula for a compound gives a very basic glimpse into the composition of the compound. For this reason, the empirical formula is a very good start in an experimental analysis to determine the formula for a new or unknown compound. However, would determining the empirical formula for an unknown substance be enough to identify that substance? Because more than one compound can have the same empirical formula, it is often necessary to know the molecular formula in order to distinguish between compounds.

Applications of Empirical and Molecular Formulas

Suppose a thief has robbed a home but left behind some tracks on the floor. The residue is analyzed and found to contain an unknown compound. An initial analysis reveals that the compound contains carbon, hydrogen, and oxygen. Further analysis reveals that its empirical formula is CH_2O. Could the investigators conclude that the unknown substance is formaldehyde, which has the molecular formula $CH_2O(g)$, and that the thief is someone who comes in contact with formaldehyde? This would be a hasty decision. Formaldehyde is one of several compounds that have the same empirical formula, as shown in **Table 6.1**. These compounds have the same *relative* amount of each element. However, their molecular formulas differ, as the compounds are composed of different *actual* amounts of each element.

Although the compounds in **Table 6.1** have the same proportions of the same elements, their chemical properties are quite different. For example, formaldehyde is toxic to humans, whereas acetic acid is the main component in vinegar, which we eat on salads and french fries. Ribose is an important biological compound that is found in Vitamin B. It is also a component the genetic material, DNA (*deoxyribo*nucleic acid). Erythrose is an intermediate in the conversion from glucose to ribose. Glucose is the main source of energy for most animals. During intense exercise, as shown in **Figure 6.7**, muscles breakdown glucose into lactic acid which causes sore muscles.

Suggested **Investigation**

Inquiry Investigation 6-B,
Chemical Analysis
Simulation

Figure 6.7 This runner's muscles are using glucose so rapidly that lactic acid builds up before it can be broken down into carbon dioxide and water.

Table 6.1 Six Compounds with the Empirical Formula CH_2O

Name	Empirical Formula	Molecular Formula	Whole-Number Multiple	M (g/mol)	Use or Function
Formaldehyde	CH_2O	CH_2O	1	30.03	Is used as a disinfectant and biological preservative
Acetic acid	CH_2O	$C_2H_4O_2$	2	60.06	Is used to produce acetate polymers; is a component of vinegar (5% solution)
Lactic acid	CH_2O	$C_3H_6O_3$	3	90.09	Causes milk to sour; forms in muscles during exercise
Erythrose	CH_2O	$C_4H_8O_4$	4	120.12	Forms during sugar metabolism
Ribose	CH_2O	$C_5H_{10}O_5$	5	150.15	Is a component of many nucleic acids and vitamin B_2
Glucose	CH_2O	$C_6H_{12}O_6$	6	180.18	Is a major nutrient for energy in cells

CH_2O $C_2H_4O_2$ $C_3H_6O_3$ $C_4H_8O_4$ $C_5H_{10}O_5$ $C_6H_{12}O_6$

Compounds with the Same Empirical Formula and Different Molecular Formulas

In the thief scenario, you compared compounds that have the same relative amount of each element, but different actual amounts of each element. Such compounds share the same empirical formula, but have different molecular formulas. As a result, they tend to have very different properties. Consider nitrogen dioxide, $NO_2(g)$, and dinitrogen tetroxide, $N_2O_4(g)$.

Nitrogen dioxide, is an orange-brown gas with a sharp, biting odour. In high concentrations, it is extremely toxic and corrosive. Recall that nitrogen dioxide is a common product in internal combustion engine exhaust, and it is released from power plants that burn fossil fuels to generate electrical power. Nitrogen dioxide is one of the main components of smog. It is the component that is responsible for making a visible smog haze above polluted urban centres, as shown in **Figure 6.8 (A)**.

Note that the ratio of nitrogen to oxygen in this gas is the simplest possible, so its empirical formula is also its molecular formula. That is, one molecule of this gas contains one nitrogen atom and two oxygen atoms.

Dinitrogen tetroxide has the same empirical formula as nitrogen dioxide, but a different molecular formula. As a result, its properties differ. For example, dinitrogen tetroxide in conjunction with hydrazine, $H_2N_4(\ell)$, make an important rocket propellant, as shown in **Figure 6.8 (B)**, because they burn on contact, without the need for a separate ignition source. Like nitrogen dioxide, it is very toxic and corrosive in high concentrations. The dinitrogen tetroxide molecule contains two nitrogen atoms and four oxygen atoms.

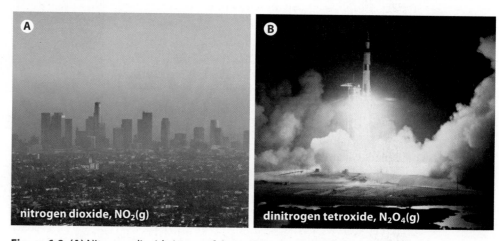

nitrogen dioxide, $NO_2(g)$

dinitrogen tetroxide, $N_2O_4(g)$

Figure 6.8 (A) Nitrogen dioxide is one of the main components of smog. **(B)** Dinitrogen tetroxide is used as a rocket propellant. The two compounds have the same empirical formulas but different molecular formulas, chemical properties, and applications.

Learning Check

7. State the ratio of carbon to hydrogen in one molecule of methane, $CH_4(g)$.

8. Both the percentage composition and the molecular formula for a compound provide information about the proportions of the elements in the compound. How are these proportions similar, and how are they different?

9. What information does a molecular formula tell you about a compound that an empirical formula does not?

10. When is the molecular formula for a compound the same as its empirical formula?

11. Why is it important to know the molecular formula for a compound? Use the chemicals in **Figure 6.8** as examples.

12. The following formulas contain only nitrogen and oxygen: NO, N_2O, NO_2, N_2O_4, and N_2O_5. Which formulas represent the empirical and molecular formulas for the same compound? Explain your answer.

Determining the Empirical Formula

You can determine the empirical formula for a compound from the percentage composition. For example, assume that a compound is 50.91% zinc, 16.04% phosphorus, and 33.15% oxygen. Because only relative amounts of the elements are needed for an empirical formula, *any amount of the compound can be used*. Therefore, assume that you have a 100 g sample. Using the percentage composition, 100 g of the compound contains 50.91 g of zinc, 16.04 g of phosphorus, and 33.15 g of oxygen. Next, determine the amount in moles of each element. You can find the amount of an element by dividing the mass by the molar mass of the element, as shown below.

Go to **scienceontario** to find out more

$$n_{Zn} = \frac{50.81 \text{ g}}{65.38 \text{ g/mol}} = 0.777 \text{ mol (Zn)}$$

$$n_P = \frac{16.04 \text{ g}}{30.97 \text{ g/mol}} = 0.518 \text{ mol (P)}$$

$$n_O = \frac{33.15 \text{ g}}{16.00 \text{ g/mol}} = 2.072 \text{ mol (O)}$$

These values give you the ratio of the amounts of each element in the substance in moles, or 0.777 mol Zn : 0.518 mol P : 2.072 mol O. This ratio is the same as the subscripts in an empirical formula. Therefore, you could write the empirical formula as:

$$Zn_{0.777}P_{0.518}O_{2.072}$$

However, you cannot have fractions of an atom. You can obtain whole number subscripts in one or two more steps. First, divide each subscript by the smallest subscript, or 0.518.

$$Zn_{\frac{0.777}{0.518}}P_{\frac{0.518}{0.518}}O_{\frac{2.072}{0.518}}$$

$$Zn_{1.5}P_1O_4$$

In many calculations, the subscripts are all whole numbers at this step. However, the subscript for zinc is 1.5. In cases like this, you can multiply all subscripts by a whole number that will make the decimal subscript into a whole number. To complete this calculation, multiply all of the subscripts by 2. Your empirical formula becomes

$$Zn_{1.5 \times 2}P_{1 \times 2}O_{4 \times 2}$$

$$Zn_3P_2O_8$$

Because this formula contains a metal, Zn, and two non-metals, P and O, the substance must be an ionic compound which has a polyatomic ion for the non-metal. Therefore, the empirical formula is the same as the formula for the compound. However, P_2O_8 is not a familiar polyatomic ion but it has the same ratio of elements as the phosphate ion. Therefore, your formula is $Zn_3(PO_4)_2$.

Rules for Determining Empirical Formulas

1. Convert percentage composition data into mass data by assuming that the total mass of the sample is 100 g.

2. Determine the number of moles of each element in the sample by dividing the mass by the molar mass of each element.

3. Convert the number of moles of each element into whole numbers that become subscripts in the empirical formula by dividing each amount in moles by the smallest amount.

4. If the subscripts are not yet whole numbers, determine the least common multiple that will make the decimal values into whole numbers. Multiply all subscripts by this least common multiple. Use these numbers as subscripts to complete the empirical formula.

Determining the Empirical Formula for a Compound with More Than Two Elements from Percentage Composition

Problem

Determine the empirical formula for a compound that is found by analysis to contain 27.37% sodium, 1.200% hydrogen, 14.30% carbon, and 57.14% oxygen.

What Is Required?

You need to determine the empirical formula for the sample that contains sodium, hydrogen, carbon, and oxygen.

What Is Given?

You know the composition of the compound: 27.37% sodium, 1.200% hydrogen, 14.30% carbon, and 57.14% oxygen.

Plan Your Strategy	Act on Your Strategy
Assume that the mass of the sample is 100 g.	In a 100 g sample of the compound, there will be 27.37 g sodium, 1.200 g of hydrogen, 14.30 g carbon, and 57.14 g oxygen.
Determine the molar masses of sodium, hydrogen, carbon, and oxygen using the periodic table	The molar mass of sodium is 22.99 g/mol. The molar mass of hydrogen is 1.01 g/mol. The molar mass of carbon is 12.01 g/mol. The molar mass of oxygen is 16.00 g/mol.
Convert each mass to moles	$n_{Na} = \dfrac{27.37 \text{ g}}{22.99 \text{ g/mol}} = 1.190518 \text{ mol}$ $n_{H} = \dfrac{1.200 \text{ g}}{1.01 \text{ g/mol}} = 1.188119 \text{ mol}$ $n_{C} = \dfrac{14.30 \text{ g}}{12.01 \text{ g/mol}} = 1.190674 \text{ mol}$ $n_{O} = \dfrac{57.14 \text{ g}}{16.00 \text{ g/mol}} = 3.57125 \text{ mol}$
Divide all the mole amounts by the lowest mole amount.	$\dfrac{1.190518}{1.188119} : \dfrac{1.188119}{1.188119} : \dfrac{1.190674}{1.188119} : \dfrac{3.57125}{1.188119}$ ratios: 1:1:1:3 Therefore, the empirical formula is $Na_1H_1C_1O_3$ or $NaHCO_3$

Check Your Solution

Work backward. Calculate the percentage composition of $NaHCO_3$.

The molar mass of $NaHCO_3$ is:

$$M_{NaHCO_3} = 1M_{Na} + 1M_{H} + 1M_{C} + 3M_{O}$$
$$= 1(22.99 \text{ g/mol}) + 1(1.01 \text{ g/mol}) + 1(12.01 \text{ g/mol}) + 3(16.00 \text{ g/mol}) = 84.01 \text{ g/mol}$$

mass percent of Na: $\dfrac{22.99 \text{ g/mol}}{84.01 \text{ g/mol}} \times 100\% = 27.37\%$

mass percent of H: $\dfrac{1.01 \text{ g/mol}}{84.01 \text{ g/mol}} \times 100\% = 1.202\%$

mass percent of C: $\dfrac{12.01 \text{ g/mol}}{84.01 \text{ g/mol}} \times 100\% = 14.30\%$

mass percent of O: $\dfrac{3(16.00 \text{ g/mol})}{84.01 \text{ g/mol}} \times 100\% = 57.14\%$

The percentage composition calculated from the empirical formula closely matches the given data. Therefore, the empirical formula $NaHCO_3$ is reasonable. The percentages do not add to 100% due to rounding.

Determining the Empirical Formula from Mass Data when a Ratio Term Is Not Close to a Whole Number

Problem

Determine the empirical formula for a compound that contains 69.88 g of iron and 30.12 g of oxygen.

What Is Required

You need to determine the empirical formula from mass data.

What Is Given?

You know the mass data of the compound: 69.88 g iron and 30.12 g oxygen

Plan Your Strategy	Act on Your Strategy
Convert each mass to moles using the molar mass.	$\dfrac{69.88 \text{ g Fe}}{55.85 \text{ g/mol}} = 1.2512$ mol Fe $\dfrac{30.12 \text{ g O}}{16.00 \text{ g/mol}} = 1.882$ mol O
Divide all the mole amounts by the lowest mole amount.	Fe:O $= \dfrac{1.2512}{1.2512} : \dfrac{1.882}{1.2512} = 1:1.5$ The incomplete formula is $Fe_1O_{1.5}$.
Determine the least common multiple that will make all of the subscripts whole numbers. Multiply all the subscripts by this whole-number multiple to determine the empirical formula.	The smallest whole-number multiple that makes all the subscripts small whole numbers is 2. Therefore, the empirical formula is Fe_2O_3.

Check Your Solution

Work backward. Determine the percentage composition. Assume there is a 100 g sample and compare the number of grams of each substance.

$M_{Fe_2O_3} = 2M_{Fe} + 3M_O = 2(55.85 \text{ g/mol}) + 3(16.00 \text{ g/mol}) = 159.7 \text{ g/mol}$

Determine the percentage composition:

mass percent of Fe $= \dfrac{111.70 \text{ g/mol}}{159.7 \text{ g/mol}} \times 100\% = 69.94\%$

mass percent of O $= \dfrac{48.00 \text{ g/mol}}{159.7 \text{ g/mol}} \times 100\% = 30.06\%$

In a 100 g sample, there is 69.94 g of iron and 30.06 g of oxygen. This closely matches the given data. The empirical formula is reasonable.

Practice Problems

Determine the empirical formulas for the compounds with the following percentage compositions.

31. 80.04% carbon and 19.96% hydrogen

32. 25.5 g of magnesium and 74.5 g of chlorine

33. 40.0% copper, 20.0% sulfur, and 40.0% oxygen

34. 26.61% K, 35.38% Cr, and 38.01% O

35. 17.6 g of hydrogen and 82.4 g of nitrogen

36. 46.3 g of lithium and 53.7 g of oxygen

37. 15.9% boron, and the rest is fluorine

38. 60.11% sulfur, and the rest is chlorine

39. 11.33% carbon, 45.29% oxygen, and the rest is sodium

40. 56.36% oxygen, and the rest is phosphorus

Determining the Molecular Formula

Because the molecular formula is so closely related to the empirical formula, you can determine the molecular formula for a compound using the empirical formula and the molar mass of the compound, as shown in the Sample Problem below. Recall that the empirical formula is multiplied by a whole number to get the molecular formula. For example, the empirical formula for benzene, $C_6H_6(\ell)$, is CH, as shown in **Figure 6.9**. This empirical formula is multiplied by the whole number 6 to get the molecular formula.

To determine the molecular formula for a compound, use the method you learned in Chapter 5 to determine the molar mass of the empirical formula. The molar mass of the actual compound must be determined experimentally using a device such as a mass spectrometer. Because it must be determined experimentally, the molar mass of the actual compound is usually given in problems where you determine the molecular formula for a compound.

To determine the molecular formula from percentage composition data, you determine the empirical formula for the compound as you did in previous problems. Then you use the molar masses of the empirical formula and the molecular formula to find the whole-number multiple, x, which relates the empirical formula to the molecular formula. The Sample Problem on the following page demonstrates how to solve this type of problem.

Figure 6.9 Benzene has six CH units joined together. The empirical formula for benzene is CH and the molecular formula is $C_6H_6(\ell)$.

Sample Problem

Molecular Formula from Empirical Formula and Molar Mass

Problem

A compound with the empirical formula CH was analyzed using a mass spectrometer. Its molar mass was found to be 78 g/mol. Determine the molecular formula.

What Is Required

You need to determine the the molecular formula.

What Is Given?

You know the empirical formula: CH
You know the molar mass of the actual compound: 78 g/mol

Plan Your Strategy	Act on Your Strategy
Determine the molar mass of CH.	$M_{CH} = M_C + M_H$ $\qquad = 12.01 \text{ g/mol} + 1.01 \text{ g/mol}$ $\qquad = 13.02 \text{ g/mol}$
To find the whole-number multiple, x, divide the experimentally determined molar mass by the molar mass of the empirical formula. To determine the molecular formula, multiply each subscript of the empirical formula by the whole-number multiple.	$M_{\text{actual compound}} = x \times (M_{CH})$ $78 \text{ g/mol} = x \times (13.02 \text{ g/mol})$ $x = \dfrac{78 \text{ g/mol}}{13.02 \text{ g/mol}}$ $x = 6$ Therefore, the molecular formula is $C_6H_6(\ell)$.

Check Your Solution

Work backward by calculating the molar mass of $C_6H_6(\ell)$.

$(6 \times 12.01 \text{ g/mol}) + (6 \times 1.01 \text{ g/mol}) = 78.12 \text{ g/mol}$

The calculated molar mass matches the molar mass that is given in the problem.

Molecular Formula from Percentage Composition and Molar Mass

Problem
Chemical analysis indicates that a compound is 28.64% sulfur and 71.36% bromine. The molar mass of the compound is 223.94 g/mol. Determine the molecular formula.

What Is Required
You need to find the molecular formula for the compound composed of sulfur and bromine.

What Is Given?
You know the percentage composition: 28.64% sulfur and 71.36% bromine
You know the molar mass of the compound: 223.94 g/mol

Plan Your Strategy	Act on Your Strategy
Assume the mass of the sample is 100 g.	In 100 g of the compound, there will be 28.64 g of sulfur and 71.36 g of bromine.
Find the molar mass of sulfur. Find the molar mass of bromine.	The molar mass of sulfur is 32.07 g/mol. The molar mass of bromine is 79.90 g/mol.
Convert each mass to moles.	$\dfrac{28.64 \text{ g S}}{32.07 \text{ g/mol}} = 0.8930 \text{ mol S}$ $\dfrac{71.36 \text{ g Br}}{79.90 \text{ g/mol}} = 0.8931 \text{ mol Br}$
Divide all the mole amounts by the lowest number of moles.	$\text{S:Br} = \dfrac{0.8930}{0.8930} : \dfrac{0.8931}{0.8930} = 1{:}1$
Determine the least common multiple that will make all of the subscripts whole numbers.	Because these subscripts are already whole numbers, you do not need to determine the least common multiple. The empirical formula is SBr.
Find the molar mass of the empirical formula.	$M_{SBr} = 1M_S + 1M_{Br}$ $= 1(32.07 \text{ g/mol}) + 1(79.90 \text{ g/mol}) = 111.97 \text{ g/mol}$
To find the whole-number multiple, x, divide the experimental molar mass by the molar mass of the empirical formula.	$x = \dfrac{\text{experimental molar mass}}{\text{empirical formula molar mass}} = \dfrac{223.94 \text{ g/mol}}{111.97 \text{ g/mol}} = 2$
To determine the molecular formula, multiply each subscript of the empirical formula by the whole-number multiple.	The molecular formula is S_2Br_2.

41. The empirical formula for glucose is $CH_2O(s)$. The molar mass of glucose is 180.18 g/mol. Determine the molecular formula for glucose.

42. The empirical formula for xylene is $C_4H_5(\ell)$, and its molar mass is 106 g/mol. What is the molecular formula for xylene?

43. The empirical formula for 1,4-butanediol is $C_2OH_5(\ell)$. Its molar mass is 90.14 g/mol. What is its molecular formula?

44. The empirical formula for styrene is $CH(\ell)$, and its molar mass is 104 g/mol. What is its molecular formula?

45. Calomel is a compound that was once popular for treating syphilis. It contains 84.98% mercury and 15.02% chlorine. It has a molar mass of 472 g/mol. What is its empirical formula?

46. The molar mass of caffeine is 194 g/mol. Determine whether the molecular formula for caffeine is $C_4H_5N_2O(s)$ or $C_8H_{10}N_4O_2(s)$.

47. An unknown compound contains 42.6% oxygen, 32% carbon, 18.7% nitrogen, and the remainder hydrogen. Using mass spectrometry, its molar mass was determined to be 75.0 g/mol. What is the molecular formula for the compound?

Continued on next page ❯

48. A compound that contains 6.44 g of boron and 1.80 g of hydrogen has a molar mass of approximately 28 g/mol. What is its molecular formula?

49. The molar mass of a compound is 148.20 g/mol. Its percentage composition is 48.63% carbon, 21.59% oxygen, 18.90% nitrogen, and the rest hydrogen.
   a. Find the empirical formula for the compound.
   b. Find its molecular formula.

50. Estradiol is the main estrogen compound that is found in humans. Its molar mass is 272.38 g/mol. The percentage composition of estradiol is 79.4% carbon, 11.7% oxygen, and 8.9% hydrogen. Determine whether its molecular formula is the same as its empirical formula. If not, what is each formula?

How to Determine Empirical and Molecular Formulas

The steps required to determine the empirical and molecular formulas for compounds are very similar. The flowchart in **Figure 6.10** illustrates these steps.

Figure 6.10 Use this flowchart as a guide to help you determine the empirical and molecular formulas for compounds.

Describe *How is the integer x related to the empirical and molecular formulas?*

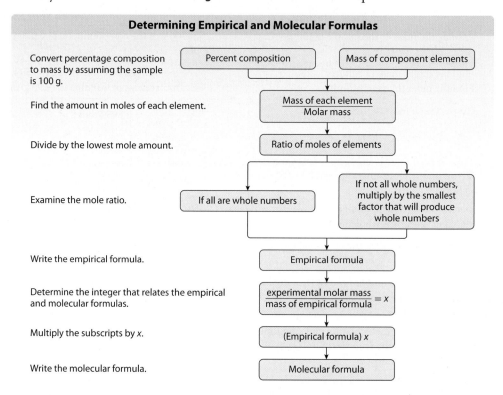

Determining Empirical and Molecular Formulas

Convert percentage composition to mass by assuming the sample is 100 g.

Find the amount in moles of each element.

Divide by the lowest mole amount.

Examine the mole ratio.

Write the empirical formula.

Determine the integer that relates the empirical and molecular formulas.

Multiply the subscripts by x.

Write the molecular formula.

- Percent composition
- Mass of component elements
- $\dfrac{\text{Mass of each element}}{\text{Molar mass}}$
- Ratio of moles of elements
- If all are whole numbers
- If not all whole numbers, multiply by the smallest factor that will produce whole numbers
- Empirical formula
- $\dfrac{\text{experimental molar mass}}{\text{mass of empirical formula}} = x$
- (Empirical formula) x
- Molecular formula

Activity | 6.2 | Exploring Formulas Using Models

This activity will give you the opportunity to explore relationships between empirical and molecular formulas by building molecular models.

Materials
- molecular modelling kit
- reference sources

Procedure

1. Construct a table to record the names, molecular formulas, empirical formulas, and molar masses of ethylene, butane, and cyclohexane.

2. Research to find the molecular structure for ethylene, butane, and cyclohexane. Then build each molecule using the molecular modelling kit.

3. Examine each molecule you built. Record its molecular formula and empirical formula in your table.

4. Calculate and record the molar mass of each compound.

Questions

1. Compare and contrast the three molecules, based on their molecular and empirical formulas.

2. What is the relationship between the actual molar mass value of each molecule and the molar mass calculated for its empirical formula?

Hydrates and Their Chemical Formulas

A *hydrate* is a compound that has a specific number of water molecules bound to each formula unit. Often, when a crystal forms from a water solution, water molecules are trapped in the crystal in a specific arrangement. An example is the calcium sulfate, $CaSO_4(s)$, in gypsum, the white powder that is used to make drywall. When gypsum crystals form, as shown in **Figure 6.11**, each formula unit of calcium sulfate incorporates two water molecules into its structure. The chemical formula for gypsum is represented as $CaSO_4 \cdot 2H_2O(s)$. This formula means that there are two water molecules for every formula unit of calcium sulfate.

Figure 6.11 Gypsum crystals form when each formula unit of calcium sulfate incorporates two water molecules. Gypsum is a mineral that is used for making drywall and plaster of Paris.

Using Hydrates and Anhydrous Compounds

When chemists work with ionic compounds in the solid state, they need to know whether these compounds are hydrates or anhydrous compounds (without water molecules). The water molecules in the crystal structure of a hydrated ionic compound usually do not interfere with the chemical activity of the compound. However, the water molecules are part of the crystal structure and, therefore, add mass to the solid.

For example, consider a 1 g sample of magnesium sulfate heptahydrate, $MgSO_4 \cdot 7H_2O(s)$, and a 1 g sample of anhydrous magnesium sulfate, $MgSO_4(s)$. Which sample do you think contains more magnesium atoms? If your answer is the anhydrous form, you are correct. The 1 g sample of the hydrate contains seven moles of water for every mole of magnesium sulfate. The anhydrous form, on the other hand, does not contain any water molecules, so there is more magnesium sulfate in the sample. As a result, there is more magnesium in the 1 g sample of the anhydrous form than in the 1 g sample of the hydrate.

Analyzing Hydrates

What analytical methods do chemists use to determine how many water molecules are attached to each formula unit in a hydrate? The simplest method is to convert the hydrated form of the compound to the anhydrous form by heating it and driving off the water molecules. The difference between the initial mass of the sample (the hydrate) and the final mass of the sample (the anhydrous form) is the mass of the water. As the Sample Problem on the following page shows, this is enough information for a chemist to determine both the percent by mass of water in a hydrate and the chemical formula for the hydrate.

Suggested Investigation

Inquiry Investigation 6-C, Determining the Chemical Formula for a Hydrate

---

Learning Check

13. Compare and contrast empirical and molecular formulas.

14. List the steps required to determine the empirical formula from mass data.

15. How is the molar mass usually determined for an unknown compound when you are trying to determine its molecular formula?

16. Explain the difference between a hydrate and the anhydrous form.

17. Why is it important for chemists to know if they are using a hydrate or the anhydrous form when they are performing investigations?

18. How can you determine the mass of water in a hydrate?

Determining the Formula for a Hydrate

Problem
A 50.0 g sample of barium hydroxide, $Ba(OH)_2 \cdot xH_2O(s)$ contains 27.2 g of $Ba(OH)_2(s)$.
Calculate the percent by mass of water in $Ba(OH)_2 \cdot xH_2O(s)$, and find the value of x.

What Is Required?
You need to calculate the percent by mass of water in the hydrate of barium hydroxide.
You need to determine how many water molecules are bonded to each formula unit of
$Ba(OH)_2(s)$ to form the hydrate.

What Is Given?
You know the formula for the sample: $Ba(OH)_2 \cdot xH_2O(s)$.
You know the mass of the sample: 50.0 g
You know the mass of the $Ba(OH)_2(s)$ in the sample: 27.2 g

Plan Your Strategy	Act on Your Strategy
Find the mass of the water in the hydrate by finding the difference between the mass of the barium hydroxide and the total mass of the sample. Divide by the total mass of the sample, and multiply by 100%.	percent by mass of water in $Ba(OH)_2 \cdot xH_2O$ $$= \frac{(\text{total mass of sample}) - (\text{mass of } Ba(OH)_2 \text{ in sample})}{\text{total mass of sample}} \times 100\%$$ $$= \frac{50.0 \text{ g} - 27.2 \text{ g}}{50.0 \text{ g}} \times 100\% = 45.6\%$$
Find the amount in moles of barium hydroxide in the sample. Then find the amount in moles of water in the sample. To find out how many water molecules bond to each formula unit of barium hydroxide, divide each answer by the amount in moles of barium hydroxide.	$n_{Ba(OH)_2} = \frac{m_{Ba(OH)_2}}{M_{Ba(OH)_2}} = \frac{27.2 \text{ g}}{171.35 \text{ g/mol}} = 0.159 \text{ mol } Ba(OH)_2$ $$n_{H_2O} = \frac{50.0 \text{ g} - 27.2 \text{ g}}{18.02 \text{ g/mol}} = 1.27 \text{ mol } H_2O$$ $$\frac{1.27 \text{ mol } H_2O}{0.159 \text{ mol } Ba(OH)_2} = \frac{8 \text{ mol } H_2O}{1 \text{ mol } Ba(OH)_2}$$ Therefore, the formula for the hydrate is $Ba(OH)_2 \cdot 8H_2O(s)$.

Check Your Solution
The percent by mass of water in $Ba(OH)_2 \cdot 8H_2O(s)$ is $\frac{144.16 \text{ g/mol}}{315.51 \text{ g/mol}} \times 100\% = 45.691\%$

Using the given mass data, the percent by mass of water in the hydrate of $Ba(OH)_2(s)$ is

$$\frac{(50.0 \text{ g} - 27.2 \text{ g/mol})}{50.0 \text{ g}} \times 100\% = 45.6\%. \quad \text{The answer is reasonable.}$$

For the compounds in questions 51 to 54, calculate the percent by mass of water.

51. $MgSO_3 \cdot 6H_2O(s)$

52. $LiCl \cdot 4H_2O(s)$

53. $CaSO_4 \cdot 2H_2O(s)$

54. $Na_2CO_3 \cdot 10H_2O(s)$

55. List the following hydrates in order, from greatest to least percent by mass of water: $CaCl_2 \cdot 2H_2O(s)$, $MgSO_4 \cdot 7H_2O(s)$, $Ba(OH)_2 \cdot 8H_2O(s)$, $Mn(SO_4)_2 \cdot 2H_2O(s)$.

56. A 3.34 g sample of a hydrate, $SrS_2O_3 \cdot xH_2O(s)$, contains 2.30 g of $SrS_2O_3(s)$. Find the value of x.

57. A hydrate of zinc chlorate, $Zn(ClO_3)_2 \cdot xH_2O(s)$, contains 21.5% zinc by mass. Find the value of x.

58. Determine the formula for the hydrate of chromium(III) nitrate that is 40.50% water by mass.

59. The mass of a sample of a hydrate of magnesium iodide is 1.628 g. It is heated until it is anhydrous and its mass is 1.072 g. Determine the formula for the hydrate.

60. A chemist needs 1.28 g of sodium hypochlorite, $NaOCl(s)$, for an experiment, but she only has sodium hypochlorite pentahydrate, $NaOCl \cdot 5H_2O(s)$. How many grams of the hydrate should she use?

Section Summary

- The empirical formula shows only the relative amounts of the elements, not the actual amounts.

- The molecular formula for a compound shows the actual number of each type of atom in the compound.

- The molecular formula is a whole-number multiple of the empirical formula. It can be determined from the empirical

formula and molar mass of the compound or from the percentage composition and molar mass of the compound.

- The water content of a hydrate can be determined by measuring the mass of a sample before and after heating.

- Chemists need to know whether an ionic substance is a hydrate or in an anhydrous form in order to use the correct molecular formula in a calculation.

Review Questions

1. **K/U** Explain why you do not need to know the actual mass of a substance when you are determining the empirical formula for the substance from its percentage composition.

2. **K/U** Most analytical instruments require a very small sample of a substance—much smaller than 100 g—to determine the composition of the substance. Why is 100 g the most convenient mass to use when determining the empirical formula?

3. **T/I** Determine the empirical formula for a compound that contains 78.77% tin and 21.23% oxygen.

4. **T/I** Determine the empirical formula for a compound that contains 20.24% aluminum and 79.76% chlorine.

5. **T/I** Determine the empirical formula for a compound that is 24.74 % K, 34.76 % Mn, and the rest O.

6. **T/I** Determine the empirical formula for the compound that is represented by this pie graph, and give the graph an appropriate title.

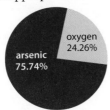

7. **T/I** A compound that contains 22.35 g of lead and 7.65 g of chlorine has a molar mass of 278.11 g/mol. What is its empirical formula?

8. **C** Draw a Venn diagram to show the similarities and differences between the empirical formula and the molecular formula for a compound. Provide an example of a compound to show the differences.

9. **T/I** Turquoise, $CuAl_6(PO_4)_4(OH)_8 \cdot 5H_2O(s)$, is one of the most valuable non-transparent gemstones. It is made from a hydrate of copper aluminum phosphate. What percent by mass is the anhydrous component of the mineral?

10. **T/I** The formula for a hydrate of zinc sulfate is $ZnSO_4 \cdot xH_2O(s)$. If 1 mol of anhydrous zinc sulfate is 56.14% of the mass of 1 mol of the hydrate, what is the value of x?

11. **T/I** Ikaite, $CaCO_3 \cdot xH_2O(s)$, is a hydrate of a calcium carbonate that is found in stalagmites and stalactites, the limestone pillar formations that often form in underground caves. If 1 mol of anhydrous calcium carbonate is 48.08% of the mass of ikaite, what is the value of x?

12. **A** Imagine that you are a lawyer and you are representing an Olympic athlete who has been charged with taking anabolic steroids. The prosecutor presents, as forensic evidence, the empirical formula for the banned substance that was found in the athlete's urine sample. How might you deal with this evidence as the lawyer for the defence?

13. **T/I** Tetrafluoroethene, $C_2F_4(g)$, is the monomer found in the polymer polytetrafluorethene, more commonly known as Teflon™. Teflon™ is used as a coating on frying pans and other cookware.
    a. What is the empirical formula for Teflon™?
    b. How does the molar mass of Teflon™ compare with the molar mass of its empirical formula?

14. **T/I** EDTA, which is short for ethylenediaminetetraacetic acid, is a chemical used in the textile industry to prevent metal impurities from changing the colours of dyed products. The empirical formula for EDTA is $C_5H_8NO_4$. The molar mass of EDTA is 292.24 g/mol. What is its molecular formula?

15. **T/I** Acetone is an organic solvent that can mix with water and most organic solvents. Its empirical formula is C_3H_6O. Its molar mass is 58.08 g/mol. What is its molecular formula?

16. **T/I** An organic fuel has the empirical formula C_4H_4O. The molar mass of the fuel is 408.48 g/mol. What is the molecular formula for the fuel?

QUIRKS & QUARKS

with BOB MCDONALD

CBC ⊕

THIS WEEK ON QUIRKS & QUARKS

Bubbling Methane in the Beaufort Sea

The world's need for fuel is growing, and the existing sources are not enough to sustain this need. Natural gas is an important fuel for heating, transportation, and energy generation. Countries around the world are exploring for new sources of natural gas to supply their growing energy needs.

Scott Dallimore is a geotechnical engineer and researcher at Natural Resources Canada. Bob McDonald interviewed Dallimore to find out if gas hydrates might be a new source of natural gas.

Pingo-like Features

Under the surface of the Beaufort Sea are some very strange geological formations, which look like mounds rising from the seabed. These formations have been named pingo-like features, after similar formations found on land. No one is really sure how the pingo-like features came to be there. Dallimore believes that they formed when methane gas and water rose up through the surface of the seabed.

The pingo-like features contain methane hydrate. The methane-ice froze more than 10 000 years ago, when the land was above

water and covered in permafrost. When the ice crystals formed, methane gas was trapped in the voids of the ice crystal. If you put a flame to a small sample of methane hydrate, the methane trapped in the ice burns and the ice melts. Burning methane is a combustion reaction, as shown in the equation below:

$$CH_4(g) + 2O_2(g) \rightarrow CO_2(g) + 2H_2O(g)$$

There are several hundred pingo-like features offshore from Tuktoyaktuk in the Northwest Territories. Some are a few hundred metres in diameter. Because the ocean water is warming from climate change, the methane hydrate ice is thawing. As a result, the pingo-like features are releasing natural gas.

A Possible Energy Source

On the sea floor, the natural gas is under extreme pressure. When the natural gas is brought to sea level, its volume increases 160 times. These pingo-like features may be the largest source of natural gas on Earth. But can the natural gas in the pingo-like features be harnessed for fuel?

▶ Related Career

Natural Resources Canada is responsible for making and enforcing policies related to our natural resources. Employees deal with water, minerals, trees, and animals, as well as protected areas, such as parks and reserves. These workers monitor the health of the environment, developing public education programs, and rehabilitating wildlife and wild spaces.

Go to **scienceontario** to find out more

Methane hydrate burns as the methane gas is released from the ice.

The pingo-like features on the sea floor off Tuktoyaktuk are one of the natural wonders of Canada. They may also be a valuable new source of natural gas.

QUESTIONS

1. Research and summarize the properties, risks, and benefits of methane hydrates.

2. Find out more about the current research into the use of methane hydrates for energy.

3. Jobs that involve protecting our natural resources involve a range of expertise and technologies. Describe one job and one technology that are related to methane hydrate research.

CHEMISTRY Connections

Mass Spectrometer: Chemical Detective

Suppose that a forensic scientist needs to identify the inks in a document to test for possible counterfeiting. The scientist can analyze the inks using a mass spectrometer, such as the one shown at the right. A mass spectrometer breaks the compounds in a sample of an unknown substance into smaller fragments. The fragments are then separated according to their masses, and the exact composition of the sample can be determined. Mass spectrometry is one of the most important techniques for studying unknown substances.

face of magnetic pole

vacuum chamber

❶ Electron bombardment
A beam of high-energy electrons bombards a vaporized sample, knocking electrons from the atoms in the sample and forming positive ions.

positive ions

electron beam

detector

Data Analysis

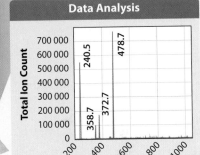

Mass-to-Charge Ratio (m/Z)

❺ Data analysis
A data system generates a graphic display of the results. The lines are located at the mass-to-charge ratios that correspond to the components in the ink sample. A similar analysis can be performed with a different ink sample. The samples can then be compared to determine whether they originated from the same pen.

❷ Particle acceleration
The positive ions are accelerated by an electric field created between two metal grids. The beam of accelerated ions moves toward the next chamber of the mass spectrometer.

vapour entry

❹ Ion detection
A detector measures the deflection and the total ion count.

❸ Ion deflection
The ions in the vacuum chamber are deflected by a magnetic field. The amount of deflection depends on the mass-to-charge ratio of the ions. The greater the mass-to-charge ratio, the less the ions are deflected.

Connect to Society

Research a case in which a mass spectrometer was used to distinguish between different inks. Then write a summary of the procedure and results.

Inquiry INVESTIGATION

6-A

Skill Check

✓ Initiating and Planning

✓ Performing and Recording

✓ Analyzing and Interpreting

✓ Communicating

Safety Precautions

- Do not perform this investigation unless you are wearing welder's goggles to shield your eyes from the bright white light.

- Do not look directly at the burning magnesium.

- Do not put a hot crucible on your work area or the balance.

- Use EXTREME CAUTION when you are near an open flame.

Materials

- 8 cm strip of magnesium, Mg(s), ribbon

- small square of sandpaper or steel wool

- opaque protective shield that filters the UV part of the electromagnetic spectrum

- paper towel

- electronic balance

- clean, dry crucible with lid

- Bunsen burner secured to a retort stand

- retort stand

- ring clamp

- clay triangle

- welder's goggles

- flint lighter

- crucible tongs

- ceramic pad

Teacher Demonstration

Determining the Percentage Composition of Magnesium Oxide

When magnesium metal is heated over a flame, a combustion reaction takes place between the magnesium and the oxygen in air to form magnesium oxide:

$$2Mg(s) + O_2(g) \rightarrow 2MgO(s)$$

In this investigation, your teacher will react a pre-measured mass of pure magnesium metal with oxygen in the surrounding air until all the magnesium is converted into magnesium oxide powder. Then your teacher will measure the mass of the magnesium oxide that was produced to determine the percentage composition of magnesium oxide.

Pre-Lab Questions

1. Why do you think you cannot look directly at the magnesium when it is burning?

2. How is experimental error reduced by keeping the lid on the crucible when the magnesium is burning?

3. When the combustion process is finished, can the product be thrown in the garbage? Explain.

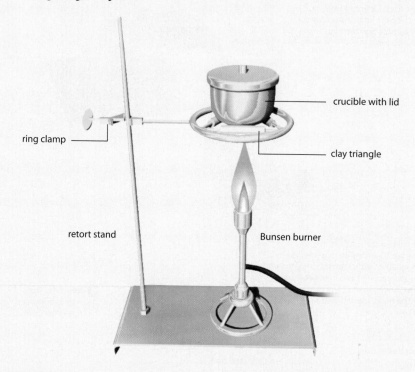

Question

What is the percentage composition of magnesium oxide?

Procedure-Teacher

1. Have students copy the Observations Table into their laboratory notebooks to record data as it is measured.

2. Assemble the equipment as shown in the diagram. Use a clamp to secure the Bunsen burner to a retort stand.

3. Set up the opaque protective shield so that students do not look directly at the burning magnesium, but they see the light on the wall behind you. Protect your eyes with the welder's goggles when the magnesium is burning.

4. Brush both sides of an 8 cm long magnesium strip with sandpaper or steel wool to remove the oxide coating. Wipe the sanded magnesium strip with a dry paper towel to remove any residue. Allow students to see the magnesium strip before you ignite it.

5. Measure and record the mass of the empty crucible and lid. Add the magnesium strip to the crucible. Measure and record the mass of the crucible, lid, and magnesium. Have students record the masses as you measure them.

    Caution: You must wear the welder's goggles while the magnesium is burning. Make sure students do not look directly at the burning magnesium. The shield should be in place to protect their eyes while the magnesium burns.

6. With the lid off, place the crucible with the magnesium on the clay triangle. Heat the crucible with a hot flame. Using the crucible tongs, hold the lid of the crucible nearby. As soon as the magnesium ignites, carefully cover the crucible with the lid. Continue heating for 1 min.

7. Using the crucible tongs, briefly lift the side of the crucible to let more oxygen into the crucible. Try not to let smoke escape the crucible.

8. Heat the crucible until the magnesium ignites again. Repeat this process of heating the magnesium and lifting and closing the crucible lid until the magnesium no longer ignites. Heat 4 to 5 min more with the lid off.

9. Using the crucible tongs, transfer the crucible to the ceramic pad on your work area to cool.

10. With the lid on, leave the crucible and contents to cool completely, for at least 15 min. When the crucible and lid are cool enough to touch, measure the mass. Make sure students record this mass in their Observations Table. Allow students to see the contents of the crucible.

11. Dispose of the magnesium oxide as required.

Procedure-Students

1. Copy the Observations Table into your laboratory notebook. Carefully watch as the demonstration is performed. Record data and observations in your table.

Observations Table

Mass of clean, empty crucible and lid (g)	
Mass of crucible, lid, and magnesium (g)	
Mass of crucible, lid, and magnesium oxide (g)	
Mass of magnesium oxide produced (g)	
Mass of oxygen that must have reacted (g)	
Percent mass of magnesium in magnesium oxide $\dfrac{m_{(Mg)}}{m_{(MgO)}} \times 100\%$	

Analyze and Interpret

1. What mass of magnesium was used in the reaction?

2. What mass of magnesium oxide was produced?

3. Calculate the mass of oxygen that reacted with the magnesium.

Conclude and Communicate

4. Use the data to calculate the percentage composition of magnesium oxide.

5. Calculate the percentage error (PE) using the difference between the experimental mass percent (EP) of magnesium and the actual mass percent (AP) of magnesium.

6. Suppose that some magnesium oxide smoke escaped during the investigation. Would the Mg:O ratio have increased, decreased, or remained unchanged? Explain your answer using sample calculations.

7. How would the value you calculated for the percentage composition of magnesium oxide have been affected if all the magnesium in the crucible had not reacted? Support your answer with sample calculations.

Extend Further

8. **INQUIRY** Based on the percent error, evaluate the design of the investigation. Was this a good investigation for determining the percentage composition of magnesium oxide? What changes might improve the Procedure?

9. **RESEARCH** Magnesium is commonly used in fireworks because it produces a bright white light when it burns. Research other uses of magnesium that are based on this property.

Inquiry
INVESTIGATION
Skill Check
✓ **Performing and Recording**
✓ **Analyzing and Interpreting**
✓ **Communicating**

6-B

Safety Precautions

Materials

- sample of a dry mixture of clean sand and sample of $C_xH_{2x}O_x$
- distilled water
- beaker (250 mL or larger)
- electronic balance
- graphite pencil
- filter paper
- funnel
- Erlenmeyer flask (at least 250 mL, for support)
- graduated cylinder
- glass stirring rod
- wash bottle
- watch glass

Chemical Analysis Simulation

The empirical formula CH_2O is common to several different compounds, many of which are white granular solids that dissolve readily in water. Before chemists can analyze and identify compounds, they often have to separate and isolate them. In this investigation, you will be given a dry mixture that contains clean sand and a mystery compound, $C_xH_{2x}O_x$. The only other information you will have is how many moles of the mystery compound are present in the sample. You will use mass measurements and filtration to determine the composition of the mystery compound.

Pre-Lab Questions

1. Read the steps in the Procedure. Draft a flowchart as you read, to record what you expect to happen at each stage of the separation. Write the contents of the mixture and separated fractions inside the boxes in the flowchart (for example, dry sand and $C_xH_{2x}O_x$, wet sand only). Write the steps in the Procedure (for example, dissolve crystal, filter, wash and rinse) along the branch lines. You may also want to make up a set of possible results to help you anticipate what all the steps will be. Create a table and insert possible results for the mass of the sample, the mass of the filter paper, the mass of the filter paper and sand (after drying), and the mass of the sand.

2. The filtration will separate the mixture into two fractions. Even though you can work very carefully to determine the dry mass of chemical in each fraction, you will need to recover only one of the fractions very precisely. Why?

3. Which fraction do you expect to be easier to separate and recover more precisely? Why?

Questions

What compound, with empirical formula CH_2O, is in the sand/$C_xH_{2x}O_x$ mixture?

Procedure

1. Prepare a results table with the following headings: Mass of Sample (g), Mass of Filter Paper (g); Mass of Dry Sand + Filter Paper (after filtration) (g); Mass of Sand (g); Mass of $C_xH_{2x}O_x$ (g).

2. Measure the mass of the beaker. Add the dry sand mixture to the beaker, and re-measure the mass of the beaker. The mass of the sample is the difference between these masses. Record the three masses in your results table.

3. Using the graphite pencil, write your name on the filter paper. Measure the mass of the filter paper, in grams, to two decimal places. Record this mass in your table.

4. To prepare the filtration apparatus, fold the filter paper and set it inside the funnel. Set the funnel on top of the Erlenmeyer flask.

5. Add approximately 50 mL of water to the beaker. Use the glass rod to stir the contents of the beaker until all the crystals have dissolved.

6. While the sand is suspended in the solution, slowly pour the solution into the funnel. Be careful not to splash, spill, or go over the top of the filter paper.

7. After you have transferred most of the solution from the beaker to the funnel, use the wash bottle to carefully rinse and transfer any remaining sample that may be stuck to the beaker. It is critical for all the sand to be transferred to the filter paper in the funnel.

8. After the solution has passed through the filter paper, wash and rinse the sand residue with the wash bottle to ensure that no trace of residue remains in the beaker.

9. Carefully remove the filter paper with the sand residue. Leave the filter paper in a draft-free location to air-dry on the watch glass, or place in a drying oven.

10. When the filter paper and sand residue are completely dry (the next day), measure the post-filtration mass. Record your results in your table.

11. Calculate the mass of the sand by subtracting the mass of the filter paper from the mass of the filter paper and sand (after filtration). Record the mass of the sand in your table.

12. Calculate the mass of $C_xH_{2x}O_x$ by subtracting the mass of the sand from the mass of the original sample. Record the mass of $C_xH_{2x}O_x$ in your table.

Analyze and Interpret

1. Calculate the molar mass of the $C_xH_{2x}O_x$. Show all your steps.

2. Using the molar mass and the empirical formula, determine the molecular formula for the compound. Show all your steps.

Conclude and Communicate

3. Refer to **Table 6.1** to identify your sample (the white granular compound). Record the identity of your sample.

4. If your filter paper tore during the filtration, describe what you would do to salvage the results. Write your description as steps in a procedure.

Extend Further

1. **INQUIRY** If you had to do this investigation again, what would you do the same? What would you do differently?

2. **RESEARCH** You used sand and filter paper as filters in this investigation. Filters are useful in chemistry because they are designed to block certain substances while allowing others to pass through. For example, filters can be used to remove solid substances that are suspended in fluids. Use text and Internet resources to research how filtration is used to monitor and remove environmental contaminants in water and air.

Safety Precautions

- Heat the hydrate at only a low to medium temperature.
- Use EXTREME CAUTION when you are near an open flame

Materials

- 3 to 5 g of hydrated copper(II) sulfate, $CuSO_4 \cdot xH_2O(s)$
- 400 mL beaker (if hot plate is used) or porcelain evaporating dish (if Bunsen burner is used)
- glass stirring rod
- electronic balance
- scoop
- tongs
- hot plate or Bunsen burner
- ceramic pad or hot pad

Determining the Chemical Formula for a Hydrate

A hydrate can often be converted to an anhydrous ionic compound by heating. This property makes hydrates useful for determining percentage composition and formulas experimentally. In this investigation, you will determine the chemical formula for the hydrate of copper(II) sulfate, $CuSO_4 \cdot xH_2O(s)$. There is an advantage to using copper(II) sulfate: a noticeable colour change occurs. Hydrated copper(II) sulfate is blue, whereas the anhydrous compound is very pale blue.

Prelab Questions

1. How will you know when the reaction is complete?

2. Would your results differ if you used more or less hydrated copper(II) sulfate?

3. If you used a test tube with a small opening instead of a beaker or evaporating dish, how might it affect your results?

Question

What is the chemical formula for the hydrate of copper(II) sulfate, $CuSO_4 \cdot xH_2O(s)$?

Prediction

Predict what change(s) will occur when you heat the hydrate of copper(II) sulfate.

Procedure

Note: If you are using a hot plate as your heat source, use the 400 mL beaker. If you are using a laboratory burner or Bunsen burner, use the porcelain evaporating dish.

1. Make a table like the one below to record your observations.

Observations Table

Mass of empty beaker or evaporating dish	
Mass of beaker or evaporating dish + hydrated copper(II) sulfate	
Mass of beaker or evaporating dish + anhydrous copper(II) sulfate	

2. Measure the mass of the beaker (or evaporating dish) and stirring rod. Record the mass in your table.

3. Add 3 to 5 g of hydrated copper(II) sulfate to the beaker.

4. Measure the mass of the beaker, stirring rod, and hydrated copper(II) sulfate. Record the mass in your table.

5. If you are using a hot plate, heat the beaker with the hydrated copper(II) sulfate until the crystals lose their blue colour. You may need to stir occasionally with the stirring rod. Be sure to keep the heat at a medium setting. Otherwise, the beaker may break.

6. When you see the colour change, stop heating the beaker. Turn off or unplug the hot plate. Remove the beaker with the beaker tongs. Allow the beaker and crystals to cool on the ceramic pad or hot pad.

7. Measure the mass of the beaker with the white crystals. Record this mass in your table.

8. Return the anhydrous copper(II) sulfate to your teacher when you are finished. Do not put it in the sink or in the garbage.

Analyze and Interpret

1. **a.** Calculate the percent by mass of water in your sample of hydrated copper(II) sulfate. Show your calculations clearly.

   b. Do you expect the percent by mass of water that you calculated to be similar to the percent by mass that other groups calculated? Explain.

2. **a.** On the board, write the mass of your sample of hydrated copper(II) sulfate, the mass of the anhydrous copper(II) sulfate, and the percent by mass of water that you calculated.

   b. How do your results compare with other groups' results?

Conclude and Communicate

3. Based on your observations, determine the chemical formula for $CuSO_4 \cdot xH_2O(s)$.

4. Suppose that you heated a sample of a hydrated ionic compound in a test tube. What might you expect to see inside, near the mouth of the test tube? Explain.

5. You calculated the percent by mass of water in your sample of hydrated copper(II) sulfate.

   a. Using your observations, calculate the percentage composition of the hydrated copper(II) sulfate.

   b. Do you think it is more useful to have the percent by mass of water in a hydrate or the percentage composition, assuming that you know the formula for the associated anhydrous ionic compound? Explain your answer.

6. Compare the formula for hydrated copper(II) sulfate that you obtained with the formulas that other groups obtained. Are there any differences? How might these differences have occurred?

7. Suppose that you did not completely convert the hydrate to the anhydrous compound. Explain how this would affect

   a. the calculated percent by mass of water in the compound.

   b. the chemical formula you determined.

8. Suppose that you heated the hydrate too quickly and some of it was lost as it spattered out of the container. Explain how this would affect

   a. the calculated percent by mass of water in the compound.

   b. the chemical formula you determined.

9. Suggest a source of error (not already mentioned) that would result in a value of x that is

   a. higher than the actual value

   b. lower than the actual value

Extend Further

10. **INQUIRY** How would you design an investigation to determine whether a compound is a hydrate or is anhydrous? Write the steps you would include in your procedure.

11. **RESEARCH** Copper sulfate is often used in educational chemistry kits to grow crystals. Explain what properties of copper sulfate make it ideal for this use. Research other uses of copper sulfate that depend on the same properties.

Section 6.1 | Chemical Proportions and Percentage Composition

Elements in a compound are always present in the same proportions by mass. These proportions are represented by the percentage composition of a compound, which is the mass percents of all the elements that make up the compound.

KEY TERMS
law of definite proportions
mass percent
percentage composition

KEY CONCEPTS
- The law of definite proportions states that the elements in a chemical compound are always present in the same proportions by mass.
- The same elements can form different compounds when combined in different whole-number ratios.
- The mass percent of an element in a compound is the mass of the element expressed as a percentage of the total mass of the compound.
- The percentage composition of a compound is the percent by mass of each element in a compound.
- Percentage composition can be calculated using either the mass data for a substance or the chemical formula for the substance.

Section 6.2 | Empirical and Molecular Formulas

The molecular formula for a compound is the whole-number multiple of the empirical formula. The molecular formula can be determined from mass data or percentage composition data.

KEY TERMS
empirical formula
molecular formula
ratio

KEY CONCEPTS
- The empirical formula shows only the relative amounts of the elements, not the actual amounts.
- The molecular formula for a compound shows the actual number of each type of atom in the compound.
- The molecular formula is a whole-number multiple of the empirical formula. It can be determined from the empirical formula and molar mass of the compound, from the percentage composition and molar mass of the compound, or from mass data and the molar mass of the compound.

Knowledge and Understanding

Select the letter of the best answer below.

1. The law of definite proportions states that
   a. elements in the same compound are sometimes found in fixed mass proportions
   b. elements in the sample compound are always present in the same proportions by mass
   c. the source of a chemical affects its composition
   d. water found on the Moon is chemically different from water found on Earth
   e. none of the above

2. A molecular formula can provide more information than an empirical formula about
   a. which elements make up the molecule
   b. the relative amounts of the elements in the molecule
   c. the relative masses of the elements in the molecule
   d. the actual amounts of the elements in the molecule
   e. the mass of the sample

3. Which formula is an example of an empirical formula?
   a. CH_2O
   b. $C_2H_4O_2$
   c. $C_3H_6O_3$
   d. $C_4H_8O_4$
   e. $C_5H_{10}O_5$

4. Which statement about compounds that have the same empirical formula is true?
   a. The compounds have the same properties.
   b. The compounds have the same molar mass.
   c. The compounds have the same physical state.
   d. The compounds are composed of the same elements.
   e. The compounds react with other compounds similarly.

5. While calculating the empirical formula for a compound, you get the element ratio, $Fe_{1.33}O_{1.00}$. Which number is the least common multiple that makes these subscripts small whole numbers?
   a. 1
   b. 2
   c. 3
   d. 4
   e. 5

6. Which compound is an example of a hydrate?
   a. $Na_2CO_3(s)$
   b. $Ag_2CrO4(s)$
   c. $Cu(NO_3)_2(s)$
   d. $(NH_4)_3PO_4(s)$
   e. $CaCl_2 \cdot 2H_2O(s)$

7. Which is the formula for cobalt(II) chloride hexahydrate?
   a. $CoCl_2 \cdot 6H_2O(s)$
   b. $CaCl_2 \cdot 2H_2O(s)$
   c. $CuSO_4 \cdot 5H_2O(s)$
   d. $NaCO_3 \cdot 10H_2O(s)$
   e. $(NH_4)_2C_2O4 \cdot H_2O(s)$

8. Why does a chemist need to know if a compound that is being used in an investigation is in the form of a hydrate or an anhydrous compound?
   a. Anhydrous compounds react much faster because they contain no water.
   b. The water in hydrates must be considered in mass calculations.
   c. Hydrates usually chemically combine with different compounds than anhydrous compounds.
   d. Hydrates produce much more heat during the chemical reaction because of the water in the compound.
   e. Hydrates usually do not react because the water requires too much energy for the reaction to start automatically.

9. Which statement about anhydrous compounds is true?
   a. Any compound can exist as an anhydrous compound.
   b. An anhydrous compound is formed when the water molecules in a hydrate are driven off by heat.
   c. All anhydrous compounds are ionic compounds that do not contain water molecules in their structure.
   d. An anhydrous compound can never become a hydrate again.
   e. Anhydrous compounds are the ionic compounds formed on land, not in the oceans.

Answer the questions below.

10. State, as a ratio, the relative amounts of the elements in each compound.
    a. $H_2O_2(\ell)$
    b. $C_2H_4O_2(\ell)$
    c. $Na_3PO_4(s)$
    d. $AgNO_3(s)$

11. When does a molecular formula not provide more information than the empirical formula?

12. The following table summarizes the different ways in which the composition of a compound can be understood. Some of the information is missing. Copy and complete the table.

Understanding the Composition of a Compound

Type of Composition	Information Already Known	Analytical Information Required	Information Provided about the Composition of the Substance		
			Elements	Relative Amounts	Actual Amounts
?	Percentage composition	?	?	Yes	?
Percentage composition	Pure substance	?	?	?	?
?	Empirical formula	Molar mass	?	?	?

13. Explain why the empirical formula for a compound might also be referred to as the simplest formula for the compound.

14. A chemist wants to determine the molecular formula for a compound. What information must the chemist have?

15. Do all pure samples of a compound have the same percentage composition? Explain your answer.

16. Suppose that you are given the molar mass of a substance and the elements that make up the substance. Can you determine the molecular formula for the substance? Explain your answer.

17. Can an ionic compound be described by a molecular formula? Explain.

18. Explain why a molecular formula, rather than an empirical formula, is essential for identifying a specific compound.

Thinking and Investigation

19. What is the empirical formula for a compound that contains 58.3% magnesium and 41.7% chlorine?

20. What is the empirical formula for a compound that is found by analysis to contain 0.78% hydrogen, 61.98% bromine, and 37.23% oxygen?

21. Galactose is a simple sugar found in milk. The empirical formula for galactose is CH_2O. Chemical analysis shows that the molar mass of galactose is 180 g/mol. Determine the molecular formula for galactose.

22. The empirical formula for tartaric acid, a common acid found in grapes, is $C_2H_3O_3$. Its molar mass is determined by mass spectrometry to be 150.087 g/mol. What is the molecular formula for tartaric acid?

23. Arachidic acid is a saturated fatty acid found in peanut oil. A sample was found to contain 60.06 mg carbon, 10.08 mg hydrogen, and 8.00 mg oxygen. Its molar mass is 312.53 g/mol. What is its molecular formula?

24. The empirical formula for a chlorofluorohydrocarbon is $C_4H_4Cl_3F_2$. Its molar mass is 589.29 g/mol. What is its molecular formula?

25. Analysis of a chemical used in photographic developing fluid indicated that it contains 29.09% oxygen and 65.45% carbon. The remainder is hydrogen. Chemists determined that the molar mass is 110.0 g/mol. What is the molecular formula for the compound?

26. The molecular formula for 2-butene, or β-butylene, is $C_4H_8(g)$. What is the empirical formula for β-butylene?

27. A mystery compound has a molar mass of 68.23 g/mol. It can exist as a pentahydrate. Calculate the percent by mass of the anhydrous compound, compared with the hydrated form of the compound.

28. The anhydrous form of a compound is 38.95% of the mass of its heptahydrate form. Calculate its molar mass.

29. Determine the formula and name for a hydrate that is 14.8% water and 85.2% barium chloride.

30. Three naturally occurring iron compounds are pyrite, $FeS_2(s)$ hematite, $Fe_2O_3(s)$ and siderite, $FeCO_3(s)$. Which contains the greatest percent by mass of iron?

31. The stimulant in coffee, tea, and colas is caffeine, $C_8H_{10}N_4O_2(s)$. Determine its percentage composition.

32. The formulas for hydrates are written with a special notation. Write the formula for each of the following hydrates, and calculate the percent by mass of water in the hydrate.

   a. barium hydroxide octahydrate

   b. sodium carbonate decahydrate

   c. cobalt(II) chloride hexahydrate

   d. iron(III) phosphate tetrahydrate

   e. calcium chloride dihydrate

33. Chemical analysis reveals that a sample of dioxin contains 14.29 g of carbon, 1.20 g of hydrogen, and 9.52 g of oxygen. Dioxin has a molar mass of 84.07 g/mol. What is its molecular formula?

34. A classmate tells you that his data from an investigation show that the molecular formula for a compound is 2.5 times its empirical formula. Could this be correct? Explain your answer.

Communication

35. Using a flowchart, list the steps involved in finding the percentage composition of a compound.

36. Develop a short presentation for your class that explains why moles are used to determine the percentage composition for a compound, even though mass percent values are calculated in grams.

37. Describe the procedure you would use to determine the formula for a hydrate. Provide the reason for each step in your procedure.

38. Summarize your learning in this chapter using a graphic organizer. To help you, the Chapter 6 Summary lists the Key Terms and Key Concepts. Refer to Using Graphic Organizers in Appendix A to help you decide which graphic organizer to use.

Application

39. Two ores that are used as sources of iron are hematite, Fe_2O_3, and magnetite, Fe_3O_4. Determine which ore provides the greater percent of iron per kilogram.

40. **BIG IDEAS** Relationships in chemical reactions can be described quantitatively. For example, in a class investigation, Jana heated powdered zinc with oxygen in a crucible. She collected the following mass data before and after the reaction. (Assume that all the zinc was used up in the reaction.)

Experimental Data

Trial	Mass of Crucible (g)	Mass of Crucible + Zn(s) (g) (before reaction)	Mass of Crucible + Contents (g) (after reaction and cooling)
1	0.45	1.60	1.88
2	0.50	1.65	1.93
3	0.45	1.70	1.90
4	0.40	1.75	2.08

a. What is the percentage composition in the final sample in each trial?

b. Which trials would you keep? Which trial would you discard for likely error? Why?

c. What would you suggest as the likely reason for the error in the discarded trial?

d. What are x and y equal to in the empirical formula for the product, Zn_xO_y?

41. Nickel is used in many consumer products, including stainless steel, rechargeable batteries, magnets, and guitar strings. About 30% of the world's current nickel production comes from Sudbury, Ontario.

a. Three of the mineral compounds in which nickel is found are pentlandite, $(Fe,Ni)_9S_8(s)$, millerite, $NiS(s)$, and nickeline, $NiAs(s)$ (Note: $(Fe,Ni)_9$ is a mixture of iron and nickel, totalling 9 ions. For your calculations, assume that the mixture contains a ratio of 5 Fe ions : 4 Ni ions.). Suppose that these compounds are in three separate formations. You are a geochemist who must decide which nickel formation to mine first. You have approximately equal samples of pentlandite, millerite, and nickeline. Find the mass percent of nickel in each sample, and list the three samples in order from highest to lowest percentage of nickel. Which formation would you mine first?

b. Nickel mining in the Sudbury area has had some negative environmental effects. Research what these effects have been and what has been done to lessen these effects.

42. **T/I** Use the information in the table of naturally occurring minerals to calculate the mass of a sample of each hydrate if it contained as much magnesium as a 10.00 g sample of anhydrous magnesium sulfate contains.

Known hydrates of magnesium sulfate

Formula	Chemical Name	Mineral Name
$MgSO_4$	Anhydrous magnesium sulfate	Does not occur naturally
$MgSO_4 \cdot H_2O$	Magnesium sulfate monohydrate	Kieserite
$MgSO_4 \cdot 2H_2O$	Magnesium sulfate dihydrate	Sanderite
$MgSO_4 \cdot 4H_2O$	Magnesium sulfate tetrahydrate	Starkeyite
$MgSO_4 \cdot 5H_2O$	Magnesium sulfate pentahydrate	Pentahydrite
$MgSO_4 \cdot 6H_2O$	Magnesium sulfate hexahydrate	Hexahydrite
$MgSO_4 \cdot 7H_2O$	Magnesium sulfate heptahydrate	Epsomite
$MgSO_4 \cdot 11H_2O$	Magnesium sulfate elevenhydrate	Meridianiite

Select the letter of the best answer below.

1. **K/U** An empirical formula is similar to percentage composition because it provides information about
   a. the proportions of the elements in a substance
   b. the number of atoms of each element
   c. the mass of each element
   d. the structure of the substance
   e. the molar mass of the substance

2. **K/U** What information does an empirical formula sometimes, but not always, provide?
   a. the elements in the substance
   b. the mass amounts of each element
   c. the relative amounts of each element
   d. the actual amounts of each element
   e. the structure of the substance

3. **K/U** Which of the following is an empirical formula?
   a. C_2H_4
   b. C_6H_{10}
   c. C_2H_2
   d. H_2O_2
   e. $Na_6Cr_2H_7$

4. **K/U** Which of the following statements is false?
   a. The subscripts in an empirical formula give the smallest whole-number ratio of moles of elements in the compound.
   b. The molecular formula gives the actual number of atoms of each element in a molecule, or formula unit, of a substance.
   c. The molecular formula is a whole-number multiple of the empirical formula.
   d. The empirical formula and molecular formula can never be the same.
   e. None of the above are false.

5. **T/I** The percentage composition of butanoic acid is 54.5% carbon, 9.1% hydrogen, and 36.4% oxygen. The molar mass of butanoic acid is 88.1 g/mol. What is the molecular formula for butanoic acid?
   a. $C_3H_4O_3$
   b. C_2H_4O
   c. $C_5H_{12}O$
   d. $C_4H_8O_2$
   e. CH_4O_3

6. **T/I** Propane, $C_3H_8(g)$, is a fuel often used in gas grills. Which is the percentage composition of propane?
   a. 10.11% carbon, 89.89% hydrogen
   b. 81.68% carbon, 18.32% hydrogen
   c. 54.85% carbon, 45.15% hydrogen
   d. 62.50% carbon, 37.50% hydrogen
   e. 37.50% carbon, 62.50% hydrogen

7. **T/I** An unknown compound at a crime scene has the following chemical analysis: 48.64% carbon, 8.16% hydrogen, and 43.20% oxygen. Which of the following is the correct empirical formula?
   a. CHO
   b. C_2HO_2
   c. C_3H_8O
   d. $C_2H_4O_3$
   e. $C_3H_6O_2$

8. **T/I** An unknown compound is determined to have the following composition: 40.68 g of carbon, 5.08 g of hydrogen, and 54.24 g of oxygen. If its molar mass is 118.1 g/mol, what is the molecular formula for the compound?
   a. $C_2H_3O_2$
   b. $C_4H_6O_4$
   c. CH_2O_2
   d. $C_4H_3O_8$
   e. $C_5H_3O_5$

9. **T/I** You measure out two 50.0 g samples: one of calcium chloride dihydrate and one of anhydrous calcium chloride. Which of the following statements is true?
   a. The anhydrous form will have a lower percent by mass of calcium, but a higher percent by mass of water than the hydrate.
   b. The hydrate will have a higher percent by mass of calcium than the anhydrous form.
   c. The hydrate will have a higher percent by mass of calcium and a higher percent by mass of water than the anhydrous form.
   d. The anhydrous form will have a higher percent by mass of calcium than the hydrate.
   e. The two samples will contain the same percent by mass of calcium.

10. **K/U** Which of the following statements is true?
    a. The formula for a hydrate consists of the formula for a compound and the number of water molecules associated with one formula unit.
    b. Anhydrous compounds are formed when hydrates are heated.
    c. The name of a hydrate consists of the compound name and the word hydrate, with a prefix indicating the number of water molecules in 1 mol of the compound.
    d. all of the above are true
    e. none of the above are true

Use sentences and diagrams as appropriate to answer the questions below.

11. **C** You read a statement in a book that says different pure samples of a given compound can have different percentage compositions. Write a letter to the editor explaining why this is incorrect.

12. **C** Using a table, compare empirical and molecular formulas. Provide examples.

13. **C** Draw a concept map to relate the following terms: molar mass of an element, molar mass of a compound, percentage composition, empirical formula, and molecular formula. Provide an example for each term.

14. **T/I** The graph below shows the percentage composition of a compound that contains barium, carbon, and oxygen. Determine the empirical formula and give the graph an appropriate title.

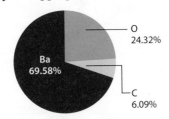

O
24.32%

Ba
69.58%

C
6.09%

15. **T/I** Which compound has the larger percent by mass of sulfur: $H_2SO_3(aq)$ or $H_2S_2O_8(aq)$?

16. **T/I** The graph below shows the percentage composition of a solid that contains oxygen and nitrogen. What is the empirical formula for the solid?

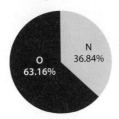

N
36.84%

O
63.16%

17. **T/I** What is the empirical formula for a compound that contains 33.3% sodium, 20.3% nitrogen, and 46.4% oxygen?

18. **T/I** A hydrate of magnesium sulfate, $MgSO_4 \cdot xH_2O(s)$, contains 9.87% magnesium by mass. Find x.

19. **T/I** A compound contains 49.98 g of carbon and 10.47 g of hydrogen. The molar mass of the compound is 116.24 g/mol. Determine its molecular formula.

20. **T/I** A liquid, composed of 46.68% nitrogen and 53.32% oxygen, has a molar mass of 60.01 g/mol. What is the molecular formula for the liquid?

21. **T/I** Iron reacts with oxygen in the air to form rust, $Fe_2O_3 \cdot 4H_2O(s)$. If 65.2 g of rust forms, determine the percent by mass of iron(III) oxide, $Fe_2O_3(s)$, in the rust.

22. **T/I** A chloride of silicon contains 20.9% silicon.
   a. What is the empirical formula for the compound?
   b. By chemical analysis, the molar mass is determined to be 269 g/mol. What is the molecular formula for the compound?

23. **T/I** An oxide of nitrogen contains 30.45 g of nitrogen and 69.55 g of oxygen.
   a. What is the empirical formula for the compound?
   b. By analysis, the molar mass is determined to be 92.02 g/mol. What is the molecular formula?

24. **A** Natron is the name of a mixture of salts that was used by the ancient Egyptians to dehydrate corpses before mummification. Natron is composed of $Na_2CO_3(s)$, $NaHCO_3(s)$, $NaCl(s)$, and $CaCl_2(s)$. The $Na_2CO_3(s)$ absorbs water from tissues to form $Na_2CO_3 \cdot 7H_2O(s)$.
   a. Name the compound $Na_2CO_3 \cdot 7H_2O(s)$.
   b. Calculate the percent by mass of water in $Na_2CO_3 \cdot 7H_2O(s)$.
   c. What mass of anhydrous $Na_2CO_3(s)$ would have been required to dessicate (remove all the water from) an 80 kg body that was 78% water by mass?

25. **A** A mining company has two possible sources of copper: chalcopyrite, $CuFeS_2(s)$, and chalcocite, $Cu_2S(s)$. Which ore would yield the greater quantity of copper? (Assume that the mining conditions and the extraction of copper from the ore are identical for both ores.) Explain your answer.

Self-Check

If you missed question …	1	2	3	4	5	6	7	8	9	10	11	12	13	14	15	16	17	18	19	20	21	22	23	24	25
Review section(s)…	6.2	6.2	6.2	6.2	6.2	6.1	6.2	6.2	6.2	6.2	6.1	6.2	6.1, 6.2	6.2	6.1	6.2	6.2	6.2	6.1	6.2	6.2	6.2	6.2	6.2	6.1

Chemical Reactions and Stoichiometry

Specific Expectations

In this chapter, you will learn how to . . .

- D1.1 **analyze** processes in the home, the workplace, and the environmental sector that involve the use of chemical quantities and calculations (7.1, 7.2, 7.3)

- D1.2 **assess**, on the basis of research, the importance of quantitative accuracy in industrial chemical processes and the potential impact on the environment if the quantities used are not accurate (7.1)

- D2.1 **use** appropriate terminology related to quantities in chemical reactions (7.1, 7.2, 7.3)

- D2.6 **solve** problems related to quantities in chemical reactions by performing calculations that involve percentage yield and limiting reactants (7.2, 7.3)

- D2.7 **conduct** an inquiry to **determine** the actual yield, theoretical yield, and percentage yield of the products in a chemical reaction, **assess** the effectiveness of the procedure, and **suggest** sources of experimental error (7.3)

- D3.4 **explain** the quantitative relationships that are expressed in a balanced chemical equation, using appropriate units of measure (7.1)

Textile mills use a variety of chemical compounds to process the textiles. Chemical solutions often contain excess amounts of reactants to ensure that all the chemical reactions occur and that all the textiles are processed. Unfortunately, several compounds that are used in textile processing have ended up in Canada's waterways, ground water, and landfills. Two of these compounds, nonylphenol ethoxylates (NPEs) and nonylphenol (NP), are toxic to aquatic wildlife. However, there is good news. Since 1998, textile mills have reduced their use of these toxic compounds, as well as the overall amount of water pollution they cause.

Using excess amounts of chemical compounds can be harmful to the environment. Stoichiometric calculations are helpful for determining the correct amounts of reactants to use in industrial and laboratory reactions.

Comparing Mole Relationships

Baking soda and sodium bicarbonate are common names for sodium hydrogen carbonate, $NaHCO_3(s)$. When baking soda is heated, such as in baking, it decomposes to produce sodium carbonate, $Na_2CO_3(s)$, carbon dioxide, $CO_2(g)$, and water, $H_2O(g)$. In this activity, you will observe the decomposition of baking soda and calculate the mole relationship for the reactant and principal product in this reaction.

Safety Precautions

- Wear safety eyewear throughout this activity.
- Use EXTREME CAUTION when you are near an open flame.
- Tie back loose hair and clothing.
- Allow the crucible to cool after the reaction, before touching it.

Materials

- baking soda, $NaHCO_3(s)$
- electronic balance
- crucible
- clay triangle
- retort stand with iron ring
- Bunsen burner secured to a retort stand
- flint lighter
- crucible tongs

Procedure

1. Read steps 2 to 4 in this Procedure, and make a data table to record your observations. Give your table an appropriate title.

2. Determine the mass of the dry empty crucible. Place 2 to 3 g of baking soda in the crucible, and determine the mass. Calculate the mass of the baking soda. Record the data in your table.

3. Set the crucible containing the baking soda on a clay triangle supported by an iron ring, which is attached to a retort stand. Heat the crucible gently with a Bunsen burner for 5 or 6 min. Then increase the flame for an additional 3 or 4 min. Be sure that the crucible is in the flame of the burner. During the reaction, record your observations.

4. Allow the crucible and its contents to cool to room temperature. Determine the mass of the crucible and its contents. Calculate the mass of the product, sodium carbonate, formed in the reaction. Record your data.

Questions

1. Write a balanced chemical equation for the decomposition of sodium hydrogen carbonate.

2. Calculate the amount in moles of sodium hydrogen carbonate used and the amount in moles of sodium carbonate produced.

3. Determine the following ratio:

$$\frac{\text{sodium hydrogen carbonate (mol)}}{\text{sodium carbonate (mol)}}$$

4. Compare the ratio from question 3 to the same ratio using the coefficients from the balanced chemical equation. Describe how the ratios compare.

What Is Stoichiometry?

Key Terms

stoichiometry

mole ratio

stoichiometry the study of the quantitative relationships among the amounts of reactants used and the amounts of products formed in a chemical reaction

Stoichiometry is the study of the quantitative relationships among the amounts of reactants used and the amounts of products formed in a chemical reaction. The basic tool of stoichiometry is a balanced chemical equation. A balanced chemical equation is essential for making calculations and predictions related to quantities in a chemical reaction. A balanced chemical equation is much like a cooking recipe. The outcome of a chemical reaction or a cooking recipe depends on the quantities of reactants or starting ingredients. Suppose that you are making a turkey sandwich. **Figure 7.1** shows how you might express your sandwich recipe as an equation.

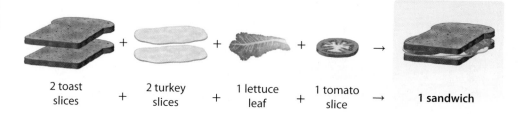

| 2 toast slices | + | 2 turkey slices | + | 1 lettuce leaf | + | 1 tomato slice | → | **1 sandwich** |

Figure 7.1 A turkey sandwich might consist of two toast slices, two turkey slices, one lettuce leaf, and one tomato slice.

Identify how much of each ingredient you would need to make four turkey sandwiches.

A balanced chemical equation gives the same kind of information as a recipe about the quantities of reactants that are needed to carry out a chemical reaction. Using the correct quantities of reactants prevents environmental problems, like the one shown in **Figure 7.2**, that are caused when excessive quantities of harmful chemicals enter ecosystems.

Aquatic wildlife are sensitive to changes in their environment. When chemicals are dumped into waterways, many changes occur that affect the wildlife. As shown below, chemicals that are toxic to wildlife can result in the death of large numbers of individuals. Some chemicals are directly toxic to organisms, while others act indirectly by changing the abiotic conditions of an aquatic ecosystem, such as the acidity and temperature of the water. When large numbers of organisms die, the decomposition of their bodies depletes the water of oxygen. Such oxygen depletion leads to the death of more aquatic organisms.

Go to **scienceontario** to find out more

Figure 7.2 Substances from chemical processes can be toxic to the environment. Quantitative accuracy in industrial chemical processes and responsible disposal of the wastes prevents environmental problems from occurring.

Particle Ratios in a Balanced Chemical Equation

The coefficients in front of the chemical formulas in a balanced chemical equation represent the relative numbers of particles involved in the chemical reaction, as shown in **Figure 7.3**. From the balanced chemical equation, you know that two molecules of hydrogen and one molecule of oxygen react to form two molecules of water.

$$2H_2(g) \quad + \quad O_2(g) \quad \rightarrow \quad 2H_2O(g)$$

Figure 7.3 The coefficients in a chemical equation represent the numbers of particles involved in the chemical reaction. In this reaction, four hydrogen atoms combine with two oxygen atoms to form two molecules of water.

Interpret How many atoms of hydrogen and oxygen are needed to produce 10 molecules of water?

At the molecular level, the ratio of the components in this reaction is

$$\text{2 molecules of } H_2(g) : \text{1 molecule of } O_2(g) : \text{2 molecules of } H_2O(g)$$

Suppose that you want to produce twice as many molecules of water. You simply multiply the ratio of the reactants by 2 to get

$$\text{4 molecules of } H_2(g) : \text{2 molecules of } O_2(g) : \text{4 molecules of } H_2O(g)$$

What if you want to produce 20 molecules of water? How many molecules of oxygen do you need? You know that you need one molecule of oxygen for every two molecules of water. In other words, the number of molecules of oxygen that you need is one half the number of molecules of water that you want to produce. This value can be determined mathematically by equating the known particle ratio for oxygen to water, as shown below.

$$\frac{x}{20 \text{ molecules of } H_2O} = \frac{1 \text{ molecule of } O_2}{2 \text{ molecules of } H_2O}$$

Solve for the unknown, x, to determine the number of oxygen molecules you need.

$$x = 20 \text{ molecules of } H_2O \times \frac{1 \text{ molecule of } O_2}{2 \text{ molecules of } H_2O} = 10 \text{ molecules of } O_2$$

Therefore, 10 oxygen molecules are needed to produce 20 water molecules, as shown in **Figure 7.4**.

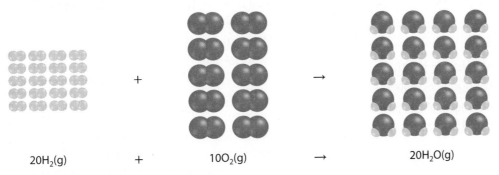

$$20H_2(g) \quad + \quad 10O_2(g) \quad \rightarrow \quad 20H_2O(g)$$

Figure 7.4 If you compare the chemical equation and molecules in this figure with the ones in **Figure 7.3**, you will see that each coefficient is multiplied by 10.

The Sample Problem on the next page shows you how to use ratios of coefficients to determine the number of particles that are produced in a chemical reaction. Notice that ratios of coefficients are used for individual elements, such as hydrogen and oxygen in the above example, and for large molecules, like octane in the Sample Problem on the next page.

Using Ratios of Coefficients in a Balanced Chemical Equation

Problem

The combustion of octane, $C_8H_{18}(g)$, is represented by the following balanced equation:

$$2C_8H_{18}(g) + 25O_2(g) \rightarrow 16CO_2(g) + 18H_2O(g)$$

If 450 molecules of water are produced, how many molecules of carbon dioxide, CO_2, are produced?

What Is Required?

You need to find the number of molecules of carbon dioxide that are produced when 450 molecules of water are produced.

What Is Given?

You know the balanced chemical equation:

$$2C_8H_{18}(g) + 25O_2(g) \rightarrow 16CO_2(g) + 18H_2O(g)$$

You know the number of water molecules that are formed: 450

Plan Your Strategy	Act on Your Strategy
Use the balanced chemical equation to determine the ratio of coefficients for carbon dioxide molecules to water molecules.	The balanced chemical equation is $$2C_8H_{18} + 25O_2(g) \rightarrow 16CO_2(g) + 18H_2O(g)$$ The ratio of coefficients is 16 molecules of CO_2 : 18 molecules of H_2O.
Equate the known ratio of coefficients for carbon dioxide to water to the unknown ratio. Then solve for the unknown.	$$\frac{x}{450 \text{ molecules of } H_2O} = \frac{16 \text{ molecules of } CO_2}{18 \text{ molecules of } H_2O}$$ $$x = 450 \text{ molecules of } H_2O \times \frac{16 \text{ molecules of } CO_2}{18 \text{ molecules of } H_2O}$$ $$= 400 \text{ molecules of } CO_2$$

Check Your Solution

The units are correct. The ratio 400 : 450 is equivalent to the ratio 16 : 18. The answer is reasonable.

Write ratios of coefficients for the equations in questions 1–4 and answer the remaining questions.

1. $2Mg(s) + O_2(g) \rightarrow 2MgO(s)$

2. $2NO(g) + O_2(g) \rightarrow 2NO_2(g)$

3. $Ca(s) + 2H_2O(\ell) \rightarrow Ca(OH)_2(s) + H_2(g)$

4. $2C_2H_6(g) + 7O_2(g) \rightarrow 4CO_2(g) + 6H_2O(g)$

5. How many molecules of nitrogen, $N_2(g)$, produce 10 molecules of ammonia, $NH_3(g)$, in the following reaction?
$$N_2(g) + 3H_2(g) \rightarrow 2NH_3(g)$$

6. Aluminum reacts with chlorine gas to form aluminum chloride:
$$2Al(s) + 3Cl_2(g) \rightarrow 2AlCl_3(s)$$
How many molecules of aluminum chloride form when 155 atoms of aluminum react with an excess of chlorine gas?

7. How many formula units of calcium chloride are produced by 6.7×10^{23} molecules of hydrochloric acid in the following reaction?
$$Ca(OH)_2(aq) + 2HCl(aq) \rightarrow CaCl_2(s) + 2H_2O(\ell)$$

8. How many formula units of magnesium chloride are produced by 7.7×10^{24} molecules of hydrochloric acid in this reaction?
$$Mg(OH)_2(aq) + 2HCl(aq) \rightarrow MgCl_2(s) + 2H_2O(\ell)$$

9. The combustion of ethanol, $C_2H_5OH(\ell)$, is represented by the following equation:
$$C_2H_5OH(\ell) + 3O_2(g) \rightarrow 2CO_2(g) + 3H_2O(\ell)$$
How many molecules of oxygen, $O_2(g)$, produce 1.81 $\times 10^{24}$ molecules of carbon dioxide, $CO_2(g)$, if an excess of ethanol is present?

10. Iron reacts with chlorine gas to form iron(III) chloride. How many atoms of iron react with three molecules of chlorine?

Mole Ratios in a Balanced Chemical Equation

Recall the reaction of hydrogen to produce water, $2H_2(g) + O_2(g) \rightarrow 2H_2O(g)$, from which you obtained the ratio of molecules:

2 molecules of $H_2(g)$: 1 molecule of $O_2(g)$: 2 molecules of $H_2O(g)$

Previously, you multiplied each term in the ratio by 2 and by 10, and still had a correct ratio. It is more useful to multiply each term by 6.02×10^{23}, which is the numerical value of the Avogadro constant.

$2(6.02 \times 10^{23})$ molecules of $H_2(g)$: $1(6.02 \times 10^{23})$ molecule of $O_2(g)$: $2(6.02 \times 10^{23})$ molecules of $H_2O(g)$

As you know, 6.02×10^{23} molecules is one mole. Therefore, the ratio becomes:

2 mol of $H_2(g)$: 1 mol of $O_2(g)$: 2 mol of $H_2O(g)$

The ratio of the amounts in moles of any two substances in a balanced chemical equation is called the **mole ratio**.

mole ratio the ratio of the amounts (in moles) of any two substances in a balanced chemical equation

Learning Check

1. How much of each ingredient described in **Figure 7.1** is needed to make five turkey sandwiches?

2. Why are balanced chemical equations necessary for solving stoichiometric problems?

3. What do the coefficients in the following balanced chemical equation represent?
$$CH_4(g) + 2O_2(g) \rightarrow CO_2(g) + 2H_2O(g)$$

4. What relationships can be determined from a balanced chemical equation?

5. Why are coefficients, not subscripts, used in mole ratios?

6. Consider the following chemical equation:
$$2C_2H_6(g) + 7O_2(g) \rightarrow 4CO_2(g) + 6H_2O(g)$$
   a. Write the mole ratio for ethane, $C_2H_6(g)$, and carbon dioxide.
   b. Write the mole ratio for ethane and oxygen, $O_2(g)$.
   c. Write the mole ratio for carbon dioxide and water.

Using Mole Ratios

Mole ratios can be manipulated to solve problems. For example, consider the mole ratio below:

1 mol $O_2(g)$: 2 mol $H_2O(g)$

This mole ratio can be used to predict the amount of water, in moles, that will be produced if a certain amount of oxygen, $O_2(g)$, reacts according to the following chemical equation:

$$2H_2(g) + O_2(g) \rightarrow 2H_2O(g)$$

Suppose that you want to know the amount of water that is produced by 3.2 mol of oxygen. You know that you obtain 2 mol of water for every 1 mol of oxygen. Therefore, you can use this ratio in the following proportion:

$$\frac{n_{H_2O}}{3.2 \text{ mol } O_2} = \frac{2 \text{ mol } H_2O}{1 \text{ mol } O_2}$$

$$n_{H_2O} = 3.2 \text{ mol } O_2 \times \frac{2 \text{ mol } H_2O}{1 \text{ mol } O_2} = 6.4 \text{ mol } H_2O$$

The Sample Problem on the next page illustrates how the mole ratios from a balanced chemical equation are used to solve problems.

Using Mole Ratios in a Balanced Chemical Equation

Problem
What amount in moles of copper(II) oxide, CuO(s), forms when 0.0045 mol of malachite, $Cu_2(CO_3)(OH)_2(s)$, decomposes completely according to the following equation?

$$Cu_2(CO_3)(OH)_2(s) \rightarrow CO_2(g) + H_2O(g) + 2CuO(s)$$

What Is Required?
You need to determine the amount in moles of copper(II) oxide that is produced.

What Is Given?
You know the balanced chemical equation for the reaction:

$$Cu_2(CO_3)(OH)_2(s) \rightarrow CO_2(g) + H_2O(g) + 2CuO(s)$$

You know the amount of malachite: 0.0045 mol

Malachite

Plan Your Strategy	Act on Your Strategy
Use the balanced chemical equation to write the mole ratio of copper(II) oxide to malachite.	$Cu_2(CO_3)(OH)_2(s) \rightarrow CO_2(g) + H_2O(g) + 2CuO(s)$ The ratio of copper(II) oxide to malachite is 2 mol CuO: 1 mol $Cu_2(CO_3)(OH)_2$
Equate the known mole ratio for copper(II) oxide to malachite with the unknown mole ratio. Then solve for the unknown to determine the amount of copper(II) oxide.	$\dfrac{n_{CuO}}{0.0045 \text{ mol } Cu_2(CO_3)(OH)_2} = \dfrac{2 \text{ mol CuO}}{1 \text{ mol } Cu_2(CO_3)(OH)_2}$ $n_{CuO} = 0.0045 \text{ mol } Cu_2(CO_3)(OH)_2 \times \dfrac{2 \text{ mol CuO}}{1 \text{ mol } Cu_2(CO_3)(OH)_2}$ $= 0.0090 \text{ mol CuO}$

Check Your Solution
The ratio of copper(II) oxide to malachite is 2:1. Accordingly, the amount of copper(II) oxide calculated is twice the amount of malachite.

Practice Problems

11. What amount in moles of silver chromate, $Ag_2CrO_4(s)$, is produced from 0.50 mol of silver nitrate, $AgNO_3(aq)$?

$$2AgNO_3(aq) + Na_2CrO_4(aq) \rightarrow$$
$$Ag_2CrO_4(s) + 2NaNO_3(aq)$$

12. What amount in moles of water forms when 6.00 mol of carbon dioxide is consumed in the following reaction?

$$2NH_3(g) + CO_2(g) \rightarrow NH_2CONH_2(s) + H_2O(g)$$

13. Calculate the amount in moles of ammonia, $NH_3(g)$, that is needed to prepare 22 500 mol of the fertilizer ammonium sulfate, $(NH_4)_2SO_4(s)$.

$$2NH_3(g) + H_2SO_4(aq) \rightarrow (NH_4)_2SO_4(s)$$

14. Calculate the amount in moles of oxygen that is needed to react with 2.4 mol of ammonia to produce poisonous hydrogen cyanide, HCN(g).

$$2NH_3(g) + 3O_2(g) + 2CH_4(g) \rightarrow$$
$$2HCN(g) + 6H_2O(g)$$

15. What amount in moles of fluorine, $F_2(g)$, yields 2.35 mol of xenon tetrafluoride, $XeF_4(s)$?

$$Xe(g) + 2F_2(g) \rightarrow XeFe_4(s)$$

16. These equations show two possible reactions:

$$2N_2(g) + O_2(g) \rightarrow 2N_2O(g)$$
$$N_2(g) + 2O_2(g) \rightarrow 2NO_2(g)$$

a. What amount in moles of oxygen reacts with 93.5 mol of nitrogen to form dinitrogen monoxide, $N_2O(g)$?

b. What amount in moles of nitrogen dioxide, $NO_2(g)$ forms in the other reaction?

17. What amount in moles of oxygen reacts with 11.3 mol of propane gas, $C_3H_8(g)$, during the combustion of propane?

$$C_3H_8(g) + 5O_2(g) \rightarrow 3CO_2(g) + 4H_2O(g)$$

18. What amount in moles of phosphorus produces 6.45 mol of tetraphosphorus hexoxide, $P_4O_6(s)$?

$$P_4(s) + 3O_2(g) \rightarrow P_4O_6(s)$$

19. Silver tarnishes when it is exposed to small amounts of hydrogen sulfide, $H_2S(g)$, in the air.
$$4Ag(s) + 2H_2S(g) + O_2(g) \rightarrow 2Ag_2S(s) + 2H_2O(\ell)$$
How many molecules of hydrogen sulfide react with 1.7 mol of silver?

20. When heated, magnesium hydrogen carbonate, $Mg(HCO_3)_2(s)$, decomposes and forms magnesium carbonate, $MgCO_3(s)$, carbon dioxide and water, vapour. What amount in moles of water is produced from 7.24×10^5 mol of magnesium hydrogen carbonate?

Activity 7.1 Chalk It Up to Molar Relationships

Common chalk (calcium carbonate), $CaCO_3(s)$, reacts with hydrochloric acid according to the following balanced equation:

$$CaCO_3(s) + 2HCl(aq) \rightarrow CO_2(g) + H_2O(\ell) + CaCl_2(aq)$$

In this activity, you will carry out the reaction between chalk and hydrochloric acid and then determine the molar relationships.

Safety Precautions

- Wear safety eyewear throughout this activity.
- Wear gloves throughout this activity.
- Do not inhale vapour from hydrochloric acid.
- Be careful when using hydrochloric acid.
- If acid gets on skin, flush with plenty of water.

Materials

- 1.0 g piece of chalk, $CaCO_3(s)$
- 40 mL of 1 mol/L hydrochloric acid, $HCl(aq)$
- balance (with precision to 0.01 g)
- 2 beakers (100 mL)
- graduated cylinder

Procedure

1. Measure and record the mass of an empty 100 mL beaker.
2. Your teacher will give you a piece of chalk, with a mass of approximately 1.0 g. Place the chalk in the beaker. Measure and record the mass of the beaker and the chalk.
3. Fill another 100 mL beaker with 40 mL of 1 mol/L hydrochloric acid. Place this beaker on the balance, next to the beaker containing the chalk. Measure and record the total mass of both beakers and their contents.
4. Remove the beakers from the balance. Slowly add hydrochloric acid from the second beaker to the beaker containing the chalk, a bit at a time, until all the chalk has disappeared and the solution produces no more bubbles.
5. Measure and record the total mass of both beakers and their contents. Dispose of the reacted chemicals and clean up your workarea as directed by your teacher.

Questions

1. Calculate the amount in moles of calcium carbonate used and the amount in moles of carbon dioxide produced. (**Hint:** Find the difference between the total mass of both beakers and their contents before and after the reaction. This difference represents the mass of carbon dioxide gas produced.)

2. According to the balanced chemical equation, how many formula units of calcium carbonate react to form one molecule of carbon dioxide? What amount in moles of each compound is involved in the reaction?

3. Based on the balanced chemical equation and the amount of chalk you used, what amount in moles of carbon dioxide was expected to form?

4. How does the amount of carbon dioxide you calculated in question 3 relate to the amount of carbon dioxide that was actually produced? Explain your results.

Mass Relationships in Chemical Equations

The mole ratio of any two substances in any balanced chemical equation consists of simple whole numbers. However, the mass ratio of any two substances in an actual chemical reaction rarely consists of whole numbers. Consider the reaction of hydrogen gas with oxygen: $2H_2(g) + O_2(g) \rightarrow 2H_2O(g)$

In this reaction, the mole ratio of hydrogen to oxygen is 2 mol $H_2(g)$: 1 mol $O_2(g)$.

Recall, from Chapter 5, that you can find the molar mass (M) of an element in the periodic table. The equation $m = n \times M$ can be used to find the mass of each compound in a reaction, as follows:

$$m_{H_2} = n_{H_2} \times M_{H_2} = 2 \text{ mol} \times \frac{2.02 \text{ g}}{\text{mol}} = 4.04 \text{ g} \qquad m_{O_2} = n_{O_2} \times M_{O_2} = 1 \text{ mol} \times \frac{32.00 \text{ g}}{\text{mol}} = 32.00 \text{ g}$$

Therefore, the mass ratio of hydrogen to oxygen is 4.04 g $H_2(g)$: 32.00 g $O_2(g)$.

Stoichiometric Mass Calculations

Go to **scienceontario** to find out more

If you know the amount in moles, number of particles, or mass in grams of any substance in a chemical reaction, you can calculate the amount, number of particles, or mass of any other substance in the reaction. This allows you to calculate the quantities of reactants required for a specific chemical reaction or to predict how much product will form in terms of moles, molecules, or grams. The following Sample Problems show how to use stoichiometry to determine these values.

Sample Problem

Mass Stoichiometry: Reactant to Product

Problem

Some scientists propose generating oxygen by photosynthesis during a mission to Mars. Photosynthesis is a process that uses energy from sunlight to drive a long series of reactions in which carbon dioxide and water are combined to form glucose and oxygen.

$$6CO_2(g) + 6H_2O(\ell) \rightarrow C_6H_{12}O_6(s) + 6O_2(g)$$

This reaction could help to eliminate carbon dioxide from the spacecraft, while producing breathable oxygen. An astronaut produces an average of 1.00×10^3 g of carbon dioxide each day. What mass of oxygen, per astronaut, would need to be produced by photosynthesis each day?

What Is Required?

You need to find the mass of oxygen produced from 1.00×10^3 g of carbon dioxide.

What Is Given?

You know the balanced chemical equation: $6CO_2(g) + 6H_2O(\ell) \rightarrow C_6H_{12}O_6(s) + 6O_2(g)$
You know the mass of carbon dioxide: 1.00×10^3 g

Plan Your Strategy	Act on Your Strategy
Write the balanced equation for the reaction and the mole ratio of oxygen to carbon dioxide.	$6CO_2(g) + 6H_2O(\ell) \rightarrow C_6H_{12}O_6(s) + 6O_2(g)$ The ratio of oxygen to carbon dioxide is 6 mol O_2: 6 mol CO_2.
Calculate the molar masses, M, of oxygen and carbon dioxide.	$M_{O_2} = 2M_O = 2(16.00 \text{ g/mol}) = 32.00 \text{ g/mol}$ $M_{CO_2} = 1M_C + 2M_O$ $\quad = 1(12.01 \text{ g/mol}) + 2(16.00 \text{ g/mol}) = 44.01 \text{ g/mol}$
Convert the mass of carbon dioxide into an amount in moles using the molar mass of carbon dioxide and $n = \frac{m}{M}$.	$n = \frac{m}{M}$ $\quad = \frac{1.00 \times 10^3 \text{g}}{44.01 \text{ g/mol}} = 22.722 \text{ mol } CO_2$
To solve for the amount of oxygen in moles, use the mole ratio of oxygen to carbon dioxide from the balanced equation, as well as the amount of carbon dioxide in moles.	$\frac{n_{O_2}}{22.722 \text{ mol } CO_2} = \frac{6 \text{ mol } O_2}{6 \text{ mol } CO}$ $n_{O_2} = 22.722 \text{ mol } CO_2 \times \frac{6 \text{ mol } O_2}{6 \text{ mol } CO_2} = 22.722 \text{ mol } O_2$
Convert the amount of oxygen in moles into the mass of oxygen using the molar mass of oxygen and the equation $m = n \times M$.	$m = n \times M$ $\quad = 22.722 \text{ mol} \times 32.00 \text{ g/mol} = 727 \text{ g } O_2$ Therefore, 727 g of oxygen is theoretically produced.

Check Your Solution

An alternative method for solving stoichiometric problems, called the factor-label method, can be used to check the solution. To use this method, set up the equation so that all the terms cancel, except for the required term.

$$m_{O_2} = 1.00 \times 10^3 \text{ g } CO_2 \times \frac{1 \text{ mol } CO_2}{44.01 \text{ g } CO_2} \times \frac{6 \text{ mol } O_2}{6 \text{ mol } CO_2} \times \frac{32.00 \text{ g } O_2}{1 \text{ mol } O_2} = 727 \text{ g } O_2$$

Mass Stoichiometry: Reactant to Reactant

One disadvantage of using photosynthesis to produce oxygen for a mission to Mars is that photosynthesis requires water, which is not readily available in space. What mass of water is required to remove the carbon dioxide from one astronaut's exhaled breath each day, using photosynthesis? Recall that an astronaut produces an average of 1.00×10^3 g of carbon dioxide each day and the reaction for photosynthesis is

$$6CO_2(g) + 6H_2O(\ell) \rightarrow C_6H_{12}O_6(s) + 6O_2(g)$$

Astronaut Working in Space

What Is Required?

You need to find the mass of water that is required to react with 1.00×10^3 g of carbon dioxide.

What Is Given?

You know the balanced chemical equation:

$$6CO_2(g) + 6H_2O(\ell) \rightarrow C_6H_{12}O_6(s) + 6O_2(g)$$

You know the mass of carbon dioxide: 1.00×10^3 g

Plan Your Strategy	Act on Your Strategy
Write the balanced equation for the reaction and the mole ratio of water to carbon dioxide.	$6CO_2(g) + 6H_2O(\ell) \rightarrow C_6H_{12}O_6(s) + 6O_2(g)$ The ratio of water to carbon dioxide is 6 mol H_2O: 6 mol CO_2.
Calculate the molar masses, M, of carbon dioxide and water.	$\begin{aligned} M_{CO_2} &= M_C + 2M_O \\ &= 1(12.01 \text{ g/mol}) + 2(16.00 \text{ g/mol}) \\ &= 44.01 \text{ g/mol} \end{aligned}$ $\begin{aligned} M_{H_2O} &= 2M_H + 1M_O \\ &= 2(1.01 \text{ g/mol}) + 1(16.00 \text{ g/mol}) \\ &= 18.02 \text{ g/mol} \end{aligned}$
Convert the mass of carbon dioxide into an amount in moles using the molar mass of carbon dioxide and $n = m/M$.	$\begin{aligned} n = \frac{m}{M} &= \frac{1.00 \times 10^3 \text{ g}}{44.01 \text{ g/mol}} \\ &= 22.722 \text{ mol } CO_2 \end{aligned}$
To determine the amount of water in moles, use the mole ratio of water to carbon dioxide from the balanced equation, as well as the amount of carbon dioxide in moles.	$\dfrac{n_{H_2O}}{22.722 \text{ mol } CO_2} = \dfrac{6 \text{ mol } H_2O}{6 \text{ mol } CO_2}$ $n_{H_2O} = 22.722 \text{ mol } CO_2 \times \dfrac{6 \text{ mol } H_2O}{6 \text{ mol } CO_2}$ $= 22.722 \text{ mol } H_2O$
Convert the amount of water into the mass of water using the molar mass of water and the equation $m = n \times M$.	$\begin{aligned} m = n \times M &= 22.722 \text{ mol} \times 18.02 \text{ g/mol} \\ &= 409 \text{ g } H_2O \end{aligned}$ Therefore, 409 g of water is required each day to remove the carbon dioxide from one astronaut's exhaled breath.

Check Your Solution

The units are correct. Because carbon dioxide and water have a mole ratio of 6:6 or 1:1, and the molar mass of water is a little less than half of the molar mass of carbon dioxide, the mass of water should be less than 500 g. The answer is reasonable.

Using the factor-label method to check the answer,

$$m_{H_2O} = 1.00 \times 10^3 \text{ g } CO_2 \times \frac{1 \text{ mol } CO_2}{44.01 \text{ g } CO_2} \times \frac{6 \text{ mol } H_2O}{6 \text{ mol } CO_2} \times \frac{18.02 \text{ g } H_2O}{1 \text{ mol } H_2O} = 409 \text{ g } H_2O$$

21. The production of acetic acid, $CH_3COOH(\ell)$, is represented by the following chemical equation:

$$CH_3OH(\ell) + CO(g) \rightarrow CH_3COOH(\ell)$$

Calculate the mass of acetic acid that is produced by the reaction of 6.0×10^4 g of carbon monoxide with sufficient methanol, $CH_3OH(\ell)$.

22. Calculate the mass of silver nitrate, $AgNO_3(aq)$, that must react with solid copper to provide 475 kg of copper nitrate, $Cu(NO_3)_2(aq)$.

$$Cu(s) + 2AgNO_3(aq) \rightarrow 2Ag(s) + Cu(NO_3)_2(aq)$$

23. What mass of oxygen is produced if 22.7 mol of carbon dioxide is consumed in a controlled photosynthesis reaction?

$$6CO_2(g) + 6H_2O(\ell) \rightarrow C_6H_{12}O_6(s) + 6O_2(g)$$

24. Sodium phosphate, $Na_3PO_4(aq)$, is an all-purpose cleaner that can be used to clean walls before painting. It is often referred to as trisodium phosphate, or TSP, and it must be handled with care because it is corrosive. It is prepared by the following reaction:

$$3NaOH(aq) + H_3PO_4(aq) \rightarrow Na_3PO_4(aq) + 3H_2O(\ell)$$

What amount in moles of TSP is produced if 14.7 g of sodium hydroxide reacts with phosphoric acid, $H_3PO_4(aq)$?

25. What mass of hydrogen is produced when 3.75 g of aluminum reacts with sulfuric acid, $H_2SO_4(aq)$?

$$2Al(s) + 3H_2SO_4(aq) \rightarrow 3H_2(g) + Al_2(SO_4)_3(aq)$$

26. Nitrogen monoxide, $NO(g)$, reacts with oxygen gas to form nitrogen dioxide, $NO_2(g)$. What mass of nitrogen dioxide is produced from 2.84 g of nitrogen monoxide?

27. Iron(III) oxide, $Fe_2O_3(s)$, reacts with carbon monoxide to form solid iron and carbon dioxide in the following reaction:

$$Fe_2O_3(s) + 3CO(g) \rightarrow 2Fe(s) + 3CO_2(g)$$

What mass (in grams) of carbon dioxide is produced from 12.4 g of iron(III) oxide?

28. Methane, $CH_4(g)$, reacts with sulfur, $S_8(s)$, to produce carbon disulfide, $CS_2(\ell)$, and hydrogen sulfide, $H_2S(g)$. Carbon disulfide is often used in the production of cellophane.

$$2CH_4(g) + S_8(s) \rightarrow 2CS_2(\ell) + 4H_2S(g)$$

What mass of methane is required if 4.09 g of hydrogen sulfide is produced?

29. The addition of concentrated hydrochloric acid to manganese(IV) oxide, $MnO_2(s)$, produces chlorine gas, $Cl_2(g)$.

$$4HCl(aq) + MnO_2(s) \rightarrow$$
$$MnCl_2(aq) + Cl_2(g) + 2H_2O(\ell)$$

What mass of manganese(IV) oxide is needed to react with 8.65×10^{-2} g of hydrochloric acid?

30. Aluminum carbide, $Al_4C_3(s)$, is a yellow powder that reacts with water, $H_2O(\ell)$, to produce aluminum hydroxide, $Al(OH)_3(s)$, and methane, $CH_4(g)$. Write a balanced chemical equation for the reaction and determine the mass of water required to react with 14.0 g of aluminum carbide.

Stoichiometry and Reactions in the Laboratory

In this section, you have learned how to do stoichiometric calculations using a balanced chemical equation. Stoichiometric calculations are based on the assumption that all the substances occur in an exact mole ratio, as shown in the chemical equation. However, reactants often are not present in the exact ratio. Furthermore, just as you might not always get exactly a dozen cookies from a chocolate-chip cookie recipe, as shown in **Figure 7.5**, the amount of final product that is predicted by stoichiometry is not always produced in a laboratory. In the next two sections, you will learn how to predict how much product will form in a given chemical equation.

Figure 7.5 The amount of final product is not always what is predicted when making cookies and when performing chemical reactions in the laboratory.

Section Summary

- The coefficients of a balanced chemical equation can be used to represent the relative amounts (in moles) of particles (atoms, ions, molecules, or formula units).

- Stoichiometric calculations are used to predict the amounts of reactants used or products formed in a chemical reaction.

- A mole ratio from a balanced chemical equation relates the amount in moles of one reactant or product to the amount in moles of another reactant or product.

- The amount in moles of any substance can be converted to number of particles or mass units, such as grams.

Review Questions

1. **K/U** What important chemical information about the reactants and products in a reaction is obtained from the coefficients of a balanced chemical equation?

2. **K/U** Why is a balanced chemical equation needed for stoichiometric calculations?

3. **K/U** Determine all the possible mole ratios for each balanced chemical equation.
   a. $4Al(s) + 3O_2(g) \rightarrow 2Al_2O_3(s)$
   b. $3Fe(s) + 4H_2O(\ell) \rightarrow Fe_3O_4(s) + 4H_2(g)$
   c. $2HgO(s) \rightarrow 2Hg(\ell) + O_2(g)$
   d. $2SO_2(g) + O_2(g) \rightarrow 2SO_3(g)$
   e. $CaO(s) + H_2O(\ell) \rightarrow Ca(OH)_2(s)$

4. **T/I** The oxidation of aluminum is represented by the following chemical equation:
   $$4Al(s) + 3O_2(g) \rightarrow 2Al_2O_3(aq)$$
   What mass of oxygen is required to oxidize 25 mol of aluminum?

5. **T/I** The reaction of nitrogen gas with hydrogen gas is represented by the following chemical equation:
   $$N_2(g) + 3H_2(g) \rightarrow 2NH_3(g)$$
   What mass (in grams) of nitrogen reacts with 6.0 g of hydrogen?

6. **T/I** A student says that 1.0 g of magnesium reacts with 1.0 g of chlorine, $Cl_2(g)$, according to this equation:
   $$Mg(s) + Cl_2(g) \rightarrow MgCl_2(s)$$
   Using mathematical calculations, explain why the student's reasoning is incorrect.

7. **A** Iron ore, $Fe_2O_3(s)$, is treated with carbon monoxide, $CO(g)$, to extract and purify the iron. This reaction is represented by the following unbalanced equation:
   $$\_\_\_Fe_2O_3(s) + \_\_\_CO(g) \rightarrow \_\_\_Fe(s) + \_\_\_CO_2(g)$$
   a. Balance the chemical equation.
   b. Calculate the minimum mass of carbon monoxide that must be ordered by a refining company for every metric tonne of iron ore that is processed.

8. **C** Complete a flowchart to show how you would use mole ratios to determine the unknown amount of a substance that reacts with a known amount of another substance.

9. **T/I** The neutralization reaction of hydrobromic acid, $HBr(aq)$, and calcium hydroxide, $Ca(OH)_2(aq)$, is represented by the following balanced chemical equation:
   $$2HBr(aq) + Ca(OH)_2(aq) \rightarrow CaBr_2(aq) + 2H_2O(\ell)$$
   Copy and complete this table to show all the quantity ratios that are implied by the balanced chemical equation.

Neutralization Reaction

	2HBr(aq)	Ca(OH)$_2$(aq)	CaBr$_2$(aq)	2H$_2$O(ℓ)
Amount (mol)				
Number of Units				
Mass (g)				

10. **T/I** When heated, the orange crystals of ammonium dichromate, $(NH_4)_2Cr_2O_7(s)$, slowly decompose to form green chromium(III) oxide, $Cr_2O_3(s)$. Colourless nitrogen gas and water vapour are given off.
   $$(NH_4)_2Cr_2O_7(s) \rightarrow Cr_2O_3(s) + N_2(g) + 4H_2O(g)$$
   a. How many formula units of chromium(III) oxide are produced from the decomposition of 7.0 g of ammonium dichromate?
   b. How many formula units of ammonium dichromate are needed to produce 2.75 g of water vapour?

11. **C** A neighbour over-fertilizes his lawn. Fertilizer that cannot be absorbed by the plants runs into a nearby lake. The excess fertilizer causes algal bloom. Research algal bloom and write a letter to your neighbour explaining why using the correct chemical quantities is important for the environment.

12. **C** Gardening often involves the use of fertilizers, pesticides, and herbicides (organic and synthetic). Write a brief public service announcement explaining why it is important to use correct chemical quantities to grow a healthy garden and to protect the surrounding environment.

Limiting and Excess Reactants

Key Terms

stoichiometric amount

limiting reactant

excess reactant

Consider again the turkey sandwich example in **Figure 7.1**. The recipe for a turkey sandwich can be written as an equation:

2 toast slices + 2 turkey slices + 1 lettuce leaf + 1 tomato slice → 1 turkey sandwich

A certain number of each ingredient is necessary to make a turkey sandwich according to the recipe. However, there is always the possibility that there will be a short supply of at least one ingredient. When an ingredient runs out, the production of turkey sandwiches by the recipe stops. In other words, the ingredient that runs out first limits the quantity of sandwiches that can be produced. For example, suppose that you have four toast slices, six turkey slices, three lettuce leaves, and three tomato slices, as shown in **Figure 7.6**. Because each sandwich requires two slices of toast, you can make only two sandwiches, even though you have enough of the other ingredients to make three turkey sandwiches. The number of slices of toast limits the number of sandwiches you can make. The other ingredients are in excess and are left over.

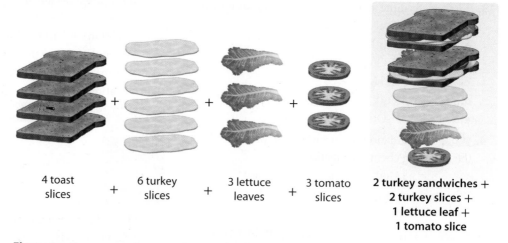

| 4 toast slices | + | 6 turkey slices | + | 3 lettuce leaves | + | 3 tomato slices | | **2 turkey sandwiches +**
2 turkey slices +
1 lettuce leaf +
1 tomato slice |

Figure 7.6 Once you make two turkey sandwiches, there is no more toast to make additional sandwiches. Toast is the "limiting ingredient." The other ingredients are in excess and are left over.

Limiting and Excess Reactants in Chemical Reactions

stoichiometric amount
the exact molar amount of a reactant or product, as predicted by a balanced chemical equation

limiting reactant
a reactant that is completely consumed during a chemical reaction, limiting the amount of product that is produced

excess reactant
a reactant that remains after a reaction is over

If the reactants in a chemical reaction are present in amounts that correspond exactly to the mole ratios from the balanced chemical equation, they are said to be present in **stoichiometric amounts**. If the reactants are present in stoichiometric amounts, ideally the reaction stops when no trace of the reactants are left. However, in actual chemical reactions, one reactant is usually in shorter supply than the other reactants. In other words, it is rare for the reactants in a chemical reaction to be present in amounts that correspond exactly to the mole ratios from the balanced chemical equation. In most reactions, there is at least one leftover reactant when the chemical reaction stops.

Consider a candle burning in a room. There is an unlimited amount of oxygen in the air, so the reaction proceeds until the wax is gone. However, if a candlesnuffer, like the one in **Figure 7.7,** is put over the burning candle, the amount of oxygen is limited. The combustion reaction stops when the supply of oxygen is gone, even though there is wax available. The reactant that limits or stops a reaction, such as the oxygen in **Figure 7.7,** is called the **limiting reactant** or limiting reagent. The limiting reactant determines the amount of product that is formed. The reactants that are left over, such as the candle wax, are called **excess reactants**.

The Limiting Reactant Forms Less Product

The limiting reactant is not necessarily the reactant that is present in the smaller amount. It is the reactant that forms the smaller amount of product. For example, consider the chemical reaction that produces water:

$$2H_2(g) + O_2(g) \rightarrow 2H_2O(g)$$

If oxygen is present in excess and 2 mol of hydrogen is available, then 2 mol of water is produced. However, if hydrogen is in excess and 2 mol of oxygen is available, then 4 mol of water is produced. In each situation, the limiting reactant is the reactant that forms the smaller amount of product. Similarly, the limiting reactant is not necessarily the reactant that has the lower mass. It is the reactant that produces the lower mass of product.

Identifying the Limiting Reactant in a Chemical Reaction

It is important to identify the limiting and excess reactants. This is because the amount of limiting reactant that is available for a chemical reaction determines the amount of product that is formed and the amount of excess reactant that is left over. To identify the limiting reactant, you need to determine which reactant yields the smaller amount of product in a chemical reaction. The activity below demonstrates the importance of considering the limiting reactant.

Figure 7.7 There was excess oxygen available for the burning candle until a candlesnuffer was used to limit the oxygen.

Suggested Investigation

Inquiry Investigation 7-A, Limiting and Excess Reactants

Activity 7.2 Identify the Limiting Item

Suppose that you have been hired by a furniture company. Your job is to put together kits for making kitchen chairs. Each kit contains all the parts that are needed to assemble one kitchen chair. The equation for one kitchen chair is given below.

Procedure

1. Assume that you have 36 frames, 128 legs, 256 leg braces, 100 hardware packages, and 1000 assembly manuals. How many complete chair kits can you make?

2. Determine the item that will limit the number of complete chair kits you can make.

3. Determine the items you have in excess amounts.

4. Calculate how much of each excess item remains after you make the chair kits.

Questions

1. You have 36 chair frames. Why are the chair frames not the limiting item, even though they are present in the smallest quantity?

2. Does an item that is available in excess affect the quantity of complete chair kits that you can make? Explain your answer.

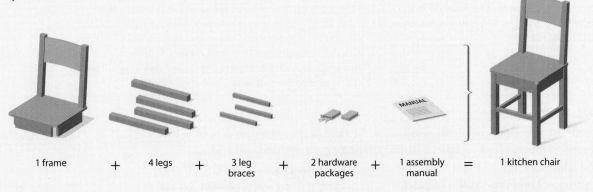

1 frame   +   4 legs   +   3 leg braces   +   2 hardware packages   +   1 assembly manual   =   1 kitchen chair

Identifying the Limiting Reactant in Problems

When solving problems, it is important to identify which reactant is limiting. The Sample Problem below demonstrates how to identify the limiting reactant.

Fertilizer

Sample Problem

Identifying the Limiting Reactant

Problem

The chemical compound ammonia is prepared from its elements according to the following chemical equation:

$$N_2(g) + 3H_2(g) \rightarrow 2NH_3(g)$$

The production of ammonia is an important industrial process in the manufacture of fertilizers. If 4.20 g of nitrogen gas reacts with 0.750 g of hydrogen gas, which is the limiting reactant?

What Is Required?

You need to determine whether nitrogen gas or hydrogen gas is the limiting reactant.

What Is Given?

You know the balanced chemical equation:

$$N_2(g) + 3H_2(g) \rightarrow 2NH_3(g)$$

You know the mass of nitrogen: 4.20 g

You know the mass of hydrogen: 0.750 g

Plan Your Strategy	Act on Your Strategy
Calculate the molar masses, M, of nitrogen and hydrogen.	$M_{N_2} = 2M_N = 2\left(\dfrac{14.01 \text{ g N}}{1 \text{ mol}}\right) = 28.02 \text{ g/mol}$ $M_{H_2} = 2M_H = 2\left(\dfrac{1.01 \text{ g H}}{1 \text{ mol}}\right) = 2.02 \text{ g/mol}$
Convert the masses of nitrogen and hydrogen into amounts (in moles) using $n = \dfrac{m}{M}$.	$n_{N_2} = \dfrac{m_{N_2}}{M_{N_2}} = \dfrac{4.20 \text{ g}}{28.02 \text{ g/mol}} = 0.149\ 89 \text{ mol } N_2(g)$ $n_{H_2} = \dfrac{m_{N_2}}{M_{N_2}} = \dfrac{0.750 \text{ g}}{2.02 \text{ g/mol}} = 0.371\ 29 \text{ mol } H_2(g)$
Calculate the amount of ammonia that is produced by the given amount of nitrogen and the given amount of hydrogen.	$n_{NH_3} = 0.149\ 89 \text{ mol } N_2 \times \dfrac{2 \text{ mol } NH_3}{1 \text{ mol } N_2} = 0.300 \text{ mol } NH_3$ $n_{NH_3} = 0.371\ 29 \text{ mol } H_2 \times \dfrac{2 \text{ mol } NH_3}{3 \text{ mol } H_2} = 0.248 \text{ mol } NH_3$
Compare the amounts of ammonia that are produced by nitrogen and hydrogen to determine the limiting reactant.	The given amount of hydrogen produces less ammonia than the given amount of nitrogen. Therefore, the limiting reactant is hydrogen gas. Notice that there is more hydrogen gas than nitrogen gas, in terms of moles. Hydrogen gas is the limiting reactant, however, because 3 mol of hydrogen gas is needed to react with 1 mol of nitrogen gas.

Check Your Solution

According to the balanced chemical equation, the ratio of nitrogen to hydrogen is 1:3. The ratio of nitrogen to hydrogen, based on the amounts calculated, is 0.15:0.37. Divide this ratio by 0.15 to get 1:2.47. For each mole of nitrogen, there is only 2.47 mol of hydrogen. However, 3 mol is required by stoichiometry. Therefore, hydrogen gas is the limiting reactant.

31. Hydrogen fluoride, HF(g), is a highly toxic gas. It is produced according to the following balanced chemical equation:

$$CaF_2(s) + H_2SO_4(aq) \rightarrow 2HF(g) + CaSO_4(s)$$

Determine the limiting reactant when 1.00 g of calcium fluoride, $CaF_2(s)$, reacts with 15.5 g of sulfuric acid, $H_2SO_4(aq)$.

32. An ester is an organic compound that forms when a carboxylic acid reacts with an alcohol. Esters often are used as essences or scents. One such ester is methyl salicylate, $C_8H_8O_3(aq)$, which is oil of wintergreen. It is formed by the reaction of salicylic acid, $C_7H_6O_3(aq)$, and methanol, $CH_3OH(aq)$, as shown below:

$$C_7H_6O_3(aq) + CH_3OH(aq) \rightarrow C_8H_8O_3(aq) + H_2O(\ell)$$

If 100.11 g of salicylic acid and 90.4 g of methanol are used to produce oil of wintergreen, which is the limiting reactant?

33. Acetylene, $C_2H_2(g)$, is used in welding. It forms when calcium carbide, $CaC_2(s)$, reacts with water, as shown below:

$$CaC_2(s) + 2H_2O(\ell) \rightarrow Ca(OH)_2(aq) + C_2H_2(g)$$

If 5.50 mol of calcium carbide reacts with 3.75 mol of water, which is the limiting reactant?

34. Nickel(II) chloride, $NiCl_2(aq)$, reacts with sodium phosphate, $Na_3PO_4(aq)$, according to the following balanced chemical equation:

$$3NiCl_2(aq) + 2Na_3PO_4(aq) \rightarrow$$
$$Ni_3(PO_4)_2(s) + 6NaCl(aq)$$

If 10.0 g of each reactant is used, which is the limiting reactant?

35. Copper metal reacts with nitric acid, $HNO_3(aq)$, as follows:

$$3Cu(s) + 8HNO_3(aq) \rightarrow$$
$$3Cu(NO_3)_2(aq) + 2NO(g) + 4H_2O(\ell)$$

If 2.5 g of copper reacts with 25.0 g of nitric acid, which reactant is in excess?

36. Lithium reacts with oxygen to form lithium oxide, $Li_2O(s)$.

$$4Li(s) + O_2(g) \rightarrow 2Li_2O(s)$$

When 20.0 g of lithium metal reacts with 30.0 g of oxygen gas, which reactant is limiting and which reactant is in excess?

37. Chlorine gas is used in the textile industry to bleach fabric. Excess chlorine is removed by a reaction with sodium thiosulfate, $Na_2S_2O_3(aq)$, as shown below:

$$Na_2S_2O_3(aq) + 4Cl_2(g) + 5H_2O(\ell) \rightarrow$$
$$2NaHSO_4(aq) + 8HCl(aq)$$

If 42.5 g of sodium thiosulfate and 175 g of chlorine gas react with excess water, which is the limiting reactant?

38. Acrylonitrile, $C_3H_3N(g)$, is prepared by the reaction of propylene, $C_3H_6(g)$, with nitric oxide, $NO(g)$.

$$4C_3H_6(g) + 6NO(g) \rightarrow$$
$$4C_3H_3N(g) + 6H_2O(g) + N_2(g)$$

If 126 g of propylene reacts with 175 g of nitric oxide, which is the limiting reactant?

39. Insoluble silver carbonate, $Ag_2CO_3(s)$, forms in the following balanced chemical reaction:

$$2AgNO_3(aq) + K_2CO_3(aq) \rightarrow Ag_2CO_3(s) + 2KNO_3(aq)$$

What mass of silver nitrate, $AgNO_3(aq)$, reacts with 25.0 g of potassium carbonate, $K_2CO_3(aq)$, if there is at least 5.5 g of silver nitrate in excess?

Learning Check

7. Explain what is meant by the term "stoichiometric amount."

8. Identify the limiting reactant and excess reactant in each situation. Describe any assumptions you made.
   a. A pilot flame flickers in a gas fireplace.
   b. Vinegar is used to remove deposits in a kettle.
   c. A peeled potato turns brown while sitting on a kitchen counter.

9. Four slices of toast, four slices of turkey, two lettuce leaves, and one slice of tomato are available to make turkey sandwiches. Based on **Figure 7.1**, which ingredient is the limiting ingredient?

10. Is the limiting reactant always the compound that is present in the smaller amount? Explain your answer.

11. Why are reactants present in excess amounts not considered when determining the product yield by stoichiometric calculations?

12. When a small quantity of phosphorus, $P_4(s)$, reacts with oxygen gas in open air, which reactant do you think is in excess? Explain your answer.

Identifying the limiting reactant is crucial for predicting the amount of product that is formed in a chemical reaction, as shown in the following Sample Problem.

Sample Problem

Stoichiometry Using a Limiting Reactant

Problem

The thermite reaction, shown on the right, is a reaction of powdered aluminum with iron(III) oxide, $Fe_2O_3(s)$. This reaction produces so much heat that the iron formed is actually molten (liquid). The balanced chemical equation is

$$2Al(s) + Fe_2O_3(s) \rightarrow Al_2O_3(s) + 2Fe(\ell)$$

If 113.00 g of aluminum powder is mixed with 279.50 g of iron(III) oxide, what mass of molten iron forms?

What Is Required?

You need to find the mass of molten iron that forms.

What Is Given?

You know the balanced chemical equation:

$$2Al(s) + Fe_2O_3(s) \rightarrow Al_2O_3(s) + 2Fe(\ell)$$

You know the mass of aluminum powder: 113.00 g

You know the mass of iron(III) oxide: 279.50 g

Thermite Reaction

Plan Your Strategy	Act on Your Strategy
Calculate the molar masses, M, of aluminum, iron(III) oxide, and iron.	$M_{Al} = 26.98$ g/mol $M_{Fe_2O_3} = 2M_{Fe} + 3M_O$ $\quad\quad = 2(55.85 \text{ g/mol}) + 3(16.00 \text{ g/mol})$ $\quad\quad = 159.70$ g/mol $M_{Fe} = 55.85$ g/mol
Convert the masses of aluminum and iron(III) oxide into amounts (in moles) using the equation $n = \dfrac{m}{M}$.	$n_{Al} = \dfrac{m_{Al}}{M_{Al}} = \dfrac{113.00 \text{ g}}{26.98 \text{ g/mol}} = 4.18829$ mol Al $n_{Fe_2O_3} = \dfrac{m_{Fe_2O_3}}{M_{Fe_2O_3}} = \dfrac{279.50 \text{ g}}{159.70 \text{ g/mol}} = 1.75016$ mol Fe_2O_3
Calculate the amount of iron that forms by the given amount of aluminum and the given amount of iron(III) oxide.	$n_{Fe} = 4.18829 \text{ mol Al} \times \dfrac{2 \text{ mol Fe}}{2 \text{ mol Al}} = 4.18829$ mol Fe(s) $n_{Fe} = 1.75016 \text{ mol Fe}_2\text{O}_3 \times \dfrac{2 \text{ mol Fe}}{1 \text{ mol Fe}_2\text{O}_3} = 3.50032$ mol Fe(s)
Compare the amounts of iron that form by aluminum and iron(III) oxide to determine the limiting reactant.	Iron(III) oxide produces less iron than aluminum does. Therefore, the limiting reactant is iron(III) oxide.
Determine the mass of iron that forms using $m = n \times M$ and the amount of iron that the limiting reactant forms.	$m = n \times M$ $\quad = 3.50 \text{ mol} \times 55.85 \text{ g/mol}$ $\quad = 195.49$ g Fe Therefore, 195.49 g of molten iron is formed.

Check Your Solution

The amount in moles of iron(III) oxide is less than half the amount of aluminum. The mole ratio is 2 mol of aluminum to 1 mol of iron(III) oxide. Iron(III) oxide is the limiting reactant.

40. The formation of water is represented by the following equation:
$$2H_2(g) + O_2(g) \rightarrow 2H_2O(g)$$
   a. What is the limiting reactant if 4 mol of oxygen reacts with 16 mol of hydrogen?
   b. What amount (in moles) of water is produced in this reaction?

41. Silver nitrate, $AgNO_3(aq)$, reacts with iron(III) chloride, $FeCl_3(aq)$, to produce silver chloride, $AgCl(s)$, and iron(III) nitrate, $Fe(NO_3)_3(aq)$.
$$3AgNO_3(aq) + FeCl_3(aq) \rightarrow$$
$$3AgCl(s) + Fe(NO_3)_3(aq)$$
   a. If a solution containing 18.00 g of silver nitrate is mixed with a solution containing 32.4 g of iron(III) chloride, which is the limiting reactant?
   b. What amount in moles of iron(III) nitrate is produced in this reaction?

42. Barium sulfate, $BaSO_4(s)$, forms in the following reaction:
$$Ba(NO_3)_2(aq) + Na_2SO_4(aq) \rightarrow$$
$$BaSO_4(s) + 2NaNO_3(aq)$$
   If 75.00 g of barium nitrate, $Ba(NO_3)_2(aq)$, reacts with 100.00 g of sodium sulfate, $Na_2SO_4(aq)$, what mass of barium sulfate is produced?

43. Zinc oxide, $ZnO(s)$, is formed by the reaction of zinc sulfide, $ZnS(s)$, with oxygen.
$$2ZnS(s) + 3O_2(g) \rightarrow 2ZnO(s) + 2SO_2(g)$$
   If 16.7 g of zinc sulfide reacts with 6.70 g of oxygen, what mass of zinc oxide is produced?

44. The following balanced chemical equation represents the reaction of calcium carbonate, $CaCO_3(s)$, with hydrochloric acid:
$$CaCO_3(s) + 2HCl(aq) \rightarrow$$
$$CaCl_2(aq) + CO_2(g) + H_2O(\ell)$$
   If 155 g of calcium carbonate reacts with 245 g of hydrochloric acid, what mass of calcium chloride, $CaCl_2(s)$, is produced?

45. The reaction of aluminum hydroxide, $Al(OH)_3(aq)$, with hydrochloric acid produces water and aluminum chloride, $AlCl_3(s)$.
$$3HCl(aq) + Al(OH)_3(aq) \rightarrow 3H_2O(\ell) + AlCl_3(s)$$
   What mass of aluminum chloride is produced when 8.0 g of hydrochloric acid reacts with an equal mass of aluminum hydroxide?

46. The reaction between solid white phosphorus, $P_4(s)$, and oxygen gas produces solid tetraphosphorus decoxide, $P_4O_{10}(s)$. Determine the mass of tetraphosphorus decoxide that is formed when 25.0 g of solid white phosphorus and 50.0 g of oxygen are combined.

47. A solution containing 14.0 g of silver nitrate, $AgNO_3(aq)$, is added to a solution containing 4.83 g of calcium chloride, $CaCl_2(aq)$. Find the mass of silver chloride, $AgCl(s)$, produced.

48. The reaction between solid sodium and iron(III) oxide, $Fe_2O_3(s)$, is one in a series of reactions that occurs when an automobile air bag inflates.
$$6Na(s) + Fe_2O_3(s) \rightarrow 3Na_2O(s) + 2Fe(s)$$

   If 100.0 g of solid sodium and 100.0 g of iron(III) oxide are used in this reaction, what mass of solid iron will be produced?

49. Manganese(III) fluoride, $MnF_3(s)$, is formed by the reaction of manganese(III) iodide, $MnI_2(s)$, with fluorine gas.
$$2MnI_2(s) + 13F_2(g) \rightarrow 2MnF_3(s) + 4IF_5(\ell)$$
   a. If 1.23 g of manganese(III) iodide reacts with 25.0 g of fluorine, what mass of manganese(III) fluoride is produced?
   b. Which reactant is in excess? How much of this reactant remains at the end of the reaction?

50. Silver nitrate, $AgNO_3(aq)$, reacts with calcium chloride, $CaCl_2(aq)$, in the following reaction:
$$2AgNO_3(aq) + CaCl_2(aq) \rightarrow$$
$$2AgCl(s) + Ca(NO_3)_2(aq)$$
   There are 7.000 mol of each reactant present.
   a. What is the mass of the excess reactant?
   b. What is the mass of the limiting reactant?
   c. What are the masses of each product that forms?

Applications of Stoichiometry and Limiting Reactants

Stoichiometry, chemical reactions, and limiting reactants are not topics that are relevant only in chemistry class or in the laboratory. You benefit from stoichiometry in your daily life, although you might not realize it. The photographs in **Figure 7.8** show some examples of quantitative chemistry in everyday applications. The activity below also highlights the importance of quantities in chemical reactions in our society.

Figure 7.8 Many processes in the home, workplace, and environment involve the use of chemical quantities and calculations. For example, concentrated herbicides **(A)** must be properly diluted before they are applied to invasive species. Hairdressers **(B)** mix chemicals to colour, straighten, and curl hair.

Activity 7.3 Stoichiometric Applications

Processes in the home and in the workplace often involve the use of chemical quantities and calculations. When you mix cleaning solutions in your home, you often mix them in required proportions according to instructions on the product labels. These proportions are determined by chemists and are designed to give the best results. Pharmacists and other medical professionals must mix medication doses, such as chemotherapy, using stoichiometry to obtain a mixture that achieves the desired results without harming patients. Gardeners, farmers, and nursery workers often mix fungicides, pesticides, herbicides, and fertilizers with water or other ingredients to achieve the correct proportions. If these products are not mixed correctly, the plants might die and damage to the surrounding environment can occur.

Procedure

1. Choose one of the following products that involve the use of stoichiometry and limiting and excess reactants.
   - a specific type of pesticide, herbicide, or fungicide (organic or synthetic) that is used in home gardens or in commercial operations
   - a pharmaceutical product (either an over-the-counter medication or a prescribed medication)
   - a household cleaning solution (organic or synthetic)
   - a consumer product, such as hair colouring, hair relaxer, or permanent wave mixture
   - a product of your choice (approved by your teacher)
2. Use the Internet, product instruction sheets, pharmaceutical inserts, or other reliable resources to research information about your topic.

Questions

1. Answer the following questions as you complete your research:
   a. What is the purpose of your product?
   b. What chemical quantities or calculations are required to ensure safe use of your product?
   c. What are the possible consequences if your chosen product is not mixed correctly?
   d. When your product is being applied or used, what are the limiting and excess reactants?
2. Prepare a short presentation of your findings to share with the class. Consider questions that your classmates might ask about the topic as you prepare your presentation.

Limiting Reactants and Product Formed

In this section, you were introduced to the concept of the limiting reactant in a chemical reaction. You learned that once you have identified the limiting reactant, you can use stoichiometric calculations to predict how much product will be formed in a chemical reaction. However, sometimes the amount of product that is actually formed is quite different from the amount you predicted. In the next section, you will learn why.

Section Summary

- A limiting reactant is a reactant that is completely consumed during a chemical reaction, and therefore limits or stops the reaction. Reactants that remain after the reaction stops are called excess reactants.

- To identify the limiting reactant, the amount of product that is produced from each reactant is determined. Then the amounts from both reactants are compared to determine which reactant produces the smaller amount of product.

- Determining the limiting reactant is necessary for all stoichiometric calculations that are used to determine the amount of product that forms.

- In chemical processes, limiting and excess reactants must be managed to ensure that the reactants produce the products safely and efficiently.

Review Questions

1. **K/U** Calcium reacts with water. If a small piece is dropped into a beaker of water, which reactant is in excess during this reaction?

2. **T/I** If 47.2 mol of lead(II) oxide, $PbO(s)$, reacts with 6 mol of oxygen according to the equation below, which is the limiting reactant?
$$6PbO(s) + O_2(g) \rightarrow 2Pb_3O_4(g)$$

3. **T/I** If 72.15 g of pentane, $C_5H_{12}(g)$, reacts with 6.9 mol of oxygen, according to the equation below, which is the limiting reactant?
$$C_5H_{12}(g) + 8O_2(g) \rightarrow 5CO_2(g) + 6H_2O(g)$$

4. **T/I** In an experiment, 57.4 g of iron(III) chloride in solution reacts with 45.3 g of sodium hydroxide in solution, as shown below.
$$FeCl_3(aq) + 3NaOH(aq) \rightarrow Fe(OH)_3(s) + 3NaCl(aq)$$
   a. Which reactant is the limiting reactant?
   b. How much of the excess reactant remains after the reaction?
   c. How much of each product forms?

5. **T/I** Solid sodium metal reacts with chlorine gas to form table salt.
$$2Na(s) + Cl_2(g) \rightarrow 2NaCl(s)$$
What is the minimum mass of chlorine gas that is required to consume 2.25 g of solid sodium?

6. **K/U** Define the term "limiting reactant," and give a specific example from your own experience.

7. **C** Using a numbered list of steps, explain the process of identifying the limiting reactant in a chemical reaction.

8. **C** A candle is burning at the bottom of a container. The container is then covered until the candle is almost extinguished. When the cover is removed, the candle flame recovers and burns normally again. Use stoichiometry and the concept of the limiting reactant to explain your observations.

9. **T/I** Copper reacts with nitric acid, $HNO_3(aq)$, as follows:
$$3Cu(s) + 8HNO_3(aq) \rightarrow$$
$$3Cu(NO_3)_2(aq) + 2NO(g) + 4H_2O(\ell)$$
What mass of nitrogen monoxide, $NO(g)$, is produced when 50.0 g of copper reacts with 150.0 g of nitric acid?

10. **T/I** The following chemical equation represents the reaction of silver nitrate, $AgNO_3(aq)$, with sodium chloride:
$$AgNO_3(aq) + NaCl(aq) \rightarrow AgCl(s) + NaNO_3(aq)$$
Different amounts of silver nitrate are added to a fixed amount of sodium chloride. For each mass of silver nitrate, the mass of sliver chloride, $AgCl(s)$, precipitate is determined and plotted on the graph below.

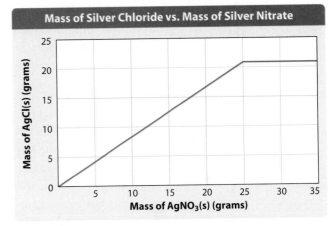

Mass of Silver Chloride vs. Mass of Silver Nitrate

   a. Why does the graph level off after 25 g of silver nitrate is added to the sodium chloride?
   b. What amount in moles of sodium chloride is the fixed amount?

11. **A** You pour household vinegar on mineral deposits on a kitchen faucet. Some of the deposits wash away. You pour more vinegar on the deposits and they all wash way. Explain what occurred using the terms "limiting reactant" and "excess reactant."

Reaction Yields

Key Terms

theoretical yield
actual yield
competing reaction
percentage yield

The highest grade that a student can earn on a test is 100 percent. However, most students do not receive a grade of 100 percent. The actual grade earned is usually less than the theoretical highest grade of 100 percent. The actual percentage grade on a test is calculated using the following equation:

$$\text{percentage grade} = \frac{\text{points earned}}{\text{maximum possible points}} \times 100\%$$

Calculating your test grade is similar to calculating chemical reaction yields.

Theoretical Yield and Actual Yield

theoretical yield the amount of product that is predicted by stoichiometric calculations

actual yield the actual amount of product that is recovered after a reaction is complete

competing reaction a reaction that occurs along with the principal reaction and that involves the reactants and/or products of the principal reaction

The stoichiometric calculations you have learned so far in this chapter allow you to calculate the amount of product that forms, in theory, in a chemical reaction. This is called the **theoretical yield** of the reaction. However, the theoretical yield is not always the same as the amount of product that actually forms in a chemical reaction. Just as most students are unlikely to answer 100 out of 100 questions on a test correctly, most reactions do not produce the theoretical amount of product. The actual amount of product that forms in a chemical reaction conducted in a laboratory or in industry is called the **actual yield**.

Competing Reactions (Reactants) That Affect Yield

Reactions do not yield as much product as expected for a variety of reasons. For example, the actual yield might be affected by a competing reaction. A **competing reaction** is a reaction that occurs along with the principal reaction and that involves the reactants and/or products of the principal reaction. An example of a competing reaction occurs when hydrocarbons are burned as fuel. Depending on the availability of oxygen and other circumstances, a hydrocarbon gas, such as propane, may burn in both a complete combustion reaction (the principal reaction) and an incomplete combustion reaction (the competing reaction) according to the following chemical equations:

$$C_3H_8(g) + 5O_2(g) \rightarrow 3CO_2(g) + 4H_2O(g) \text{ (complete combustion)}$$

$$2C_3H_8(g) + 7O_2(g) \rightarrow 6CO(g) + 8H_2O(g) \text{ (incomplete combustion)}$$

Because the competing reaction uses the same reactants as the principal reaction, the principal reaction has a lower yield.

One product of the incomplete combustion reaction above is carbon monoxide, $CO(g)$, an odourless toxic gas. A carbon monoxide detector, like the one shown in **Figure 7.9**, alerts people if the gas is present in their home.

Figure 7.9 Propane can produce carbon monoxide during incomplete combustion. Because carbon monoxide is poisonous, detectors are used in homes to warn people if the gas is present.

Competing Reactions (Products) That Affect Yield

A competing reaction does not always involve the reactants in the principal reaction. It can also involve the products. For example, phosphorus reacts with chlorine and forms phosphorus trichloride, as shown in **Figure 7.10**. Some of the phosphorus trichloride then reacts with chlorine and forms phosphorus pentachloride, resulting in less actual yield. The chemical equations for these reactions are

$$2P(s) + 3Cl_2(g) \rightarrow 2PCl_3(\ell)$$

$$PCl_3(\ell) + Cl_2(g) \rightarrow PCl_5(s)$$

In the first reaction, phosphorus and chlorine gas produce phosphorus trichloride. In the second reaction, some of the phosphorus trichloride is converted into phosphorus pentachloride. The second reaction is the competing reaction. It causes the actual yield of phosphorus trichloride to be less than the theoretical yield.

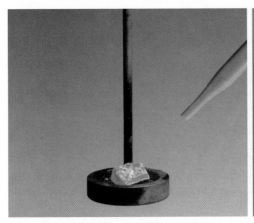 

Figure 7.10 When phosphorus reacts with chlorine gas, a competing reaction lowers the actual yield of phosphorus trichloride.

Other Factors That Affect Yield

Competing reactions are not the only things that can affect the actual product yield. Reaction rates can also affect product yield. For example, if a reaction is extremely slow, the reaction might not have gone to completion at the time the product yield is measured. Because the reaction is incomplete, the product yield will be lower than predicted.

Reaction rates can be affected by factors such as surface area, temperature, pressure, and reaction vessel conditions. Typically, larger surface areas of the reactants and higher reaction temperatures and pressures result in more frequent collisions by the particles that make up the reactants. Increasing the collision rate of the particles of the reactants increases the rate at which product forms. A reaction vessel can hamper the collisions of particles of the reactants because of factors such as a rough, pitted, or dirty surface, or an odd shape. As a result, fewer products will be formed. Another factor that reduces actual product yield is the purity of the reactants. If theoretical yield calculations are made on the assumption that the reactants are 100 percent pure when, in fact, they are not, the actual yield will be lower than expected. For example, if you use household vinegar instead of pure acetic acid in a reaction, your actual product yield will be much lower than expected because household vinegar is only about 5 percent acetic acid. The remaining 95 percent is water.

The laboratory techniques used to collect the final product also affect product yield. For example, if a product is slightly soluble in a filtrate, some of the product will remain dissolved in the filtrate. Furthermore, if the product is soluble in water, some of the product might dissolve in the water used to clean the filter paper. Other laboratory techniques can also cause product loss, such as the product clinging to the glassware, stirring rods, and other equipment used in the investigation.

Summary of Factors That Affect Product Yield Summary

There are many factors that can affect the actual product yield. **Table 7.1** summarizes several of those factors. When you perform the Investigations 7-B and 7-C, you will observe some of these factors.

Table 7.1 Factors That Affect the Actual Yield of a Reaction

Factor	Description
Competing reaction	• Competing reactions involving the principal reactants and/or products produce multiple products and reduce the yield of the desired product.
Reaction rate	• A slow reaction does not go to completion and reaction products are collected too soon. • The surface area of the reactants is small, which reduces the probability of collisions of the particles of the reactants. • Environmental conditions of the reaction, such as temperature and pressure, slow particle movement and reduce the probability of reactant particle collisions. • Reaction vessel conditions, such as rough, pitted, or dirty surfaces or an odd-shaped vessel, impede reactant particle collisions.
Purity of the reactant	• Impure reactants contain contaminants. • Incorrect theoretical yield calculations are based on impure reactants and the mass of the impurity is not considered.
Laboratory techniques	• Improper lab techniques reduce actual product yield. • A slightly soluble product results in some of the product staying in the filtrate rather than being collected on the filter paper. • A slightly soluble product dissolves in the water used to rinse the product on the filter paper; some of the product is washed away during the rinsing process. • Some of the product clings to the reaction vessel, filter paper, stirring rods, spatulas, and other equipment used in the investigation and this mass is not recorded.

Learning Check

13. Explain the difference between actual yield and theoretical yield.

14. Which is usually higher, the actual yield or the theoretical yield? Explain your answer.

15. Explain how laboratory techniques may reduce the actual yield.

16. Draw and label a diagram showing how a reaction vessel can influence the actual yield.

17. Explain how impure reactants can affect the actual yield.

18. Use an analogy to explain how a competing reaction can reduce the actual yield.

Calculating Percentage Yield

percentage yield the actual yield of a reaction, expressed as a percentage of the theoretical yield

Chemists often need to know the efficiency of a chemical reaction, such as when they are doing a cost analysis for a chemical process or when they are ordering materials for a chemical process. The **percentage yield** of a chemical reaction is the actual yield expressed as a percentage of the theoretical yield:

$$\text{percentage yield} = \frac{\text{actual yield}}{\text{theoretical yield}} \times 100\%$$

In this equation, the theoretical yield is the amount of product that is determined by theoretical stoichiometric calculations, and the actual yield is the amount of product that is actually recovered in the reaction. The percentage yield can be calculated using either the amount in moles or mass. The following Sample Problems demonstrate how to use this equation.

Calculating Percentage Yield

Problem

The following chemical equation represents the production of ethanol, $C_2H_5OH(aq)$, by the fermentation of glucose, $C_6H_{12}O_6(aq)$:

$$C_6H_{12}O_6(aq) \rightarrow 2C_2H_5OH(aq) + 2CO_2(g)$$

If 20.0 g of glucose reacts but only 1.40 g of ethanol is produced, what is the percentage yield of the reaction?

glucose

What Is Required?

You need to find the percentage yield of the reaction.

What Is Given?

You know the balanced chemical equation:

$$C_6H_{12}O_6(aq) \rightarrow 2C_2H_5OH(aq) + 2CO_2(g)$$

You know the actual yield of ethanol: 1.40 g

You know the mass of glucose: 20.0 g

Plan Your Strategy	Act on Your Strategy
Find the molar masses of glucose and ethanol.	$M_{C_6H_{12}O_6} = 6M_C + 12M_H + 6M_O$ $\quad = 6(12.01 \text{ g/mol}) + 12(1.01 \text{ g/mol}) + 6(16.00 \text{ g/mol}) = 180.18 \text{ g/mol}$ $M_{C_2H_5OH} = 2M_C + 6M_H + 1M_O$ $\quad = 2(12.01 \text{ g/mol}) + 6(1.01 \text{ g/mol}) + 1(16.00 \text{ g/mol}) = 46.08 \text{ g/mol}$
Determine the actual yield amount in moles of ethanol.	$n = \dfrac{m}{M} = \dfrac{1.40 \text{ g}}{46.08 \text{ g/mol}} = 0.03038 \text{ mol } C_2H_5OH$
To determine the theoretical yield, convert the mass of glucose in grams to amount in moles. Then, use the mole ratio to find the yield of ethanol in moles.	$n_{C_6H_{12}O_6} = 20.0 \text{ g} \left(\dfrac{1 \text{ mol } C_6H_{12}O_6}{180.18 \text{ g } C_6H_{12}O_6} \right) = 0.111 \text{ mol } C_6H_{12}O_6$ $n_{C_2H_5OH} = 0.111 \text{ mol } C_6H_{12}O_6 \left(\dfrac{2 \text{ mol } C_2H_5OH}{1 \text{ mol } C_6H_{12}O_6} \right) = 0.222 \text{ mol } C_2H_5OH$
Calculate the percentage yield of ethanol.	$\text{percentage yield} = \dfrac{\text{actual yield}}{\text{theoretical yield}} \times 100\% = \dfrac{0.03038 \text{ mol}}{0.222 \text{ mol}} \times 100\%$ $\quad = 13.7\%$ Therefore, the percentage yield is 13.7%.

Check Your Solution

To check your answer, calculate the percentage yield using mass and the factor-label method.

$$\text{theoretical yield}_{C_2H_5OH} = (20.0 \text{ g } C_6H_{12}O_6) \times \left(\frac{1 \text{ mol } C_6H_{12}O_6}{180.18 \text{ g } C_6H_{12}O_6} \right) \times \left(\frac{2 \text{ mol } C_2H_5OH}{1 \text{ mol } C_6H_{12}O_6} \right) \times \left(\frac{46.08 \text{ g } C_2H_5OH}{1 \text{ mol } C_2H_5OH} \right)$$

$$= 10.2 \text{ g } C_2H_5OH$$

$$\text{actual percentage yield}_{C_2H_5OH} = \frac{1.40 \text{ g}}{10.2 \text{ g}} \times 100\% = 13.7\%$$

The two methods produce the same answer.

Predicting the Actual Yield Using Percentage Yield

Problem

Ammonium nitrate, $NH_4NO_3(s)$, is a compound that is used to make fertilizer, like the fertilizer shown on the right. It is produced in a chemical reaction that is represented by the following balanced equation:

$$NH_3(g) + HNO_3(aq) \rightarrow NH_4NO_3(s)$$

Suppose that 4.950 kg of ammonia, $NH_3(g)$, is available. What mass in grams of ammonium nitrate is produced, if the reaction is only 89.5% efficient?

What Is Required?

You need to calculate the mass in grams of ammonium nitrate, that forms in the reaction.

What Is Given?

You know the percentage yield of ammonium nitrate: 89.5%

You know the mass of ammonia: 4.950 kg

You know the balanced chemical equation:

$NH_3(g) + HNO_3(aq) \rightarrow NH_4NO_3(s)$

Plan Your Strategy	Act on Your Strategy
Convert the mass of ammonia from kilograms to grams.	$(4.950 \text{ kg NH}_3) \times \left(\dfrac{1000 \text{ g}}{1 \text{ kg}} \right) = 4.950 \times 10^3 \text{ g NH}_3$
Find the molar masses of ammonia and ammonium nitrate.	$M_{NH_3} = 1M_N + 3M_H = 1(14.01 \text{ g/mol}) + 3(1.01 \text{ g/mol}) = 17.04 \text{ g/mol}$ $M_{NH_4NO_3} = 2M_N + 4M_H + 3M_O$ $= 2(14.01 \text{ g/mol}) + 4(1.01 \text{ g/mol}) + 3(16.00 \text{ g/mol}) = 80.06 \text{ g/mol}$
Find the amount of ammonia in moles.	$n = \dfrac{m}{M} = \dfrac{4.950 \times 10^3 \text{ g}}{17.04 \text{ g/mol}}$ $= 290.493 \text{ mol NH}_3$
Calculate the theoretical yield of ammonium nitrate in moles.	$n_{NH_4NO_3} = 290.493 \text{ mol NH}_3 \times \left(\dfrac{1 \text{ mol NH}_4NO_3}{1 \text{ mol NH}_3} \right)$ $= 290.493 \text{ mol NH}_4NO_3$
Calculate the theoretical yield of ammonium nitrate in grams.	$m = n \times M = 290.493 \text{ mol} \times 80.06 \text{ g/mol}$ $= 2.3257 \times 10^4 \text{ g NH}_4NO_3$
Use the percentage yield equation to solve for the unknown.	$\text{percentage yield} = \dfrac{\text{actual yield}}{\text{theoretical yield}} \times 100\%$ $\text{actual yield} = \dfrac{\text{percentage yield} \times \text{theoretical yield}}{100\%}$ $= \dfrac{(89.5\%)(2.3257 \times 10^4 \text{ g NH}_4NO_3)}{100\%}$ $= 2.08 \times 10^4 \text{ g NH}_4NO_3$ Therefore, the actual yield of ammonium nitrate is $2.08 \times 10^4 \text{ g NH}_4NO_3$.

Check Your Solution

The actual yield should be about 90% of the theoretical yield, so the answer is reasonable.

51. During an investigation, calcium carbide, $CaC_2(s)$, reacted with excess water to make calcium hydroxide, $Ca(OH)_2(aq)$, and acetylene, $C_2H_2(g)$.

$$CaC_2(s) + 2H_2O(\ell) \rightarrow Ca(OH)_2(aq) + C_2H_2(g)$$

The data table for this investigation is given below.

Data Table

Mass of Calcium Carbide That Reacted	2.38 g
Mass of Acetylene That Was Produced	0.77 g

What was the theoretical yield and percentage yield of acetylene?

52. Suppose that 0.250 mol of potassium carbonate, $K_2CO_3(s)$, reacts with excess hydrochloric acid as follows:

$$K_2CO_3(s) + 2HCl(aq) \rightarrow$$
$$H_2O(\ell) + CO_2(g) + 2KCl(aq)$$

a. Calculate the theoretical yield of potassium chloride.

b. Calculate the percentage yield of water if 0.189 mol of water is produced.

53. Phosphoric acid, $H_3PO_4(aq)$, is neutralized by potassium hydroxide, $KOH(aq)$, according to the following reaction:

$$H_3PO_4(aq) + 3KOH(aq) \rightarrow K_3PO_4(aq) + 3H_2O(\ell)$$

If 49.0 g of potassium phosphate, $K_3PO_4(aq)$, is recovered after 49.0 g of phosphoric acid reacts with 49.0 g of potassium hydroxide, what is the percentage yield of the reaction?

54. The reaction of glucose, $C_6H_{12}O_6(s)$, with sulfuric acid, $H_2SO_4(\ell)$, produces carbon as follows:

$$C_6H_{12}O_6(s) + 2H_2SO_4(\ell) \rightarrow$$
$$6C(s) + 6H_2O(\ell) + 2H_2SO_4(aq)$$

a. If 20.8 g of glucose reacts with excess sulfuric acid, what is the theoretical yield, in grams, of carbon?

b. If the percentage yield is 72.0%, what mass of carbon is produced?

55. Calcium chloride, $CaCl_2(aq)$, is mixed with silver nitrate, $AgNO_3(aq)$, to form calcium nitrate, $Ca(NO_3)_2(aq)$, and silver chloride, $AgCl(s)$.

$$CaCl_2(aq) + 2AgNO_3(aq) \rightarrow$$
$$Ca(NO_3)_2(aq) + 2AgCl(s)$$

If this reaction has an 81.5% yield, what mass of silver chloride is produced when 21.2 g of calcium chloride is added to excess silver nitrate?

56. The following reaction has a 68% yield.

$$AlCl_3(aq) + 4NaOH(aq) \rightarrow$$
$$NaAlO_2(aq) + 3NaCl(aq) + 2H_2O(\ell)$$

Calculate the actual mass of sodium chloride that is recovered if 18.2 g of aluminum chloride, $AlCl_3(aq)$, reacts with 16.00 g of sodium hydroxide.

57. Ethyl butanoate, $C_6H_{13}O_2(\ell)$, is an organic ester that has the flavour and scent of pineapple. It is prepared as follows:

$$C_4H_9O_2(\ell) + C_2H_6O(\ell) \rightarrow C_6H_{13}O_2(\ell) + H_2O(\ell)$$

During an investigation, 0.573 mol of butanoic acid, $C_4H_9O_2(\ell)$, reacts with excess ethanol, $C_2H_6O(\ell)$. What mass of ethyl butanoate is produced if this reaction has a 92.0% yield?

58. An impure sample of barium hydroxide, $Ba(OH)_2(aq)$, has a mass 0.540 g. It is dissolved in water and then treated with excess sulfuric acid, $H_2SO_4(aq)$. This results in the formation of a precipitate of barium sulfate, $BaSO_4(s)$.

$$H_2SO_4(aq) + Ba(OH)_2(aq) \rightarrow BaSO_4(s) + 2H_2O(\ell)$$

The barium sulfate is filtered, and any remaining sulfuric acid is washed away. Then the barium sulfate is dried and its mass is measured to be 0.62 g. What mass of barium hydroxide was in the original (impure) sample?

59. Iron pyrite, $FeS_2(s)$, reacts with oxygen as shown in the reaction below:

$$4FeS_2(s) + 11O_2(g) \rightarrow 2Fe_2O_3(s) + 8SO_2(g)$$

a. In a laboratory, 5.000 kg of an impure mineral, which contains 45.3% iron pyrite, reacts with oxygen. Calculate the mass of iron(III) oxide, $Fe_2O_3(s)$, that forms. Assume that all the pyrite reacts.

b. Suppose that the reaction has a 78.0% yield, due to an incomplete reaction. How many grams of iron(III) oxide is produced?

60. Sodium oxide, $Na_2O(s)$, reacts with water to form the base sodium hydroxide.

$$Na_2O(s) + H_2O(\ell) \rightarrow 2NaOH(aq)$$

If this reaction has a 91% yield, what mass of sodium hydroxide is obtained when 0.483 mol of sodium oxide reacts with excess water?

The cost of manufactured items is primarily based on the cost of producing the items. Many manufacturers aim to produce items as inexpensively as possible, while maintaining the quality of the items. They try to ensure that they use enough reactants to produce a high yield. At the same time, they must try to ensure that they do not waste reactants or introduce chemical waste into the environment. Wasting reactants unnecessarily increases the cost of producing the items. Introducing chemical waste into the environment can create environmental damage that is expensive to clean up and can sometimes not be undone.

Suppose that you work for a company that wants to test iron(III) oxide, $Fe_2O_3(s)$, as a pigment for cosmetics. You want to produce iron(III) oxide from steel wool.

Procedure

Note any visible differences you can see in the two different grades of steel wool in the photographs below, and then answer the questions.

Questions

1. Steel wool contains iron and small amounts of carbon. If the steel wool is held over a Bunsen burner, the iron in the steel wool combines with oxygen in the air and forms iron(III) oxide. Write a balanced chemical equation for this reaction.

2. How would you determine the actual and theoretical percentage yields of iron(III) oxide for each grade of steel wool?

3. While conducting the reaction described in question 1, you discover that oxygen is a limiting factor. What could you do to overcome this limiting factor and increase your percentage yield?

4. Suppose that you can buy steel wool from two different companies. The steel wool from one company contains less iron and more impurities. The steel wool from the other company is almost pure iron but is more expensive. Describe at least two factors that you must consider before choosing which steel wool you will use to produce your iron(III) oxide. Explain your reasoning.

The Importance of Actual Percentage Yield

Suggested Investigation

Inquiry Investigation 7-C, Finding the Percentage Yield of a Double Displacement Reaction

You have learned how to predict the amount of product that forms from a certain amount of reactant. It is important to maximize the percentage yield of any chemical process, but there are several reasons why the maximum yield may not be produced. The amount that is predicted by stoichiometric calculations may be reduced because of impure reactants or because of how the product is collected.

It is often useful to know the percentage yield of a chemical reaction. If you know that the percentage yield of a reaction is about 14 percent, you can calculate the amounts of reactants that are needed to produce a desired mass of product. It is important to know the percentage yield of a large chemical process because of the cost of production. A small difference in the percentage yield of a product could mean a loss of thousands of dollars for a chemical company.

Section Summary

- Stoichiometric calculations are used to determine the theoretical yield, which is the maximum theoretical amount of product that can be produced from a given amount of reactants in a chemical reaction.

- The actual yield is determined through experimentation and is the actual amount of product that is produced in a chemical reaction.

- The percentage yield is the ratio of the actual yield to the theoretical yield, expressed as a percent.

Review Questions

1. **K/U** Does it matter which units you use, grams or moles, when calculating the percentage yield of a reaction? Explain.

2. **K/U** Briefly compare and contrast the terms "theoretical yield" and "actual yield" using an analogy of your own choice.

3. **K/U** Explain why the reaction rate can reduce the actual yield of product.

4. **T/I** Silver nitrate, $AgNO_3(aq)$, reacts with potassium bromide, $KBr(aq)$, to produce silver bromide, $AgBr(s)$.
   $$AgNO_3(aq) + KBr(aq) \rightarrow AgBr(s) + KNO_3(aq)$$
   If 14.64 g of silver bromide is obtained when 14.00 g of silver nitrate reacts with excess potassium bromide, what is the percentage yield?

5. **T/I** Sodium phosphate, $Na_3PO_4(aq)$, reacts with barium nitrate, $Ba(NO_3)_2(aq)$, as represented by:
   $$2Na_3PO_4(aq) + 3Ba(NO_3)_2(aq) \rightarrow$$
   $$Ba_3(PO_4)_2(s) + 6NaNO_3(aq)$$
   If 5.00 g of sodium phosphate reacts with 10.90 g of barium nitrate and 7.69 g of solid precipitate is recovered, what is the percentage yield?

6. **K/U** Zinc reacts with sulfuric acid, $H_2SO_4(aq)$, to yield zinc sulfate, $ZnSO_4(aq)$, and hydrogen gas. What assumption is usually made about the purity of the reactants, such as the zinc?

7. **C** Use a flowchart to describe the steps you would take to find the percentage yield of a reaction, given the mass of the two reactants and the mass of the product.

8. **C** You find a website that describes an investigation to produce carbon dioxide gas using the following equation:
   $$CH_3COOH(aq) + NaHCO_3(s) \rightarrow$$
   $$NaCH_3COO(aq) + CO_2(g) + H_2O(\ell)$$
   The procedure states that household white vinegar, which is dilute acetic acid, $CH_3COOH(aq)$, could be used instead of pure acetic acid. Write a posting for the site explaining why the product yield will be low when vinegar is used instead of pure acetic acid.

9. **A** Explain why percentage yield is important to companies that produce chemical compounds for profit.

10. **A** Explain how each laboratory technique given below could be changed to increase the yield.
    a. stirring the reaction material with a wooden splint
    b. transferring a reactant from one beaker to another
    c. drying a mixture in an evaporating dish over a laboratory gas burner
    d. adding aqueous reactants from a graduated cylinder to the reaction beaker

11. **T/I** Sodium iodide, $NaI(s)$, and chlorine gas undergo a single displacement reaction, as represented by the following equation:
    $$2NaI(s) + Cl_2(g) \rightarrow I_2(s) + 2NaCl(s)$$
    a. If 4.0 g of sodium iodide and 4.0 g of chlorine gas react, what are the theoretical yields of both products?
    b. If the percentage yield of sodium chloride is 67%, what is the actual yield of sodium chloride?
    c. Would it be appropriate to assume that the yield of the other product, iodine, $I_2(s)$, is also 67%? Explain.

12. **T/I** To dehydrate bluestone, which is a form of copper(II) sulfate pentahydrate, $CuSO_4 \cdot 5H_2O(s)$, it was heated over a Bunsen burner. The following equation represents the dehydration reaction:
    $$CuSO_4 \cdot 5H_2O(s) \rightarrow CuSO_4(s) + 5H_2O(\ell)$$
    Three masses were measured, before and after the reaction, as shown in the data table below. Use these masses to calculate the theoretical yield, actual yield, and percentage yield of both copper sulfate, $CuSO_4(s)$, and water.

Data Table

Mass of Empty Crucible and Lid	15.146 g
Mass of Crucible and Copper(II) Sulfate Pentahydrate	19.273 g
Mass of Crucible and Copper Sulfate	18.059 g

Safety Precautions

- Wear safety eyewear throughout this investigation.

- Wear a lab coat or apron throughout this investigation.

- Wear protective gloves when performing this investigation.

- Tie back loose hair and clothing.

- The reaction mixture may get hot. Do not hold the beaker as the reaction proceeds.

- Copper(II) chloride is toxic. Do not ingest. Do not inhale the vapours.

- If you get copper(II) chloride on your skin, flush with plenty of water.

Materials

- 2.0 g of copper(II) chloride, $CuCl_2(s)$

- 0.50 g of aluminum foil, $Al(s)$

- distilled water

- graduated cylinder

- spatula

- electronic balance

- 100 mL beaker

- stirring rod

Limiting and Excess Reactants

In this investigation, you will observe a limiting reactant in a single displacement reaction. In the reaction, 0.50 g of solid aluminum, $Al(s)$, will react with 2.0 g of copper (II) chloride, $CuCl_2(s)$, dissolved in water. After the reaction, you will determine which compound was the limiting reactant and which compound was the excess reactant.

Pre-Lab Questions

1. What is the balanced chemical equation for the single displacement reaction described in the introduction of this investigation?

2. When a chemical reaction is in progress, what observations indicate that a chemical change is occurring?

3. What observations might indicate that a chemical reaction is complete.

Question

How can you use observations and calculations to determine the limiting reactant in the chemical reaction of aluminum foil and aqueous copper(II) chloride?

When aluminum and copper(II) chloride react, a single displacement reaction occurs.

Procedure

1. Prepare a data table to record your observations and measurements. Give your table an appropriate title.

2. Observe the physical properties of the copper(II) chloride and the aluminum foil. Write your observations in your data table.

3. Measure and record the mass of both reactants. Place the copper(II) chloride in a beaker. Add 50 mL of distilled water to the beaker. Stir the contents of the beaker with a stirring rod, and record your observations.

4. Tear the aluminum foil into small pieces. Then add it to the copper(II) chloride solution. Record the initial colour of the solution, and describe the appearance of the metal at the beginning of the reaction.

5. Record your observations as the reaction proceeds. Pay particular attention to any colour changes and any changes in the appearance of the metal. Stir occasionally with the stirring rod.

6. When the reaction is complete, ask your teacher for proper disposal instructions. Do not pour anything down the drain.

Analyze and Interpret

1. **a.** What colour was the copper(II) chloride solution before the aluminum foil was added?

   b. What colour was the solution after the reaction was complete? Did all the copper(II) chloride react? Explain.

   c. Describe the appearance of the aluminum foil. Did all the aluminum foil react? How do you know?

   d. Based on the colour of the metal and the balanced chemical equation, what is the solid that formed during the chemical reaction?

2. **a.** Use the masses in your data table to determine the limiting reactant in this chemical reaction. Which reactant was in excess?

   b. Compare your observations and your calculations. Do your calculations support your observations? Explain.

Conclude and Communicate

3. Summarize your findings. Are stoichiometric calculations an effective way to predict limiting and excess reactants? Explain your answer.

4. Write a conclusion to explain how your experimental observations did or did not support your theoretical calculations.

Extend Further

5. **INQUIRY** Magnesium and hydrochloric acid react as follows:

$$Mg(s) + 2HCl(aq) \rightarrow MgCl_2(aq) + H_2(g)$$

   a. Examine the equation. What would you expect to observe if magnesium was the limiting reactant? What would you expect to observe if the hydrochloric acid was the limiting reactant?

   b. Suppose that you have a piece of magnesium and a beaker that contains hydrochloric acid of unknown concentration. Design a procedure you could use to determine which reactant is the limiting reactant. State the safety precautions that you should take. If your teacher approves, carry out your procedure.

6. **RESEARCH** Use text or Internet sources to obtain the material data safety sheets (MSDSs) for copper(II) chloride and aluminum chloride. Examine the information about first aid measures, accidental release measures, handling and storage, exposure controls, and personal protection. Write a brief summary of your findings.

Safety Precautions

- Wear safety eyewear throughout this investigation.

- Wear a lab coat or apron throughout this investigation.

- Tie back loose hair and clothing

- Copper(II) chloride dihydrate is toxic. Do not ingest. Do not inhale the vapours.

- Do not inhale hydrochloric acid vapours.

- If you get copper(II) chloride dihydrate or hydrochloric acid solution on your skin, flush with plenty of water.

Materials

- 5.00 g of copper(II) chloride dihydrate, $CuCl_2 \cdot 2H_2O(s)$

- distilled water

- 1.00 g piece of steel wool, $Fe(s)$

- 10 mL of 1 mol/L hydrochloric acid, $HCl(aq)$

- two 250 mL beakers

- waterproof grease marker

- electronic balance

- stirring rod

- wash bottle with distilled water

- drying oven or heat lamp

Finding the Percentage Yield of a Single Displacement Reaction

The percentage yield of any reaction can be affected by several different factors. One of these factors is the impurity of the reactants. In this investigation, you will conduct a single displacement reaction in which the iron, $Fe(s)$, in steel wool reacts with aqueous copper(II) chloride, $CuCl_2(aq)$, to produce aqueous iron(II) chloride, $FeCl_2(aq)$, and a precipitate. You will then determine the percentage yield of the precipitate in the reaction and determine the effect of an impure reactant on this yield.

Pre-Lab Questions

1. What is the balanced chemical equation for the reaction in this investigation?

2. Which reactant is considered impure in this investigation?

3. How might an impure reactant affect the percentage yield of a reaction? Explain.

4. Why is it important to flush your skin with plenty of water if it comes in contact with copper(II) chloride dihydrate or hydrochloric acid?

Question

What is the percentage yield of the precipitate in the reaction of iron and copper(II) chloride when steel wool and aqueous copper(II) chloride are used as reactants?

Procedure

1. Obtain a clean, dry 250 mL beaker. Mark the beaker with a grease marker so that you can identify it later.

2. Copy the following data table into your notebook.

Data Table

Material	Mass (grams)
Empty labelled beaker	
Copper(II) chloride dihydrate	
Iron (steel wool)	
Beaker containing clean, dry copper	

3. Measure the mass of the beaker, and record the mass in your data table.

4. Measure 5.00 g of copper(II) chloride dihydrate, and place it in the beaker. Add 50 mL of distilled water. Stir to dissolve the solid blue material. (Note: When a copper(II) chloride dihydrate is dissolved in water, it is aqueous copper(II) chloride.)

5. Measure 1.00 g of iron (steel wool) and record the mass in your data table.

6. Unravel the iron (steel wool) over a piece of paper to catch all the pieces. Then add the steel wool to the solution in the beaker. Allow the reaction mixture to sit until all the iron has reacted. It should take 2 to 3 min for the reaction to go to completion.

7. When the reaction is complete and no iron is visible, carefully pour the solution into a 250 mL beaker by decanting the liquid. To decant the liquid, slowly pour the liquid down the length of a stirring rod, as shown in the diagram below. Decanting the liquid in this way prevents the liquid from flowing down the side of the reaction beaker instead of flowing into the second beaker. Decant the liquid carefully to prevent any precipitate from escaping from the reaction beaker.

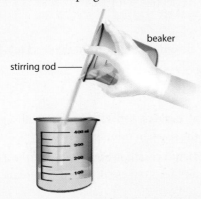

stirring rod ——

beaker

8. Rinse the precipitate several times with distilled water, using a wash bottle. Decant the water each time.

9. Clean the precipitate with hydrochloric acid by adding approximately 10 mL of 1 mol/L hydrochloric acid. Decant the hydrochloric acid, and perform a final distilled water rinse of the precipitate. Try to decant as much liquid as possible from the beaker without losing any precipitate.

10. Place your labelled reaction beaker, containing the cleaned precipitate, in a drying oven overnight.

11. On the following day, determine the mass of the beaker with the dry precipitate.

12. Dispose of the precipitate as instructed by your teacher.

Analyze and Interpret

1. **a.** Identify the precipitate that was formed in the reaction.

   b. Which reactant was the limiting reactant, and which was the excess reactant?

   c. Calculate the theoretical yield of the precipitate, in grams, using the mass of the steel wool that you measured in Procedure step 5.

   d. Compare the mass of the precipitate you collected (the actual yield) with the expected theoretical yield. Describe your observations.

2. Calculate the percentage yield of this reaction.

Conclude and Communicate

3. Suggest reasons why the percentage yield of this reaction was not 100%.

4. Explain how the impurity of the iron affected the actual yield.

5. Review the laboratory techniques that you used in this investigation. Did these laboratory techniques affect your actual percentage yield? Explain.

Extend Further

6. **INQUIRY** Design an investigation that uses the same balanced chemical equation as this investigation but improves your percentage yield. State the safety precautions that you should take. With your teacher's approval, conduct your investigation. Calculate the actual percentage yield, and compare it with the percentage yield you obtained in this investigation.

7. **RESEARCH** Conduct research to find the percent by mass of iron in steel wool. Do you think your percentage yield would have been different if you had used pure iron in this investigation? Explain your reasoning.

Skill Check

Initiating and Planning

✓ Performing and Recording

✓ Analyzing and Interpreting

✓ Communicating

Safety Precautions

- Wear safety eyewear throughout this investigation.

- Wear a lab coat or apron throughout this investigation.

- Tie back loose hair and clothing.

- If you get chemical compounds from this lab on your skin, flush with plenty of cold water.

- **Caution:** Calcium chloride is corrosive.

Materials

- 2.00 g of sodium carbonate, $Na_2CO_3(s)$

- distilled water

- 1.00 g of calcium chloride, $CaCl_2(s)$

- graduated cylinder

- two 250 mL beakers

- waterproof grease marker

- electronic balance, accurate to two decimal places

- stirring rod

- filter paper

- funnel

- retort stand with ring clamp

- Erlenmeyer flask

- wash bottle with distilled water

- drying oven or heat lamp

Finding the Percentage Yield of a Double Displacement Reaction

The percentage yield of a reaction is affected by different factors, including reaction procedures and laboratory techniques. In this investigation, you will conduct a double displacement reaction in which a solution of sodium carbonate, $Na_2CO_3(aq)$, reacts with a solution of calcium chloride, $CaCl_2(aq)$, to produce aqueous sodium chloride, $NaCl(aq)$, and a solid precipitate. You will then determine the percentage yield of calcium carbonate, $CaCO_3(s)$, in the reaction and determine the effects of reaction procedures and laboratory techniques on this yield.

Pre-Lab Questions

1. Explain, in your own words, what a double displacement reaction is.

2. What is the balanced chemical equation for the reaction in this investigation?

3. List three ways in which the actual product yield can be reduced during the filtration of a precipitate.

Question

What is the percentage yield of calcium carbonate from the reactants sodium carbonate and calcium chloride?

Procedure

1. Obtain two clean, dry 250 mL beakers. Label the beakers with a grease marker, using the names of the reactant solutions.

2. Copy the following data table into your notebook. Measure the masses of the empty labelled beakers, and record these masses in your data table.

Data Table

Material	Mass (g)
Sodium carbonate	
Empty labelled beaker	
Beaker and sodium carbonate	
Sodium carbonate	
Calcium chloride	
Empty labelled beaker	
Beaker and calcium chloride	
Calcium chloride	
Product	
Clean, dry, labelled filter paper	
Filter paper and dry calcium carbonate precipitate	

3. Measure 2.00 g of sodium carbonate, and place it in the corresponding labelled beaker. Measure and record the mass of the beaker with the sodium carbonate. Add 50 mL of distilled water to the beaker. Stir to dissolve the solid material. If the stirring rod is removed from the beaker, rinse it with distilled water, and allow the water to flow into the beaker so that you do not lose the reactant.

4. Measure 1.00 g of calcium chloride, and place it in the other labelled beaker. Measure and record the mass of the beaker with the calcium chloride. Add 50 mL of distilled water to the beaker. Stir to dissolve the solid material. If the stirring rod is removed from the beaker, rinse it as described in step 3.

5. When both solids have dissolved, gently pour the entire contents of the sodium carbonate beaker, a portion at a time, into the calcium chloride beaker while stirring constantly.

6. Observe and record all the changes that occur in the reaction beaker.

7. When the reaction is complete (about 2 min), label a piece of filter paper. Measure the mass of the filter paper, and record the mass in your data table. Set up the funnel, filter paper, and flask as shown in step 3 in the diagram.

8. Gently swirl the reaction beaker to suspend the precipitate. Carefully pour the entire contents, a portion at a time, onto the filter paper.

9. Rinse the reaction beaker several times with distilled water, using a wash bottle, and pour the rinse through the filter paper to capture all the precipitate.

10. Rinse the precipitate on the filter paper with more distilled water from the wash bottle to be sure that all the dissolved materials have been removed.

11. When the filtrate stops dripping from the funnel, carefully remove the filter paper from the funnel and place it in a drying oven overnight.

12. After the precipitate has dried, measure the mass of the filter paper and precipitate. Record the mass in your data table.

13. Dispose of the precipitate as instructed by your teacher.

Analyze and Interpret

1. **a.** What is the solid precipitate that formed in this reaction?
   b. Which reactant was the limiting reactant, and which was the excess reactant?
   c. Calculate the theoretical yield of the precipitate, in grams, using the actual recorded masses.
   d. Compare the mass of the precipitate you collected (the actual yield) with the expected theoretical yield. Describe any differences.

2. Calculate the percentage yield of this reaction.

Conclude and Communicate

3. How might reaction procedures and laboratory techniques have affected the percentage yield of this reaction? Explain your reasoning.

Extend Further

4. **INQUIRY** Design new steps for this investigation to improve the laboratory techniques and possibly improve the percentage yield. State the safety precautions that you should take. With your teacher's permission, redo this investigation using your new steps. Calculate the actual percentage yield, and compare it with the original actual percentage yield.

5. **RESEARCH** Sodium carbonate is commonly called soda ash or washing soda. Research sodium carbonate as a "green" alternative for cleaning clothes. Write a short summary of your findings.

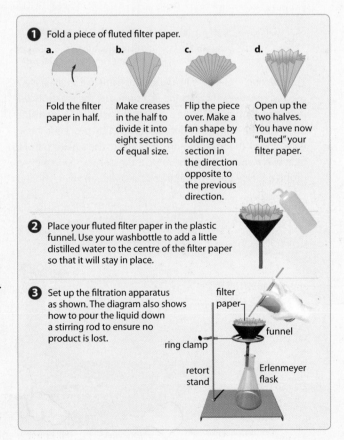

1 Fold a piece of fluted filter paper.

a. Fold the filter paper in half.

b. Make creases in the half to divide it into eight sections of equal size.

c. Flip the piece over. Make a fan shape by folding each section in the direction opposite to the previous direction.

d. Open up the two halves. You have now "fluted" your filter paper.

2 Place your fluted filter paper in the plastic funnel. Use your washbottle to add a little distilled water to the centre of the filter paper so that it will stay in place.

3 Set up the filtration apparatus as shown. The diagram also shows how to pour the liquid down a stirring rod to ensure no product is lost.

filter paper

funnel

ring clamp

retort stand

Erlenmeyer flask

Case Study

Industrial-sized Safety

Preventing Environmental Harm from Large-scale Chemical Reactions

Scenario

Suppose that you are learning about different types of disasters in a world issues class. In class, you discussed natural disasters, such as earthquakes and volcanic eruptions. You also discussed human-caused industrial disasters, such as accidents at industrial chemical plants.

Potential for Explosion

One industrial disaster occurred on April 12, 2004, at a chemical plant in Georgia, U.S.A. Employees had just begun the production of a large-scale batch of triallyl cyanurate (TAC) in the 7500 L chemical reactor like the ones shown below.

TAC is a chemical that is used to produce resins, plastics, coatings, and adhesives. The main reaction uses allyl alcohol and cyanuric chloride to generate TAC and hydrogen chloride, according to the following balanced equation:

$$3C_3H_6O(\ell) \ + \ C_3N_3Cl_3(s) \ \rightarrow \ C_{12}N_3H_{15}O_3(s) \ + \ 3HCl(g)$$

| allyl alcohol | cyanuric chloride | triallyl cyanurate (TAC) | hydrogen chloride |

The reaction was tested in a laboratory and in a small-scale reactor. However, this was the first time that the reaction was being performed in a large-scale reactor with large quantities.

In this chemical reaction, as the reaction proceeds, heat is produced. In the laboratory tests and in the small reactors, heat was removed from the reaction vessel and there were no problems. However, in the large-scale reaction vessel, heat did become a problem. At 9:30 P.M. on the first day of operation, the large-scale reaction vessel overheated. The pressure in the reactor built up and eventually a gasket burst. A vapour cloud of allyl alcohol, and possibly hydrogen chloride, was released into the atmosphere. Allyl alcohol is toxic to humans and wildlife, especially aquatic life, and it can damage crops.

Within minutes, the employees evacuated the site and called 911. Then 14 minutes after the explosion, nearby residents complained of an odour. After nine hours, emergency personnel used an overhead water spray to force the escaping vapour cloud down. Eventually, the area was sealed off. However, environmental damage had been done. Contaminated water run-off killed fish and other organisms in two nearby creeks. Vegetation up to 0.8 km downwind of the chemical plant was chemically burned.

For humans, the effects of allyl alcohol exposure include severe eye and respiratory irritation. In cases of prolonged exposure, lung, liver, and kidney complications can occur. More than 100 families and businesses were evacuated because of the toxic allyl alcohol vapour cloud, and 154 people were decontaminated and checked at a local hospital.

This is the site in which the accident occurred in Georgia, U.S.A. The large tank on the trailer contains one of the reactants. The reaction vessels are behind the trailer under the awning.

"Scaling Up" from Laboratory to Industry

As this chemical accident demonstrates, the scale of a chemical reaction matters. The quantities of chemicals used for industrial processes are on the order of tonnes rather than grams. Hundreds or thousands of litres are used, rather than just a few millilitres. Before a chemical process is used in an industrial chemical plant, chemists test the reaction using small quantities in a laboratory. In this way, the reaction is more easily controlled. In addition, large quantities of materials are not wasted while chemists perfect and optimize the reaction conditions. When the optimal conditions are determined, chemical engineers take over. They optimize the process for large-scale production.

Safety is important when performing chemical reactions. In a laboratory, even a small spill or gas release can have a negative effect on nearby people or the environment. However, when a chemical reaction is "scaled up" from a laboratory to a chemical plant, special safety considerations are required. Extra caution is needed because the potential for dangerous situations, including serious harm to the environment, is far greater.

Many chemical reactions, like the reaction at the Georgia plant, release heat. In a laboratory, where experiments are carried out in glass test tubes and flasks, the quantities are small. Thus, the amount of heat produced usually does not create a problem. When small amounts of heat are produced, the heat is usually absorbed by the equipment and surroundings. Inside a large, closed reaction vessel, like the ones shown in the photograph, more heat is produced because more chemicals are used. Moreover, there are limited ways for the excess heat to escape. Chemical engineers must plan for the safe removal of heat. Otherwise, the contents of the reaction vessel can get too hot, and accidents can occur, like the one in Georgia.

One scenario that chemists and chemical engineers must avoid is a "runaway" reaction. In this situation, the temperature increase caused by a chemical reaction becomes uncontrolled and self-accelerates. As the contents of the reaction vessel become hotter, the reaction speeds up and produces even more heat. Furthermore, as the temperature increases, the pressure inside the reaction vessel also increases. The increases in temperature and pressure can lead to an explosion. In a large industrial reaction, there must be a way to control the reaction and a way to remove the excess heat and pressure. The accident described earlier demonstrates the importance of planning for these control devices and what can occur if they are not included.

Research and Analyze

1. Conduct research to identify three industrial chemical accidents that had an impact on the environment. What role, if any, did quantitative error play in these accidents?

2. Research the many tools that are available to help emergency personnel handle chemical incidents quickly and effectively. For example, one tool, which is shown in the photograph below, is the Palmtop Emergency Action for Chemicals (PEAC) tool. Research the PEAC tool online and record your findings.

3. Analyze the circumstances that led to the three industrial chemical accidents you identified in question 1. What steps could have prevented each accident? What steps could have been taken to limit the impact of each accident on the environment?

Take Action

1. **Plan** In a group, discuss the issue of industrial chemical disasters and runaway reactions. What can be done to prevent runaway reactions? What safety measures should be taken? How can the surrounding environment be protected? Share the results of the research and analysis you conducted for questions 1 to 3 above.

2. **Act** Prepare a multimedia presentation about runaway reactions and industrial chemical disasters to present to your class.

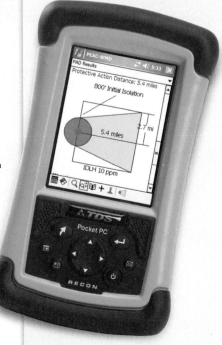

Chapter 7 | SUMMARY

Section 7.1   What Is Stoichiometry?

A balanced chemical equation describes what occurs in a chemical reaction. It is used to predict the relationships among the quantities of reactants and products in a chemical reaction.

KEY TERMS
mole ratio
stoichiometry

KEY CONCEPTS
- The coefficients of a balanced chemical equation can be used to represent the relative amounts (in moles) of particles (atoms, ions, molecules, or formula units).
- Stoichiometric calculations are used to predict the amounts of reactants used or products formed in a chemical reaction.
- A mole ratio from a balanced chemical equation relates the amount in moles of one reactant or product to the amount in moles of another reactant or product.
- The amount in moles of any substance can be converted to number of particles or mass units, such as grams.

Section 7.2   Limiting and Excess Reactants

Reactants are usually not present in exact ratios, as represented by a balanced chemical equation. One reactant usually limits the amount of product that is produced.

KEY TERMS
excess reactant
limiting reactant
stoichiometric amount

KEY CONCEPTS
- A limiting reactant is a reactant that is completely consumed during a chemical reaction, and therefore limits or stops the reaction. Reactants that remain after the reaction stops are called excess reactants.
- To identify the limiting reactant, the amount of product that is produced from each reactant is determined. Then the amounts from both reactants are compared to determine which reactant produces the smaller amount of product.
- Determining the limiting reactant is necessary for all stoichiometric calculations that are used to determine the amount of product that forms.
- In industrial chemical processes, limiting and excess reactants must be managed to ensure that the reactants produce the products safely and efficiently.

Section 7.3   Reaction Yields

The actual yield of product that is recovered from a chemical reaction is usually less than the theoretical yield that is predicted by means of calculations. There are several factors that can lead to percentage yield that is less than 100 percent.

KEY TERMS
actual yield
competing reaction
percentage yield
theoretical yield

KEY CONCEPTS
- Stoichiometric calculations are used to determine the theoretical yield, which is the maximum theoretical amount of product that can be produced from a given amount of reactants in a chemical reaction.
- The actual yield is determined through experimentation and is the actual amount of product that is produced in a chemical reaction.
- The percentage yield is the ratio of the actual yield to the theoretical yield, expressed as a percent.

Knowledge and Understanding

Select the letter of the best answer below.

1. Stoichiometry is the study of
   a. qualitative relationships in chemical changes
   b. quantitative relationships in chemical changes
   c. the relationship between the amount in moles of a substance and the mass of the substance
   d. quantitative relationships in physical changes
   e. chemical and physical changes that occur in measured amounts

2. Suppose that you are using a recipe to make spiced hot cocoa. For the recipe, you need 1 cup of milk, $\frac{3}{4}$ tablespoon of sweetened cocoa, 1 tablespoon of whipped cream, $\frac{1}{8}$ teaspoon of cinnamon, and $\frac{1}{8}$ teaspoon of cloves. You have 100 mL of milk, 16 tablespoons of sweetened cocoa, 4 tablespoons of whipped cream, and 14 teaspoons of each of the spices. Which ingredient is the limiting ingredient?
   a. cinnamon   d. milk
   b. cloves   e. whipped cream
   c. cocoa

3. Which kind of information is given by the coefficients in a chemical equation?
   a. the actual yield of the product
   b. the physical states of the compounds
   c. the amounts (in moles) of the compounds
   d. the limiting reactant in the chemical equation
   e. the excess reactant in the chemical equation

4. If the reactants in a chemical reaction are present in stoichiometric amounts, which scenario will most likely occur?
   a. The percentage yield of the reaction will be very low.
   b. The reaction will proceed until all the reactants are consumed.
   c. The reaction will proceed until the excess reactant is gone.
   d. The reaction will proceed until the limiting reactant is gone.
   e. The actual yield of the product will be greater than the theoretical yield.

5. Which factor should result in a higher percentage yield?
   a. impure reactants
   b. low reaction pressure
   c. low reaction temperature
   d. no competing reaction
   e. poor laboratory techniques

6. In the equation for percentage yield, which ratio is used?
   a. actual yield to theoretical yield
   b. theoretical yield to actual yield
   c. grams of reactants to grams of products
   d. grams of products to grams of reactants
   e. grams of limiting reactant to grams of excess reactant

7. A 2.5 g lump of zinc is dropped into a solution of hydrochloric acid, producing zinc chloride, $ZnCl_2$(aq), and a gas. The hydrochloric acid is in excess. Which equation would you use to calculate the theoretical yield of the reaction?
   a. $Zn(s) + 2HCl(aq) \rightarrow ZnCl_2(aq) + H_2(g)$
   b. $Zn(s) + 2HCl(aq) \rightarrow ZnCl(aq) + H_2(g)$
   c. $3Zn(s) + HCl(aq) \rightarrow 3ZnH_2(aq) + Cl_2(g)$
   d. $Zn(s) + HCl(aq) \rightarrow ZnCl_2(aq) + H_2(g)$
   e. $2Zn(s) + 2HCl(aq) \rightarrow 2ZnCl(aq) + H_2(g)$

8. What is the mole ratio for this reaction, when balanced?
$$KClO_3(s) \rightarrow KCl(s) + O_2(g)$$
   a. 2:3:2
   b. 2:2:4
   c. 3:3:2
   d. 1:2:3
   e. 2:2:3

Answer the questions below.

9. Compare and contrast a cooking recipe and a balanced chemical equation.

10. Identify the limiting reactant and excess reactant in each situation.
    a. Baking soda is added to vinegar, and some of the baking soda falls to the bottom of the cup.
    b. Vinegar is poured over hard water deposits, and some of the deposits remain after a few minutes.
    c. An antacid tablet bubbles and eventually disappears when placed in a glass of water.

11. What information is given by the coefficients in mole ratios?

12. If the limiting reactant in a chemical reaction is impure, what will happen to the percentage yield of the product?

13. Explain, in general terms, how the application of stoichiometry benefits manufacturers of industrial chemicals.

14. Consider a chemical reaction that involves two reactants. If you know the amount (in moles) of one of the reactants, briefly explain how you would find the amount (in moles) required for the other reactant.

15. Outline the steps you would take to determine how many molecules are produced in a chemical reaction, if you know the number of molecules of the limiting reactant.

Thinking and Investigation

16. A student was working with the following chemical reaction and made a statement that "this reaction requires three times as much hydrogen as nitrogen to proceed."

$$N_2(g) + 3H_2(g) \rightarrow 2NH_3(g)$$

What must be added or changed in the student's statement to make it true?

17. The following unbalanced chemical equation represents the reaction of aluminum with bromine, to form aluminum bromide, $AlBr_3(s)$:

$$\_\_Al(s) + \_\_Br_2(g) \rightarrow \_\_AlBr_3(s)$$

a. Balance the chemical equation.

b. What amount (in moles) of bromine reacts with 5.0 mol of aluminum?

c. What amount (in moles) of aluminum bromide is produced if 5.0 mol of aluminum is used?

18. What are the theoretical yield and percentage yield if 14.59 g of iron(II) chloride, $FeCl_2(aq)$, and hydrogen gas are produced when 6.51 g of iron reacts with excess hydrochloric acid?

19. What amount (in moles) of sodium oxide, $Na_2O(s)$, is produced when 4.5 mol of solid sodium completely reacts with oxygen gas?

20. Solid silver chromate, $Ag_2CrO_4(s)$, forms when excess potassium chromate, $K_2CrO_4(s)$, is added to a solution that contains 0.500 g of silver nitrate, $AgNO_3(aq)$. Determine the theoretical yield of silver chromate. Calculate the percentage yield if 0.455 g of silver chromate is produced.

21. Solid copper wire is placed in a colourless solution of silver nitrate, $AgNO_3(aq)$, as shown below.

If the reaction goes to completion, which of the two reactants is the limiting reactant? Explain.

22. Consider the following reaction:

$$Li_3N(s) + 3H_2O(\ell) \rightarrow NH_3(g) + 3LiOH(aq)$$

a. Will 2.39 g of lithium nitride, $Li_3N(s)$, react completely with 2.90 g of water? Explain.

b. What mass of lithium hydroxide, $LiOH(aq)$, would you expect to be produced if the quantities given in part a are used?

23. In an investigation, students were asked to decompose potassium chlorate, $KClO_3(s)$, with heat and to determine the percentage yield. The decomposition equation for potassium chlorate is given below.

$$2KClO_3(s) \rightarrow 2KCl(s) + 3O_2(g)$$

Design a procedure that could be used to determine the percentage yields of potassium chloride, $KCl(s)$, and oxygen gas.

24. An investigation was performed to determine the percentage yield of the following chemical reaction:

$$2AgNO_3(aq) + CaCl_2(aq) \rightarrow 2AgCl(s) + Ca(NO_3)_2(aq)$$

The results were recorded in the data table below. Use the results to calculate the percentage yield of the silver chloride, $AgCl(s)$, precipitate.

Data Table

What Was Measured	Mass
Mass of clean, dry filter paper	1.27 g
Mass of empty beaker	124.75 g
Mass of beaker and $AgNO_3(s)$	126.23 g
Mass of dry precipitate and filter paper	2.36 g

Communication

25. Develop an analogy for the concept of limiting and excess reactants. Illustrate the analogy using diagrams or photographs.

26. Examine the following balanced chemical equation:

$$2A + B \rightarrow 3C + D$$

Using a graphic organizer, explain how you would calculate the mass (in grams) of C that can be obtained when a given mass of A reacts with a given amount (in moles) of B. Assume that you know the molar masses of A and C. Include proper units. For simplicity, assume that A is the limiting reactant, but do not forget to show how you would determine the limiting reactant.

27. A chemical company wants to create a short radio advertisement that shows how consideration of percentage yield in the company's manufacturing process helps to reduce the company's impact on the environment. Write a script for an advertisement that would be about 30 seconds long.

28. Summarize your learning in this chapter using a graphic organizer. To help you, the Chapter 7 Summary lists the Key Terms and Key Concepts. Refer to Using Graphic Organizers in Appendix A to help you decide which graphic organizer to use.

Application

29. Analyze a process in your home that uses chemical quantities. Briefly describe the process, and explain how it uses chemical quantities.

30. Choose an industrial chemical process that occurs in Canada. Research the importance of quantitative accuracy in this process and the potential impact on the environment if accurate quantities are not used.

31. Hydrogen is proposed by some scientists as a viable fuel alternative in automobiles, instead of fossil fuels such as gasoline (which contains octane $C_8H_{18}(g)$), propane, $C_3H_8(g)$, and methane, $CH_4(g)$. Energy is released from hydrogen by burning it directly or by using a hydrogen fuel cell. In a hydrogen fuel cell, hydrogen gas reacts with oxygen gas to produce electrical energy and water. The chemical equations for different fuels are shown in the table below.

Chemical Equations for Different Fuels

Fuel	Reaction
Hydrogen	$2H_2(g) + O_2(g) \rightarrow 2H_2O(\ell) + $ energy
Gasoline	$2C_8H_{18}(g) + 25O_2(g) \rightarrow 18H_2O(g) + 16CO_2(g) + $ energy
Propane	$C_3H_8(g) + 5O_2(g) \rightarrow 4H_2O(g) + 3CO_2(g) + $ energy
Methane	$CH_4(g) + 2O_2(g) \rightarrow 2H_2O(g) + CO_2(g) + $ energy

a. Compare the mole ratios of each fuel listed in the table to oxygen, in each of the four reactions, by completing the table below.

Mole Ratios

Type of Fuel	Coefficient			
	Fuel	O_2	H_2O	CO_2
Hydrogen	2	1	2	---
Gasoline				
Propane				
Methane				

b. Many scientists infer that climate change and global warming are caused by excess carbon dioxide in the atmosphere. What are some possible benefits for the environment if hydrogen is used as a fuel, instead of fossil fuels?

32. Suppose that you have been asked to design an investigation based on the following equation:
$$2AgNO_3(aq) + CaCl_2(aq) \rightarrow 2AgCl(s) + Ca(NO_3)_2(aq)$$
You will use the precipitate, silver chloride, $AgCl(s)$, to determine the percentage yield of the reaction. Research both reactants for their reactivity, cost, disposal as hazardous waste, and other health and safety concerns. Choose one of the two reactants to be present in excess, based on your findings.

33. Burning any fossil fuel in air, rather than in pure oxygen, $O_2(g)$, means that the oxygen is impure. One of the contaminants is nitrogen gas, $N_2(g)$, which makes up over 70% of air. Conduct research to determine which nitrogen compounds might be created in the high heat of combustion and what environmental impact they might have.

34. Large petroleum deposits exist in the province of Alberta. These deposits are often called tar sands. Recently, however, some media have referred to them as oil sands. To yield gasoline and other fossil fuels, these deposits must be heavily refined in a costly process that produces carbon dioxide, a greenhouse gas.

a. Discuss possible differences in the public perception of the terms "oil sands" and "tar sands."

b. Explain how the low percentage yield of fossil fuels from these sands could have an impact on the environment.

35. **BIG IDEAS** Relationships in chemical reactions can be described quantitatively. For example, the masses of reactants and products are often used in stoichiometric calculations, but masses are not the only quantities that can be used. Review your chemistry notebook and this textbook. What is another type of quantity that can be used in stoichiometric calculations?

36. **BIG IDEAS** The efficiency of chemical reactions can be determined and optimized by applying an understanding of quantitative relationships in such reactions. For example, air and water pollution can be caused by dumping excess reactants from chemical reactions into waterways and the air. Excess solid wastes also create pollution and disposal issues. Write a letter to a local politician, describing environmental problems that are caused by using excessive amounts of chemical compounds and why industrial chemical companies must be encouraged to use correct stoichiometric amounts.

Select the letter of the best answer below.

1. **K/U** What is the mole ratio for this balanced chemical equation?
$$N_2(s) + 3H_2(g) \rightarrow 2NH_3(g)$$
   a. 1:1:1
   b. 1:3:2
   c. 1:2:3
   d. 2:5:2
   e. 2:5:4

2. **K/U** What is the mole ratio for the following chemical equation, when balanced?
$$P(s) + O_2(g) \rightarrow P_2O_5(s)$$
   a. 1:1:1
   b. 4:5:2
   c. 2:1:1
   d. 2:5:2
   e. 2:5:4

3. **K/U** What is the first step when solving any stoichiometry problem?
   a. ensuring that the equation is balanced
   b. determining the mole ratios
   c. calculating the amount of reactant
   d. identifying the limiting reactant
   e. calculating the theoretical yield

4. **K/U** What is the best description of a limiting reactant in a chemical reaction?
   a. the product that is consumed completely
   b. the product that is not consumed completely
   c. the reactant that is not consumed completely
   d. the reactant that is dissolved in solution
   e. the reactant that is consumed completely

5. **K/U** The percentage yield of a reaction is
   a. the actual amount of product that is obtained during the reaction
   b. the maximum amount of product that can be obtained under ideal conditions
   c. a comparison between the actual yield and the theoretical yield
   d. the ratio of the reactants, expressed as a percent
   e. the ratio of the products, expressed as a percent

6. **T/I** A reaction has a theoretical yield of 9.10 g and an actual yield of 1.72 g. What is its percentage yield?
   a. 18.9%
   b. 15.7%
   c. 18%
   d. 81.1%
   e. 19.0%

7. **K/U** Which situation would most likely result in the highest actual yield?
   a. Competing reactions occur.
   b. The beaker is rough and pitted.
   c. The product clings to the stirring rod.
   d. The product is soluble in filtrate wash.
   e. The reaction occurs at a high pressure.

8. **K/U** Which best describes a competing reaction?
   a. a reaction that yields more product than the desired reaction
   b. a reaction that occurs along with the principal reaction and involves the reactants and/or products of the principal reaction
   c. a reaction that is used to slow down the principal reaction
   d. a reaction that is used to speed up the principal reaction
   e. a reaction that has a higher theoretical yield than the principal reaction

9. **T/I** A student determines that the percentage yield of an experiment is 82.5%. If the final amount of product was measured to be 5.20 g, what was the theoretical yield?
   a. 6.30 g
   b. 4.95 g
   c. 9.91 g
   d. 2.53 g
   e. 32.76 g

Use sentences and diagrams, as appropriate, to answer the questions below.

10. **K/U** Define the terms "limiting reactant" and "excess reactant" in your own words.

11. **A** Describe a real-life scenario in which there are limiting reactants.

12. **T/I** If 20.00 g of sodium hydroxide is mixed with a sufficient amount of iron(III) nitrate, $Fe(NO_3)_3(aq)$, a clear solution and a yellow solid precipitate are produced. What mass of precipitate is produced?

13. **T/I** Carbon combines with oxygen to form several different oxides, including carbon monoxide and carbon dioxide.
   a. The equation for one of these reactions is given below.
$$2C(s) + O_2(g) \rightarrow 2CO(g)$$
   What amount (in moles) of oxygen reacts with 9.35×10^{-2} mol of carbon to form carbon monoxide?
   b. Carbon can also react with oxygen gas as follows:
$$C(s) + O_2(g) \rightarrow CO_2(g)$$
   What amount (in moles) of oxygen reacts with 4.5 mol of carbon to form carbon monoxide?

14. **C** Using a graphic organizer, outline how the Avogadro constant and molar mass are used in

stoichiometric calculations. Use a sample calculation to illustrate your answer.

15. **T/I** Chromite, $FeCr_2O_4(s)$, is an important ore of chromium. To extract chromium, chromite is reacted with carbon to produce ferrochrome, $FeCr_2(s)$, as follows:
$$2C(s) + FeCr_2O_4(s) \rightarrow FeCr_2(s) + 2CO_2(g)$$
What mass of chromite reacts with 10.0 kg of carbon?

16. **T/I** Solid zinc metal and iron(III) chloride, $FeCl_3(aq)$, are used to obtain pure iron metal. The reaction is only 55.000% efficient, however, and yields 150.00 g of iron metal.
   a. Write the balanced chemical equation.
   b. Determine the theoretical yield of iron metal in this reaction.
   c. Determine the minimum mass of zinc metal that must be used to obtain 150.00 g of iron metal.

17. **C** Solid sodium hydrogen carbonate, $NaHCO_3(s)$, decomposes when heated to form sodium carbonate, $Na_2CO_3(s)$, carbon dioxide gas, and one other product. Develop a procedure that outlines the steps that could be used to determine the percentage yield of sodium carbonate.

18. **C** Use an analogy to explain the term "excess reactant."

19. **T/I** In an investigation, magnesium ribbon was burned over a watch glass with oxygen as follows:
$$2Mg(s) + O_2(g) \rightarrow 2MgO(s)$$

   a. Based on the photograph, infer reasons for why this reaction had a relatively low percentage yield?

b. Use the data in the table below to calculate the percentage yield of this reaction.

Data Table

Mass of watch glass and crucible tongs	273.527 g
Mass of watch glass, crucible tongs, and 10 cm piece of magnesium ribbon	273.567 g
Mass of watch glass, crucible tongs, and ash, $MgO(s)$	273.564 g

20. **T/I** If 2.0 g of hydrogen sulfide, $H_2S(g)$, reacts with 5.0 g of sodium hydroxide what mass of the excess reactant is present when the reaction is complete?
$$H_2S(g) + 2NaOH(aq) \rightarrow Na_2S(aq) + 2H_2O(\ell)$$

21. **C** During an investigation, a student used a stirring rod to stir a solid precipitate that was drying, prior to being weighed. Explain to the student how this might affect the actual yield of the reaction.

22. **T/I** Under the right conditions, aluminum oxidizes and corrodes as represented by:
$$4Al(s) + 3O_2(g) \rightarrow 2Al_2O_3(s)$$
Calculate the amount (in moles) of aluminum oxide, $Al_2O_3(s)$, that is produced when 10.0 g of aluminum is used.

23. **T/I** The following equation represents the combustion of methane, $CH_4(g)$, with oxygen:
$$CH_4(g) + 2O_2(g) \rightarrow CO_2(g) + 2H_2O(g)$$
Determine the limiting reactant if 20.0 g of methane reacts with 20.0 g of oxygen.

24. **T/I** Nitrogen can combine with oxygen to form several different oxides of nitrogen. The following equations represent two of these reactions:
$$2N_2(g) + O_2(g) \rightarrow 2N_2O(g)$$
$$N_2(g) + O_2(g) \rightarrow 2NO(g)$$
Write the mole ratios for each reaction, and use them to discuss how the limiting reactant in each reaction is identified.

25. **T/I** Sulfuric acid, $H_2SO_4(aq)$, forms when sulfur dioxide, $SO_2(s)$, reacts with oxygen and water. What amount (in moles) of sulfuric acid is produced from 12.5 mol of sulfur dioxide?

Self-Check

If you missed question …	1	2	3	4	5	6	7	8	9	10	11	12	13	14	15	16	17	18	19	20	21	22	23	24	25
Review section(s)…	7.1	7.1	7.1	7.2	7.3	7.1	7.3	7.1	7.1	7.2	7.2	7.1	7.1	7.1	7.1	7.3	7.2	7.2	7.3	7.2	7.3	7.1	7.2	7.1	7.1

Modelling Copper Recovery in the Electronics Industry

What do laptop computers, blinking novelty brooches, like the one below and musical greeting cards have in common? They all contain printed circuit boards (PCBs). Printed circuit boards are an inexpensive way to make electronic devices. Instead of using individual wires to conduct electric current in a device, PCBs use copper pathways.

To make a PCB, the electric circuit is printed on a non-conductive plate that is coated with a layer of electrically conductive copper, as shown in the photo below. Special unreactive ink is used to print the electric circuit on the plate. The plate is then submerged in an etching fluid to remove all of the copper except the copper underneath the ink outline. The unreactive ink is then carefully removed, leaving a shiny, conductive copper pathway on the plate. Finally, electrical components, such as batteries, light-emitting diodes, and switches, are added to the circuit.

The used etching fluid and copper are considered waste. However, it would be irresponsible to simply dump the used etching fluid, untreated, into the environment. In addition,

copper is an expensive commodity, so it would not be cost-effective to dispose of it as waste. Instead, the copper is recovered from the etching fluid and used again.

As part of a student-corporate challenge, you are selected to design a process to recover copper from etching waste. You must design a procedure to determine the most efficient method for recovering 10 g of copper metal from a copper(II) nitrate solution. Because nitric acid can be used to etch copper plates, you will use a copper(II) nitrate solution to model the recovery of copper from etching waste. You must use a minimum amount of copper(II) nitrate in your process.

On the basis of your experimental results, you will create a presentation to promote your method of copper recovery from the etching fluid to the production manager at an electronics company. You must demonstrate that your process is a cost-effective and an efficient method of copper recovery.

How can you extract 10 g of copper metal from copper(II) nitrate with the highest percentage yield?

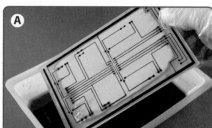

The outline of the circuit board is printed on a copper coated plate.

After the circuit board is etched, electric components, such as batteries, diodes, and switches, are added to the circuit board.

A power source, such as a small battery, light-emitting diodes, and an on/off switch are attached to a circuit board to make the lights on the brooch work.

Safety Precautions

- Wear safety eyewear, gloves, and a lab apron or other protective clothing throughout this investigation.
- Tie back loose hair and clothing.
- Use EXTREME CAUTION when you are near an open flame.
- Wash your hands after handling the materials in this investigation.

Initiate and Plan

1. Write a general equation for the displacement of copper metal from copper(II) nitrate solution. Use the metal activity series and a list of available metals from your teacher to write equations for specific single displacement reactions that you could investigate. Your experimental design should include three different reactions using three different metals.

2. Calculate the mass of each metal and the mass of copper(II) nitrate you would need to recover 10 g of copper metal. Record this information in a table. Remember to give your table a title.

3. Because copper(II) nitrate is the only source of copper in this reaction, it should be the limiting reactant. How will you use this information in your experimental design?

4. For each reaction, draw a flow chart that shows what goes into the reaction vessel and what comes out of the reaction vessel. Remember to show the physical state of each chemical compound in each chemical reaction. Use the flow charts to help you decide how to calculate your percentage yield.

5. Write a step-by-step procedure for your chemical process. Go to Scientific Inquiry in Appendix A for help with writing your procedure. You can also refer to the investigations in this book as models. Remember to list the materials and safety precautions, as well as each step in the procedure. Include instructions for separating, rinsing, and drying the solid product at the end of the reaction before determining its mass. Identify the independent, dependent, and controlled variables for each reaction in your investigation.

6. You will find it helpful to run each reaction twice: the first time, to determine percentage yield; the second time, with altered amounts, to compensate for percentage yield and to aim for an actual yield of 10 g of copper metal. Note: each reaction should be allowed to continue for 24 hours before the final product is analyzed.

7. Have your teacher approve your procedure before you begin.

Perform and Record

1. Conduct your investigation. Record detailed observations of each reaction as it proceeds.

2. Make coloured diagrams or take photographs of the contents of the reaction vessels to illustrate what is occurring as the reactions proceed. The drawings and photographs can be used as visual aids in your presentation.

3. Make sure that you collect all of the information that you need to complete the actual yield and percentage yield calculations.

Analyze and Interpret

1. Calculate the theoretical yield, actual yield, and percentage yield for each reaction. Display the percentage yield for each reaction in a bar graph.

2. Based on your results, what can you conclude about the recovery of copper from a copper(II) nitrate solution?

3. Identify possible sources of error in your procedure, and suggest possible improvements to your experimental design.

4. Using experience gathered from your investigation, infer the importance of quantitative accuracy in industrial chemical processes, such as copper recovery.

Communicate Your Findings

Prepare an electronic slide show, a webcast, or a poster board presentation to promote your method of copper recovery. Be sure to include your experimental design, procedure diagrams or photographs, qualitative observations, results, conclusions, and recommendations for experimental design improvements.

Assessment Criteria

Once you complete your project, ask yourself these questions. Did you…

☑ **K/U** determine which metals displace copper?

☑ **C** state correct balanced chemical equations for each reaction?

☑ **C** draw the flow charts neatly and correctly?

☑ **T/I** consider factors such as how quickly and completely compounds in different physical states react?

☑ **A** control appropriate variables and use equipment and materials safely, accurately, and effectively?

☑ **T/I** adjust your calculations correctly to obtain 10 g of copper in each case?

☑ **K/U** calculate yields correctly?

☑ **C** correctly draw the bar graph?

☑ **T/I** identify sources of error and make suggestions for improvements?

☑ **C** organize your presentation in a clear and logical manner using appropriate scientific vocabulary?

BIG IDEAS

- Relationships in a chemical reaction can be described quantitatively.
- The efficiency of a chemical reaction can be determined and optimized by applying an understanding of the quantitative relationships in the reaction.

Overall Expectations

In this unit, you learned how to…

- **analyze** processes in the home, workplace, and environment that involve chemical quantities and calculations, and **assess** the importance of quantitative accuracy in industrial chemical processes
- **investigate** quantitative relationships in chemical reactions, and solve related problems
- **demonstrate** an understanding of the concept of the mole and its significance to the quantitative analysis of chemical reactions

Chapter 5	The Mole: A Chemist's Counter

KEY IDEAS

- A mole is the amount of substance that contains as many particles (atoms, molecules, ions, or formula units) as exactly 12 g of the isotope carbon-12. It is the SI base unit that is used to measure the amount of a substance.
- Chemists use the mole to count very small particles, such as atoms, molecules, ions, and formula units.
- The Avogadro constant allows chemists to convert back and forth between the number of individual particles and the amount in moles of particles.
- The mass of one atom of an element in atomic mass units has the same numerical value as the mass of one mole of atoms of the same element in grams.

- Molar mass is an important tool for converting back and forth between the mass and the amount in moles of an element or a compound.
- Conversions can be made between the number of individual particles of a substance and the mass of the substance, using both the Avogadro constant and the molar mass.
- All of these conversions are important when calculating and measuring the amounts of substances in a chemical reaction.

Chapter 6	Proportions in Chemical Compounds

KEY IDEAS

- The law of definite proportions states that the elements in a chemical compound are always present in the same proportions by mass.
- The same elements can form different compounds when the atoms are combined in different whole-number ratios.
- The percentage composition of a compound is the mass percents of all the elements in the compound. It can be calculated using either the mass data or chemical formula for the compound.
- The empirical formula for a compound shows only the relative amounts of the elements, not the actual amounts. It can be determined using the percentage composition

of the compound (from chemical analysis) and the molar mass of each element (from the periodic table).

- The molecular formula for a compound shows the actual number of each type of atom in the compound. It is a whole-number multiple of the empirical formula, and it can be determined from the empirical formula and molar mass of the compound or from the percentage composition and molar mass of the compound.
- Information that is needed to determine the composition of a compound can be obtained through chemical analysis.

KEY IDEAS

- The coefficients of a balanced chemical equation can be used to represent amounts in moles of particles (atoms, ions, molecules, or formula units).

- Stoichiometric calculations are used to predict the amounts of reactants used or products formed in a chemical reaction.

- Mole ratios are determined from the coefficients of a balanced chemical equation and are used in stoichiometric calculations. A mole ratio relates the number of moles of one reactant or product to the number of moles of another reactant or product in a particular chemical equation.

- A limiting reactant is a reactant that is completely consumed during a chemical reaction. Reactants that remain after the reaction stops are called excess reactants.

- Stoichiometric calculations can be used to determine the theoretical yield, which is the maximum amount of product(s) that can be produced from given amounts of reactants in a chemical reaction.

- The actual yield is determined through experimentation and is the actual amount of product that is produced in a chemical reaction.

- The percentage yield is the ratio of the actual yield to the theoretical yield, expressed as a percentage.

Knowledge and Understanding

Select the letter of the best answer below.

1. What is the molar mass of sodium phosphate, $Na_3PO_4(aq)$?
   a. 69.95 g/mol
   b. 99.04 g/mol
   c. 117.96 g/mol
   d. 136.94 g/mol
   e. 163.94 g/mol

2. How many formula units of sodium chloride are present in a 3.500 mol sample?
   a. 2.108×10^{-24} formula units
   b. 5.811×10^{-24} formula units
   c. 2.108×10^{23} formula units
   d. 3.500×10^{23} formula units
   e. 2.108×10^{24} formula units

3. According to the law of definite proportions, how do elements always combine?
   a. It depends on the source of the substance.
   b. with the addition of heat
   c. in a constant ratio by mass
   d. in a constant ratio by volume
   e. in a way that always produces a precipitate

4. Which formula could be an empirical formula?
   a. CO_2
   b. C_4H_8
   c. $H_{36}O_{18}$
   d. $Na_4S_2O_8$
   e. $C_{24}H_{44}O_{22}$

5. Magnesium is an element in Epsom salts and many other bath salts, as shown below. The molar mass of magnesium is 24 g. What is the mass of only one atom of magnesium?
   a. 2.0×10^{-23} g
   b. 4.0×10^{-23} g
   c. 1.0×10^{23} g
   d. 2.0×10^{23} g
   e. 4.0×10^{23} g

6. Which statement about the percentage composition of a pure substance is correct?
   a. It is constant for elements, but not compounds.
   b. It depends on the amount in moles.
   c. It is always greater than 80 percent.
   d. It depends on the size of the sample used.
   e. It is the combination of the percent masses of all elements in the compound.

7. When predicting the amount of product that is produced in a given reaction, which do you use?
   a. mass ratios
   b. mole ratios
   c. atomic ratios
   d. molecular ratios
   e. formula unit ratios

8. Which substance contains the greatest amount (in moles)?
   a. 4 g of methane, $CH_4(g)$
   b. 8 g of hydrogen, $H_2(g)$
   c. 28 g of nitrogen, $N_2(g)$
   d. 35 g of chlorine, $Cl_2(g)$
   e. 64 g of oxygen, $O_2(g)$

Answer the questions below.

9. Explain the difference between the terms "molecular molar mass" and "formula unit molar mass" when referring to substances such as carbon dioxide, $CO_2(g)$, and sodium chloride, $NaCl(s)$.

10. What information do you need if you want to determine the empirical formula of a chemical compound?

11. What information does an empirical formula tell you about the molecule?

12. Ammonia, $NH_3(g)$, is used to process textiles, such as the nylon in the jacket shown in the photograph. Nitrogen and hydrogen are used to synthesize ammonia.

  a. Write the balanced chemical equation for the synthesis of ammonia.

  b. If 10.0 g of nitrogen gas is used, what amount in moles of hydrogen gas is required to complete the reaction? What is the mass of the hydrogen gas in grams?

13. Glucose, $C_6H_{12}O_6(s)$, and ribose, $C_5H_{10}O_5(s)$, are sugars. What are the empirical formulas for these sugars? Use these sugars as examples to describe the difference between empirical and molecular formulas.

14. The percentage yield of the following reaction is only 72 percent.

$$CaO(s) + 2HCl(aq) \rightarrow CaCl_2(aq) + H_2O(\ell)$$

If 250 mol of calcium chloride, $CaCl_2$ (aq), is required, what mass (in kilograms) of solid calcium oxide, $CaO(s)$, is required?

15. Methane, $CH_4(g)$, is a major component of natural gas. In a complete combustion reaction, 0.530 mol of methane reacts with 0.497 mol of oxygen gas as follows:

$$CH_4(g) + 2O_2(g) \rightarrow CO_2(g) + 2H_2O(g)$$

  a. What is the limiting reactant?

  b. What is the theoretical yield of water (in grams) in this reaction?

16. Pure iodine is to be prepared according to the following chemical equation.

$$K_2Cr_2O_7(aq) + 6NaI(aq) + 7H_2SO_4(aq) \rightarrow Cr_2(SO_4)_3(aq) + 3I_2(s) + 7H_2O(\ell) + 3Na_2SO_4(aq) + K_2SO_4(aq)$$

If the percentage yield is 86.5%, what mass in grams of potassium dichromate would be needed to obtain 38.5 g of iodine?

Thinking and Investigation

17. A student conducted an investigation to study the concept of the mole. The student found that 25 dry macaroni pasta pieces had a mass of 8.2 g.

  a. What is the mass of 1 mol of pasta pieces?

  b. About 100 pasta pieces is enough for a meal for one person. What amount in moles is 100 pasta pieces? What mass, in grams, is this amount of pasta?

  c. What amount of pasta in moles is required to feed the world's population (6 billion people) three meals a day for one week?

  d. What is the mass of the amount of pasta you calculated in part c?

18. A bright orange compound that easily ignites and burns is analyzed in a laboratory. A technician finds that a 10.0 g sample contains 2.7 g of potassium, 3.5 g of chromium, and 3.8 g of oxygen. What is the percent composition of this compound?

19. Sodium sulfate, $Na_2SO_4(s)$, is a chemical compound that is used to manufacture detergents. Calculate the percentage composition of sodium sulfate.

20. An unknown compound was analyzed in a laboratory, and the following data were obtained:

Data Table

Molar Mass g/mol	Element	Percentage Composition
798.7	lead	51.89%
	sulfur	16.06%
	oxygen	32.05%

a. Determine the empirical formula and the molecular formula for the compound.

b. Name the molecular compound.

21. The elements that are shown below were studied in a laboratory.

Antimony	Sulfur

When a sample of pure antimony was heated with excess sulfur, antimony sulfide was formed. The excess sulfur was then removed. The mass data for the investigation are given in the following table. Complete the data table, and calculate the empirical formula for this newly formed compound.

Mass Data for Chemical Reaction

What Was Measured	Mass (g)
Initial sample of pure antimony	3.653
Antimony sulfide formed	5.096
Sulfur consumed by the reaction	

22. A 56.4 g sample of a compound with a molecular molar mass of 210 g/mol was found to be composed of 12.35 g of sodium 25.77 g of carbon, 1.07 g of hydrogen, and an unknown mass of oxygen. What is the molecular formula for this compound?

23. a. Determine the empirical formula of a compound with the following composition by mass: 18.0% C, 2.5% H, 63.5% I, and 16.0% O

b. If this compound has a molar mass of 400 g/mol, what is its molecular formula?

24. An unknown substance that contained carbon, hydrogen, and oxygen underwent a combustion reaction in the presence of an excess of oxygen. The products of the reaction were analyzed and the following data were obtained:

Mass Data from Combustion Analysis

What Was Measured	Mass
Original sample	42.705 g
Carbon dioxide collected	62.59 g
Water collected	25.62 g
Molar mass of compound (from mass spectrometer)	150.13 g/mol

a. Calculate the percentage composition and the empirical formula for the compound.

b. What is the molecular formula for the compound?

25. A group of students conducted an investigation to verify the percentage composition of copper(II) sulfate pentahydrate, $CuSO_4 \cdot 5H_2O(s)$. The students removed the water from the hydrate by heating it in an open ceramic crucible with a gas burner.

a. Calculate the theoretical percentage composition of water in the molecule.

b. Explain how each of the following would affect the students' calculation of the percentage composition of water in the molecule:

i. heating the material for a short time;

ii. spilling some of the initial copper(II) sulfate pentahydrate on the electronic balance and not noticing;

iii. inadvertently removing some of the reaction material when stirring it with the crucible tongs

26. After completing an investigation to isolate a precipitate from a reaction mixture using filter paper, a student became frustrated. Her consistent attempts to improve her lab techniques resulted in very little change in the percentage yield. Choose any simple precipitation reaction as an example, and provide a balanced chemical equation. Using your understanding of percentage yield and the techniques for collecting a precipitate in a chemical reaction, describe at least three reasons why the student will never be able to achieve a percentage yield near 100 percent.

27. Imagine that you are a chemistry teacher. You must prepare an investigation to show a single displacement reaction of solid aluminum foil with a blue aqueous copper(II) sulfate solution, $CuSO_4 \cdot 5H_2O(aq)$. The purpose of the investigation is to show the production of solid copper as the blue copper(II) sulfate is totally consumed. You have already told students to dissolve 1 g of copper(II) sulfate pentahydrate to make their solution. What minimum mass of aluminum foil must students use in order to have at least 0.25 g of aluminum remaining after the reaction has gone to completion?

Communication

28. Imagine that you are a chemistry teacher and your students are having trouble learning how to solve problems involving limiting reactants and percentage yield. You decide to develop a step-by-step guide for your students to follow. You will start with the chemical equation:

__Al(s) + __CuO(aq) → __Al₂O₃(aq) + __Cu(s)

You will choose masses of aluminum and copper(II) oxide for the reactants and a mass of copper metal that is recovered. Write out a bulleted list of steps that will guide students through the process of balancing the equation, finding the limiting reactant, finding the expected yield of product, and finding the percentage yield. Beside each bullet, write that step of the example.

29. List three conversion factors that are used in molar conversions, and provide an example of each. Use a graphic organizer to summarize how these conversion factors are used when completing a stoichiometry problem.

30. You are given a dry compound that might be a hydrated compound. Write a procedure outlining the steps you would take to determine whether or not the compound is a hydrate. As part of your step-by-step procedure, include how to determine the number of water molecules in the chemical formula for a hydrate.

31. **BIG IDEAS** The efficiency of chemical reactions can be determined and optimized by applying an understanding of quantitative relationships in such reactions. Create a graphic organizer that shows how the concept of the mole, percent composition, empirical and molecular formulas, and stoichiometry are used by chemical engineers to improve the efficiency of chemical reactions and reduce their negative impact on the environment and society.

32. Drive-through windows are criticized because they promote excess idling of car engines. The idling produces carbon dioxide, a greenhouse gas that is associated with climate change. Carbon dioxide is one product of the combustion of octane, $C_8H_{18}(g)$, in gasoline.
   a. Write the balanced equation for the combustion of octane.
   b. Calculate the amount in moles and the mass in grams of carbon dioxide that is produced by every 10.0 g of octane.

33. The fertilizer industry produces many different synthetic fertilizers, with different percentage compositions. Although these fertilizers benefit society by increasing the quality and quantity of food crops, they have a major impact on the environment. Research one of these fertilizers.
   a. Perform a risk/benefit analysis to determine whether or not its benefits to society outweigh its risk to the environment.
   b. Develop a presentation in the format of your choice that summarizes your findings.

Application

34. A high interest rate, expressed as a percentage, is important for the banking industry. It allows the banking industry to make a profit lending money. Explain why a high percentage yield from an industrial chemical process is equally important for the chemical industry.

35. A chemist analyzed an unknown compound and determined that it was composed of 2.2 percent hydrogen, 26.7 percent carbon, and 71.1 percent oxygen. What is the empirical formula for this compound?

36. When copper, is heated in the presence of oxygen gas, as shown below, an oxide of copper forms. If 8.24 g of copper reacts with 2.08 g of oxygen, what is the empirical formula for the oxide of copper? What is the name of this copper oxide?

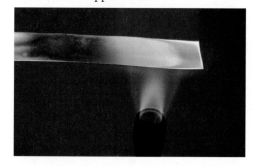

37. Acetone is a liquid solvent that works well as a nail polish remover. Acetone contains 62.0 percent carbon, 10.4 percent hydrogen, and 27.5 percent oxygen. The molecular molar mass of acetone is 58.1 g/mol. What is the molecular formula for acetone?

38. The pain reliever Aspirin® (acetylsalicylic acid, $C_9H_8O_4(s)$) is produced by reacting salicylic acid, $C_7H_6O_3(aq)$, with acetic anhydride, $C_4H_6O_3(aq)$. Acetic acid, $CH_3COOH(aq)$ is the other product of this reaction:

$C_7H_6O_3(aq) + C_4H_6O_3(aq) \rightarrow$
$$C_9H_8O_4(s) + CH_3COOH(aq)$$

a. If the process is 100 percent efficient, what mass of salicylic acid is required to produce 150 kg of Aspirin®?

b. What mass of salicylic acid is required if the actual percentage yield is 80 percent Aspirin®?

c. If the manufacturing process used 189 kg of salicylic acid and 129 kg of acetic anhydride, what is the expected theoretical yield of Aspirin®?

d. The actual yield of the process described in part c was 186 kg of Aspirin®. What is the percentage yield for this process?

39. Glauber's salt is a form of hydrated sodium sulfate, $Na_2SO_4 \cdot xH_2O(s)$, that is used medicinally as a mild laxative. Chemical analysis of this hydrate shows that it is 44.1 percent sodium sulfate. Determine the formula and name of this hydrate.

40. **BIG IDEAS** Silver reacts with sulfur-containing compounds in the air to form silver sulfide, $Ag_2S(s)$, or tarnish, as shown in the photograph below. One way to remove silver sulfide from silver metal is to react it with aluminum, as shown in the following equation:

$2Al(s) + 3Ag_2S(s) + 6H_2O(\ell) \rightarrow$
$$6Ag(s) + 2Al(OH)_3(s) + 3H_2S(aq)$$

What mass of silver is recovered for every gram of aluminum consumed?

41. Propane, $C_3H_8(g)$, can be used as an alternative fuel in vehicles, if the gasoline engine is exchanged for an engine that uses propane. Gasoline is a mixture of carbon compounds, including octane, $C_8H_{18}(\ell)$.

a. Write the complete combustion reactions for propane and octane.

b. Look at the fuel to oxygen ratio for both fuels, and infer why one of these fuels is more likely to produce dangerous carbon monoxide, $CO(g)$, in its exhaust.

42. Chemistry is used on the International Space Station to overcome many problems. For example, to improve environmental air quality, water from human respiration is removed from the air by running it through a layer of solid lithium oxide, $Li_2O(s)$:

$Li_2O(s) + H_2O(g) \rightarrow 2LiOH(s)$

a. The space station uses prepackaged 25.0 kg cartridges of lithium oxide. How much water does each cartridge remove? If the environmental control unit on the space station reports that 29.5 kg of water should be removed from the air, how many lithium oxide cartridges need to be used?

b. Lithium hydroxide, the same compound that is produced when water is removed from the air, can remove the excess carbon dioxide exhaled by the astronauts:

$LiOH(s) + CO_2(g) \rightarrow LiHCO_3(s)$

Each lithium hydroxide cartridge that is used to remove carbon dioxide from the air has a lithium hydroxide mass of 32.0 kg. If the process for removing the carbon dioxide has a percentage yield of 88 percent, what mass of carbon dioxide is removed by each canister of lithium hydroxide?

Select the letter of the best answer.

1. **K/U** The mole is a unit of
   a. mass
   b. volume
   c. weight
   d. quantity
   e. atomic mass

2. **T/I** How many phosphate ions, PO_4^{-3}, are in 3.40×10^{23} formula units of calcium phosphate, $Ca_3(PO_4)_2(s)$?
   a. 2
   b. 3.40×10^{23}
   c. 6.80×10^{23}
   d. 1.02×10^{24}
   e. 6.02×10^{23}

3. **T/I** Which value is the molar mass of zinc sulfate heptahydrate, $ZnSO_4 \cdot 7H_2O(s)$, closest to?
   a. 161 g/mol
   b. 288 g/mol
   c. 182 g/mol
   d. 240 g/mol
   e. 312 g/mol

4. **T/I** What is the mass of a molecule, called a buckyball, that contains 60 carbon atoms?
   a. 1.20×10^{-21} g
   b. 4.34×10^{-26} g
   c. 1.66×10^{-24} g
   d. 8.36×10^{-20} g
   e. 9.96×10^{-23} g

5. **K/U** Which of the following statements about the simplest formula and molecular formula of a compound are correct?
   I. The simplest formula may be the same as the molecular formula.
   II. The molecular formula is n(simplest formula), where $n = 1, 2, 3,....$
   III. The simplest formula can never be the same as the molecular formula.
   IV. The simplest formula can be called the empirical formula.
   a. II and IV
   b. I, II, and IV
   c. I, II, and III
   d. II and III
   e. I and IV

6. **K/U** Which formula is an empirical formula?
   a. $C_2H_4(g)$
   b. $H_2O_2(\ell)$
   c. $C_2H_2(g)$
   d. $Na_2Cr_2O_7(s)$
   e. $C_2O_4H_2(aq)$

7. **T/I** When determining the empirical formula for a compound, a student obtained the ratio $C_{3.2}H_6O$. What is the correct empirical formula for the compound?
   a. C_3H_6O
   b. $C_7H_{13}O_2$
   c. $C_{16}H_{30}O_5$
   d. $C_{16}H_{30}O_6$
   e. $C_{32}H_{60}O_{10}$

8. **T/I** A hydrocarbon is converted completely to water vapour and carbon dioxide gas. If the mole ratio of water to carbon dioxide is 1.33:1.00, what might the hydrocarbon be?
   a. CH_4
   b. C_2H_2
   c. C_2H_6
   d. C_3H_4
   e. C_3H_8

9. **T/I** When 1.84 g of strontium nitrate is mixed with excess copper(II) sulfate in an aqueous solution, a precipitate of strontium sulfate forms.
   $Sr(NO_3)_2(aq) + CuSO_4(aq) \rightarrow SrSO_4(s) + Cu(NO_3)_2(aq)$
   If 1.50 g of strontium sulfate is recovered by filtration, what is the percentage yield?
   a. 70.5%
   b. 93.9%
   c. 92.0%
   d. 107%
   e. 83.8%

10. **K/U** How many atoms are in 1 mol of neon gas?
    a. 3.00×10^{23}
    b. 5.00×10^{23}
    c. 6.02×10^{23}
    d. 12.00×10^{23}
    e. 20.18×10^{23}

Use sentences, diagrams, and calculations, as appropriate, to answer the questions below.

11. **T/I** Ammonium nitrate, $NH_4NO_3(s)$, produces dinitrogen monoxide, $N_2O(g)$ and liquid water when it decomposes. Determine the mass of water produced from the decomposition of 25.0 g of solid ammonium nitrate.

12. **C** A classmate does not understand what the law of definite proportions means. Develop a short lesson plan that you could use to explain this law.

13. **T/I** A 2.800 g sample of a compound containing only carbon and hydrogen is subjected to combustion in an excess of oxygen. The water vapour that forms during the complete combustion reaction has a mass of 3.6 g. In a separate experiment, the molar mass of the compound is found to be 84 g/mol. Determine the empirical formula and molecular formula of the compound.

14. **T/I** The actual percentage yield of a chemical reaction that produces a precipitate is much lower than predicted. What are three possible explanations for this?

15. **T/I** When reacted with sulfur, copper can form either copper(I) sulfide, $Cu_2S(s)$, or copper(II) sulfide, $CuS(s)$. A student places a measured amount of copper with an excess of sulfur in a crucible in a fume hood. She heats the mixture for several minutes, until the copper has reacted with the sulfur and any excess sulfur has combined with oxygen gas to form sulfur dioxide gas, $SO_2(g)$, which escapes from the crucible. Use the student's data to determine the formula for the copper sulfide produced.

Mass Data from Reaction of Copper with Excess Sulfur

Object and Contents	Mass (g)
Empty crucible + lid	21.55 g
Crucible + copper + lid before reaction	22.83 g
Crucible + lid + copper sulfide product	23.17 g

16. **A** Phenolphthalein is commonly used in quantitative analyses. The empirical formula for the form of phenolphthalein is $C_5H_3O(aq)$. The molar mass of this form of phenolphthalein is 316.31g/mol. Determine the molecular formula for this form of phenolphthalein.

17. **T/I** The following equation represents the synthesis of vanadium(V) oxide, $V_2O_5(s)$, directly from its elements:
$$4V(s) + 5O_2(g) \rightarrow 2V_2O_5(s)$$
If 1.06×10^{23} atoms of vanadium are reacted with 9.43 g of oxygen gas, what mass of vanadium oxide is produced?

18. **C** Develop a diagram that shows why a balanced chemical equation is needed to solve a stoichiometric problem.

19. **K/U** Acetylene, $C_2H_2(g)$, consists of equal numbers of carbon atoms and hydrogen atoms, as shown in the figure. Explain why acetylene is not 50 percent carbon by mass.

20. **A** During a crime scene investigation, chemists discover the presence of a gas that is not normally found at that location. Analysis shows that the gas is a compound that contains nitrogen and oxygen. Explain what additional information could be used to identify the gas.

21. **A** A chemist decides to determine the empirical formula of an artificial sweetener. She determines that the sweetener contains 60.85 percent carbon, 5.64 percent hydrogen, 8.69 percent nitrogen, and 24.82 percent oxygen. What is the empirical formula of the sweetener?

22. **T/I** Using an example, explain why you do not need to consider excess reactants when carrying out stoichiometric calculations.

23. **T/I** Yeast can act on a sugar, such as glucose, $C_6H_{12}O_6(s)$ to produce ethyl alcohol, $C_2H_5OH(\ell)$, and carbon dioxide, according to the following reaction:
$$C_6H_{12}O_6(s) \rightarrow 2C_2H_5OH(\ell) + 2CO_2(g)$$
If 223 g of ethyl alcohol are recovered after 1.63 kg of glucose react, what is the percentage yield of the reaction?

24. **C** Write, in point form, an experimental procedure to determine the simplest formula of zinc chloride, $ZnCl_x(s)$. In addition to standard glassware and equipment, assume that solid zinc, and concentrated hydrochloric acid solution are available. Be sure to indicate any measurements that you would need to record.

25. **A** Researchers are investigating whether it is possible to extract water from vehicle exhaust. The complete combustion of each litre of gasoline (assume pure octane, $C_8H_{18}(\ell)$, density = 0.703 g/mL) yields about one litre of liquid water (density = 1.00 g/mL). Using calculations, determine whether this idea is feasible.

Self-Check

If you missed question …	1	2	3	4	5	6	7	8	9	10	11	12	13	14	15	16	17	18	19	20	21	22	23	24	25
Review section(s)…	5.1	5.1	6.3	5.2	6.2	6.2	6.2	6.2	6.2	5.1	7.1	6.1	6.3	7.3	6.2	5.2	7.2	7.1	6.1	6.2	6.3	7.2	7.3	6.2	7.1

UNIT 4 Solutions and Solubility

BIG IDEAS

- Properties of solutions can be described qualitatively and quantitatively, and can be predicted.

- Living things depend for their survival on the unique physical and chemical properties of water.

- People have a responsibility to protect the integrity of Earth's water resources.

Overall Expectations

By the end of this course, you will

- **analyze** the origins and effects of water pollution, and a variety of economic, social, and environmental issues related to drinking water

- **investigate** qualitative and quantitative properties of solutions, and **solve** related problems

- **demonstrate** an understanding of qualitative and quantitative properties of solutions

Unit 4 Contents

Chapter 8
Solutions and Their Properties

What factors affect solubility, and how can the properties of solutions be described?

Chapter 9
Reactions in Aqueous Solutions

How are qualitative analysis and quantitative analysis used to describe the composition of a solution, and how do humans use and affect the world's water supply?

Chapter 10
Acids and Bases

What are acids and bases, and how is titration used to find the concentrations of solutions of acids and bases?

Fishing and boating are popular activities that allow Canadians to experience nature, while enjoying the company of friends and family. Aboriginal peoples and the first settlers used Ontario's numerous lakes, rivers, and streams as a source of food and transportation. Today, these waterways support commercial and recreational fishing, agriculture, and industry. In addition, the Great Lakes provide over eight million Canadians with drinking water.

Water is essential to life, mainly because many substances can dissolve in it. Fish depend on the oxygen dissolved in water to breathe. All living things rely on water to transport dissolved ions inside their bodies. However, many undesirable substances can also dissolve in water. These substances can pollute lakes, rivers, and ground water. Some pollution comes from above, in the form of acids and other substances dissolved in rain or snow. Other pollution comes from industrial effluent, discharge from towns and cities, and run-off from farms. This unit is about solutions—particularly those involving water—and the chemical reactions between dissolved substances.

As you study this unit, look ahead to the **Unit 4 Project** on pages 484 to 485, which gives you an opportunity to demonstrate and apply your new knowledge and skills. Keep a planning folder so you can complete the project in stages as you progress through the unit.

Safety in the Chemistry Laboratory and Classroom

- Follow all the safety rules that are provided by your teacher and by the labels on chemicals and equipment.
- All materials that are used in the science classroom must be safely handled and properly disposed of.

- Safety equipment, such as safety eyewear, a laboratory coat or apron, and protective gloves, should be worn when indicated.
- Immediately wash off any chemicals that come in contact with your skin.
- Inform your teacher if an accident occurs.

1. The following safety symbols are included in the instructions for an investigation that involves chemicals. What does each symbol mean?

a. ⬡ **b.** ⬡ **c.** ⬡

2. When learning about chemical safety, you may encounter the acronyms MSDS and WHMIS.
  a. What do the acronyms stand for?
  b. What information can you find in a chemical's MSDS?
  c. What is WHMIS used for?
  d. Where are the MSDSs located in your classroom?

3. Suppose that you are working in a classroom laboratory and your laboratory partner drops a beaker containing a corrosive chemical. The beaker breaks, and some of the chemical splashes on your lab partner's arm. Describe what you should do in response to this accident. Include any instructions that you would give to your laboratory partner.

Classification of Matter

- Matter is anything that has mass and volume.
- All matter is made up of pure substances and mixtures.
- A pure substance is a substance that is made up of only one type of particle. A pure substance may be either an element or a compound. An element is made up of only one kind of atom. A compound is made up of atoms of two or more kinds of elements, joined together by chemical bonds.

- A mixture is made up of two or more substances that are not chemically combined.
- A mixture may be a heterogeneous mixture or a homogeneous mixture. A heterogeneous mixture has a varied composition. A homogeneous mixture has a uniform composition throughout.

4. Classify each of the following as an element, a compound, or a mixture.
  a. salt water
  b. aluminum foil
  c. water
  d. stainless steel
  e. copper wire
  f. table sugar

5. If a sample of pure water is vaporized, what is the smallest particle in the vapour?
  a. a hydrogen atom, $H(g)$
  b. a hydrogen ion, $H^+(g)$
  c. an electron
  d. a molecule, $H_2O(g)$
  e. a compound

6. Explain how compounds and mixtures differ, in terms of how they are broken down into simpler substances.

7. Give three examples of heterogeneous mixtures and three examples of homogeneous mixtures.

Volume Measurement and Unit Conversions

- The volume of a regular solid can be calculated using a geometric formula. The volume of an irregular solid can be measured using water displacement.
- The volume of a liquid can be measured using a graduated cylinder or a graduated pipette.
- The volume of a gas is equal to the size of the container that holds the gas.

- The SI unit of volume is the cubic metre, m^3. However, the units that are most commonly used in a laboratory are the millilitre, mL, and the litre, L. The decilitre, dL, is sometimes used in non-laboratory situations.
- When using the SI (metric) system, a unit of volume can be easily converted into another unit of volume by multiplying by an appropriate conversion factor.

8. Part of a graduated cylinder, calibrated in millilitres, is shown. A student is making an error when reading the volume of the liquid in the graduated cylinder.

a. Is the student likely to record the volume of the liquid as greater or smaller than the actual volume? Explain.

b. What should the student do to correctly read the volume of the liquid?

c. What is the volume of the liquid?

9. Make each of the following conversions.
  a. 26.8 mL to litres   **c.** 125 mL to decilitres
  b. 0.355 L to millilitres   **d.** 1.00 dL to millilitres

10. Describe two methods you could use to measure the volume of a marble.

Classification of Chemical Reactions

- The four main types of chemical reactions are synthesis reactions, decomposition reactions, single displacement reactions, and double displacement reactions.
- A synthesis reaction is a reaction in which two or more elements or compounds combine chemically to form a new compound.
- A decomposition reaction is a reaction in which a single compound breaks down into two or more simpler substances.

- A single displacement reaction is a reaction in which one element or ion takes the place of another element or ion in a compound.
- A double displacement reaction is a reaction in which elements or ions in two compounds change places.

11. Which equation represents a double displacement reaction?
  a. $CuO(s) + H_2(g) \rightarrow Cu(s) + H_2O(g)$
  b. $Zn(s) + H_2SO_4(aq) \rightarrow ZnSO_4(aq) + H_2(g)$
  c. $2C_2H_6(g) + 7O_2(g) \rightarrow 4CO_2(g) + 6H_2O(g)$
  d. $CuSO_4(aq) + Fe(s) \rightarrow FeSO_4(aq) + Cu(s)$
  e. $AgNO_3(aq) + NaCl(aq) \rightarrow NaNO_3(aq) + AgCl(s)$

12. An electric current was passed through a container of water, causing a chemical reaction. The products of the reaction were caught in inverted test tubes, as shown in the photograph on the right. What type of reaction occurred? Explain your answer.

13. Describe the four types of chemical reactions, using analogies that involve dance partners.

- Chemists express the amount of a substance in moles (mol). One mole of a substance contains $6.022\,141\,99 \times 10^{23}$ particles of the substance. This number is called the *Avogadro constant*.
- The molar mass of a compound is the sum of the molar masses of the atoms in the formula of the compound.
- The amount of a substance (in moles) is its mass (in grams) divided by its molar mass (in grams/mole).

- The coefficients in a balanced chemical equation give the ratio between the reactants and products, when expressed in moles.
- Stoichiometry calculations use the relative quantities in moles of the reactants and products in a reaction, as given by the coefficients in the balanced chemical equation.

14. What is the molar mass of caffeine, $C_8H_{10}N_4O_2(s)$?
Molar masses: C = 12, H = 1, N = 14, O = 16
  a. 43 g/mol
  b. 147 g/mol
  c. 178 g/mol
  d. 194 g/mol
  e. 248 g/mol

15. What mass of carbon dioxide is present in 0.50 mol of the gas?
  a. 0.50 g
  b. 14 g
  c. 20 g
  d. 22 g
  e. 44 g

16. In a chemical reaction, 1.0 mol of aluminum reacted with 1.5 mol of chlorine gas to form 1.0 mol of aluminum chloride. Which chemical equation correctly shows these amounts?
  a. $Al_2(s) + Cl_3(g) \rightarrow Al_2Cl_3(s)$
  b. $Al_2(s) + 3Cl(g) \rightarrow Al_2Cl_3(s)$
  c. $2Al(s) + 3Cl_2(g) \rightarrow 2AlCl_3(s)$
  d. $Al(s) + 3Cl(g) \rightarrow AlCl_3(s)$
  e. $2Al(s) + 3Cl_2(g) \rightarrow Al_2Cl_3(s)$

17. Sodium chloride, $NaCl(s)$, has a molar mass of 58.5 g/mol. How would you find the amount in moles of sodium chloride in a 100-g sample?
  a. divide the molar mass by 100 g
  b. divide 100 g by the molar mass
  c. divide 100 g by Avogadro's constant
  d. multiply Avogadro's constant by the molar mass
  e. multiply 100 g by the molar mass

18. When aqueous solutions of lead(II) nitrate and potassium iodide are mixed, a chemical reaction takes place, as shown below.

The chemical equation for this reaction is
$Pb(NO_3)_2(aq) + 2KI(aq) \rightarrow PbI_2(s) + 2KNO_3(aq)$
  a. What is the name of the yellow substance? What is its formula?
  b. What type of chemical reaction occurred?
  c. If 0.2 mol of lead(II) nitrate is mixed with 0.3 mol of potassium iodide, which reactant is the limiting reactant? What amount in moles of the solid precipitate will form?

19. Ethene burns in oxygen, as shown by the following balanced chemical equation:
$C_2H_4(g) + 3O_2(g) \rightarrow 2CO_2(g) + 2H_2O(g)$
If 3.0 mol of carbon dioxide gas is formed in the reaction, what amount of oxygen gas was required?
  a. 1.5 mol
  b. 2.0 mol
  c. 3.0 mol
  d. 4.5 mol
  e. 6.0 mol

- Acids are compounds that increase the concentration of hydrogen ions, $H^+(aq)$, when dissolved in water.
- Bases are compounds that increase the concentration of hydroxide ions, $OH^-(aq)$, when dissolved in water.
- Acids and bases share some properties, such as the ability to conduct electric current. Other properties of acids and bases, such as taste and texture, are different.

- The pH scale ranges from 0 to 14. Acidic solutions have a pH less than 7, neutral solutions have a pH equal to 7, and basic solutions have a pH greater than 7.
- A pH indicator shows the acidity or basicity of a compound by changing colour within a small range of pH values.
- A neutralization reaction occurs when an acid and a base react to form a salt and water. A neutralization reaction is a form of double displacement reaction.

20. Which property is characteristic of acidic solutions?
   a. They feel slippery.
   b. They taste bitter.
   c. They react with carbonates to form carbon dioxide gas.
   d. They turn red litmus paper blue.
   e. They have a pH greater than 7.

21. Complete each chemical equation to illustrate ionization or dissociation in solution. Identify each compound as an acid or a base.
   a. $HCl(aq) \rightarrow$
   b. $NaOH(aq) \rightarrow$
   c. $Ca(OH)_2(aq) \rightarrow$
   d. $HClO_3(aq) \rightarrow$

22. How does the pH of an acidic solution change as a base is added to the solution?

23. Which of these household liquids would turn red litmus paper blue? Explain your choice(s).

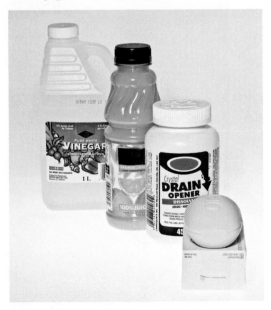

24. Which equation shows an acid-base neutralization reaction?
   a. $H_2O(\ell) + CO_2(g) \rightarrow H_2CO_3(aq)$
   b. $Mg(s) + 2HCl(aq) \rightarrow MgCl_2(aq) + H_2(g)$
   c. $HNO_3(aq) + KOH(aq) \rightarrow KNO_3(aq) + H_2O(\ell)$
   d. $NH_3(g) + H_2O(\ell) \rightarrow NH_4^+(aq) + OH^-(aq)$
   e. $2NaOH(aq) + MgCl_2(aq) \rightarrow$
   $$2NaCl(aq) + Mg(OH)_2(s)$$

25. Which tool would be the least accurate for finding the pH of a solution?
   a. pH paper
   b. litmus paper
   c. pH meter
   d. universal indicator
   e. pH probe

26. Classify each property listed below as being a property of an acid, a base, or both. You could use a Venn diagram to classify the properties.
   a. increases hydrogen ions in solution
   b. increases hydroxide ions in solution
   c. slippery
   d. conducts electric current
   e. tastes bitter
   f. tastes sour
   g. can be corrosive
   h. can be found in household materials
   i. has a pH less than 7

27. Equal amounts of 1.0 mol/L hydrochloric acid, $HCl(aq)$, and 1.0 mol/L sodium hydroxide, $NaOH(aq)$, are mixed. What is the pH of the products? Explain your answer.

CHAPTER
8

Solutions and Their Properties

Specific Expectations

In this chapter, you will learn how to...

- E2.1 **use** appropriate terminology related to aqueous solutions and solubility (8.1)

- E2.2 **solve** problems related to the concentration of solutions by **performing** calculations (8.3)

- E2.3 **prepare** solutions of a given concentration by **dissolving** a solid solute in a solvent or by **diluting** a concentrated solution (8.4)

- E3.1 **describe** the properties of water and **explain** why these properties make water such a good solvent (8.2)

- E3.2 **explain** the process of formation for solutions that are produced by dissolving ionic and molecular compounds in water, and for solutions that are produced by dissolving non-polar solutes in non-polar solvents (8.2)

- E3.3 **explain** the effects of changes in temperature and pressure on the solubility of solids, liquids, and gases (8.2)

The seawater that makes up the world's oceans is a solution composed mostly of water and sodium chloride. Sodium chloride, which is often called *table salt*, is a very important compound. People use it in many different ways, including for de-icing roads, softening water, and seasoning food.

Much of the salt used in Canada is obtained from salt mines. In fact, the salt mine in Goderich, Ontario is the largest in the world. The Goderich mine, shown in the photograph, is more than 500 m underground and extends about 5 km under Lake Huron. Where did that salt come from and how did it get there?

Throughout Earth's history, changing climates have affected the depths and locations of oceans and inland seas. The warm climate during the Silurian period (from about 445 to 415 million years ago) melted glaciers and much of the ice at the poles. The sea level rose, and large areas of land were covered by seawater. At the end of the Silurian period, the sea level dropped, leaving large basins of seawater in many low-lying areas of land. The seawater in these basins evaporated and large deposits of salt crystals built up. Today, the Goderich mine extracts salt from one of these deposits.

Layered Liquids

Some liquids dissolve in one another, but others do not. How can you determine whether two liquids dissolve in each other?

Safety Precautions

- Wear safety eyewear, gloves, and a lab apron throughout this activity.
- Lugol's solution contains iodine, which can stain clothing and the skin.
- Follow your teacher's chemical disposal instructions.

Materials

- water
- mineral oil
- dropper bottle that contains food colouring
- dropper bottle that contains Lugol's solution
- test tube
- stopper

Procedure

1. Pour water into the test tube until it is about one-quarter full. Then add an equal volume of mineral oil. Record your observations.

2. Add two or three drops of food colouring. Record your observations.

3. Put the stopper in the test tube. Gently shake the test tube. Record your observations.

4. Remove the stopper, and add two or three drops of Lugol's solution to the test tube. Stopper the test tube, and gently shake it again. Record your observations.

Questions

1. Did the liquids used in step 1 dissolve in each other? Explain your answer.

2. In which liquid(s) does food colouring dissolve? Explain how you know.

3. Did Lugol's solution colour the oil, the water, or both?

4. Lugol's solution contains iodine molecules, $I_2(aq)$, as well as $I^-(aq)$ and $I_3^-(aq)$ ions. Iodine molecules have an intense colour, but $I^-(aq)$ ions are colourless. In which liquid(s) do iodine molecules dissolve? How do you know?

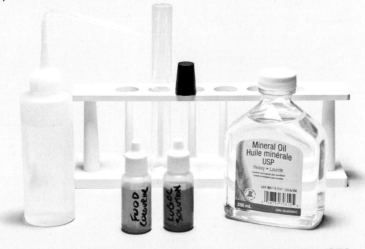

Classifying Solutions

Key Terms

solution

solvent

solute

aqueous solution

solubility

saturated solution

unsaturated solution

supersaturated solution

You encounter a great variety of solutions in your daily life. For example, you likely know that the tea in **Figure 8.1** is a solution. However, are you aware that the stainless steel in the spoon is also a solution? A **solution** is a homogeneous mixture of two or more substances. *Homogeneous* means that the mixture has a uniform composition throughout. Once a solution has been fully mixed, the substances in it remain evenly distributed throughout the solution. Solutions are homogeneous even at the microscopic scale. All samples of a given solution, no matter how small, contain the same substances in the same proportions.

For example, each drop in a cup of tea contains the same proportions of water and dissolved substances that give the coffee its colour and flavour. In another pot of tea, however, the proportions will likely be somewhat different. If the tea is stronger, the proportions of the substances from the tea leaves will be greater. But, again, each drop in a cup of the stronger tea will have the same proportions of water and other substances as any other drop of tea in that cup. Similarly, the proportions of metals in solution in a stainless steel spoon may vary from one manufacturer to another. However, these proportions are consistent throughout any particular stainless steel spoon.

Figure 8.1 This cup contains two solutions. The coffee is an aqueous solution, and the stainless steel spoon is a solution of metals.

solution a homogeneous mixture of two or more substances

solvent the component of a solution that is present in the greatest amount

solute a substance that is dissolved in a solvent

aqueous solution a solution that contains water

The substance that is present in the greatest amount in a solution is called the **solvent**. Any other substance in the solution is a **solute**. In a cup of coffee, water is the solvent and all the different substances from the coffee beans are solutes. Solutions that contain water, such as coffee and tea, are called **aqueous solutions**. This term comes from the Latin word *aqua*, meaning water. Aqueous solutions may be coloured or colourless, but they are always clear. Not all solutions have water as a solvent. In a stainless steel spoon, for example, the solvent and the solutes are metals.

Types of Solutions

A solution can be formed from a solvent and a solute in any state: solid, liquid, or gas. Thus, nine different types of solutions are possible. These different types of solutions are shown in **Table 8.1**.

Table 8.1 Examples of Solutions in Different States of Matter

Solid Dissolved in Solid	Solid Dissolved in Liquid	Solid Dissolved in Gas
The structural steel that is used to construct buildings usually contains from 0.2 to 1.5% carbon dissolved in iron. 	On average, 1 L of seawater contains about 19 g of chloride ions, 11 g of sodium ions, and 5 g of other solutes, including magnesium, sulfate, calcium, and potassium ions. 	Molecules of naphthalene or paradichlorobenzene separate from the surface of a mothball to form a solution with air. The vapour from a mothball is toxic to moths and many other organisms. 
Liquid Dissolved in Solid	**Liquid Dissolved in Liquid**	**Liquid Dissolved in Gas**
Liquid mineral spirits or liquid toluene are dissolved in solid wax to make the wax easier to apply. 	Antifreeze liquids are mixed with water in car radiators to form solutions with low freezing points. 	Humidity results from water dissolving in the air. A hygrometer is a tool that measures humidity. 
Gas Dissolved in Solid	**Gas Dissolved in Liquid**	**Gas Dissolved in Gas**
Most ice contains a small amount of dissolved air. 	When a bottle of a carbonated beverage is opened, some of the dissolved carbon dioxide bubbles out of the solution. 	Natural gas is a solution of methane gas with ethane, nitrogen, carbon dioxide, and other gases dissolved in it. 

Solubility and Saturation

solubility the maximum amount of solute that will dissolve in a given quantity of solvent at a specific temperature

Substances vary in how readily they will dissolve in a solvent. The **solubility** of a substance is defined in terms of the maximum amount of solute that will dissolve in a given quantity of solvent at a specific temperature. Solubility in water is often stated in terms of the mass, in grams, of solute that will dissolve in a decilitre (100 mL) of water at 20°C. For example, the solubility of table salt, NaCl(s), in water is 35.9 g/100 mL at 20°C, whereas the solubility of oxygen, O_2(g), is only 0.0009 g/100 mL at 20°C.

Qualitatively, solubility can be described using the terms described in the bulleted points below. **Figure 8.2** applies these terms to the solubilities of three ionic compounds:

- A solute is described as *soluble* if more than 1 g will dissolve in 100 mL of solvent. Thus, table salt is soluble in water.

- A solute is described as *sparingly* (or *slightly*) *soluble* if it has a solubility between 0.1 g and 1 g per 100 mL of solvent.

- A solute is described as *insoluble* if less than 0.1 g of a solute will dissolve in 100 mL of solvent. In chemistry, the term "insoluble" should not be taken literally. It does not necessarily mean that no solute at all will dissolve. For example, oxygen is described as being insoluble in water, but the small amount that does dissolve is tremendously important to life in oceans, lakes, and rivers.

Figure 8.2 At 20°C, sodium chloride is soluble in water (35.9 g/100 mL), calcium hydroxide is sparingly soluble in water (0.183 g/100 mL), and calcium carbonate is insoluble (about 0.005 g/100 mL).

Saturated, Unsaturated, and Supersaturated Solutions

saturated solution a solution that cannot dissolve more solute

unsaturated solution a solution that could dissolve more solute

supersaturated solution a solution that contains more dissolved solute than a saturated solution at the same temperature

A **saturated solution** cannot dissolve any more solute. To determine whether a solution is saturated, you can compare the amount of dissolved solute per unit volume of the solution with the solubility of the solute. Therefore, the presence of solute that will not dissolve in a solution indicates that the solution is saturated. For example, dissolving 36 g of table salt in 100 mL of water at 25°C forms a saturated solution. If 40 g of table salt is added to 100 mL of water at 25°C, then 4 g of the salt will remain undissolved. In contrast, an **unsaturated solution** of any solute will dissolve more of the same solute if it is added to the solution. A **supersaturated solution** contains more dissolved solute than a saturated solution at the same temperature. A change in temperature can cause a saturated solution to become supersaturated. For example, many saturated solutions of a solid dissolved in water become supersaturated when cooled. Conversely, most saturated solutions of a gas in a liquid become supersaturated when heated.

Supersaturated solutions are unstable. If a crystal of solute is added to a supersaturated liquid solution of the solute, the excess solute will precipitate, leaving the solution saturated. Similarly, a disturbance in a supersaturated solution of a vapour in air can cause the vapour to condense into droplets.

1. Some sunscreens contain finely powdered zinc oxide, suspended in a lotion or ointment. Explain why these sunscreens are not solutions of zinc oxide.

2. Dry air contains about 78 percent nitrogen, 21 percent oxygen, and 0.9 percent argon, plus much smaller percentages of several other gases. Which gas is the solvent in air? Explain why.

3. Give three examples of aqueous solutions: one with a solid solute, one with a liquid solute, and one with a gas solute.

4. Water that condenses from air in the gas tank of a car tends to collect at the bottom of the tank or at the lowest point in the fuel line. What does this information indicate about the solubility of water in gasoline?

5. Can an insoluble compound form a saturated solution? Explain your reasoning.

6. When you open a can of soft drink, is the soft drink a supersaturated solution? Explain your reasoning.

Activity 8.1 Unsaturated, Saturated, and Supersaturated Solutions

In this activity, you will make qualitative observations of the properties of unsaturated, saturated, and supersaturated solutions of sodium thiosulfate pentahydrate, $Na_2S_2O_3 \cdot 5H_2O(s)$.

Safety Precautions

- Wear safety eyewear throughout this activity.
- When you are heating a test tube, keep the mouth of the test tube pointed away from you and other students.
- Tie back loose hair and clothing.
- Wash your hands after completing this activity.
- Use EXTREME CAUTION when you are near an open flame.

Materials

- 2 mL of distilled water
- 15 g of sodium thiosulfate pentahydrate, $Na_2S_2O_3 \cdot 5H_2O(s)$
- 10 mL graduated cylinder
- test tube
- scoopula
- test-tube rack
- stirring rod
- wooden test-tube holder
- Bunsen burner secured to a retort stand
- beaker of cold water
- watch or timer

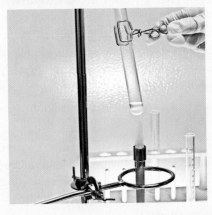

Procedure

1. Use the 10 mL graduated cylinder to pour 2 mL of distilled water into the test tube. Add about a quarter of a scoopula of sodium thiosulfate pentahydrate crystals. Place the test tube in a test-tube rack. Stir to dissolve the crystals.

2. Gradually add more sodium thiosulfate pentahydrate crystals to the solution, stirring as you add them, until no more dissolve. Feel the side of the test tube as you stir.

3. Add the rest of the sodium thiosulfate pentahydrate, except for a few crystals. Using the test-tube holder, gently heat the solution over the Bunsen burner. Do not boil the solution. Move the test tube from side to side in the flame so that the heat is not concentrated on the bottom of the test tube. Note how many of the added crystals dissolve.

4. Place the warm test tube into a beaker of cold water, and allow it to cool for about 3 min. Record the appearance of the cooled solution.

5. Add one or two crystals of sodium thiosulfate to the cooled solution. Feel the side of the test tube again. Record your observations.

6. Dispose of the solution and clean the equipment as directed by your teacher.

Questions

1. At what step in the Procedure did the solution become saturated? Explain your reasoning.

2. Explain the high solubility of sodium thiosulfate pentahydrate in water. (**Hint:** Think about the formula of this compound.)

3. Explain what happened in step 5.

4. Suppose that you are given a clear solution in a test tube along with a small sample of the solid solute. How could you determine whether the solution is unsaturated, saturated, or supersaturated?

Section Summary

- A solution is a completely homogeneous mixture. It can be formed from a solute and a solvent in any initial state: solid, liquid, or gas.

- The solvent in a solution is the substance present in the greatest amount. Solutes are anything dissolved in a solvent.

- The solubility of a substance is stated in terms of the maximum amount of solute that dissolves in a given quantity of solvent at a specific temperature.

- A solution can be classified as unsaturated, saturated, or supersaturated, depending on the amount of dissolved solute per unit volume of solution.

Review Questions

1. **K/U** A solution can have more than one solute.
   a. Give an example of a solution that has more than one solute.
   b. Explain why a solution can have many solutes, but only one solvent.

2. **K/U** Explain why stainless steel, seawater, and air are all considered solutions.

3. **A** Some processes for making coffee leave tiny particles of ground coffee beans suspended in the coffee when it is served. Can these particles be considered a solute if they are evenly distributed throughout the cup of coffee? Explain.

4. **K/U** Explain the difference between a solvent and a solute.

5. **K/U** Distinguish between the terms "insoluble" and "unsaturated." Give an example to illustrate each term.

6. **A** In a decorative lava lamp, a clear glass jar contains coloured water and a mixture of wax and carbon tetrachloride. Heat from a light bulb in the base of the lamp melts and expands the wax mixture, causing it to form blobs that change shape as they slowly float to the top of the lamp. The wax blobs then cool and sink back down again.
   a. Identify the two solutions in a lava lamp.
   b. Explain why a lava lamp would not work if the solutions dissolved in each other.

7. **K/U** A liquid can be a pure substance, a solution of two liquids, a solution of a gas in a liquid solvent, or a solution of a solid in a liquid solvent. Give one example of each type of liquid.

8. **T/I** Suppose that you are given two unlabelled beakers. One beaker contains water, and the other beaker contains a solution of ethanol and water. How could you safely determine which beaker contains just water?

9. **T/I** If a jar of honey sits for a long time, sugar may crystallize in the jar. Describe two possible causes for this crystallization.

10. **A** Describe two processes that could cause water vapour in the atmosphere to condense and form clouds.

11. **C** Draw a flowchart that shows a procedure for determining whether an aqueous solution is unsaturated, saturated, or supersaturated.

12. **A** About half of all the sodium chloride that is processed worldwide is used to melt snow and ice on roads. Explain how the solubility of sodium chloride is related to environmental concerns about this use.

13. **C** Two beakers in a laboratory are full of white crystals. One beaker contains sodium chloride, and the other beaker contains sucrose. The labels have come off the beakers, however, so you are not sure which is which. A chemistry handbook lists the solubility of sodium chloride in water as 36 g/100 mL at 25°C and the solubility of sucrose in water as 200 g/100 mL at 25°C. Write a procedure that describes how to safely identify the two chemicals.

14. **K/U** Explain why all supersaturated solutions are unstable.

15. **T/I** At 25°C, 35 g of solute A makes a saturated solution when dissolved in 100 mL of water, and 26 g of solute B makes a saturated solution when dissolved in 80 mL of water. Which solute is more soluble? Explain your reasoning.

Factors That Affect Solubility and Rate of Dissolving

Why do different substances have a wide range of solubilities? This range can be explained in terms of the forces that act between the particles of the substances in solutions. This section describes how these forces and other factors determine the solubilities of substances in different types of solutions.

Solubility and Forces Between Particles

The formation of most solutions depends on the relative strength of three categories of forces:

1. forces that attract particles of the solute to each other

2. forces that attract particles of the solute to particles of the solvent

3. forces that attract particles of the solvent to each other

When a solution forms, particles of the solute are attracted to particles of the solvent. All the intermolecular forces between the particles of the solute are broken, as well as some of the intermolecular forces between the particles of the solvent. However, when gases are mixed together, gaseous solutions form readily (unless the gases react chemically with each other). For such solutions, the forces of attraction between the particles of the gases are small enough that their influence on solubility is negligible.

Solubility in Water

Most of the solutions you will study in this unit are aqueous solutions. Water is a very good solvent. The polar nature of water molecules enables water to dissolve a wide range of solutes. **Figure 8.3 (A)** shows the polar oxygen-hydrogen bonds and the V-shape of a water molecule, which were described in Section 2.3. Oxygen atoms are more *electronegative* than hydrogen atoms. As a result, the shared electrons in the covalent bonds in a water molecule are displaced toward the oxygen atom. This displacement gives the oxygen atom a small net negative charge and leaves the hydrogen atoms with a net positive charge. Both of the O-H bonds in a water molecule are polar and act as dipoles.

There is a strong attraction between the oxygen atom of one water molecule and the hydrogen atoms of adjacent water molecules. This attraction is an example of **hydrogen bonding**. Hydrogen bonding is a type of dipole-dipole attraction between molecules. **Figure 8.3 (B)** illustrates the hydrogen bonding between water molecules. A hydrogen bond can form between a hydrogen atom on one molecule and a highly electronegative atom, such as oxygen, fluorine, or nitrogen, on another molecule.

Key Terms

hydrogen bonding
hydration
rate of dissolving

Suggested Investigation

Plan Your Own Investigation 8-A, Investigating Solutes and Solvents

hydrogen bonding the attraction between a hydrogen atom on one molecule and a very electronegative atom on another molecule

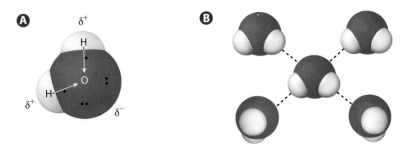

Figure 8.3 (A) The hydrogen atoms in a water molecule have a net positive charge, whereas the oxygen atom has a net negative charge. **(B)** The attraction between the positive and negative charges on the polarized bonds in water molecules creates hydrogen bonds between adjacent molecules. The hydrogen bonds are shown by the dashed lines.

Solubility of Ionic Compounds in Water

Most ionic compounds are soluble in water. The attraction between the ions in a soluble ionic compound (the solute) and the dipoles on the water molecules (the solvent) is great enough to pull ions away from the surface of the ionic compound. **Figure 8.4** shows what happens when sodium chloride, NaCl(s), dissolves in water. The positive end (a hydrogen atom) of a dipole on a water molecule attracts Cl^- ions, and the negative end (the oxygen atom) attracts Na^+ ions. Water molecules surround each ion, in a process called **hydration**. This process causes the ions in a soluble ionic compound to separate and disperse through the water when the compound dissolves.

The attraction between the ions in some ionic compounds is *greater* than the attraction between the ions and dipoles on water molecules. These strongly bound ionic compounds will not dissolve in water.

The chemical formula of an ionic compound in solution does not indicate that the solution contains solute ions bonded to each other. Instead, the formula represents the overall composition of the solute. For example, NaCl(aq) indicates that the dissolved solute consists of equal numbers of sodium and chlorine ions. These ions have separated and distributed evenly throughout the solution.

hydration the process in which water molecules surround the molecules or ions of a solute

Go to **scienceontario** to find out more

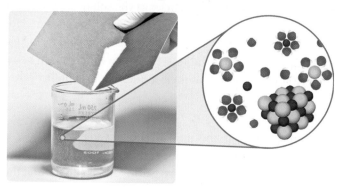

Figure 8.4 When a soluble ionic compound, such as sodium chloride, is added to water, the ions separate and several water molecules surround each ion. During hydration, water molecules (shown in red) surround sodium ions (shown in blue) and chloride ions (shown in yellow).

Solubility of Molecular Compounds in Water

Dipole-dipole attraction can also occur between molecules of any polar compound, as shown in **Figure 8.5**. This attraction is much weaker than the attraction between ions in an ionic compound. However, dipole-dipole attraction is often strong enough to hold polar molecules together as a solid crystal. For example, sucrose (table sugar) readily forms crystals because of dipole-dipole attraction among its polar molecules.

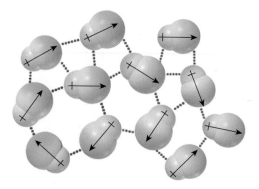

Figure 8.5 The arrows in the molecules show the molecule's dipoles. The part of the molecule near the tail of the arrow has a partial positive charge, whereas the part of the molecule near the head of the arrow has a partial negative charge. The dotted lines show the dipole-dipole interactions between the molecules.

Solubility of Polar Compounds in Water

Most polar compounds dissolve in water. The dipole-dipole attraction between the polar solute molecules is generally much weaker than the hydrogen bonds between the solute molecules and the water molecules. As shown in **Figure 8.6**, a molecule of sucrose, $C_{12}H_{22}O_{11}(s)$, has eight polar —OH groups that can form hydrogen bonds with water molecules. Consequently, sucrose is very soluble in water. The attraction between sucrose molecules and water molecules is stronger than the attraction of sucrose molecules to each other. Thus, water molecules readily separate sucrose molecules from the solid crystals, as shown in **Figure 8.7 (A)**. The sucrose molecules in solution are then surrounded by water molecules, which are attracted to the —OH groups.

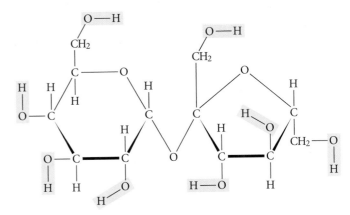

Figure 8.6 The molecular structure of sucrose includes eight polar —OH groups.

Non-polar molecules do not dissolve in water. These molecules are only weakly attracted to the water molecules. Thus, the attraction is far too weak to break the hydrogen bonds between the water molecules. As shown in **Figure 8.7 (B)**, water and a non-polar compound cannot dissolve each other.

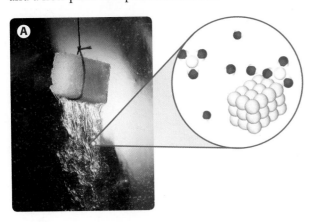

Figure 8.7 (A) Sucrose is a polar molecule so it dissolves readily in water. Sucrose molecules (shown in white) are hydrated by water molecules (shown in red). **(B)** The non-polar molecules in vegetable oil are insoluble in water.

Conductivity of Aqueous Solutions

Conductivity tests can indicate whether a compound dissolved in water is ionic or molecular. An ionic compound dissociates into separate ions when it dissolves. These ions are free to move, so they can carry charge to electrodes immersed in the solution. The molecules of most molecular compounds remain intact when they dissolve. The intact molecules are neutral, so they are not attracted to either positive or negative electrodes. As shown in **Figure 8.8**, solutions of ionic compounds can conduct an electric current, but solutions of most molecular compounds conduct little or no current.

Figure 8.8 A conductivity tester can distinguish between ionic and most molecular substances in aqueous solution.

Explain *Why does the needle of the conductivity tester not move when the sucrose solution is tested?*

Predicting Whether an Ionic Compound Is Soluble in Water

What factors determine which ionic compounds will be soluble in water? When water dissolves a solute, it must separate the particles of the solute. Ionic substances are held together by the attraction between ions with opposite charges. When the attraction between the cations (positively charged ions) and anions (negatively charged ions) in a substance is strong, they are more difficult to separate, making the substance less soluble in water.

The force of attraction between ions with opposite charges depends on two factors: the amount of charge on each ion, and the size of each ion. The data for magnesium compounds in **Table 8.2** demonstrate the effects of these two factors. The first two lines of **Table 8.2** show the effect of ion charge on solubility. An oxide ion has a greater charge than a fluoride ion and compounds containing oxide ions are less soluble than compounds containing fluoride ions. The last four lines of **Table 8.2** show the effect of ionic radius on solubility. The greater the ionic radius, the greater the solubility.

Table 8.2 The Effects of Ionic Charge and Radius on the Solubility of Magnesium Compounds

Anion Symbol and Relative Ionic Radius	Formula of Compound	Solubility at 25°C (g/100 mL)
O^{2-}	MgO	0.0006
F^-	MgF_2	0.008
Cl^-	$MgCl_2$	54
Br^-	$MgBr_2$	101
I^-	MgI_2	148

Effect of ion charge (brackets the first two rows)

Effect of ion size (arrow down the right side)

Ion Charge

The force of attraction between a cation and an anion is directly proportional to the amount of charge on each ion. The greater the charge on each ion, the less soluble the compound will be. For example, the oxide ion, O^{2-}, has twice the charge of the fluoride ion, F^-. Thus, the ionic bond in an oxide is stronger than the ionic bond in a fluoride with the same cation. This stronger bond makes most oxides insoluble.

Ion Size

As ion size increases, so does solubility. Compounds with larger ions are *usually* more soluble than compounds with smaller ions. The force of attraction between opposite charges decreases as the distance between the charges increases. The greater the radii of the ions in a compound are, the greater the separation of the charges. For example, silver nitrate, $AgNO_3(s)$, is very soluble in water, but silver chloride, $AgCl(s)$, is insoluble. Both anions have a single charge, but the NO_3^- ion is much larger than the Cl^- ion. Thus, the attraction between the Ag^+ and NO_3^- ions is weaker than the attraction between the Ag^+ and Cl^- ions. This weaker bond makes $AgNO_3(s)$ more soluble in water.

Solubility Guidelines for Ionic Compounds

Considering the combined effects of ion charge and ion size, you can predict that an ion with a small charge and a large radius should form an ionic compound that is soluble in water. The box below and **Table 8.3** summarize the solubility guidelines for ionic compounds.

Solubility Guidelines

1. The hydrogen ion, ammonium ion, and all Group 1 (alkali metal) ions form soluble compounds with nearly all anions.

2. Nitrate and acetate ions form soluble compounds with nearly all cations.

3. Chloride, bromide, and iodide ions form compounds that have low solubility with silver, lead(II), mercury(I), copper(I), and thallium cations only.

4. The fluoride ion forms compounds that have low solubility with magnesium, calcium, barium, and lead(II) cations only.

5. The sulfate ion forms compounds that have low solubility with calcium, strontium, barium, and lead(II) cations only.

6. The sulfide ion forms soluble compounds with only the ions listed in guideline 1 and the Group 2 cations.

7. The hydroxide ion forms soluble compounds with only the cations listed in guideline 1, as well as strontium, barium, and thallium cations.

8. Phosphate, carbonate, and sulfite ions form compounds that have low solubility with all cations except those listed in guideline 1.

Table 8.3 Solubility of Common Ionic Compounds in Water

	Anion +	Cation →	Solubility of Compound*
1.	Most	Alkali metal ions: Li^+, K^+, Rb^+, Cs^+, Fr^+	Soluble
	Most	hydrogen ion, H^+	Soluble
	Most	ammonium ion, NH_4^+	Soluble
2.	nitrate, NO_3^-	Most	Soluble
	ethanoate (acetate), CH_3COO^-	Ag^+	Low solubility
		Most others	Soluble
3.	chloride, Cl^- bromide, Br^- iodide, I^-	Ag^+, Pb^{2+}, Hg_2^{2+}, Cu^+, Tl^+	Low solubility
		All others	Soluble
4.	fluoride, F^-	Mg^{2+}, Ca^{2+}, Ba^{2+}, Pb^{2+}	Low solubility
		Most others	Soluble
5.	sulfate, SO_4^{2-}	Ca^{2+}, Sr^{2+}, Ba^{2+}, Pb^{2+}	Low solubility
		All others	Soluble
6.	sulfide, S^{2-}	Alkali ions and H^+, NH_4^+, Be^{2+}, Mg^{2+}, Ca^{2+}, Sr^{2+}, Ba^{2+}	Soluble
		All others	Low solubility
7.	hydroxide, OH^-	Alkali ions and H^+, NH_4^+, Sr^{2+}, Ba^{2+}, Tl^+	Soluble
		All others	Low solubility
8.	phosphate, PO_4^{3-} carbonate, CO_3^{2-} sulfite, SO_3^{2-}	Alkali ions and H^+, NH_4^+	Soluble
		All others	Low solubility

*Compounds listed as soluble have solubilities of at least 1 g/100 mL of water at 25°C and 100 kPa.

Predicting Whether a Molecular Compound Is Soluble in Water

Size also affects the solubility of a molecular compound. Methanol, $CH_3OH(\ell)$, and ethanol, $CH_3CH_2OH(\ell)$, are completely soluble in water because hydrogen bonds form between water and the $-OH$ group in a methanol or ethanol molecule. The rest of a methanol or ethanol molecule is non-polar, but the effect of the $-OH$ group predominates when the non-polar part of the molecule is relatively small. **Table 8.4** shows how the size of the non-polar portion of a molecule affects solubility.

Table 8.4 Solubility and Molecule Size

Increasing size of non-polar portion of molecule →

Name of Compound	Methanol	Ethanol	Propanol	Butanol	Pentanol
Chemical Formula	CH_3OH	CH_3CH_2OH	$CH_3CH_2CH_2OH$	$CH_3(CH_2)_3OH$	$CH_3(CH_2)_4OH$
Solubility at 25°C and 100 kPa	Soluble	Soluble	Soluble	9 g/100 mL	3 g/100 mL

"Like Dissolves Like"

Go to **scienceontario** to find out more

As described above, the force of attraction between a non-polar molecule and a water molecule is much weaker than the hydrogen bonding between water molecules. As a result, most non-polar molecules, such as grease and wax, are insoluble in water. However, non-polar solvents can dissolve non-polar molecules because the forces of attraction between the solute and solvent molecules are similar.

Thus, polar solvents tend to dissolve ionic and polar molecules, and non-polar solvents tend to dissolve non-polar molecules. These findings are often summarized as "like dissolves like," which means that a solute will probably dissolve in a solvent if the solute and the solvent have the same polar or non-polar nature. **Table 8.5** lists the solute-solvent combinations and their likelihood of forming a solution.

Table 8.5 Predictions Using the Generalization "Like Dissolves Like"

		SOLUTE	
		Polar or Ionic	**Non-polar**
SOLVENT	**Polar**	Usually soluble	Usually insoluble
	Non-polar	Usually insoluble	Usually soluble

"Like Dissolves Like" in the Solubility of Gases

The generalization "like dissolves like" explains why some gases are soluble in water but others are not. **Figure 8.9** shows molecular models for several gases. Hydrogen chloride, $HCl(g)$, and ammonia, $NH_3(g)$, are soluble in water because they are polar molecules. In contrast, carbon dioxide, $CO_2(g)$, is relatively insoluble because its molecules are non-polar.

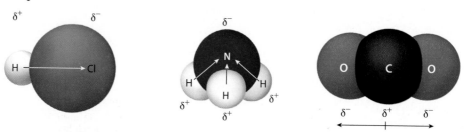

Figure 8.9 The bonds and arrangement of atoms in hydrogen chloride and ammonia cause the molecules to be polar. However, carbon dioxide is non-polar because both ends of the molecule have a partial negative charge.

Molecules That Have Both Polar and Non-Polar Components

As described earlier, sucrose and alcohol molecules contain both polar and non-polar bonds. Many other molecular compounds also have both polar and non-polar components in their molecules. **Figure 8.10** shows the structure of acetic acid, $CH_3COOH(\ell)$, which includes a polar O-H bond. To reflect this structure, the formula for acetic acid is commonly written as CH_3COOH instead of $C_2H_4O_2$. The $-OH$ group forms a hydrogen bond with water. This hydrogen bond makes acetic acid soluble in water. Acetic acid also has a non-polar $-CH_3$ group, which enables it to dissolve in non-polar solvents, such as benzene, $C_6H_6(\ell)$, and carbon tetrachloride, $CCl_4(\ell)$. Similarly, soap and detergent molecules have both polar and non-polar bonds, so these molecules can dissolve in water and in oil and grease.

Figure 8.10 The molecular structure of acetic acid shows its polar O-H bond and its non-polar $-CH_3$ group.

Temperature and Solubility

As mentioned in Section 8.1, changes in temperature can make a saturated solution become supersaturated. This effect shows that the solubility of many substances changes with temperature. At higher temperatures, solvent particles have greater energy, resulting in more frequent and more energetic collisions with a solute. As you will learn, pressure can also affect solubility. Therefore, solubilities are often stated at *standard ambient temperature and pressure* (SATP), which is 25°C and 100 kPa.

Effect of Temperature on the Solubility of Solids

For a given solid solute and liquid solvent, solubility usually *increases* with temperature. For example, boiling water is better than warm water for making tea or coffee because the hot water dissolves more of the flavour compounds and other substances from tea leaves or ground coffee beans. For example, the solubility in water of caffeine (present in coffee and many types of tea) increases from 2.2 g/100 mL at 25°C to 40 g/100 mL at 100°C.

The solubility of a compound should always be specified at a certain temperature. A graph that shows the relationship between the solubility of a solute and the temperature of the solvent is called a *solubility curve*. The solubility curves of several ionic substances are shown in **Figure 8.11**.

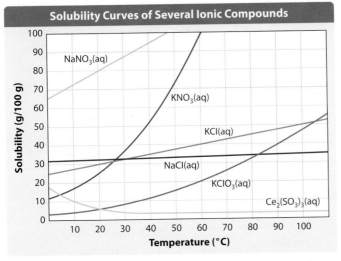

Figure 8.11 caption graph: **Solubility Curves of Several Ionic Compounds**
Solubility (g/100 g) vs Temperature (°C); curves labelled NaNO₃(aq), KNO₃(aq), KCl(aq), NaCl(aq), KClO₃(aq), Ce₂(SO₃)₃(aq)

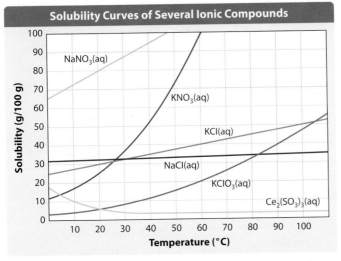

Suggested Investigation

Inquiry Investigation 8-B, Plotting a Solubility Curve

Figure 8.11 The solubility of most ionic substances in water increases with temperature. In this graph, the solubility is given in grams of solute per 100 g of *solvent*. This is different from other expressions of solubility, which are given in grams of solute per 100 mL of *solution*.

Analyze *According to the graph, what is the solubility of potassium chloride at 80°C?*

Effect of Temperature on the Solubility of Liquids and Gases

When two liquids at the same temperature mix to form a solution, there is usually little or no energy exchanged between the particles of the liquids. For this reason, the solubility of one liquid in another liquid is usually *not* affected by temperature. Similarly, the solubility of one gas in another gas is usually *not* affected by temperature.

The solubility of a gas in a liquid, however, depends on both temperature and pressure. There is a large change in the kinetic energy of gas molecules when they enter or leave a solution. Molecules in the gaseous state have greater kinetic energy than the same molecules dissolved in a solvent. Increasing the temperature of a solution that contains dissolved gas molecules provides energy for the gas molecules to escape. Thus, the solubility of most gases in most liquid solvents *decreases* with an increase in temperature. This decrease in solubility causes a carbonated drink to go "flat" much more quickly at room temperature than it does when cold. In aqueous solution, carbon dioxide reacts with water to form carbonic acid:

$$H_2O(\ell) + CO_2(g) \rightleftharpoons H_2CO_3(aq)$$

The double arrow in the equation above indicates that the reaction is reversible, which means that it can proceed in either direction. When a carbonated drink goes flat, it loses some of the carbonic acid. The drink then tastes less tart and is much less fizzy.

Activity 8.2 — The Effect of Temperature on Dissolved Carbon Dioxide

In this activity, you will observe how temperature affects dissolved carbon dioxide gas in soda water.

Safety Precautions

- Wear safety eyewear throughout this activity.
- If using a hot plate, do not touch the hot surface.
- Do not unplug the hot plate by pulling on the cord.
- If using a Bunsen burner, make sure that there are no flammable materials nearby. Tie back loose hair and clothing.
- Use EXTREME CAUTION when you are near an open flame.

Materials

- refrigerated can of soda water
- universal indicator solution
- two 100 mL beakers
- balance
- thermometer
- retort stand
- hot plate (or Bunsen burner secured to the retort stand)
- wire gauze square
- one or two ring clamps

Procedure

1. Open a can of cold soda water. Pour about 50 mL of the soda water into each of two 100 mL beakers. Make the amounts in the beakers as equal as you can. Note the rate at which bubbles form in the soda water.

2. Add a few drops of universal indicator to both beakers. Record the colour of the solutions, and estimate the pH.

3. Measure and record the mass of each beaker of soda water. Measure and record the temperature of the soda water in each beaker.

4. Heat one beaker to about 50°C. Place the wire gauze under the beaker to protect it during heating. Use a ring clamp to secure the beaker to a retort stand. (If using a Bunsen burner, use the second ring clamp to hold the wire gauze.) Compare the rate at which bubbles form in the heated beaker with the rate at which they form in the unheated beaker. Record any change in the colour of the soda water in each beaker. Estimate the pH of the soda water in each beaker.

5. Allow the heated soda water to cool. Record any change in its colour, and estimate its pH.

6. Measure and record the mass of each beaker of soda water. Calculate any change in mass by comparing the final and initial masses.

Questions

1. Which sample of soda water lost the most mass? Explain your observation.

2. Did the soda water become more or less acidic when it was heated? Explain why there was a change in acidity.

Environmental Effects of Increased Temperatures

All the important gases in the atmosphere become less soluble in water as the temperature of the water increases. Because of this relationship, *heat* (or *thermal*) *pollution* and the warming of Earth can have serious consequences. Many manufacturing processes and electricity-generating plants, such as the one shown in **Figure 8.12**, release waste heat into the environment. Commonly, water is pumped from a nearby lake or river and passed through a heat-exchange system. Then the heated water is returned to its source. This transfer of waste heat to the environment is a form of heat pollution. The warmer water decreases the solubility of oxygen. Many of the organisms living in the water become more susceptible to disease, and some may even die from a shortage of oxygen.

Figure 8.12 The transfer of waste heat to a body of water can decrease the concentration of oxygen that is dissolved in the water.

Pressure and Solubility

Pressure is force per unit area. Pressure has very little effect on the solubility of a liquid or a solid, but it has a significant effect on the solubility of a gas. The solubility of a gas in a liquid is directly proportional to the pressure of *that particular* gas above the liquid, but it is not affected by the pressure of any other gas. For example, the solubility of carbon dioxide in water is directly proportional to the pressure of carbon dioxide above the water, as shown in **Figure 8.13**. Injecting a different gas to increase the pressure in a bottle or can would have no effect on the solubility of the carbon dioxide in the soft drink. The pressure of the carbon dioxide in a fresh, sealed bottle or can is more than 10 000 times the pressure of the carbon dioxide in the atmosphere. This large increase in the pressure of the carbon dioxide greatly increases its solubility in water.

Figure 8.13 (A) A soft drink is bottled with a high pressure of carbon dioxide to increase the solubility of this gas in the drink. **(B)** The pressure of carbon dioxide in the atmosphere is much lower than the pressure in the bottle, which causes the gas to come out of solution.

Pressure, Solubility, and Scuba Diving

The solubility of gases is vitally important to scuba divers. Most recreational divers, such as the people shown in **Figure 8.14**, breathe compressed air. The deeper that a scuba diver descends, the greater the water pressure is. A valve system automatically regulates the pressure of the air that the diver breathes to match the pressure of the surrounding water. A scuba diver breathes air at a greater pressure underwater than at the surface of the water. Therefore, more of the gases in the air, especially nitrogen, dissolve in the diver's blood. When the diver surfaces at the end of a dive, he or she must move to the surface slowly. Slow surfacing allows the dissolved nitrogen to come out of the blood solution gradually. If the diver surfaces too quickly, the effect is similar to opening a soft drink. Bubbles will form in the blood, leading to a painful, and sometimes fatal, condition known as the "bends." Even after a diver is out of the water, the amount of nitrogen dissolved in the diver's blood is higher than normal. Therefore, people should not fly within 24 hours of completing a dive. Air pressure in airplanes is lower than sea-level air pressure, and a diver could get the bends if he or she flies too soon after a dive.

Figure 8.14 Scuba divers wear gas tanks during their dives. The tanks contain air at very high pressures.

Learning Check

7. Benzene, $C_6H_6(\ell)$, is an industrial solvent for fats and oils. Would you expect benzene to be soluble in water? Explain your answer.

8. Explain why the solubilities of compounds that contain fluoride ions are different from the solubilities of compounds that contain the other halides listed in **Table 8.3**.

9. Consider the solubilities of potassium chlorate, $KClO_3(s)$; methanol, $CH_3OH(\ell)$; and nitrogen, $N_2(g)$. Which substance's solubility in water is *least* affected by temperature? Explain your answer.

10. Explain why the taste of a warm soft drink differs from the taste of the same soft drink at a cooler temperature.

11. Use the solubility curves in **Figure 8.11** to predict whether a solution of 50 g of $KNO_3(s)$ in 100 g of water at 40°C is saturated, unsaturated, or supersaturated.

12. Why are many species of fish less likely to be found in warm surface water than in cooler water?

Factors That Affect the Rate of Dissolving

The **rate of dissolving** is a measure of *how quickly* a solute dissolves in a solvent. The rate of dissolving is primarily of concern when preparing a solution with a solid solute and a liquid solvent. The rate of dissolving should not be confused with solubility, which is a measure of the *amount* of solute that dissolves in a given volume of solvent at a particular temperature. Cooks, bakers, laboratory technicians, chemical engineers, and other people who work with solutions use a variety of techniques and equipment to increase the rate of dissolving.

The factors that affect the rate of dissolving are related to the number of collisions between solute particles and solvent particles in a given period of time. The three most important factors are agitation or mixing, temperature, and surface area.

rate of dissolving a measure of how quickly a solute dissolves in a solvent

Agitation or Mixing

Stirring or shaking a mixture increases the rate of dissolving by increasing the number of collisions between solute particles and solvent particles. For example, stirring sugar (a solid solute) makes it dissolve faster in tea or coffee. Similarly, when making juice or lemonade from a liquid concentrate, shaking or stirring the liquids increases their rate of dissolving. Kitchens, laboratories, and chemical plants use a wide variety of stirring machines.

Temperature

At higher temperatures, solvent particles have increased kinetic energy and they collide with solute particles more frequently. Most solid solutes dissolve faster in a solvent at higher temperatures.

Surface Area

A given mass of a solid solute has a greater surface area when the individual grains or crystals of the solute are smaller. For example, the surface area of a sugar cube is much smaller than the total surface area of the same mass of powdered sugar. Increasing the surface area of a solute increases the amount of solute that is in direct contact with the solvent. Thus, the greater area increases the rate of collisions with the solvent molecules, making the solute dissolve more rapidly. Grinding a solid solute increases the rate at which it will dissolve. **Figure 8.15** shows a simple technique that chemists and non-scientists around the world use for grinding solids.

Figure 8.15 A mortar (the bowl) and pestle (the rounded stick) are used to grind solids into powder to increase the solid's rate of dissolving.

Section Summary

- Intermolecular forces affect solubility.
- Water is a good solvent because it can dissolve a wide range of solutes.
- Ionic compounds that contain ions with relatively large charges or small radii tend to be insoluble in water.
- Polar solvents tend to dissolve ionic and polar molecules. Non-polar solvents tend to dissolve non-polar molecules.
- Solubility is affected by temperature and pressure.
- The rate of dissolving of a solid solute in a liquid solvent is affected by temperature, the surface area of the solute, and agitation.

Review Questions

1. **K/U** Describe the process of hydration when a sugar molecule dissolves.

2. **K/U** Water is called "the universal solvent."
   a. Explain how the bonding in water makes it an excellent solvent.
   b. Is "universal solvent" an accurate description of water? Explain your answer.

3. **A** Vitamin A is soluble in fats but not in water. This vitamin is found in foods such as yellow fruit and green vegetables. People who have little access to meat frequently show signs of vitamin A deficiency even when their diet contains a good supply of the necessary fruit and vegetables. Why do these people show signs of vitamin A deficiency?

4. **K/U** Explain why calcium hydroxide, $Ca(OH)_2(s)$, is insoluble in water, but sodium hydroxide, $NaOH(s)$, is soluble.

5. **T/I** Is iodine, $I_2(s)$, more likely to dissolve in water or in carbon tetrachloride, $CCl_4(\ell)$? Explain.

6. **A** Explain why the solubility of a gas in olive oil is similar to the solubility of the same gas in animal fat. How could this information be useful to a researcher who is evaluating the safety of a new anesthetic gas? Anesthetics are medicines that are given to patients before surgery to keep them from feeling pain during the procedure.

7. **T/I** Food colouring is often added to ice cream, candies, and icing for birthday cakes. Are the molecules in food colouring more likely to be polar or non-polar? Explain your answer.

8. **A** A scuba diver who is experiencing the "bends" as a result of nitrogen bubbles in the blood may be treated in a recompression chamber. The air pressure in this sealed chamber can be increased to many times the atmospheric pressure. Explain how a recompression chamber can be used to remove the nitrogen bubbles safely.

9. **T/I** Use the solubility curves in **Figure 8.11** to predict whether each solution of potassium chlorate, $KClO_3(aq)$, is saturated, unsaturated, or supersaturated.
   a. 15 g of $KClO_3(aq)$ in 100 g of water at 40°C
   b. 10 g of $KClO_3(aq)$ in 50 g of water at 65°C
   c. 8 g of $KClO_3(aq)$ in 50 g of water at 50°C

10. **T/I** Use **Table 8.3** to predict the solubility of the following ionic compounds in water:
    a. copper(II) sulfate, $CuSO_4(s)$
    b. magnesium hydroxide, $Mg(OH)_2(s)$

11. **C** Explain why sodium chloride, $NaCl(s)$, is insoluble in benzene, $C_6H_6(\ell)$. In your explanation, refer to the forces of attraction between the various particles.

12. **C** Sketch a graph with three curves to show approximately how the solubilities in water of a gas, a liquid, and a solid vary with temperature.

13. **K/U** Distinguish between solubility and rate of dissolving.

14. **K/U** Suppose you are dissolving a solid in a liquid.
    a. List three factors that affect the rate of dissolving.
    b. Identify a factor that affects both solubility and rate of dissolving.

15. **A** The molecular structure of methanol is shown below. Use this structure to explain why methanol is soluble in both water and gasoline.

$$H-\overset{\overset{\displaystyle H}{|}}{\underset{\underset{\displaystyle H}{|}}{C}}-O-H$$

16. **T/I** Boiling is an effective method for sterilizing water. The water will remain sterile if it is covered while it cools to room temperature. Water that is sterilized in this way is not suitable for filling a fish tank, however. Explain why.

Concentrations of Solutions

Benzoic acid, $C_6H_5COOH(s)$, is commonly used as a food preservative and as an ingredient in cosmetics, germicides, and medications for treating fungus infections. It occurs naturally in cranberries, such as the ones shown in **Figure 8.16**, and in most other types of berries. However, benzoic acid is moderately toxic. If too large a quantity is consumed, it causes abdominal pain, nausea, and vomiting. When benzoic acid is added to food, the mass is typically no more than 0.1 percent of the mass of the food. The World Health Organization recommends that people limit their daily consumption of benzoic acid to 5 mg per kilogram of body mass.

Benzoic acid is not the only chemical that can be dangerous if the *concentration* is too great. The proportions of many chemicals in food, medicines, and the body often determine whether the chemicals are harmful or beneficial.

Concentration is the quantity of solute per unit quantity of solution or unit quantity of solvent. This section describes several of the most commonly used measures of concentration. All of these measures express concentration in terms of unit quantities of solution.

Key Terms

concentration

concentrated

dilute

percent (m/v)

percent (m/m)

percent (v/v)

parts per million (ppm)

parts per billion (ppb)

molar concentration

concentration the ratio of the quantity of solute to the quantity of solvent or the quantity of solution

Figure 8.16 The cranberry is a North American fruit. Aboriginal peoples mixed cranberries with ground meat. The benzoic acid in the berries preserved the meat and allowed it to be stored for a long time. The concentration of benzoic acid was high enough to be toxic to micro-organisms, but not high enough to be toxic to humans.

Expressing Concentration

Qualitatively, a solution can be described as **concentrated** or **dilute**. A concentrated solution contains many particles of solute per unit quantity of solution, whereas a dilute solution contains relatively few solute particles. Concentration is a *ratio* of the quantity of solute to the quantity of solution. A concentrated solution contains a high ratio of solute to solution, whereas a dilute solution contains a low ratio of solute to solution.

Most calculations that involve solutions require quantitative measures of concentration. These measures enable you to compare the concentrations of different solutions and to find the amount of a substance in a given volume of solution. The simplest quantitative measures of concentration are stated as a quantity of solute per unit quantity of solution. For example, a solution with a concentration of 10 g of solute per litre of solution is clearly a bit more dilute than a solution with a concentration of 11 g of the same solute per litre of solution.

concentrated having a high ratio of solute to solution

dilute having a low ratio of solute to solution

Calculating Percent Concentrations

Ratios of solute to solution are commonly expressed as percents. These measures of concentration are referred to as *percent concentrations*. Each percent concentration is a ratio multiplied by 100. These ratios always refer to the *solution as a whole*—they are not ratios of solute to solvent.

There are three common methods for expressing percent concentration:

- mass/volume percent, or percent (m/v)
- mass percent, or percent (m/m)
- volume percent, or percent (v/v)

Mass/Volume Percent

A mass/volume percent, or **percent (m/v)**, expresses the mass of solute dissolved in a volume of *solution* as a percent:

percent (m/v) a ratio of the mass of solute to the volume of solution, expressed as a percent

$$\text{percent (m/v)} = \frac{\text{mass of solute [in grams]}}{\text{volume of solution [in millilitres]}} \times 100\%$$

This formula expresses the concentration of a solution as a percentage relative to a concentration of 1 g/mL. The units for percent (m/v) concentration are % g/mL, but these units are usually not indicated. The notation "% (m/v)" is used instead. The Sample Problem below demonstrates how to calculate a percent (m/v) concentration.

Sample Problem

Finding Percent (m/v) Concentration

Problem

An intravenous solution for a patient was prepared by dissolving 17.5 g of glucose in distilled water to make 350 mL of solution. Find the percent (m/v) concentration of the solution. (Treat the zero in 350 mL as a significant figure.)

What Is Required?

You need to calculate the percent (m/v) concentration.

What Is Given?

You know the mass of the dissolved solute: 17.5 g
You know the volume of the solution: 350 mL

Plan Your Strategy	Act on Your Strategy
Write the formula for percent (m/v) concentration.	$\text{percent (m/v)} = \dfrac{\text{mass of solute [in grams]}}{\text{volume of solution [in millilitres]}} \times 100\%$
Substitute the given data to calculate the concentration.	$\text{percent (m/v)} = \dfrac{17.5 \text{ g}}{350 \text{ mL}} \times 100\% = 5.00\%$ The concentration of glucose in the intravenous solution was 5.00% (m/v).

Check Your Solution

The answer appears reasonable since the number of grams of solute is much less than the number of millilitres of solution. The answer is a percent and has three significant digits, the same as the given information.

Using Percent (m/v) Concentration to Find Mass

Problem

Sucrose sugar syrups can have percent concentrations that are greater than 100% (m/v).
Find the mass of sucrose in 475 mL of 166% (m/v) sugar syrup.

What Is Required?

You need to calculate the mass of sucrose.

What Is Given?

You know the concentration of the solution: 166% (m/v)
You know the volume of the solution: 475 mL

Plan Your Strategy	Act on Your Strategy
Write the formula for percent (m/v) concentration.	$\text{percent (m/v)} = \dfrac{\text{mass of solute [in grams]}}{\text{volume of solution [in millilitres]}} \times 100\%$
Rearrange the formula to solve for the mass of solute.	$\dfrac{\text{percent (m/v)}}{100\%} = \dfrac{\text{mass of solute [in grams]}}{\text{volume of solution [in millilitres]}}$ $\text{mass of solute [in grams]} = \dfrac{\text{percent (m/v)}}{100\%} \times \text{volume of solution [in millilitres]}$
Substitute the known values to calculate the mass of sucrose, m.	$m = \dfrac{166\% \text{ g/mL}}{100\%} \times 475 \text{ mL} = 788 \text{ g}$ The mass of sucrose in the sugar syrup is 788 g.

Check Your Solution

The mass in grams of solute is greater than the volume in millilitres of solvent, which
is reasonable since the concentration is greater than 100% (m/v). The answer has three
significant digits, the same as the given information.

1. A pharmacist adds 20.0 mL of distilled water to 30.0 g of a powdered medicine. The volume of the solution formed is 25.0 mL. What is the percent (m/v) concentration of the solution?

2. A solution contains 21.4 g of sodium nitrate, $NaNO_3(s)$, dissolved in 250 mL of solution. Find the percent (m/v) concentration of the solution.

3. A chemist slowly evaporated 1.80 L of a 1.75% (m/v) solution of calcium nitrate, $Ca(NO_3)_2(aq)$. What mass of solute should the chemist obtain?

4. What mass of potassium permanganate must be dissolved to make 2.0 L of a 4.0% (m/v) solution?

5. A chemist measured 25.00 g of water and bubbled hydrogen chloride gas into the water. The resulting solution had a mass of 26.68 g and a volume of 25.2 mL. Determine the percent (m/v) concentration of the solution.

6. A student carefully evaporated all the water from an 80.0 mL salt solution. She found that the mass of the residue from the sample was 1.40 g. Calculate the percent (m/v) concentration of the salt solution.

7. A household bleach has a concentration of 4.60% (m/v) of sodium hypochlorite, $NaOCl(aq)$. What mass of sodium hypochlorite does a 2.84 L container of this bleach contain?

8. Ringer's solution contains 0.86% (m/v) $NaCl(aq)$, 0.03% (m/v) $KCl(aq)$, and 0.033% (m/v) $CaCl_2(aq)$. Calculate the mass of each of these compounds in a 300 mL bag of Ringer's solution.

9. What volume of 5.0% (m/v) solution of sodium chloride, $NaCl(aq)$, can be made using 40 g of $NaCl(s)$?

10. How would you prepare 400 mL of a 3.5% (m/v) solution of sodium acetate?

Mass Percent

The concentration of a solution that contains a solid solute dissolved in a liquid solvent can be expressed as a percent ratio of the mass of the solute to the mass of the solution. This method for expressing concentration is called mass percent, or **percent (m/m)**.

To calculate the percent, the masses of the solute and solution must be expressed in the same units.

$$\text{percent (m/m)} = \frac{\text{mass of solute}}{\text{mass of solution}} \times 100\%$$

Mass percent values are sometimes referred to as *weight/weight* or *(w/w)* percents. If you see a concentration expressed as percent (w/w), treat it as percent (m/m).

The concentration of a solid solution is usually expressed as a mass percent. For example, sterling silver consists of 92.5% silver and 7.5% copper (or sometimes a mixture of copper and other metals), where the percents are mass ratios. Sterling silver that is made with 7.5% copper could be described as a 7.5% (m/m) solution of copper dissolved in silver. Stainless steel, shown in **Figure 8.17**, is a solution of chromium and nickel dissolved in iron. Solid solutions of two or more metals, such as sterling silver or stainless steel, are often called *alloys*. The Sample Problem below demonstrates a calculation using percent (m/m) concentration.

Figure 8.17 Cooking pans are often made with 18/8 stainless steel, which contains 18% (m/m) chromium and 8% (m/m) nickel in iron.

Explain Why is iron considered to be the solvent in stainless steel?

Sample Problem

Using Percent (m/m) Concentration to Find Mass

Problem

Find the mass of pure silver in a sterling silver ring that has a mass of 6.45 g.

What Is Required?

You need to find the mass of pure silver in the ring.

What Is Given?

You know the mass of the ring: 6.45 g
You know the concentration of silver in sterling silver alloys: 92.5% (m/m)

Plan Your Strategy	Act on Your Strategy
Write the formula for percent (m/m).	$\text{percent (m/m)} = \dfrac{\text{mass of solute}}{\text{mass of solution}} \times 100\%$
Rearrange the equation to solve for the mass of solute.	$\text{percent (m/m)} \times \text{mass of solution} = \text{mass of solute} \times 100\%$ $\text{mass of solute} = \dfrac{\text{percent (m/m)} \times \text{mass of solution}}{100\%}$
Substitute the given data to find the mass of silver.	$\text{mass of solute} = \dfrac{92.5\% \times 6.45 \text{ g}}{100\%} = 5.97 \text{ g}$ The mass of pure silver in the ring is 5.97 g.

Check Your Solution

The answer is reasonable because the mass of silver is close to the mass of the ring.

11. Calculate the percent (m/m) concentration of a solution that contains 11 g of pure sodium hydroxide in 75 g of solution.

12. A physiotherapist makes a footbath solution by dissolving 120 g of magnesium sulfate (Epsom salts), $MgSO_4(s)$, in 3.00 kg of water. Calculate the percent (m/m) of magnesium sulfate in the solution. (**Hint**: Remember to use the mass of *solution*.)

13. How much chromium, nickel, and iron would you need to make a 500 kg batch of 18/8 stainless steel, which is steel made with 18% (m/m) chromium and 8% (m/m) nickel in iron?

14. When evaporated, a sample of a solution of silver nitrate, $AgNO_3(aq)$, leaves a residue with a mass of 3.47 g. The original sample had a mass of 43.88 g. Calculate the percent (m/m) concentration of the silver nitrate solution.

15. A solution is made by dissolving 12.9 g of carbon tetrachloride, $CCl_4(\ell)$, in 72.5 g of benzene, $C_6H_6(\ell)$. Calculate the percent (m/m) concentration of carbon tetrachloride in the solution.

16. Since pure gold is quite soft, gold jewellery is usually made with an alloy. An 18 karat gold alloy contains 75% (m/m) gold. How much of this alloy can a jeweller make with 8.00 g of pure gold?

17. Surgical steel is an iron alloy that is easy to clean and sterilize. It contains 12 to 14% (m/m) chromium. Calculate the minimum mass of chromium in a 40 g instrument made from surgical steel.

18. Pure iron is a relatively soft metal. Adding carbon to iron makes a type of steel that is much stronger than pure iron. Calculate the percent (m/m) concentration of carbon in a 5.0 kg steel bar that contains 85 g of carbon.

19. A technician who was monitoring the health of a lake analyzed a sample of water from the lake. The sample had a mass of 155 g and contained 1.12 mg of dissolved oxygen. Calculate the percent (m/m) concentration of oxygen in the sample.

20. A mining company in Sudbury reported mining 6.91×10^5 t of ore, from which it extracted 1.68×10^3 t of nickel, 1.6×10^4 t of copper, and 1.6 t of platinum. What is the percent (m/m) concentration of each metal in the ore?

Volume Percent

When two liquids are mixed to form a solution, it is easier to measure their volumes than their masses. Volume percent, or **percent (v/v)**, is the ratio of the volume of solute to the volume of solution, expressed as a percent:

percent (v/v) a ratio of the volume of solute to the volume of solution, expressed as a percent

$$\text{percent (v/v)} = \frac{\text{volume of solute}}{\text{volume of solution}} \times 100\%$$

Solutions that are sold in pharmacies and hardware stores are often labelled with volume percent concentrations, as shown in **Figure 8.18**. The next Sample Problem demonstrates a calculation using percent (v/v) concentration. When doing these calculations, the volumes used must have the same units. Note that the volume of solution produced by dissolving one liquid in another is usually not equal to the sum of the volumes of the two separate liquids.

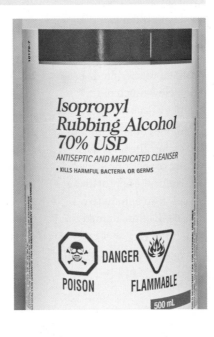

Figure 8.18 This bottle of rubbing alcohol is 70% (v/v) isopropyl alcohol, $C_3H_8O(\ell)$. USP stands for *United States Pharmacopeia*, which is a document that establishes medicine standards in the United States, Canada, and other countries.

Using Percent (v/v) Concentration to Find Volume

Problem

Acetic acid, $CH_3COOH(\ell)$, is a liquid at room temperature. How much pure water should be added to 15.0 mL of pure acetic acid to make a 5.00% (v/v) solution of acetic acid? Assume that the total volume of the solution equals the sum of the volumes of the water and the acetic acid.

What Is Required?

You need to calculate the volume of the solvent, water.

What Is Given?

You know the concentration of the acetic acid: 5.00% (v/v); You know the volume of the solute: 15.0 mL

Plan Your Strategy	Act on Your Strategy
Write the formula for percent (v/v).	$\text{percent (v/v)} = \dfrac{\text{volume of solute}}{\text{volume of solution}} \times 100\%$
Rearrange the equation to solve for the volume of the solution.	$\text{percent (v/v)} \times \text{volume of solution} = \text{volume of solute} \times 100\%$ $\text{volume of solution} = \dfrac{\text{volume of solute} \times 100\%}{\text{percent (v/v)}}$
Substitute the given data, and calculate the volume of the solution.	$\text{volume of solution} = \dfrac{15.0 \text{ mL} \times 100\%}{5.00\%} = 300 \text{ mL}$
Subtract the volume of the solute from the volume of the solution.	$\text{volume of solvent} = 300 \text{ mL} - 15.0 \text{ mL} = 285 \text{ mL}$ Thus, 285 mL of pure water should be added to make a 5.00% (v/v) solution.

Check Your Solution

The answer is reasonable. The volume of water that should be added is less than the volume of the solution, but much more than the volume of the acetic acid. The answer has three significant digits, the same as the given information.

Practice Problems

21. The rubbing alcohol that is sold in pharmacies is usually a 70% (v/v) aqueous solution of isopropyl alcohol. What volume of isopropyl alcohol is present in a 500 mL bottle of this solution?

22. If 80 mL of ethanol is diluted with water to a final volume of 500 mL, what is the percent (v/v) concentration of ethanol in the solution?

23. A particular brand of windshield washer fluid contains 40% (v/v) methanol. How much pure methanol does a 4.0 L container of this fluid contain?

24. A concentrated solution of engine coolant contains 75% (v/v) ethylene glycol in water. The label tells consumers to use a 1:1 mixture of the concentrate with water in their cars. Determine the approximate volume of pure ethylene glycol in an automotive cooling system that contains 6.0 L of the diluted solution.

25. Describe how to prepare a 5.00% (v/v) solution with 50.0 mL of pure ethylene glycol.

26. Gasoline sold in Ontario must contain at least 5.0% (v/v) ethanol. How much ethanol is a driver likely to get when buying 30 L of gasoline?

27. A vending machine mixes a liquid flavour concentrate with water in a ratio of 1:10 to make coffee. Determine the percent (v/v) concentration of the flavour concentrate in the drink.

28. Your teacher has 4.0 L of a 15% (v/v) solution of sulfuric acid in water. What will the volume of the solution be if it is diluted to 10% (v/v)?

29. Two concentrations of the same cleaning chemical are mixed: 6.0 L of a 75% (v/v) solution and 14.0 L of a 25% (v/v) solution. What is the concentration of the resulting solution?

30. You need 125 mL of white vinegar, which has a concentration of 5.0% (v/v) of acetic acid. You are out of white vinegar. However, you do have pickling vinegar with a concentration of 8.5% (v/v) of acetic acid. How much pickling vinegar should you dilute to substitute for the white vinegar?

Very Small Concentrations

Very dilute solutions have concentrations that are much less than 1% (m/m). For such solutions, it is often convenient to express concentrations in terms of **parts per million (ppm)** or **parts per billion (ppb)**. A concentration of 1 ppb is extremely dilute. For example, one part per billion of water in a full swimming pool that is 10 m long, 5 m wide, and 2 m deep is only 0.1 mL. However, there are some chemicals that can have serious health effects at a concentration of 1 ppb.

Both parts per million and parts per billion are fractions, with the mass of the solute divided by the mass of the *solution*:

parts per million (ppm)
a ratio of solute to solution $\times 10^6$

parts per billion (ppb)
a ratio of solute to solution $\times 10^9$

Parts per million

$$\text{ppm} = \frac{\text{mass of solute}}{\text{mass of solution}} \times 10^6$$

Parts per billion

$$\text{ppb} = \frac{\text{mass of solute}}{\text{mass of solution}} \times 10^9$$

Sample Problem

Calculating Concentration in ppm

Problem
Health Canada's guideline for the maximum mercury content in commercial fish is 0.5 parts per million (ppm). When a 1.6 kg salmon was tested, it was found to contain 0.6 mg of mercury. Would this salmon be safe to eat?

What Is Required?
You need to find the concentration of mercury in the salmon, in parts per million, and then determine if this concentration is less than Health Canada's guideline.

What Is Given?
You know the allowable level of mercury: 0.5 ppm
You know the mass of the salmon: 1.6 kg
You know the amount of mercury in the salmon: 0.6 mg

Plan Your Strategy	Act on Your Strategy
Write the formula for ppm.	$\text{ppm} = \dfrac{\text{mass of solute}}{\text{mass of solution}} \times 10^6$
Substitute the given data. Express both masses in grams so that you can cancel the units.	$\text{ppm} = \dfrac{0.6 \text{ mg}}{1.6 \text{ kg}} \times 10^6 = \dfrac{6 \times 10^{-4} \text{ g}}{1.6 \times 10^3 \text{ g}} \times 10^6 = 0.4$
Compare the calculated concentration to Heath Canada's guideline.	The concentration of mercury is 0.4 ppm, which is less than the maximum in the guideline. Therefore, the salmon would be safe to eat.

Check Your Solution
The answer appears to be reasonable. The units divided correctly, and the answer has one significant digit, the same as the number of significant digits in the mass of mercury.

31. A sample of lake water has a mass of 310 g and contains 2.24 mg of dissolved oxygen. Calculate the oxygen concentration in parts per million.

32. The agricultural use of the pesticide DDT has been banned in Canada since 1969 because of its effect on wildlife. In 1967, the average concentration of DDT in trout taken from Lake Simcoe, in Ontario, was 16 ppm. Today, the average concentration is less than 1 ppm. What mass of DDT is present in a 2.2 kg smallmouth bass contaminated with 16 ppm of DDT?

33. Dry air contains about 0.000 07% (m/m) helium. Express this concentration in parts per million.

34. A fungus that grows on peanuts produces aflatoxin, a potentially deadly toxin. A quality control inspector tests a 100 g sample from a shipment of peanuts to check that it contains no more than 25 ppb of aflatoxin. What mass of aflatoxin would the sample contain if the concentration is 25 ppb?

35. A sample of water contains one atom of lead for every million water molecules. Calculate the concentration of lead, in parts per million, in this sample.

36. The concentration of chlorine in swimming pools is generally kept in the range from 1.4 to 4.0 mg/L. A pool contains 3.0 ppm of chlorine. Is this concentration within the acceptable range? Show your work, and explain your reasoning. (**Hint:** 1 L of water has a mass of 1 kg.)

37. Water supplies that contain more than 500 ppm of dissolved calcium carbonate, $CaCO_3(aq)$, are considered unacceptable for most domestic purposes. What is the maximum mass of calcium carbonate that would be acceptable in a 250 mL sample of tap water?

38. Since 1991, house paint produced in Canada must contain less than 600 ppm of lead. What is the maximum mass of lead permitted in a can that contains 7.0 kg of paint?

39. Cadmium is a highly toxic metal. The average level of cadmium in the blood of Canadians is about 0.35 ppb. At this level, what mass of cadmium would be present in 1.5 kg of blood?

40. Find the concentration, in parts per million, of a solution that contains 0.1 g of solute per litre.

Molar Concentration

molar concentration
the amount (in moles) of solute dissolved in 1 L of solution

Percent concentrations are easy to determine from simple mass and volume measurements. However, calculations for reactions often involve the amount (in moles) of reactants and products. For this reason, a measure of concentration based on the amount of solute is particularly useful. **Molar concentration** (or *molarity*) is the amount (in moles) of solute dissolved in 1 L of *solution*.

$$\text{molar concentration} = \frac{\text{amount of solute [in moles]}}{\text{volume of solution [in litres]}}$$

The symbol for molar concentration is c. Thus, the formula for molar concentration can be written as

$$c = \frac{n}{V}$$

where n is the amount of solute in moles, and V is the volume of solution in litres. The units for molar concentration are moles per litre (mol/L).

In the past, the symbol "M" was often used for mol/L. However, IUPAC and other international standards recommend writing "mol/L" for clarity. The symbol M is used for other quantities. Writing "mol/L" also helps you check the units when you perform a calculation.

Molar concentration is especially useful to chemists because it is directly related to the number of solute particles in a solution. Given the molar concentration and volume of a solution, you can easily find the amount (in moles) of dissolved solute. Molar concentrations are the most convenient form for stoichiometry calculations that involve reactions in solutions, as you will see in **Chapters 9** and **10**.

Molar Concentrations of Ions

An extra step is often needed for calculations of the concentrations of ions when an ionic compound dissolves in water. A solution of sodium chloride contains the same amount of sodium and chloride ions because $NaCl(aq)$ dissociates to give equal numbers of these ions:

$$NaCl(aq) \rightarrow Na^+(aq) + Cl^-(aq)$$

However, a solution of calcium chloride has twice as many chloride ions as calcium ions:

$$CaCl_2(aq) \rightarrow Ca^{2+}(aq) + 2Cl^-(aq)$$

Thus, the molar concentration of chloride ions is twice the molar concentration of calcium ions. You can use the chemical formula of an ionic substance to determine the numbers of ions formed and the ratio of the molar concentrations of the different ions. The analogy shown in **Figure 8.19** will help you visualize the dissociation of ionic compounds that contain more than one ion of the same type.

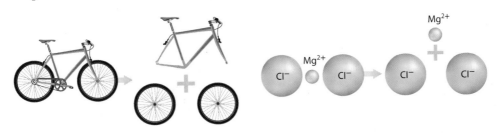

Figure 8.19 A bicycle contains two wheels, just as each formula unit of magnesium chloride contains two chloride ions.

Describe *another object with components that can be separated in the same ratio as the ions in magnesium chloride.*

Sample Problem

Calculating Molar Concentrations

Problem
A student dissolved 0.212 mol of iron(III) chloride, $FeCl_3(s)$, to make a 175 mL solution. Find the molar concentration of the solution and the concentrations of the ions in the solution.

What Is Required?
You need to calculate the concentration of iron(III) chloride.
You must also determine the concentration of iron(III) ions and chloride ions.

What Is Given?
You know the amount of iron(III) chloride in the sample: 0.212 mol
You know the volume of the solution: 175 mL

Plan Your Strategy	Act on Your Strategy
Write the formula for molar concentration and substitute the known values to find the molar concentration of iron(III) chloride.	$c = \dfrac{n}{V} = \dfrac{0.212 \text{ mol}}{175 \text{ mL}} = \dfrac{0.212 \text{ mol}}{0.175 \text{ L}} = 1.2114 \text{ mol/L}$
Write the chemical equation for the dissociation of iron(III) chloride.	$FeCl_3(aq) \rightarrow Fe^{3+}(aq) + 3Cl^-(aq)$
Use the coefficients in the chemical equation to find the molar concentrations of the ions.	$c_{Fe^{3+}} = \dfrac{1.2114 \text{ mol } FeCl_3}{L} \times \dfrac{1 \text{ mol } Fe^{3+}}{1 \text{ mol } FeCl_3} = 1.21 \text{ mol/L } Fe^{3+}$ $c_{Cl^-} = \dfrac{1.2114 \text{ mol } FeCl_3}{L} \times \dfrac{3 \text{ mol } Cl^-}{1 \text{ mol } FeCl_3} = 3.63 \text{ mol/L } Cl^-$ The concentrations of $FeCl_3(aq)$ and $Fe^{3+}(aq)$ are both 1.21 mol/L, and the concentration of $Cl^-(aq)$ is 3.63 mol/L.

Check Your Solution
The concentration of iron(III) chloride has the correct units and significant digits. The concentration of chloride ions is three times the concentration of iron(III) ions, which reflects the formula $FeCl_3$.

Calculating Mass from Molar Concentration

Problem

To replace lost fluids and electrolytes, patients often receive an infusion of an intravenous solution immediately after surgery. The most commonly used intravenous solution, normal saline, contains 0.154 mol/L of sodium chloride, NaCl(aq). This concentration is close to the concentration that is normally found in the bloodstream. However, for some patients, the intake of sodium must be carefully monitored. Calculate the mass of sodium in a 500 mL bag of normal saline solution. Assume that the volume of the solution is accurate to three significant figures.

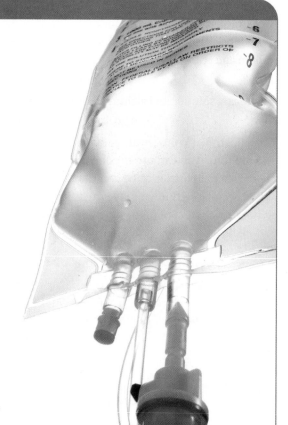

What Is Required?

You need to calculate the mass of sodium in the bag of solution.

What Is Given?

You know the molar concentration of sodium chloride: 0.154 mol/L
You know the volume of the solution: 500 mL
You know the molar mass of sodium according to the periodic table: 22.99 g/mol

Plan Your Strategy	Act on Your Strategy
Determine the molar concentration of sodium ions in the solution.	The chemical formula is NaCl. There is 1 mol of $Na^+(aq)$ for each mole of NaCl(aq). Therefore, the concentration of sodium ions is 0.154 mol/L.
Rearrange the concentration formula to isolate n, the amount of sodium.	$c = \dfrac{n}{V}$ $n = cV$
Substitute the known values to calculate n.	$n = (0.154 \text{ mol/\cancel{L}})(0.500 \, \cancel{L})$ $= 0.0770 \text{ mol}$
Write the formula to find the mass of sodium in the solution.	$m = nM$ M is the molar mass of sodium.
Look up the molar mass of sodium.	$M = 22.99 \text{ g/mol}$
Calculate the mass, m, of sodium in the solution.	$m = nM$ $= (0.0770 \, \cancel{\text{mol}})(22.99 \text{ g/}\cancel{\text{mol}})$ $= 1.77 \text{ g}$ The mass of sodium in a 500 mL bag of normal saline solution is 1.77 g.

Check Your Solution

The mass is reasonable because 0.154 mol/L is a low concentration, so the mass of solute in solution should be a small amount. The units are correct because the question asked to find a mass, and grams is a unit of mass. The number of significant digits is correct because values given in the problem each have three significant digits.

41. Find the molar concentration of each saline solution.
   a. 0.60 mol NaCl(s) dissolved in 0.40 L of solution
   b. 0.90 g NaCl(s) dissolved in 100 mL of solution

42. What volume of 0.25 mol/L solution can be made using 14 g of sodium hydroxide?

43. Calculate the molar concentration of each solution.
   a. 14 g of copper(II) sulfate, $CuSO_4(s)$, dissolved in 70 mL of solution
   b. 5.07 g of sucrose, $C_{12}H_{22}O_{11}(s)$, dissolved in 23.6 mL of solution
   c. 1.1 g of calcium nitrate, $Ca(NO_3)_2(s)$, dissolved in 70 mL of solution

44. At 20°C, a saturated solution of calcium sulfate, $CaSO_4(aq)$, has a concentration of 0.0153 mol/L. A student takes 65 mL of this solution and evaporates it. What mass of solute should be left in the evaporating dish?

45. Find the mass of solute in each aqueous solution.
   a. 28 mL of 0.045 mol/L calcium hydroxide, $Ca(OH)_2(aq)$
   b. 50 mL of 4.0 mol/L acetic acid, $CH_3COOH(aq)$
   c. 5.31 L of 0.675 mol/L ammonium phosphate, $(NH_4)_3PO_4(aq)$

46. Calculate the molar concentrations of the ions in each solution.
   a. 18 g of sodium sulfate, $Na_2SO_4(s)$, dissolved in 210 mL of solution

b. 15 g of ammonium phosphate, $(NH_4)_3PO_4(s)$, dissolved in 98 mL of solution

c. 20 mg of calcium phosphate, $Ca_3(PO_4)_2(s)$, dissolved in 1.7 L of solution

47. A student dissolves 28.46 g of silver nitrate, $AgNO_3(s)$, in water to make 580 mL of solution. Find the molar concentration of the solution.

48. Formalin is an aqueous solution that is made by dissolving formaldehyde gas, HCHO(g), in water. A saturated formalin solution has a concentration of about 37% (m/v). This concentration is used to preserve biological specimens. Calculate the molar concentration of 37% (m/v) formalin.

49. What volume of a 0.555 mol/L aqueous solution contains 12.8 g of sodium carbonate, $Na_2CO_3(aq)$?

50. Zinc oxide, ZnO(s), has a solubility of 0.16 mg/100 mL in water at 30°C. Find the molar concentration of a saturated solution of zinc oxide at 30°C.

Different Measures for Different Solutions

Table 8.6 summarizes various ways that concentration can be measured. In the next section, you will find out how to prepare a solution with a specific concentration.

Suggested Investigation

Plan Your Own Investigation 8-C, Determining the Concentration of a Solution

Table 8.6 Measures of Concentration

Type of Concentration	Formula	Common Application
Concentration as a Percent • mass/volume percent	$\text{percent (m/v)} = \dfrac{\text{mass of solute [in grams]}}{\text{volume of solution [in millilitres]}} \times 100\%$	• intravenous solutions, such as a saline drip
• mass percent	$\text{percent (m/m)} = \dfrac{\text{mass of solute}}{\text{mass of solution}} \times 100\%$	• concentration of metals in an alloy
• volume percent	$\text{percent (v/v)} = \dfrac{\text{volume of solute}}{\text{volume of solution}} \times 100\%$	• solutions prepared by mixing liquids
Very Small Concentrations • parts per million	$\text{ppm} = \dfrac{\text{mass of solute}}{\text{mass of solution}} \times 10^6$	• safety limits for contaminants, such as mercury or lead in food or water
• parts per billion	$\text{ppb} = \dfrac{\text{mass of solute}}{\text{mass of solution}} \times 10^9$	
Molar Concentration	$\text{molar concentration} = \dfrac{\text{amount of solute [in moles]}}{\text{volume of solution [in litres]}}$ $c = \dfrac{n}{V}$	• solutions used as reactants

Section Summary

- The concentration of a solution can be stated in terms of the ratio of the volumes or masses of the solute and the solution, using percent (m/v), percent (m/m), or percent (v/v).

- Small concentrations can be expressed as parts per million (ppm) or parts per billion (ppb).
- Molar concentration is calculated by dividing moles of solute by litres of solution.

Review Questions

1. **C** Use diagrams to explain the difference between a concentrated solution and a dilute solution.

2. **C** Use a Venn diagram to compare the terms "solubility" and "concentration."

3. **T/I** A 50 g sample of seawater is found to contain 0.02 g of sodium chloride.
   a. State the concentration of sodium as a mass percent.
   b. Express the concentration of sodium in parts per million.

4. **T/I** Calcium carbonate, $CaCO_3(aq)$, may be naturally present in household water supplies. Suppose that a toilet tank holds 6.0 L of water, and the water contains 90 ppm of calcium carbonate. What mass of calcium carbonate is in the water in the tank?

5. **A** Aldrin and dieldrin are pesticides that used to be allowed for the control of soil insects. In Ontario, the maximum allowable total concentration of aldrin plus dieldrin in drinking water is 0.7 ppb. If a 250 mL sample of drinking water is found to contain 0.0001 mg of aldrin and dieldrin, does the concentration exceed the standard? Explain.

6. **T/I** A researcher distilled an 85.1 mL sample of a solution of liquid hydrocarbons. The distillation process separated out 20.3 mL of hexane. Find the percent (v/v) concentration of hexane in the solution.

7. **T/I** Phosphoric acid, $H_3PO_4(aq)$, can be used to remove rust. Find the molar concentration of an 85% (m/v) solution of phosphoric acid in water.

8. **A** The concentration of a certain ionic compound in aqueous solution is 0.186 mol/L. It is the only solute in the solution. The concentration of potassium ions in the same solution is 0.558 mol/L. Explain why the concentrations of ions in the solution are different. What might the ionic compound be?

9. **T/I** Since the Industrial Revolution, the atmospheric concentration of carbon dioxide has increased from 280 parts per million to 380 parts per million. The mass of Earth's atmosphere is estimated to be 5.3×10^{18} kg. What mass of carbon dioxide has been added to Earth's atmosphere?

10. **T/I** Use this solubility curve to help you determine the molar concentration of a saturated solution of potassium chlorate, $KClO_3(aq)$, at 30°C.

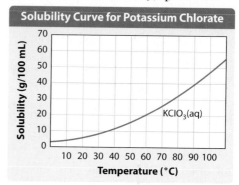

11. **A** The Canadian Ambient Air Quality Objective for ground-level ozone is 82 ppb. What is the maximum mass of ozone, $O_3(g)$, allowed per cubic metre of air? The density of air is 1.2 kg/m³.

12. **T/I** One teaspoon of table salt is added to the water in a swimming pool. The swimming pool is 10 m long, 5.0 m wide, and 2.0 m deep. One teaspoon of table salt has a mass of 4.5 g. Calculate the concentration (in parts per billion) of table salt in the water. Assume 10 m has two significant digits.

13. **A** The concentration of dissolved iron(II) ions, $Fe^{2+}(aq)$, in a sample of ground water is 7.2×10^{-5} mol/L. Is this concentration an acceptable level if the recommended maximum level is 300 ppb? Show your work, and explain your reasoning.

14. **T/I** A pharmacist dilutes a 10% (m/v) saline solution until the final volume is four times the initial volume. Find the molar concentration of sodium chloride in the diluted solution.

15. **A** A researcher who was studying toxic algae measured the concentration of phytoplankton as 5×10^{-4}% (m/v). Phytoplankton are microscopic plants.
   a. Explain why phytoplankton suspended in seawater is not a solution.
   b. Discuss whether mass/volume concentration is a valid measure for mixtures that are not solutions.

Preparing Solutions in the Laboratory

What do the effectiveness of a medicine, the safety of a chemical reaction, the cost of an industrial process, and the taste of a soft drink have in common? They all depend on a solution with a known concentration being carefully made. A solution with an accurate, known concentration is called a **standard solution**.

Preparing a Standard Aqueous Solution

There are two ways that an aqueous solution with a known concentration of solid solute can be prepared. You can dissolve a measured mass of pure solute in water and then dilute the solution to a known volume, or you can dilute a standard solution by adding a known volume of additional water.

Preparing a Standard Aqueous Solution from a Solid Solute

The basic procedure for preparing a standard aqueous solution with a solid solute is as follows:

1. Measure a mass of solute using a balance.
2. Dissolve the solute in water.
3. Add more water to dilute the solution to the required volume.
4. Mix the solution thoroughly.
5. Transfer the solution into a clean, dry, WHMIS-labelled container.

To make up a solution with a specific volume, chemists use a piece of special glassware called a volumetric flask. A **volumetric flask** is a pear-shaped glass container with a flat bottom and long neck. A *graduation mark* on the neck indicates the exact level to which the flask should be filled. Volumetric flasks are available in a variety of sizes, as shown in **Figure 8.20**. These flasks measure a fixed volume of solution to ±0.1 mL at a specified temperature, usually 20°C. If you were performing an experiment in which significant digits and error were important, the volume of solution in a 500 mL volumetric flask would be recorded as 500.0 mL ± 0.1 mL. **Table 8.7** on the next page describes how you can use a volumetric flask to prepare a standard aqueous solution.

Key Terms

standard solution
volumetric flask

---

standard solution a solution with a known concentration

volumetric flask glassware that is used to make a liquid solution with an accurate volume

Figure 8.20 Volumetric flasks are used for making solutions but not for storing them.

Identify *the features that the various flasks have in common.*

Table 8.7 Using a Volumetric Flask to Prepare a Standard Aqueous Solution

1. Place the known mass of solute in a clean beaker. Use distilled water to dissolve the solute completely.	
2. Rinse a clean volumetric flask of the required volume with a small quantity of distilled water. Discard the rinse water. Repeat the rinsing several times.	
3. Transfer the solution from the beaker to the volumetric flask using a funnel.	
4. Using a wash bottle, rinse the beaker with distilled water, and pour the rinse water into the volumetric flask. Repeat this rinsing several times.	
5. Using a wash bottle or a beaker, add distilled water to the volumetric flask until the level is just below the graduation mark. Then remove the funnel from the volumetric flask.	
6. View the neck of the volumetric flask straight on from the side, so that the graduation mark looks like a line, not an ellipse. Add distilled water, drop by drop, until the bottom of the *meniscus* (the curved surface of the solution) appears to touch the graduation mark.	

Calculating the Concentration of a Diluted Solution

A *stock solution* can be diluted to prepare a standard aqueous solution. A stock solution is usually a concentrated solution that is diluted before it is used. The key to understanding dilution is realizing that adding more solvent to a solution does not remove or add any particles of solute. The amount of solute is exactly the same before and after dilution, as illustrated in **Figure 8.21**.

The formula for molar concentration can be rearranged to give an expression for n, the amount of solute (in moles):

$$c = \frac{n}{V} \qquad n = cV$$

Thus, n, the amount of solute (in moles), is the product of c, the concentration (in moles per litre), and V, the volume of the solution (in litres). Since diluting a solution does not change n, the product of the molar concentration and volume before dilution equals the product of the molar concentration and volume after dilution:

$$c_1 V_1 = c_2 V_2$$

where c_1 and V_1 are the molar concentration and volume of the concentrated solution, and c_2 and V_2 are the molar concentration and volume of the diluted solution.

This dilution equation can be used to calculate quantities and concentrations when a stock solution is diluted. Often, the concentration of a stock solution is known to only two or three significant figures. Higher precision is usually possible when using a solid solute to prepare a standard solution.

Figure 8.21 When a solution (**A**) is diluted, the volume of solvent increases (**B**), but the number of solute particles remains the same.

Sample Problem

Diluting a Concentrated Solution

Problem
Your teacher has a stock solution of 12 mol/L hydrochloric acid. A class experiment requires 2.0 L of 0.10 mol/L hydrochloric acid. What volume of concentrated solution should be used to make the dilute solution for the experiment?

What Is Required?
You need to find the volume (V_1) of concentrated solution to be diluted.

What Is Given?
You know the initial concentration, c_1: 12 mol/L
You know the diluted concentration, c_2: 0.10 mol/L
You know the volume of the diluted solution, V_2: 2.0 L.

Plan Your Strategy	Act on Your Strategy
Write the dilution equation.	$c_1 V_1 = c_2 V_2$
Divide both sides of the equation by c_1 to isolate V_1.	$V_1 = \dfrac{c_2 V_2}{c_1}$
Substitute the known quantities to calculate V_1, the volume of stock solution required.	$V_1 = \dfrac{0.10 \text{ mol/L} \times 2.0 \text{ L}}{12 \text{ mol/L}} = 0.017 \text{ L}$ Therefore, 0.017 L, or 17 mL, of the stock solution should be diluted to make the required dilute solution.

Check Your Solution
The units and number of significant digits are correct. The answer is reasonable: only a small volume of concentrated solution is needed because the dilute solution is much less concentrated than the stock solution.

51. Suppose that you are given a stock solution of 1.50 mol/L ammonium sulfate, $(NH_4)_2SO_4(aq)$. What volume of the stock solution do you need to use to prepare each of the following solutions?
  a. 50.0 mL of 1.00 mol/L $(NH_4)_2SO_4(aq)$
  b. 2.00×10^2 mL of 0.800 mol/L $(NH_4)_2SO_4(aq)$
  c. 250 mL of 0.300 mol/L $NH_4^+(aq)$

52. What is the concentration of the solution that is obtained by diluting 60.0 mL of 0.580 mol/L potassium hydroxide to each of the following volumes?
  a. 350 mL
  b. 180 mL
  c. 3.00 L

53. What volume of a 1.60 mol/L stock solution of calcium chloride, $CaCl_2(aq)$, would you use to make 0.500 L of a 0.300 mol/L solution?

54. Water is added to 100 mL of 0.15 mol/L sodium nitrate, $NaNO_3(aq)$, to make 700 mL of diluted solution. Calculate the molar concentration of the diluted solution.

55. A solution is made by diluting 25 mL of 0.34 mol/L calcium nitrate, $Ca(NO_3)_2(aq)$, solution to 100 mL. Calculate the following concentrations for the solution:
  a. the concentration of calcium nitrate
  b. the concentration of nitrate ions

56. A laboratory stockroom has a stock solution of 90% (m/v) sulfuric acid, $H_2SO_4(aq)$. If a technician dilutes 50 mL of the stock solution to a final volume of 300 mL, what will be the new mass/volume percent concentration? (**Hint:** The dilution formula can be used for concentration expressed in any units, provided that the units remain the same.)

57. What volume of 1.25 mol/L potassium iodide solution can you make with 125 mL of 3.00 mol/L potassium iodide solution?

58. Hydrochloric acid is available as a stock solution with a concentration of 10 mol/L. If you need 1.0 L of 5.0 mol/L hydrochloric acid, what volume of stock solution should you measure out? Approximately how much distilled water will you need to make the dilution?

59. Write a procedure you could use to make each aqueous solution using a solid solute.
  a. 0.50 L of 0.25 mol/L silver nitrate, $AgNO_3(aq)$
  b. 125 mL of 0.350 mol/L potassium carbonate, $K_2CO_3(aq)$
  c. 4.00×10^2 mL of 0.200 mol/L potassium permanganate, $KMnO_4(aq)$

60. Outline a procedure for making each aqueous solution by diluting a stock solution.
  a. 0.50 L of 1.0 mol/L sodium hydroxide, $NaOH(aq)$, using 17 mol/L sodium hydroxide
  b. 150 mL of 0.300 mol/L ammonia, $NH_3(aq)$, using 6.0 mol/L ammonia
  c. 1.75 L of 0.0675 mol/L ammonium bromide, $NH_4Br(aq)$, using 0.125 mol/L ammonium bromide

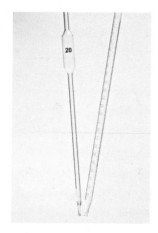

Figure 8.22 Volumetric and graduated pipettes accurately measure volumes.

Preparing a Standard Aqueous Solution by Diluting a Solution

Many of the solutions that are used in a chemistry laboratory are prepared by diluting a concentrated stock solution. You can use a graduated cylinder or a pipette to measure the volume of stock solution that will be diluted. A pipette is an instrument for measuring and transporting a volume of liquid. A pipette is more accurate than a graduated cylinder. There are different types and sizes of pipettes, as shown in **Figure 8.22**. The two most common types are the *volumetric* (or transfer) pipette and the *graduated* (or Mohr) pipette. A volumetric pipette, like a volumetric flask, has a single graduation mark on its stem. Volumetric pipettes are used for measuring common volumes, such as 5, 10, 25, or 50 mL. A graduated pipette can be used to measure any volume within the range of its graduation markings. Suction bulbs or pipette pumps are used for drawing liquid up into the pipette, as described in **Table 8.8**.

After a stock solution is measured with a pipette, the measured solution is placed in a volumetric flask. Water is then added to the flask, similar to the way water is added when preparing a standard aqueous solution using a solid solute. However, if a concentrated acid is being diluted, special safety precautions must be followed, as described on page 388.

Table 8.8 Using a Volumetric Pipette to Measure the Volume of a Stock Solution for Dilution

1. Make sure that the outside of the pipette, especially the tip, is dry. If not, wipe it with a paper towel.	
2. Squeeze the pipette bulb, and then place it over the top of the pipette. If using a pipette pump, place it over the top of the pipette.	
3. Rinse the pipette as follows. Place the tip of the pipette below the surface of the stock solution. Release the bulb carefully to draw up some liquid until the pipette is about half full. Remove the pipette bulb, invert the pipette, and drain the liquid into a beaker for waste. Repeat this rinsing two or three times.	
4. Fill the pipette with stock solution so that the level is past the graduation mark, but do not allow stock solution into the pipette bulb.	
5. Remove the pipette bulb, and quickly seal the top of the pipette with your finger or thumb.	
6. Remove the tip of the pipette from the stock solution. Lift your finger slightly, and let stock solution drain out slowly until the meniscus reaches the graduation mark. Wipe the tip of the pipette with a piece of paper towel.	
7. Move the pipette to the container into which you want to transfer the stock solution. Touch the tip of the pipette against the inside of the container, and release your finger to allow the liquid to drain. A small volume of liquid will remain inside the pipette. The pipette has been calibrated to allow for this volume of liquid. *Do not* force this liquid from the pipette.	

Suggested Investigation

Inquiry Investigation 8-D, Preparing and Diluting a Standard Solution

13. What is the advantage of preparing a standard solution from a solid solute rather than a stock solution?

14. Name two pieces of glassware that are used by chemists to prepare a solution with a known concentration.

15. Briefly describe two different ways to make 1.000 L of an aqueous solution with a known concentration.

16. Explain why the amount of solute is the same before and after dilution, but the concentration is less.

17. You are about to use a pipette to add 10.0 mL of 0.5 mol/L $CuSO_4(aq)$ to a beaker, when you see that the beaker contains a few drops of water. You previously rinsed the beaker using distilled water, but you did not dry the beaker. The amount of copper sulfate is important in the reaction you will be performing. Should you dry the beaker now or not worry about it? Explain your answer.

18. You need to make a solution that contains lead ions in a concentration of 0.002 g/L. The limit of accuracy of the balance you have is about 1 mg. How would you prepare the solution?

Safety Considerations When Diluting Acids

Some concentrated acids are dangerous. When diluting a concentrated acid, teachers, chemists, and lab technicians must follow strict safety procedures. A material safety data sheet (MSDS) is available for every hazardous chemical. An MSDS lists the properties of a chemical and the procedures for handling it safely.

A concentrated acid should always be diluted in a fume hood, such as the one shown in **Figure 8.23**, because breathing in the fumes causes acid to collect in the air passages and lungs. Rubber gloves are necessary to protect the hands, and a lab coat is necessary to protect clothing. Even a small splash of a concentrated acid will make a hole in fabric. Safety eyewear is also essential.

Mixing a strong, concentrated acid with water is a highly *exothermic* process—it releases a lot of heat. A concentrated acid is denser than water. When poured into water, such acids sink into the water and dissolve in solution. The heat that is generated is spread throughout the solution. As a result, this procedure is relatively safe. However, adding water to a concentrated acid is *not* safe. If water were added to a concentrated acid, the water would float on top of the solution. The heat generated could easily boil the solution at the acid-water boundary and splatter highly corrosive liquid. Also, the sudden change in temperature could crack the glassware, causing a dangerous spill. The rule when diluting concentrated acids is to add acid to water. You might remember this more easily with this memory aid: "add acid to water, like you oughter."

Figure 8.23 Proper safety precautions, such as working in a fume hood, must be taken whenever you use a concentrated acid.

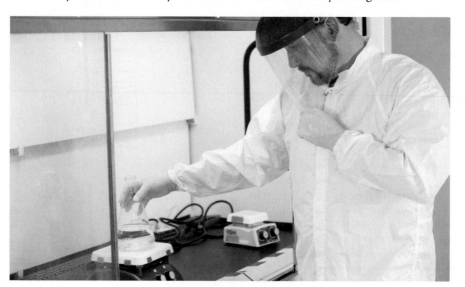

QUIRKS & QUARKS

with BOB MCDONALD

CBC ⊛

THIS WEEK ON QUIRKS & QUARKS

Antimony: The New Lead

Worldwide efforts to lower the level of lead in the atmosphere have been quite successful over the last 30 years. But now some scientists are worried about the level of another element: the metalloid antimony, Sb(s). The chemical processes in the human body do not use antimony. Antimony is toxic, with effects similar to those of arsenic. Dr. William Shotyk, an environmental geochemist at the University of Heidelberg in Germany, has been measuring the levels of lead and antimony in Europe and the Arctic. He has discovered that, while the level of lead has been dropping, the level of antimony has been increasing. Bob McDonald interviewed Dr. Shotyk to learn about antimony contamination in the environment.

Antimony, a shiny metalloid, is used as a catalyst to make PET, a plastic often used for drink bottles. A small amount of antimony remains in the plastic after manufacturing.

Useful but Dangerous

Antimony is present in many everyday products, including computers, batteries, and some types of glass. However, much of the antimony that is produced globally is used as a flame retardant in textiles and as a catalyst for making plastics. Some of these textiles and plastic products are incinerated when they are thrown out. Incineration releases tiny airborne particles of antimony compounds, which can travel thousands of kilometres. Dr. Shotyk thinks that these particles contribute to the rising level of antimony in arctic ice.

As well, Dr. Shotyk found that antimony leaches from polyethylene terephthalate (PET) bottles, which are commonly used for water and other drinks. The longer the water is in the bottle, the more antimony that is dissolved in the water. To eliminate this hazard, Japan is now using titanium as a catalyst for producing PET plastic bottles. Titanium is non-toxic and insoluble in water.

QUESTIONS

1. Describe the origins of antimony pollution.

2. Research the industrial life cycle of antimony. How is antimony produced? What processes and products is it used for? Can antimony be recycled?

3. Scientific research often involves a team of people, with different skills and education. List three careers related to geochemistry. Describe the skills and education required for each of these careers.

Section Summary

- A standard solution can be made from a solid solute by using a volumetric flask.

- Stock solutions can be used to produce more dilute solutions.

- When a solution is diluted, the volumes and concentrations before and after dilution are related by the dilution equation: $c_1V_1 = c_2V_2$.

- The safe procedure for diluting a concentrated acid is to add the acid to water.

Review Questions

1. **K/U** Distinguish between a standard solution and a stock solution.

2. **K/U** Why should a hot solution never be poured into a volumetric flask?

3. **T/I** A solution is prepared by dissolving 25.4 g of copper(II) sulfate in a 1 L volumetric flask. Determine the molar concentration of this solution.

4. **C** Your class needs 4.0 L of aqueous sulfuric acid, $H_2SO_4(aq)$, solution with a concentration of 0.10 mol/L. Outline a procedure that your teacher could use to make this solution from a stock solution of 12 mol/L aqueous sulfuric acid. Include any safety procedures that must be followed.

5. **T/I** What volume of 0.250 mol/L solution could you make with 55.9 g of potassium chloride, KCl(s)?

6. **T/I** Distilled water and 6.00 mol/L hydrochloric acid solution are mixed to produce 750 mL of a solution with a concentration of 2.00 mol/L. What volume of the hydrochloric acid solution is needed? Estimate the volume of water added.

7. **T/I** What mass of sodium acetate, $NaCH_3COO(s)$, is needed to make 40.0 mL of a solution with a concentration of 1.25 mol/L?

8. **T/I** A chemical company accidentally released 475 L of a 5.50 mol/L solution of a toxic compound into a nearby stream. The stream flows into a small pond. What volume of water would the pond have to contain to dilute the compound to a safe concentration of 0.35 μmol/L? (1 μmol = 10^{-6} mol)

9. **A** A student makes the following errors while preparing a solution. Describe how each error affects the concentration of the resulting solution. Explain what the student should do to avoid the error.

   a. The student dissolves the solute in a beaker and transfers the solution to a volumetric flask, but forgets to rinse the beaker and add the rinse to the volumetric flask.

   b. The student looks at the graduation mark on the volumetric flask from above, as shown.

10. **A** A tanker car, carrying concentrated sulfuric acid, $H_2SO_4(aq)$, jumps the train track. At the scene, the firefighters find that acid is leaking from the tanker car. Would it be safe for the firefighters to use water from their hoses to dilute the acid? What precautions should they take?

11. **C** A stock solution of 16 mol/L nitric acid, $HNO_3(aq)$, is available. Describe a safe procedure for using this stock solution to make 2.0 L of 4.0 mol/L aqueous nitric acid solution.

12. **T/I** Hydrogen peroxide solution, $H_2O_2(aq)$, is available with a concentration of 1.667 mol/L. When diluted to a concentration of 0.25 mol/L, the solution is used as a disinfectant. What volume of water should be added to 100 mL of the concentrated hydrogen peroxide solution to prepare the disinfectant?

13. **A** Describe how each of the following undesirable properties would affect the concentration of a standard solution, made from a solid solute.

   a. The solid contains unreactive impurities.

   b. The solid absorbs water vapour from the air.

   c. The solid slowly decomposes during storage.

14. **C** Potassium hydrogen phthalate (KHP) is a solid that is used to make a standard solution, which reacts with bases. Outline a procedure for making 250.0 mL of a 0.1000 mol/L aqueous solution of KHP. The molar mass of KHP is 204.2 g/mol.

Safety Precautions

- Wear safety eyewear, a laboratory apron, and gloves throughout this investigation.

- Avoid contact with iodine. It can discolour your skin.

- Mineral oil is flammable. Keep it away from open flames.

- Wash your hands after performing this investigation.

Suggested Materials

- mineral oil
- water
- table salt (sodium chloride), $NaCl(s)$
- iodine, $I_2(s)$
- sugar (sucrose), $C_{12}H_{22}O_{11}(s)$
- petroleum jelly
- paraffin wax, $C_{25}H_{52}(s)$
- glycerol, $C_3H_5(OH)_3(\ell)$
- scoopula
- 12 test tubes
- test-tube rack
- stoppers to fit the test tubes

*Go to **Organizing Data in a Table** in **Appendix A** for help with constructing data tables.*

Investigating Solutes and Solvents

In this investigation, you will make a hypothesis about the types of substances that dissolve in two different solvents: water, which is polar, and mineral oil, which is non-polar. Then you will design and perform an experiment with various solutes to test your hypothesis.

Pre-Lab Questions

1. Sketch the structure of a water molecule. Label your diagram to explain why a water molecule is polar.

2. Mineral oil is a solution of hydrocarbons, which are compounds that contain only hydrogen and carbon atoms. Why are hydrocarbons non-polar?

3. If you perform an investigation to compare the solubility of table salt in water with its solubility in mineral oil, which variables should be kept constant?

4. All of the substances used in this investigation are commonly found around the home. Why must you still wear gloves while handling these substances?

Question

What hypothesis can you form about solubility and the nature of solutes and solvents?

Prediction

Consider these six solutes: table salt, iodine, sugar, petroleum jelly, paraffin wax, and glycerol. Make a table to list your predictions about which of these solutes will dissolve in the two solvents, water and mineral oil. For each combination of solute and solvent, put "yes" if you think the solute will dissolve, "no" if you think it will not dissolve, or "?" if you are uncertain. Justify your predictions.

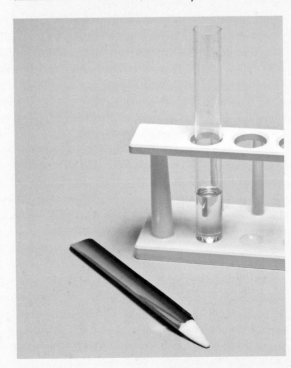

Plan and Conduct

1. Write a procedure for your investigation, and describe how you will control the experimental variables. Be sure to list relevant safety precautions in your procedure. Note that this investigation can be performed using small volumes of solvent.

2. Make a table to record your observations. Make sure that your table has spaces to record information about all the solutes and solvents that you will use.

3. Make a list of the solutes, classifying each one as ionic, polar, or non-polar. If you are uncertain about how to classify a solute, consider it as having both polar and non-polar properties.

4. Obtain your teacher's approval for your procedure, and then carry out your investigation. When you are finished, dispose of the materials and clean the equipment as directed by your teacher.

Analyze and Interpret

1. List the substances that dissolved in water.

2. List the substances that dissolved in mineral oil.

3. List any substance that dissolved in both water and mineral oil.

4. If you changed your mind about classifying a solute as ionic, polar, or non-polar because of your investigation, write a sentence describing the evidence for the new classification.

5. Describe any sources of error in your investigation. How could you change your procedure or materials to improve your investigation?

Conclude and Communicate

6. Are substances that dissolve in water ionic, polar, or non-polar? Explain your reasoning.

7. Are substances that dissolve in mineral oil ionic, polar, or non-polar? Explain your reasoning.

Extend Further

8. **INQUIRY** Outline a procedure for determining how readily an aqueous solution of a soap dissolves the substances you identified as insoluble in water.

9. **RESEARCH** Research the structure of the compounds that are used to make soaps and detergents. Why are these substances soluble in water? How are they able to dissolve grease? What is the difference between a soap and a detergent?

Safety Precautions

- Wear safety eyewear and a laboratory apron throughout this investigation.

- If you are using a Bunsen burner, check that there are no flammable solvents nearby. Tie back loose hair and clothing.

- Use EXTREME CAUTION when you are near an open flame.

- If you are using a hot plate, be careful not to touch the hot surface.

- Do not unplug the hot plate by pulling on the cord.

Materials

- 9.0 g of potassium nitrate, $KNO_3(s)$

- 20 mL of distilled water

- 300 mL of tap water

- balance

- large test tube

- graduated pipette

- stirring wire

- thermometer

- 400 mL beaker

- hot plate *or* Bunsen burner secured to the retort stand

- wire gauze square

- retort stand with test-tube clamp, thermometer clamp, and one or two ring clamps

*Go to **Constructing Graphs** in Appendix A for help with graphing.*

Plotting a Solubility Curve

A solubility curve is a graph that shows the relationship between temperature and the solubility of a solute. The solubility curve can be used to determine the solubility of a solute at any temperature in the range that it covers. In this investigation, you will create a series of saturated solutions to determine the solubility of potassium nitrate, $KNO_3(s)$, at various temperatures.

Pre-Lab Questions

1. Read the entire investigation, and identify the independent variable. On which axis should you plot this variable?

2. Why is it important to measure the mass of solute accurately in the second step of the Procedure?

3. Do you expect potassium nitrate to be more or less soluble in water at higher temperatures? Justify your answer.

4. Why is a retort stand required for this investigation even if you are using a hot plate and not a Bunsen burner?

Question

How does temperature affect the solubility of potassium nitrate in water?

Prediction

Sketch a solubility curve that shows the relationship you expect for the solubility of a typical solid dissolved in water at different temperatures. Label both axes of your graph.

Procedure

1. Prepare a data table, like the one shown below, in your notebook. Have rows for up to six sets of data.

Solubility of Potassium Nitrate

Mass of $KNO_3(s)$:_____ g

Data Table

Temperature (°C)	Volume of Water (mL)	Concentration of $KNO_3(s)$ (g/mL)	Solubility of $KNO_3(s)$ (g/100 mL)

2. Place a large test tube on the balance, and zero the balance. Add about 9.0 g of potassium nitrate to the test tube. Measure and record the mass of the potassium nitrate.

3. Pour about 300 mL of tap water into the beaker. Place the beaker on a wire gauze square and a hot plate, or use a ring clamp and a wire gauze square to support the beaker over a Bunsen burner. Use a ring clamp to secure the beaker to the retort stand.

4. Use a graduated pipette to add 10.0 mL of distilled water to the test tube. Use a retort stand with a test-tube clamp to hold the test tube in the water bath. Place a thermometer and stirring wire in the test tube. Use a thermometer clamp to position the thermometer in the test tube. Make sure that the thermometer bulb is below the surface of the solution and that the bottom of the test tube does not touch the beaker. Check the diagram below to make sure that you have the apparatus properly assembled.

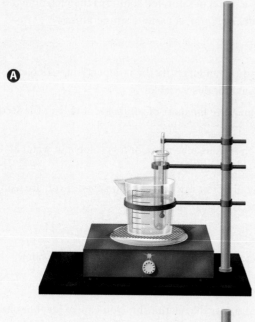

A

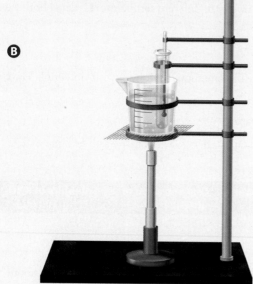

B

Securely clamp the thermometer, the test tube, and the beaker when using either a hot plate (**A**) or a Bunsen burner (**B**).

5. While stirring, heat the contents of the test tube to about 50 to 55°C. Heat and stir the solution until all the solute *just* dissolves. Turn off the heat.

6. Stir the solution as it cools. At the first sign of crystals forming in the solution (they may appear as a faint haze), read the temperature and record it in your table.

7. Use a graduated pipette to add 2.0 mL of distilled water to the test tube. Stir the solution. If necessary, warm the solution to re-dissolve the solute.

8. Repeat steps 6 and 7 until you have recorded the temperature at which crystals form with 20 mL of water in the test tube. You may have to stop sooner if there is not enough time to complete six temperature measurements.

9. Dispose of the potassium nitrate solution and clean the equipment as directed by your teacher.

Analyze and Interpret

1. Calculate the concentration data for the third column of your table by dividing the mass of potassium nitrate by the volume of water in the test tube. Multiply the result by 100 to determine the solubility, in grams per 100 mL, to enter in the fourth column of your table.

2. Plot a graph of the results, with solubility on the *y*-axis and temperature on the *x*-axis. Label the axes. Draw the best smooth curve between the points. (Do not simply join the points.) Give your graph an appropriate title.

Conclude and Communicate

3. Describe how the solubility of potassium nitrate in water varies with temperature.

4. Compare your graph with the sketch you made to predict the relationship between the solubility of a solute and the temperature of water. Explain any differences between your predicted graph and your experimental results.

5. Use your graph to estimate the solubility of potassium nitrate at
   a. 45°C   **b.** 60°C   **c.** 25°C

Extend Further

6. INQUIRY If 25 mL of a saturated solution of potassium nitrate at 50°C is cooled to 30°C, what mass of solute should crystallize? Outline an experimental procedure to check your prediction and, with your teacher's permission, carry out your procedure.

7. RESEARCH Use the Internet or other resources to research the effects of heat transfers from power plants, industrial processes, and cities to nearby bodies of water. Describe how these heat transfers affect the solubility of pollutants and other substances in the bodies of water.

Safety Precautions

- Wear safety eyewear and a laboratory apron throughout this investigation.
- Your investigation must include appropriate safety measures.
- Obtain your teacher's permission before proceeding with your investigation.

Suggested Materials

- aqueous solution (Your teacher will let you know what the solute is.)
- any apparatus that is available in the laboratory

Go to **Organizing Data in a Table** in **Appendix A** for help with constructing a data table.

Determining the Concentration of a Solution

Your teacher will give you a solution that was made by dissolving a solid solute in water. You will design and perform an investigation to determine the concentration of the solution.

Pre-Lab Questions

1. Which two measurements must be made to determine the concentration of a solute in a solution?
2. Write the formula for the mass percent concentration of a solution.
3. Write the formula for the molar concentration of a solution.
4. What safety measures will you take during this investigation?

Question

How can you determine the concentration of a solution?

Plan and Conduct

1. Design a procedure that will allow you to measure the quantities needed to calculate the concentration of a solution. Be sure to list relevant safety precautions in your procedure.
2. Design a data table for your results. Include a space for the name of the solute in your solution. Give your table an appropriate title.
3. When your teacher approves your procedure, proceed with your investigation.
4. Dispose of your solution and clean the equipment as directed by your teacher.

Analyze and Interpret

1. Express the concentration of the solution you analyzed as a mass percent and as a molar concentration. Show your calculations.

Conclude and Communicate

2. List at least two possible sources of error in your measurements.
3. List at least two ways that you could improve your procedure.
4. Some compounds partially decompose when they are heated, producing a gas and another solid. If your solution contained such a compound, how would it affect the results of your investigation?

Extend Further

5. **INQUIRY** Copper(II) nitrate decomposes on heating to form copper(II) oxide, nitrogen dioxide gas, and oxygen gas. If you were given an aqueous solution of copper(II) nitrate, how could you determine its concentration?
6. **RESEARCH** Use the Internet or other resources to learn how the properties of a solution are related to the properties and concentration of the solute. Describe several different techniques for determining the concentration of a solution.

Safety Precautions

- Wear safety eyewear and a laboratory apron throughout this investigation.

- Copper(II) sulfate pentahydrate is poisonous. Wash your hands at the end of this investigation.

- If you spill any solution on your skin, flush your skin immediately with a lot of cool water.

- Do not unplug the projector by pulling on the cord.

Materials

- copper(II) sulfate pentahydrate, $CuSO_4 \cdot 5H_2O(s)$

- distilled water

- standard solutions of copper(II) sulfate pentahydrate in large test tubes, labelled 1 to 6

- electronic balance

- scoopula

- beaker

- stirring rod

- funnel

- 100 mL volumetric flask

- pipette

- pipette bulb or pipette pump

- large test tube

- marker pen or pencil

- medicine dropper

- overhead projector or light box

- paper towel

Preparing and Diluting a Standard Solution

Copper(II) sulfate pentahydrate, $CuSO_4 \cdot 5H_2O(s)$, is moderately toxic. It is sometimes used to prevent the growth of fungi and algae. In this investigation, you will prepare a standard solution of copper(II) sulfate pentahydrate. Then you will dilute the solution to a specific concentration. You will check the concentration by comparing the colour intensity of the diluted solution with the colour intensities of solutions with known concentrations.

Pre-Lab Questions

1. Determine the following quantities. Show your calculations.
   a. the molar mass of copper(II) sulfate pentahydrate
   b. the amount (in moles) of copper(II) sulfate pentahydrate that you will need to make 100 mL of a 0.500 mol/L solution
   c. the mass of copper(II) sulfate pentahydrate that you will need to make 100 mL of a 0.500 mol/L solution

2. Write a procedure that you can use to prepare 100 mL of a standard aqueous solution of 0.500 mol/L copper(II) sulfate pentahydrate.

3. Why is washing your hands after doing this investigation particularly important?

Question

How can you use colour to compare the concentrations of solutions?

Procedure

1. Show your answers to the Pre-Lab Questions to your teacher. Once your teacher has approved the procedure you designed, use it to prepare the standard solution.

2. Calculate the volume of a 0.500 mol/L copper(II) sulfate pentahydrate solution that you will need to prepare 100 mL of 0.0500 mol/L solution.

3. Choose an appropriate pipette for the volume you calculated. Transfer this volume of the standard solution to a clean 100 mL volumetric flask.

4. Make the diluted standard solution of 0.0500 mol/L copper(II) sulfate pentahydrate solution.

5. Label a large test tube. Transfer some of your diluted (0.0500 mol/L) solution to the large test tube. Compare your solution with the standard solutions provided by your teacher. The depth of the solution must be the same in all the test tubes. Use a medicine dropper to adjust the volume of solution in your labelled test tube, if necessary.

6. Compare the colour intensities and concentrations of the standard solutions provided by your teacher. Note any pattern. You can compare the colour intensities by looking down through the test tubes while holding them over a bright, diffuse light source, as shown in the photograph. Wrap each test tube with a small piece of paper towel to stop light from entering the side.

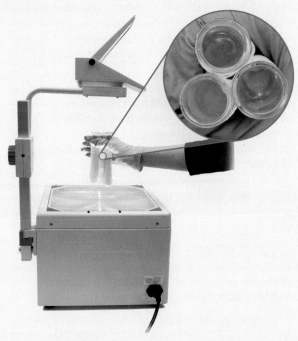

7. Compare the colour of your diluted solution with the colours of the standard solutions. Identify the standard solution that most closely matches the diluted solution you prepared.

8. Dispose of the solutions and clean the equipment as directed by your teacher.

Analyze and Interpret

1. Suppose that you had used the same volumetric flask for preparing the standard solution in step 1 and the dilution in steps 3 and 4. How would the concentration of the dilution have been affected if you had not rinsed the flask before using it again in step 3? Explain your reasoning.

2. What sources of error could have affected the concentration of the solutions you prepared? How large of an effect do you think these errors had?

3. Why was it important, in step 5, to make sure that all the solutions were the same depth?

Conclude and Communicate

4. Based on your observations, how is the appearance of a coloured solution related to its concentration? Explain your reasoning.

5. Estimate the concentration of your diluted solution, based on comparing it with the standard solutions. Does your estimate indicate any errors in your preparation of the diluted solution?

Extend Further

6. **INQUIRY** Suppose that you are given an aqueous solution of copper(II) sulfate pentahydrate with an unknown concentration. You observe the colour of the solution to be less intense than the colour of a standard 0.500 mol/L solution. You have 250 mL of the standard solution available. Write a procedure that uses the technique of dilution to estimate the unknown concentration.

7. **RESEARCH** Use the library or Internet sources to find out how the intensity of the colour of a solution can be measured quantitatively. Describe one method.

Section 8.1 | Classifying Solutions

Every solution is a homogeneous mixture, with a composition that can be described both qualitatively and quantitatively.

KEY TERMS

aqueous solution	solution
saturated solution	solvent
solubility	supersaturated solution
solute	unsaturated solution

KEY CONCEPTS

- A solution is a completely homogeneous mixture. It can be formed from a solute and a solvent in any initial state: solid, liquid, or gas.
- The solvent in a solution is the substance present in the greatest amount. Solutes are anything dissolved in a solvent.
- The solubility of a substance is stated in terms of the maximum amount of solute that dissolves in a given quantity of solvent at a specific temperature.
- A solution can be classified as unsaturated, saturated, or supersaturated, depending on the amount of dissolved solute per unit volume of solution.

Section 8.2 | Factors That Affect Solubility and Rate of Dissolving

The key factor that determines whether a solute is soluble in a solvent is the nature of the bonds within and between the solute and the solvent

KEY TERMS

hydration	rate of dissolving
hydrogen bonding	

KEY CONCEPTS

- Intermolecular forces affect solubility.
- Water is a good solvent because it can dissolve a wide range of solutes.

- Ionic compounds that contain ions with relatively large charges or small radii tend to be insoluble in water.
- Polar solvents tend to dissolve ionic and polar molecules. Non-polar solvents tend to dissolve non-polar molecules.
- Solubility is affected by temperature and pressure.
- The rate of dissolving a solid solute in a liquid solvent is affected by temperature, the surface area of the solute, and agitation.

Section 8.3 | Concentrations of Solutions

The concentration of a solution can be expressed quantitatively in several ways.

KEY TERMS

concentrated	parts per million (ppm)
concentration	percent (m/m)
dilute	percent (m/v)
molar concentration	percent (v/v)
parts per billion (ppb)	

KEY CONCEPTS

- The concentration of a solution can be stated in terms of the ratio of the volumes or masses of the solute and the solution, using percent (m/v), percent (m/m), or percent (v/v).
- Small concentrations can be expressed as parts per million (ppm) or parts per billion (ppb).
- Molar concentration is calculated by dividing moles of solute by litres of solution.

Section 8.4 | Preparing Solutions in the Laboratory

A standard solution can made from either a solid solute or a stock solution.

KEY TERMS

standard solution	volumetric flask

KEY CONCEPTS

- A standard solution can be made from a solid solute by using a volumetric flask.

- Stock solutions can be used to produce more dilute solutions.
- When a solution is diluted, the volumes and concentrations before and after dilution are related by the dilution equation: $c_1V_1 = c_2V_2$.
- The safe procedure for diluting a concentrated acid is to add the acid to water.

Knowledge and Understanding

Select the letter of the best answer below.

1. When a teaspoon of table salt is added to a glass of water, the salt and the water are, in order,
   a. the solute and the solution.
   b. the solvent and the solution.
   c. the solute and the solvent.
   d. the solvent and the solute.
   e. the solution and the solute.

2. A student adds a crystal of copper(II) sulfate to an aqueous solution of the same solute. After some time, the student observes that the size of the crystal does not appear to have changed. Which statement is probably true?
   a. The original solution was unsaturated.
   b. The original solution was saturated.
   c. The original solution was supersaturated.
   d. Copper(II) sulfate is slightly soluble in water.
   e. Copper(II) sulfate is insoluble in water.

3. Which substance has intact molecules when dissolved in water?
   a. copper(II) sulfate
   b. sodium hydroxide
   c. ammonium chloride
   d. sucrose
   e. potassium nitrate

4. Which compound is most soluble in water?
   a. barium sulfate
   b. calcium sulfate
   c. lead sulfate
   d. magnesium sulfate
   e. strontium sulfate

5. Which volume of 4.00×10^{-2} mol/L solution of barium nitrate, $Ba(NO_3)_2(aq)$, contains 2.00×10^{-2} mol of nitrate ions?
   a. 4.00 L
   b. 2.00 L
   c. 1.00 L
   d. 500 mL
   e. 250 mL

6. Which factor has the *least* effect on the solubility of an ionic compound in water?
   a. the temperature of the solvent
   b. the number of ions in each formula unit of the compound
   c. the ionic radius of the anions and cations in the solute
   d. the charge on the ions in the solute
   e. the pressure of gas above the solution

7. Which of the following is most likely to increase the solubility of carbon dioxide gas in a liquid solvent?
   a. increasing the temperature of the carbon dioxide
   b. increasing the temperature of the solvent
   c. lowering the air pressure above the solvent
   d. increasing the pressure of the carbon dioxide above the solvent
   e. using a solvent with polar molecules, such as water

8. What is the symbol for molar concentration?
   a. M
   b. C
   c. c
   d. m
   e. V

Answer the questions below.

9. Mixing 2 mL of linseed oil and 4 mL of turpentine makes a binder for oil paint. Which liquid is the solvent?

10. Why must you shake oil and vinegar salad dressing before using it?

11. How does pressure affect the solubility of solids, liquids, and gases that dissolve in water?

12. Sulfur dioxide, $SO_2(g)$, is produced by volcanoes and by various industrial processes. The solubility of sulfur dioxide in water is 9.4 g/100 mL at 25°C.
    a. Is sulfur dioxide classified as soluble or insoluble in water at 25°C? Give reasons for your answer.
    b. Is sulfur dioxide likely to be a polar molecule or a non-polar molecule? Explain your reasoning.

13. Would you expect the following substances to be soluble or insoluble in water? Briefly explain your reasoning for each.
    a. potassium chloride, KCl(s)
    b. carbon tetrachloride, $CCl_4(\ell)$
    c. sodium sulfate, $Na_2SO_4(s)$

14. What is the key difference between percent concentration and molar concentration?

15. How does a volumetric pipette differ from a graduated pipette?

16. What safety precautions should be taken when making a dilute solution from highly concentrated hydrochloric acid?

Thinking and Investigation

17. Ammonia is a gas at room temperature and pressure, but it can be liquefied easily. What substances would you expect to be soluble in liquid ammonia? Explain your reasoning.

18. Isopropyl (rubbing) alcohol is sometimes sold with a concentration of 90% (v/v).

   a. What volume of isopropyl alcohol does a 50 mL bottle of this concentration contain?

   b. What volume of 70% (v/v) isopropyl alcohol can you make by diluting 50 mL of 90% (v/v) isopropyl alcohol?

19. Naval brass, an alloy of copper with zinc and tin, resists corrosion by salt water. A 6.0 kg ship's bell made of this alloy contains 234 g of zinc and 60 g of tin. Determine the percent (m/m) concentration of each metal in the bell.

20. Ammonium nitrate is a commonly used fertilizer. It is very soluble in water. In Ontario, the maximum allowable concentration of nitrate ions, $NO_3^-(aq)$, in drinking water is 10 ppm. What is the maximum amount of nitrate ions in 250 mL of water that meets this standard?

21. Determine the molar concentration of distilled water. Assume that the mass of 1 L of water is 1000 g.

22. If young children have an illness that causes diarrhea and vomiting, they can quickly become seriously ill from dehydration and the loss of sodium and other vital chemicals. Rehydration solutions contain sodium along with chemicals to help the intestine absorb the sodium. This table lists the solutes in a rehydration solution that is designed for children.

Composition of Pediatric Rehydration Solution	
Solute	Concentration (g/100 mL)
sodium chloride, $NaCl(aq)$	0.205
dextrose, $C_6H_{12}O_6(aq)$	2.5
potassium citrate, $K_3C_6H_5O_7(aq)$	0.204
sodium citrate, $Na_3C_6H_5O_7(aq)$	0.086

   a. Calculate the mass/volume percent concentration of potassium citrate in the rehydration solution.

   b. Calculate the approximate concentration of sodium citrate in parts per million. State any assumptions you make to do this calculation.

   c. Calculate the molar concentration of dextrose.

   d. Calculate the molar concentration of sodium ions, $Na^+(aq)$, from sodium citrate.

   e. Calculate the total molar concentration of sodium ions in the solution.

23. A teacher wants to dilute 200 mL of 12 mol/L hydrochloric acid to make a 1 mol/L solution. Determine the approximate volume of water that is required for the dilution. State any assumptions you make.

24. Potassium alum, $KAl(SO_4)_2 \cdot 12H_2O(aq)$, can be used to stop bleeding from a small cut. The solubility of potassium alum at various temperatures is given below.

Solubility of Potassium Alum	
Temperature (°C)	Solubility (g/100 mL)
0	4
10	10
20	15
30	23
40	31
50	49
60	67
70	101
80	135

   a. Plot a graph of solubility versus temperature.

   b. Use your graph to estimate the solubility of potassium alum at 67°C.

   c. Use your graph to estimate the temperature at which 120 g of potassium alum will form a saturated solution in 100 mL of water.

25. Use the solubility curve below to determine

   a. the mass of sodium chloride that will dissolve in 1.0 L of water at 80°C.

   b. the minimum temperature that is required to dissolve 24 g of potassium nitrate in 40 g of water.

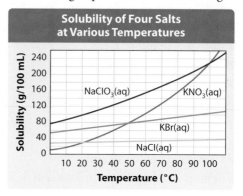

Communication

26. Write a procedure you could use to determine how agitation and the surface area of a solid solute affect the rate of dissolving.

27. **BIG IDEAS** Properties of solutions can be described qualitatively and quantitatively, and can be predicted.

a. Use diagrams to explain why the size and charge of the ions in a compound are key factors in determining whether the compound is soluble in water.

b. Describe how you can predict whether a covalent compound will be soluble in water.

28. A water molecule can form hydrogen bonds with other molecules.

a. Draw a diagram to show how hydrogen bonds form in pure water.

b. Explain how hydrogen bonds enable water to dissolve many ionic compounds and polar covalent compounds.

c. Explain how hydrogen bonds prevent most non-polar compounds from dissolving in water.

29. Draw a flowchart that shows how you could predict the solubility of an ionic compound.

30. Your teacher gives you three test tubes. Each test tube contains a clear, colourless liquid. One liquid is an aqueous solution of an ionic substance. Another liquid is an aqueous solution of a molecular substance. The third liquid is distilled water. Outline a procedure you could use to determine which liquid is which.

31. At 20°C, the solubility of oxygen gas in water is more than twice the solubility of nitrogen gas. A student in your class analyzed the concentration of dissolved gases in the water of an unpolluted pond. The concentration of oxygen was much less than twice the concentration of nitrogen. Prepare an explanation of these findings that the student could present to your class.

32. Draw diagrams to explain "like dissolves like" in terms of intermolecular forces. Use the concept of intermolecular forces to explain why oil does not dissolve in water.

33. Summarize your learning in this chapter using a graphic organizer. To help you, the Chapter 8 Summary lists the Key Terms and Key Concepts. Refer to Using Graphic Organizers in Appendix A to help you decide which graphic organizer to use.

Application

34. A crystalline mineral called cinnabar can be used to make an intense reddish-orange artist's pigment. Cinnabar consists of mercury(II) sulfide, HgS(s).

a. Predict the solubility of cinnabar pigment in water. Justify your prediction.

b. Predict the solubility of cinnabar pigment in oil. Justify your prediction.

c. Based on your predictions, would you classify paint that contains cinnabar pigment as a solution? Explain your reasoning.

35. The concentration of gold in seawater is very low and varies from one area to another. As a result, estimates of this concentration vary a great deal. One of the higher estimates is that 1000 t of seawater contains about 0.1 g of dissolved gold.

a. Express this concentration in parts per billion.

b. Express the concentration in moles per litre.

c. Comment on the economics of extracting gold from the oceans.

36. The table below gives the solubility of three different solids in water and mineral oil.

Solubilities of the Mixture Components		
Solid	Soluble in Water?	Soluble in Mineral Oil?
A	Yes	No
B	No	Yes
C	No	No

Suppose that you have a sample that contains a mixture of all three solids. Explain how you can separate the mixture to obtain a pure sample of each solid.

37. Sodium chloride is present in most types of soil. In some areas that have low rainfall but some ground water, salt deposits can form on the surface of the soil. Explain why such salt deposits occur.

38. Use the Internet or library resources to research homogenized milk. Describe how raw milk is homogenized. Explain why homogenized milk is not a solution. How is the mixture of substances in milk classified?

39. Radon is an invisible, odourless, radioactive gas. It is emitted from soil, rock, and building materials that contain traces of uranium. Use text or Internet resources to answer the following questions.

a. Which method of reporting radon concentration is commonly used?

b. What are the health risks of inhaling radon gas?

c. What measures can be taken to reduce the concentration of radon gas inside a home?

Select the letter of the best answer below.

1. **K/U** What is the effect of increased temperature on a substance dissolving in water?
   a. Most solid solutes become less soluble, but some are more soluble.
   b. All gases are more soluble.
   c. Most solid solutes are more soluble.
   d. The rate of dissolving decreases.
   e. Most solutions of solid solutes will become saturated.

2. **K/U** Which one of the following compounds will form an aqueous solution that conducts electricity?
   a. $I_2(s)$
   b. $CaCl_2(s)$
   c. $C_6H_{12}O_6(s)$
   d. $CH_3CH_2COH(\ell)$
   e. $C_6H_6(\ell)$

3. **K/U** Select the substance that is most soluble in water at 25°C.
   a. $O_2(g)$
   b. $AgCl(s)$
   c. $Mg_3(PO_4)_2(s)$
   d. $CuS(s)$
   e. $NH_4CH_3COO(s)$

4. **K/U** Which of the following changes would increase the amount of carbon dioxide gas dissolved in one litre of water?
   a. stirring the solution
   b. increasing the temperature of the gas
   c. decreasing the temperature of the solution
   d. decreasing the pressure of carbon dioxide
   e. adding another gas, such as nitrogen

5. **K/U** Which action will help you dissolve more sugar in a glass of iced tea?
   a. chilling the tea in a refrigerator
   b. stirring with a spoon
   c. increasing air pressure
   d. allowing some water to evaporate
   e. putting the tea in an insulated cup

6. **T/I** Which of the following solutions contains the greatest amount (in moles) of ions?
   a. 0.1 mol/L $C_6H_{12}O_6(aq)$
   b. 0.1 mol/L $NaCl(aq)$
   c. 0.1 mol/L $MgCl_2(aq)$
   d. 0.05 mol/L $Al_2(SO_4)_3(aq)$
   e. 0.05 mol/L $Na_3PO_4(aq)$

7. **T/I** What volume of 6.00% (v/v) $H_2SO_4(aq)$ is required to obtain 15.0 mL of pure sulfuric acid?
   a. 37.5 mL
   b. 90.0 mL
   c. 250 mL
   d. 600 mL
   e. 900 mL

8. **T/I** What is the concentration of a solution containing 1.00 g of sodium chloride dissolved in 430 mL of water?
   a. 2.23×10^{-3} mol/L
   b. 3.98×10^{-2} mol/L
   c. 0.136 mol/L
   d. 7.19 mol/L
   e. 7.36 mol/L

9. **T/I** What mass of potassium chromate, $K_2CrO_4(s)$ (molar mass 194.2 g/mol), must be dissolved in 2.00 L to prepare a 0.100 mol/L solution?
   a. 9.71 g
   b. 19.4 g
   c. 38.8 g
   d. 194 g
   e. 388 g

10. **T/I** What is the molar concentration of a solution prepared by adding 250 mL of 3.00 mol/L hydrochloric acid to 750 mL of water?
    a. 0.250 mol/L
    b. 0.750 mol/L
    c. 1.00 mol/L
    d. 1.25 mol/L
    e. 1.33 mol/L

Use sentences and diagrams as appropriate to answer the questions below.

11. **K/U** Give an example of a solution that has more than one solute, when the solution is in the following states:
    a. solid.
    b. liquid.
    c. gas.

12. **K/U** What will happen if you try to add more solid solute to a supersaturated solution?

13. **K/U** Explain why small amounts of insoluble substances can be dissolved in water.

14. **K/U** Use your understanding of the structure of the water molecule to explain why water is such a good solvent.

15. **T/I** Predict whether each compound is soluble in water. Briefly explain your reasoning.
    a. $KNO_3(s)$
    b. $BaSO_4(s)$
    c. $CaCO_3(s)$
    d. $Al(OH)_3(s)$

16. [A] Nail polish remover contains acetone, $CH_3COCH_3(\ell)$. Use the structure of acetone, shown below, to explain why it is soluble in water and a good solvent for many molecular compounds.

17. [T/I] Sodium nitrate is sometimes added to processed meats as a preservative. Use **Figure 8.11** to estimate the solubility of sodium nitrate at 25°C.

18. [T/I] In jewellery, the karat is used as the unit representing the proportion of gold. Pure gold is 24 karat gold. What is the percent (m/m) gold present in a 14 karat ring?

19. [T/I] Your blood contains about 0.72% (v/v) white blood cells (leukocytes), which are essential for fighting infections. Assuming you have a blood volume of about 4.0 L, what volume of pure leukocytes is present?

20. [T/I] Symptoms of mercury poisoning appear after a person has accumulated more than about 20 mg of mercury in their body. If a 50 kg person has 20 mg of mercury in their body, what is the concentration expressed in ppm?

21. [T/I] Each year, outbreaks of "red tide," a sudden growth in toxic algae, shut down parts of Canada's shellfish industry by contaminating popular shellfish like clams and oysters with dangerous toxins. The increased concentration of phytoplankton colours the seawater. A researcher measured the concentration of phytoplankton as 5×10^{-4}% (m/v). What volume of seawater contains 1.0 g of phytoplankton?

22. [T/I] Concentrated nitric acid, $HNO_3(aq)$, is available as a stock solution with a concentration of 15.9 mol/L. If you need 3.00 L of 5.00 mol/L nitric acid, what volume of stock solution should you measure out? Approximately how much distilled water will you need to make the dilution?

23. [A] A technician in a research laboratory was asked to prepare a standard solution of 1.00 mol/L $Na^+(aq)$. The chemicals available include NaCl(s) and 6.00 mol/L NaOH(aq). Which should the technician use to make the standard solution? Explain your choice and outline how the solution should be prepared.

24. [C] An aqueous solution of potassium permanganate, $KMnO_4(aq)$, is used as an antiseptic and a fungicide. Describe how to make 250 mL of 0.100 mol/L potassium permanganate.

25. [A] You prepare two aqueous solutions for an experiment, each with the same mass of solute, and each with the same volume. Solution A is silver nitrate, $AgNO_3(aq)$, and solution B is potassium chromate, $K_2CrO_3(aq)$.
  a. Which solution has greater molar concentration? Explain your answer.
  b. Write the name and formula of the precipitate formed when the solutions are mixed together.

26. [T/I] Aqueous ammonia, $NH_3(aq)$, with a concentration of 14.6 mol/L, is sold by chemical supply companies. A certain brand of household glass cleaner is a 4.0 mol/L aqueous solution of ammonia. What volume of the concentrated ammonia solution is needed to prepare 750 mL of the glass cleaner?

27. [A] A solution of sodium hydroxide can absorb carbon dioxide from the air and then react to precipitate sodium carbonate, $Na_2CO_3(s)$.
  a. Explain how this reaction would affect the concentration of a standard 1.00 mol/L solution of sodium ions, $Na^+(aq)$, prepared from 6.00 mol/L aqueous sodium hydroxide.
  b. How could you prepare a more accurate standard solution of sodium ions?

28. [C] Water that percolates through a landfill produces leachate, a solution of substances from the waste in the landfill. The bottom of a landfill is constructed to prevent leachate from contaminating the ground water. Prepare a presentation that explains why a mixture that contains both ionic and non-polar compounds presents a challenge for containing leachate.

Self-Check

If you missed question …	1	2	3	4	5	6	7	8	9	10	11	12	13	14	15	16	17	18	19	20	21	22	23	24	25	26	27	28
Review section(s)…	8.2	8.2	8.2	8.2	8.2	8.3	8.3	8.3	8.3, 8.4	8.4	8.1	8.1	8.1	8.2	8.2	8.2	8.2	8.3	8.3	8.3	8.3	8.4	8.4	8.4	8.4, 8.3, 8.2	8.4	8.4	8.2

Reactions in Aqueous Solutions

Specific Expectations

In this chapter, you will learn how to . . .

- E1.1 **analyze** the origins and cumulative effects of pollutants that enter our water systems, and **explain** how these pollutants affect water quality (9.3)

- E1.2 **analyze** economic, social, and environmental issues related to the distribution, purification, or use of drinking water (9.4)

- E2.5 **write** balanced net ionic equations to represent precipitation and neutralization reactions (9.1)

- E2.6 **use** stoichiometry to **solve** problems involving solutions and solubility (9.2)

- E2.8 **conduct** an investigation to determine the concentrations of pollutants in your local treated drinking water, and **compare** the results to commonly used guidelines and standards (9.3)

- E3.4 **identify**, using a solubility table, the formation of precipitates in aqueous solutions (9.2)

Life exists almost everywhere on Earth, even under extreme conditions. Worms and crabs live in darkness on the sea floor, at enormous pressure, near hot volcanic vents called *black smokers*. Mosses grow on rocks in the extreme cold of northern Canada. Some types of bacteria can live only in airless conditions, while other types thrive in scalding hot springs. All known forms of life have one thing in common—water. The ability of water to dissolve many different compounds allows it to carry the nutrients that are required by the vast range of life forms on Earth. However, water can also carry harmful compounds, both natural substances and pollutants. In this chapter, you will learn more about how to represent reactions in solutions and how to measure the reactants and products in these reactions. You will also learn about the sources of contaminants in water, the health problems that these contaminants can cause, and some techniques that can be used to purify contaminated water.

Easy Tests on Hard Water

Tap water often contains ions that react with soap to form insoluble substances. In *hard water*, the concentrations of these ions are high enough to interfere with the formation of a lather. *Soft water* contains relatively few of these ions, so it readily forms a lather.

Materials

- distilled water
- tap water
- 0.02 mol/L sodium chloride, NaCl(aq)
- 0.01 mol/L calcium chloride, CaCl₂(aq)
- 0.01 mol/L magnesium chloride, MgCl₂(aq)
- liquid hand soap
- 5 large test tubes with stoppers
- marker or grease pencil
- test-tube rack
- medicine dropper

Safety Precautions

- Wear safety eyewear and a lab coat or apron throughout this activity.
- Flush any spills on skin or clothing with plenty of cool water. Inform your teacher immediately.

Procedure

1. Label the five test tubes as follows: distilled water, tap water, NaCl(aq), CaCl₂(aq), MgCl₂(aq). Place the test tubes in a test-tube rack.

2. Pour each liquid into the corresponding test tube until each test tube is about half filled. Make sure that the level of the solution is the same in all the test tubes.

3. Add one drop of liquid soap to the test tube that contains distilled water. Seal the test tube with a stopper. Place your thumb over the stopper, and shake the test tube for 10 s. Record your observations.

4. Repeat step 3 for each of the other liquids.

5. Dispose of the liquids as directed by your teacher.

Questions

1. Rank the solutions based on the lather they produced.

2. Compare the concentrations of chloride ions, Cl⁻(aq), in the three solutions with known concentrations. Could chloride ions cause hard water? Explain.

3. Would you rate your tap water as hard or soft? Explain your reasoning.

4. Which of the ions in the solutions could cause hard water? Justify your answer.

Net Ionic Equations and Qualitative Analysis

Key Terms

spectator ion

ionic equation

net ionic equation

precipitate

qualitative analysis

flame test

Flowerpot Island, shown in **Figure 9.1**, is a tourist attraction off the Bruce Peninsula in Ontario. The island is composed of a rock called dolomite, which is a mixture of calcium carbonate, $CaCO_3(s)$, and magnesium carbonate, $MgCO_3(s)$. How dolomite rock forms is not well understood. It could be the product of reactions with calcium, magnesium, and carbonate ions, which are commonly present in water. Adding a solution of sodium carbonate to any solution that contains calcium ions and magnesium ions will precipitate calcium carbonate and magnesium carbonate.

Figure 9.1 The dolomite on Flowerpot Island was probably formed from ions in an aqueous solution.

As shown in **Figure 9.2**, some reactions in aqueous solutions cause spectacular colour changes. The reaction shown in **Figure 9.2** is a double displacement reaction between silver nitrate and sodium chromate. A chemical reaction between two aqueous solutions that contain ions is *always* a double displacement reaction. Water dissociates ionic substances into their component ions, allowing reactant ions to mix and react more readily. However, water is not a reactant in the chemical equation, and neither are some of the ions in the solution. Non-reacting ions in an aqueous solution are called **spectator ions**. Spectator ions are usually ions that form soluble compounds. As described in Section 8.2, these ions often have a single charge and a large radius.

spectator ion an ion that is present in a solution but not involved in a chemical reaction

Figure 9.2 In the reaction shown here, the reactants are a colourless solution of silver nitrate, $AgNO_3(aq)$, and a yellow solution of sodium chromate , $Na_2CrO_4(aq)$. The products are a red precipitate of silver chromate, $Ag_2CrO_4(s)$, and a colourless solution of sodium nitrate, $NaNO_3(aq)$.

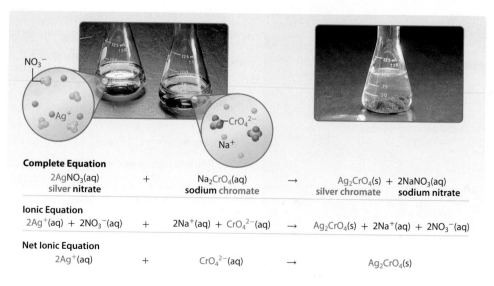

Complete Equation

$$2AgNO_3(aq) + Na_2CrO_4(aq) \rightarrow Ag_2CrO_4(s) + 2NaNO_3(aq)$$

silver **nitrate**   sodium **chromate**   silver **chromate**   **sodium nitrate**

Ionic Equation

$$2Ag^+(aq) + 2NO_3^-(aq) + 2Na^+(aq) + CrO_4^{2-}(aq) \rightarrow Ag_2CrO_4(s) + 2Na^+(aq) + 2NO_3^-(aq)$$

Net Ionic Equation

$$2Ag^+(aq) + CrO_4^{2-}(aq) \rightarrow Ag_2CrO_4(s)$$

Writing Ionic and Net Ionic Equations

The following equation represents the reaction shown in **Figure 9.2** in terms of intact compounds:

$$2AgNO_3(aq) + Na_2CrO_4(aq) \rightarrow Ag_2CrO_4(s) + 2NaNO_3(aq)$$

This representation is not fully accurate because soluble ionic substances dissociate into ions in water.

An **ionic equation** replaces the formulas of soluble ionic compounds with the ions that these compounds form in water. Recall that the solubility guidelines in **Table 8.3** indicate whether a substance is soluble in water. Silver nitrate, sodium chromate, and sodium nitrate are all soluble. Each of these compounds dissociates into its respective ions in water. Since silver chromate is insoluble, it is shown as an intact compound in the ionic equation:

$$2Ag^+(aq) + 2NO_3^-(aq) + 2Na^+(aq) + CrO_4^{2-}(aq) \rightarrow Ag_2CrO_4(s) + 2Na^+(aq) + 2NO_3^-(aq)$$

Note that the total charge on the left side of the ionic equation equals the total charge on the right side. Also note that sodium ions, $Na^+(aq)$, and nitrate ions, $NO_3^-(aq)$, appear on both sides of the equation. These ions are spectator ions in the reaction. You can cancel any term that appears on both sides of an equation:

$$2Ag^+(aq) + \cancel{2NO_3^-(aq)} + \cancel{2Na^+(aq)} + CrO_4^{2-}(aq) \rightarrow Ag_2CrO_4(s) + \cancel{2Na^+(aq)} + \cancel{2NO_3^-(aq)}$$
$$2Ag^+(aq) + CrO_4^{2-}(aq) \rightarrow Ag_2CrO_4(s)$$

Cancelling the spectator ions leaves the **net ionic equation**. A net ionic equation shows only the ions that react and the insoluble product(s), or **precipitates**, of the reaction.

Benefits of Net Ionic Equations

In the example above, the net ionic equation shows that silver ions react with chromate ions to form silver chromate. A solution of an ionic compound always contains both cations and anions. However, the net ionic equation indicates that the source of the silver ions does not matter. The silver ions could be from silver nitrate or from any other soluble silver salt, such as silver acetate, $AgCH_3COO(aq)$. Similarly, sodium chromate could be replaced with ammonium chromate, $(NH_4)_2CrO_4(aq)$. Mixing aqueous solutions of silver acetate and ammonium chromate would produce the same precipitate as the mixture shown in **Figure 9.2**. Changing the spectator ions, from sodium to ammonium or from nitrate to acetate, does not change the reaction because these ions do not participate in the chemical reaction, just as the spectators in **Figure 9.3** do not participate in the game they are watching.

ionic equation a chemical equation in which soluble ionic substances are written in dissociated form

net ionic equation an ionic equation that does not include spectator ions

precipitate an insoluble product in a reaction

Figure 9.3 Spectators are present at a sporting event but do not take part in the event, just as spectator ions are present in a reaction but do not take part in it.

Writing Net Ionic Equations

Use the following rules to write a net ionic equation for a reaction in an aqueous solution.

Rules for Writing a Net Ionic Equation

1. Write the complete chemical equation for the reaction.

2. Rewrite the soluble ionic compounds as ions. For example, show ammonium chloride as $NH_4^+(aq)$ and $Cl^-(aq)$, instead of $NH_4Cl(aq)$.

3. Leave insoluble ionic compounds as formula units. For example, zinc sulfide is insoluble, so you write it as $ZnS(s)$, not $Zn^{2+}(aq)$ and $S^{2-}(aq)$.

4. Leave molecular compounds as molecular formulas since these compounds produce relatively few ions in an aqueous solution. For example, write aqueous carbon dioxide as $CO_2(aq)$ and water as $H_2O(\ell)$.

5. Write all acids as formula units, except for the six strong acids listed below. Since these strong acids ionize almost completely in water, write them as ions:
   - Write hydrochloric acid, HCl, as $H^+(aq)$ and $Cl^-(aq)$.
   - Write hydrobromic acid, HBr, as $H^+(aq)$ and $Br^-(aq)$.
   - Write hydroiodic acid, HI, as $H^+(aq)$ and $I^-(aq)$.
   - Write sulfuric acid, H_2SO_4, as $2H^+(aq)$ and $SO_4^{2-}(aq)$.
   - Write nitric acid, HNO_3, as $H^+(aq)$ and $NO_3^-(aq)$.
   - Write perchloric acid, $HClO_4$, as $H^+(aq)$ and $ClO_4^-(aq)$.

6. Cancel out the spectator ions. Keep only covalent compounds, the ions that react, and the precipitates that form in the reaction. Any gas that is involved in the reaction must appear in the net ionic equation.

7. Check that both the charges and the atoms are balanced in the net ionic equation.

As you learned in Section 4.2, a double displacement reaction always produces a precipitate, a gas, or water. The next two Sample Problems describe how to write net ionic equations for reactions that produce a precipitate and for reactions that produce water. A reaction between an acid and a base that produces water is called a *neutralization reaction*.

Writing net ionic equations is helpful for distinguishing the important reactants and products in a reaction. Read the Sample Problems and then use what you learned to complete the Practice Problems on page 410.

Sample Problem

Net Ionic Equation for a Reaction That Forms a Precipitate

Problem

What substance will precipitate when an aqueous solution of sodium sulfide, $Na_2S(aq)$, is mixed with an aqueous solution of silver nitrate, $AgNO_3(aq)$? Write the net ionic equation for the reaction.

What Is Required?

You need to identify the precipitate in the reaction between an aqueous solution of sodium sulfide and an aqueous solution of silver nitrate. Then you need to write the net ionic equation for the reaction.

What Is Given?

You know that a precipitate forms when sodium sulfide reacts with silver nitrate.

Solubility guidelines are listed in **Table 8.3** in Section 8.2.

Plan Your Strategy	Act on Your Strategy
Use the solubility guidelines to identify the precipitate.	The exchange of ions in the reaction gives two possibilities for the precipitate: $NaNO_3(s)$ or $Ag_2S(s)$. According to the solubility guidelines, all ionic compounds that contain either sodium or nitrate ions are soluble. However, most sulfides are insoluble. Thus, the precipitate must be silver sulfide, $Ag_2S(s)$.
Write the complete chemical equation for the reaction.	$Na_2S(aq) + 2AgNO_3(aq) \rightarrow 2NaNO_3(aq) + Ag_2S(s)$
Write $Na_2S(aq)$, $AgNO_3(aq)$, and $NaNO_3(aq)$ as ions. Leave $Ag_2S(s)$ as a formula unit since this ionic compound is insoluble.	$2Na^+(aq) + S^{2-}(aq) + 2Ag^+(aq) + 2NO_3^-(aq) \rightarrow 2Na^+(aq) + 2NO_3^-(aq) + Ag_2S(s)$
Cancel the spectator ions on both sides of the equation.	$2\cancel{Na^+(aq)} + S^{2-}(aq) + 2Ag^+(aq) + \cancel{2NO_3^-(aq)} \rightarrow \cancel{2Na^+(aq)} + \cancel{2NO_3^-(aq)} + Ag_2S(s)$
Write the net ionic equation.	$2Ag^+(aq) + S^{2-}(aq) \rightarrow Ag_2S(s)$

Check Your Solution

The net ionic equation is balanced, including the charges on the ions. The solubility guidelines indicate that silver sulfide is insoluble in water.

Sample Problem

Net Ionic Equation for a Neutralization Reaction

Problem

Write the net ionic equation for the double displacement reaction between aqueous hydrobromic acid, HBr(aq), and aqueous potassium hydroxide, KOH(aq).

What Is Required?

You need to write the net ionic equation for the reaction.

What Is Given?

You know that the reaction between hydrobromic acid and potassium hydroxide is a double displacement reaction.

Plan Your Strategy	Act on Your Strategy
Write the complete chemical equation for the reaction.	$HBr(aq) + KOH(aq) \rightarrow KBr(aq) + H_2O(\ell)$
Write $HBr(aq)$, $KOH(aq)$, and $KBr(aq)$ as ions. Leave $H_2O(\ell)$ as a formula unit since very few water molecules ionize.	$H^+(aq) + Br^-(aq) + K^+(aq) + OH^-(aq) \rightarrow K^+(aq) + Br^-(aq) + H_2O(\ell)$
Identify the spectator ions, and cancel them on both sides of the equation.	$H^+(aq) + \cancel{Br^-(aq)} + \cancel{K^+(aq)} + OH^-(aq) \rightarrow \cancel{K^+(aq)} + \cancel{Br^-(aq)} + H_2O(\ell)$
Write the net ionic equation.	$H^+(aq) + OH^-(aq) \rightarrow H_2O(\ell)$

Check Your Solution

The net ionic equation is balanced, including the charges on the ions.

This net ionic equation can also be used to represent all neutralization reactions between other strong acids and strong bases. These reactions are discussed in detail in Chapter 10.

1. Write the net ionic equation for this reaction:
   $$Ba(ClO_3)_2(aq) + Na_3PO_4(aq) \rightarrow$$
   $$Ba_3(PO_4)_2(s) + NaClO_3(aq)$$

2. Write the net ionic equation for this reaction:
   $$Na_2SO_4(aq) + Sr(OH)_2(aq) \rightarrow SrSO_4(s) + NaOH(aq)$$

3. Write the net ionic equation for this reaction:
   $$MgCl_2(aq) + NaOH(aq) \rightarrow Mg(OH)_2(s) + NaCl(aq)$$

4. Barium sulfate, $BaSO_4(s)$, is used in some types of paint as a white pigment and as a filler. Barium sulfate precipitates when an aqueous solution of barium chloride, $BaCl_2(aq)$, is mixed with an aqueous solution of sodium sulfate, $Na_2SO_4(aq)$. Write the complete chemical equation and the net ionic equation for this reaction.

5. Identify the precipitate and the spectator ions in the reaction that occurs when an aqueous solution of sodium sulfide is mixed with an aqueous solution of iron(II) sulfate. Write the net ionic equation.

6. Identify the spectator ions in the reaction between each pair of aqueous solutions. Then write the net ionic equation for the reaction.

   a. ammonium phosphate and zinc sulfate
   b. lithium carbonate and nitric acid
   c. sulfuric acid and barium hydroxide

7. When aqueous solutions of sodium iodide and lead(II) nitrate are mixed, a bright yellow precipitate of lead(II) iodide forms. Write a net ionic equation to represent this reaction.

8. A chemical reaction can be represented by the following net ionic equation:
   $$2Al^{3+}(aq) + 3Cr_2O_7^{2-}(aq) \rightarrow Al_2(Cr_2O_7)_3(s)$$
   Suggest two aqueous solutions that could be mixed to cause this reaction.

9. Iron(III) ions, $Fe^{3+}(aq)$, can be precipitated from a solution by adding potassium hydroxide, $KOH(aq)$. Write the net ionic equation for the reaction between iron(III) nitrate, $Fe(NO_3)_3(aq)$, and potassium hydroxide. Identify the spectator ions.

10. Complete and balance each equation. Then write the corresponding net ionic equation.
    a. $Pb(NO_3)_2(aq) + Na_2CO_3(aq) \rightarrow$
    b. $Co(CH_3COO)_2(aq) + (NH_4)_2S(aq) \rightarrow$

Activity 9.1 Reactions in Aqueous Solutions

Observe three reactions between aqueous solutions, and write the net ionic equation for each reaction.

Safety Precautions

- Wear chemical safety goggles throughout this activity.
- Wear a lab coat or apron throughout this activity.
- Hydrochloric acid and sodium hydroxide are corrosive. Flush any spills off skin or clothing immediately.
- Wash your hands after completing this activity.

Materials

- 0.1 mol/L sodium hydroxide, NaOH(aq), in a dropper bottle
- 0.1 mol/L ammonium chloride, $NH_4Cl(aq)$, in a dropper bottle
- 0.1 mol/L magnesium chloride, $MgCl_2(aq)$, in a dropper bottle
- 0.1 mol/L hydrochloric acid, HCl(aq), in a dropper bottle
- phenolphthalein indicator in a dropper bottle
- six-well plate
- toothpicks

Procedure

1. Add a few drops of ammonium chloride to one of the wells in the six-well plate. Then add a few drops of sodium hydroxide to the same well. Stir the mixture with a toothpick. Cautiously smell the liquid mixture by wafting the air that is just above the mixture toward your nose. Record your observations.

2. Add a few drops of magnesium chloride to another well. Then add a few drops of sodium hydroxide to the same well. Stir the mixture with a clean toothpick. Record your observations.

3. Add five drops of hydrochloric acid to one of the wells. Add one drop of phenolphthalein to the same well. Then add sodium hydroxide, drop by drop, until you observe a change.

4. Dispose of the solutions as instructed by your teacher.

Questions

1. Write a chemical equation for the reaction you observed in step 1. Identify the spectator ions in this reaction and write the net ionic equation for this reaction. Finally, explain how the net ionic equation corresponds to your observations.

2. Repeat question 1 for the reaction you observed in step 2.

3. Repeat question 1 for the reaction you observed in step 3.

Qualitative Analysis

Qualitative analysis identifies substances in a sample. You can often identify whether certain ions are in a sample by observing the colour in a flame test, the colour of a solution, or the formation of a precipitate. Qualitative analysis can tell you what ions are present in a solution. Quantitative analysis, which you will learn about in the next section, tells you *how much* of a given ion is present in a solution.

Flame Tests

Many metal ions produce a distinctive colour when they are heated. Thus, one way to test for the presence of metal ions is to heat a small sample of a solid, or a drop of a solution, in a flame and observe the colour. This type of qualitative analysis is called a **flame test**. The flame colours of some common ions are listed in **Table 9.1**. Fireworks, shown in **Figure 9.4**, are a dramatic demonstration of the various colours that are produced when metal ions are heated.

Table 9.1 Flame Colours of Some Metal Ions

Ion	Symbol	Colour
lithium	Li^+	Crimson red
sodium	Na^+	Yellow-orange
potassium	K^+	Lavender
cesium	Cs^+	Blue
calcium	Ca^{2+}	Reddish-orange
strontium	Sr^{2+}	Bright red
barium	Ba^{2+}	Yellowish-green
copper	Cu^{2+}	Bluish-green
lead	Pb^{2+}	Bluish-white

The Bunsen burner was invented by the German chemist Robert Wilhelm Eberhard Bunsen (1811–1899). It produces a clean, hot flame that can be used to heat samples of chemicals. Flame tests can be performed using a Bunsen burner and a clean wire loop, made from either platinum or an alloy of nickel and chromium. To test an aqueous solution, the wire is dipped into the solution. To test a solid, the wire can be moistened with hydrochloric acid or nitric acid to help the solid stick to the wire.

As shown in **Figure 9.5**, the wire is placed in the flame of the Bunsen burner. The electrons in the atoms of the sample absorb energy from the flame. The electrons then re-emit some of the energy as visible light. Since the arrangement of the electrons within the atoms determines the colours of the light that the electrons emit, some elements produce characteristic colours. The bright yellow-orange light that is emitted by some streetlights comes from sodium atoms heated by an electric current. Bunsen used light from heated samples to discover the elements cesium and rubidium. The wavelengths of light emitted by different elements are so unique that astronomers analyze light from distant stars to determine which elements are present in those stars.

Flame tests are usually very sensitive—the characteristic colours of metal cations can be seen using tiny samples. Because platinum is very expensive and nickel-chromium alloys produce a trace of orange colour in the flame, a wooden splint is sometimes used instead of the wire loop.

qualitative analysis analysis that identifies elements, ions, or compounds in a sample

flame test qualitative analysis that uses the colour that a sample produces in a flame to identify the metal ion(s) in the sample

Figure 9.4 Fireworks are a spectacular demonstration of the different colours of light that are given off by metal ions when they are heated.

Analyze *Which metal ions could produce the colours in this fireworks display?*

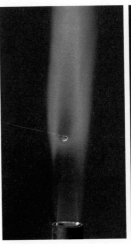

Figure 9.5 These photographs show flame tests of strontium and copper. Notice that the colour of the copper flame is greener than a typical Bunsen burner flame.

Go to **scienceontario** to find out more

Colours of Ions in Solutions

Flame tests can be used to identify only certain metallic ions. However, aqueous solutions of the ionic compounds of certain cations and anions also have characteristic colours. Therefore, the colour of a solution can help to identify some of the ions in the solution. For example, most aqueous solutions that contain aqueous copper(II) ions are blue. **Table 9.2** lists the colours of aqueous solutions of some common ions.

Table 9.2 Colours of Some Common Ions in Aqueous Solutions

	Ion	Symbol	Colour
Cations	chromium(II) copper(II)	$Cr^{2+}(aq)$ $Cu^{2+}(aq)$	Blue
	chromium(III) copper(I) iron(II) nickel(II)	$Cr^{3+}(aq)$ $Cu^+(aq)$ $Fe^{2+}(aq)$ $Ni^{2+}(aq)$	Green
	iron(III)	$Fe^{3+}(aq)$	Pale yellow
	cobalt(II) manganese(II)	$Co^{2+}(aq)$ $Mn^{2+}(aq)$	Pink
Anions	chromate	$CrO_4^{2-}(aq)$	Yellow
	dichromate	$Cr_2O_7^{2-}(aq)$	Orange
	permanganate	$MnO_4^-(aq)$	Purple

Suggested Investigation

Inquiry Investigation 9-A, Qualitative Analysis

Precipitation Reactions

Another way to identify an unknown ion in a solution is to add a known reactant to the solution and observe whether a precipitate forms. Then the solubility guidelines can be used to infer which ion must have been present in the unknown solution. **Figure 9.6** shows how a series of precipitation reactions can be used to identify different ions in a solution. Each time a precipitate forms, ions are removed from the solution. After a precipitate has been filtered out, other reactants can be added to the *filtrate* (the filtered solution). The colour of the solution after each reaction can help to identify the ions that are still present. A flame test may also be used on a precipitate after it has been rinsed.

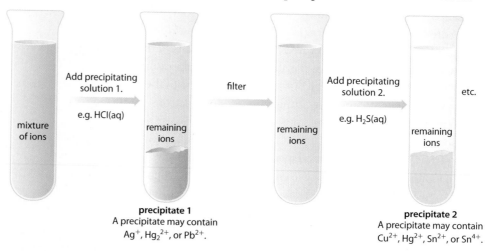

Figure 9.6 A series of precipitation reactions can be used to identify ions in a solution.

Explain Why could a precipitate caused by adding hydrochloric acid, HCl(aq), contain silver ions but not copper ions?

1. Give two specific examples of substances that are never shown as ions in a net ionic equation.

2. Explain how a chemical equation differs from an ionic equation.

3. After you have written a net ionic equation, why should you check to make sure that it is balanced for charges as well as for atoms?

4. When an aqueous solution of ammonium phosphate is mixed with an aqueous solution of sodium carbonate, all the ions are spectator ions. Explain why.

5. Describe three different qualitative analysis tests.

6. If an aqueous solution is a certain colour, will a sample of the solution cause the same colour in a flame test? Explain your answer.

Activity 9.2 Identifying Unknown Aqueous Solutions

In this activity, you will interpret observations of flame tests, solution colours, and precipitation reactions to identify dissolved metal ions.

Procedure

Examine the observations listed in the table below, and then answer the questions.

Observations from Testing a Solution of Unknown Metal Ions

Test	Observation
1. Solution colour	The solution is colourless.
2. Addition of sodium hydroxide, NaOH(aq), to the solution	A white precipitate is produced. When the mixture is filtered, the filtrate is colourless.
3. Flame test on the precipitate from test 2	The flame colour is red.
4. Addition of sodium sulfate, Na$_2$SO$_4$(aq), to the filtrate from test 2	A second white precipitate is produced. When the mixture is filtered, the filtrate is colourless.
5. Flame test on the precipitate from test 4	The flame colour is red, but a different red than the flame colour in test 3.

Questions

1. List all the ions that cause a red flame and produce a precipitate in the presence of hydroxide ions.

2. List all the cations that could cause a red flame and produce a precipitate in the presence of sulfate ions.

3. If all traces of the two metal cations are removed from the solution in test 4, what might the flame colour be when a sample of the solution is tested? Explain your prediction.

4. List the solution colours and precipitation reactions you would expect to observe in tests to identify the metal ions in solutions that contain the following cations:

   a. Na$^+$(aq) only

   b. Cu^{2+}(aq) only

   c. Na$^+$(aq) and Ag$^+$(aq)

   d. Cu^{2+}(aq) and Ag$^+$(aq)

Qualitative Analysis and Quantitative Analysis

Qualitative analysis is only useful for determining which ions are present in a solution. Sometimes, the identity of the ions is all the information that a chemist needs. But other times, chemists want to know the amount or concentration of the ions present. To find the concentration of the ions, chemists need to perform quantitative analysis, which is described in the next section.

If chemists need to perform both qualitative analysis and quantitative analysis on the same solution, they often do qualitative analysis first. However, they can do both kinds of analysis at the same time when doing precipitation reactions. To do so, they must keep accurate records of the amounts of unknown solution and reagent used and carefully measure the amount of precipitate formed by the reaction.

Section Summary

- A net ionic equation omits the spectator ions and shows only the ions that react and the product(s) of the reaction.

- You can use observations of flame tests, solution colours, and the formation of precipitates to identify ions in a sample.

Review Questions

1. **K/U** What is a spectator ion? What characteristics does a spectator ion often have?

2. **K/U** Identify the spectator ions in each reaction.
   a. $3CuCl_2(aq) + 2(NH_4)_3PO_4(aq) \rightarrow$
   $$Cu_3(PO_4)_2(s) + 6NH_4Cl(aq)$$
   b. $2Al(NO_3)_3(aq) + 3Ba(OH)_2(aq) \rightarrow$
   $$2Al(OH)_3(s) + 3Ba(NO_3)_2(aq)$$
   c. $2NaOH(aq) + MgCl_2(aq) \rightarrow$
   $$2NaCl(aq) + Mg(OH)_2(s)$$

3. **T/I** Write a net ionic equation for each reaction in question 2.

4. **T/I** An aqueous solution of copper(II) sulfate is mixed with an aqueous solution of sodium carbonate.
   a. State the name and formula for the precipitate that forms.
   b. Write the net ionic equation for the reaction.
   c. Identify the spectator ions.

5. **T/I** For each of the following net ionic equations, list two soluble ionic compounds that can be mixed together in solution to produce the reaction represented by the equation. (**Note:** There are many correct answers.)
   a. $3Ba^{2+}(aq) + 2PO_4{}^{3-}(aq) \rightarrow Ba_3(PO_4)_2(s)$
   b. $Mg^{2+}(aq) + 2OH^-(aq) \rightarrow Mg(OH)_2(s)$
   c. $2Al^{3+}(aq) + 3Cr_2O_7{}^{2-}(aq) \rightarrow Al_2(Cr_2O_7)_3(s)$

6. **K/U** Explain why there are many correct answers for question 5.

7. **C** Draw a flowchart that summarizes how to write net ionic equations for double displacement reactions.

8. **K/U** What is the difference between qualitative analysis and quantitative analysis?

9. **A** Why might a chemist need to carry out qualitative analysis on a solution?

10. **A** Lithium carbonate is the active ingredient in some anti-depression medications. What tests could you perform to confirm the presence of lithium carbonate, $Li_2CO_3(s)$, in a tablet?

11. **T/I** Limewater is a solution of calcium hydroxide, $Ca(OH)_2(aq)$. It can be used to test for the presence of carbon dioxide. When carbon dioxide is bubbled through limewater, a milky-white precipitate is produced.

   a. Write a chemical equation and a net ionic equation to show what happens when carbon dioxide is bubbled through limewater.
   b. Is this test an example of qualitative or quantitative analysis? Explain your answer.

12. **T/I** An ion in a solution forms a yellow precipitate when sodium iodide, $NaI(aq)$, is added to the solution. The precipitate produces a blue-white colour when it is heated in a flame.

   a. Suggest a formula for the ion and a formula for the precipitated compound.
   b. Write a net ionic equation to represent the reaction.

13. **T/I** All the solutions below have the same concentration. Use **Table 9.2** to infer what ion causes the colour in each solution. How much confidence do you have in your inferences? How could you check your inferences?

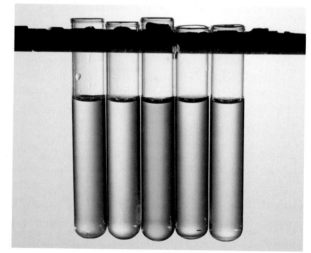

14. **A** To answer the following questions, refer to the solubility guidelines in Section 8.2.
   a. What aqueous solution will precipitate $Pb^{2+}(aq)$ ions but not $Cu^+(aq)$ or $Mg^{2+}(aq)$ ions?
   b. What aqueous solution will precipitate $Cu^+(aq)$ ions but not $Mg^{2+}(aq)$ ions?
   c. Using your answers to parts a and b, outline a procedure that would allow you to precipitate the $Pb^{2+}(aq)$ ions, followed by $Cu^+(aq)$ ions, and then $Mg^{2+}(aq)$ ions.

Solution Stoichiometry

Cookbook recipes involve chemistry. If you bake a cake, you need to assemble the ingredients, which are the reactants, in the correct quantities. If you add too little baking powder, the cake will not rise properly, as shown in **Figure 9.7**. If you add too much baking powder, the cake may be too crumbly. Similarly, for some chemical reactions, the proportions of the reactants are critical. Chemists and chemical engineers use stoichiometry calculations to determine the quantities they need for chemical reactions. In **solution stoichiometry**, known volumes and concentrations of reactants or products are used to determine the volumes, concentrations, or masses of other reactants or products.

Key Terms

solution stoichiometry

quantitative analysis

solution stoichiometry
stoichiometry that is applied to substances in solutions

quantitative analysis
an analysis that determines how much of a substance is in a sample

Figure 9.7 Accurate measurement is important in both cooking and chemistry. A cake made from improperly measured ingredients may not look good and may not taste good either!

Solution stoichiometry is often used in **quantitative analysis**, which involves determining how much of a substance is present in a sample. To solve solution stoichiometry problems, you need to use the same strategies you learned in Chapter 7. The rules below outline how to apply these strategies to reactants and products in solutions.

Rules for Solving Solution Stoichiometry Problems

1. Always use a balanced equation: either the balanced chemical equation or the net ionic equation. If a precipitate forms as the result of the reaction, the net ionic equation may be easier to use.

2. The concentration in moles per litre, c, the amount of a given substance in moles, n, and the volume of the solution in litres, V, are related by the following equation:

$$c = \frac{n}{V}$$

Thus, the amount of a substance is given by the equation $n = c \times V$.

3. Use the coefficients in the balanced net ionic equation (or the balanced chemical equation) to write the known mole ratio of the substances. Equate the known mole ratio to the mole ratio of the other substances in the reaction, which includes an unknown quantity. Then solve for the unknown quantity.

Concentration of a Solution from the Mass of the Precipitate

Problem

A student carefully measured 100 mL of a silver nitrate solution, $AgNO_3(aq)$, of unknown concentration. The student poured the solution into a beaker, added a coil of copper, and left the mixture for several days. When the reaction was complete, the student carefully scraped the silver from the copper wire and filtered the solution to obtain all the silver. The dry precipitate had a mass of 1.65 g. What was the molar concentration of the silver nitrate solution?

What Is Required?

You need to find the molar concentration, c, of the silver nitrate solution.

What Is Given?

You know the volume of the silver nitrate solution: 100 mL
You know the mass of silver precipitated: 1.65 g

Plan Your Strategy	Act on Your Strategy
Write the chemical equation for the single displacement reaction.	$Cu(s) + 2AgNO_3(aq) \rightarrow Cu(NO_3)_2(aq) + 2Ag(s)$
Look up the molar mass of silver in the periodic table, and use it to calculate the amount in moles of silver precipitated.	$M = 107.87 \text{ g/mol}$ amount of Ag $= 1.65 \text{ g} \times \dfrac{1 \text{ mol}}{107.87 \text{ g}} = 0.015\,296 \text{ mol}$
Equate the mole ratios and cross multiply to solve for n, the amount in moles of silver nitrate.	$\dfrac{2 \text{ mol AgNO}_3}{2 \text{ mol Ag}} = \dfrac{n}{0.015\,296 \text{ mol Ag}}$ $n = \dfrac{2 \text{ mol AgNO}_3 \times 0.015\,296 \text{ mol Ag}}{2 \text{ mol Ag}} = 0.015\,296 \text{ mol AgNO}_3$
Calculate the concentration of silver nitrate.	$c = \dfrac{n}{v} = \dfrac{0.015\,296 \text{ mol}}{100 \text{ mL}} = \dfrac{0.015\,296 \text{ mol}}{0.100 \text{ L}} = 0.153 \text{ mol/L}$ The molar concentration of the silver nitrate solution was 0.153 mol/L.

Check Your Solution

The units for amount and concentration are correct. The answer has three significant digits and seems reasonable.

Concentration of Ions from the Mass of the Precipitate

Problem

When excess aqueous lead(II) nitrate, $Pb(NO_3)_2(aq)$, was added to 125 mL of a solution of sodium iodide, $NaI(aq)$, a bright yellow precipitate of lead(II) iodide formed. The dry precipitate had a mass of 4.13 g. What was the concentration of iodide ions in the solution of sodium iodide?

What Is Required?

You need to find the initial concentration of iodide ions in the sodium iodide solution.

What Is Given?

You know the volume of the sodium iodide solution: 125 mL
You know the mass of lead(II) iodide that was precipitated: 4.13 g

Plan Your Strategy	Act on Your Strategy
Write the net ionic equation for the double displacement reaction.	$Pb(NO_3)_2(aq) + 2NaI(aq) \rightarrow PbI_2(s) + 2NaNO_3(aq)$ $Pb^{2+}(aq) + \cancel{2NO_3^-(aq)} + \cancel{2Na^+(aq)} + 2I^-(aq) \rightarrow PbI_2(s) + \cancel{2Na^+(aq)} + \cancel{2NO_3^-(aq)}$ $Pb^{2+}(aq) + 2I^-(aq) \rightarrow PbI_2(s)$
Calculate the molar mass of lead(II) iodide, and use it to find the amount in moles of precipitate.	$M = M_{Pb} + 2M_I = 207.2 \text{ g/mol} + 2(126.90 \text{ g/mol})$ $\quad = 461.0 \text{ g/mol}$ amount of $PbI_2 = 4.13 \text{ g} \times \dfrac{1 \text{ mol}}{461.0 \text{ g}} = 0.008\,958\,8 \text{ mol}$
Equate the mole ratios and solve for n, the amount in moles of iodide ions, $I^-(aq)$.	$\dfrac{2 \text{ mol I}^-}{1 \text{ mol PbI}_2} = \dfrac{n}{0.008\,958\,8 \text{ mol PbI}_2}$ $n = \dfrac{2 \text{ mol I}^- \times 0.008\,958\,8 \text{ mol PbI}_2}{1 \text{ mol PbI}_2} = 0.017\,918 \text{ mol I}^-$
Calculate the concentration of iodide ions.	$c = \dfrac{n}{V} = \dfrac{0.017\,918 \text{ mol}}{125 \text{ mL}} = \dfrac{0.017\,918 \text{ mol}}{0.125 \text{ L}} = 0.143 \text{ mol/L}$ The concentration of iodide ions was 0.143 mol/L.

Check Your Solution

The units for amount and concentration are correct. The answer has three significant digits and appears to be reasonable.

Practice Problems

11. If 8.5 g of pure ammonium phosphate, $(NH_4)_3PO_4(s)$, is dissolved in distilled water to make 400 mL of solution, what are the concentrations (in moles per litre) of the ions in the solution?

12. A strip of zinc metal was placed in a beaker that contained 120 mL of a solution of copper(II) nitrate, $Cu(NO_3)_2(aq)$. The mass of the copper produced was 0.813 g. Find the initial concentration of the solution of copper(II) nitrate.

13. When 75.0 mL of silver nitrate, $AgNO_3(aq)$, was treated with excess ammonium carbonate, $(NH_4)_2CO_3(aq)$, 2.47 g of dry precipitate was recovered. Write the net ionic equation for the reaction, and calculate the concentration of the original silver nitrate solution.

14. When an excess of sodium sulfide, $Na_2S(aq)$, was added to 125 mL of 0.100 mol/L iron(II) nitrate, $Fe(NO_3)_2(aq)$, a black precipitate formed. Identify the precipitate, and calculate the maximum mass of precipitate that can be collected from the reaction.

15. What mass of silver chloride, $AgCl(s)$, can be precipitated from 75 mL of 0.25 mol/L silver nitrate, $AgNO_3(aq)$, by adding excess magnesium chloride, $MgCl_2(aq)$?

16. What mass of bromine gas can be collected by bubbling excess chlorine gas through 850 mL of a 0.350 mol/L solution of sodium bromide, $NaBr(aq)$?

17. What mass of strontium carbonate, $SrCO_3(s)$, can be precipitated from 50.0 mL of 0.165 mol/L strontium nitrate, $Sr(NO_3)_2(aq)$, by adding excess sodium carbonate, $NaCO_3(aq)$?

18. Before it was banned in the 1970s due to its non-selective toxicity, thallium(I) sulfate, $Tl_2SO_4(s)$, was the active ingredient in some pesticides. A chemist measured 100.0 mL of a solution of thallium(I) sulfate and added excess aqueous potassium iodide to precipitate yellow thallium(I) iodide, $TlI(s)$. The mass of the dry precipitate was 2.45 g. Find the molar concentration of the thallium(I) sulfate solution.

19. A sample of a substance known to contain chloride ions was dissolved in distilled water in a 1 L volumetric flask. Then 25.00 mL of this solution was treated with excess silver nitrate, $AgNO_3(aq)$. The precipitate of silver chloride, $AgCl(s)$, was filtered and dried. The mass of the dry precipitate was 0.765 g.
a. Calculate the concentration of chloride ions.
b. If the original substance was sodium chloride, $NaCl(s)$, what mass of it was dissolved in the volumetric flask?

20. Food manufacturers sometimes add calcium acetate, $Ca(CH_3COO)_2(s)$, to sauces as a thickening agent. When analyzed, a 250 mL solution of calcium acetate was found to contain 0.200 mol of acetate ions.
a. Find the molar concentration of the calcium acetate solution.
b. What mass of calcium acetate was dissolved to make the solution?

Limiting Reactant Problems

Suggested Investigation

Plan Your Own Investigation 9-B, Determining the Mass Percent Composition of a Mixture

The amount of any product in a chemical reaction is determined by the amount of the reactant that is completely consumed. In industrial chemical processes, the most expensive reactant is usually the limiting reactant, in order to minimize costs. In Chapter 7, you learned how to solve limiting reactant problems by using information about the reactants to determine which reactant is limiting. In solution stoichiometry, you usually find the amount of a reactant in moles, given the volume and concentration of the solution, and then find the limiting reactant.

Learning Check

7. Write an equation for molar concentration, c. Explain what each variable in the equation represents.

8. Why is a balanced chemical or net ionic equation essential for a stoichiometric calculation?

9. What information do you need if you want to convert a mass/volume percent concentration to a molar concentration?

10. What is a limiting reactant?

11. For some solutions of ionic compounds, the molar concentration of each ion is equal to the molar concentration of the compound. What can you infer about the charges on the ions in these solutions?

12. Explain what stoichiometric calculations for solutions have in common with the stoichiometric calculations in Chapter 7.

Sample Problem

Limiting Reactant and Mass of Precipitate

Problem

In one process for water purification, aluminum sulfate, $Al_2(SO_4)_3$(aq), reacts with calcium hydroxide, $Ca(OH)_2$(aq), to form a precipitate of aluminum hydroxide, $Al(OH)_3$(s). Bacteria and dirt particles that were suspended in the impure water stick to the precipitate as it settles out of solution. Find the mass of aluminum hydroxide that precipitates when 20.0 mL of 0.0150 mol/L aqueous aluminum sulfate is mixed with 30.0 mL of 0.0185 mol/L aqueous calcium hydroxide.

What Is Required?

You need to find the mass of the aluminum hydroxide precipitate.

What Is Given?

You know the volume of the aluminum sulfate solution: 20.0 mL
You know the concentration of the aluminum sulfate solution: 0.0150 mol/L
You know the volume of the calcium hydroxide solution: 30.0 mL
You know the concentration of the calcium hydroxide solution: 0.0185 mol/L

Plan Your Strategy	Act on Your Strategy
Write the chemical equation for the reaction.	$Al_2(SO_4)_3$(aq) + $3Ca(OH)_2$(aq) → $2Al(OH)_3$(s) + $3CaSO_4$(aq)
Calculate the amount in moles of each reactant.	$n = cV$ amount of $Al_2(SO_4)_3$(aq) = 0.0150 mol/L × 0.0200 L $= 3.00 \times 10^{-4}$ mol amount of $Ca(OH)_2$(aq) = 0.0185 mol/L × 0.0300 L $= 5.55 \times 10^{-4}$ mol
To allow for the mole ratio of the reactants, divide the amount of each reactant by its coefficient in the chemical equation. The smaller result identifies the limiting reactant.	$\dfrac{\text{amount of } Al_2(SO_4)_3}{\text{coefficient}} = \dfrac{3.00 \times 10^{-4} \text{ mol}}{1}$ $= 3.00 \times 10^{-4}$ mol $\dfrac{\text{amount of } Ca(OH)_2}{\text{coefficient}} = \dfrac{5.55 \times 10^{-4} \text{ mol}}{3}$ $= 1.85 \times 10^{-4}$ mol $Ca(OH)_2$(aq) is the limiting reactant because it is the smaller amount.

Use the mole ratio of calcium hydroxide and aluminum hydroxide and the amount of calcium hydroxide to find n, the amount of precipitate.	$\dfrac{3 \text{ mol Ca(OH)}_2}{2 \text{ mol Al(OH)}_3} = \dfrac{5.55 \times 10^{-4} \text{ mol Ca(OH)}_2}{n}$ $n = \dfrac{2 \text{ mol Al(OH)}_3 \times 5.55 \times 10^{-4} \text{ mol Ca(OH)}_2}{3 \text{ mol Ca(OH)}_2} = 3.70 \times 10^{-4} \text{ mol Al(OH)}_3$
Calculate the molar mass of aluminum hydroxide, and use it to find the mass, m, of the precipitate.	$M_{Al(OH)_3} = 26.98 \text{ g/mol} + 3(16.00 \text{ g/mol}) + 3(1.01 \text{ g/mol}) = 78.01 \text{ g/mol}$ $m = nM = 3.70 \times 10^{-4} \text{ mol} \times 78.01 \text{ g/mol} = 0.0289 \text{ g}$ The mass of aluminum hydroxide precipitate is 0.0289 g.

Check Your Solution

The mass is in grams and appears reasonable. It has the correct number of significant digits.

Sample Problem

Finding the Minimum Volume for a Complete Reaction

Problem

A kidney stone is a hard mass that can form in the kidneys or urinary tract. The most common type of kidney stone contains primarily calcium oxalate, $CaC_2O_4(s)$. A chemist wants to react 60.0 mL of 0.135 mol/L sodium oxalate, $Na_2C_2O_4(aq)$, with 0.226 mol/L calcium chloride, $CaCl_2(aq)$, to precipitate calcium oxalate. What is the minimum volume of calcium chloride solution required? What mass of calcium oxalate will be precipitated?

What Is Required?

You need to find the volume of calcium chloride solution that is needed for a complete reaction and the mass of calcium oxalate that will be precipitated.

What Is Given?

You know the volume, V_1, of the sodium oxalate solution: 60.0 mL
You know the concentration, c_1, of the sodium oxalate solution: 0.135 mol/L
You know the concentration, c_2, of the calcium chloride solution: 0.226 mol/L

Plan Your Strategy	Act on Your Strategy
Write the chemical equation for the reaction.	$Na_2C_2O_4(aq) + CaCl_2(aq) \rightarrow CaC_2O_4(s) + 2NaCl(aq)$
Find n_1, the amount of sodium oxalate.	$n_1 = c_1 \times V_1 = 0.135 \text{ mol/L} \times 0.0600 \text{ L} = 0.008\ 10 \text{ mol}$
Use the mole ratio of the reactants to find n_2, the amount of calcium chloride solution that is required for a complete reaction.	$\dfrac{1 \text{ mol Na}_2C_2O_4}{1 \text{ mol CaCl}_2} = \dfrac{0.008\ 10 \text{ mol Na}_2C_2O_4}{n_2}$ $n_2 = \dfrac{1 \text{ mol CaCl}_2 \times 0.008\ 10 \text{ mol Na}_2C_2O_4}{1 \text{ mol Na}_2C_2O_4} = 0.008\ 10 \text{ mol CaCl}_2$
Use the known amount and concentration of the calcium chloride solution to find V_2, the volume that is needed.	$V_2 = \dfrac{n_2}{c_2} = \dfrac{0.008\ 10 \text{ mol}}{0.226 \text{ mol/L}} = 0.035\ 84 \text{ L} = 35.8 \text{ mL}$
Use mole ratios to find $n_{CaC_2O_4}$, the amount of calcium oxalate that will be precipitated.	$\dfrac{1 \text{ mol Na}_2C_2O_4}{1 \text{ mol CaC}_2O_4} = \dfrac{0.00810 \text{ mol Na}_2C_2O_4}{n_3}$ $n_{(CaC_2O_4)_2} = \dfrac{1 \text{ mol CaC}_2O_4 \times 0.008\ 10 \text{ mol Na}_2C_2O_4}{1 \text{ mol Na}_2C_2O_4} = 0.008\ 10 \text{ mL CaC}_2O_4$
Find the molar mass of calcium oxalate.	$M_{CaC_2O_4} = M_{Ca} + 2M_C + 4M_O$ $= 40.08 \text{ g/mol} + 2(12.01 \text{ g/mol}) + 4(16.00 \text{ g/mol}) = 128.1 \text{ g/mol}$
Use the amount of calcium oxalate and the molar mass to calculate the mass, m, of the precipitate.	$m = nM = 0.008\ 10 \text{ mol} \times 128.1 \text{ g/mol} = 1.04 \text{ g}$ Therefore, 1.04 g of calcium oxalate will be precipitated.

Check Your Solution

The volume is in millilitres, and it appears reasonable compared with the volume of the other reactant. The mass of the precipitate also seems reasonable.

21. Lead(II) sulfide, PbS(s), is a black, insoluble substance. Calculate the maximum mass of lead(II) sulfide that will precipitate when 6.75 g of sodium sulfide, $Na_2S(s)$, is added to 250 mL of 0.200 mol/L lead(II) nitrate, $Pb(NO_3)_2(aq)$.

22. Silver chromate, $Ag_2CrO_4(s)$, is a brick-red insoluble substance that is used to stain neurons so they can be viewed under a microscope. Silver chromate can be formed by the reaction between silver nitrate, $AgNO_3(aq)$, and potassium chromate, $K_2CrO_4(aq)$, as shown in the photograph below. Calculate the mass of silver chromate that forms when 25.0 mL of 0.125 mol/L silver nitrate reacts with 20.0 mL of 0.150 mol/L sodium chromate.

23. Mercury compounds are poisonous, but mercury ions can be removed from a solution by precipitating insoluble mercury (II) sulfide, HgS(s). Determine the minimum volume of 0.0783 mol/L sodium sulfide, $Na_2S(aq)$, that is needed to precipitate all the mercury ions in 75.5 mL of 0.100 mol/L, $Hg(NO_3)_2(aq)$.

24. What is the minimum mass of sodium carbonate, $NaCO_3(s)$, that is needed to precipitate all the barium ions from 50.0 mL of 0.125 mol/L barium nitrate, $Ba(NO_3)_2(aq)$?

25. What is the maximum mass of lead(II) iodide, $PbI_2(s)$, that can precipitate when 40.0 mL of a 0.345 mol/L solution of lead(II) nitrate, $Pb(NO_3)_2(aq)$, is mixed with 85.0 mL of a 0.210 mol/L solution of potassium iodide, KI(aq)? Why might the actual mass precipitated be less?

26. Carbonates react with dilute hydrochloric acid to generate carbon dioxide gas. What volume of 2.00 mol/L hydrochloric acid is needed to react with 3.35 g of calcium carbonate?

27. A 15.8 g strip of zinc metal was placed in 100.0 mL of silver nitrate, $AgNO_3(aq)$. When the reaction was complete, the strip of zinc had a mass of 13.1 g. What was the concentration of the silver nitrate solution?

28. Vinegar is an aqueous solution of acetic acid, $CH_3COOH(aq)$. What volume of 1.07 mol/L aqueous sodium hydroxide will completely react with 25.0 mL of 0.833 mol/L household vinegar?

29. Before toothpaste was invented, people sometimes used calcium carbonate, $CaCO_3(s)$, to clean their teeth. What mass of calcium carbonate can be precipitated by reacting 80.0 mL of a 0.100 mol/L solution of sodium carbonate, $Na_2CO_3(aq)$, with 50.0 mL of a 0.100 mol/L solution of calcium chloride, $CaCl_2(aq)$?

30. Barium chromate, $BaCrO_4(s)$, is an insoluble yellow solid. Determine the concentration of barium ions in a solution made by mixing 50.0 mL of a 0.150 mol/L solution of barium nitrate, $Ba(NO_3)_2(aq)$, with 50.0 mL of a 0.120 mol/L solution of potassium chromate, $K_2CrO_4(aq)$.

Solution Stoichiometry versus Other Types of Stoichiometry

The purpose of solution stoichiometry is the same as the purpose of other types of stoichiometry: to determine the quantities involved in chemical reactions. Both kinds of stoichiometry can be used to calculate quantities of reactants or products, to identify limiting reactants and excess reactants, and to determine reaction yields. Solution stoichiometry differs only because the calculations involve concentrations and volumes. However, if you have the concentration and volume of a solution in a stoichiometry problem, you can find the amount (in moles) of a reactant or product involved in the given reaction. Once you find the amount of a substance, you can solve the problem like any other stoichiometry problem.

Section Summary

- In solution stoichiometry, known volumes and concentrations of reactants or products are used to determine the volumes, concentrations, or masses of other reactants or products.

- The solution to a stoichiometry problem should always include a balanced chemical equation or the net ionic equation.

Review Questions

1. **K/U** Which solution has the greater concentration of chloride ions: 0.10 mol/L magnesium chloride, $MgCl_2(aq)$, or 0.15 mol/L sodium chloride, $NaCl(aq)$? Explain your reasoning.

2. **T/I** Calculate the molar concentration of iodide ions in each aqueous solution.
   a. 15.0 g of potassium iodide dissolved in 200 mL of solution
   b. 12.0 g of calcium iodide dissolved in 180 mL of solution

3. **T/I** What is the minimum volume of 0.220 mol/L calcium chloride, $CaCl_2(aq)$, that is needed to precipitate all the silver ions in 110 mL of 0.166 mol/L silver nitrate, $AgNO_3(aq)$?

4. **T/I** Lead(II) acetate is a poisonous compound. It is used as a colour additive in hair dyes. What volume of 1.25 mol/L lead(II) acetate, $Pb(CH_3COO)_2(aq)$, contains 0.500 mol of lead(II) ions, $Pb^{2+}(aq)$?

5. **T/I** A piece of iron was added to a beaker that contained 0.585 mol/L copper(II) sulfate, $CuSO_4(aq)$. The solid copper that precipitated was dried, and its mass was found to be 5.02 g. Some unreacted iron remained in the beaker. Calculate the minimum volume of the copper(II) sulfate solution.

6. **T/I** To generate hydrogen gas, a teacher added 25.0 g of mossy zinc to 220 mL of 3.00 mol/L hydrochloric acid in an Erlenmeyer flask.
   a. What mass of hydrogen gas was generated?
   b. After the reaction, what was the concentration of zinc chloride, $ZnCl_2(aq)$, in the flask?

7. **A** A type of stomach medication is a tablet that contains a mixture of 1.00 g of sodium hydrogen carbonate, $NaHCO_3(s)$, and 1.00 g of citric acid, $H_3C_6H_5O_7(s)$. When dropped into water, the chemicals in the tablet react to produce carbon dioxide gas:
$$3NaHCO_3(aq) + H_3C_6H_5O_7(aq) \rightarrow$$
$$3CO_2(g) + 3H_2O(\ell) + Na_3C_6H_5O_7(aq)$$
   a. Which substance is in excess?
   b. What mass of this substance remains unreacted when the tablet is dropped into a glass of water?

8. **T/I** When 50 mL of 0.20 mol/L sodium sulfate, $Na_2SO_4(aq)$, was mixed with 80 mL of 0.10 mol/L lead(II) acetate, $Pb(CH_3COO)_2(aq)$, a white precipitate formed. Identify the precipitate, and calculate the maximum mass of dry solid that can be collected.

9. **T/I** To measure the concentration of copper(II) sulfate in the water discharged from an industrial plant, a chemist measured 600 mL of the water and then added excess aqueous sodium sulfide. When dried, the precipitate of copper(II) sulfide, $CuS(s)$, had a mass of 0.125 g. Calculate the molar concentration of copper ions in the water sample.

10. **T/I** Mixing solutions of calcium chloride, $CaCl_2(aq)$, and potassium carbonate, $K_2CO_3(aq)$, will cause calcium carbonate, $CaCO_3(s)$, to precipitate. Suppose that you have the following solutions available: 0.500 mol/L $CaCl_2(aq)$ and 1.00 mol/L $K_2CO_3(aq)$. What volume of each solution should be mixed together to form 10.0 g of calcium carbonate?

11. **C** Suppose that you are given a white powder, known to contain a mixture of an alkali metal carbonate and magnesium carbonate, $Mg(CO_3)_2(s)$. Outline the procedure you would use to identify the unknown alkali metal and determine the percent by mass of the alkali metal carbonate in the mixture.

12. **A** Lead poisoning can have long-lasting effects. One of the most effective treatments for lead poisoning is the ion called $EDTA^{4-}$, which stands for ethylenediaminetetraacetate. $EDTA^{4-}$ ions bond to lead(II) ions in a 1:1 ratio. A doctor determines that a child's blood has a dangerously high concentration of 1.0×10^{-5} mol/L of lead(II) ions. The doctor estimates that the child's total blood volume is about 1.6 L. Find the minimum volume of a 0.025 mol/L solution of $EDTA^{4-}$ ions that is needed to treat the child.

13. **A** Suppose that you need to determine the concentration of copper(II) ions in a sample of waste water by precipitating the metal ion. Outline a procedure you could use, and describe factors that could affect the accuracy of your measurements.

Water Quality

Key Terms

fresh water

ground water

surface water

hard water

soft water

maximum allowable
concentration (MAC)

---

fresh water water that
is not salty

ground water water
that seeps through
the ground below the
surface

surface water water on
the surface of the land

Water is the most abundant substance on Earth's surface. However, naturally available water is never pure because it can dissolve so many substances. Some of the dissolved substances are beneficial to human health, but others are harmful. Many of the harmful substances are pollutants such as pesticides, exhaust gases from vehicles, and by-products of industrial processes.

Over 97 percent, by mass, of Earth's water is in the oceans. Ocean water contains dissolved substances—mainly salts—that make it unsuitable for drinking or irrigation unless it is first treated. So, less than 3 percent of the water on Earth is **fresh water**. Most of this fresh water is in the form of polar ice. A large amount of Earth's **ground water** is not easily accessible because it is in remote locations or too far below Earth's surface. **Surface water**, such as lakes, rivers, and reservoirs, is a tiny fraction of the water on Earth—less than 0.02 percent, as shown in **Figure 9.8**.

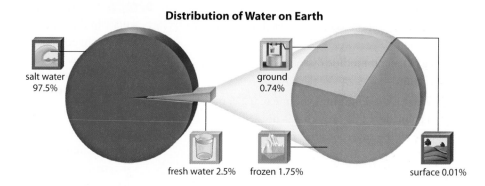

Distribution of Water on Earth

salt water
97.5%

ground
0.74%

fresh water 2.5%   frozen 1.75%   surface 0.01%

Figure 9.8 Very little of the water on Earth is fresh water, and very little fresh water is available for people to drink.

Interpret *About what percentage of the water on Earth is unavailable for human consumption?*

Canada, which has less than 1 percent of the world's population, has 22 percent of the world's source of fresh water. So, in most regions of Canada, the *quality* of the water is usually of greater concern than the quantity available. Water quality is affected by the substances that have dissolved in it. The sources of these substances can be divided into three broad categories: naturally occurring materials, pollutants, and treatments used to improve the water for human consumption. Many of the ions that are found in water come from more than one of these categories of sources.

Harmful Substances from Natural Sources

The water cycle on Earth circulates water from the land to the sky, and back again. The natural supply of fresh water comes from rain and snow, which contains dissolved gases from the atmosphere. When carbon dioxide gas in the atmosphere dissolves in water, the resulting solution behaves like a dilute solution of carbonic acid:

$$CO_2(g) + H_2O(\ell) \longrightarrow H_2CO_3(aq)$$

As this slightly acidic water filters through soil and rock, it dissolves certain compounds from these materials. Ions that leach into ground water in this way include calcium, magnesium, iron(II), iron (III), carbonate, and sulfate. Most of the ions are largely harmless to plants and animals, and calcium is often beneficial. However, in some parts of the world, ions such as arsenic and fluoride may be present in concentrations that are high enough to be harmful to human health.

Harmful Arsenic Ions in Drinking Water

Arsenic is found naturally in the ground water beneath river deltas, where minerals containing arsenic compounds have been deposited as sediment. In these locations, arsenic is present as ions, such as the arsenate ion, $AsO_4^{3-}(aq)$.

In most of Canada, the concentration of arsenic in ground water is less than 5 ppb, which is considered to be safe. However, in Bangladesh, which is mainly delta land, as many as 35 million people have drinking water that contains more than 50 ppb of arsenic. Long-term exposure to arsenic in drinking water can cause cancer and may be linked to diabetes and other medical problems.

Risks and Benefits of Fluoride Ions in Drinking Water

Fluorine is the 17th most abundant element in Earth's crust, and all water supplies contain some fluoride ions. Too many fluoride ions leads to stained teeth, called *dental fluorosis*, shown in **Figure 9.9**. But if the concentration of fluoride ions in drinking water is less than 1 ppm, people benefit from healthier teeth without staining. Fluoride ions form strong ionic bonds. When fluoride ions replace some of the anions that are present in tooth enamel, the enamel becomes more resistant to decay. In regions where the natural concentration of fluoride ions is relatively low, many municipalities add fluoride ions to the water supply to boost the concentration to a level that will help prevent tooth decay.

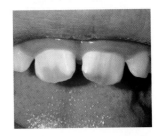

Figure 9.9 Fluoride ions prevent tooth decay, but too much fluoride can cause brown stains on teeth.

Activity 9.3 Removing Phosphate Ions from Drinking Water

The solubility of phosphate ions, $PO_4^{3-}(aq)$, is similar to the solubility of arsenate ions, $AsO_4^{3-}(aq)$. Phosphate ions can enter ground water as a result of pollution from fertilizers. You will model the precipitation of arsenate ions and phosphate ions using sodium phosphate, $Na_3PO_4(aq)$. Which reaction(s) could be effective for removing arsenate and phosphate ions from drinking water?

Safety Precautions

- Wear chemical safety goggles throughout this activity.
- Wear a lab coat or apron throughout this activity.
- If you spill any calcium hydroxide solution on your skin, wash it off immediately with plenty of cool water.
- When you have completed this activity, wash your hands.

Materials

- 0.1 mol/L sodium phosphate, $Na_3PO_4(aq)$
- 0.1 mol/L aluminum sulfate, $Al_2(SO_4)_3(aq)$, in a dropper bottle
- saturated solution of calcium hydroxide, $Ca(OH)_2(aq)$, in a dropper bottle
- 0.1 mol/L iron(III) sulfate, $Fe_2(SO_4)_3(aq)$, in a dropper bottle
- 4 test tubes with stoppers
- test-tube rack

Procedure

1. Design a table for recording the results of adding three different solutions to a solution of sodium phosphate.

2. Pour a few millilitres of sodium phosphate solution into each of four test tubes. Make sure that the volume of sodium phosphate is about the same in all the test tubes.

3. Add a few drops of aluminum sulfate solution to the first test tube. Cover the test tube with a stopper, and shake the test tube while holding the stopper down. Describe the precipitate that forms. Continue adding aluminum sulfate solution and shaking the test tube until no more precipitate forms. Place the test tube into the test-tube rack for later observation.

4. Repeat step 3 for each of the other two solutions.

5. Shake the fourth test tube, without adding anything to the sodium phosphate solution.

6. Compare the precipitates. Note which precipitates, if any, did not settle at the bottom of the test tube. Also compare the relative total volumes in the test tubes to determine which required the largest volume of reactant to cause complete precipitation.

Questions

1. Based on your observations, which precipitate should be relatively easy to separate by filtration? Explain.

2. Compare the volumes of the reactants you needed to precipitate all the phosphate ions.

3. What can you conclude from your observations of the fourth test tube?

4. Write the net ionic equation for each reaction.

5. What factors would you consider if you were choosing one of the three reactants in this activity to precipitate arsenate ions or phosphate ions from drinking water?

Calcium and Magnesium Ions Can Cause Hard Water

Ground water always contains ions dissolved from the surrounding rocks. The rocks shown in **Figure 9.10** are limestone, which is primarily calcium carbonate, $CaCO_3(s)$. Calcium ions enter into solution by reacting with the carbonic acid that is present in rainwater according to the following net ionic equation:

$$CaCO_3(s) + 2H^+(aq) + CO_3^{2-}(aq) \longrightarrow Ca^{2+}(aq) + 2HCO_3^-(aq)$$

If your water supply has a large concentration of calcium ions, you may notice that you have difficulty forming a lather with soap. Soap reacts with dissolved calcium and magnesium ions to form a scum of insoluble substances. Water with high concentrations of dissolved calcium and magnesium ions is called **hard water**. Water with relatively low concentrations of these ions is called **soft water** and lathers readily.

Figure 9.10 Many of the rocks in southern Ontario are limestone, which makes most of the water in this region hard.

Figure 9.11 The lime scale in this water pipe was caused by hard water.

Another sign of hard water is calcium carbonate deposits (often called lime scale) that build up inside water pipes, kettles, and humidifiers, as shown in **Figure 9.11**. You can use ordinary white vinegar to remove lime scale from the inside of a kettle or humidifier. Vinegar is an aqueous solution of acetic acid, $CH_3COOH(aq)$. Acetic acid reacts with the calcium and magnesium carbonates in the scale to form soluble compounds. (Remember that all acetates are soluble.) The following equation shows the reaction between calcium carbonate and acetic acid:

$$CaCO_3(s) + 2CH_3COOH(aq) \longrightarrow Ca(CH_3COO)_2(aq) + H_2O(\ell) + CO_2(g)$$

Learning Check

13. Is the water in your home hard or soft? What observations support your opinion?

14. What ions are you likely to find in greater concentrations in hard water than in soft water?

15. Why does the hardness of water differ from place to place throughout the world?

16. In Bangladesh, water from relatively shallow wells is likely to contain a greater concentration of arsenic than surface water or water from deep wells. Suggest a reason for this difference.

17. Why is a chemical treatment that precipitates phosphate ions from an aqueous solution likely to remove arsenate ions from ground water?

18. What are the signs of dental fluorosis? If you were a dentist and saw these signs in a patient, what advice would you offer?

Harmful Water Pollutants from Human Activities

Human activities, such as manufacturing, farming, transportation, and garbage disposal, can lead to the pollution of water systems. Sources of pollutants are often classified as either point source or non-point source:

- A *point source* of pollution has a single source with a specific location, as shown in **Figure 9.12 (A)**. Examples include a wrecked tanker that is leaking oil or a pulp mill that discharges effluent into a river.

- A *non-point source* of pollution does not come from a single, easily defined location. A non-point source may involve substances spread over large areas, such as pesticides and fertilizers from farmland or golf courses, as shown in **Figure 9.12 (B)**. A non-point source can also be a combination of thousands or even millions of small point sources, such as the exhaust from cars or the mercury from compact fluorescent lamps (CFLs).

The cumulative effect of many small point sources of pollution on Ontario's lakes and ground water is a major problem. In 2009, Ontario introduced the Cosmetic Pesticide Act to protect the environment from harmful lawn and garden pesticides. Yet, pollutants including lead, mercury, nitrates, and phosphates remain a problem in Canada and around the world.

Figure 9.12 (A) Point source of pollution: Waste water from a factory can quickly pollute a body of water. **(B)** Non-point source of pollution: Run-off from farms can carry fertilizer and pesticides into nearby waterways.

Effects and Sources of Lead Pollution

Exposure to lead can cause a variety of medical problems, including abdominal pain, kidney failure, nerve damage, and brain disorders. Babies and children are particularly susceptible. Lead in fresh water rarely comes from natural sources. Much of the lead in municipal drinking water comes from old water pipes. Until the 1950s, lead pipe was commonly used for the underground connection from a water main to a home because lead lasts longer than iron when buried in the ground. At that time, the potential health effects of the small amount of lead that leaches from a lead pipe were not known. Many municipalities now have programs to replace underground lead pipes.

Lead is released into the environment by some industrial processes, especially ore smelting, the manufacture and recycling of car batteries, and the production of some types of plastic. Over the last 30 years, the concentrations of lead measured in the air and the oceans have dropped dramatically—more than 75 percent in some tests. This decrease is largely the result of a worldwide effort to phase out the use of leaded gasoline. In 1990, Canada banned the use of leaded gasoline in on-road vehicles.

Effects and Sources of Mercury Pollution

Mercury and most mercury compounds are highly toxic. They affect the central nervous system, producing symptoms such as tremors, irritability, insomnia, numbness, and tunnel vision. Mercury can also damage the liver and kidneys. Some mercury exposure does come from natural sources. For example, volcanoes may be responsible for about half of the mercury in the atmosphere. However, much of the mercury deposited in the Great Lakes comes from emissions from coal-fired power plants. Other major sources of mercury emissions include gold mines, cement plants, and smelters for non-ferrous metals.

Effects and Sources of Nitrate and Phosphate Pollution

Livestock waste and nitrogen-based fertilizers are the greatest sources of nitrate ion, $NO_3^-(aq)$, pollution in Canada. Nitrate fertilizer is applied to farmland to increase crop yields, but the soluble fertilizer easily finds its way into lakes and rivers. There, it increases the growth of plants and algae, as shown in **Figure 9.13**. Phosphate ions, $PO_4^{3-}(aq)$, also promote excessive growth of aquatic plants. When bacteria decompose the remains of these plants, dissolved oxygen is removed from the water. The decreased oxygen supply puts stress on fish, especially in the summer when warmer water naturally contains less dissolved oxygen. In the 1960s, Lake Erie experienced many *algal blooms*: rapid growths of large quantities of algae. **Figure 9.14** shows how algal blooms happen and how they affect an aquatic wetland.

Figure 9.13 Nitrate and phosphate pollution can cause excessive plant growth in lakes and rivers.

A high concentration of nitrate ions in drinking water is harmful to babies who are less than three months old. Bacteria in a baby's digestive system convert nitrate ions to nitrite ions, $NO_2^-(aq)$. When the nitrite ions enter the bloodstream, they bond to hemoglobin, leaving less hemoglobin available to carry oxygen in the bloodstream. The baby's tissues can become starved for oxygen, causing the lips and fingertips to become blue. This condition is called *blue-baby syndrome*. Older babies are less susceptible to this condition since they have more acid in their stomach. Stomach acid inhibits the bacteria that convert nitrate ions to nitrite ions.

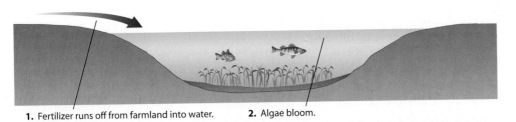

1. Fertilizer runs off from farmland into water.   **2.** Algae bloom.

3. Submerged plants die due to reduced light.   **4.** Algae and other plants die.

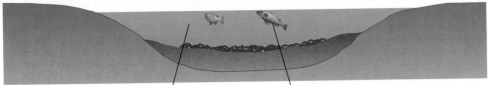

5. Bacteria use oxygen during decomposition.   **6.** Oxygen levels in the water drop too low for fish to survive.

Figure 9.14 Although algal blooms appear to help plant growth at first, the overall environmental effect is negative because of the loss of both plant and animal life.

How Airborne Pollution Contributes to Water Pollution

Motor vehicles, refineries, and many factories release carbon dioxide, sulfur dioxide, and nitrogen oxide gases. These non-metallic oxides dissolve in rainwater and contribute to the formation of acid rain. The greater the acidity, the greater the amounts of various compounds dissolved from soil and rocks by the rainwater. For example, aluminum is the most abundant metal in Earth's crust. Acidified rainwater leaches more aluminum ions into ground water and surface water. The dissolved aluminum ions can harm fish, since these ions impede the extraction of oxygen from water in the gills.

Leachates from Plastics

Drinking water may contain dissolved substances from unexpected sources. For example, polycarbonates are hard, clear plastics that are commonly used to make water bottles and other containers, like the containers shown in **Figure 9.15**. Polycarbonates are made using a chemical called bisphenol A, often abbreviated as BPA. A polycarbonate bottle can leach BPA into the water it contains. BPA is known to trigger biological changes like those caused by estrogen, the female sex hormone. Recent research suggests that BPA exposure is linked to breast cancer and heart disease, and may be a factor in several other serious disorders.

Canada was the first country to ban BPA from baby bottles and to restrict its use for lining infant formula cans. However, BPA is still used for making a variety of products, including kitchen containers, and for lining food and beverage cans. Unless processed in a properly contained landfill or recycling centre, discarded plastic containers can leach BPA and other pollutants into the environment.

Figure 9.15 Bisphenol A can leach from polycarbonate containers, such as these water cooler jugs.

Drinking Water Standards

maximum allowable concentration (MAC) a drinking water standard for a substance that is known or suspected to affect health when above a certain concentration

Ideally, drinking water should be clear and colourless, and it should not have an unpleasant taste or odour. More critically, drinking water must not contain disease-causing organisms or unsafe concentrations of compounds that could affect health. The absence of detectable micro-organisms, such as *E. coli* and coliform bacteria, is one example of a standard for drinking water quality. Ontario has standards for drinking water quality that specify the **maximum allowable concentration (MAC)** of many substances and micro-organisms that are known to affect human health. For example, drinking water must contain no *E. coli* bacteria and no more than five coliform bacteria per 100 mL.

In Canada, the federal, provincial, and territorial governments co-operate to produce guidelines for drinking water. These guidelines are the basis for the provincial and territorial standards. The guidelines and standards are often adjusted in response to new research about possible hazards. When insufficient data are available to establish a MAC with reasonable certainty, an *interim* maximum acceptable concentration (IMAC) is stated. The guidelines for drinking water also include aesthetic objectives (AOs), which suggest limits for substances that affect the taste, odour, or colour of drinking water but are not health hazards. **Table 9.3** lists some of the standards for chemicals in drinking water.

Suggested **Investigation**

Plan Your Own Investigation 9-C, Testing Drinking Water

Table 9.3 Maximum Concentrations of Selected Ions and Compounds in Ontario's Drinking Water

Ions or Compounds (Common Sources)	Maximum Allowable Concentration (mg/L)	Interim Maximum Allowable Concentration (mg/L)	Aesthetic Objective (mg/L)
arsenic, $As^{3+}(aq)$, and $As^{5+}(aq)$ (some types of soil and rock; mining activities)		0.025	
cadmium, $Cd^{2+}(aq)$ (some types of batteries)	0.005		
iron, $Fe^{2+}(aq)$ and Fe^{3+} (some types of rock; iron water mains)			0.3
lead, $Pb^{2+}(aq)$ (lead-alloy solder; lead water pipes; old house paint*)	0.01		
mercury, $Hg_2^{2+}(aq)$ and $Hg^{2+}(aq)$ (fluorescent lamps; some batteries; some types of fish, such as tuna)	0.001		
chloride, $Cl^-(aq)$ (water treatment)			250
fluoride, $F^-(aq)$ (some rocks; water treatment)	1.5		
nitrate, $NO_3^-(aq)$, and nitrite, $NO_2^-(aq)$ (fertilizer; animal waste)	10.0		
Alachlor (herbicide)		0.005	
Benzene (component of gasoline)	0.005		
Dioxin and furan (burning of waste, especially plastics)		0.000 000 015	

*Lead is no longer permitted in water-supply piping, and lead in new house paint must not exceed 0.06 percent (m/m). **Source:** Ontario Ministry of the Environment

Section Summary

- Only a tiny fraction of the water on Earth is readily available fresh water.

- Drinking water is obtained from surface water and ground water, and always contains dissolved substances from the environment.

- The dissolved substances in fresh water can include naturally occurring materials, pollutants, and chemicals added for water treatments.

- Drinking water standards specify the maximum allowable concentrations of substances that are known to affect human health.

Review Questions

1. **K/U** The substances that are dissolved in fresh water can be divided into three broad categories. List these categories, and give an example of a substance and a source for each.

2. **K/U** Describe two ways that you could test for hard water.

3. **A** List two point sources of pollution and two non-point sources of pollution that may affect your local water supply.

4. **K/U** How might sediment deposited by a river affect the quality of local ground water?

5. **K/U** List four household substances that can pollute the ground water if disposed of improperly.

6. **A** Describe a process that could be used to remove arsenic from drinking and irrigation water. Suggest reasons why this process is not used on a large scale in Bangladesh.

7. **T/I** Road salt is applied on ice-covered roads to improve driving conditions. When the ice melts, it dissolves the road salt. Is the run-off that contains the dissolved road salt a point source of pollution? Explain your reasoning.

8. **T/I** Why should the well water on a farm be tested regularly? Which contaminants would be of most concern?

9. **C** Which ions or compounds in **Table 9.3** have the lowest acceptable concentration? Draft a letter to the editor of your local newspaper, explaining the source of these ions or compounds. In your letter, describe what people can do to ensure that this type of pollution does not occur.

10. **T/I** Use **Table 9.3** to decide which of the following are unacceptable in a sample of drinking water. Which of the following should be of most concern? Explain your reasoning.
    - iron ions, 0.35 mg/L
    - chloride ions, 200 ppm
    - benzene, 0.000 007 g/L

11. **K/U** Excessive plant growth in a body of water is called an *algal bloom*.
    a. What substances cause algal blooms?
    b. List two sources of this type of pollution.
    c. Why might fish in a water system with this type of pollution be at risk, especially on a hot summer day?

12. **C** Volatile organic compounds (VOCs) include gasoline fumes and vapours from solvents. The table below lists the main sources of VOC emissions in Canada. Use the data in the table to construct a pie chart.

Sources of VOC Emissions in Canada*

Source	Percentage (%)
Combustion of fossil fuels for transportation	44
Production and distribution of petroleum and natural gas	26
Commercial and consumer products (e.g., solvents, paints)	12
Combustion of wood for home heating	8
Other sources	10

*Does not include sources such as agricultural animals and forest fires. **Source:** Environment Canada

13. **A** Crawford Lake is a small lake on the Niagara Escarpment in southern Ontario. This lake is unusually deep for its size. The rock cap of the escarpment is limestone, which contains both calcium carbonate, $CaCO_3(s)$, and magnesium carbonate, $Mg(CO_3)_2(s)$.
    a. Is the water in Crawford Lake more likely to be hard or soft? Explain your reasoning.
    b. Below a depth of 15 m, the water in Crawford Lake contains very little dissolved oxygen. Suggest why the shape of the lake causes this lack of oxygen.

14. **A** Herbicides can increase crop yields by preventing the growth of weeds. Discuss the risks and benefits of using a herbicide that is slow to break down into less potent chemicals.

Water Treatment

Key Terms

temporary hardness

permanent hardness

desalination

potable water

reverse osmosis

Ontario is fortunate to have abundant fresh water. However, Ontario also has examples of how pollution can make fresh water unfit for drinking. In May 2000, one of the wells that supply drinking water for the town of Walkerton was contaminated by run-off from a nearby farm. This run-off contained *E. coli* bacteria from animal manure. Water from the well was not tested or processed properly. As a result, the bacteria killed seven people and made about 2500 other people seriously ill.

In 1962, Reed International in Dryden, Ontario, began using a mercury cell process to make bleaching chemicals for its pulp and paper plant, shown in **Figure 9.16**. The company dumped waste water that contained mercury from this process directly into the Wabigoon-English River system. Within a few years, residents of the two closest communities downstream, Grassy Narrows and Whitedog, were showing signs of mercury poisoning from eating fish from the river. In 1970, commercial fishing on the river was stopped and some tourism businesses were shut down because of the extensive mercury contamination. Research published in 2010 found that mercury pollution is still having effects on the health of the people in Grassy Narrows and Whitedog. Even children born 30 years after the plant closed are showing clear signs of mercury poisoning.

In this section, you will explore chemical and physical methods for making water safe to drink.

Figure 9.16 Between 1962 and 1970, Reed International dumped over 9000 kg of mercury into the Wabigoon-English River system.

Treating Water Hardness

As described in Section 9.3, hard water contains dissolved calcium and magnesium carbonates. Limestone, for example, reacts with rainwater that contains dissolved carbon dioxide:

$$CaCO_3(s) + H_2O(\ell) + CO_2(aq) \rightleftharpoons Ca(HCO_3)_2(aq)$$

temporary hardness
hardness that can be removed from water by boiling

The double arrow in the chemical equation means that the reaction is *reversible*. When you read the equation from *right to left*, it describes dissolved calcium hydrogen carbonate forming calcium carbonate precipitate, water, and carbon dioxide gas. This reverse reaction takes place when hard water is heated. Calcium hydrogen carbonate and magnesium hydrogen carbonate are the main causes of **temporary hardness** in water.

Temporary Hardness and Permanent Hardness

Temporary hardness can be removed from water by boiling. Boiling drives most of the dissolved carbon dioxide out of the solution. Without dissolved carbon dioxide, the calcium carbonate that forms cannot react to form calcium hydrogen carbonate. So, the calcium carbonate precipitates out of the solution. **Figure 9.17** shows calcium carbonate and magnesium carbonate deposits that have been precipitated by heating hard water. While boiling water to remove temporary hardness is a simple solution, it is not practical for large volumes of water. The inside of large hot water boilers, for example, would soon become encrusted with calcium carbonate and magnesium carbonate deposits.

Permanent hardness is usually caused by dissolved calcium sulfate, $CaSO_4(aq)$, and magnesium sulfate, $MgSO_4(aq)$, which cannot be removed by boiling. However, permanent hardness can be removed by chemical methods. The least expensive water-softening chemical is washing soda, which is the common name for hydrated sodium carbonate, $Na_2CO_3 \cdot 10H_2O$. The addition of sodium carbonate to hard water precipitates insoluble magnesium and calcium carbonate. The following equation represents the reaction between calcium sulfate and sodium carbonate:

$$CaSO_4(aq) + Na_2CO_3(s) \rightarrow CaCO_3(s) + Na_2SO_4(aq)$$

However, the addition of sodium carbonate makes water basic. Ion-exchange water softeners use a system that avoids making the water basic.

Figure 9.17 Deposits of calcium carbonate and magnesium carbonate on water-heating elements can reduce the heater's efficiency.

Ion-Exchange Water Softeners

An ion-exchange water softener, like the one in **Figure 9.18**, exchanges the ions that cause hard water, principally calcium and magnesium, for sodium ions. Dissolved sodium compounds do not form a precipitate when the water is heated, and they do not form insoluble substances with soap.

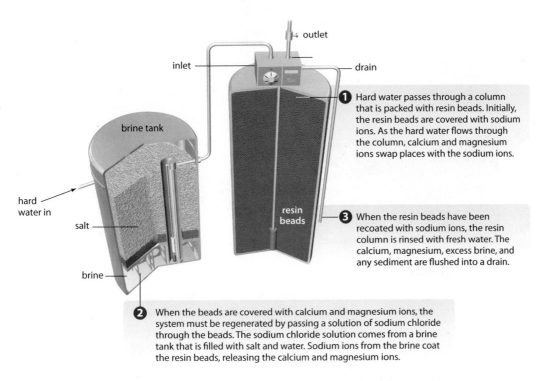

① Hard water passes through a column that is packed with resin beads. Initially, the resin beads are covered with sodium ions. As the hard water flows through the column, calcium and magnesium ions swap places with the sodium ions.

③ When the resin beads have been recoated with sodium ions, the resin column is rinsed with fresh water. The calcium, magnesium, excess brine, and any sediment are flushed into a drain.

② When the beads are covered with calcium and magnesium ions, the system must be regenerated by passing a solution of sodium chloride through the beads. The sodium chloride solution comes from a brine tank that is filled with salt and water. Sodium ions from the brine coat the resin beads, releasing the calcium and magnesium ions.

Figure 9.18 Water softened by an ion-exchange softener contains a high concentration of sodium ions. People on a low-sodium diet should avoid drinking water from a sodium ion water softener.

Explain the benefits and disadvantages of a home water softener.

Desalination

Most of the water on Earth's surface is too salty to drink. Dissolved ions regulate the flow of substances in and out of our cells. When someone drinks salt water, the ion balance is upset and water flows out of the cells. The cells become dehydrated. The more salt water that a person drinks, the worse the dehydration of the cells becomes. Excess salt is also harmful to many plants. The process of obtaining fresh water from salt water is called **desalination**. Water that is safe to drink is called **potable water**.

desalination the process of obtaining fresh water from salt water

potable water water that is safe to drink

Distillation

Most of the world's desalination plants are located in the Middle East. The majority of Middle East desalination plants distill ocean water, like the plant shown in **Figure 9.19**. Ocean water is heated and the vapour, which is free from dissolved substances, is condensed. Desalination plants in the Middle East use local supplies of relatively cheap oil to heat the water. Elsewhere in the world, however, energy costs usually make thermal desalination plants impractical. Even oil-rich countries, such as Saudi Arabia, are now developing desalination plants that use processes requiring less energy and that can be powered partly by solar energy.

Figure 9.19 The Jubail desalination plant, on the Arabian Gulf in Saudi Arabia, is the largest desalination plant in the world. Some of the energy that is needed to boil the water comes from waste heat from an adjacent electrical power plant.

Go to **scienceontario** to find out more

Reverse Osmosis

A more energy-efficient process for producing potable water is **reverse osmosis.** Osmosis is the natural tendency of a solvent, such as water, to move through a semi-permeable membrane to make the concentrations of solutes on both sides of the membrane equal. Water can pass through the semi-permeable membrane, but solutes, such as salt, cannot. If a semi-permeable membrane separates two aqueous solutions with different concentrations, water will flow from the more dilute solution into the more concentrated solution until there is no longer a difference in concentration. In reverse osmosis, high pressure is applied to the more concentrated solution to force water through the semi-permeable membrane in the opposite direction.

reverse osmosis the process of using high pressure to force water from a concentrated solution through a semi-permeable membrane to get a less concentrated solution

Reverse Osmosis Desalination Plants

Large reverse osmosis desalination plants, such as the one shown in **Figure 9.20**, are usually built in coastal areas where ocean water can be easily taken into the plant. Only part of the water taken in is purified. The rest, which contains dissolved salt and other impurities, is returned to the ocean with no added chemicals and no thermal pollution.

The largest seawater desalination plant in North America is located on Tampa Bay in Florida. This plant uses reverse osmosis to produce up to 100 million litres of potable water daily.

Figure 9.20 High pressure forces water across a semi-permeable membrane inside these vessels. The water that passes through the membranes is completely free of salt and other ions.

Water Supply Treatment

Birth and death records can provide an overview of the health of a population. The first collection of such data was made in the middle of the 17th century in London, England. The average life expectancy for a Londoner was then about 27 years. The graph in **Figure 9.21** shows data for life expectancy in North America. Many health scientists have concluded that the increasing life expectancy, since around 1800, is due mostly to the widespread improvement in the quality of water supplies, especially in towns and cities.

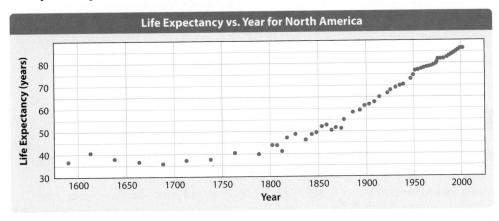

Figure 9.21 Life expectancy in North America has increased steadily since 1800, partly because of improvements in water treatment.

Analyze *Which century had the greatest increase in life expectancy?*

Municipal Water Treatment

Go to **scienceontario** to find out more

Figure 9.22 gives an overview of the processes that are used by municipalities to treat drinking water. In general, water enters a treatment plant from a surface source, such as a lake, river, or reservoir, or from a ground water source, such as a well or spring. On its way to the storage tower, the water is filtered, treated with chemicals to remove suspended particles, and treated with chlorine or ozone to kill harmful bacteria. After treatment, the water is tested to make sure that it is safe and meets current water standards.

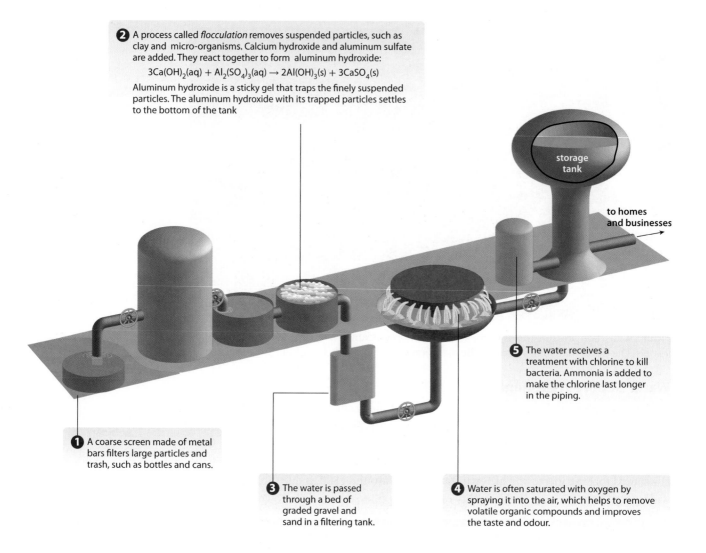

2 A process called *flocculation* removes suspended particles, such as clay and micro-organisms. Calcium hydroxide and aluminum sulfate are added. They react together to form aluminum hydroxide:

$$3Ca(OH)_2(aq) + Al_2(SO_4)_3(aq) \rightarrow 2Al(OH)_3(s) + 3CaSO_4(s)$$

Aluminum hydroxide is a sticky gel that traps the finely suspended particles. The aluminum hydroxide with its trapped particles settles to the bottom of the tank

storage tank

to homes and businesses

5 The water receives a treatment with chlorine to kill bacteria. Ammonia is added to make the chlorine last longer in the piping.

1 A coarse screen made of metal bars filters large particles and trash, such as bottles and cans.

3 The water is passed through a bed of graded gravel and sand in a filtering tank.

4 Water is often saturated with oxygen by spraying it into the air, which helps to remove volatile organic compounds and improves the taste and odour.

Figure 9.22 Municipalities use a combination of physical and chemical processes to purify water.

Learning Check

19. How does temporary water hardness differ from permanent hardness?

20. Why must salt be added, from time to time, to a home water softener?

21. Figure 9.19 shows a seawater desalination plant located next to an oil-burning electrical generating plant. What are the advantages of this arrangement?

22. Why are there are no large commercial desalination plants in Canada?

23. What evidence suggests that water treatment is an enormous benefit to society?

24. What are the basic steps in treating water for a municipal water supply?

Waste-Water Treatment

In most towns and cities, the water that is used in homes and businesses is collected by a system of pipes and sewers that carries it to a waste-water treatment plant, like the one shown in **Figure 9.23**. The goal of municipal waste-water treatment is to remove solids, chemicals, and dangerous bacteria from sewage. The waste water can then be released back into the environment.

Figure 9.23 Waste-water treatment plants remove possible pollutants from sewage so that those pollutants cannot enter and harm the environment.

Waste water (or sewage) first passes through screens that are made of closely spaced metal bars. These screens remove large solids and garbage that could block the flow through the machinery. The subsequent treatment of sewage often has three stages, as shown in **Figure 9.24**. Depending on their needs, municipalities use one, two, or all three of these stages.

Suggested **Investigation**

Inquiry Investigation 9-D, Treating Waste Water

- *Primary treatment* in a holding tank removes solid materials by sedimentation and by skimming scum from the surface of the water. The addition of calcium hydroxide and aluminum sulfate causes a sticky aluminum hydroxide precipitate to form:

$$3Ca(OH)_2(aq) + Al_2(SO_4)_3(aq) \rightarrow 2Al(OH)_3(s) + 3CaSO_4(s)$$

The precipitate settles slowly, removing suspended particles and some bacteria.

- *Secondary treatment* uses natural micro-organisms that feed on organic matter in the sewage. These bacteria convert organic material into carbon dioxide, water, and nitrogen compounds.

- *Tertiary treatment* involves chemical precipitation of nitrogen, phosphorus, and organic compounds.

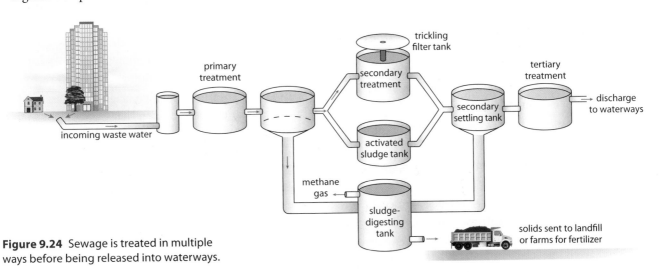

Figure 9.24 Sewage is treated in multiple ways before being released into waterways.

Section Summary

- Temporary water hardness is due to dissolved calcium carbonate and magnesium carbonate, and can be removed by boiling the water.

- Permanent water hardness is usually due to dissolved calcium sulfate and magnesium sulfate, and cannot be removed by boiling the water.

- Potable water can be produced from seawater or polluted water by various technologies, including distillation and reverse osmosis.

- Municipal water treatment plants process water to meet provincial standards, thus ensuring that the water is safe to drink.

- Waste-water treatment plants reduce pollutants and micro-organisms to levels that are low enough for the water to be safely returned to the environment.

Review Questions

1. **K/U** Hard water scale is caused by the build-up of calcium carbonate and magnesium carbonate.
   a. List the appliances in your home that are most likely to be affected by hard water scale.
   b. How can you remove hard water scale from an appliance?

2. **T/I** Why is lime scale more likely to form on hot-water pipes than on cold-water pipes?

3. **T/I** When people dump chemicals and waste into waterways, it is harmful to the environment. However, people also dump chemicals into water when they are purifying it for use. Explain the difference between these two situations.

4. **T/I** How many sodium ions are exchanged for each calcium ion in a water softener? Explain your answer.

5. **T/I** Energy from the Sun can be used to heat water, especially in areas that receive plenty of sunshine. Why do you think the desalination plant shown in **Figure 9.19** uses oil, rather than the Sun's energy?

6. **T/I** What by-products are produced by a distillation desalinization plant?

7. **A** Could a reverse-osmosis filter system be used to soften hard water? If so, would it have any advantage over an ion-exchange water softener?

8. **K/U** Briefly describe the physical processes that are used to clean municipal waste water.

9. **K/U** Aeration is the process of dissolving oxygen in water. Describe how aeration can help to purify water.

10. **K/U** Which water purification processes at a water treatment plant do not take place in nature?

11. **A** Due to the cost, not all municipalities use tertiary treatment to purify waste water. What would be the potential threat to aquatic life if water was discharged into a lake without tertiary treatment?

12. **C** The following diagram shows a simple solar distillation apparatus. Draw a sketch that explains how this apparatus produces potable water.

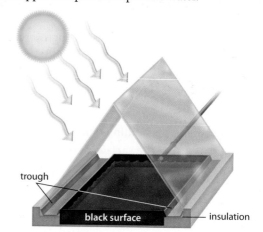

trough
black surface — insulation

13. **A** Pickling is a food preservation technique that uses salt to dehydrate food to make the food inhospitable to bacteria. The food is then stored in an acid solution that further inhibits bacterial growth. How does a concentrated salt solution help the pickling process?

14. **K/U** Name two different household wastes, and the substances they contain that should be removed at a waste-water treatment plant.

15. **T/I** An advertisement for a magnetic water conditioner claims that it can alter calcium ions so they cannot cause lime scale. Outline an investigation you could perform to test whether this claim is true.

16. **A** Lead(II) ions can be present in water that has passed through lead pipes or through pipes joined using lead-tin solder.
   a. Identify a compound that could be used to precipitate lead from an aqueous solution.
   b. Write the net ionic equation for the reaction.

Safety Precautions

- Wear safety eyewear and a lab coat or apron throughout this investigation.

- Hydrochloric acid and sulfuric acid are corrosive. Flush any spills on your skin with plenty of cool water. Inform your teacher immediately.

- Be careful not to contaminate the dropper bottles. The tip of a dropper should not make contact with either the plate or another solution. Put the cap back onto the bottle immediately after use.

- Part 2 of this investigation requires an open flame.

- Use EXTREME CAUTION when you are near an open flame.

- Tie back loose hair and clothing.

- Follow your teacher's disposal instructions.

- When you have completed this investigation, wash your hands.

Materials

- 12-well plate, 24-well plate, or spot plate

- toothpicks

- Bunsen burner secured to a retort stand

- heat-resistant pad

- wooden splints that have been soaked in water

Go to **Organizing Data in a Table** in **Appendix A** for help with making a data table.

Qualitative Analysis

In this investigation, you will apply your knowledge of flame tests, solution colours, chemical reactions, and solubilities to identify ions in solutions.

Pre-Lab Questions

1. In Part 2, why must you use wooden splints that have been soaked in water rather than dry splints?

2. Consider four different aqueous solutions, each containing one of the following types of cations:
   $Na^+(aq)$, $Ag^+(aq)$, $Ca^{2+}(aq)$, and $Cu^{2+}(aq)$.
   a. Describe the colour of each solution.
   b. Which type of cation will give a reddish-orange colour in a flame test?

Questions

What cations are in unknown solutions A, B, C, and D?

What anions are in unknown solutions X, Y, and Z?

Procedure

Part 1: Using Acids to Identify Cations

Reagents

- 4 dropper bottles of unknown solutions, labelled A, B, C, and D, each containing one of the following types of cations: sodium, $Na^+(aq)$; silver, $Ag^+(aq)$; calcium, $Ca^{2+}(aq)$; and copper(II), $Cu^{2+}(aq)$

- labelled dropper bottle containing dilute hydrochloric acid, $HCl(aq)$

- labelled dropper bottle containing dilute sulfuric acid, $H_2SO_4(aq)$

1. Read steps 2 and 3 below. Design a suitable table for recording your observations.

2. Place one or two drops of each unknown solution (A, B, C, or D) into its own well or spot. Then add one or two drops of hydrochloric acid to each unknown solution, and stir the mixture with a clean toothpick. Record your observations.

3. Repeat step 2 but, this time, test each unknown solution with one or two drops of sulfuric acid. Record your observations.

Part 2: Using Flame Tests to Identify Cations

Reagents

- labelled dropper bottle containing sodium ions, $Na^+(aq)$

- labelled dropper bottle containing silver ions, $Ag^+(aq)$

- labelled dropper bottle containing calcium ions, $Ca^{2+}(aq)$

- labelled dropper bottle containing copper(II) ions, $Cu^{2+}(aq)$

- unknown solutions A, B, C, and D from Part 1

Note: Your teacher may demonstrate this part of the investigation.

4. Read steps 5 to 9. Design a table to record your observations.

5. Observe and record the appearance of each known solution. If you think that you can identify any of the unknown solutions, record your identification. (Refer to Table 9.2.)

6. Set up the Bunsen burner on the heat-resistant pad. Light the burner. Adjust the air supply to produce a hot flame with a blue cone.

7. Wet one end of a soaked wooden splint with a few drops of the solution containing sodium ions. Carefully hold the splint so that the wet tip is just in the Bunsen burner flame, near the blue cone. You may need to hold the splint in this position for as long as 30 s to allow the solution to vaporize and mix with the flame. Do not let the splint burn. Record the colour of the flame.

8. For a control, repeat step 7 with distilled water. Record your observations. You can use the other end of the splint for a second test. Dispose of the used splints in the container that your teacher provides.

9. Repeat the flame test for each of the other known solutions. Then test each of the unknown solutions.

Part 3: Identifying Anions
Reagents

- 3 dropper bottles of unknown solutions, labelled X, Y, and Z, each solution containing one of the following types of anions: sulfate, SO_4^{2-}(aq); carbonate, CO_3^{2-}(aq); and iodide, I^-(aq)

- labelled dropper bottle containing barium ions, Ba^{2+}(aq)

- labelled dropper bottle containing silver ions, Ag^+(aq)

- labelled dropper bottle containing hydrochloric acid, HCl(aq)

10. Place one or two drops of each unknown solution (X, Y, or Z) into its own well or spot. Then add one or two drops of barium ion solution to each unknown solution. Stir each mixture with a clean toothpick. Record your observations.

11. Add a drop of hydrochloric acid to any well or spot where you observed a precipitate in step 10. Stir the mixture, and record your observations.

12. Repeat step 11 but, this time, test each unknown solution with one or two drops of silver ion solution. Record the colour of any precipitate that forms.

Analyze and Interpret

1. You tested four cations in Part 1.
   a. Which of the four cations should form a precipitate with hydrochloric acid? Refer to the solubility guidelines in Table 8.3 (in Section 8.2), if necessary.

Write the net ionic equation for the precipitation reaction.
   b. Do your observations confirm your prediction? Explain.
   c. Which of the cations should form a precipitate with sulfuric acid? Write the net ionic equations for these reactions.
   d. Do your observations confirm your predictions? Explain.
   e. Which cations should form soluble chloride compounds?
   f. Which cations should form soluble sulfate compounds?

2. Which cation is not colourless in aqueous solutions?

3. Based on your analysis so far, tentatively identify each unknown solution.

4. Use your observations of the flame tests to confirm or refute the identifications you made in question 5. If you are not sure about your identifications, compare your observations and analysis with other students' results. If necessary, repeat some of your tests.

5. Which anions should form precipitates with barium ions? Write the net ionic equations for the reactions.

6. Which precipitate should react when hydrochloric acid is added? Give reasons for your prediction.

7. Which anions should form precipitates with silver ions? Write net ionic equations for the reactions.

8. Tentatively identify each anion. Check your observations against the results you obtained when you added hydrochloric acid.

Conclude and Communicate

9. Identify the unknown cations and anions in this investigation. Explain why you do, or do not, have confidence in your identifications. What could you do to verify your conclusions?

10. What can you conclude from your control test in Procedure step 8?

Extend Further

11. **INQUIRY** Design an investigation to determine whether a solution contains potassium and/or calcium ions.

12. **RESEARCH** Atomic absorption spectroscopy (AAS) can be used to carry out both qualitative and quantitative analyses. Use text and Internet resources to research AAS. Briefly describe how it is used in analytical work.

Safety Precautions

- Wear safety eyewear throughout this investigation.

- Wear a lab coat or apron throughout this investigation.

Suggested Materials

- mixture of strontium chloride, $SrCl_2(s)$, and sodium chloride, $NaCl(s)$

- distilled water

- 1.0 mol/L sodium sulfate, $Na_2SO_4(aq)$

- electronic balance

- other equipment, as needed

*Go to **Scientific Inquiry** in **Appendix A** for information about planning an investigation.*

Determining the Mass Percent Composition of a Mixture

In this investigation, you will be given a sample of white powder that contains strontium chloride, $SrCl_2(s)$, and sodium chloride, $NaCl(s)$, in unknown proportions. These substances are soluble in water. You will plan and then carry out a procedure to determine the percent by mass of strontium chloride in the mixture.

Pre-Lab Questions

1. A solution of sodium sulfate, $Na_2SO_4(aq)$, is added to an aqueous solution that contains dissolved strontium chloride and sodium chloride. What ionic substance do you predict will precipitate?

2. Write the chemical equation and the net ionic equation for the precipitation reaction in question 1.

3. How does the amount of precipitate relate to the amount of strontium chloride in the reaction in question 1?

4. Write a formula to express the percent by mass of strontium chloride in a mixture that contains strontium chloride and sodium chloride.

5. What mass measurements are required to calculate the percent by mass of strontium chloride in the mixture described in question 4?

6. The sodium chloride used in this investigation is identical to the table salt used to season food. Why should you not use the sodium chloride from this investigation for cooking?

Question

How can a precipitation reaction be used to find the percent by mass of strontium chloride in a mixture?

Plan and Conduct

1. Write a procedure for your investigation. Include the mass measurements you will need to make, and describe the steps you will need to follow to precipitate one of the two substances in the mixture. Explain how you will obtain a dry sample of the precipitate. List the required equipment and necessary safety precautions.

2. Obtain your teacher's approval for your procedure, and then carry out your investigation.

Analyze and Interpret

1. Calculate the amount (in moles) of precipitate. Show your calculations.

2. Which ion in the precipitate was also present in the mixture? Which substance in the mixture contained this ion?

3. Calculate the mass of the substance you identified in question 2 that was present in the mixture you were given. Show your calculations.

4. Calculate the percentages by mass of strontium chloride and sodium chloride in the mixture. Show your work.

Conclude and Communicate

5. Write a sentence to explain how each of the following experimental errors would affect the calculated percentage of strontium chloride in a mixture.

    a. After the mixture was dissolved in water, not enough sodium sulfate was added.

    b. Some of the precipitate was left in a beaker, so it was not collected on the filter paper.

    c. The precipitate was not completely dry when its mass was measured.

6. Describe another possible experimental error, other than those in question 5. Explain how this error would affect the calculated percentage of strontium chloride in a mixture.

7. In this investigation, you were given a solution of sodium sulfate to precipitate one substance from the mixture. Write the name and formula for another solution you could have used to form the same precipitate.

8. Suggest improvements you could make to your procedure.

9. **INQUIRY** Strontium chlorate, $Sr(ClO_3)_2(s)$, forms white crystals, which cause a red colour when added to fireworks. Sodium phosphate, $Na_3PO_4(s)$, is a white powder that is sold in hardware stores as a cleaner. Both compounds are soluble in water. Write a procedure to determine the percent by mass of each compound in a mixture of the two compounds. Include the relevant chemical equations.

10. **RESEARCH** Barium is in the same group in the periodic table as strontium. Use text or Internet resources to research the properties of barium sulfate, $BaSO_4(s)$, that are useful in medical tests. Is barium sulfate soluble in water? Why is the solubility of barium sulfate important in medical tests on patients?

Safety Precautions

- Wear safety eyewear throughout this investigation.

- Wear a lab coat or apron throughout this investigation.

- Do not drink water from a beaker in the laboratory.

- Your investigation must include appropriate safety measures.

- Obtain your teacher's approval before proceeding with your investigation.

Suggested Materials

- tap water
- drinking-water test kit
- 50 mL beaker

Testing Drinking Water

In this investigation, you will determine the concentrations of substances dissolved in tap water by using a drinking-water test kit.

Pre-Lab Questions

1. What is one way that you can determine the pH of a solution?

2. What determines whether water is classified as being hard or soft?

3. You are testing normal tap water in this investigation. You probably drink the same water regularly when you are at home. Why should you not drink any of the water when doing this investigation?

Question

How does the quality of your drinking water compare to the standards set by the government?

Plan and Conduct

1. Read the manufacturer's directions on the drinking-water test kit.

2. Write a procedure that you can follow to test your school's tap water. Include a table that you can use to record your data.

3. After you teacher approves your procedure, begin your investigation.

4. Dispose of your water and other materials as directed by your teacher.

Analyze and Interpret

1. Is your drinking water acidic (pH less than 7) or basic (pH greater than 7)? Why do you think it has this property?

2. Classify your drinking water as hard or soft. How do the local rocks and soil account for its hardness or softness?

3. The test-kit manufacturer describes some of the tests as "semi-quantitative." Suggest a reason why these tests give only a rough estimate of the concentration of substances in the sample.

Conclude and Communicate

4. Compare your test results with the maximum allowable concentrations listed in **Table 9.3**.

Extend Further

5. **INQUIRY** What other ions in drinking water are regulated by the Ontario government? Describe some qualitative tests you could do to identify these ions.

6. **RESEARCH** Mercury poisoning can be very dangerous, so the level of mercury in drinking water is carefully monitored. However, people can consume mercury in different ways. Use the Internet to research different ways that people consume mercury. How can people reduce their exposure to all forms of mercury?

Safety Precautions

- Wear safety eyewear and a laboratory apron throughout this investigation.
- When you have finished this investigation, wash your hands.

Materials

- small foam or wax paper cup
- paper clip
- pea gravel
- sand
- 50 mL sample of "waste water"
- 1 mL quicklime (calcium oxide), $CaO(s)$
- 0.5 mL aluminum sulfate, $Al_2(SO_4)_3(s)$
- 0.5 mL powdered activated charcoal
- pH test strip
- 250 mL beaker
- filter stand with funnel holder
- 2 beakers (100 mL)
- wire gauze with no ceramic centre
- plastic teaspoon
- stirring rod
- filter paper
- funnel

Go to *Organizing Data in a Table* in *Appendix A* for help with making a data table.

Treating Waste Water

In this investigation, you will investigate methods for cleaning waste water. You will treat a sample of "waste water" that has physical properties similar to those of sewage, but is safe to handle. You will evaluate the properties of the treated water before returning it to the environment.

Pre-Lab Questions

1. Which part of the Procedure for this investigation is similar to the primary treatment at a waste-water plant?

2. In step 3, why is gravel placed in the cup before sand? Why are layers of gravel and sand used, rather than mixing the gravel and sand together?

3. Which steps of the process used in a waste-water treatment plant are not modelled in this investigation?

4. Why should you check with your teacher before pouring your treated water down the drain?

Question

Can a sample of dirty water be cleaned using simple physical and chemical methods?

Procedure

1. Prepare a table for recording your observations of the solids in the waste water and the clarity, colour, and odour of the waste water at each step of the treatment.

2. Use a straightened paper clip to poke six to eight holes in the bottom of the foam or wax-paper cup. Skip this step if the cup already has holes in the bottom.

3. Place about 1 cm of gravel in the bottom of the cup. Then add about 3 cm of sand over the gravel. Finally, add another 1 cm of gravel on top of the sand.

4. Your teacher will give you about 50 mL of "waste water" in a 250 mL beaker. In the table you prepared, record your observations of the untreated water.

5. Set up the filter stand with the support ring over a clean, dry 100 mL beaker. Place the wire gauze on the support ring, and then place the sand and gravel filtration cup on the wire gauze, as shown in the photograph below.

6. Carefully and slowly pour the waste water into the cup. In the table you prepared, record your observations of the water that passes through the cup.

7. Use the plastic spoon to add about 1 mL of quicklime and about 0.5 mL of aluminum sulfate to the filtered water. Stir the mixture to dissolve the added chemicals. Record your observations of the water.

8. Add about 0.5 mL of powdered activated charcoal to the beaker of water, and stir to distribute the charcoal evenly. Leave the beaker for a few minutes. While waiting, prepare the filter paper as described in step 9. Record your observations of the water.

9. Fold a piece of filter paper in half and then in half again. Open one of the edges of the filter paper to form a small cone. Place the cone in the funnel. Moisten the paper with clean tap water to help the cone stay in place. Place a clean, dry 100 mL beaker under the funnel.

10. Filter the solution by carefully pouring the liquid into the filter-paper cone. Leave as much of the charcoal and any other solids in the beaker as you can.

11. Record your observations of the water. Measure and record the pH of the water using a pH test strip.

12. Check with your teacher before pouring the treated water into the sink because you are returning it to the local water system. Do not allow any solids to go down the drain. Dispose of the solids you collected and the filter paper as directed by your teacher.

Analyze and Interpret

1. What is the advantage of using sand and gravel, rather than filter paper, for the initial filtration?

2. What was the main effect of adding quicklime and aluminum sulfate in step 7?

3. What was the main effect of adding activated charcoal in step 8?

Conclude and Communicate

4. The acceptable range for the pH of drinking water is from 6.5 to 8.5. Was the pH of the water you treated in this range? If not, suggest further treatment that could achieve an acceptable pH.

5. The aesthetic quality of water refers to properties such as colour, odour, and clarity. Summarize the change in each of these properties after the sand and gravel filtration, and after the addition of quicklime and aluminum sulfate. Why are quicklime and aluminum sulfate added after the sand/gravel filtration, rather than before?

| **Extend Further** |

6. **INQUIRY** A petri dish that contains agar can be used to test for the presence of micro-organisms. Micro-organisms feed on the agar and multiply. Design an investigation to compare the number of micro-organisms in tap water with the number of micro-organisms in the water you treated in this investigation. (Note: Do not perform your investigation without first getting your teacher's approval.)

7. **RESEARCH** How is water purified in nature? Use text and Internet resources to find out how evaporation by the Sun, transpiration by plants, and filtration by rocks and soil remove many impurities. Summarize your findings in a visual form, suitable for either a notice board or a computer presentation.

Case Study

What's Going Down the Drain?

Treating Pharmaceutical Chemicals in Waste Water

Scenario

You have just seen today's newspaper, and the headline below catches your eye. The article describes research showing that thousands of chemicals found in prescription medications enter into municipal waste water every day. The chemicals make their way into the municipal system when they are disposed of by pouring them down sinks and drains or by excreting them from the body in urine. The article states that, for the most part, the effects of mixtures of these pharmaceutical chemicals in the environment are poorly understood or even completely unknown. There is growing concern among some scientists that these waste products are forming a whole new class of pollutants that could have dangerous effects on the environment and on human health over the long term.

This information about the increase in pharmaceutical waste entering sewage systems across Canada troubles you, so you decide to address your Municipal Council and ask council members about water treatment in your area. Your brief presentation will urge the municipality to check procedures for measuring the quantities of pharmaceutical wastes in treated water, thus ensuring the community's ground water and drinking water supply are protected.

To prepare for your presentation to Council, you do some research on how municipal waste is treated. When pharmaceutical wastes enter the sewage and waste-water system, they end up at the municipal waste-water treatment plant. Some waste-water plants use three separate processes to treat waste water. The primary treatment involves the separation and removal of solids and some bacteria.

The secondary treatment introduces micro-organisms to the water, to consume the organic matter and pathogens found in the waste water. The water then moves to a tank where the micro-organisms and any remaining solids settle out. In some cases, a third stage occurs, in which chemicals are added to precipitate out nitrogen, phosphorus, and organic compounds. The treated water is then released back out into the ecosystem, where it enters into the region's source water.

Treatment of waste water varies from municipality to municipality, depending on the following factors:

- the size of the population
- the economic feasibility of building, operating, and maintaining the waste-water treatment facility
- the availability of trained staff to operate and monitor the treatment processes

Many municipalities use only the primary process to treat the waste water. Thus, the effluent released back into the environment can include human waste, solids and other debris, nutrients, pathogens, potentially toxic chemicals, ingredients from personal care and household products, and pharmaceutical chemicals. In Canada, the government does not currently regulate the treatment of pharmaceutical chemicals in waste-water treatment plants, so these chemicals are not targeted for removal.

You decide to prepare a bulletin to distribute throughout your community, stating facts about pharmaceutical waste, contamination, waste-water treatment, and water quality. You are hoping that other community members will come forward and give presentations that summarize their concerns as well, so that Council will understand how important the issue is to the community.

SUNDAY TIMES | August 7, 2011

METRO

Tap Water Found to be Contaminated

By GEORGE SMITHERS
Associated Press

Lorem ipsum dolor sit amet, consectetuer adipiscing elit, sed diam nonummy nibh euismod tincidunt ut laoreet dolore magna aliquam erat volutpat. Ut wisi enim ad minim veniam, quis nostrud exerci tation ullamcorper suscipit lobortis nisl ut aliquip ex ea commodo consequat. Duis autem vel eum iriure dolor in hendrerit in

Lorem ipsum dolor sit amet, consectetuer adipiscing elit, sed diam nonummy nibh euismod tincidunt ut laoreet dolore magna aliquam erat volutpat. Ut wisi enim ad minim veniam, quis nostrud exerci tation ullamcorper suscipit lobortis nisl ut aliquip ex ea commodo consequat. Duis autem vel eum iriure dolor in hendrerit in vulputate velit esse molestie consequat, vel illum dolore eu feugiat nulla facilisis at vero eros et accumsan et iusto odio dignissim qui blandit praesent luptatum zzril delenit augue

Do You Know What's in Your Water?

- Pure water doesn't exist in nature. Water is the universal solvent; therefore, it is always found with some type of solutes dissolved in it. These solutes can be naturally occurring (such as metals found in the soil), or they can be the result of human activities (such as contaminants added to source water from medicines, household wastes, and industrial operations). Not every substance dissolved in water is harmful to humans or the environment. In fact, some dissolved substances are beneficial or even necessary for living things.

- Sources of pharmaceutical waste that can contaminate the sewage entering into treatment plants include prescription medications, and non-biodegradable personal care items such as scented skin-care products.

- Medications that mimic animal hormones can have a negative impact because they may alter the action of hormones in animals. One known impact is the growth of female sex organs in male fish. Sources of these chemicals include medicines such as birth control pills, as well as pesticides and industrial waste.

- Waste-water treatment plants are not designed to detect and treat most pharmaceutical waste. This means that the water being released from the treatment plant into the environment still contains these potentially toxic chemicals.

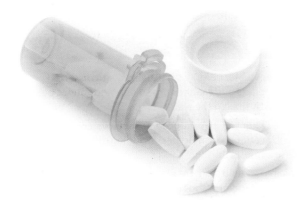

Pharmaceutical products, such as the medicine shown here, can be life savers for many people. However, pharmaceutical wastes can enter the sewage system from homes or businesses. Sewage is treated at waste-water treatment plants and is then released back into the environment. But the pharmaceutical wastes stay dissolved in the water because there are no specific processes for removing them.

Research and Analyze

1. Which level of government regulates pollutants in the water treatment system? Research which department is responsible and what that department is doing, if anything, to monitor pharmaceutical waste in municipal waste water.

2. There have been reports of cancer medications and other strong medications being found in trace amounts in tap water that has been treated. The chronic effects of drinking these toxins are not currently known. Research the types of studies that are underway to determine levels of pharmaceutical waste in treated water. What are the main topics under study and what waste products are causing the most concern?

3. Upgrading water and waste-water facilities across the country to a level that will ensure full treatment of pharmaceutical waste will require enormous amounts of money. Furthermore, the techniques to treat these contaminants are still being developed. Analyze some economic, social, and environmental issues related to upgrading water and waste water facilities across the country to a level that will ensure treatment of pharmaceutical waste. You may wish to use a risk-benefit analysis chart (See Analyzing STSE Issues in Appendix A for information about this type of analysis.)

Take Action

1. **Plan** In a group, discuss the issue of pharmaceutical waste in the water supply in municipalities in Canada. What are the key issues to consider when thinking about how to ensure that drinking water is safe for consumption? Share the results of the research and analysis you conducted for questions 1 to 3 above.

2. **Act** Create a script for a television commercial about your concerns for ensuring that source water is protected from pharmaceutical waste. The commercial will air on your local cable television network. It will be directed at community members and will present some possible solutions for addressing the issue. Ensure that you create a product that will enable individuals to understand some of the major economic, social, and environmental issues related to drinking water.

Section 9.1 | Net Ionic Equations and Qualitative Analysis

Net ionic equations are a useful model for representing reactions. Qualitative tests can help to identify elements in an unknown compound.

KEY TERMS

flame test
ionic equation
net ionic equation

precipitate
qualitative analysis
spectator ion

KEY CONCEPTS

- A net ionic equation omits the spectator ions and shows only the ions that react and the product(s) of the reaction.
- You can use observations of flame tests, solution colours, and the formation of precipitates to identify ions in a sample.

Section 9.2 | Solution Stoichiometry

Solution stoichiometry applies the principles of stoichiometry calculations to the volumes, concentrations, and masses involved in reactions that take place in solutions.

KEY TERMS

quantitative analysis
solution stoichiometry

KEY CONCEPTS

- In solution stoichiometry, known volumes and concentrations of reactants or products are used to determine the volumes, concentrations, or masses of other reactants or products.
- The solution to a stoichiometry problem should always include a balanced chemical equation or the net ionic equation.

Section 9.3 | Water Quality

Water naturally contains dissolved ions and compounds, which affect its suitability for drinking and irrigation.

KEY TERMS

fresh water
ground water
hard water
maximum allowable concentration (MAC)
soft water
surface water

KEY CONCEPTS

- Only a tiny fraction of the water on Earth is readily available fresh water.
- Drinking water is obtained from surface water and ground water, and always contains dissolved substances from the environment.
- The dissolved substances in fresh water can include naturally occurring materials, pollutants, and chemicals added for water treatments.
- Drinking water standards specify the maximum allowable concentrations of substances that are known to affect human health.

Section 9.4 | Water Treatment

Water treatment processes can kill micro-organisms and reduce certain dissolved substances to safe levels, both in drinking water and in waste water returned to the environment.

KEY TERMS

desalination
permanent hardness
potable water

reverse osmosis
temporary hardness

KEY CONCEPTS

- Temporary water hardness is due to dissolved calcium carbonate and magnesium carbonate, and can be removed by boiling the water.

- Permanent water hardness is usually due to dissolved calcium sulfate and magnesium sulfate, and cannot be removed by boiling the water.
- Potable water can be produced from seawater or polluted water by various technologies, including distillation and reverse osmosis.
- Municipal water treatment plants process water to meet provincial standards, thus ensuring that the water is safe to drink.
- Waste-water treatment plants reduce pollutants and micro-organisms to levels that are low enough for the water to be safely returned to the environment.

Knowledge and Understanding

Select the letter of the best answer below.

1. When a solution of potassium sulfate, K_2SO_4(aq), is mixed with a solution of lead(II) nitrate, $Pb(NO_3)_2$(aq), a reaction takes place. What are the spectator ions in this reaction?
   a. K^+(aq) and $SO_4{}^{2-}$(aq)
   b. K^+(aq) and $NO_3{}^-$(aq)
   c. Pb^{2+}(aq) and $NO_3{}^-$(aq)
   d. Pb^{2+}(aq) and $SO_4{}^{2-}$(aq)
   e. K^+(aq), Pb^{2+}(aq), $SO_4{}^{2-}$(aq), and $NO_3{}^-$(aq)

2. Which ion promotes the growth of plant life in lakes and streams?
   a. chloride, Cl^-(aq)
   b. phosphate, $PO_4{}^{3-}$(aq)
   c. arsenic(V), As^{5+}(aq)
   d. carbonate, $CO_3{}^{2-}$(aq)
   e. sulfate, $SO_4{}^{2-}$(aq)

3. Which ion causes hardness in water?
   a. sodium, Na^+(aq)
   b. nitrate, $NO_3{}^-$(aq)
   c. magnesium, Mg^{2+}(aq)
   d. chloride, Cl^-(aq)
   e. hydrogen carbonate, $HCO_3{}^-$(aq)

4. Which process could be used to make drinking water from seawater?
   a. aeration
   b. filtration through sand and gravel
   c. filtration through activated carbon
   d. distillation
   e. ion exchange

5. Which solution contains the same number of ions per litre as 1.0 mol/L $MgCl_2$(aq)?
   a. 1.5 mol/L sodium chloride, NaCl(aq)
   b. 0.5 mol/L lead(II) sulfate, $PbSO_4$(aq)
   c. 1.0 mol/L aluminum chloride, $AlCl_3$(aq)
   d. 2.0 mol/L potassium chloride, KCl(aq)
   e. 3.0 mol/L hydrochloric acid, HCl(aq)

6. The temporary hardness in a sample of tap water was removed by boiling the water. Which compound does the resulting precipitate contain?
   a. sodium carbonate, Na_2CO_3(s)
   b. calcium carbonate, $CaCO_3$(s)
   c. calcium hydrogen carbonate, $Ca(HCO_3)_2$(s)
   d. calcium chloride, $CaCl_2$(s)
   e. calcium sulfate, $CaSO_4$(s)

7. What is the net ionic equation for the following reaction?
   $$Ca(NO_3)_2(aq) + Na_2S(aq) \rightarrow CaS(s) + 2NaNO_3(aq)$$
   a. Ca^{2+}(aq) + $2NO_3{}^-$(aq) + $2Na^+$(aq) + S^{2-}(aq) $\rightarrow CaS(s) + 2NaNO_3(aq)$
   b. Ca^{2+}(aq) + $2NO_3{}^-$(aq) + $2Na^+$(aq) + S^{2-}(aq) $\rightarrow CaS(s) + 2Na^+(aq) + 2NO_3{}^-(aq)$
   c. Ca^{2+}(aq) + S^{2-}(aq) $\rightarrow CaS(s)$
   d. Na^+(aq) + $NO_3{}^-$(aq) $\rightarrow NaNO_3(aq)$
   e. $Ca(NO_3)_2(aq) + Na_2S(aq) \rightarrow CaS(s) + 2NaNO_3(aq)$

8. Which solution has the deepest colour?
   a. $Pb(NO_3)_2$(aq)
   b. $NaNO_3$(aq)
   c. $Mg(NO_3)_2$(aq)
   d. KNO_3(aq)
   e. $Cu(NO_3)_2$(aq)

Answer the questions below.

9. A precipitation reaction can be written as a complete chemical equation or as a net ionic equation.
   a. Distinguish between a complete chemical equation and a net ionic equation.
   b. List two advantages of a net ionic equation.

10. Identify the spectator ions in this skeleton equation:
    $$NaCl(aq) + CuSO_4(aq) \rightarrow CuCl_2(s) + Na_2SO_4(aq)$$

11. Write the net ionic equation for each reaction.
    a. $Cu(s) + Fe(NO_3)_2(aq) \rightarrow Cu(NO_3)_2(aq) + Fe(s)$
    b. $Na_2SO_4(aq) + Sr(OH)_2(aq) \rightarrow 2NaOH(aq) + SrSO_4(s)$
    c. $KOH(aq) + HI(aq) \rightarrow KI(aq) + H_2O(\ell)$

12. Chemists use qualitative analysis to identify ions in a solution. Identify an ion that could be present in each solution.
    a. a blue solution that gives a bluish-green flame test
    b. a colourless solution that gives a reddish-orange flame test
    c. a purple solution that gives a yellow-orange flame test

13. Why might a chemist do a qualitative analysis on a solution? Give an example. Why might a chemist do a quantitative analysis? Give an example.

14. Suggest one benefit and one drawback of removing phosphate ions from waste water using a chemical treatment.

15. State two ways in which water in the environment can become polluted.

16. Describe two hazards resulting from excess nitrate ions in water.

Thinking and Investigation

17. What is the maximum amount (in moles) of barium sulfate that can be precipitated by reacting 0.10 L of 0.10 mol/L aqueous barium chloride, $BaCl_2(aq)$, with 0.20 L of 0.10 mol/L aqueous sodium sulfate, $Na_2SO_4(aq)$?

18. State the concentrations of magnesium ions, $Mg^{2+}(aq)$, and nitrate ions, $NO_3^-(aq)$, in a 1.5 mol/L solution of magnesium nitrate, $Mg(NO_3)_2(aq)$. Explain your reasoning.

19. The burning sensation of "heartburn" is caused by excess stomach acid, HCl(aq), rising into the esophagus. The active ingredient in several liquid over-the-counter antacid medications is magnesium hydroxide, $Mg(OH)_2(s)$.

  a. Why do the instructions on a bottle of one of these antacid medications tell you to shake the bottle before pouring a dose?

  b. Write the chemical equation and the net ionic equation for the reaction between magnesium hydroxide and hydrochloric acid.

20. Identical volumes of two solutions are prepared using the same masses of sodium carbonate and magnesium carbonate. Which solution has the greater concentration of carbonate ions? Show your reasoning.

21. Magnesium is an essential element for many types of plants. An old gardening book suggests using 15 g of Epsom salts dissolved in 4.0 L of water to provide enough magnesium for a small lawn. Epsom salts are crystals of magnesium sulfate heptahydrate, $MgSO_4 \cdot 7H_2O(s)$. Find the concentration of magnesium ions, $Mg^{2+}(aq)$, in this solution.

22. Determine the molar concentrations of the ions in each aqueous solution.

  a. 62.8 g of sodium nitrate, $NaNO_3(s)$, dissolved in 125 mL of water

  b. 25.0 g of calcium acetate, $Ca(CH_3COO)_2(s)$, dissolved in 180 mL of water

  c. 12.6 g of ammonium phosphate, $(NH_4)_3PO_4(s)$, dissolved in 200 mL of water

23. Calcium ions, $Ca^{2+}(aq)$, are important for regulating heartbeat. The concentration of calcium ions in the blood of a healthy person is about 1.2×10^{-3} mol/L. In one patient, the concentration of calcium ions was found to be 3.0 mg/100 mL. What mass of calcium carbonate could raise this patient's concentration of calcium ions to that of a healthy person? The blood volume of this patient is about 5.0 L.

24. A strip of zinc metal was placed in a beaker that contained 110 mL of a solution of lead(II) nitrate, $Pb(NO_3)_2(aq)$. After a few days the replacement reaction was complete, and the solid lead was removed from the metal strip and dried. The mass of dried lead was 2.18 g. Write the net ionic equation for the reaction, and calculate the molar concentration of the lead(II) nitrate solution.

25. In a quantitative analysis experiment, 125 mL of aqueous copper(II) bromide was treated with excess aqueous sodium fluoride. When dried, the black precipitate had a mass of 18.5 g. Write the net ionic equation for the reaction, and calculate the concentration of copper(II) bromide.

26. Magnesium chloride reacts with sodium hydroxide in an aqueous solution according to the following skeleton equation:

$MgCl_2(aq) + NaOH(aq) \rightarrow Mg(OH)_2(s) + NaCl(aq)$

  a. Write the balanced chemical equation for this reaction.

  b. Write the net ionic equation.

  c. Find the maximum mass of magnesium hydroxide that will precipitate if 80.0 mL of 0.312 mol/L magnesium chloride is mixed with 100 mL of 0.315 mol/L sodium hydroxide.

27. Use **Table 9.1** to identify the substances in the photograph below. Are you certain about your answers? Explain why or why not.

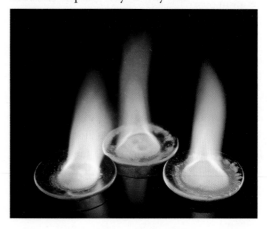

28. Suppose that you are given two test tubes, labelled A and B. Each test tube is filled with a clear, colourless, aqueous solution. One solution contains calcium chloride, and the other solution contains potassium nitrate. Outline two tests you would perform to identify the contents of each test tube. Describe the observations you would expect for each solution.

Communication

29. **BIG IDEAS** People have a responsibility to protect the integrity of Earth's water resources. Design a leaflet you could distribute to homes in your area, suggesting ways that people could reduce the volume of water they use and reduce the pollution in the waste water from their homes.

30. **BIG IDEAS** Living things depend for their survival on the unique physical and chemical properties of water. Some lifeboats are equipped with manually operated reverse-osmosis equipment that can produce potable water from seawater. Use text and Internet sources to research reverse-osmosis equipment. Draw a diagram to show how it works.

31. **C** A research chemist was investigating the chemical properties of bismuth trichloride, $BiCl_3(s)$. When excess water was added to 100 g of bismuth trichloride, a precipitate formed. Make a table that the chemist could use to determine which of the following reactions took place:

   I. $BiCl_3(s) + H_2O(\ell) \longrightarrow$
   $$BiOCl(s) + 2H^+(aq) + 2Cl^-(aq)$$

   II. $2BiCl_3(s) + H_2O(\ell) \longrightarrow$
   $$Bi_2OCl_4(s) + 2H^+(aq) + 2Cl^-(aq)$$

   III. $2BiCl_3(s) + 3H_2O(\ell) \longrightarrow$
   $$Bi_2O_3(s) + 6H^+(aq) + 6Cl^-(aq)$$

32. Use text or Internet resources to research the effectiveness, costs, and by-products of desalination plants. Then make a table to compare the advantages and disadvantages of the distillation and reverse-osmosis processes.

33. New processes, using equipment such as membrane bioreactors, rotating biological contactors, and ultraviolet disinfectors, are being developed for treating sewage. Use text or Internet resources to learn about several of these new processes. Then draft a letter to a town council recommending two processes that the council should consider when expanding or overhauling the town's aging waste-water treatment plant. Give reasons for your recommendations.

34. Summarize your learning in this chapter using a graphic organizer. To help you, the Chapter 9 Summary lists the Key Terms and Key Concepts. Refer to Using Graphic Organizers in Appendix A to help you decide which graphic organizer to use.

Application

35. Chlorine is toxic to fish. Even at low concentrations, it can damage their gills. An aqueous solution of sodium thiosulfate can be added to tap water to remove dissolved chlorine:
   $$2Na_2S_2O_3(aq) + Cl_2(aq) \longrightarrow 2NaCl(aq) + Na_2S_4O_6(aq)$$
   Suppose that you are setting up an aquarium that holds 60 L of water. You use a test strip to estimate that your tap water contains 10 mg/L chlorine. You use a chart on the bottle of sodium thiosulfate solution to determine that you should add 60 mL of the solution to the water in the aquarium. If this amount removes all the dissolved chlorine, what is the molar concentration of the sodium thiosulfate solution?

36. Urea, $(NH_2)_2CO(s)$, is widely used as a fertilizer. It is also excreted in the urine of mammals.
   a. Identify at least two different ways that urea could pollute the environment.
   b. Use text or Internet resources to find out how bacteria in the soil convert urea to nitrite and nitrate ions. Write the chemical equations for the three reactions in this conversion.

37. Lead from shotgun shells could be a source of lead exposure for hunters, especially in First Nations communities that rely heavily on hunting wild game for food. Describe how the level of lead could build up in the body of someone who regularly eats wild game. What evidence would you collect to support your hypothesis?

38. In some municipalities, waste water and storm water are carried in the same sewers. After a prolonged heavy rain, the volume of storm water can be too much for a waste-water treatment plant to handle. As a result, some waste water has to be released without being treated. How would a lake that received untreated water from a treatment plant be affected?

39. The concentration of dissolved chlorine in tap water can be determined by adding excess potassium iodide, $KI(aq)$, and then using aqueous sodium thiosulfate to precipitate the resulting iodide ions. The net ionic equation for the first reaction is
   $$2I^-(aq) + Cl_2(aq) \longrightarrow 2Cl^-(aq) + I_2(aq)$$
   a. In Ontario, the aesthetic objective for chlorine in tap water is a maximum concentration of 250 mg/L. Find the minimum volume of a 0.001 00 mol/L solution of potassium iodide that should be used to test a 500 mL sample of water to meet this objective.
   b. Why might more than the minimum amount of potassium iodide solution be added?

Chapter 9 | SELF-ASSESSMENT

Select the letter of the best answer below.

1. **K/U** Zinc nitrate, $Zn(NO_3)_2(aq)$, reacts with sodium hydroxide, $NaOH(aq)$, to form a precipitate. Identify the spectator ion(s) in the reaction.
   a. $NO_3^-(aq)$
   b. $Na^+(aq)$ and $OH^-(aq)$
   c. $Na^+(aq)$ and $NO_3^-(aq)$
   d. $OH^-(aq)$
   e. $Zn^{2+}(aq)$ and $OH^-(aq)$

2. **K/U** What is the net ionic equation for the reaction between nitrous acid, $HNO_2(aq)$, and potassium hydroxide, $KOH(aq)$?
   a. $HNO_2(aq) + KOH(aq) \rightarrow KNO_2(aq) + H_2O(\ell)$
   b. $HNO_2(aq) + OH^-(aq) \rightarrow NO_2^-(aq) + H_2O(\ell)$
   c. $H^+(aq) + NO_2^-(aq) + K^+(aq) + OH^-(aq)$
      $\rightarrow KNO_2(aq) + H_2O(\ell)$
   d. $K^+(aq) + NO_2^-(aq) \rightarrow KNO_2(s)$
   e. $H^+(aq) + OH^-(aq) \rightarrow H_2O(\ell)$

3. **K/U** Which ion gives a characteristic colour in a flame test?
   a. lithium, $Li^+(aq)$
   b. carbonate, $CO_3^{2-}(aq)$
   c. zinc(II), $Zn^{2+}(aq)$
   d. sulfate, $SO_4^{2-}(aq)$
   e. ammonium, $NH_4^+(aq)$

4. **T/I** What is the maximum mass of silver chloride , $AgCl(s)$, that can be precipitated from 0.050 L of 0.20 mol/L silver nitrate, $AgNO_3(aq)$, by adding excess magnesium chloride, $MgCl_2(aq)$?
   a. 0.010 g
   b. 0.72 g
   c. 0.95 g
   d. 1.4 g
   e. 1.7 g

5. **T/I** What is the minimum volume of 0.15 mol/L $Na_2S(aq)$ solution that is needed to precipitate all the lead ions from 75 mL of 0.10 mol/L $Pb(NO_3)_2(aq)$ solution?
   a. 75 mL
   b. 50 mL
   c. 25 mL
   d. 15 mL
   e. 10 mL

6. **T/I** What volume of 0.25 mol/L $(NH_4)_2SO_4(aq)$ solution contains 1.5 mol of ammonium ions?
   a. 0.17 L
   b. 0.38 L
   c. 1.5 L
   d. 3.0 L
   e. 6.0 L

7. **K/U** Which of the following is a point source of pollution?
   a. a coal-burning power plant
   b. urban roads
   c. urban lawns and gardens
   d. croplands
   e. cottages bordering a lake

8. **K/U** Which of the following would have very little effect on oxygen levels in a lake?
   a. controlling the run-off of fertilizers from agricultural fields
   b. banning the use of pesticides
   c. banning phosphate detergents
   d. tertiary treatment of municipal sewage
   e. controlling drainage from poultry, livestock, and hog farms

9. **K/U** Hard water contains a relatively large concentration of
   a. sodium ions, $Na^+(aq)$
   b. hydroxide ions, $OH^-(aq)$
   c. lead(II) ions, $Pb^{2+}(aq)$
   d. aluminum ions, $Al^{3+}(aq)$
   e. calcium ions, $Ca^{2+}(aq)$

10. **T/I** Which type of municipal sewage treatment is incorrectly classified?
   a. use of natural micro-organisms: secondary
   b. skimming of scum from the surface of the water in holding tanks: primary
   c. removal of solids by sedimentation: primary
   d. chemical precipitation: secondary
   e. conversion of organic material into carbon dioxide, water, and nitrogen compounds: secondary

Use sentences and diagrams, as appropriate, to answer the questions below.

11. **C** Suppose that you are given two unlabelled solutions. One contains silver nitrate, and the other contains calcium nitrate. Compare and contrast the observations you would make while doing qualitative analysis tests to identify the solutions.

12. (T/I) Aqueous sodium phosphate and aqueous calcium hydroxide react in a double displacement reaction.

a. Complete the chemical equation for the reaction:
$$Na_3PO_4(aq) + Ca(OH)_2(aq) \rightarrow$$

b. Write the net ionic equation for this reaction.

13. (T/I) Tin(II) chloride, $SnCl_2(aq)$, and potassium phosphate, $K_3PO_4(aq)$, are mixed.

a. State the name and formula for the precipitate that forms.

b. Identify the spectator ions in the reaction.

c. Write the net ionic equation for the reaction.

14. (T/I) What two cations may be responsible for the colour of the solution shown below? What test could you do to determine which of the two ions is present?

15. (C) Make a poster that shows how to determine the molar concentration of lead(II) nitrate, $Pb(NO_3)_2$, in a solution by precipitating the lead ions. Identify the solution you would use to precipitate the ions, and write the equation for the reaction.

16. (A) The compound cisplatin, $Pt(NH_3)_2Cl_2(s)$, is used to treat various cancers. It can be prepared by reacting aqueous solutions of potassium tetrachloroplatinate, $K_2PtCl_4(aq)$, and ammonia, $NH_3(aq)$:
$$K_2PtCl_4(aq) + 2NH_3(aq) \rightarrow 2KCl(aq) + Pt(NH_3)_2Cl_2(s)$$
What is the maximum mass of cisplatin that could be obtained from 150 mL of 0.250 mol/L ammonia solution?

17. (T/I) The fertilizer ammonium sulfate is manufactured by bubbling ammonia gas into a sulfuric acid solution:
$$2NH_3(g) + H_2SO_4(aq) \rightarrow (NH_4)_2SO_4(aq)$$

a. What amount (in moles) of ammonium sulfate is formed when 4.0 mol of ammonia gas reacts with the sulfuric acid solution?

b. What mass of ammonia gas is needed to manufacture 10×10^3 kg of ammonium sulfate fertilizer?

18. (A) A flask that contained 25.0 mL of 6.0 mol/L sulfuric acid shattered when it was accidentally knocked onto the floor. Find the minimum mass of sodium hydrogen carbonate, $NaHCO_3(s)$, that would be needed to neutralize the spilled acid. The reaction that occurs is:
$$H_2SO_4(aq) + 2NaHCO_3(s) \rightarrow$$
$$2CO_2(g) + 2H_2O(\ell) + Na_2SO_4(aq)$$

19. (K/U) Distinguish between ground water and surface water. What is the source of your drinking water?

20. (K/U) List three sources of surface water.

21. (K/U) How does temporary water hardness differ from permanent water hardness? Which dissolved substances account for each type of hardness?

22. (T/I) Why is it harder to rinse off soap with soft water than with hard water?

23. (A) About 40 years ago, much of the water that was used by people who lived in Bangladesh came from surface sources. However, many children died from diarrhea as a result of drinking the water. What was probably present in the water to make the children sick?

24. (C) Briefly compare and contrast two methods that are used to desalinate water. What are the advantages and disadvantages of each method?

25. (C) Substances can enter waterways from different sources. Make a table that shows three naturally occurring substances that could be present in water from a well and three types of water pollution that result from human activities.

Self-Check

If you missed question …	1	2	3	4	5	6	7	8	9	10	11	12	13	14	15	16	17	18	19	20	21	22	23	24	25
Review section(s)…	9.1	9.1	9.1	9.2	9.2	9.2	9.3	9.3	9.4	9.4	9.1	9.1	9.1	9.1	9.2	9.2	9.2	9.2	9.3	9.3	9.4	9.3	9.3	9.4	9.3

CHAPTER 10

Acids and Bases

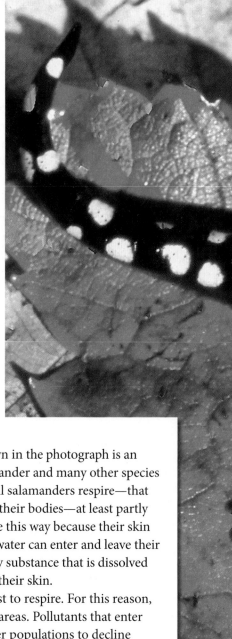

Specific Expectations

In this chapter, you will learn how to . . .

- E2.1 **use** appropriate terminology related to aqueous solutions and solubility (10.1)

- E2.2 **solve** problems related to the concentration of solutions by **performing** calculations involving moles, and **express** the results in various units (10.2)

- E2.7 **determine** the concentration of an acid or a base in a solution, **using** the acid-base titration technique (10.2)

- E3.5 **explain** the Arrhenius theory of acids and bases (10.1)

- E3.6 **explain** the difference between strong acids and weak acids, and between strong bases and weak bases, in terms of degree of ionization (10.1)

The somewhat shiny creature that is shown in the photograph is an amphibian called a *salamander*. This salamander and many other species of salamanders can be found in Ontario. All salamanders respire—that is, exchange oxygen and carbon dioxide in their bodies—at least partly through their skin. Salamanders can respire this way because their skin is permeable, which means that gases and water can enter and leave their body through their skin. Unfortunately, any substance that is dissolved in water can also enter their body through their skin.

Salamanders must keep their skin moist to respire. For this reason, many salamanders live in or near wetland areas. Pollutants that enter wetland areas are causing many salamander populations to decline because dissolved pollutants pass into the salamanders' bodies. Acid rain is another major factor that affects salamander populations. Acid rain changes the pH of the water in wetlands, which kills salamander eggs. In fact, if the pH of the water drops below 6, more than 60 percent of the salamanders in the eggs that are laid in the water will die. Furthermore, the salamanders that hatch from the eggs that survive are likely to be deformed.

Luckily for salamanders and other wetland wildlife, acid rain has decreased in recent years. The level of the air pollutants that cause acid rain has dropped dramatically since 1980, and the Canadian and United States' governments are still working to reduce these pollutants.

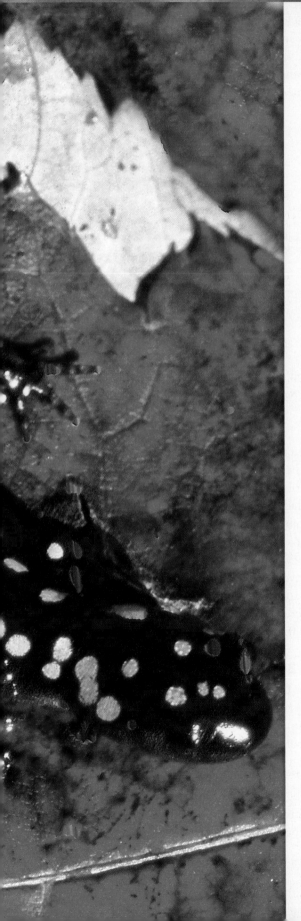

"Acid Rain" in a Petri Dish

How can you investigate some of the properties of sulfur dioxide gas, $SO_2(g)$, and model the formation of acid rain?

Safety Precautions

- Wear safety eyewear throughout this activity.
- Wear gloves and a lab coat or apron throughout this activity.
- After using a dropper, place it back into its container immediately to avoid contaminating the reagent.
- Sulfuric acid is corrosive. Flush any spills on skin or clothing with plenty of cool water and inform your teacher immediately.
- Sulfur dioxide may affect people with bronchitis or asthma. If you have asthma or other breathing difficulties, ask to be excused during this activity.

Materials

- blue litmus paper
- distilled water
- bromocresol green in a dropper bottle
- 0.5 mol/L sodium thiosulfate solution, $Na_2SO_3(aq)$, in a dropper bottle
- 2 mol/L sulfuric acid, $H_2SO_4(aq)$, in a dropper bottle
- petri dish with a cover
- sheet of white paper

Procedure

1. Place a clean petri dish on a sheet of white paper or another white background. Moisten a strip of blue litmus paper with distilled water. Arrange the litmus paper in the petri dish so that one end is near the edge of the dish and the other end is close to the centre.

2. Place small and large drops of bromocresol green at various distances from the centre of the petri dish. Record the colour of the drops.

3. Place two drops of sodium thiosulfate solution in the centre of the petri dish. Then add two drops of sulfuric acid to the sodium thiosulfate. The reaction that occurs will generate sulfur dioxide gas. **Quickly place the cover over the dish**, and observe for about 2 min. Record any changes in the litmus paper and the drops of bromocresol green.

4. When you finish, place the petri dish in the fumehood without opening it.

Questions

1. Blue litmus paper turns red in an acid. Bromocresol green turns yellow in an acid. Gases such as sulfur dioxide can react with water. What can you conclude about sulfur dioxide gas from your results?

2. In step 3, which drops do you think were more acidic? Explain.

3. Did the smaller or larger drops of bromocresol green change colour first? Suggest a reason for what you observed.

Arrhenius Acids and Bases

Key Terms

Arrhenius theory of acids
and bases

ionization

dissociation

acid-base indicator

universal indicator

pH scale

strong acid

weak acid

strong base

weak base

Every day, you use, see, and even eat many acids and bases. Citrus fruits, such as oranges and lemons, contain citric acid. Household cleaners, such as soap and ammonia, contain bases. There are even acids and bases inside the human body. These acids and bases are important for normal body functions. **Figure 10.1** shows a variety of common items that have acidic or basic properties.

Acidic substances

Basic substances

Figure 10.1 Many foods and household items contain acids or bases.

Identify *What properties do lemons and grapes have in common?*

Acid-base Theories

Compounds and solutions can be categorized as acidic, basic, or neutral. What determines whether a substance is acidic or basic? Scientists have developed a few different theories to explain acids and bases. Each of these theories is useful in different situations, but none of them completely explains every acid-base interaction.

An early acid-base theory was developed by a Swedish chemist and physicist named Svanté Arrhenius, shown in **Figure 10.2**. Arrhenius's theory was part of his doctoral thesis in 1884. His examiners found his theory hard to accept, and Arrhenius was just barely awarded his degree. However, he received a Nobel Prize for this theory in 1903, and today his theory is recognized as a fundamental concept in chemistry. You will learn about other acid-base theories in higher-level chemistry classes, In this chapter, however, the focus will be on the **Arrhenius theory of acids and bases**.

Arrhenius theory of acids and bases an acid-base theory stating that acids ionize to produce hydrogen ions in solution and bases dissociate to produce hydroxide ions in solution

Figure 10.2 Svanté Arrhenius (1859–1927) was only 25 years old when he proposed his theory of acids and bases. Scientists still use his theory today.

The Arrhenius Theory of Acids and Bases

The Arrhenius theory of acids and bases uses the concept of ions in solution to explain the nature of acids and bases. According to Arrhenius, an acid must contain a hydrogen atom that can become a hydrogen ion, $H^+(aq)$, in solution. For example, hydrogen chloride, $HCl(g)$, is a molecular substance that forms ions when dissolved in water:

$$HCl(g) \xrightarrow{\text{in water}} H^+(aq) + Cl^-(aq)$$

Similarly, a base must contain a hydroxyl group, $-OH$, which is a source of hydroxide ions, $OH^-(aq)$, in an aqueous solution. For example, sodium hydroxide, $NaOH(s)$, is an ionic substance that breaks apart into ions when dissolved in water:

$$NaOH(s) \xrightarrow{\text{in water}} Na^+(aq) + OH^-(aq)$$

Arrhenius Acids and Bases
- An acid is a substance that ionizes in water to produce one or more hydrogen ions, $H^+(aq)$.
- A base is a substance that dissociates in water to form one or more hydroxide ions, $OH^-(aq)$.

Acids are molecular compounds that are held together by covalent bonds. When acids dissolve in water, they *form* ions and therefore undergo **ionization**. Conversely, most bases are made up of ions because they are ionic compounds. When a base is dissolved in water, the ions in the base break apart, or dissociate. Therefore, the base undergoes **dissociation**.

The Arrhenius theory explains the properties of acids and bases. Acids have characteristic properties because acids form hydrogen ions in solution. For example, the reaction between an acid and zinc and the sour taste of an acid are both due to the presence of hydrogen ions. Bases have characteristic properties because most bases dissociate to form hydroxide ions in solution.

The Arrhenius theory also explains neutralization reactions. Consider the reaction between hydrochloric acid and sodium hydroxide:

$$HCl(aq) + NaOH(aq) \rightarrow NaCl(aq) + H_2O(\ell)$$

In this reaction, and in all reactions between Arrhenius acids and bases, the hydrogen ions from the acid combine with the hydroxide ions from the base to form water. Thus, a base and an acid can neutralize each other.

ionization the process of forming ions

dissociation the process in which ions break apart when dissolved in solution

Properties of Acids and Bases

Acids and bases are substances that have been used for thousands of years for a wide variety of applications. The extensive use of acids and bases is in large part due to the characteristic physical and chemical properties that each of these substances has. **Table 10.1** summarizes some important properties of acids and bases, and provides some common examples of substances that are acidic or basic. When reading through **Table 10.1**, make note of the differences between acids and bases. One important difference, which is commonly used to identify a substance as being acidic or basic, is the substance's pH, which can be estimated with pH indicators.

Table 10.1 Properties of Acids and Bases

Property	Acids	Bases
Taste*	Sour	Bitter
Texture*	No characteristic texture	Slippery to the touch
Electrical conductivity in aqueous solution	Conduct electricity	Conduct electricity
pH	Less than 7	Greater than 7
Indicator colours • Litmus • Phenolphthalein	Red Colourless	Blue Pink
Corrosion	Corrode tissues and metals	Corrode tissues but not metals
Reactions • With metals • With carbonates	React with active metals to produce hydrogen gas React with carbonates to produce carbon dioxide gas	No reaction No reaction
Common examples	Citrus fruits, vinegar, carbonated drinks, vitamin C	Soap, baking soda, oven cleaner, household ammonia

*Never taste any chemical in the laboratory and do not touch chemicals without wearing protective gloves.

The pH Scale

Go to **Logarithms and Calculating pH** in Appendix A for information about calculating the pH of a solution.

Advertisements for soaps, shampoos, and skin creams often use the terms "pH balanced" and "pH controlled." Gardeners and farmers monitor the pH of the soil, because it can affect plant growth. Some plants, such as the hydrangeas shown in **Figure 10.3**, respond visibly to the acidity or basicity—also known as alkalinity—of the soil in which they are planted. If the pH of your blood becomes too high or too low, your blood will lose its ability to transport oxygen and you will become very sick. Clearly, pH values can be important, but what does the term "pH" actually mean?

The **pH scale** is used to describe the acidity or basicity of a solution based on the concentration of hydrogen ions in solution. Acids produce hydrogen ions in solution. Therefore, an acidic solution has a hydrogen ion concentration that is greater than its hydroxide ion concentration and has a pH that is less than 7. Bases produce hydroxide ions in solution. Thus, a basic solution has a hydroxide ion concentration that is greater than its hydrogen ion concentration and has a pH that is greater than 7. Neutral solutions, such as pure water, have equal concentrations of hydrogen ions and hydroxide ions. The pH of a neutral solution is 7. **Figure 10.4** shows the pH values of various common substances.

Acid-base Indicators and pH

Acids and bases react differently with **acid-base indicators**, such as litmus and bromocresol green. This property is exploited when using acid-base indicators to determine if a substance is acidic or basic. How acid-base indicators work is based on a characteristic colour they have at certain pH values. An acid-base indicator has one colour in a solution when the pH is below a certain level and a noticeably different colour when the pH is above this level.

Some acid-base indicators, such as litmus, can be used to distinguish between acids and bases because they change colour around pH 7. Other indicators can be used to measure pH values that are greater or lower than 7, depending on the pH at which they change colour. **Table 10.2** lists some common indicators and the pH ranges they can measure. Often, something called a **universal indicator** is used. This type of acid-base indicator contains a mixture of chemicals that changes colour throughout the range of pH values from 1 to 14.

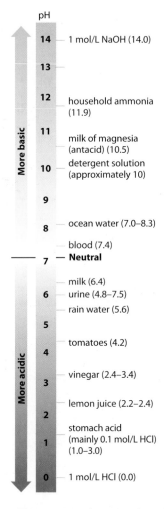

pH

More basic

14 — 1 mol/L NaOH (14.0)

13 —

12 — household ammonia (11.9)

11 — milk of magnesia (antacid) (10.5)

10 — detergent solution (approximately 10)

9 —

8 — ocean water (7.0–8.3)

— blood (7.4)

7 — **Neutral**

— milk (6.4)

6 — urine (4.8–7.5)

— rain water (5.6)

5 —

— tomatoes (4.2)

4 —

— vinegar (2.4–3.4)

3 —

— lemon juice (2.2–2.4)

2 —

— stomach acid (mainly 0.1 mol/L HCl) (1.0–3.0)

1 —

More acidic

0 — 1 mol/L HCl (0.0)

Figure 10.4 The pH scale has values between 0 and 14. The higher the pH of a substance, the more basic the substance is.

Figure 10.3 The blue flowers on a hydrangea plant indicate that the soil is acidic. When the soil is more basic, the flowers are pink.

Table 10.2 Common Indicators

Indicator	pH Range	Colour Change as pH Increases
methyl violet	0.0 to 1.6	yellow to blue
orange IV	1.4 to 2.8	red to yellow
methyl orange	3.2 to 4.4	red to yellow
bromocresol green	3.8 to 5.4	yellow to blue
methyl red	4.8 to 6.0	red to yellow
bromothymol blue	6.0 to 7.6	yellow to green to blue
phenol red	6.6 to 8.0	yellow to red
phenolphthalein	8.2 to 10.0	colourless to pink
indigo carmine	11.4 to 13.0	blue to yellow

Suggested Investigation

Inquiry Investigation 10-A, Physical and Chemical Properties of Acids and Bases

Learning Check

1. According to the Arrhenius theory of acids and bases, what characterizes an acid? What characterizes a base?

2. Acids and bases are commonly found in the home. They can be identified by their properties.
   a. Name two foods that contain acidic substances.
   b. Name two household items that contain basic substances.

3. Use **Table 10.1** to create a graphic organizer, such as a Venn diagram, that compares the properties of acids and bases.

4. Identify each substance as an acid or an base.
   a. $HBr(aq)$ d. $HClO_4(aq)$ g. $Sr(OH)_2(aq)$
   b. $KOH(aq)$ e. $Ca(OH)_2(aq)$ h. $CsOH(aq)$
   c. $H_3PO_4(aq)$ f. $HNO_3(aq)$

5. When one drop of phenolphthalein is added to a clear colourless solution, the solution becomes pink. If another sample of the solution is tested with a piece of blue litmus paper, what observation would you expect? Explain your answer.

6. For each of the following, identify whether the hydrogen ion concentration is higher or lower than the hydroxide ion concentration.
   a. a solution with a pH of 4
   b. lemon juice
   c. a solution of sodium hydroxide

Activity 10.1 Determining the pH of an Unknown Solution with Indicators

The pH values of solutions can be determined in different ways. One way to determine a pH value is to use an indicator.

Safety Precautions

- Wear safety eyewear throughout this activity.
- Wear gloves and a lab coat or apron throughout this activity.
- Solutions of acids and bases can be toxic and corrosive or caustic. Flush any spills on skin or clothing with plenty of cool water. Inform your teacher immediately.
- If any solutions get into your eyes, flush your eyes at an eye-wash station for 15 min and inform your teacher.

Materials

- 4 indicator solutions: methyl orange, methyl red, bromothymol blue, and phenolphthalein in dropper bottles
- 4 solutions of unknown pH in dropper bottles
- spot plate or 4 small test tubes

Procedure

1. Place two drops of methyl orange indicator in four wells on the spot plate (or in four small test tubes).

2. Add five drops of each unknown solution to the indicator.

3. Record the colours.

4. Repeat steps 1 to 3 with the other three indicators.

Questions

1. Using **Table 10.2**, estimate the pH of each of the four solutions.

2. Your teacher will use a pH meter to determine the pH of the solutions. How do the results you obtained using indicators compare with the results your teacher obtained using a pH meter?

3. With the permission of your teacher, repeat this activity using a universal indicator, such as pH paper. Which method, a combination of indicators or a universal indicator, is more accurate? Explain your answer.

Strong and Weak Acids

Go to **scienceontario** to find out more

Household vinegar contains approximately 1.0 mol/L acetic acid, CH₃COOH(aq). The pH of household vinegar is about 2.4 and can be eaten and handled without causing harm. In comparison, 1.0 mol/L hydrochloric acid, HCl(aq), has a pH of 0. Hydrochloric acid is corrosive and must be handled with caution. The difference is based on the strengths of the acids.

Strong Acids

A solution that contains a high concentration of ions conducts electric current better than a solution that contains a low concentration of ions, as shown in **Figure 10.5**. Any solution of hydrochloric acid contains many more hydrogen ions than a solution with the same concentration of acetic acid. In fact, *all* the molecules of hydrochloric acid ionize in water, whereas relatively few molecules of acetic acid ionize.

Figure 10.5 The brightness of a conductivity tester is a clue to the concentration of ions in a solution. **(A)** A solution of 1 mol/L hydrochloric acid, HCl(aq), contains many ions and conducts electricity very well. **(B)** A solution of 1 mol/L acetic acid, CH₃COOH(aq), contains relatively few ions and is a poor conductor of electricity.

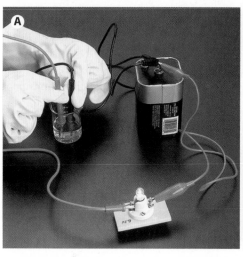
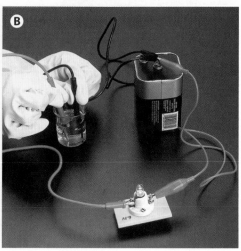

strong acid an acid that ionizes completely in water

An acid that ionizes completely in water is called a **strong acid**. For example, hydrochloric acid, HCl(aq), is a strong acid. When hydrochloric acid is dissolved in water, all the molecules ionize to form hydrogen ions, $H^+(aq)$, and chloride ions, $Cl^-(aq)$. In fact, the concentration of hydrogen ions in a dilute solution of a strong acid is equal to the concentration of the acid. There are very few strong acids. The most common strong acids are listed in **Table 10.3**.

Acid rain, which affects salamanders and other living things in the environment, is harmful because it contains strong acids. The acids in acid rain are sulfuric acid, $H_2SO_4(aq)$, and nitric acid, $HNO_3(aq)$. These acids form when sulfur dioxide, $SO_2(g)$, and nitrogen dioxide, $NO_2(g)$, react with water and oxygen in the atmosphere. The primary sources of these gases are power plants and factories that burn fossil fuels. To reduce acid rain, many factories employ different methods to remove the sulfur compounds from the gases that are emitted into the atmosphere.

Table 10.3 The Most Common Strong Acids

Name	Formula
hydrochloric acid	HCl(aq)
hydrobromic acid	HBr(aq)
hydroiodic acid	HI(aq)
perchloric acid	HClO₄(aq)
nitric acid	HNO₃(aq)
sulfuric acid	H₂SO₄(aq)

Weak Acids

Most acids are weak. A **weak acid** is an acid that does not ionize completely in water. When dissolved in water, most of the molecules in a weak acid remain intact. In an aqueous solution of a weak acid, the number of acid molecules that ionize depends on the concentration and the temperature of the solution. In most weak acids, only a small number of acid molecules ionize. For example, only about 1 in 100 acetic acid molecules ionize in a 0.1 mol/L solution at room temperature. Thus, the concentration of hydrogen ions in a solution of a weak acid is always less than the concentration of the dissolved acid. **Figure 10.6** shows how the ionization of a strong acid differs from the ionization of a weak acid. Four examples of weak acids are acetic acid, $CH_3COOH(aq)$, hydrocyanic acid, $HCN(aq)$, hydrofluoric acid, $HF(aq)$, and phosphoric acid, H_3PO_4.

weak acid an acid that ionizes very slightly in a water solution

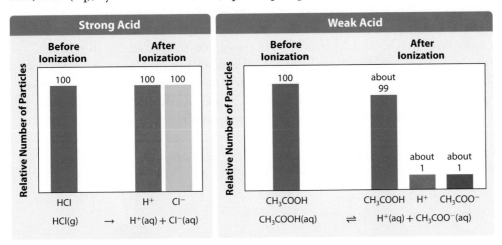

Figure 10.6 Hydrochloric acid, HCl(aq), ionizes completely in water, whereas only about 1 percent of acetic acid molecules ionize in water.

The Strong Binary Acids

Hydrochloric acid, $HCl(aq)$, hydrobromic acid, $HBr(aq)$, and hydroiodic acid, $HI(aq)$, are strong binary acids. Recall that binary acids are acids composed of hydrogen and a non-metal. As shown in **Figure 10.7**, the electronegative chlorine atom in hydrochloric acid draws electrons away from the hydrogen atom, making the hydrogen atom relatively positive. The slightly negative pole of a water molecule is strongly attracted to the hydrogen atom on the acid molecule and is able to tear it away from the chlorine atom. Hydrobromic acid and hydroiodic acid ionize completely in water in the same way.

Binary acids that contain an atom from the halogen family are all strong, with the exception of the binary acid that contains fluorine. The fluorine atom is very electronegative and has a small radius. These properties make the H-F bond so strong that hydrofluoric acid, $HF(aq)$, does not ionize completely. Although hydrofluoric acid is a weak acid, it is extremely corrosive. It can etch glass, and it is quite dangerous if spilled on the skin.

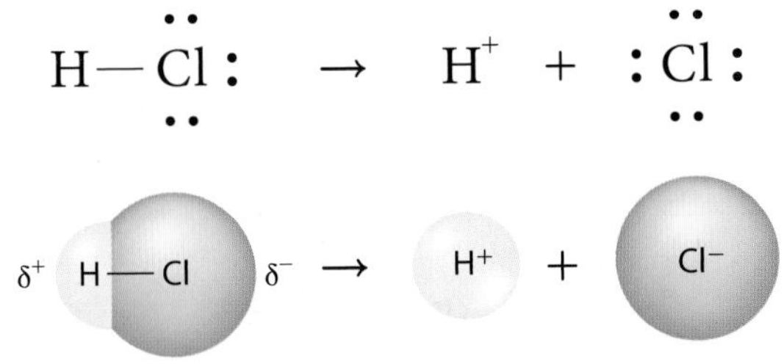

Figure 10.7 The highly polar bond in hydrochloric acid molecules causes hydrochloric acid to ionize completely in aqueous solution.

The Strong Oxoacids

Recall from Chapter 2 that oxoacids are composed of hydrogen, oxygen, and another element. The strong oxoacids are nitric acid, $HNO_3(aq)$, perchloric acid, $HClO_4(aq)$, and sulfuric acid, $H_2SO_4(aq)$. In oxoacids, the hydrogen atom that ionizes is *always* attached to an oxygen atom. The oxygen atom is electronegative and draws electrons away from the hydrogen atom. Water molecules are attracted to the resulting positive charge on the hydrogen atom. The more oxygen atoms there are in an acid molecule, the greater the polarity of the bond between the hydrogen atom that ionizes and the oxygen atom that it is attached to. Thus, you can predict that perchloric acid, $HClO_4(aq)$, is a stronger acid than chloric acid, $HClO_3(aq)$.

Acids with two hydrogen atoms that can ionize are called *diprotic acids*. (Acids with three hydrogen atoms that can ionize are called *triprotic acids*.) Sulfuric acid is the only strong diprotic acid. The high electronegativity of oxygen atoms makes each oxygen-hydrogen bond polar. As shown in **Figure 10.8**, sulfuric acid in an aqueous solution ionizes completely into hydrogen ions and hydrogen sulfate ions, $HSO_4^-(aq)$. The hydrogen sulfate ion can also act as an acid. However, the negative charge on the ion makes the hydrogen-oxygen bond much less polar, so the hydrogen is less likely to ionize to form another hydrogen ion. Thus, the hydrogen sulfate ion is a weak acid.

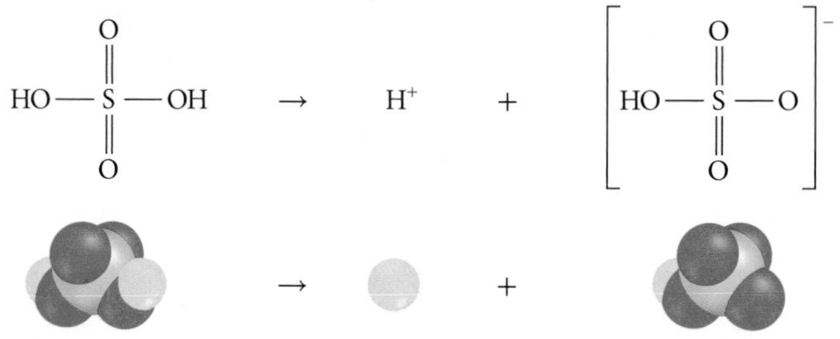

Figure 10.8 The oxygen atoms increase the polarity of both of the oxygen-hydrogen bonds in sulfuric acid. At least one hydrogen ion ionizes from each sulfuric acid molecule in water.

Strong and Weak Bases

Bases also can be classified as being strong or weak, based on their degree of dissociation in water. The characteristics of strong bases are similar to the characteristics of strong acids and the characteristics of weak bases are similar to the characteristics of weak acids.

Strong Bases

strong base a base that dissociates completely in water

A **strong base** dissociates completely in water. For example, sodium hydroxide, $NaOH(s)$, is a strong base. When sodium hydroxide is dissolved in water, it completely dissociates to form sodium ions, $Na^+(aq)$, and hydroxide ions, $OH^-(aq)$. The concentration of hydroxide ions in a dilute solution of a strong base is equal to the concentration of the base.

All hydroxides of the alkali metals (Group 1 elements) are strong bases. The alkaline earth metal (Group 2) hydroxides below beryllium in the periodic table are also strong bases. Beryllium is the exception because it is a relatively small atom and its bond with oxygen is strong. **Table 10.4** lists the strong bases that you are most likely to use in activities and investigations.

Go to **scienceontario** to find out more

Table 10.4 Some Common Strong Bases

Name	Formula
lithium hydroxide	$LiOH(aq)$
sodium hydroxide	$NaOH(aq)$
potassium hydroxide	$KOH(aq)$
calcium hydroxide	$Ca(OH)_2(aq)$
barium hydroxide	$Ba(OH)_2(aq)$

Weak Bases

Most bases are weak. A **weak base** is a base that produces relatively few hydroxide ions in water. Like a weak acid, only a small number of the particles in a weak Arrhenius base dissociate in water.

The most common weak base is ammonia, $NH_3(aq)$. However, the Arrhenius theory cannot explain why ammonia is a base since ammonia does not contain hydroxide ions. To understand why ammonia is a base, you must look at the reaction between ammonia and water. In this reaction, ammonia removes a hydrogen ion, H^+, from water, producing an ammonium ion, $NH_4^+(aq)$, and a hydroxide ion, $OH^-(aq)$:

$$NH_3(aq) + H_2O(\ell) \rightleftharpoons NH_4^+(aq) + OH^-(aq)$$

Strong and Weak versus Concentrated and Dilute

When discussing acids and bases, people often confuse the terms *strong* and *concentrated* and the terms *weak* and *dilute*. The terms *strong* and *weak* refer to the ionization or dissociation of particles in water. On the other hand, the terms *concentrated* and *dilute* refer to the amount of solute in a solvent. A concentrated solution is a solution that has a high concentration of solute. A dilute solution is a solution that has a low concentration of solute.

An example of a concentrated acid is 12.0 mol/L hydrochloric acid. An example of a dilute acid is 0.1 mol/L hydrochloric acid. Notice that both are solutions of the same acid, and this acid is a strong acid. Thus, you can have a concentrated solution of a strong acid and a dilute solution of a strong acid. You can also have a concentrated solution of a strong base and a dilute solution of a strong base.

Weak acids and weak bases can also be made into concentrated and dilute solutions. A concentrated base is a 5.0 mol/L solution of ammonia, $NH_3(aq)$. A dilute base is a 0.2 mol/L solution of ammonia. Ammonia is a weak base, so these solutions are examples of a concentrated weak base and a dilute weak base.

Figure 10.9 illustrates the differences between strong and concentrated acids and between weak and dilute acids. Similar diagrams could be drawn for bases.

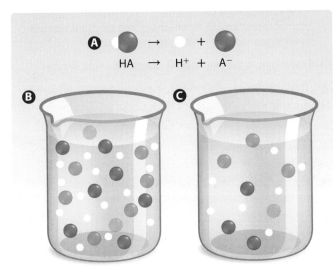

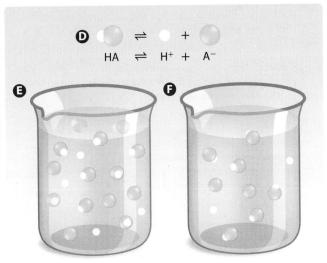

Figure 10.9 (A) When a strong acid is dissolved in water, all of its molecules ionize to form hydrogen ions and acid anions. **(B)** In a *concentrated* solution of a *strong* acid, there are many hydrogen ions and acid anions. **(C)** In a *dilute* solution of a *strong* acid, there are a few hydrogen ions and acid anions.

(D) When a weak acid is dissolved in water, only a few of its molecules ionize to form hydrogen ions and acid anions. **(E)** In a *concentrated* solution of a *weak* acid, there are many acid molecules and some hydrogen ions and acid anions. **(F)** In a *dilute* solution of a *weak* acid, there are a few acid molecules and even fewer hydrogen ions and acid anions.

7. Summarize the difference between "strong" and "concentrated" when describing a solution of an acid. Give examples to illustrate this difference.

8. The terms "concentrated" and "dilute" can be used to describe acids and bases.

  a. Give an example of a dilute solution of a strong base.

  b. Give an example of a concentrated solution of a weak acid.

9. The pH of one type of soft drink is 3.0. The soft drink contains carbonic acid, $H_2CO_3(aq)$, and phosphoric acid, $H_3PO_4(aq)$. Are these acids strong or weak? Give reasons for your answer.

10. Predict the relative strengths of the following acids: hypochlorous acid, $HClO(aq)$; chlorous acid, $HClO_2(aq)$; chloric acid, $HClO_3(aq)$; perchloric acid, $HClO_4(aq)$. Explain your reasoning.

11. Draw diagrams that show the difference between strong and weak versus concentrated and dilute bases.

12. Why are the safety warnings for investigations that use strong acids or bases much more strict than those that use weak acids or bases?

Activity   10.2   Differentiating between Weak and Strong Acids and Bases

In this activity, you will be given six unknown solutions of equal concentration. You will identify each solution as a strong acid, a strong base, a weak acid, a weak base, a neutral ionic solution, or a molecular solution.

Safety Precautions

- Wear safety eyewear throughout this activity.
- Wear gloves and a lab coat or apron throughout this activity.
- Acids and bases are often both toxic and corrosive. Wash any spills on skin or clothing with plenty of cool water, and inform your teacher immediately.
- When you have completed this activity, wash your hands.
- If any solutions get into your eyes, flush your eyes at an eye-wash station for 15 min and inform your teacher.

Materials

- unknown 0.10 mol/L molecular solution, neutral ionic solution, strong base, strong acid, weak base, and weak acid
- pH paper or pH meter
- 0.5 cm strips of magnesium, Mg(s), ribbon
- 6 beakers (50 mL)
- conductivity tester
- 6 test tubes

Procedure

1. Read steps 2 to 5 in this Procedure, and create an appropriate table to record your results.

2. Place 25 mL of each solution in a 50 mL beaker. Determine the pH of each solution, using either pH paper or a pH meter. Record your observations.

3. Determine the conductivity of each solution using the conductivity tester. Record your observations.

4. Determine the reactivity of each solution with magnesium by placing 5 mL of the solution in a test tube and adding a piece of magnesium ribbon. Record your observations. Note whether some solutions reacted more vigorously with magnesium than others.

5. Dispose of all the materials as directed by your teacher.

Questions

1. Using your data, classify each of the six unknown solutions.

2. How did you distinguish the neutral solutions from the acids and bases?

3. How did you determine which solution was the molecular solution and which was the neutral ionic solution?

4. How did you distinguish between the strong base and the weak base and between the strong acid and the weak acid?

Section Summary

- According to the Arrhenius theory, an acid is a substance that ionizes in water to produce one or more hydrogen ions, $H^+(aq)$, and a base is a substance that dissociates in water to produce one or more hydroxide ions, $OH^-(aq)$.

- Aqueous solutions of acids and aqueous solutions of bases both conduct electricity. However, the other properties of acids and bases, such as reactions with indicators and metals, are different.

- The pH scale describes the concentration of hydrogen ions in solution. Acids have pH values that are less than 7, and bases have pH values that are greater than 7.

- A strong acid or strong base ionizes or dissociates completely into ions in aqueous solution.

- Most acids and bases are weak.

Review Questions

1. **K/U** The Arrhenius theory of acids and bases describes acids and bases in terms of the ions formed when the compounds are dissolved in water.
   a. According to the Arrhenius theory, can a substance that does not contain hydrogen atoms be an acid? Explain your answer.
   b. Is every substance that contains hydrogen atoms an acid? Explain your answer.

2. **K/U** How can the properties of an acid be explained? Why does glucose, $C_6H_{12}O_6(aq)$, not have acidic properties even though 1 mol of glucose contains six times as many hydrogen atoms as 1 mol of sulfuric acid, $H_2SO_4(aq)$?

3. **C** Draw diagrams to show the difference between ionization and dissociation.

4. **K/U** What is a universal indicator? How is it useful in a laboratory?

5. **T/I** Describe two different tests you could perform in a laboratory to determine if an unknown compound is an acid.

6. **T/I** Three aqueous solutions were tested using litmus paper. The results of these tests are given in the table below. Identify each solution as neutral, acidic, or basic.

Reactions with Litmus Paper

Litmus Paper	Solution A	Solution B	Solution C
Red	Paper turns blue.	Paper stays red.	Paper stays red.
Blue	Paper stays blue.	Paper stays blue.	Paper turns red.

7. **A** A solution of household vinegar and water is sometimes used to clean glass. Marble is made up primarily of calcium carbonate, $CaCO_3(s)$. Why should vinegar never be used to clean marble tiles?

8. **K/U** The pH scale measures the acidity and basicity of solutions.
   a. What pH values correspond to acidic solutions?
   b. What pH values correspond to basic solutions?

9. **A** The pH of some common beverages are as follows: buttermilk, pH 4.5; coconut milk, pH 6.5; cranberry juice, pH 2.4; lime juice, pH 2.2; orange juice, pH 4.0.
   a. Arrange the beverages in order of increasing acidity.
   b. Predict how the taste of these beverages would compare, based on your arrangement.

10. **T/I** Explain why there are no molecules of hydrogen chloride in a 1 mol/L solution of hydrochloric acid, HCl(aq). What is present in the solution?

11. **K/U** Consider the following substances: $H_2SO_4(aq)$, $NaHCO_3(s)$, KOH(aq), HCl(aq), oxalic acid, aqueous ammonia.
    a. Which (if any) is a strong acid?
    b. Which (if any) is a weak base?

12. **K/U** What does it mean when an aqueous solution of an acid is described as "weak"?

13. **T/I** Consider 0.1 mol/L solutions of hydrochloric acid, HCl(aq); acetic acid, $CH_3COOH(aq)$; and ammonia, $NH_3(aq)$. List these solutions in order of increasing hydrogen ion concentration. Explain your reasoning.

14. **T/I** What piece of laboratory equipment could you use to distinguish between a strong base and a weak base? Describe how you would do this.

15. **A** Citric acid and most of the other acids found in nature are weak acids. Explain why this is a good thing for humans.

16. **C** Is it possible to have a concentrated solution of a weak acid? Explain your answer in a way that a student in ninth grade could understand you.

Neutralization Reactions and Acid-base Titrations

Key Terms

neutralization reaction

salt

titration

titrant

burette

end point

equivalence point

Acids and bases are found in air, soil, oceans, and waterways. Lactic acid is found in spoiled milk and in tired muscles after a strenuous workout. Acids and bases are also used widely in industry. The manufacturing of fertilizers, fabrics, soaps, plastics, pesticides, and numerous other chemicals relies on chemical reactions that involve acids and bases. **Table 10.5** lists some acids and bases that are manufactured in million tonne quantities worldwide.

Table 10.5 Ranking of Some Acids and Bases by Quantity Manufactured

Acid or Base	Uses
1. Sulfuric acid, $H_2SO_4(aq)$	Making phosphate fertilizers and car batteries
2. Ammonia, $NH_3(aq)$	Making fertilizers and explosives
3. Sodium hydroxide, $NaOH(aq)$	Making soaps and detergents; refining vegetable oil; peeling fruits and vegetables
4. Phosphoric acid, $H_3PO_4(aq)$	Making detergents, food additives, and phosphate fertilizers
5. Nitric acid, $HNO_3(aq)$	Making fertilizers, explosives, and plastics
6. Hydrochloric acid, $HCl(aq)$	Cleaning metal products; making chlorides, fertilizers, and dyes
7. Acetic acid, $CH_3COOH(aq)$	Preserving food; making vinegar, adhesives, and paints

Neutralization Reactions

The reaction between an acid and base is often called a **neutralization reaction**. In a neutralization reaction, the acid counteracts (or neutralizes) the properties of the base, and the base counteracts the properties of the acid.

During a neutralization reaction between an Arrhenius acid and an Arrhenius base, the hydrogen ions, $H^+(aq)$, from the acid combine with the hydroxide ions, $OH^-(aq)$, from the base to form water, $H_2O(\ell)$. The metal cation from the base and the anion from the acid combine to form an ionic compound called a **salt**.

In everyday language, the word *salt* is used to mean table salt, which is the ionic compound sodium chloride, $NaCl(s)$. In chemistry, however, a salt can be made of different cations and anions. For example, the neutralization reaction between hydrochloric acid, $HCl(aq)$, and potassium hydroxide, $KOH(aq)$, produces an aqueous solution of the salt potassium chloride, $KCl(aq)$, as shown in **Figure 10.10**.

neutralization reaction a reaction between an acid and a base

salt a compound composed of a metal cation from a base and an anion from an acid

$$HCl(aq) + KOH(aq) \rightarrow HOH(\ell) + KCl(aq)$$

Figure 10.10 The cation from the base and the anion from the acid form a salt. The formula for water, $H_2O(\ell)$, can be written as $HOH(\ell)$ to see more clearly how the hydrogen ion from the acid combines with the hydroxide ion from the base to form water.

The reaction between any aqueous solution of a strong acid and any aqueous solution of a strong base forms a neutral salt. If the molar amounts are balanced—that is, if there are equal numbers of aqueous hydrogen ions and aqueous hydroxide ions—all the acid and all the base will be neutralized, leaving a solution with a pH of 7. However, reactions that involve equal amounts of a weak acid and/or a weak base usually produce a solution with a pH that is not 7 because the ions react with water. Such reactions are still called neutralization reactions even though the solution at the end of the reaction is not neutral. Using the term *neutralization reaction* in this situation may be confusing because the term *neutral* in everyday language implies that everything is balanced and unreactive.

Calculations That Involve Neutralization Reactions

When solutions of an acid and a base undergo a neutralization reaction, the concentration of one can be determined if the concentration of the other is known, and if the volumes of both have been accurately measured. You can apply the solution stoichiometry techniques you learned in Chapter 9 to solve problems that involve neutralization reactions. Recall that you need to begin with the appropriate balanced chemical equation. This equation tells you the molar ratios of the reactants and products. For example, you could determine the concentration of an aqueous solution of potassium hydroxide by measuring the volume of a solution that neutralizes a measured volume of hydrochloric acid with a known concentration.

Sample Problem

Determining the Concentration of a Base in a Neutralization Reaction

Problem

A technician performed three trial reactions to determine the concentration of a solution of potassium hydroxide, KOH(aq). In each trial, a 0.1250 mol/L solution of hydrochloric acid, HCl(aq), was used to neutralize a 25.00 mL sample of the potassium hydroxide solution. The average volume of hydrochloric acid required was 32.86 mL. Determine the concentration of the potassium hydroxide solution.

What Is Required?

You need to determine the concentration of the potassium hydroxide solution.

What Is Given?

You know each of the following:

Volume of KOH(aq): 25.00 mL

Volume of HCl(aq): 32.86 mL

Concentration of HCl(aq): 0.1250 mol/L

Plan Your Strategy	Act on Your Strategy
Write the balanced chemical equation for the reaction.	$HCl(aq) + KOH(aq) \rightarrow KCl(aq) + H_2O(\ell)$
Use the formula $n = cV$ to determine the amount of hydrochloric acid. Remember to convert the volume in mL to L.	$n = cV$ $n = 0.1250 \text{ mol/L} \times 32.86 \text{ mL}$ $\quad = 0.1250 \text{ mol/}\cancel{L} \times 0.032\,86\,\cancel{L}$ $\quad = 4.1075 \times 10^{-3} \text{ mol}$
Determine the amount of potassium hydroxide needed to neutralize the hydrochloric acid.	The OH^- in KOH(aq) reacts with the H^+ in HCl(aq) in a 1:1 ratio, so the amount of KOH(aq) is 4.1075×10^{-3} mol.
Substitute values into the formula for molar concentration to calculate the concentration of the potassium hydroxide solution. Remember to convert the volume in mL to L.	$c = \dfrac{n}{V}$ $c = \dfrac{4.1075 \times 10^{-3} \text{ mol}}{0.025\,00 \text{ L}} = 0.1643 \text{ mol/L}$ The concentration of the potassium hydroxide solution is 0.1643 mol/L.

Check Your Solution

The balanced equation shows that hydrochloric acid and potassium hydroxide react in a 1:1 ratio. The volume of potassium hydroxide required for neutralization was less than the volume of hydrochloric acid. Therefore, the concentration of potassium hydroxide must be greater than the concentration of hydrochloric acid. The answer is reasonable.

1. Hydrochloric acid was slowly added to an Erlenmeyer flask that contained 50.0 mL of 1.50 mol/L sodium hydroxide, NaOH(aq), and a pH meter. The pH meter read 7.0 after the addition of 35.3 mL of hydrochloric acid. Calculate the concentration of the hydrochloric acid.

2. What volume of 0.400 mol/L sodium hydroxide, NaOH(aq), is needed to neutralize 26.8 mL of 0.504 mol/L sulfuric acid, H_2SO_4(aq), completely? **Hint:** Sulfuric acid loses two hydrogen ions during this neutralization reaction.

3. A 25.00 mL sample of a nitric acid solution, HNO_3(aq), is neutralized by 18.55 mL of a 0.1750 mol/L sodium hydroxide, NaOH(aq). What is the concentration of the nitric acid solution?

4. What volume of 1.25 mol/L hydrobromic acid, HBr(aq), will neutralize 75.0 mL of 0.895 mol/L magnesium hydroxide, $Mg(OH)_2$(aq)?

5. A solution of sodium hydroxide was prepared by dissolving 4.0 g of sodium hydroxide, NaOH(s), in 250 mL of water. It was found that 20.0 mL of the sodium hydroxide solution neutralizes 25.0 mL of vinegar. Determine the concentration of acetic acid, CH_3COOH(aq), in the sample of vinegar. Assume that acetic acid is the only acidic substance in the vinegar.

6. Phosphoric acid, H_3PO_4(aq), is a triprotic acid. If 15.0 mL of phosphoric acid completely neutralizes 38.5 mL of 0.150 mol/L sodium hydroxide, NaOH(aq), what is the concentration of the phosphoric acid?

7. The acidity of a water sample can be measured by a neutralization reaction with a solution of sodium hydroxide, NaOH(aq). What is the concentration of hydrogen ions in a water sample if 100 mL of the sample is neutralized by the addition of 8.0 mL of 2.5×10^{-3} mol/L sodium hydroxide?

8. Citric acid, $H_3C_6H_5O_7$(aq), is a weak, triprotic acid that occurs naturally in many fruits and vegetables, especially the citrus fruits from which it gets its name. What volume of 0.165 mol/L sodium hydroxide, NaOH(aq), will completely react with 40.0 mL of 0.120 mol/L citric acid? For this calculation, assume that all the hydrogen ions are released by the citric acid.

9. Phosphoric acid, H_3PO_4(aq), is a weak, triprotic acid. When phosphoric acid reacts with a base, different salts can be prepared, depending on how many hydrogen ions are replaced by cations. For example, potassium hydrogen phosphate, K_2HPO_4(aq), can be prepared in an aqueous solution by adding just enough potassium hydroxide, KOH(aq), to replace two hydrogen ions:
H_3PO_4(aq) + 2KOH(aq) → K_2HPO_4(aq) + $2H_2O(\ell)$
What volume of 0.185 mol/L potassium hydroxide should be added to 80.0 mL of 0.137 mol/L phosphoric acid to form a solution of potassium hydrogen phosphate?

10. What volume of 0.150 mol/L calcium hydroxide, $Ca(OH)_2$(aq), is needed to completely neutralize 20 mL of 0.185 mol/L sulfuric acid, H_2SO_4(aq)?

Acid-base Titration Can Determine the Concentration of a Solution

titration a procedure that is used to determine the concentration of a solution by reacting a known volume of that solution with a measured volume of a solution that has a known concentration

titrant in a titration, the solution with a known concentration

burette a clear tube with volume markings along its length and a tap at the bottom

To determine the concentration of an acid or a base, chemists perform a neutralization reaction while doing a procedure called **titration**. In a titration, the concentration of a solution is determined by reacting a known volume of that solution with a measured volume of a solution with a known concentration.

For example, suppose that you wanted to find the concentration of an acid solution. You would gradually add a basic solution with a known concentration to an accurately measured volume of the acid solution to find the volume of basic solution that would completely react. Then you would use stoichiometry to calculate the concentration of the acid. In a titration, the solution with the known concentration is the **titrant**.

Special glassware is used in a titration experiment. A sample of the solution with an unknown concentration is drawn into a volumetric pipette or a graduated pipette. The titrant is poured into a **burette**. A burette is a long, narrow graduated tube, with a tap on the bottom end. The burette is used to measure the volume of titrant that is added to the sample. The procedure on pages 468 to 469 describes how to do a titration and how to use these pieces of glassware during the titration.

Acid-base Indicators and Titration

In a titration of hydrochloric acid with aqueous potassium hydroxide, the two clear, colourless solutions react to form another clear, colourless solution. The temperature of the solution rises because the reaction is exothermic, but there is no other visible sign that a reaction took place. To know when a neutralization reaction is complete, chemists often use an acid-base indicator.

Figure 10.11 shows the colour change of phenolphthalein, a common indicator that is used in titrations. Phenolphthalein is colourless between pH 0 and pH 8. It turns pink between pH 8 and pH 10, and it is red in more basic solutions. This pH range may seem too large to be useful for a titration, but the colour change is quite abrupt. A single drop of titrant is usually enough to cause a vivid colour change. The point when the indicator changes colour is called the **end point** of the titration.

The aim of a titration is to know when the amount of titrant that has been added to the sample is just enough to react with all the acid or base the sample contains. This point is referred to as the **equivalence point**. Ideally, the end point and equivalence point should coincide—that is, the indicator should be in the middle of its colour change at the pH of the equivalence point.

Phenolphthalein is often used for titrating a strong acid with a strong base, even though the equivalence point, which is at pH 7.0, is not in phenolphthalein's range. However, near the equivalence point, the pH of a titrated solution changes *very* rapidly. A fraction of a drop of titrant beyond the equivalence point will place the pH of the solution in the range where phenolphthalein changes colour.

Phenolphthalein is also perfect for titrating a weak acid with a strong base. The salt solution formed is mildly basic, so phenolphthalein changes colour very close to the equivalence point. However, methyl orange is more suitable for titrating a weak base with a strong acid because the solution at the equivalence point is mildly acidic.

end point the point at which the indicator in a titration changes colour

equivalence point the point at which the amount of titrant is just enough to react with all of the reactant in the sample

Go to **scienceontario** to find out more

Figure 10.11 Like all indicators, phenolphthalein has a distinct colour change. The solution on the left is acidic, and the solution on the right is basic.

Suggested Investigation

Inquiry Investigation 10-B, The Concentration of Acetic Acid in Vinegar

Inquiry Investigation 10-C, The Percent (m/m) of Ascorbic Acid in a Vitamin C Tablet

Learning Check

13. What are the products of a neutralization reaction?

14. How does the pH of the equivalence point in a titration between a strong acid and a strong base compare with the pH of the equivalence point in a titration involving either a weak base or a weak acid?

15. Why must you know the concentration of one of the solutions used during titration?

16. Explain why a solution that is produced by a neutralization reaction may not have a pH of 7.

17. Explain the difference between the end point and the equivalence point of a titration.

18. A student performed a titration between hydrochloric acid and a weak base. The student used a pipette to add hydrochloric acid to the reaction flask and then added a few drops of phenolphthalein as the indicator. At the equivalence point, the solution in the reaction flask was acidic. Explain how the choice of indicator could cause an error in determining the concentration of hydrochloric acid.

Procedure for an Acid-base Titration
The following steps describe how to prepare for and perform a titration.

Rinsing the Volumetric or Graduated Pipette
Rinse a pipette with the solution whose volume you are measuring. This will ensure that any drops remaining inside the pipette will form part of the measured volume.

1. Put the pipette bulb on the pipette, as shown in **Figure A**. Place the tip of the pipette into a beaker of distilled water.
2. Relax your grip on the bulb to draw up a small volume of distilled water.
3. Remove the bulb, and discard the water by letting it drain out.
4. Pour a sample of the solution with the unknown concentration into a clean, dry beaker.
5. Rinse the pipette by drawing several millilitres of the solution with the unknown concentration from the beaker into the pipette. Coat the inner surface with the solution, as shown in **Figure B**. Discard the rinse. Rinse the pipette twice in this way. The pipette is now ready to be filled with the solution that has the unknown concentration.

Filling the Pipette
6. Place the tip of the pipette below the surface of the solution with the unknown concentration.
7. Hold the suction bulb loosely on the end of the glass stem. Use the suction bulb to draw the solution up to the point shown in **Figure C**.
8. As quickly and smoothly as you can, slide the bulb off the glass stem and place your index finger over the end.
9. Roll your finger slightly away from end of the stem to let the solution slowly drain out.
10. When the bottom of the meniscus aligns with the etched mark, as in **Figure D**, press your finger back over the end of the stem. This will prevent more solution from draining out.
11. Touch the tip of the pipette to the side of the beaker to remove any clinging drops. The measured volume inside the pipette is now ready to transfer to an Erlenmeyer flask.

Transferring the Solution
12. Place the tip of the pipette against the inside glass wall of the flask, as shown in **Figure E**. Let the solution drain slowly, by removing your finger from the stem.
13. After the solution drains, wait several seconds and then touch the tip to the inside wall of the flask to remove any drops on the end. Note: Do not remove the small amount of solution shown in **Figure F**.

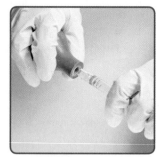

Figure A Squeeze the pipette bulb as you put it on the stem of the pipette.

Figure B Cover the ends of the pipette so that none of the solution spills out as you rock the pipette back and forth to coat its inner surface with solution.

Figure C Start with more of the unknown solution than you need. You will drain out the excess solution in the next two steps.

Figure D Always read the volume of the solution at the bottom of the meniscus.

Figure E Draining the pipette with the tip against the wall of the flask will prevent splashing.

Adding the Indicator

14. Add two or three drops of the indicator to the flask and its contents. Do not add too much indicator. Using more indicator does not make the colour change easier to see. Also, most indicators are weak acids. Too much indicator can change the amount of base needed for the neutralization. You are now ready to prepare the apparatus for the titration.

Rinsing the Burette

15. To rinse the burette, close the tap and add about 10 mL of distilled water from a wash bottle.

16. Tip the burette to one side, and roll it gently back and forth so that the water comes in contact with all the inner surfaces.

17. Hold the burette over a sink. Let the water drain out, as shown in **Figure G**. While you do this, check that the tap does not leak. Make sure that the tap turns smoothly and easily.

18. Rinse the burette twice, with 5 to 10 mL of the titrant. Remember to open the tap to rinse the lower portion of the burette. Discard the rinse solution each time.

Filling the Burette

19. Assemble a retort stand and burette clamp to hold the burette. Place a funnel in the top of the burette, and put a beaker under the burette.

20. With the tap closed, add the solution until it is above the zero mark. Remove the funnel. Carefully open the tap. Drain the solution into the beaker until the bottom of the meniscus is at or below the zero mark.

21. Touch the tip of the burette against the beaker to remove any clinging drops. Check that the part of the burette below the tap is filled with solution and contains no air bubbles. **Figure H** shows the air bubbles that you should avoid.

22. Find the initial burette reading using a meniscus reader, as shown in **Figure I**. Record the initial volume to the nearest 0.05 mL.

Titrating the Unknown Solution

23. Replace the beaker with the Erlenmeyer flask that contains the solution you want to titrate. Place a sheet of white paper under the flask to help you see the colour change.

24. Add titrant from the burette to the Erlenmeyer flask by opening the tap, as shown in **Figure J**. You may start by adding the titrant quickly, but slow down when you start to see a colour change in the solution in the flask.

25. At first, the colour change will disappear as you mix the solution in the flask. Add a small amount of titrant, and swirl thoroughly before adding any more. Stop adding titrant when the solution in the Erlenmeyer flask has a persistent colour change. If you are using phenolphthalein as an indicator, stop when the solution is a faint pink colour.

26. Use the meniscus reader to read the final volume. Record this volume, and subtract the initial volume from it to find the volume of the titrant needed to reach the end point.

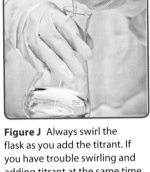

Figure J Always swirl the flask as you add the titrant. If you have trouble swirling and adding titrant at the same time, use a magnetic stirrer or have your laboratory partner swirl the flask as you add the titrant.

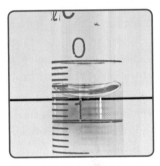

Figure I Hold the meniscus reader so that the line is under the meniscus.

Figure F A small amount of solution will always remain in the tip of the pipette. Do not remove this.

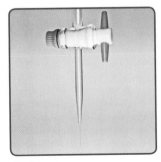

Figure G The tap is fully open when the handle on the tap is parallel to the burette and the solution inside the burette comes out quickly.

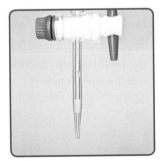

Figure H Do NOT start a titration if you have air bubbles like these in the tip of the burette. They will cause errors in your measurements.

Section Summary

- A neutralization reaction between an acid and a base in aqueous solution forms a salt and water.

- An acid-base titration is a quantitative technique in which a neutralization reaction is used to determine the concentration of one solution.

- The end point of a titration occurs when the indicator changes colour.

- An indicator must be chosen to change colour near the equivalence point of the titration, when equal amounts of acid and base have reacted.

Review Questions

1. **C** Write a general word equation that describes all neutralization reactions between an Arrhenius acid and an Arrhenius base.

2. **K/U** Write a chemical equation for each neutralization reaction in aqueous solution.
   a. sulfuric acid with potassium hydroxide
   b. hydroiodic acid with magnesium hydroxide.

3. **K/U** Why might the equivalence point in a titration be different from the end point?

4. **T/I** Before using a pipette to draw up a standard solution of an acid, a student rinses the pipette with distilled water but not with the acid solution. How will the concentration of the acid determined by the titration be affected? Explain your answer.

5. **T/I** During a titration, a student uses the pipette bulb to force out the final drop of the basic solution from the pipette. How will this mistake affect the calculated concentration of the basic solution?

6. **A** Explain why some people rinse their hair with vinegar after washing it with shampoo.

7. **K/U** Explain why phenolphthalein, which changes colour in the pH range of 8.2 to 10.0, is used as the indicator for a titration that forms a solution with pH 7 at the equivalence point.

8. **T/I** What amount of calcium hydroxide, $Ca(OH)_2(aq)$, will be neutralized by 1 mol of hydrochloric acid?

9. **A** Heartburn is a condition that is caused when fluid from the stomach moves up into the esophagus, causing irritation. Some people use milk of magnesia, which contains magnesium hydroxide, $Mg(OH)_2(s)$, to relieve the symptoms of heartburn. Explain why this medicine works.

10. **T/I** What volume of 0.996 mol/L barium hydroxide, $Ba(OH)_2(aq)$, is needed to neutralize 25.0 mL of 1.70 mol/L nitric acid, $HNO_3(aq)$?

11. **K/U** Explain how a salt is formed during a neutralization reaction.

12. **C** Describe how you would design and perform a titration in which you use 0.250 mol/L sulfuric acid, $H_2SO_4(aq)$, to determine the concentration of a strontium hydroxide, $Sr(OH)_2(aq)$, solution. Include the equation for the reaction, as well as an outline of the calculations you would make.

13. **T/I** Methanoic acid, $HCOOH(aq)$, is a weak monoprotic acid. A 25.00 mL sample of methanoic acid was titrated with a standard solution of 0.1004 mol/L sodium hydroxide, $NaOH(aq)$. Three trials were conducted. The average volume of sodium hydroxide solution that was required to reach the end point was 16.32 mL. What is the concentration of the methanoic acid solution?

14. **T/I** Suppose that you titrate 25.0 mL of 0.100 mol/L sodium hydroxide, $NaOH(aq)$, with two different acids. In the first titration, you use 0.150 mol/L hydrochloric acid, $HCl(aq)$, which is a strong acid. In the second titration, you use 0.150 mol/L acetic acid, $CH_3COOH(aq)$, which is a weak acid. How will the volume of acid used in each titration compare? Explain your answer.

15. **K/U** Explain how the meaning of the term *salt* differs in chemistry and in everyday language.

16. **T/I** Suppose that you are going to use the solutions in the bottles shown for a titration. Which solution will you put in the burette? Which solution will you measure with a graduated pipette? Explain your answers.

CHEMISTRY Connections

Acid-base Reactions on the Rise

Have you ever watched a vinegar baking soda volcano erupt? The bubbles of carbon dioxide, $CO_2(g)$, result from a decomposition reaction that quickly follows an acid-base reaction between the vinegar $CH_3COOH(aq)$, an acid, and baking soda, $NaHCO_3(aq)$, a base, as shown below.

Acid-base Reaction

$$CH_3COOH(aq) + NaHCO_3(aq) \rightarrow$$
$$NaCH_3COO(aq) + H_2CO_3(aq)$$

Decomposition

$$H_2CO_3(aq) \rightarrow CO_2(g) + H_2O(\ell)$$

The release of carbon dioxide as a result of the chemical reaction between an acid and a base, as shown in the photograph to the right, is part of the reason why baked goods rise. An ingredient that causes a batter to rise when baked is called a *leavening agent*. The two main chemical leavening agents are baking soda and baking powder.

BAKING SODA The chemical name for baking soda is sodium hydrogen carbonate. When used in cooking, baking soda reacts with mildly acidic liquids, forming carbon dioxide bubbles. Mildly acidic liquids include vinegar, molasses, honey, citrus juice, and buttermilk.

Baking soda must be mixed with other dry ingredients and added last to a batter so that the release of carbon dioxide is uniform throughout the batter. This acid-base reaction happens quickly. If baking soda is the only leavening agent in a recipe, the batter must be baked immediately before the bubbles have a chance to escape. Baking causes the bubbles to expand, and the batter rises. As the batter firms, the bubbles are trapped, as shown below.

Baking traps the bubbles that are formed during the reaction between an acid and a base, resulting in a light, airy cake.

BAKING POWDER If a recipe does not include an acidic liquid, baking powder is used. Most baking powder is a mixture of baking soda and two dry acids. One of the acids reacts with the baking soda when it dissolves in the batter, and the other acid reacts with the baking soda when it is heated.

Like baking soda, baking powder is mixed with other dry ingredients and added last to a batter. However, batter does not have to be baked immediately.

Sometimes, a batter made with mildly acidic liquid(s) includes both baking powder and baking soda. Excess acid can disrupt the action of the baking powder. The baking soda helps to neutralize the acid, and the baking powder provides a reliable source of carbon dioxide.

Carbon dioxide forms bubbles when baking soda, a base, is added to vinegar, an acid.

Connect to Society

Research to find out more about the chemical properties and relative costs of baking soda and baking powder. Create a chart summarizing your results. Based on your findings, write a recommendation to a recipe developer for an industrial bakery, suggesting which of the two bases would be best to include in recipes for mass-produced baked goods.

Safety Precautions

- Wear safety eyewear throughout this investigation.
- Wear gloves and a lab coat or apron throughout this investigation.
- Acids and bases are corrosive to the skin and may damage the eyes.
- Inform your teacher if any spill occurs so that it can be properly cleaned up.
- Flush any spill of acid or base on your skin immediately, with plenty of cold water.
- Sodium hydroxide can cause blindness if it gets in the eye. Wash your eyes at an eye-wash station for 15 min if any gets into your eye. Inform your teacher.

Materials

- 1 mol/L hydrochloric acid, HCl(aq)
- 1 mol/L acetic acid, $CH_3COOH(aq)$
- 1 mol/L sodium hydroxide, NaOH(aq)
- 1 mol/L ammonia, $NH_3(aq)$
- mossy zinc, Zn(s)
- sodium carbonate, Na_2CO_3(s)
- phenolphthalein indicator in a dropper bottle
- paper towel
- 4 beakers (50 mL)
- wax pencil or labels
- conductivity tester
- 12-well plate
- scoopula
- disposable pipettes
- forceps

Physical and Chemical Properties of Acids and Bases

In this investigation, you will examine the electrical conductivity and pH of acids and bases, as well as some chemical reactions. You will compare the properties of two different acids with the same concentration. You will also compare the properties of two different bases with the same concentration.

Pre-Lab Questions

1. What is an electrolyte?
2. Write the chemical equation for each reaction.
   a. hydrochloric acid, HCl(aq), with sodium hydroxide, NaOH(aq)
   b. acetic acid, $CH_3COOH(aq)$, with sodium hydroxide, NaOH(aq)
3. Do you need to wash off acid that spills on a gloved hand? Explain your answer.

Question

Which properties of weak and strong acids and bases are similar, and which properties are different?

Predictions

- Predict how the electrical conductivity of hydrochloric acid will compare with the electrical conductivity of acetic acid, if both have the same concentration.
- Predict how the electrical conductivity of hydrochloric acid will compare with the electrical conductivity of sodium hydroxide, if both have the same concentration.
- Predict how the pH of hydrochloric acid will compare with the pH of acetic acid, if both solutions have the same molar concentration.

Procedure

1. Copy the following table into your notebook.

Observations

Property \ Solution	HCl(aq)	$CH_3COOH(aq)$	NaOH(aq)	$NH_3(aq)$
Conductivity				
pH				
Reaction with Zn(s)				
Reaction with Na_2CO_3(s)				
Colour with phenolphthalein				
Number of drops to change colour	+ NaOH(aq)	+ $NH_3(aq)$	+ $CH_3COOH(aq)$	+ HCl(aq)

Part 1: Electrical Conductivity and pH

Your teacher may demonstrate this part of the investigation.

2. Label a clean, dry 50 mL beaker for each of the four solutions to be tested.

3. Pour hydrochloric acid to a depth of about 1 cm in the corresponding labelled beaker. Test the conductivity of the solution, and record your observations in your table. Rinse the electrodes of the tester with distilled water, and dry them with a paper towel.

4. Use universal pH paper to measure the pH of the solution. Record the pH in your table. Do not discard the solution.

5. Repeat steps 3 and 4 for the solutions of acetic acid, sodium hydroxide, and aqueous ammonia.

Part 2: Chemical Tests

6. Place a clean, dry 12-well plate on a piece of paper. Label one column of three wells with the formula for each of the four solutions you will test: $HCl(aq)$, $CH_3COOH(aq)$, $NaOH(aq)$, $NH_3(aq)$. Label the other edge of the 12-well plate with the chemical tests you will perform: $+ Zn(s)$, $+ Na_2CO_3(s)$, neutralization.

7. Pour the hydrochloric acid solution from step 4 into the first two wells in the corresponding column, keeping the last well reserved for the neutralization reaction. Pour each of the three other solutions into two wells in the corresponding column, keeping the last well empty.

8. Add a small piece of mossy zinc to each well in the row labelled "+Zn." Compare the reactions of all four solutions, and record your observations.

9. Use a scoopula to add a small amount of sodium carbonate, about the size of the head of a match, to each well in the row labelled "+Na_2CO_3(s)." Record your observations.

10. Use a disposable pipette to put 10 drops of hydrochloric acid in the first well in the third row of the well plate. Using a different pipette for each solution, put 10 drops of acetic acid in the second well, 10 drops of sodium hydroxide in the third well, and 10 drops of aqueous ammonia in the fourth well. Save these pipettes for step 12.

11. Add one drop of phenolphthalein indicator to each of these four wells. Record the colour of each solution.

12. Add sodium hydroxide, one drop at a time, to the well containing hydrochloric acid until you see a colour change. Record the number of drops you added. Similarly, add drops of aqueous ammonia to the well containing acetic acid, drops of acetic acid to the well containing sodium hydroxide, and drops of hydrochloric acid to the well containing aqueous ammonia. Record the number of drops required to change the colour of each solution.

13. Use forceps to remove any unreacted zinc from the wells in the first row of the plate. Rinse the zinc with tap water, and dry it with a paper towel. Return the zinc to your teacher. Dispose of the solutions in the beakers and the well plate, and clean the equipment as instructed by your teacher.

Analyze and Interpret

1. Classify each solution you tested as a strong or weak conductor of electricity.

2. List the solutions in order, from lowest pH to highest pH.

3. Which solution(s) reacted with zinc?

4. Which solution(s) reacted with sodium carbonate?

5. Compare the number of drops required to neutralize the 10-drop samples of acids and bases.

Conclude and Communicate

6. Review your prediction about how the electrical conductivity of hydrochloric acid would compare with the electrical conductivity of acetic acid, if both solutions had the same molar concentration. Explain your observations.

7. Review your prediction about how the electrical conductivity of hydrochloric acid would compare with the electrical conductivity of sodium hydroxide, if both solutions had the same molar concentration. Explain your observations.

8. Evaluate your prediction about how the pH of hydrochloric acid would compare with the pH of acetic acid if both solutions had the same molar concentration. Explain your observations.

9. Summarize the properties of acids and bases by making a table.

10. **INQUIRY** A gas was produced in some of the reactions you observed. Identify the reactants in each of these reactions, predict the gas that was likely produced, and write balanced equations for the reactions that you think occurred. Write a procedure to identify the gas.

11. **RESEARCH** Identify the substance that is used in oven cleaners and drain cleaners by visiting a store, looking around your home, or using Internet resources. What are the advantages of using this substance? What are the disadvantages of using it?

Safety Precautions

- Wear safety eyewear throughout this investigation.

- Wear gloves and a lab coat or apron throughout this investigation.

- Acids and bases are corrosive to the skin and may damage your eyes.

- Flush any spill of acid or base on your skin immediately, with plenty of cold water.

- Tie back loose hair and clothing.

- Sodium hydroxide can cause blindness if it gets in the eye. Wash your eyes at an eye-wash station for 15 min if any gets into your eye. Inform your teacher.

Materials

- one kind of vinegar
- distilled water
- phenolphthalein indicator in a dropper bottle
- standardized sodium hydroxide solution, NaOH(aq)
- 3 beakers (100 mL)
- labels or wax pencil
- 10 mL pipette with a suction bulb or a pipette pump
- 3 Erlenmeyer flasks (250 mL)
- retort stand
- burette clamp
- 50 mL burette
- funnel
- meniscus reader
- sheet of white paper

The Concentration of Acetic Acid in Vinegar

Vinegar is an aqueous solution that may contain several different compounds. The most important compound in vinegar is acetic acid, $CH_3COOH(aq)$. Different kinds of vinegar contain different concentrations of acetic acid. In this investigation, you will use titration to determine the concentration of acetic acid in a sample of vinegar.

Pre-Lab Questions

1. Explain why the beakers used for transferring the titration solutions must be clean and dry.

2. Explain why you should rinse the burette with sodium hydroxide solution and then discard the rinse solution before filling the burette to do this titration.

3. You will use phenolphthalein as the indicator in this titration. Why should you place vinegar, rather than sodium hydroxide, in the Erlenmeyer flask?

4. Near the end point of a titration, a partial drop on the tip of the burette can be washed into the Erlenmeyer flask using distilled water. Why will washing with distilled water not affect the end point of the titration?

5. Explain why you should not take any vinegar leftover from this investigation home to use in your kitchen.

Question

What is the concentration of acetic acid in a sample of vinegar?

Procedure

1. Record the following information in your notebook.
   - concentration of standardized sodium hydroxide solution, NaOH(aq)
   - type of vinegar
   - volume of pipette

2. Copy the following table into your notebook to record your burette readings.

Burette Readings

	Trial 1	Trial 2	Trial 3
Final Reading (mL)			
Initial Reading (mL)			
Volume of NaOH(aq) Added (mL)			

3. Review the titration procedure on pages 468 to 469.

4. Pour about 40 mL of vinegar into a clean, dry, labelled beaker.

5. Rinse the pipette with distilled water and then with vinegar.

6. Use the pipette to transfer 10 mL of vinegar into the first Erlenmeyer flask. Add approximately 50 mL of water. Then add two or three drops of phenolphthalein indicator.

7. Pour about 60 mL of standardized sodium hydroxide solution into a clean, dry, labelled beaker.

8. Set up a retort stand, burette clamp, burette, funnel, and meniscus reader. Rinse the burette with distilled water and then with the standard sodium hydroxide solution from the beaker. Discard the rinse solution. Using the funnel, fill the burette with the sodium hydroxide solution. Remove the funnel. Make sure that no air bubbles are in the burette, below the tap.

9. Place a sheet of white paper under the Erlenmeyer flask. Run sodium hydroxide into the Erlenmeyer flask while swirling the contents. You have reached the end point of the titration when the pink colour spreads through the solution rather than fading as you swirl the flask. If you are not sure whether you have reached the end point, record the burette reading. Then add just one drop, or part of a drop, by rinsing it into the Erlenmeyer flask with distilled water from a wash bottle. If you go past the end point, the solution will become very pink.

10. Repeat step 8 twice, for a total of three trials. Record your observations for each trial.

11. Dispose of the titrated solutions as instructed by your teacher. Rinse the pipette and burette with distilled water. Leave the burette tap in the open position.

Analyze and Interpret

1. Average the two closest burette readings. Average all three readings if they agree within ±0.25 mL.

2. Write the chemical equation for the reaction of acetic acid with sodium hydroxide.

3. Calculate the amount (in moles) of sodium hydroxide that you needed to reach the equivalence point. Then calculate the concentration of acetic acid in the vinegar. List any assumptions you made for your calculation.

4. Find the molar mass of acetic acid. Then calculate the mass of acetic acid in the volume of vinegar you used.

5. The density of vinegar is 1.01 g/mL. Calculate the mass of the vinegar you used. Then find the percent (m/v) of acetic acid in the vinegar.

Conclude and Communicate

6. Different kinds of vinegars contain different percentages of acetic acid. Compare your results with other students' results. What can you conclude about the concentration of the vinegar you and the rest of the class used?

7. **INQUIRY** Solid drain cleaners that are sold for home use contain sodium hydroxide mixed with other ingredients. Outline a procedure to determine the percent (m/m) of sodium hydroxide, NaOH(s), in a sample of solid drain cleaner, using a titration. (Do *not* perform the procedure.)

8. **RESEARCH** Use the Internet to learn about titrations of substances that are not acids and bases. How are these titrations similar to titrations of acids and bases?

Safety Precautions

- Wear safety eyewear throughout this investigation.

- Wear a lab coat or apron throughout this activity.

- Sodium hydroxide solution is caustic to the skin and may damage your eyes.

- Do not eat the vitamin C tablet; it may be contaminated.

- Flush any spill of base on your skin with plenty of cold water immediately.

Materials

- one commercial vitamin C tablet

- standardized solution of sodium hydroxide, NaOH(aq), approximately 0.100 mol/L

- dropper bottle containing phenolphthalein

- wash bottle with distilled water

- three 250 mL Erlenmeyer flasks

- labels or wax pencil

- balance

- mortar and pestle

- 50 mL burette

- retort stand

- burette clamp

- funnel

- meniscus reader

- sheet of white paper

*Go to **Measurement** in **Appendix A** to learn more about calculating percentage error.*

The Percent (m/m) of Ascorbic Acid in a Vitamin C Tablet

Vitamin C, or ascorbic acid ($HC_6H_7O_6$), is an essential vitamin responsible for the synthesis of collagen. It is found in a variety of foods including citrus fruits, leafy green vegetables, cabbage, and tomatoes. Prolonged deficiency of vitamin C results in scurvy, a serious disease. In spite of the many dietary sources of vitamin C, many people choose to supplement the daily adult requirement of ascorbic acid, about 60 mg, with a vitamin C supplement, in the belief that it will ward off the common cold, or at least reduce its intensity. Review the titration procedure on pages 468 to 469 before beginning this investigation. You will determine the percent (m/m) of ascorbic acid in a vitamin C tablet.

Pre-Lab

1. Why is it better to find the mass of the vitamin C tablet by adding the powder to an Erlenmeyer flask rather than using a weigh boat to measure out a quantity of powder?

2. Ascorbic acid is a monoprotic acid. The hydrogen atom responsible for the acidic behaviour of ascorbic acid is underlined in the formula $\underline{H}C_6H_7O_6$.

   a. Write a balanced chemical equation for the reaction of ascorbic acid with sodium hydroxide, NaOH(aq).

   b. What volume of 0.105 mol/L sodium hydroxide is required to neutralize 255 mg of ascorbic acid? Use the balanced equation obtained above.

   c. A 2.18 g vitamin C tablet was dissolved in water. It took 28.45 mL of 0.124 mol/L sodium hydroxide to neutralize the ascorbic acid in the tablet. Calculate the percent (m/m) of ascorbic acid in the vitamin C tablet.

3. The safety precautions warn about being careful while handling sodium hydroxide. Why is there no similar warning for handling ascorbic acid?

Procedure

1. Use a balance to determine the mass of a vitamin C tablet.

2. Record the following data in your notebook:
   - the mass of the vitamin C tablet
   - the concentration of the standardized solution of sodium hydroxide

3. Copy the data table below. You will be doing at least two titrations, but may do a third if time permits.

Titration Data Table

	Trial 1	Trial 2	Trial 3
Mass of Erlenmeyer Flask with Vitamin C Tablet Powder (g)			
Mass of Empty Erlenmeyer Flask (g)			
Mass of Vitamin C Tablet Powder (g)			
Final Burette Reading (mL)			
Initial Burette Reading (mL)			
Volume of NaOH(aq) Added (mL)			

4. Label two 250 mL Erlenmeyer flasks using labels or a wax pencil. Find the mass of each flask.

5. Using a mortar and pestle, crush the vitamin C tablet. Transfer approximately half of the powdered tablet to each of the two labelled Erlenmeyer flasks. Reweigh the flasks to determine the mass of powder in each flask.

6. Add about 75 mL of water to each flask and swirl to dissolve the ascorbic acid. Do not be concerned if the tablet does not dissolve completely. Some tablets contain insoluble fillers and binders. Add two or three drops of phenolphthalein to each flask and swirl to mix.

7. Pour about 60 mL of standardized sodium hydroxide solution into a clean, dry, labelled beaker.

8. Set up a retort stand, burette clamp, burette, funnel, and meniscus reader. Rinse the burette with distilled water and then with the standard sodium hydroxide solution. Discard the rinse solution. Using the funnel, fill the burette with the sodium hydroxide solution. Remove the funnel. Make sure that no air bubbles are in the burette, below the tap.

9. Place a sheet of white paper under the Erlenmeyer flask. Run sodium hydroxide into the Erlenmeyer flask while swirling the contents. You have reached the end point of the titration when the pink colour spreads through the solution rather than fading as you swirl the flask. If you are not sure whether you have reached the end point, record the burette reading. Then add just one drop, or part of a drop, by rinsing it into the Erlenmeyer flask with distilled water from a wash bottle. If you go past the end point, the solution will become quite pink.

10. Repeat step 9 with the second Erlenmeyer flask.

11. If time permits, obtain another tablet for further trials. Prepare the vitamin C solution as you did earlier.

12. Dispose of the titrated solutions as instructed by your teacher. Rinse the mortar, pestle and burette with distilled water. Leave the burette tap in the open position.

Analyze and Interpret

1. For each trial, determine the volume of sodium hydroxide solution required to neutralize the ascorbic acid.

2. Use the balanced chemical equation for the reaction of ascorbic acid with sodium hydroxide to determine the amount (in moles) of ascorbic acid present in each sample.

3. Determine the molar mass of ascorbic acid and use this to calculate the mass of ascorbic acid in each sample.

4. Determine the percent (m/m) of ascorbic acid in each sample.

5. Average your results to determine the percent (m/m) of ascorbic acid in a vitamin C tablet.

Conclude and Communicate

6. Your teacher will provide you with the manufacturer's claim for the mass of ascorbic acid per tablet. Calculate your percentage error as described in Appendix A.

7. Explain why it is not necessary to know the exact volume of water added to the powdered ascorbic acid.

8. What did you assume about the reactivity of any fillers or binders in the tablet with respect to sodium hydroxide?

9. Some people take mega doses of vitamin C. How does the body deal with excess vitamin C?

Extend Further

10. **INQUIRY** It is possible to purchase "liquid" vitamin C at a health food store.
   a. What's incorrect about the term "liquid"?
   b. Design an experiment to determine the concentration of ascorbic acid in a commercially available "liquid" vitamin C supplement.

11. **RESEARCH** Some people believe that taking large doses of vitamin C will prevent illness, such as the common cold. Use the Internet to learn more about vitamin C "megadosage." Compare the amount of vitamin C recommended by Health Canada to the amount of vitamin C in a megadosage diet. Describe any evidence that supports the claim that large doses of vitamin C are beneficial to one's health. Finally, describe any negative effects of large amounts of vitamin C.

Section 10.1 | Arrhenius Acids and Bases

Arrhenius acids and bases have different properties and can be classified as being strong or weak.

KEY TERMS
- acid-base indicator
- Arrhenius theory of acids and bases
- dissociation
- ionization
- pH scale
- strong acid
- strong base
- universal indicator
- weak acid
- weak base

KEY CONCEPTS
- According to the Arrhenius theory, an acid is a substance that ionizes in water to produce one or more hydrogen ions, $H^+(aq)$, and a base is a substance that dissociates in water to produce one or more hydroxide ions, $OH^-(aq)$.
- Aqueous solutions of acids and aqueous solutions of bases both conduct electricity. However, the other properties of acids and bases, such as reactions with indicators and metals, are different.
- The pH scale describes the concentration of hydrogen ions in solution. Acids have pH values that are less than 7, and bases have pH values that are greater than 7.
- A strong acid or strong base ionizes or dissociates completely into ions in aqueous solution.
- Most acids and bases are weak.

Section 10.2 | Neutralization Reactions and Acid-base Titrations

Acids and bases react in neutralization reactions. A neutralization reaction can be used to determine the concentration of an acid or a base experimentally, if the concentration of one solution is known.

KEY TERMS
- burette
- end point
- equivalence point
- neutralization reaction
- salt
- titrant
- titration

KEY CONCEPTS
- A neutralization reaction between an acid and a base in aqueous solution forms a salt and water.
- An acid-base titration is a quantitative technique in which a neutralization reaction is used to determine the concentration of one solution.
- The end point of a titration occurs when the indicator changes colour.
- An indicator must be chosen to change colour near the equivalence point of the titration, when equal amounts of acid and base have reacted.

Chapter 10 | REVIEW

Knowledge and Understanding

Select the letter of the best answer below.

1. Suppose that you have two clear, 1.0 mol/L solutions. You know that each solution could be a strong acid, a weak acid, a strong base, or a weak base. Which material(s) must you use to identity the solutions correctly?
   a. only litmus paper
   b. only a conductivity tester
   c. only phenolphthalein
   d. litmus paper and a conductivity tester
   e. litmus paper and phenolphthalein

2. How would you describe a solution with a pH of 5.6?
   a. slightly acidic
   b. very acidic
   c. neutral
   d. slightly basic
   e. very basic

3. Which chemical formula represents a salt?
   a. $C_6H_6(\ell)$
   b. $H_3PO_4(aq)$
   c. $Na_3PO_4(s)$
   d. $CH_3CH_2OH(\ell)$
   e. $KOH(s)$

4. Which statement about all acid-base indicators is true?
   a. They can be used to identify acids and bases.
   b. They can be mixed together to make a universal indicator.
   c. They change colour in response to changes in pH.
   d. They are synthetic.
   e. They are toxic.

5. Which reaction is a neutralization reaction?
   a. $Mg(OH)_2(aq) + 2HCl(aq) \rightarrow MgCl_2(aq) + 2H_2O(\ell)$
   b. $Ca(OH)_2(aq) + 2NaCl(aq) \rightarrow CaCl_2(aq) + 2NaOH(aq)$
   c. $Ca(NO_3)_2(aq) + 2KOH(aq) \rightarrow Ca(OH)_2(s) + 2KNO_3(aq)$
   d. $H_2SO_4(aq) + Mg(s) \rightarrow MgSO_4(aq) + H_2(g)$
   e. $AgNO_3(aq) + NaCl(aq) \rightarrow AgCl(s) + NaNO_3(aq)$

6. Which solution is a weak acid?
   a. $HNO_3(aq)$
   b. $HF(aq)$
   c. $HCl(aq)$
   d. $HBr(aq)$
   e. $HI(aq)$

7. Which statement about the titrant in a titration is true?
   a. It is a base.
   b. It is an acid.
   c. It is measured with a pipette.
   d. It is poured into the burette.
   e. It has an unknown concentration.

8. The acid-base indicator bromocresol green changes colour in the pH range of 3.8 (yellow) to 5.4 (blue). If bromocresol green turns blue when added to a solution, which statement is true?
   a. The solution is definitely acidic.
   b. The solution is definitely basic.
   c. The pH of the solution is between 0 and 3.8.
   d. The pH of the solution is between 3.8 and 5.4.
   e. The pH of the solution is 5.4 or greater.

Answer the questions below.

9. Write the name and chemical formula for two aqueous solutions that will turn phenolphthalein indicator red.

10. Write the name and formula for each type of acid.
    a. monoprotic acid
    b. diprotic acid
    c. triprotic acid

11. List two observations that can be explained using the Arrhenius theory.

12. Write a balanced chemical equation for each neutralization reaction.
    a. aqueous calcium hydroxide with hydrochloric acid
    b. phosphoric acid with aqueous strontium hydroxide
    c. aqueous sodium hydroxide with sulfuric acid

13. Explain why many foods contain acids but very few foods contain bases.

14. Perchloric acid, $HClO_4(aq)$, and chlorous acid, $HClO_2(aq)$, are both monoprotic acids, and they are both made of the same three elements. However, perchloric acid is a strong acid and chlorous acid is a weak acid. What causes the difference in the strengths of these acids?

15. Complete the Frayer model below to define the term *Arrhenius base*.

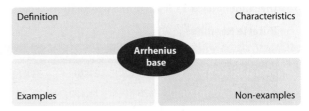

16. Explain why the process of titration is considered to be a type of quantitative analysis.

Thinking and Investigation

17. Suppose that you have to do a titration to determine the concentration of a strong acid, using a strong base as the titrant. However, you do not have an appropriate indicator to use for the titration. Describe what you could do to find the equivalence point during the titration.

18. Hypochlorous acid, HClO(aq), is an industrial disinfectant. Write the chemical equation for the ionization of hypochlorous acid in water.

19. Suppose that you are given three test tubes containing different clear, colourless solutions. Each solution contains an electrolyte. You are told that one solution is acidic, another is basic, and the third is neutral. Describe tests you could perform to identify each solution.

20. Methanoic acid, HCOOH(aq), is a monoprotic acid that is used to process latex, which is tapped from rubber trees, into natural rubber. A 25.00 mL sample of methanoic acid is neutralized by adding 25.68 mL of 0.0985 mol/L sodium hydroxide, NaOH(aq). What is the concentration of the sample?

21. What colour would you expect phenolphthalein to turn if it was added to each substance?
   a. window cleaner
   b. vinegar
   c. vitamin C
   d. dish detergent

22. Two samples of a solution with an unknown pH were tested with indicators. Methyl orange turned yellow when added to the first sample. Bromothymol blue turned yellow when added the second sample. Use **Table 10.2** to estimate the pH of the solution. What colour would you expect phenolphthalein to turn if it was added to a third sample of the same solution?

23. What amount of calcium hydroxide can be neutralized by 1 mol of hydrochloric acid?

24. A student did a titration with an unknown concentration of hydrochloric acid and 0.50 mol/L sodium hydroxide, NaOH(aq). The data that the student collected are in the table below. For each trial, the student used 25.0 mL of the hydrochloric acid.

Burette Readings

	Trial 1	Trial 2	Trial 3
Final Reading (mL)	11.15	21.20	31.15
Initial Reading (mL)	0.25	11.15	21.20

a. What volume of sodium hydroxide did the student use for each trial?

b. What data should the student use to calculate the concentration of the hydrochloric acid? Explain your answer.

c. Use the student's data to calculate the concentration of the hydrochloric acid.

25. Suppose that you have the following six unknown solutions, all of equal concentration: a strong base, a strong acid, a weak base, a weak acid, a neutral ionic solution, and a neutral molecular solution. Design a procedure to identify each solution. Summarize the results you would expect in a data table.

26. Arrange the steps below in the proper order for the process of titrating a strong acid of unknown concentration with a strong base of known concentration.
   a. put the acid in an Erlenmeyer flask
   b. record final burette reading
   c. fill burette with base
   d. add phenolphthalein to the flask
   e. measure the acid with a pipette
   f. record initial burette reading
   g. add base to the flask

Communication

27. The reaction between any acid and any base is called a neutralization reaction.
   a. What does the word "neutralization" mean in the context of an acid-base reaction? Compare and contrast this meaning with the everyday meaning of "neutralization"?
   b. If equal amounts of aqueous solutions of an acid and a base react, is the solution neutral? Explain your answer.

28. The two solutions for a titration experiment must be put in clean, dry beakers. The Erlenmeyer flask in which the solutions react must be thoroughly rinsed with distilled water, but it does not have to be dry. Compare and contrast these procedures.

29. How is a 1.0 mol/L solution of hydrochloric acid different from a 1.0 mol/L solution of acetic acid, $CH_3COOH(aq)$? Suppose that you added a strip of magnesium metal to each acid. Would you observe any differences in the reactions? Explain your answer so that Grade 9 students could understand it.

30. Draw three diagrams to show the reaction between hydrobromic acid, HBr(aq), and potassium hydroxide, KOH(aq). Show the arrangement of the atoms before the compounds are dissolved in water to make an acid and a base, the arrangement of the atoms in the individual solutions before the reaction, and the arrangement of the atoms after the reaction happens.

31. What is being modelled in the diagram below? Explain your answer.

32. Evaluate the following statement: "A concentrated acid is more dangerous than a strong acid."

33. Some words have a more specific definition in science than they do in everyday language. Make a graphic organizer that shows the difference between the scientific definition and the everyday definition of a word in this chapter.

34. Summarize your learning in this chapter using a graphic organizer. To help you, the Chapter 10 Summary lists the Key Terms and Key Concepts. Refer to Using Graphic Organizers in Appendix A to help you decide which graphic organizer to use.

Application

35. **BIG** IDEAS Properties of solutions can be described qualitatively and quantitatively, and can be predicted. Orange IV is an indicator that changes from red (at pH 1.4) to yellow (at pH 2.8). Methyl red also changes from red to yellow, but at different pH values. Methyl red is red at pH 4.8 and yellow at pH 6.0. A solution with an unknown pH is yellow in orange IV and red in methyl red. What is the approximate pH of the solution?

36. Suppose that you accidentally got some soap in your mouth while taking a shower. Explain why this was an unpleasant experience.

37. In the food industry, potatoes are soaked in a solution of sodium hydroxide for a few minutes and then sprayed with water to remove the skin. Suppose that you are responsible for preparing the solution, which needs to be about 3.75 mol/L sodium hydroxide, NaOH(aq). A 500 L vat is going to be used for soaking the potatoes.

  a. What mass of solid sodium hydroxide do you need to make the solution?

  b. To check the concentration of the sodium hydroxide solution, you titrate a 25.0 mL sample with 3.02 mol/L hydrochloric acid, using phenolphthalein indicator. You need 30.46 mL of hydrochloric acid to reach the end point. What is the concentration of the sodium hydroxide solution?

  c. The sodium hydroxide solution must be in the range of 10 to 20 percent by mass. Is the solution you prepared acceptable? Explain your reasoning.

38. Suppose that you are titrating an acid and a base using phenolphthalein indicator. Does it matter whether you put the acid or the base in the Erlenmeyer flask? Explain your answer.

39. In an acid-base reaction, 4.85 g of hydrogen sulfide, $H_2S(g)$, is bubbled into 100.0 mL of 0.110 mol/L sodium hydroxide, NaOH(aq).
$2NaOH(aq) + H_2S(g) \rightarrow Na_2S(aq) + 2H_2O(\ell)$
What mass of sodium sulfide is formed?

40. A solution is prepared by dissolving 1.00 g of a base with an unknown composition (represented by the generic formula MOH) in 100.0 mL of water. A 25.0 mL sample of this solution is then neutralized by 15.85 mL of 2.63 mol/L hydrochloric acid. What is the molar mass of MOH?

41. The active ingredient in Aspirin™ tablets is acetylsalicylic acid, $C_9H_8O_4(s)$. Acetylsalicylic acid, commonly called ASA, is a monoprotic acid. To determine the percentage of ASA in a tablet, a student began by measuring the mass of one tablet. The mass was 0.370 g. The student then crushed the tablet and dissolved it in water. Finally, she titrated the solution with 0.0750 mol/L sodium hydroxide, NaOH(aq), using an indicator. The titration required 18.50 mL of 0.0750 mol/L sodium hydroxide to reach the end point. What percentage of ASA was in the tablet?

42. **BIG** IDEAS People have a responsibility to protect the integrity of Earth's water resources. Use the library or the Internet to research acid rain. Learn which pollutants cause acid rain and what chemical reactions produce the acid in acid rain. Finally, study the ways governments and factories are working to control acid rain. Present your findings to your class as an oral presentation.

Select the letter of the best answer below.

1. **K/U** Which of the following is a common substance that contains acetic acid?
   a. lemon juice
   b. vinegar
   c. glass cleaner
   d. bleach
   e. oven cleaner

2. **T/I** A student is titrating a weak base with a strong acid. Which indicator should the student use?
   a. methyl orange (pH range of 3.2 to 4.4)
   b. phenolphthalein (pH range of 8.2 to 10.0)
   c. litmus (pH range of 5.0 to 8.0)
   d. phenol red (pH range of 6.6 to 8.0)
   e. indigo carmine (pH range of 11.4 to 13.0)

3. **T/I** A student is doing a titration to determine the concentration of an acid. The student puts a base with a known concentration in the burette. Then while doing the titration, the student washes the base down the side of the Erlenmeyer flask with a small amount of distilled water. How will the student's results be affected?
   a. The calculated concentration of the acid will be higher than the actual concentration.
   b. The calculated concentration of the acid will be lower than the actual concentration.
   c. The concentration of the acid will not be affected.
   d. The concentration of the base will decrease during the titration.
   e. The concentration of the base will increase during the titration.

4. **K/U** The acid-base indicator methyl red changes from red (pH 4.6) to yellow (pH 6.0). When this indicator is added to an aqueous solution, it turns yellow. Which statement is true?
   a. The solution must be acidic.
   b. The solution must be basic.
   c. The pH is between 0 and 6.0.
   d. The pH is between 4.0 and 6.0.
   e. The pH is greater than 6.0.

5. **T/I** Bromothymol blue has a pH range of 6.0 to 7.6. Phenolphthalein has a pH range of 8.2 to 10.0. When a strong acid completely reacts with a strong base, the pH of the solution is 7. Why is phenolphthalein used, instead of bromothymol blue, for titrations of strong acids and strong bases?

a. A lot more bromothymol blue is needed to see a colour change.
b. Bromothymol blue is rare and expensive.
c. The colour change of phenolphthalein is more dramatic.
d. Phenolphthalein is non-toxic and non-flammable.
e. Bromothymol blue does not change colour in the correct pH range.

6. **K/U** According to the Arrhenius theory, how does hydrogen chloride, $HCl(g)$, act when it is dissolved in water?
   a. as a solute
   b. as a source of $H^+(aq)$
   c. as a source of $Cl^-(aq)$
   d. as a source of $OH^-(aq)$
   e. as a solvent

7. **K/U** Which of the following pairs consists of two weak acids?
   a. sulfuric acid, $H_2SO_4(aq)$, and carbonic acid, $H_2CO_3(aq)$
   b. nitric acid, $HNO_3(aq)$, and hydrosulfuric acid, $H_2S(aq)$
   c. hydrofloric acid, $HF(aq)$, and hypofluorous acid, $HOF(aq)$
   d. hydrobromic acid, $HBr(aq)$, and phosphoric acid, $H_3PO_4(aq)$
   e. perchloric acid, $HClO_4(aq)$, and acetic acid, $CH_3COOH(aq)$

8. **K/U** Approximately 12 mol of a substance is dissolved in 1 L of water. Analysis of a sample of the solution indicates that the solution has a hydroxide ion, $OH^-(aq)$, concentration of 12 mol/L. What is the solution?
   a. a concentrated strong acid
   b. a concentrated strong base
   c. a dilute weak acid
   d. a dilute strong base
   e. a concentrated weak base

9. **T/I** What amount of magnesium hydroxide, $Mg(OH)_2(s)$, can be neutralized by 2.0 mol of hydrochloric acid?
   a. 0.50 mol
   b. 1.0 mol
   c. 1.5 mol
   d. 2.0 mol
   e. 4.0 mol

10. **T/I** What volume of 0.200 mol/L sodium hydroxide, $NaOH(aq)$, would you need to neutralize 80.0 mL of 0.400 mol/L acetic acid, $CH_3COOH(aq)$?
    a. 20.0 mL
    b. 40.0 mL
    c. 80.0 mL
    d. 160 mL
    e. 240 mL

Use sentences and diagrams, as appropriate, to answer the questions below.

11. **A** Identify two common household liquids that react together in a neutralization reaction.

12. **K/U** List three properties for each type of solution.
  a. acidic solution
  b. basic solution
  c. neutral solution

13. **A** A gardener wants to change the colour of the flowers on a hydrangea bush from pink to blue. The gardener chops up some orange peels and buries them in the soil around the bush. Explain why doing this will change the colour of the flowers.

14. **K/U** Which of the following substances is/are acidic, according to the Arrhenius theory?
  a. $CH_3OH(\ell)$ **c.** $C_6H_6(\ell)$
  b. $H_2S(aq)$ **d.** $H_3AsO_4(aq)$

15. **K/U** Which of the following substances is/are basic, according to the Arrhenius theory?
  a. $KOH(aq)$ **c.** $HClO(aq)$
  b. $Ca(OH)_2(aq)$ **d.** $CaO(aq)$

16. **A** The residents of a city are concerned that a new factory may be creating acid rain. Describe how the residents can test the acidity of the rain. How can they be sure that any change in acidity is due to the factory?

17. **T/I** When preparing for a titration, a student cannot remove the air bubbles in the burette, below the tap. The student decides to continue with the titration anyway. After the titration, when cleaning up, the student discovers that the air bubbles are no longer in the burette. Explain how the student's calculation of the unknown concentration will be affected.

18. **K/U** Explain the difference between a strong acid and a weak acid. Include an example of each type of acid in your explanation.

19. **C** Draw a flowchart that shows the major steps of titration. The flowchart should show which steps are repeated to complete three trials.

20. **T/I** Design an investigation to study the pH of a stream over one year. How would you collect the data? Why might the pH vary at different times of the year?

21. **K/U** The following beakers could contain a strong acid, a strong base, a weak acid, a weak base, an ionic salt solution (such as sodium chloride), or a molecular compound solution (such as glucose). Identify all the solutions that could be in each beaker.

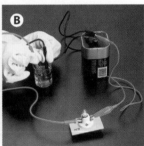

22. **T/I** During a titration, 30.0 mL of 0.0500 mol/L sodium hydroxide, $NaOH(aq)$, was added to 50.0 mL of 0.0800 mol/L hydrochloric acid. What is the pH of the solution?

23. **T/I** Aqueous ammonia, $NH_3(aq)$, is titrated with hydrobromic acid, $HBr(aq)$. Will the pH at the equivalence point be greater than, less than, or equal to 7? Explain.

24. **T/I** A titration was performed to find the concentration of a solution of hydrochloric acid. The following data were collected at the end point of the titration. What is the concentration of the acid solution? Volume of hydrochloric acid used = 18.25 mL Volume of 0.1947 mol/L sodium hydroxide used = 23.24 mL

25. **A** A student performed a titration on a drain cleaner that contained sodium hydroxide, $NaOH(aq)$. The student titrated 35.0 mL of the drain cleaner and added 50.08 mL of 0.409 mol/L hydrochloric acid to reach the equivalence point. What is the concentration of sodium hydroxide in the drain cleaner? What safety precautions should the student take?

Self-Check

If you missed question …	1	2	3	4	5	6	7	8	9	10	11	12	13	14	15	16	17	18	19	20	21	22	23	24	25
Review section(s)…	10.1	10.2	10.2	10.1	10.2	10.1	10.1	10.1	10.2	10.2	10.1, 10.2	10.1	10.1	10.1	10.1	10.1	10.2	10.1	10.2	10.1	10.1	10.2	10.2	10.2	10.2

"Eggs"-amining the Calcium Carbonate Content of Eggshells

Beginning in the 1950s, a pesticide called *DDT* was used to kill insects that can spread disease and damage crops. Because DDT degrades extremely slowly, it built up in fish, birds, and mammals. This bioaccumulation (the build-up of a chemical in living things) had unexpected environmental effects. For example, raptors such as peregrine falcons almost became extinct. Some studies showed that a concentration of DDT as low as 10 ppm caused many species of birds to produce shells that were too thin to survive incubation. DDT was banned in Canada in 1974. Today, research continues on the effects of chemicals in the environment, including their effects on birds' eggs.

How can you determine the percent by mass of calcium carbonate, CaCO₃(s), in eggshells?

Pesticides and pollutants can enter the food chain and affect wildlife, such as this peregrine falcon.

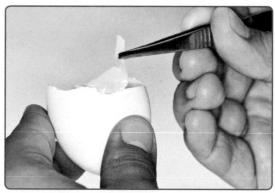

Before crushing the eggshells, you need to remove the inner membrane. You can remove it using forceps or tweezers, as shown in the photograph.

Safety Precautions

- Wear chemical safety goggles throughout this investigation.
- Wear gloves and a lab coat or apron throughout this investigation.
- Acids and bases are corrosive to the skin and can damage the eyes. If any acid or base is spilled on your skin, flush the area immediately with plenty of cold water and inform your teacher.
- Do not unplug the hot plate by pulling on the cord.

Initiate and Plan

1. In this investigation, you will add excess hydrochloric acid, HCl(aq), to react with the calcium carbonate, CaCO₃(s), in eggshells. (Your teacher will supply 1.0 mol/L hydrochloric acid.) Write the balanced chemical equation for the reaction between hydrochloric acid and calcium carbonate. **Hint:** Water and carbon dioxide are two of the products.

2. Eggshells are not pure calcium carbonate. If you decide to use 0.5 g of eggshells, you will be using less than 0.5 g of calcium carbonate. What volume of 1 mol/L hydrochloric acid will contain an excess amount in a reaction with 0.5 g of eggshells?

3. You can determine the amount of excess hydrochloric acid by titration with an aqueous solution of sodium hydroxide. (Your teacher will supply 1.0 mol/L sodium hydroxide solution.)

4. Develop a plan to determine the percent by mass of calcium carbonate in eggshells. As you develop your plan, consider the reaction you wrote in step 1, your knowledge of limiting and excess reactants, and the process of acid-base titration.
   - Write your plan in your notebook.
   - Write the procedure you will follow. Be sure to include any necessary safety precautions in your procedure.

- Make a list of all the equipment you will need. (Read steps 5 to 11 to identify any equipment you may not have thought about.)
- Make a list of all the solutions you will need, including the specific indicator you will use.
- Write the reactions that are involved in your procedure.
- Describe the calculations you will make to determine the percent by mass of calcium carbonate in eggshells.

After your teacher has approved your plan, continue with step 5.

Perform and Record

5. Prepare a data table for your results. Include space in your data table for three sets of data, even though you may only have time to perform two trials.

6. Before starting your procedure, you must make sure that your eggshells are clean and dry. Remove the inner membrane from each piece of eggshell by pulling it away from the inside of the shell with forceps or tweezers. Then dry the eggshells in an oven for 10 min. While the eggshells are drying, collect the equipment and solutions you listed in step 4.

7. Grind the eggshells into a fine powder using a mortar and pestle.

8. Accurately mass about 0.5 g of eggshell powder. Carefully transfer the powder to an Erlenmeyer flask.

9. Add several drops of ethanol, $C_2H_5OH(\ell)$, to the flask. This will help the reaction between hydrochloric acid and calcium carbonate take place. Heating the reacting mixture on a hot plate will also help to ensure that the reaction continues to completion.

10. Begin your procedure to find the percent by mass of calcium carbonate in eggshells.

11. Repeat steps 8 to 10 at least one more time. If possible, perform a third trial. Dispose of the solutions as directed by your teacher.

Analyze and Interpret

1. What amount (in moles) of hydrochloric acid did you add by pipette to the eggshell powder?

2. Calculate the amount (in moles) of hydrochloric acid that remained after reacting with the eggshells.

3. Find the amount (in moles) of hydrochloric acid that reacted.

4. Use the chemical equation for the reaction between hydrochloric acid and calcium carbonate to calculate the amount of calcium carbonate that reacted.

5. Calculate the mass of calcium carbonate in 0.5 g of eggshells.

6. Calculate the mass percent (m/m) of calcium carbonate in the eggshells.

Communicate Your Findings

7. Calculate the average percent by mass of calcium carbonate in eggshells, based on your investigation. Write a conclusion.

8. Identify possible sources of error in your procedure, and suggest improvements.

Assessment Criteria

Once you have completed your project, ask yourself these questions. Did you…

☑ **K/U** demonstrate an understanding of key concepts and skills you learned in this unit, for example, titration, concentration, limiting reactant calculations, use of indicators, and neutralization reactions?

☑ **T/I** use the equipment and solutions safely, accurately, and effectively?

☑ **C** organize your data in a clear and logical manner?

☑ **T/I** write a properly balanced equation for the reaction of an acid with calcium carbonate?

☑ **T/I** react an unknown amount of calcium carbonate with excess hydrochloric acid?

☑ **T/I** determine the amount of excess hydrochloric acid by titrating it with sodium hydroxide?

☑ **T/I** determine the percent by mass of calcium carbonate in eggshells?

☑ **T/I** identify sources of error and make suggestions for improvement?

☑ **C** write a concise conclusion, accurately using scientific vocabulary?

BIG IDEAS

- Properties of solutions can be described qualitatively and quantitatively, and can be predicted.
- Living things depend for their survival on the unique physical and chemical properties of water.
- People have a responsibility to protect the integrity of Earth's water resources.

Overall Expectations

In this unit, you learned how to…

- **analyze** the origins and effects of water pollution, and a variety of economic, social, and environmental issues related to drinking water
- **investigate** qualitative and quantitative properties of solutions, and solve related problems
- **demonstrate** an understanding of qualitative and quantitative properties of solutions

Chapter 8	Solutions and Their Properties

KEY IDEAS

- A solution is a homogeneous mixture. It can be classified based on the amount of dissolved solute per unit volume of solution.
- Whether a solute is soluble in a solvent depends on the nature of the bonds within the solute and the forces between the solute and solvent particles.
- "Like dissolves like" refers to matching the molecular properties of solute and solvent. Polar solvents tend to dissolve ionic and polar solutes. Non-polar solvents tend to dissolve non-polar solutes.

- The effect of temperature and pressure on the solubility of a solute in a solvent can be predicted.
- The concentration of a solution can be expressed quantitatively in several ways, the most important being molar concentration.
- Solutions can be made with a specific concentration using special glassware.

Chapter 9	Reactions in Aqueous Solutions

KEY IDEAS

- Net ionic equations show only the ions that react and the product of the reaction. They can be written for reactions that produce a precipitate and for neutralization reactions between acids and bases.
- Qualitative analysis uses physical and chemical tests to identify the cations and anions that are present in a sample.
- Using known volumes and concentrations of reactants or products, you can determine the concentrations, volumes, or masses of other reactants or products.

- Earth's supply of water is finite and is always returned to the environment.
- Water is such a good solvent that drinking water can include pollutants and naturally occurring materials, as well as chemicals added for water treatment.
- Water quality can be improved by using various physical and chemical processes.

Chapter 10	Acids and Bases

KEY IDEAS

- Aqueous solutions of acids and bases form specific ions in solution, which result in characteristic properties, such as reactions with indicators, carbonates, and certain metals.
- The acidity or basicity of a solution is measured using the pH scale.
- A strong acid ionizes completely in aqueous solution, increasing the concentration of hydrogen ions.
- A strong base dissociates completely into ions in aqueous solution, increasing the concentration of hydroxide ions.

- Weak acids and bases ionize incompletely. Most acids and bases are weak.
- The neutralization reaction between an acid and a base in aqueous solution forms a salt and water.
- A neutralization reaction can be used to determine the concentration of one of the solutions if the reacting volumes are measured and the concentration of the other solution is known.

Knowledge and Understanding

Select the letter of the best answer below.

1. Which curve represents the solubility of a gas in water?

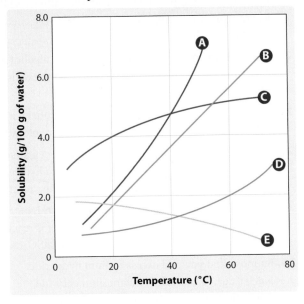

2. Which mixture contains only soluble compounds?
   a. $Fe_2(SO_4)_3$ and $Na_2Cr_2O_7$
   b. $Mg(NO_3)_2$ and $Ca(OH)_2$
   c. $MgCl_2$ and Ag_2S
   d. KBr and $CaCO_3$
   e. $(NH_4)_3PO_4$ and $Mg_3(PO_4)_2$

3. When aqueous solutions of sodium sulfate, $Na_2SO_4(aq)$, and calcium nitrate, $Ca(NO_3)_2(aq)$, are mixed, the spectator ions are
   a. $Na^+(aq)$, $Ca^{2+}(aq)$, $SO_4^{2-}(aq)$, and $NO_3^-(aq)$
   b. $Ca^{2+}(aq)$ and $SO_4^{2-}(aq)$
   c. $Na^+(aq)$ and $NO_3^-(aq)$
   d. $Na^+(aq)$ and $SO_4^{2-}(aq)$
   e. $Ca^{2+}(aq)$ and $NO_3^-(aq)$

4. Which of the following cations is/are most responsible for hard water?
   A. $Na^+(aq)$
   B. $Ca^{2+}(aq)$
   C. $Mg^{2+}(aq)$
   D. $CO_3^{2-}(aq)$

   a. A only
   b. B only
   c. D only
   d. B and C only
   e. A, B, C, and D

5. What ion exchange takes place inside a water softener?
   a. $Ca^{2+}(aq)$ replacing $Na^+(aq)$
   b. $Na^+(aq)$ replacing $Ca(aq)$
   c. $Na^+(aq)$ replacing $Cl^-(aq)$
   d. $Cl^-(aq)$ replacing $CO_3^{2-}(aq)$
   e. $Na^+(aq)$ replacing $CO_3^{2-}(aq)$

6. Which food could be identified as acidic, based on taste alone?
   a. tomatoes
   b. potatoes
   c. milk
   d. water
   e. peanuts

7. Why is acetic acid, $CH_3COOH(\ell)$, classified as a weak acid?
   a. It gives vinegar a sour taste.
   b. No acetic acid molecules ionize in water.
   c. Very few acetic acid molecules ionize in water.
   d. Acetic acid reacts slowly with a reactive metal, such as magnesium.
   e. When added to baking soda, carbon dioxide gas is formed.

8. Household vinegars are solutions of acetic acid, $CH_3COOH(\ell)$, dissolved in water, $H_2O(\ell)$. Different kinds of vinegar have different concentrations of acetic acid. How is the concentration of a vinegar usually described?
   a. mass/volume percent, % (m/v)
   b. mass percent, % (m/m)
   c. volume percent, % (v/v)
   d. parts per million, ppm
   e. molar concentration, mol/L

Answer the questions below.

9. Give an example of each type of aqueous solution.
   a. a dilute solution of a soluble compound
   b. a concentrated solution of a soluble compound
   c. a saturated solution of an insoluble compound

10. A saturated solution of potassium sulfate can be made by dissolving 12 g of solid potassium sulfate, $K_2SO_4(s)$, in 100 mL of water at a specific temperature. Explain the effect of each of the following on the solubility of a saturated solution of potassium sulfate in water.
    a. adding 15 g of solid potassium sulfate
    b. increasing the temperature of the solution
    c. adding 10 g of sodium chloride
    d. increasing the air pressure above the solution

11. What method for reporting concentration would you expect to be used in each of the following situations? Give reasons for your choice of method. In some situations, more than one method may be possible.

   a. cadmium (a toxic metal) in inexpensive jewellery

   b. a standard solution of ammonium sulfate, $(NH_4)SO_4(aq)$, used in an analytical laboratory

   c. methane gas in a coal mine

   d. tin in a pewter ornament

   e. water in rubbing alcohol

   f. atrazine (a pesticide) in fatty tissue

12. Identify the spectator ions in the following chemical equation:

$$3SnCl_2(aq) + 2K_3PO_4(aq) \rightarrow Sn_3(PO_4)_2(s) + 6KCl(aq)$$

13. List four pollutants that are found in ground water. Give a source of each pollutant.

14. Describe two experimental tests you could perform to determine whether a solution was an acid or a base.

15. Identify each substance as a strong acid, a weak acid, a strong base, or a weak base.

   a. $H_2SO_4(aq)$

   b. oxalic acid dihydrate, $(COOH)_2 \cdot 2H_2O(s)$

   c. $Be(OH)_2(aq)$

   d. $KOH(aq)$

   e. $HI(aq)$

   f. $NH_3(aq)$

16. Why are different indicators used for neutralization reactions between different acids and bases? Use specific examples to illustrate your answer.

Thinking and Investigation

17. A solution of sodium hydroxide, $NaOH(aq)$, can absorb carbon dioxide, $CO_2(g)$, from the air and then react to precipitate sodium carbonate, $Na_2CO_3(s)$.

   a. Explain how this reaction would affect the concentration of a standard 1.00 mol/L solution of sodium ions, $Na^+(aq)$, prepared from 6.00 mol/L aqueous sodium hydroxide.

   b. How could you prepare a more accurate standard solution of sodium ions?

18. Several joints, such as the hip, knee, and elbow joints, are surrounded by a liquid called synovial fluid. Synovial fluid contains dissolved gases. Why do people who suffer from rheumatism, a disease that affects joints, complain of increased pain during a low-pressure weather system?

19. Suppose that you have been given an aqueous solution containing magnesium nitrate, $Mg(NO_3)_2(aq)$, and calcium nitrate, $Ca(NO_3)_2(aq)$. What could you add to the solution to precipitate one of the cations, while leaving the other cation in solution? Write the chemical equation for the precipitation reaction.

20. Calculate the molar concentration of ethyl alcohol in a 14% (m/v) aqueous solution. Assume that the density of the solution is 1.0 g/mL.

21. If 1.00 mol/L of $NaX(aq)$ contains 4.2% solute by mass, what is the atomic mass of X? What anion could X be?

22. Identify the precipitate in the reaction between aqueous iron(III) bromide, $FeBr_3(aq)$, and potassium hydroxide, $KOH(aq)$. Write the net ionic equation for the reaction.

23. What mass of precipitate forms when 22.0 mL of 0.125 mol/L sodium hydroxide, $NaOH(aq)$, is reacted with 34.5 mL of 0.110 mol/L zinc sulfate, $ZnSO_4(aq)$?

24. Suppose that you add excess sodium hydroxide, $NaOH(aq)$, to 100 mL of magnesium chloride, $MgCl_2(aq)$. After the precipitate has been filtered and dried, you determine that it has a mass of 2.04 g.

   a. Calculate the molar concentration of the original magnesium chloride solution.

   b. What was the concentration of the chloride ions in the original solution?

25. Iron and its compounds make up 5% of Earth's crust. Many compounds that contain ferrous ions, $Fe^{2+}(aq)$, are soluble. As a result, ferrous ions are often present in water supplies, especially from wells. A simple test for iron in water is to fill a large container with a sample of the water and then leave it exposed to air for a few days. If the water contains iron, the ferrous ions will gradually change to ferric ions, $Fe^{3+}(aq)$. The water will take on the yellowish colour of ferric compounds, which are insoluble. Explain why ferric compounds are generally less soluble than ferrous compounds.

26. A student titrated nitric acid, $HNO_3(aq)$, with sodium hydroxide, $NaOH(aq)$, using phenolphthalein indicator. The following data were recorded. Use the data to calculate the concentration of nitric acid.

volume of $HNO_3(aq)$ = 25.00 mL

concentration of $NaOH(aq)$ = 0.982 mol/L

Burette Readings for the Volume of Sodium Hydroxide

Reading (mL)	Trial 1	Trial 2	Trial 3
Final reading	18.28	36.48	28.44
Initial reading	0.00	18.28	10.22

27. Potassium thiocyanate is a soluble solid. An aqueous solution of potassium thiocyanate, KSCN(aq), can be used to estimate the concentration of dissolved ferric ions in tap water. The thiocyanate ions combine with ferrous ions to form red iron(III) thiocyanate ions:

$Fe^{3+}(aq) + SCN^-(aq) \rightarrow Fe(SCN)^{2+}(aq)$

   a. Only a few drops of potassium thiocyanate solution are required to test a 5 mL sample of tap water. How would you prepare a 0.1 mol/L potassium thiocyanate solution to be used for this test?

   b. Outline a procedure you could use to estimate the concentration of ferric ions in a sample of tap water.

28. **BIG IDEAS** People have a responsibility to protect the integrity of Earth's water resources. Many homes in rural areas, where there are no sewers, have a septic system to deal with waste. Heavy solid materials settle to the bottom of the septic tank, while waste water flows through underground perforated pipes in an area called the leaching bed. Waste water from the leaching bed filters into the ground, where it is broken down by bacteria in the soil. Why should paints, solvents, bleach, and many other household products never be poured into a septic system?

29. The acid-base indicator bromocresol green changes colour in the pH range of 3.8 (yellow) to 5.4 (blue). Another acid-base indicator, methyl orange, changes colour in the pH range of 3.2 (red) to 4.4 (orange). When bromocresol green was added to a sample of an aqueous solution, it turned yellow. When methyl orange was added to another sample of the same solution, it turned red. What can you conclude about the pH of the solution?

30. Adding dilute sulfuric acid, $H_2SO_4(aq)$, to a solution that contains strontium ions produces a precipitate of strontium sulfate, $SrSO_4(s)$.

   a. What is the maximum mass of strontium sulfate that can be collected if 45.0 mL of 0.750 mol/L sulfuric acid is added to 80.0 mL of a 0.250 mol/L solution of strontium nitrate, $Sr(NO_3)_2(aq)$?

   b. Why might the measured mass of the precipitate differ from the calculated mass?

31. The concentration of dissolved chlorine in tap water can be determined by adding excess potassium iodide, KI(aq), and then using sodium thiosulfate, $Na_2S_2O_3(aq)$, to precipitate the resulting iodide ions. The net ionic equation for the first reaction is

$2I^-(aq) + Cl_2(aq) \rightarrow 2Cl^-(aq) + I_2(aq)$

   a. In Ontario, the aesthetic objective for chlorine in tap water is a maximum concentration of 250 mg/L. Find the minimum volume of a 0.001 00 mol/L solution of potassium iodide that should be used to determine if a 500 mL sample of tap water meets this objective.

   b. Why might more than the minimum volume of potassium iodide solution be added?

32. Antacids are medications that neutralize excess stomach acid. Stomach acid is primarily hydrochloric acid. Some antacids contain magnesium hydroxide, $Mg(OH)_2(aq)$. What is the maximum volume of 0.100 mol/L hydrochloric acid that can be neutralized by a three-teaspoon dose of the anatacid shown here? **Hint:** Look at the list of active ingredients to determine how much magnesium hydroxide is in a dose.

Sealed bottle. Security seal. Shake well before use.

Composition: Each mL contains: 80 mg magnesium hydroxide.

Non-medical ingredients in alphabetical order: Glycerin, menthol, peppermint oil, potassium sorbate, propylene glycol, purified water and sorbitol. **ANTACID: FOR THE RELIEF OF STOMACH UPSET: Dosage: Adults:** 1 to 4 teaspoonfuls (5–20 mL). **LAXATIVE: GENTLE RELIEF OF OCCASIONAL CONSTIPATION: Dosage: Adults:** 2 to 4 tablespoonfuls (30–60 mL). **Children 6 to 12 years:** 1 to 2 tablespoonfuls (15–30 mL). **Children 2 to 6 years:** 1 teaspoonful to 1 tablespoonful (5–15 mL). Each dose should be followed by a full glass (250 mL or 8 ounces) of water or another liquid. **Children under 2 years:** Consult a physician. Oral dosage produces bowel movement in 1/2 to 6 hours.

CAUTION: Persons suffering from kidney disease or on tetracycline should consult their physician before using this product. Overuse or extended use may cause dependence for bowel function. Do not take for more than two weeks unless directed by a physician. Do not take within 2 hours of another medicine, since the desired effect of that medicine may be reduced. Do not use in presence of fever, nausea, vomiting or abdominal pain. If symptoms persist for more than one week, consult a physician. **KEEP OUT OF THE REACH OF CHILDREN. Store between 15–30°C.**

1689-5

33. A student analyzed a certain brand of vinegar to find the concentration of acetic acid, $CH_3COOH(aq)$, that it contained. The student added the vinegar, from a burette, to three 10.00 mL samples of 0.975 mol/L NaOH(aq) until the indicator phenolphthalein turned pink. The student made the following observations in her laboratory book:

Burette Readings for the Volume of Acetic Acid

Reading (mL)	Trial 1	Trial 2	Trial 3
Final reading	11.02	22.00	33.00
Initial reading	0.00	11.02	22.00
Volume added	11.02	10.98	11.00

Calculate the molar concentration of acetic acid in the vinegar.

Communication

34. Draw a flowchart that shows how you would determine whether a substance, which could be a solid, liquid or gas, is a solution.

35. **BIG IDEAS** Properties of solutions can be described qualitatively and quantitatively, and can be predicted. Nail polish remover contains acetone, $C_3H_6O(\ell)$. Use the structure of acetone, shown below, to explain why acetone is a good solvent for many molecular compounds but is also soluble in water.

36. Salt, NaCl(s), is implicated as a cause of high blood pressure and other diseases, yet it is a common additive in prepared foods. The Canadian Food Inspection Agency regulates the salt content of prepared foods, effectively controlling the concentration of salt. Use the Internet to find the meaning of each of the following phrases, which might be seen on food packages.
  a. salt free
  b. low in salt
  c. lower in salt

37. Write a procedure that a laboratory technician could use to prepare 2.0 L of 0.25 mol/L sulfuric acid, $H_2SO_4(aq)$, using 12.0 mol/L stock solution. Include appropriate safety precautions.

38. **BIG IDEAS** Properties of solutions can be described qualitatively and quantitatively, and can be predicted. Some substances cause no problems when they are dissolved in water at low concentrations, and they may even be beneficial. However, the same substances may be harmful at higher concentrations. Describe two different examples that illustrate this.

39. Silver chloride, AgCl(s), and lead(II) chloride, $PbCl_2(s)$, are both white solids. Use the solubility data below to devise a method for estimating the percentage (m/m) of silver chloride in a mixture of silver chloride and lead(II) chloride.

Compound	Solubility (g/100 g water) at 20°C	Solubility (g/100 g water) at 100°C
AgCl(s)	0.000 15	0.0021
$PbCl_2(s)$	1.0	3.3

40. Dental toothpaste for very young children does not usually contain fluoride compounds. Write a brief paragraph that could be placed on a tube of toothpaste to explain why the toothpaste does not contain fluoride.

41. In nature, water travels through a cycle of processes from the ocean, to the atmosphere, to land, and then back to the ocean. As water travels through this cycle, which is called the water cycle, it is purified.
  a. Draw a diagram that shows the steps in the water cycle. Research the water cycle on the Internet, if necessary.
  b. Briefly describe the three processes that nature uses to purify water. Which of these processes is not commonly used at water treatment facilities?

42. Describe how you could experimentally determine the ideal soil pH in which to grow rhododendrons.

43. Copy the following table, and use it to organize the uses, advantages, and disadvantages of each type of concentration measurement.

Comparison of Concentration Measurements

Concentration Measurement	Uses	Advantages	Disadvantages
Volume/mass percent			
Mass percent			
Volume percent			
Parts per million and parts per billion			
Molar concentration			

44. The solubility of calcium hydroxide, $Ca(OH)_2(aq)$, in water at 20°C is 0.17 g/mL. Is calcium hydroxide a strong base or a weak base? Explain your answer using diagrams.

45. Sketch graphs to show the following effects of temperature and pressure on solubility. Your graphs do not have to include numerical data. Simply draw and label the axes for each graph, and then draw a curve to show the general trend.
  a. the effect of temperature on the solubility of a solid in a liquid
  b. the effect of pressure on the solubility of a solid in a liquid
  c. the effect of temperature on the solubility of a gas in a liquid
  d. the effect of pressure on the solubility of a gas in a liquid

46. Create a chemical safety poster that describes how to work with acids. The poster should include information about protective safety gear, what to do if acids spill, and how to safely dilute acids.

47. Define the term *solution* by copying and completing the chart below.

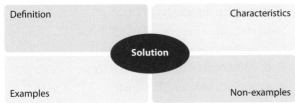

Application

48. **BIG IDEAS** People have a responsibility to protect the integrity of Earth's water resources. An accident involving an oil tanker or a leak in an underwater well can release a large volume of oil into an ocean.

a. Explain why the properties of oil and water molecules make this type of accident a serious environmental problem.

b. To help clean up a large oil spill, chemical dispersants may be used. Research what these chemical dispersants are, as well as the pros and cons of their use.

49. A 14.6 mol/L solution of aqueous ammonia, $NH_3(aq)$, is sold by many chemical supply companies. A certain brand of household glass cleaner is a 4.0 mol/L solution of aqueous ammonia. What volume of the concentrated ammonia solution is needed to prepare 750 mL of the glass cleaner?

50. Benzpyrene, $C_{20}H_{12}(s)$, is a naturally occurring carcinogen that is present in the environment and produced by cooking. A 200 g barbecued steak contains about 10 μg of benzpyrene. What is the concentration, in parts per million, of benzpyrene in the steak?

51. The list of materials for an experiment to analyze the carbohydrate content of milk includes 0.0556 mol/L aqueous glucose. How would you prepare 100 mL of the solution? The formula for glucose is $C_6H_{12}O_6(s)$.

52. **BIG IDEAS** Living things depend for their survival on the unique physical and chemical properties of water. The maximum water temperature at which trout can survive is 15°C. The maximum water temperature at which pike can survive is 24°C.

a. What does this information imply about the oxygen requirements of each type of fish? Explain your answer.

b. How could you use this information to fish in a deep lake?

53. **BIG IDEAS** People have a responsibility to protect the integrity of Earth's water resources. If industrial waste water contains mercury ions, it can be treated with sodium sulfide, $Na_2S(aq)$. Solid mercury sulfide, $HgS(s)$, can be filtered out and then buried in an approved landfill. Why is this an acceptable way to remove and dispose of mercury ions?

54. Titanium dioxide, $TiO_2(s)$, is widely used as a white pigment in paints. The largest known deposits of ilmenite ore, $FeTiO_3(s)$, are in the Allard Lake district of Québec. The sulfate process for producing titanium dioxide from ilmenite uses these reactions:

$$FeTiO_3(s) + 2H_2SO_4(aq) \rightarrow$$
$$TiOSO_4(aq) + FeSO_4(aq) + 2H_2O(\ell)$$

$$TiOSO_4(aq) + H_2O(\ell) \rightarrow TiO_2(s) + H_2SO_4(aq)$$

a. Calculate the volume of 18 mol/L sulfuric acid that is required to produce 20 000 t of titanium dioxide using the sulfate process.

b. The chloride process for refining titanium dioxide uses rutile ore, which contains a high percentage of titanium dioxide. Use text and Internet resources to learn about the chloride process. Outline the steps in this process, including the reactions involved.

c. Is the sulfate process or the chloride process less harmful to the environment? Justify your answer.

55. Hair is normally acidic. It has its maximum strength at a pH value from 4 to 5. Explain why shampooing your hair can leave it brittle and why rinsing your hair with vinegar, after shampooing it, can be beneficial.

56. Hypochlorous acid, $HOCl(aq)$, is a weak acid that is used as a bleach and a disinfectant. When 25.00 mL samples of hypochlorous acid were titrated with 1.00 mol/L sodium hydroxide, $NaOH(aq)$, an average of 19.15 mL of sodium hydroxide was needed to neutralize the acid.

a. Calculate the molar concentration of the hypochlorous acid.

b. What is the percent (m/v) of hypochlorous acid in the solution?

57. People who own ion-exchange water softeners have to buy large bags of sodium chloride pellets.

a. What are the sodium chloride pellets used for?

b. Sodium hydroxide can also be bought in pellets. Why do people not buy sodium hydroxide for their water softeners?

Select the letter of the best answer below.

1. **K/U** Which combination of observation and conclusion correctly describes what may happen when a crystal is added to an aqueous solution?
   a. The crystal dissolves; therefore the solution is saturated.
   b. The crystal remains unchanged; therefore the solution is saturated.
   c. The crystal remains unchanged; therefore the solution is unsaturated.
   d. Particles come out of solution; therefore the solution is saturated.
   e. The crystal remains unchanged; therefore the solution is supersaturated.

2. **K/U** Which mixture contains only insoluble compounds?
   a. $AgCl(s)$ and $CaBr_2(s)$
   b. $Cu(OH)_2(s)$ and $FeCO_3(s)$
   c. $Li_3PO_4(s)$ and $Pb(NO_3)_2(s)$
   d. $(NH_4)_2S(s)$ and $MgSO_4(s)$
   e. $PbI_2(s)$ and $Zn(CH_3COO)_2(s)$

3. **K/U** To confirm the presence of chloride ions in an aqueous solution, a chemist would probably add
   a. silver chloride
   b. calcium nitrate
   c. magnesium sulfide
   d. silver nitrate
   e. ammonium sulfate

4. **T/I** What is the net ionic equation for the reaction between aqueous nitric acid, $HNO_3(aq)$, and aqueous sodium hydroxide, $NaOH(aq)$?
   a. $HNO_3(aq) + NaOH(aq) \rightarrow NaNO_3(aq) + H_2O(\ell)$
   b. $H^+(aq) + NO_3^-(aq) + Na^+(aq) + OH^-(aq) \rightarrow$
      $NaNO_3(aq) + H_2O(\ell)$
   c. $H^+(aq) + NO_3^-(aq) + Na^+(aq) + OH^-(aq) \rightarrow$
      $Na^+(aq) + NO_3^-(aq) + H_2O(\ell)$
   d. $Na^+(aq) + NO_3^-(aq) \rightarrow NaNO_3(aq)$
   e. $H^+(aq) + OH^-(aq) \rightarrow H_2O(\ell)$

5. **K/U** Which of the following is a point source of pollution?
   a. urban lawns and gardens
   b. urban roads
   c. acid deposition
   d. a leaking oil well
   e. cropland

6. **K/U** The magnesium and calcium ions in hard water can be precipitated by adding
   a. sodium nitrate, $NaNO_3(aq)$
   b. sodium sulfate, $Na_2SO_4(aq)$
   c. sodium carbonate, $Na_2CO_3(aq)$
   d. sodium acetate, $NaCH_3COO(aq)$
   e. sodium chloride, $NaCl(aq)$

7. **K/U** Which substance has the lowest pH?
   a. tap water
   b. milk
   c. orange juice
   d. shampoo
   e. blood

8. **K/U** Which substance is an Arrhenius base?
   a. $CH_3COOH(\ell)$
   b. $LiOH(s)$
   c. $H_2O(\ell)$
   d. $NH_3(aq)$
   e. $HCl(aq)$

9. **T/I** Which acid is a weak acid?
   a. sulfuric acid
   b. perchloric acid
   c. acetic acid
   d. nitric acid
   e. hydrobromic acid

10. **K/U** Which equation represents an acid-base neutralization reaction?
    a. $H_2O(\ell) + SO_2(g) \rightarrow H_2SO_3(aq)$
    b. $Zn(s) + HCl(aq) \rightarrow ZnCl_2(aq) + H_2(g)$
    c. $H_3PO_4(aq) + 3NaOH(aq) \rightarrow$
       $Na_3PO_4(aq) + 3H_2O(\ell)$
    d. $NH_3(g) + H_2O(\ell) \rightarrow NH_4^+(aq) + OH^-(aq)$
    e. $KOH(aq) + CaCl_2(aq) \rightarrow 2KCl(aq) + Ca(OH)_2(s)$

11. **K/U** Predict whether each compound is soluble in water. Briefly explain your reasoning.
    a. potassium carbonate, $K_2CO_3(s)$
    b. silver sulfide, $Ag_2S(s)$
    c. lead(II) chloride, $PbCl_2(s)$
    d. iron(III) phosphate, $FePO_4(s)$

12. **T/I** Tincture of iodine is a disinfectant. It is a solution of iodine, $I_2(s)$, dissolved in ethyl alcohol. Iodine is only slightly soluble in water. Based on this information, what can you infer about the molecules of ethyl alcohol?

13. **C** Draw a flowchart for each of the following procedures. Which steps in your flowcharts are the same? How can you combine the two flowcharts into one?
    a. the steps to make an aqueous solution with a solid solute
    b. the steps to make a dilute solution from a concentrated solution

14. **T/I** Ethyl alcohol, $C_2H_5OH(\ell)$, burns with a relatively cool flame. Chefs use this property when they flambé foods. What volume of pure ethyl alcohol is present in 30 mL of a 57.14% (v/v) solution of ethyl alcohol in water?

15. **T/I** The total volume of seawater in all of Earth's oceans is about 1.35×10^{24} L. Seawater contains about 3.3% (m/m) of dissolved sodium chloride. What is the total mass (in tonnes) of dissolved sodium chloride in the oceans?

16. **A** Describe how you would safely prepare 500 mL of 1.75 mol/L sulfuric acid, $H_2SO_4(aq)$, starting from a 12.0 mol/L stock solution of sulfuric acid.

17. **C** Aqueous solutions of potassium nitrate, $KNO_3(aq)$, and calcium nitrate, $Ca(NO_3)_2(aq)$, are both clear and colourless. Make a flowchart that shows two different tests you could use to identify these solutions and the results you would observe for each test.

18. **T/I** A technician added excess aqueous sodium sulfide, $Na_2S(aq)$, to 100 mL of barium nitrate, $Ba(NO_3)_2(aq)$. When dried, the precipitate of barium sulfide had a mass of 2.675 g.

a. Write the chemical equation for the reaction.

b. Calculate the concentration, in moles per litre, of the barium nitrate solution.

19. **C** Make a poster that lists four pollutants found in ground water. Your poster should include images of the sources of each pollutant.

20. **A** Animal wastes include the soluble compound urea, $(NH_2)_2CO(aq)$. What would be the effect of discharging this compound, untreated, into an aqueous environment?

21. **A** A gardener decides to test the pH of the soil in her garden.

a. Describe how she could measure the pH.

b. She finds that the soil pH is 6.0. Is the soil acidic, basic, or neutral?

c. To grow a wider range of crops, she would like the soil pH to be about 6.5. What can she add to the soil to increase the pH?

22. **T/I** When you exercise, your muscles convert glucose into lactic acid, which is monoprotic. The structure of lactic acid is shown below:

a. Copy the structure of lactic acid, and circle the acidic hydrogen atom. Explain why the hydrogen atom is acidic.

b. Write the chemical equation for the ionization of lactic acid in water.

c. Is lactic acid a strong acid or a weak acid? Explain your answer.

23. **T/I** A student is doing a titration to determine the concentration of an acid. The student begins by reading the burette, as shown in the first diagram. When the titration is complete, the student reads the burette again, as shown in the second diagram. How will the results of the titration be affected by the student's technique? What can the student do to obtain more accurate results?

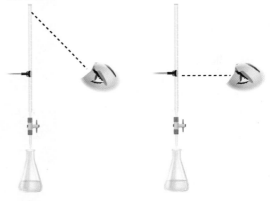

24. **A** Explain why you should never use a strong base to neutralize the spill of a strong acid.

25. **T/I** What volume of 0.500 mol/L hydrochloric acid, $HCl(aq)$, is required to react completely with 1.00 g of aluminum hydroxide, $Al(OH)_3(s)$?

Self-Check

If you missed question …	1	2	3	4	5	6	7	8	9	10	11	12	13	14	15	16	17	18	19	20	21	22	23	24	25
Review section(s)…	8.1	8.2	8.2, 9.1	9.1	9.3	8.2, 9.3, 9.4	10.1	10.1	10.1	10.2	8.2	8.2	8.4	8.3	8.3	8.4	9.1	8.3, 9.2	9.3	9.3, 9.4	10.1	10.1	10.2	8.4, 10.1, 10.2	10.2

Gases and Atmospheric Chemistry

BIG IDEAS

- Properties of gases can be described qualitatively and quantitatively, and can be predicted.

- Air quality can be affected by human activities and technology.

- People have a responsibility to protect the integrity of Earth's atmosphere.

Overall Expectations

In this unit, you will …

- **analyse** the cumulative effects of human activities and technologies on air quality, and describe some Canadian initiatives to reduce air pollution, including ways to reduce your own carbon footprint

- **investigate** gas laws that explain the behaviour of gases, and solve related problems

- **demonstrate** an understanding of the laws that explain the behaviour of gases

Unit 5 Contents

Chapter 11
Properties of Gases

How do gases differ from other states of matter, and how do these differences affect the behaviour of gases?

Chapter 12
Exploring the Gas Laws

How are the scientific laws that describe gas behaviour related?

Composed mainly of gases, the thin atmosphere that envelops our planet is essential to life. The atmosphere regulates Earth's temperatures. It helps distribute water and important nutrients on which organisms depend, and it protects all life from harmful radiation. Without the atmosphere to shield the planet from the harsh conditions of space, Earth would be a very different place. What gases make up Earth's atmosphere? How do gases behave, and what accounts for their behaviour? How do certain gases produced as a result of human activities affect air quality and health? You will find answers to questions such as these in this unit.

Pioneering studies of Earth's atmosphere represent the beginnings of scientific explorations into gases and their behaviour. These studies eventually led to the establishment of a series of gas laws that demonstrate the interrelationship among the pressure, volume, temperature, and amount of a gas. You will explore the development of these gas laws and how they are used to predict gas behaviour and gas stoichiometry in chemical reactions.

As you study this unit, look ahead to the **Unit 5 Project** on pages 582 to 583, which gives you an opportunity to demonstrate and apply your new knowledge and skills. Keep a planning folder so you can complete the project in stages as you progress through the unit.

Safety When Working with Gases in the Laboratory

- Be aware of any safety precautions that must be taken when working with gases in the laboratory. Know what precautions must be taken when handling gases that are potentially flammable, toxic, or explosive.

- Always wear a lab apron and safety eyewear when working in the laboratory with materials that could pose a safety problem. These include harmful chemicals such as acids and bases, unidentified substances, and substances that are being heated.

- Know any safety precautions before beginning an investigation. Also, make sure you understand the associated safety and WHMIS symbols that have been provided.

- Light the flame of a Bunsen burner quickly once the flow of gas begins. The air intake can be adjusted to produce a blue flame.

- When heating a substance over a flame, point the open end of the container that is being heated away from yourself and others. Never allow a container to boil dry.

- Immediately wash any part of your body that comes in contact with a laboratory chemical.

- If you are asked to smell a substance, hold the container slightly in front of and beneath your nose, and then waft the fumes toward your nostrils.

1. Which of the following gases is combustible?
   a. CO_2
   b. N_2
   c. H_2
   d. Ar
   e. CO

2. Which of the following gases has a greenish yellow color and is highly toxic?
   a. CO_2
   b. O_2
   c. N_2
   d. Cl_2
   e. CH_4

3. What is the meaning of the following WHMIS symbol?

   a. flammable gas
   b. combustible gas
   c. poisonous gas
   d. choking gas
   e. compressed gas

4. Natural gas is relatively odourless. A gas that has the smell of rotting eggs is placed in pipes and lines that carry natural gas so that gas leaks, if they occur, can be detected. Identify the formula for the gas with the rotting-egg smell.
   a. CO_2
   b. S
   c. SO_2
   d. H_2S
   e. CS_2

5. Under what conditions is toxic carbon monoxide gas generated?

6. Certain procedures that generate gases in a classroom laboratory must be performed in a fume hood. What is a fume hood, and why is it necessary for certain procedures to be performed using a fume hood?

7. Describe the precautions that must be taken when testing a gas produced using a flaming or glowing splint.

8. Why must extra precautions be taken when handling a volatile flammable liquid? Describe the precautions that you would take.

9. Describe the safety steps that must be taken if a Bunsen burner goes out unexpectedly.

10. Many activities and investigations in school laboratories involve chemical reactions that generate gases. Why is it often not safe to generate gases in a sealed, stoppered container?

- All matter is composed of particles in constant motion.
- The state of a substance is determined by the energy of its particles, the distance between particles, and the attractive forces between them.
- Particles at a higher temperature (higher kinetic energy) move faster, on average, than particles at a lower temperature (lower kinetic energy).
- Energy is involved when matter changes state. Melting, vaporizing, and sublimating (changing from the solid state to the gaseous state) involve the absorption of energy by particles. Freezing, condensing, and sublimation from the gaseous state to the solid state involve the release of energy from particles.

11. Which statement describes a substance that is in the gaseous state?
    a. Its particles cannot flow past one another.
    b. It cannot be compressed into a smaller volume.
    c. It takes the shape of the container.
    d. Its particles of matter are close together.
    e. Its particles are strongly attracted to one another.

12. Water vapour is less dense than ice because
    a. molecules in the gaseous state are in constant motion.
    b. molecules in the gaseous state have more potential energy than in solids.
    c. molecules in the gaseous state have more kinetic energy than in solids.
    d. molecules in the gaseous state have less mass.
    e. molecules in the gas phase have more empty space between them than in solids.

13. The change of state from a gas to a liquid is called
    a. melting
    b. condensation
    c. vaporization
    d. sublimation
    e. boiling

14. Which of the following has a fixed volume but no fixed shape?
    a. solid carbon dioxide dry ice at −80°C
    b. oxygen gas at room temperature
    c. liquid ammonia at 5°C
    d. water vapour at 100°C
    e. solid ice at −10°C

15. Water cools from 95°C to 25°C. In terms of energy, which of the following has occurred?
    a. potential energy has increased, kinetic energy has decreased
    b. potential energy has not changed, kinetic energy has decreased

    c. potential energy has decreased, kinetic energy has increased
    d. potential energy has increased, kinetic energy has not changed
    e. potential energy has decreased, kinetic energy has decreased

The next two questions refer to the graph below.

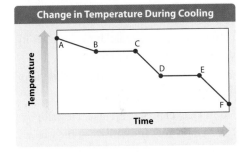

16. During which segments is this substance changing state?
    a. AB, CD, EF
    b. AB, EF
    c. BC, CD, DE
    d. BC, EF
    e. BC, DE

17. During which segments are particles of this substance experiencing a reduction in kinetic energy?
    a. BC, DE
    b. AB, EF, CD
    c. BC, CD, DE
    d. BC, EF
    e. Kinetic energy remains the same in all segments.

18. Make a table to compare solids, liquids, and gases in terms of the following properties:
    a. compressibility
    b. viscosity
    c. molecular motion
    d. distance between particles

19. A balloon is blown up and then tied. Explain each of the scenarios below, using a series of diagrams and descriptions. Use circles to represent gas molecules.
    a. The air inside the balloon is heated.
    b. The air inside the balloon is cooled.

Density Calculations and Conversions

- The density of a substance is its mass per unit volume. The mathematical expression for density is $d = \frac{m}{v}$.
- Density is related to the number of particles in a given volume of space. Particles of most substances are closer together in the solid state than in the liquid state, and farthest apart in the gaseous state. Water is an exception—the density of ice is greater than that of liquid water. However, for all substances, the density is greatest in the solid state and lowest in the gaseous state.
- For equal volumes of gases that have different densities, the gas with the higher density will weigh more. Therefore, gases of differing densities can form layers with the most dense gas at the bottom and the least dense gas at the top.

The next two questions refer to the table below.

Density of Common Gases at 0°C

Gas	Density of Gas (g/L)
Carbon Dioxide	1.799
Nitrogen	1.251
Oxygen	1.429
Helium	0.179
Argon	1.784

20. Assuming no mixing of gases, predict how the gases in the table would layer in a sealed container.

21. Air contains all the gases in the table. The density of air at 0°C is 1.293 g/L. Explain this observation.

22. A gas is trapped in an expandable syringe. Describe two ways that you could alter the density of the gas.

23. The density of solid carbon dioxide is 1.562 g/cm³. Convert this to grams per litre and compare this quantity to that of gaseous carbon dioxide. Use the particle model to illustrate the difference in density between the solid and gaseous state of carbon dioxide.

24. The concentration of gases is sometimes expressed using units of parts per million (ppm) and parts per billion (ppb). A concentration of carbon monoxide gas of 3200 ppm (3196 mg/L) can cause unconsciousness and death in less than 30 min.
    a. What is this concentration of gas in mol/L?
    b. How many molecules of carbon monoxide gas would exist in a cubic metre of this air?

25. The density of air is 1.285 g/L at 25°C.
    a. Calculate the mass of air if the volume is 2500 L.
    b. If the mass of air is 560 g, determine the corresponding volume of air.
    c. An air-filled balloon has a mass of 3.16 g and a circumference of 49.7 cm. If it is perfectly spherical, calculate the density of air. Is the air temperature above or below 25°C? Explain your answer.

26. Graph the data in the following table, using mass as the dependent variable and volume as the independent variable.

Volume (L)	Mass (g)
0.5	0.8995
2.0	3.598
3.5	6.297
5.0	8.995
6.5	11.694
8.0	14.392

a. Calculate the density of air using the slope of the line.
b. Predict the mass of air at a volume of 9.5 L.
c. Predict the volume of the air at a mass of 7.53 g.
d. Why must the temperature be stated when determining density by this method?

27. Consider the following graph.

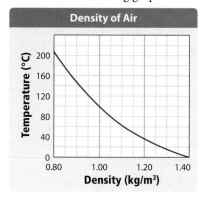

a. Describe the graphical relationship that exists between density and temperature of a gas. What is this type of mathematical relationship called?
b. Predict the density of air at 50°C and 120°C.
c. Predict the temperature of air if the density of air is 1.10 kg/m³ and 0.88 kg/m³.
d. If air was substituted with a different gas, would a similar relationship be observed?

- Chemists express the amount of a substance in moles. One mole of a substance contains $6.022\ 141\ 99 \times 10^{23}$ particles of the substance. This number is called the Avogadro constant.
- The molar mass of a compound is the sum of the molar masses of the atoms in the formula of the compound.
- The amount of a substance (in mol) is its mass (in g) divided by its molar mass (in g/mol).

- The coefficients in a balanced chemical equation give the ratio between the reactants and products, when expressed in moles.
- Stoichiometry calculations use the relative quantities (in mol) of the reactants and products in a reaction, as given by the coefficients in the balanced chemical equation.

28. Balance the following equation for the combustion of benzene: $C_6H_6(\ell) + O_2(g) \rightarrow H_2O(g) + CO_2(g)$
 a. $C_6H_6(\ell) + 9O_2(g) \rightarrow 3H_2O(g) + 6CO_2(g)$
 b. $C_6H_6(\ell) + 9O_2(g) \rightarrow 6H_2O(g) + 6CO_2(g)$
 c. $2C_6H_6(\ell) + 15O_2(g) \rightarrow 6H_2O(g) + 12CO_2(g)$
 d. $C_6H_6(\ell) + 15O_2(g) \rightarrow 3H_2O(g) + 6CO_2(g)$
 e. $2C_6H_6(\ell) + 9O_2(g) \rightarrow 6H_2O(g) + 12CO_2(g)$

29. How many molecules of molecular oxygen react with four molecules of ammonia, $[NH_3]$ (g), to form four molecules of nitrogen monoxide and six molecules of water?
 a. 2   **d.** 4
 b. 10   **e.** 5
 c. 3

30. How many grams of oxygen are needed to react completely with 200.0 g of ammonia?
 $4NH_3(g) + 5O_2(g) \rightarrow 4NO(g) + 6H_2O(g)$
 a. 469.7 g   **d.** 3.406 g
 b. 300.6 g   **e.** 2.180 g
 c. 250.0 g

31. An important reaction sequence in the industrial production of nitric acid is the following:
 $N_2(g) + 3H_2(g) \rightarrow 2NH_3(g)$
 $4NH_3(g) + 5O_2(g) \rightarrow 4NO(g) + 6H_2O(\ell)$
 Starting from 20.0 mol of nitrogen gas in the first reaction, how many moles of oxygen gas are required in the second one?
 a. 12.5 mol O_2 **d.** 50.0 mol O_2
 b. 20.0 mol O_2 **e.** 1.00×10^2 mol O_2
 c. 25.0 mol O_2

32. Each of the following reactions produces a gas. Predict the products of the reaction and write the balanced chemical equation.
 a. solid sodium hydrogen carbonate is reacted with aqueous hydrochloric acid
 b. molten sodium chloride is decomposed using electrolysis

 c. solid magnesium carbonate is decomposed with heat from a Bunsen burner
 d. aqueous hydrogen peroxide is decomposed using an aqueous sodium iodide catalyst
 e. zinc solid is reacted with aqueous hydrochloric acid
 f. liquid water is decomposed using electrolysis
 g. aqueous ammonium carbonate is heated in a test tube
 h. iron(II) sulfide is reacted with aqueous hydrochloric acid
 i. solid sulfur is burned in the presence of oxygen gas
 j. nitrogen triiodide is explosively decomposed

33. A chemical reaction conducted in a test tube is suspected of generating oxygen gas.
 a. Using laboratory equipment normally available in a high school classroom, explain how you would collect the gas.
 b. Describe one test and the subsequent result that could be conducted on the gas to confirm its identity.

34. Nitrogen monoxide, which is present in urban air pollution, immediately converts to nitrogen dioxide as it reacts with oxygen gas.
 a. Write the balanced chemical equation for the formation of nitrogen dioxide from nitrogen monoxide.
 b. What mole ratio would you use to convert from moles of oxygen gas to moles of nitrogen dioxide?

35. Octane (C_8H_{18}) is a fuel component of gasoline. When octane is combusted, carbon dioxide and water vapour are released.
 a. Write a balanced equation for the reaction.
 b. The density of gasoline is 0.77 g/cm^3. Assuming that the fuel is entirely octane, what mass of carbon dioxide is produced during the complete combustion of 40.0 L of this fuel?
 c. How many molecules of carbon dioxide were released during the reaction?

Specific Expectations

In this chapter, you will learn how to . . .

- F2.1 **use** appropriate terminology related to gases and atmospheric chemistry (11.1, 11.2, 11.3)

- F2.2 **determine**, through inquiry, the quantitative and graphical relationships between the pressure, volume, and temperature of a gas (11.2, 11.3)

- F2.3 **solve** quantitative problems by performing calculations based on Boyle's law, Charles's law, Gay-Lussac's law, the combined gas law, and the ideal gas law (11.2, 11.3)

- F3.2 **describe** the different states of matter and **explain** their differences in terms of the forces between atoms, molecules, and ions (11.1)

- F3.3 **use** the kinetic molecular theory to **explain** the properties and behaviour of gases in terms of types and degrees of molecular motion (11.1)

- F3.5 **explain** Boyle's law, Charles's law, Gay-Lussac's law, the combined gas law, and the ideal gas law (11.2, 11.3)

Since the mid-1950s, weather balloons filled with hydrogen or helium gas have carried thermometers, barometers, and humidity-measuring instruments high into the upper atmosphere to collect weather-forecasting data on which all Canadians depend. About 170 years earlier, a balloon similar to modern weather balloons ascended into the air before a crowd of 400 000 Parisian onlookers. Carrying its inventor, French scientist Jacques Charles, and his assistant, the balloon drifted 43 km over the French countryside, becoming the world's first free-floating, piloted aircraft. To advance science even further, Charles brought a thermometer and a barometer onboard the flight to take measurements of air temperature and pressure during the history-making journey.

Experiments involving gas-filled balloons have played an important role in the development of our knowledge of gases. The careful observations that scientists made enabled them to state and mathematically define scientific laws that govern the safe and efficient use of gases to this day. You will learn about some of these laws—including two stated by Jacques Charles—in this chapter.

Expanding and Shrinking Soap Bubble

Most gases are clear and colourless, so they are invisible to the eye. However, you can set up situations that let you visualize the effects of the properties of gases. In this activity, infer the properties of air by observing the behaviour of a soap bubble.

Safety Precautions

- Wear safety eyewear throughout this activity.
- Be careful to avoid contact with the rubbing alcohol. If you get any rubbing alcohol on your hands, wash them thoroughly with running water and soap and inform your teacher.

Materials

- 100 mL of soap solution
- rubbing alcohol
- petri dish that is 10 cm in diameter
- soft-drink can, empty and with tab removed
- cotton ball

Procedure

1. Add soap solution to the petri dish until it is about half-full.
2. Turn the empty soft-drink can upside down, and place the top of the can in the soap solution. Remove the can from the soap solution, and turn it upright on a flat surface.
3. Gently wrap your hands around the can, and then very gently squeeze the can. Observe any soap bubble that forms and its behaviour as you gently squeeze the can with a steady pressure and then release.
4. Gently wrap one of your hands around the can, holding it lightly. *Do not squeeze the can.* Observe any soap bubble that forms.
5. Rub your hands together vigorously for 5 to 10 seconds, and then repeat step 3. Observe any changes to the soap bubble.
6. Wet the cotton ball with rubbing alcohol. While holding the can with one hand, use the cotton ball to transfer rubbing alcohol to a small area of the outside of the can, and then remove your hand. Observe any changes to the soap bubble. (The purpose of the rubbing alcohol is to remove heat from the can. As the rubbing alcohol evaporates, heat is transferred from inside the can to outside the can.)

Questions

1. Is there a gas inside the can? How do you know?
2. Is there a gas inside the soap bubble? How do you know?
3. Explain what you think caused the behaviour of the soap bubble in step 3.
4. Explain what you think caused the behaviour of the soap bubble in steps 4, 5, and 6.
5. Infer at least three properties of gases that you observed in this activity.

Gases: Properties and Behaviour

Key Terms

kinetic molecular theory
of gases

ideal gas

On Earth, matter typically exists in three physical states: solid, liquid, and gas. All three states of a substance are composed of particles (ions, atoms, or molecules). However, the behaviour of the particles differs in each state. These differences are summarized in **Table 11.1**.

Particles that make up a solid are packed tightly and closely together and are locked into place in an organized framework. This regular arrangement of the particles explains why solids have a fixed volume and shape—the particles are unable to move past one another to alter the shape.

Particles that make up a liquid are very close together, but they are not held in fixed positions. Therefore, they have no regular arrangement. As a result, particles of a liquid can move past one another, allowing the liquid to flow and take the shape of its container.

Particles of a gas are greatly separated from each other—much more so compared to particles in liquids and solids. Thus, particles of a gas have no regular arrangement. Gas particles move freely in all directions, bouncing off each other and the walls of any container in which they may be held.

Table 11.1 Properties and Particles of the Three States of Matter

State	Properties	Particles
Solid	• Constant shape • Constant volume • Almost incompressible	• Particles are organized in a regular pattern (this is also known as having "low disorder") and they vibrate in a fixed position.
Liquid	• Variable shape • Constant volume • Almost incompressible	• Particles are less organized than in a solid and they are able to slide over and past one another.
Gas	• Variable shape • Variable volume • Compressible (can be pushed or squeezed by a force to occupy a smaller volume)	• Particles are much less organized than in other states and they bounce off each other and the walls of their container.

Changes of State and Forces between Particles

Two main factors determine the state of a substance: the forces holding the particles (ions, atoms, and molecules) together and the kinetic energy of the particles, which tends to pull them apart. If there were no forces between particles, all substances would be gases. Forces are necessary for particles to form liquids or solids. If the forces are very strong, a large amount of kinetic energy is needed to pull the particles apart. If the forces are weak, particles with smaller amounts of kinetic energy can pull away from one another. As shown in **Table 11.2**, the three types of forces that act between particles are attractions between oppositely charged ions, attractions between polar molecules, and attractions between non-polar molecules.

Table 11.2 Attractive Forces That Influence the State of Matter

	Type of Attractive Force	State of Matter	Example
Stronger Force	Between oppositely charged particles	Usually solid	Table salt, NaCl(s)
	Between polar molecules	Solid, liquid, or gas	Glucose, $C_6H_{12}O_6$(s) Ethanol, CH_3CH_2OH(ℓ) Ammonia, NH_3(g)
Weaker Force	Between non-polar molecules	Solid, liquid, or gas	Paraffin, $C_{30}H_{62}$(s) Pentane, C_5H_{12}(ℓ) Carbon dioxide, CO_2(g)

Attractive Forces between Oppositely Charged Particles

Oppositely charged particles exert a pulling force on each other due to electrostatic attraction. Ionic bonding is an example of electrostatic attraction. In an ionic bond, a positive ion is attracted to a negative ion. Ionic bonds are very strong. As a result, ionic compounds usually exist as solids at room temperature.

Attractive Forces between Polar Molecules

Attractive forces between polar molecules also occur. Although the overall charge of a polar molecule is neutral, one end of the molecule is partially positive and the other is partially negative. As a result, the polar molecule has a permanent dipole effect, as shown in **Figure 11.1**. Dipole-dipole forces between polar molecules are not as strong as ionic bonds. Thus, substances made up of polar molecules can exist as liquids and gases at room temperature. For example, ethanol is polar and exists as a liquid at room temperature. Hydrogen chloride is polar but is a gas at room temperature.

Hydrogen bonding is a specific type of dipole-dipole interaction. The strength of a hydrogen bond can vary, and this affects the state of a substance. For example, although water and ammonia are both polar molecules, the N–H hydrogen bonds in ammonia are less polar (and therefore weaker) than the O–H hydrogen bonds in water. Therefore, at room temperature, water is a liquid, while ammonia is a gas.

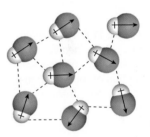

Figure 11.1 In a polar substance such as hydrogen chloride, dipole-dipole forces of attraction exist between the molecules.

Explain Why do attractive forces occur between the different regions of these hydrogen chloride molecules?

Attractive Forces between Non-polar Molecules

Attractive forces exist between non-polar molecules due to the formation of temporary dipoles. However, these forces are weak. Most small non-polar molecules, such as carbon dioxide, do not hold together long enough to maintain a solid or liquid form at room temperature, so they exist as gases at room temperature. Larger non-polar molecules, such as pentane, C_5H_{12}(ℓ), are liquids at room temperature. As a non-polar molecule increases in size, the attractive forces between the molecules of a substance also increase. When the attractive forces between the molecules are strong, more energy in the form of heat must be added to weaken them before a change in state can occur.

Learning Check

1. Using water as an example, name and distinguish between the three physical states of matter.

2. How is the state of a substance related to the attractive forces between its particles?

3. How is the state of a substance related to the arrangement of its particles?

4. Why would all substances be gases if there were no forces between particles?

5. Describe the relationship between the state of a substance and the kinetic energy of its particles.

6. Use a graphic organizer to distinguish the three types of forces between particles of substances.

Figure 11.2 This model of a particle depicts vibrational motion—the motion of something as it moves back and forth about a fixed position.

Figure 11.3 This model of a particle depicts rotational motion—the motion of something as it spins on its axis.

Figure 11.4 This model of a particle depicts translational motion—the motion of something that moves freely from one place to another.

The Kinetic Energy of Particles and Temperature of a Substance

If the particles of a substance had no kinetic energy but still had attractive forces between them, the substance would be solid. Particles must have kinetic energy to pull away from each other. Particles in a substance have three types of motion and thus three types of kinetic energy. **Figure 11.2** illustrates vibrational motion and thus vibrational kinetic energy. All particles, regardless of the state of the substance, have vibrational energy. **Figure 11.3** shows rotational motion and thus rotational kinetic energy. Particles in liquids and gases, and some solids, have rotational kinetic energy. **Figure 11.4** illustrates translational motion and thus translational kinetic energy. Only particles in liquids and gases can have translational kinetic energy.

The temperature of a substance is directly related to the average kinetic energy of its particles. To heat a substance means to add energy to the substance, which goes into the kinetic energy of its particles. As you heat a solid, the temperature increases and the particles vibrate more and more rapidly. Eventually, the particles will have enough energy to pull away from each other. However, they are still attracted to the nearby particles. Thus, in a liquid, particles are constantly sliding past one another. The strength of the forces determines how much energy is needed, and thus what the temperature must be, when particles pull away from one another.

As you heat a liquid, the temperature increases and all types of kinetic energy increase. Eventually, each particle will have enough energy to completely escape from all of the other particles in the substance as the substance becomes a gas. The strength of forces between particles determines how much energy must be added and thus what the temperature must be for the substance to become a gas.

The Distinguishing Properties of Gases

Each state of matter has physical properties that distinguish that state from another state. The following properties distinguish gases from solids and liquids.

- *Gases are compressible.* The volume of a gas decreases greatly when pressure is exerted on the gas. Similarly, the volume of a gas increases when the pressure is reduced. In contrast, the volumes of liquids and solids remain almost constant during changes in pressure. The incompressibility of solids and liquids is explained by particles in these states being unable to move independently of each other. The movement of one particle affects the movement of the other particles. Gas particles, however, are able to move independently of one another.

- *Gases expand as the temperature is increased,* if the pressure remains constant. The volumes of liquids and solids can expand with increasing temperature as well, but to a much smaller extent compared to volumes of gases.

- *Gases have very low viscosity.* The viscosity of water is approximately 55 times greater than the viscosity of air. Air and all other gases flow through pipes more freely compared to liquids. The low viscosity of gases enables them to escape quickly through small openings in their containers.

- *Gases have much lower densities* than solids or liquids. The density of water vapour is approximately 1/1000 the density of liquid water.

- *Gases are miscible.* Substances that mix completely with each other are said to be miscible. All gases are miscible. Some liquids, such as water and alcohol, are miscible. Other liquids, such as water and oil, are *immiscible* (they do not mix).

The Kinetic Molecular Theory of Gases

The **kinetic molecular theory of gases** provides a scientific model for explaining the behaviour of gases. To develop the theory, scientists defined a hypothetical substance called an ideal gas. An **ideal gas** is defined by specific characteristics related to the energy and motion of the gas particles, as shown in **Figure 11.5**. The kinetic molecular theory of gases is based on the following assumptions.

- Gas particles are in constant, random motion. Gas particles travel in straight lines until they collide with other gas particles or with the walls of their container. Therefore, an ideal gas has high *translational kinetic* energy.

- Individual gas particles are considered point masses. A *point mass* is a mass that has no volume—it takes up no space. The volume of an individual gas particle is considered negligible compared to the container holding the gas. Gas particles are considered to be extremely far apart and most of the container is thought of as empty.

- The gas particles do not exert attractive or repulsive forces on one another.

- The gas particles interact with one another and with the walls of their container only through *elastic collisions*. In an elastic collision—for example, when billiard balls collide—kinetic energy is conserved. Particles can exchange kinetic energy with one another in a collision but the total kinetic energy remains constant.

- The average kinetic energy of gas particles is directly related to temperature. The greater the temperature, the greater the average motion of the particles and the greater their average kinetic energy.

kinetic molecular theory of gases the theory that explains gas behaviour in terms of the random motion of particles with negligible volume and negligible attractive forces

ideal gas a hypothetical gas made up of particles that have mass but no volume and no attractive forces between them

Go to **scienceontario** to find out more

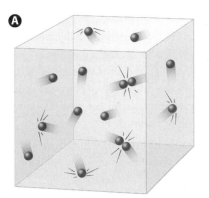 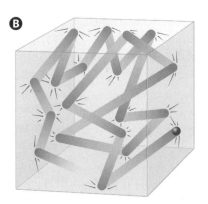

Figure 11.5 Characteristics of ideal gas particles (molecules). In **(A)**, ideal gas molecules move with random motion, colliding with one another and with the walls of their container. In **(B)**, the paths of each individual molecule follow straight lines between collisions.

Infer how the kinetic molecular theory explains the increase of the volume of a gas with an increase in temperature, and make a sketch in your notebook to illustrate your ideas.

Keep in mind that no gas is ideal. However, despite the fact that the kinetic molecular theory explains gases as if they were ideal, this important theory quite accurately describes the behaviour of real gases at ordinary temperatures and pressures. In fact, using the basic properties of an ideal gas, scientists can explain many properties of gases and can make accurate predictions about their behaviour under various circumstances and conditions. For example, the miscibility of gases can be explained by considering the large amount of space available between the molecules of a gas. The molecules of a second gas should fit readily into the spaces between the molecules of the first gas, because both types of gas molecules have negligible volume. Thus, two gases should mix evenly and completely as the molecules move about constantly. This predicted behaviour can be verified through observation.

Section Summary

- Particles of matter behave in characteristic ways in each of the three states of matter. Attractive forces between particles and the kinetic energy of the particles influence how particles behave in each state.

- Attractive forces between particles that affect the states of matter are attractions between oppositely charged particles, attractions between polar molecules, and attractions between non-polar molecules.

- The properties that distinguish gases from liquids and solids include compressibility, low viscosity, even and complete mixing, low density, and expansion as a result of an increase in temperature.

- The kinetic molecular theory of gases is a model that explains the visible properties of gases based on the behaviour of individual atoms or molecules of an ideal gas.

Review Questions

1. **K/U** What is the relationship between the strength of the attractive forces between the particles of a substance and the state of matter of the substance?

2. **K/U** How does the polarity of a molecule affect the state of a substance composed of that molecule?

3. **C** Draw a diagram that shows the attractive forces that occur for methanol, CH_3OH, and phosphine, PH_3. Based on these forces, identify the molecule that you predict would be a gas at room temperature, and explain why. (**Hint:** To draw methanol, the carbon atom is bonded to three hydrogen atoms and the oxygen atom. The fourth hydrogen atom is bonded to the oxygen.)

4. **T/I** Ammonia, $NH_3(g)$, is a gaseous compound with a boiling point of $-33.34°C$. Why is its boiling point substantially lower than the boiling point of water? Explain your answer.

5. **K/U** Describe two properties of matter in the solid or liquid state that distinguish it from matter in the gas state.

6. **T/I** A party balloon filled with helium gas is left to float in a room. Over time, the balloon falls back to the floor. Explain the behaviour of the balloon. (**Hint:** There are microscopic pores in the surface of the material that is used to make a party balloon.)

7. **K/U** For each property, explain how a gas differs from a liquid.
   a. viscosity
   b. compressibility
   c. density
   d. miscibility

8. **A** Identify a property of gases that is important in hot-air ballooning. Explain why you think it is important.

9. **T/I** Use the kinetic molecular theory to explain how a basketball is inflated.

10. **A** Which property of gases best explains each of the following situations? Explain your reasoning.
    a. A full propane tank can provide enough fuel to run a propane barbecue for several months.
    b. A carbon monoxide leak in the lower level of a building causes carbon monoxide gas to spread quickly throughout the whole building.
    c. Forced air heating is often a better choice for home heating than hot water (radiator-type) heating.

11. **A** Hand pumps are often used to fill deflated bicycle tires with air. In a hand pump, a piston is pushed through a cylinder and air is transferred to the deflated tire. How do the properties of compressibility, low resistance to flow, and even and complete mixing relate to the inflation of tires?

12. **K/U** How does an elastic collision differ from an inelastic collision? To visualize an inelastic collision, imagine throwing a ball of putty against a wall.

13. **K/U** Describe the characteristics of an ideal gas. How do real gases differ from this hypothetical model?

14. **T/I** The images below show three possible paths for a gas molecule moving inside a filled volleyball. Which of these diagrams represents the most likely path of the gas molecule? Justify your choice in terms of the kinetic molecular theory of gases.

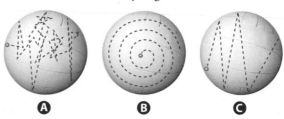

15. **C** Draw diagrams to show how the kinetic molecular theory can explain the following:
    a. why a heated gas expands to fill its container
    b. why gases can be easily compressed

Gases and Pressure Changes

If you have flown in an airplane, you may have felt discomfort in your ears during take-off or landing. As soon as your ears "popped," they probably felt better. A similar experience is common while riding up and down on an elevator. Your ears become blocked and then unblocked due to changes in atmospheric pressure. Although you are usually unaware of the effect of atmospheric pressure on your body, the atmosphere is always exerting a large amount of pressure on you from all directions.

To describe and explain the behaviour of the gases in Earth's atmosphere, as well as the behaviour of other gases, you need to understand the meanings of pressure and gas pressure. *Pressure* is the force that is exerted on an object per unit of surface area. The phrase "per unit of surface area" means the area over which the force is distributed. The equation for pressure is

$$\text{pressure} = \frac{\text{force}}{\text{area}} \quad \text{or} \quad P = \frac{F}{A}$$

The SI unit for force is the newton (N), and the unit for area is the square metre (m^2). Therefore, the corresponding unit of pressure is newtons per square metre (N/m^2). Later in this chapter, you will learn how this unit for pressure is related to other commonly used units for presssure, such as the pascal (Pa) and millimetres of mercury (mmHg).

Atmospheric Pressure

Earth's atmosphere is a spherical envelope of gases that surrounds the planet and extends from Earth's surface outward to space. The gas molecules that make up the atmosphere are pulled down toward Earth's surface by gravity, and these molecules exert pressure on all objects on Earth. Thus, the atmosphere exerts pressure on everything on Earth's surface. Earth's **atmospheric pressure** may be described as the force that a column of air exerts on Earth's surface, divided by the area of Earth's surface at the base of the column, as shown in **Figure 11.6**. The force that the column of air exerts is its weight. Unlike force, however, which is exerted in only one direction, pressure is exerted *equally in all directions*.

Early Studies of Atmospheric Pressure

People invented technologies that made use of atmospheric pressure before anyone understood how these technologies worked! For example, in 1594, the Italian scientist Galileo Galilei (1564–1642) was awarded a patent for his invention of a suction pump that used air to lift water up to the surface from about 10 m underground. (No plans or illustrations of this pump exist, but the pump operated using the kinetic energy supplied by one horse.) Although Galileo's pump functioned well, nobody—including Galileo himself—understood how water moved up the tube of the pump and why the pump could lift the water no higher than 10 m.

From 1641 to 1642, Evangelista Torricelli (1608–1647) served as Galileo's assistant. His work eventually led to an understanding of how the water moved up the tube of the pump. Torricelli hypothesized that the water rose in the tube because the surrounding air was pushing down on the rest of the water. Instead of water, he used mercury for his studies, because mercury is 13.6 times more dense than water. Thus, Torricelli could use a column of mercury that is 13.6 times shorter than the 10 m column of water.

Key Terms

atmospheric pressure

standard atmospheric pressure (SAP)

Boyle's law

---

atmospheric pressure
the force exerted on Earth's surface by a column of air over a given area

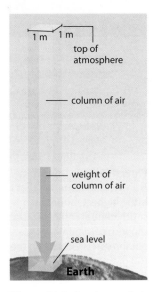

Figure 11.6 A column of air extending from sea level to the top of the atmosphere, with a cross-sectional area of 1 m^2, weighs 101 325 N. The mass of this column of air is 10 329 kg.

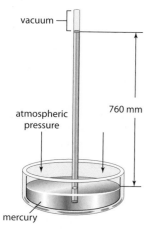

vacuum

atmospheric
pressure

760 mm

mercury

Figure 11.7 Torricelli's apparatus used mercury to test the hypothesis that underground water was being pushed up the tube of a water pump because of the air pressure acting on the surrounding water.

Predict *What would cause the level of mercury in the tube to change?*

Figure 11.8 Puy-de-Dôme is a volcanic mountain that is the highest mountain in south-central France. Today, modern communications equipment is located on the top of this mountain.

Torricelli's Hypothesis Is Confirmed

Torricelli designed an experiment with an apparatus like the one in **Figure 11.7**. He proposed that if a long tube was filled with mercury and inverted into a dish of mercury, the mercury in the tube would drop down and leave a vacuum at the closed end of the tube. Torricelli hypothesized that the pressure that the column of mercury would exert on the mercury in the dish would be equal to the pressure that the atmosphere was exerting on the surface of the mercury in the dish outside the tube. His hypothesis was verified by experiments, and the limitations of the early suction pump could now be explained. The early suction pump could not lift water more than 10 m because atmospheric pressure was *pushing* the water up, not because the pump was pulling the water up. Atmospheric pressure is about approximately the same as the pressure exerted by 10 m of water—thus the 10 m limit.

The Relationship between Atmospheric Pressure and Altitude

In 1647, the French scientist and philosopher, Blaise Pascal (1623–1662), read a letter written by Torricelli, in which he compared the atmosphere to an ocean of air. In the letter, Torricelli hypothesized that the weight of air might be greater near Earth's surface than it was at the top of mountains. The next year, Pascal designed an experiment to test Torricelli's hypothesis that atmospheric pressure decreases with altitude (distance from Earth's surface). He asked his brother-in-law, Florin Perier, to carry an apparatus like Torricelli's up and down a mountain called Puy de Dôme, shown in **Figure 11.8**. Perier measured the length of the column of mercury at the base of the mountain, during his climb of the mountain, and on the top. Perier verified that as he ascended the mountain, the column of mercury became shorter. At the top of the mountain, the column of mercury was 76 mm shorter than it was at the base of the mountain.

Figure 11.9 shows that at higher altitudes (distances from Earth's surface) the atmospheric pressure is lower and the density of air particles is less than at lower altitudes. Why is this the case? Each layer of the atmosphere exerts a force on the layer below it. Because the weight of the entire column of air exerts a force on the bottom layer, the bottom layer is the most compressed. As the altitude increases, the amount of air above that level becomes smaller and, therefore, exerts a smaller force on the air just below it. The higher layers of air are less compressed than the lower layers.

Besides demonstrating that atmospheric pressure exists and that it changes with altitude, the combined efforts of Torricelli, Pascal, and Perier resulted in an instrument for measuring atmospheric pressure: the mercury barometer. Barometers based on Torricelli's design have been in use since the mid-1600s. Although newer technologies have been developed, many mercury barometers are still used around the world today.

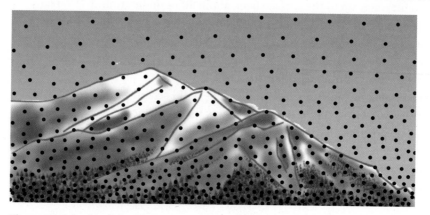

Figure 11.9 In this diagram, the dots represent air molecules. People often refer to the air "thinning" at increased altitudes. This means that there are fewer gas molecules in the air for a given volume at lower atmospheric pressure.

Units of Gas Pressure

Because mercury barometers were often used to measure atmospheric pressure, a common unit of pressure is the millimetre of mercury, or mmHg. **Standard atmospheric pressure (SAP)**, the atmospheric pressure in dry air at a temperature of 0°C at sea level, is 760 mmHg. Since standard atmospheric pressure is a common reference point, the unit *atmosphere* (atm) is also often used.

Recall that the newton per square metre (N/m²) is the SI unit for pressure. This unit is also called the pascal (Pa), in honour of Blaise Pascal. Standard atmospheric pressure is 101 325 Pa. Because this is such a large number, the unit kPa (kilopascal) is commonly used. Thus, standard atmospheric pressure is often expressed as 101.325 kPa.

Another unit often used in chemistry is the bar (b). One bar is equal to 100 kPa, and 1 atm is equal to 1.01325 bar. In the United States, pressure is often measured and expressed in pounds per square inch (psi). You may have seen the recommended pressure for bicycle and automotive tires in units of psi. Finally, in honour of Torricelli's work, standard atmospheric pressure has also been defined as 760 torr. One torr represents a column of mercury that is 1 mm high at 0°C. The following expression summarizes the various units of pressure and their equivalent values.

$$1 \text{ atm} = 760 \text{ mmHg} = 760 \text{ torr} = 101\ 325 \text{ Pa} = 101.325 \text{ kPa} = 1.01325 \text{ bar} = 14.7 \text{ psi}$$

> **standard atmospheric pressure (SAP)** atmospheric pressure in dry air at a temperature of 0°C at sea level

Different products and technologies have tended to use or report atmospheric pressure in different units. Table 11.3 lists some of the instruments that use these different units to measure pressure.

Table 11.3 Units of Pressure Used for Various Instruments

Unit of Pressure	Symbol	Examples of Instruments That Use the Unit
standard atmosphere	atm	Gas compressors, pneumatic tools (tools such as jackhammers driven by compressed gas)
millimetres of mercury	mmHg	Blood pressure meters, barometers
torr	torr	Vacuum pumps
pascal	Pa	Pressure sensors in pipelines
kilopascal	kPa	Tire inflation gauges; heating, ventilating, and air-conditioning systems
bar	bar	Pressure sensors in scuba gear, steam traps used to remove condensed water from pipes carrying steam
millibar	mb	Barometers
pounds per square inch	psi	Hydraulic pumps, tire inflation gauges

Converting among Units of Pressure

Because people in different industries report pressure using different units, it is often necessary to convert between different units of pressure. Knowing the equivalent unit values makes the conversion straightforward. For example, suppose that the atmospheric pressure in Kenora, Ontario, is measured to be 732 mmHg and you want to know what this pressure is in kilopascals. Because 760 mmHg is equivalent to 101.325 kPa, conversion of 732 mmHg to kPa is

$$732 \text{ mmHg} \times \left(\frac{101.325 \text{ kPa}}{760 \text{ mmHg}} \right) = 97.6 \text{ kPa}.$$

7. What is atmospheric pressure?

8. Explain how Torricelli's apparatus worked.

9. Convert each of the following to the indicated unit.
  a. 3.58 atm to kPa   **c.** 770 mmHg to kPa
  b. 20.5 psi to atm   **d.** 470 torr to Pa

10. If the optimum tire pressure for a bicycle is 3 bar, and your tire pressure gauge is in units of psi, develop a formula that you can use to convert these units.

11. Why must mountain climbers understand the relationship between altitude and atmospheric pressure?

12. To make a birdbath, you fill a 2 L soft-drink bottle with water and invert it in a dish of water. When the level of the water in the dish falls below the level of the water at the rim of the bottle, water flows from the bottle to refill the dish. Explain why this happens.

The Relationship between Gas Pressure and Volume

Meteorologists use weather balloons to carry instrument packages called radiosondes high into the atmosphere. The balloon is partly inflated with helium or hydrogen gas because, at the same pressure, these gases are less dense than air; thus, the balloon rises to high altitudes. As altitude increases, atmospheric pressure decreases and the balloon expands. Eventually, the balloon bursts and a parachute opens to bring the radiosonde safely back to the ground. Thus, the use of weather balloons relies on a relationship between gas pressure and volume.

Activity   11.1   Cartesian Diver

When the pressure exerted on a gas increases, the volume of the gas decreases. Similarly, when the pressure exerted on a gas decreases, the volume of the gas increases. In this activity, you will construct a device called a Cartesian diver to monitor changes in the volume of air as a result of changes in external pressure on the air.

Safety Precautions
- Ensure that the bottle cap is secured tightly before performing this activity.

Materials
- water
- 750 mL or 2 L plastic bottle with cap
- eyedropper

Procedure

1. Fill the bottle to the top with water.

2. Fill the eye dropper approximately half-full with water, and place it in the bottle with its open end down. The eye dropper should float with one end just barely above the surface of the water. If necessary, add water to the eyedropper until it is barely floating. Fasten the lid of the bottle tightly.

3. Squeeze the bottle gently. Observe the eyedropper and its contents. Experiment with the device by varying the degree of compression on the bottle.

Questions

1. When you squeezed the bottle, what happened to the contents of the eyedropper? What happened to the position of the eye dropper? Provide an explanation for the motion you observed, based on the relationship between pressure and volume of a gas.

2. When you released the bottle, what happened to the contents of the eyedropper? What happened to the position of the eyedropper? Provide an explanation for the motion you observed, based on the relationship between pressure and volume of a gas.

3. How might the results of this activity relate to the ability of submarines to surface and dive?

Observations Leading to Boyle's Law

How can meteorologists predict the altitude at which a weather balloon will burst? If the balloon is designed to burst when it reaches three times the volume to which it was inflated before its release, how can they know the altitude at which that will occur? Decisions related to these questions are based on studies of the relationship between the volume of a gas and its pressure that were first published by Irish scientist Robert Boyle in 1662.

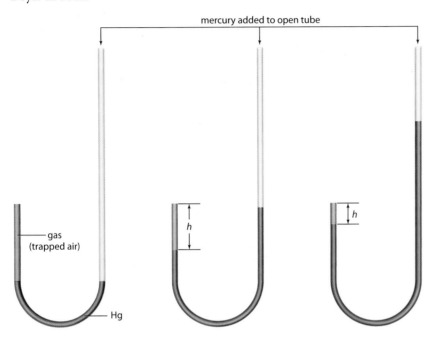

Figure 11.10 In Boyle's experiment, the left side of the U-tube is sealed and mercury has trapped air in the tube. As more mercury is added to the right side of the tube, the mercury increases the pressure on the trapped gas (air), causing its volume to decrease, demonstrated by the decreased height, h, of the column of trapped air in the tube..

Robert Boyle (1627–1691) studied the relationship between the pressure and volume of a gas, while the amount of the gas and the temperature of the gas were kept constant. By making careful measurements of the volume of a trapped gas, he described what happened when the pressure exerted on the gas was increased. **Figure 11.10** shows an apparatus like the one Boyle used. He measured the height of the column of trapped gas and the height of the column of mercury. The height of the mercury column is directly related to the pressure it exerts on the trapped gas. Therefore, Boyle was able to infer the relationship between the pressure on the air and its volume.

Boyle showed that if the temperature and the amount of gas are constant, an increase in external pressure on a gas causes the volume of the gas to decrease by the same factor. For example, at constant temperature, if the pressure of a gas doubles, the volume of the gas decreases by one-half. Similarly, if the pressure on a gas is reduced by one-half, the volume of the gas doubles. These observations led to **Boyle's law**, which states that the volume of a fixed amount of gas at a constant temperature varies inversely with the pressure.

$$V \propto \frac{1}{P}$$

Suggested **Investigation**

Inquiry Investigation 11-A, Studying Boyle's Law

Boyle's law a gas law stating that the volume of a fixed amount of gas at a constant temperature is inversely proportional to the applied (external) pressure on the gas: $V \propto \frac{1}{P}$

Developing a Mathematical Expression of Boyle's Law

A relationship in which one variable increases proportionally as the other variable decreases is known mathematically as an inverse proportion. As Boyle observed, in the case of the volume and pressure of a gas, volume, V, decreases as pressure, P, increases. Thus, the relationship can be expressed as $V \alpha \frac{1}{P}$, where α is the symbol for proportionality. The graphs in **Figure 11.11** help illustrate how the mathematical equation for Boyle's law is developed. Graph A represents volume, V, versus pressure, P, and the graph's shape is typical of an inversely proportional relationship. If the relationship is in fact an inverse proportion, you should get a straight line by plotting volume against the inverse of pressure, $\frac{1}{P}$, which is evident in Graph B.

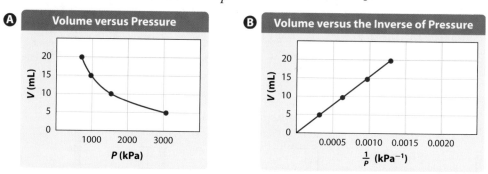

Figure 11.11 The graph for volume versus pressure (**A**) shows an inverse relationship. When you plot volume versus the inverse of pressure (**B**), you get a straight line.

You can use Graph B in **Figure 11.11** to write a linear relationship relating volume and pressure of a gas. From your study of mathematics, you know that the general expression for a straight line is $y = mx + b$, where m is the slope of the line and b is the y-intercept. This information allows you to develop Boyle's law in mathematical form by following a few steps, as shown below.

Begin with the general expression for a straight line.	$y = mx + b$
In Graph B, the y-axis represents volume, V, and the x-axis represents the inverse of pressure, $\frac{1}{P}$. Use these values to rewrite the expression.	$V = m\left(\frac{1}{P}\right) + b$
The symbol m represents the slope of the line and b is the y-intercept. From the graph, you can see that the line passes through the origin. Thus, $b = 0$.	$V = m\left(\frac{1}{P}\right)$
Multiply both sides of the equation by P. This shows that PV is equal to a constant, which is the slope of the line.	$PV = Pm\left(\frac{1}{P}\right)$ $PV = m$
Let P_1V_1 represent pressure and volume at one data point on the graph, and let P_2V_2 represent pressure and volume at a second data point. The product of pressure and volume at each point equals the constant, m.	$P_1V_1 = m$ and $P_2V_2 = m$
Because the products of P_1V_1 and P_2V_2 are equal to the same constant, they are equal to each other.	$P_1V_1 = P_2V_2$

Therefore, the mathematical expression for Boyle's law is

$$P_1V_1 = P_2V_2$$

Remember, this mathematical relationship only applies if the amount of gas and the temperature remain constant. You will use this expression of Boyle's law to understand other gas laws that you will explore in the rest of this chapter and in the next chapter.

The following Sample Problem and Practice Problems will reinforce your understanding of Boyle's law.

Using Boyle's Law to Calculate Volume

Problem

A weather balloon with a volume of 2.00×10^3 L at a pressure of 96.3 kPa rises to an altitude of 1.00×10^3 m, where the atmospheric pressure is measured to be 60.8 kPa. Assuming there is no change in temperature or amount of gas, what is the final volume of the weather balloon?

What Is Required?

You need to find the volume, V_2, after the pressure on the balloon has decreased.

What Is Given?

You know the pressure and volume for the first set of conditions and the pressure for the final set of conditions.

$P_1 = 96.3$ kPa

$V_1 = 2.00 \times 10^3$

$P_2 = 60.8$ kPa

You know the temperature does not change.

Plan Your Strategy	Act on Your Strategy
Pressure and volume are changing, at constant temperature and amount of gas. Therefore, use the equation for Boyle's law.	$P_1V_1 = P_2V_2$
Isolate the variable V_2 by dividing each side of the equation by P_2.	$P_1V_1 = P_2V_2$ $\dfrac{P_1V_1}{P_2} = \dfrac{\cancel{P_2}V_2}{\cancel{P_2}}$ $\dfrac{P_1V_1}{P_2} = V_2$
Substitute numbers and units for the known variables in the formula and solve. Make certain that the same units for pressure are used in the equation.	$V_2 = \dfrac{P_1V_1}{P_2}$ $= \dfrac{(96.3 \ \cancel{kPa})(2.00 \times 10^3 \text{ L})}{60.8 \ \cancel{kPa}}$ $= 3.17 \times 10^3$ L

According to Boyle's law, when the amount and temperature of a gas are constant, there is an inverse relationship between the pressure and volume of a gas: $V \propto \dfrac{1}{P}$

Alternative Solution

Plan Your Strategy	Act on Your Strategy
According to Boyle's law, a decrease in pressure will cause an increase in volume. Determine the ratio of the initial pressure and the final pressure that is greater than 1.	$P_1 = 96.3$ kPa $P_2 = 60.8$ kPa pressure ratio > 1 is $\dfrac{96.3 \text{ kPa}}{60.8 \text{ kPa}}$
To find the final volume, multiply the initial volume of the balloon by the ratio of the two pressures that is greater than 1.	$V_2 = V_1 \times$ pressure ratio $= (2.00 \times 10^3 \text{ L}) \times \dfrac{96.3 \ \cancel{kPa}}{60.8 \ \cancel{kPa}}$ $= 3.17 \times 10^3$ L

Check Your Solution

The units cancel out to leave the correct unit of volume, L. You would expect the volume to increase when the pressure decreases, which is represented by the value determined.

Note: Assume that the temperature and amount of gas are constant in all of the following problems.

1. 1.00 L of a gas at 1.00 atm pressure is compressed to 0.437 L. What is the new pressure of the gas?

2. A container with a volume of 60.0 mL holds a sample of gas. The gas is at a pressure of 99.5 kPa. If the container is compressed to one-quarter of its volume, what is the pressure of the gas in the container?

3. Atmospheric pressure on the peak of Mount Everest can be as low as 0.20 atm. If the volume of an oxygen tank is 10.0 L, at what pressure must the tank be filled so that the gas inside would occupy a volume of 1.2×10^3 L at this pressure?

4. If a person has 2.0×10^2 mL of trapped intestinal gas at an atmospheric pressure of 0.98 atm, what would the volume of gas be (in litres) at a higher altitude that has an atmospheric pressure of 0.72 atm?

5. Decaying vegetation at the bottom of a pond contains trapped methane gas. 5.5×10^2 mL of gas are released. When the gas rises to the surface, it now occupies 7.0×10^2 mL. If the surface pressure is 101 kPa, what was the pressure at the bottom of the pond?

6. The volume of carbon dioxide in a fire extinguisher is 25.5 L. The pressure of the gas in this can is 260 psi. What is the volume of carbon dioxide released when sprayed if the room pressure is 15 psi?

7. A 50.0 mL sample of hydrogen gas is collected at standard atmospheric pressure. What is the volume of the gas if it is compressed to a pressure of 3.50 atm?

8. A portable air compressor has an air capacity of 15.2 L and an interior pressure of 110 psi. If all the air in the tank is released, what volume will that air occupy at an atmospheric pressure of 102 kPa?

9. A scuba tank with a volume of 10.0 L holds air at a pressure of 1.75×10^4 kPa. What volume of air at an atmospheric pressure of 101 kPa was compressed into the tank if the temperature of the air in the tank is the same as the temperature of the air before it was compressed?

10. An oxygen tank has a volume of 45 L and is pressurized to 1200 psi.
    a. What volume of gas would be released at 765 torr?
    b. If the flow of gas from the tank is 6.5 L per minute, how long would the tank last?

Kinetic Molecular Theory and Boyle's Law

Go to **scienceontario** to find out more

Pressure on the walls of a gas-filled container is caused by collisions of gas molecules with the walls. Each collision of a gas molecule exerts a force on the wall. The average force exerted by all the gas molecules divided by the surface area of the container is equivalent to the pressure on the walls of the container. Examine **Figure 11.12** to see what happens when you change the external pressure on the gas. The containers have pistons that will move until the external pressure and the internal pressure are equal. If you increase the external pressure, the piston will move down, reducing the volume available to the gas molecules. The gas molecules are now closer together and collide with one another and the walls of the container more often. As the number of collisions over time increases, the average force exerted by all the molecules increases; thus, the gas pressure increases. If the temperature remains constant and no gas escapes or enters, the decrease in the volume of the container will be inversely proportional to the increase in the gas pressure.

Figure 11.12 The kinetic molecular theory can explain the relationship between pressure and volume. (d_1 and d_2 represent average distances of molecules from the container wall.)

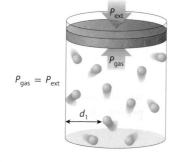
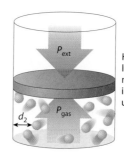

$P_{gas} = P_{ext}$

P_{ext} increases, T and n fixed

Higher P_{ext} causes lower V, which causes more collisions, increasing the pressure until $P_{gas} = P_{ext}$

Section Summary

- Atmospheric pressure is the force that a column of air exerts on Earth's surface divided by the area of Earth's surface at the base of that column.

- Atmospheric pressure decreases as altitude increases.

- Boyle's law states that the volume of gas is inversely proportional to the external pressure exerted on the gas when the temperature and amount of gas are constant. The equation for Boyle's law is $P_1V_1 = P_2V_2$.

- The relationship between pressure and volume of a gas can be explained using the kinetic molecular theory. As the external pressure on a gas increases, the volume of the gas decreases. As the volume decreases, the gas molecules become closer together, causing the frequency of collision of the molecules to increase, thus increasing the gas pressure.

Review Questions

1. **K/U** Describe how Torricelli's studies demonstrated the existence of atmospheric pressure. Include an explanation of the apparatus he used in his investigations.

2. **K/U** What is the relationship between atmospheric pressure and altitude?

3. **A** Fluid moves up a drinking straw, against gravity.
   a. Explain how fluid from a glass can rise in the straw, without anyone applying suction to drink from it.
   b. Would the use of a drinking straw be affected by changes in altitude? Explain.

4. **K/U** Explain why people who climb high mountains commonly carry bottled oxygen with them.

5. **T/I** For each of the following, determine which measurement is the higher pressure.
   a. 1.25 atm or 101.325 kPa
   b. 1.5 bar or 740 mmHg
   c. 1 bar or 105 kPa
   d. 800 mmHg or 1.25 atm

6. **T/I** A student collects hydrogen gas in a balloon fitted over the top of an Erlenmeyer flask. She records the atmospheric pressure as 98.5 kPa. Later she notices that the volume of the balloon has noticeably decreased. Initially, she hypothesizes that some of the hydrogen gas has escaped from the balloon. What data would you advise her to collect in order to confirm or refute her hypothesis? Explain your reasoning.

7. **A** One popular demonstration of gas behaviour involves putting a marshmallow in a flask and then reducing the air pressure in the flask. The marshmallow quickly swells up. How can you explain this observation?

8. **A** When scuba divers are rising after a dive, why is it important that they do not hold their breath?

9. **K/U** Describe how Robert Boyle investigated the relationship between the pressure and volume of a gas.

10. **C** A student performs an investigation to study the relationship between the pressure and volume of a gas. The experimental data are represented in the graph.

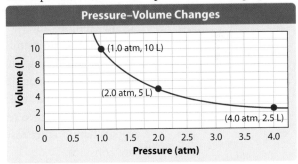

Pressure–Volume Changes

(1.0 atm, 10 L)
(2.0 atm, 5 L)
(4.0 atm, 2.5 L)

Volume (L) / Pressure (atm)

   a. Do the data support Boyle's law? Explain.
   b. How could the data be plotted so that the graph is a straight line? Draw such a graph, including x- and y-axis labels.

11. **T/I** The volume of a gas at 75 kPa is 4.0 L. What is the volume of the gas if the pressure increases to 100.0 kPa? Assume the temperature and amount of gas are constant.

12. **T/I** An air compressor tank has a volume of 60.0 L. The air from this compressor was released and found to be 2.50×10^2 L at a room air pressure of 100.0 kPa. What was the pressure of the air in the compressor tank? Assume the temperature and amount of gas are constant.

13. **T/I** A 20 L tank is filled with helium gas at a pressure of 10 000 kPa. How many balloons, each with a volume of 2.0 L, can be filled to a pressure of 100 kPa? Assume the temperature and amount of gas are constant.

14. **C** Predict what would occur if a 3 L helium balloon was taken underwater to a depth of 30 m where the pressure is 3 atm. Using kinetic molecular theory and a series of diagrams, explain any changes in volume. (Assume there is no loss of helium and the water does not change temperature.)

Gases and Temperature Changes

Key Terms

absolute zero

Charles's law

Gay-Lussac's law

So far, you have studied gas laws in terms of a fixed amount of gas at a constant temperature. However, you know that gas behaviour is affected by changes in temperature as well as changes in pressure. Scientific inquiries conducted in the late 1700s and early 1800s demonstrated connections between the temperature of a gas and its pressure and volume. These inquiries led to the development of two gas laws that complement Boyle's law. One of these laws describes the relationship between the temperature and volume of a gas at constant pressure. The other law describes the relationship between temperature and pressure when the volume of the gas is fixed.

The Relationship between Gas Volume and Temperature

During the last two decades of the 1700s, ballooning became a popular pursuit among inventors and scientists in France. Both the hot-air balloon and the hydrogen balloon were invented during this time. Two prominent French scientists, Jacques Charles (1746–1823) and Joseph Louis Gay-Lussac (1778–1850), were especially interested in ballooning. These scientists realized that there was a connection between the behaviour of the balloons and the properties and behaviour of gases in general.

Working independently, both Charles and Gay-Lussac discovered the same principle: As long as the amount of gas and the pressure on the gas are constant for a specific experiment, there is an increase in volume of a gas with an increase in temperature. Thus, the volume of a gas is proportional to the temperature of a gas. You can see this idea represented in **Figure 11.13**. The graphs show that plotting volume versus temperature results in a straight-line graph, when pressure and amount of gas are constant.

These graphs also show another consistent result that Charles and Gay-Lussac obtained in their studies. When linear plots of volume versus temperature are extrapolated to zero volume, all the lines converge at one value of temperature: −273.15°C. Using more recent technology to collect data than was available to either Charles or Gay-Lussac, scientists have found that the temperature for zero volume of a gas is, in fact, −273.15°C. In actual circumstances, no real gas can have a volume of zero, but experimental data do in fact show that volume approaches zero as temperature approaches −273.15°C.

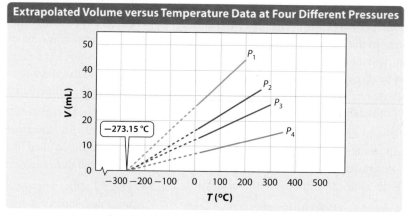

Figure 11.13 These graphs represent four experiments in which the same amount of gas was used and data were collected at four different pressures (P_1 to P_4). The solid portions of the lines represent the temperatures at which data were taken. The dashed portions represent the extrapolation of the volume versus temperature plots. All of the plots intersect at −273.15°C.

The Kelvin Temperature Scale and Absolute Zero

In 1848, twenty-five years after the death of Jacques Charles, Scottish physicist Lord Kelvin (William Thomson, 1824–1907) interpreted the significance of the extrapolated temperature at zero volume of a gas. Kelvin suggested that −273.15°C was the lowest possible temperature, or **absolute zero**. He then established a new temperature scale based on absolute zero as the starting point on the scale. The temperature scale was named the Kelvin scale in his honour.

Figure 11.14 compares the Celsius and Kelvin temperature scales. The name of a unit in the Kelvin scale is the kelvin (K). The size of the kelvin is the same as the size of a Celsius degree, but the term "degree" is not used when reporting temperatures on the Kelvin scale. As well, the starting points for these two temperature scales are different. Notice that there are no negative values on the Kelvin scale. What would happen if you tried to calculate a temperature that is twice as warm as −5°C? Mathematically, the answer would be −10°C, but this is a *colder* temperature. When mathematical manipulations are involved in studying gas behaviour, you need to convert temperatures from the Celsius scale to the Kelvin scale.

> **absolute zero** the lowest theoretical temperature, equivalent to −273.15°C; the temperature at which the volume of a gas approaches zero

For converting Celsius to kelvin: K = °C + 273.15
For converting kelvin to Celsius: °C = K − 273.15

The rounded-off value of 273 is often used as the conversion factor relating K and °C.

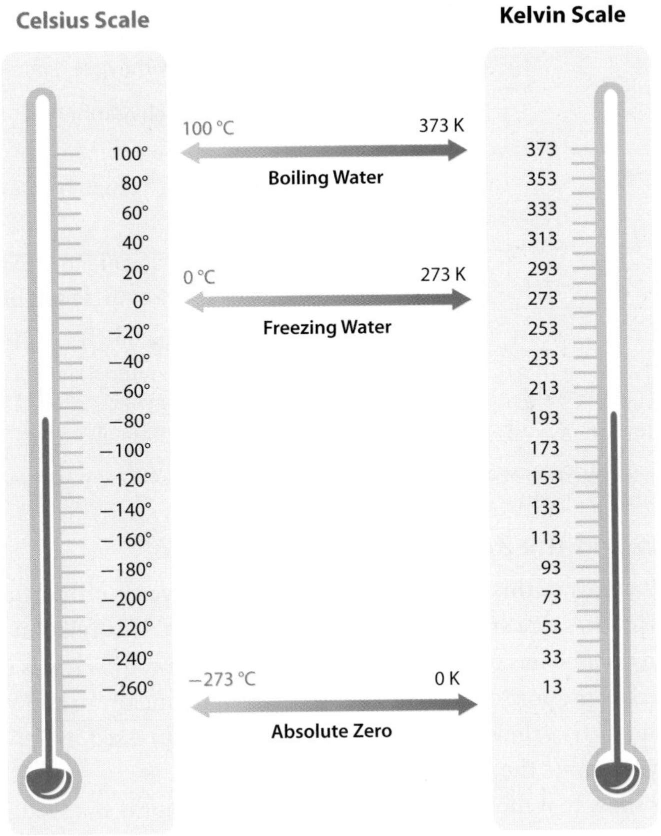

Figure 11.14 There are 273 temperature units between absolute zero and the freezing temperature of water on the Celsius and Kelvin scales. There are also 100 temperature units between the freezing and boiling temperatures of water on both scales.

Apply *If you double a Celsius temperature, by how much does the Kelvin temperature increase?*

13. What is the relationship between the temperature and volume of a gas at constant pressure and amount?

14. What is absolute zero, and what is its significance?

15. Examine the graph in **Figure 11.13**. What do all the graph lines have in common?

16. Make the following temperature conversions.
 a. 27.3°C to K   **c.** 373.2 K to °C
 b. −25°C to K   **d.** 23.5 K to °C

17. Why is it necessary to keep the pressure of a gas constant when studying the relationship between temperature and volume of a gas?

18. A teacher pours liquid nitrogen at a temperature of 77 K over a balloon. Predict the changes that would occur to the balloon.

Activity 11.2 Analyzing the Temperature-Volume Relationship of a Gas

In this activity, you will use data from the table below and the graph that you construct from them to analyze the relationship between the temperature of a gas and its volume and to infer the importance of the Kelvin temperature scale.

Volume versus Temperature Data

Temperature (°C)	Temperature (K)	Volume (cm³)	$\dfrac{\text{Volume (cm}^3)}{\text{Temperature (°C)}}$	$\dfrac{\text{Volume (cm}^3)}{\text{Temperature (K)}}$
8.0		29.5		
20.0		30.8		
30.0		32.1		
40.0		32.9		
50.0		33.9		
60.0		35.0		
70.0		36.1		

Materials
- graph paper
- ruler

Procedure

1. Copy and complete the data table. For the second column, you must calculate the Kelvin equivalent. For the last two columns, you must calculate the quotient of volume divided by temperature.

2. Draw one graph using the data from columns 1 and 3. Draw a second graph using the data from columns 2 and 3.

Questions

1. Use a Venn diagram to describe how the two graphs that you drew are similar and how they are different.

2. What is the x-intercept on each graph? What does each represent?

3. What is the y-intercept on each graph? What does each represent?

4. How do the values of $\frac{V}{T}$ (°C) compare to the values of $\frac{V}{T}$ (K)? Explain the significance of these sets of data.

5. Based on the data in this activity, what relationship seems to exist between the volume and temperature of a gas, when pressure and amount of gas remain constant? How is that relationship affected by the temperature scale that is used?

Charles's Law and the Kelvin Temperature Scale

Charles's law a gas law stating that the volume of a fixed amount of gas at a constant pressure is directly proportional to the Kelvin temperature of the gas: $V \propto T$

You learned at the start of this section that the volume of a gas is proportional to its temperature, when pressure and amount of gas are constant. This relationship between temperature and volume has become known as **Charles's law**. This law is often stated in terms of a *directly* proportional relationship between temperature and volume. This statement only holds true, however, if the temperature is expressed in Kelvin units. To understand why, examine the graphs in **Figure 11.5**.

Both graphs show that the plot of temperature versus volume is a straight line, but notice that Graph A—in which temperature is in degrees Celsius—does *not* show a direct proportion. The graph of the line does not pass through the origin, and doubling the temperature does not double the volume. Graph B *does* show a direct proportion; temperature is in kelvins, and the graph of the line passes through the origin. A temperature of 0 K corresponds to 0 mL. Doubling the temperature doubles the volume.

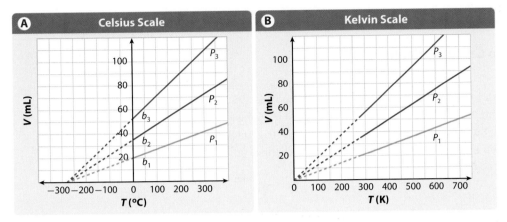

Figure 11.15 **(A)** Using the Celsius temperature scale produces straight-line graphs that have three different y-intercepts. **(B)** Using the Kelvin temperature scale produces straight-line graphs with the same y-intercept.

Analyze *Identify the x- and y-intercepts for each graph.*

Suggested**Investigation**

Inquiry Investigation 11-B, Studying Charles's Law

Developing a Mathematical Expression of Charles's Law

If temperature is expressed in Kelvin, the mathematical expression of Charles's law can be derived easily. The steps are similar to the method you used to create the mathematical expression for Boyle's law.

Begin with the general expression for a straight line.	$y = mx + b$
In a graph of volume versus temperature, let the y-axis represent volume, V, and let the x-axis represent temperature, T, in kelvins. Use these values to rewrite the expression.	$V = mT + b$
The symbol m represents the slope of the line and b is the y-intercept. Because temperature is expressed in kelvins, the line representing volume and temperature passes through the origin. Thus, $b = 0$.	$V = mT$
Divide both sides of the equation by T. This shows that $\frac{V}{T}$ is equal to a constant, m, which is the slope of the line.	$\frac{V}{T} = \frac{mT}{T}$ $\frac{V}{T} = m$
Let $\frac{V_1}{T_1}$ represent the volume and temperature at one data point on the graph and let $\frac{V_2}{T_2}$ represent volume and temperature at a second data point. The quotient of volume divided by temperature at each point equals the constant, m.	$\frac{V_1}{T_1} = m$ and $\frac{V_2}{T_2} = m$
Because the quotients of $\frac{V_1}{T_1}$ and $\frac{V_2}{T_2}$ are equal to the same constant, they are equal to each other.	$\frac{V_1}{T_1} = \frac{V_2}{T_2}$

Charles's law is expressed mathematically as

$$\frac{V_1}{T_1} = \frac{V_2}{T_2}$$

This relationship only applies if the pressure and amount of gas are kept constant and if temperature is in kelvin units. The following Sample Problems and Practice Problems will reinforce your understanding of Charles's law.

Using Charles's Law to Calculate Volume of a Gas

Problem

A balloon inflated with air in a room in which the temperature of the air is 295 K has a volume of 650 mL. The balloon is put into a refrigerator at 277 K and left long enough for the air in the balloon to reach the same temperature as the air in the refrigerator. Predict the volume of the balloon, assuming that the amount of air has not changed and the air pressure in the room and in the refrigerator are the same.

What Is Required?

You need to find the volume, V_2, of the balloon after it has been cooled to 277 K.

What Is Given?

You know the volume and temperature of the air sample for the first set of conditions and the temperature for the second set of conditions:

$V_1 = 650$ mL

$T_1 = 295$ K

$T_2 = 277$ K

Plan Your Strategy	Act on Your Strategy
Temperature and volume are changing at constant pressure and amount of gas. Therefore, use the equation for Charles's law.	$\dfrac{V_1}{T_1} = \dfrac{V_2}{T_2}$
Isolate the variable V_2 by multiplying each side of the equation by T_2 and rearranging the equation.	$\dfrac{V_1}{T_1}(T_2) = \dfrac{V_2}{\cancel{T_2}}(\cancel{T_2})$ $\dfrac{V_1 T_2}{T_1} = V_2$
Substitute numbers and units for the known variables in the formula and solve. Since the lowest number of significant digits in values in the question is two, the final volume is reported to two significant digits.	$V_2 = \dfrac{V_1 T_2}{T_1}$ $= \dfrac{(650 \text{ mL})(277 \cancel{K})}{(295 \cancel{K})}$ $= 610$ mL

According to Charles's law, when the amount and pressure of a gas are constant, there is a directly proportional relationship between the volume of the gas and its Kelvin temperature:

$V \propto T$

Alternative Solution

Plan Your Strategy	Act on Your Strategy
According to Charles's law, a decrease in temperature will cause a decrease in volume. Determine the ratio of the initial temperature and the final temperature that is less than 1.	$T_2 = 277$ K $T_1 = 295$ K temperature ratio < 1 is $\dfrac{277 \text{ K}}{295 \text{ K}}$
To find the final volume, multiply the initial volume of the balloon by the ratio of the two Kelvin temperatures that is less than 1.	$V_2 = V_1 \times$ temperature ratio $= (650 \text{ mL}) \times \dfrac{277 \cancel{K}}{295 \cancel{K}}$ $= 610$ mL

Check Your Solution

Volume units remain when the other units cancel out. Because the temperature decreases, the volume is expected to decrease. The answer represents a lower value for the volume.

Using Charles's Law to Calculate Temperature of a Gas

Problem
A birthday balloon is filled to a volume of 1.50 L of helium gas in an air-conditioned room at 294 K. The balloon is then taken outdoors on a warm sunny day and left to float as a decoration. The volume of the balloon expands to 1.55 L. Assuming that the pressure and amount of gas remain constant, what is the air temperature outdoors in kelvins?

What Is Required?
You need to find the outdoor air temperature, T_2, in K.

What Is Given?
You know the volume and temperature of the air sample for the initial set of conditions and the volume for the final set of conditions:

$V_1 = 1.50$ L

$T_1 = 294$ K

$V_2 = 1.55$ L

Plan Your Strategy	Act on Your Strategy
Temperature and volume are changing at constant pressure and amount of gas. Therefore, use the equation for Charles's law.	$\dfrac{V_1}{T_1} = \dfrac{V_2}{T_2}$
Isolate the variable T_2 by multiplying each side of the equation first by T_2 and then by $\dfrac{T_1}{V_1}$.	$\left(\dfrac{V_1}{T_1}\right)(T_2) = \left(\dfrac{V_2}{T_2}\right)(T_2)$ $\left(\dfrac{V_1}{T_1}\right)\left(\dfrac{T_1}{V_1}\right)T_2 = V_2\left(\dfrac{T_1}{V_1}\right)$ $T_2 = \dfrac{V_2 T_1}{V_1}$
Substitute numbers and units for the known variables in the formula and solve. Since the number of significant digits in values in the question is three, the final volume is reported to three significant digits.	$T_2 = \dfrac{V_2 T_1}{V_1}$ $= \dfrac{(1.55\text{ L})(294\text{ K})}{(1.50\text{ L})}$ $= 304$ K

Alternative Solution

Plan Your Strategy	Act on Your Strategy
According to Charles's law, an increase in temperature will cause an increase in volume. Determine the ratio of the initial volume and the final volume that is greater than 1.	$V_1 = 1.50$ L $V_2 = 1.55$ L volume ratio > 1 is $\dfrac{1.55\text{ L}}{1.50\text{ L}}$
To find the final temperature, multiply the initial temperature of the balloon by the ratio of the two volumes that is greater than 1.	$T_2 = T_1 \times$ volume ratio $= (294\text{ K}) \times \dfrac{1.55\text{ L}}{1.50\text{ L}}$ $= 304$ K

Check Your Solution
The unit for the answer is kelvins. When the other units cancel out, kelvins remain. Because the volume of the balloon had increased, you would expect that the temperature had increased. The answer represents an increase in temperature.

Note: Assume that the pressure and amount of gas are constant in all of the problems except question 20.

11. A gas has a volume of 6.0 L at a temperature of 250 K. What volume will the gas have at 450 K?

12. A syringe is filled with 30.0 mL of air at 298.15 K. If the temperature is raised to 353.25 K, what volume will the syringe indicate?

13. The temperature of a 2.25 L sample of gas decreases from 35.0°C to 20.0°C. What is the new volume?

14. A balloon is inflated with air in a room in which the air temperature is 27°C. When the balloon is placed in a freezer at −20.0°C, the volume is 80.0 L. What was the original volume of the balloon?

15. At a summer outdoor air temperature of 30.0°C, a particular size of bicycle tire has an interior volume of 685 cm^3. The bicycle has been left outside in the winter and the outdoor air temperature drops to −25.0°C. Assuming the tire had been filled with air in the summer, to what volume would the tire be reduced at the winter air temperature?

16. At 275 K, a gas has a volume of 25.5 mL. What is its temperature if its volume increases to 50.0 mL?

17. A sealed syringe contains 37.0 mL of trapped air. The temperature of the air in the syringe is 295 K. The sun shines on the syringe, causing the temperature of the air inside it to increase. If the volume increases to 38.6 mL, what is the new temperature of the air in the syringe?

18. A beach ball is inflated to a volume of 25 L of air in the cool of the morning at 15°C. During the afternoon, the volume changes to 26 L. What was the Celsius air temperature in the afternoon?

19. The volume of a 1.50 L balloon at room temperature increases by 25.0 percent when placed in a hot-water bath. How does the temperature of the water bath compare with room temperature?

20. Compressed gases can be condensed when they are cooled. A 5.00×10^2 mL sample of carbon dioxide gas at room temperature (assume 25.0°C) is compressed by a factor of four, and then is cooled so that its volume is reduced to 25.0 mL. What must the final temperature be (in °C)? (**Hint:** Use both Boyle's law and Charles's law to answer the question.)

Kinetic Molecular Theory and Charles's Law

Go to **scienceontario** to find out more

Applying the kinetic molecular theory to Charles's law is shown in **Figure 11.16**. The Kelvin temperature of a gas is directly proportional to the average kinetic energy of the gas molecules. An object's kinetic energy is related to its speed ($E_k = \frac{1}{2} mv^2$). As the temperature of a gas increases, the molecules move at higher speeds. As a result, they collide with the walls of the container and with one another more frequently and with greater force. Therefore, they exert a greater pressure on the walls of the container. If, however, the external pressure on the gas stays the same, the gas pressure causes the container to increase in size. As the volume of the container gets larger, the gas molecules must travel farther to collide with the walls of the container and with one another. As the collisions become less frequent, the pressure drops. The process continues until the pressure inside the container is once again equal to the external pressure.

Figure 11.16 When the temperature of a gas increases, the speed of the gas molecules increases. The gas molecules collide with the walls of the container more frequently, thus increasing the pressure. If the external pressure remains the same, the gas pushes the piston up and increases the volume of the container.

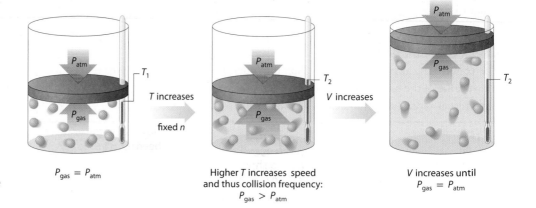

$P_{gas} = P_{atm}$

T increases

fixed n

Higher T increases speed and thus collision frequency:

$P_{gas} > P_{atm}$

V increases

V increases until

$P_{gas} = P_{atm}$

Gay-Lussac's Law: The Relationship between Temperature and Pressure

Most gases, such as those used for many industrial, commercial, and medical purposes, are stored in containers that have a fixed volume. Compressed gases such as oxygen and nitrogen are usually stored in rigid cylinders and tanks, often with gauges such as the one shown in **Figure 11.17**. You know that temperature is a measure of the average kinetic energy of the molecules making up a substance. If the temperature of a gas increases, but the volume of its container cannot increase, what happens to the pressure of the gas inside?

Extending the work of Charles, Louis Gay-Lussac discovered the relationship between temperature and pressure acting on a fixed volume of a gas. As the temperature of the gas increases, so does the pressure. In fact, when temperature is expressed in kelvins, the relationship between temperature and pressure is directly proportional. This relationship is now referred to as **Gay-Lussac's law**.

> **Gay-Lussac's law**
> a gas law stating that the pressure of a fixed amount of gas at a constant volume is directly proportional to its Kelvin temperature: $P \propto T$

Figure 11.17 Many gases are stored at high pressure in tanks such as this one. These tanks typically have gauges that monitor the pressure of the gas.

Explain *Why do compressed gas cylinders often have a pressure relief valve, which causes the release of gas when the temperature increases?*

Activity 11.3 Egg in a Bottle (Teacher Demonstration)

This activity shows the relationship between gas pressure and temperature when gas volume is constant. Your teacher will demonstrate it in order to avoid wasting food.

Materials
- peeled, hard-boiled egg
- heat-proof glass bottle (neck opening should be just a little too small for the peeled hard-boiled egg to go through)
- ice-water bath
- hot-water bath

Procedure
1. Observe as a cooled, peeled, hard-boiled egg is placed in the opening of a bottle. The egg should have its tapered-side down and be sitting on the rim of the bottle.

2. The bottle will be placed in a hot-water bath for 5 to 10 min and then transferred to a cold-water bath. Observe what happens to the egg.

Questions
1. Describe what happened to the egg when the bottle was transferred to the cold-water bath.

2. Describe the change in pressure and temperature of the air in the bottle when the bottle is in the hot-water bath and then in the cold-water bath.

3. Explain why the result you observed occurred.

4. How would the result change if a flexible container had been used instead of a bottle?

Go to **scienceontario**
to find out more

Developing a Mathematical Expression for Gay-Lussac's Law

Gay-Lussac's law states that the pressure of a fixed amount of gas, at constant volume, is directly proportional to its Kelvin temperature. The relationship can be expressed as $P \propto T$, where T is given in kelvins.

Using the general expression for a straight line ($y = mx + b$) and applying the same mathematical treatment used for Charles's law, a mathematical expression for Gay-Lussac's law is

$$\frac{P_1}{T_1} = \frac{P_2}{T_2}$$

For this equation, P_1 and T_1 represent the initial pressure and temperature conditions and P_2 and T_2 represent the final pressure and temperature conditions. The relationship applies as long as the volume and amount of a gas are constant and the temperature is expressed in kelvins.

The following Sample Problem and Practice Problems will reinforce your understanding of Gay-Lussac's law.

Sample Problem

Using Gay-Lussac's Law To Calculate Pressure of a Gas

Problem

The pressure of the oxygen gas inside a canister with a fixed volume is 5.0 atm at 298 K. What is the pressure of the oxygen gas inside the canister if the temperature changes to 263 K? Assume the amount of gas remains constant.

What is Required?

You need to find the new pressure, P_2, of the oxygen gas inside the canister resulting from a decrease in temperature:

What is Given?

You know the initial pressure of the oxygen gas in the canister, as well as the initial and final air temperatures:

$P_1 = 5.0$ atm

$T_1 = 298$ K

$T_2 = 263$ K

Plan Your Strategy	Act on Your Strategy
Temperature and pressure are changing at constant volume and amount of gas. Therefore, use the equation for Gay-Lussac's law.	$\dfrac{P_1}{T_1} = \dfrac{P_2}{T_2}$
Isolate the variable P_2 by multiplying each side of the equation by T_2	$\dfrac{P_1}{T_1}(T_2) = \dfrac{P_2}{\cancel{T_2}}(\cancel{T_2})$ $\dfrac{P_1 T_2}{T_1} = P_2$
Substitute numbers and units for the known variables in the formula and solve. Since the lowest number of significant digits in values in the question is two, the final pressure is reported to two significant digits.	$P_2 = \dfrac{P_1 T_2}{T_1}$ $= \dfrac{(5.0 \text{ atm})(263 \text{ K})}{298 \text{ K}}$ $= 4.4 \text{ atm}$

According to Gay-Lussac's law, when the amount and volume of a gas are constant, there is a directly proportional relationship between the pressure of the gas and its Kelvin temperature:

$$P \propto T$$

Alternative Solution

Plan Your Strategy	Act on Your Strategy
According to Gay-Lussac's law, a decrease in temperature will cause a decrease in pressure. Determine the ratio of the initial temperature and the final temperature that is less than 1.	$T_1 = 298$ K $T_2 = 263$ K temperature ratio < 1 is $\dfrac{263\ \text{K}}{298\ \text{K}}$
To find the final pressure, multiply the initial pressure of the gas by the ratio of the two temperatures that is less than 1.	$P_2 = P_1 \times$ temperature ratio $= (5.0\ \text{atm}) \times \dfrac{263\ \cancel{\text{K}}}{298\ \cancel{\text{K}}}$ $= 4.4$ atm

Check Your Solution
The result shows the expected decrease in pressure. With kelvin units cancelling out, the remaining unit, atm, is a pressure unit.

Practice Problems

Note: Assume that the volume and amount of gas are constant in all of the following problems.

21. A gas is at 105 kPa and 300.0 K. What is the pressure of the gas at 120.0 K?

22. The pressure of a gas in a sealed canister is 350.0 kPa at a room temperature of 298 K. The canister is placed in a refrigerator and the temperature of the gas is reduced to 278 K. What is the new pressure of the gas in the canister?

23. A propane barbeque tank is filled in the winter at −15.0°C to a pressure of 2500 kPa. What will the pressure of the propane become in the summer when the air temperature rises to 20.0°C?

24. A rubber automobile tire contains air at a pressure of 370 kPa at 15.0°C. As the tire heats up, the temperature of the air inside the tire rises to 60.0°C. What would the new pressure in the tire be?

25. A partially filled aerosol can has an internal pressure of 14.8 psi when the temperature is 20.0°C.
  a. What would the pressure in the can be, in kPa, if it were placed into an incinerator for disposal, which would have the effect of raising the temperature inside the can to 1800°C?

  b. Approximately how many times higher is that new pressure compared to standard atmospheric pressure?

26. A sealed can of gas is left near a heater, which causes the pressure of the gas to increase to 1.4 atm. What was the original pressure of the gas if its temperature change was from 20.0°C to 90.0°C?

27. Helium gas in a 2.00 L cylinder has a pressure of 1.12 atm. When the temperature is changed to 310.0 K, that same gas sample has a pressure of 2.56 atm. What was the initial temperature of the gas in the cylinder?

28. A sample of neon gas is contained in a bulb at 150°C and 350 kPa. If the pressure drops to 103 kPa, find the new temperature, in °C.

29. A storage tank is designed to hold a fixed volume of butane gas at 2.00×10^2 kPa and 39.0°C. To prevent dangerous pressure buildup, the tank has a relief valve that opens at 3.50×10^2 kPa. At what Celsius temperature does the valve open?

30. If a gas sample has a pressure of 30.7 kPa at 0.00°C, by how many degrees Celsius does the temperature have to increase to cause the pressure to double?

CHEMISTRY Connections

Health Under Pressure

You live, work, and play in air that is generally about 1 atm in pressure and contains 21% oxygen. Have you ever wondered what might happen if the pressure and the oxygen content of the air were greater? Would you recover from illness or injury more quickly? These questions are at the heart of hyperbaric medicine.

HYPERBARIC MEDICINE The prefix *hyper-* means above or excessive, and a bar is a unit of pressure equal to 100 kPa, which is roughly normal atmospheric pressure. Thus, the term hyperbaric refers to pressure that is greater than normal. Patients receiving hyperbaric therapy are exposed to pressures greater than the pressure of the atmosphere at sea level.

THE OXYGEN CONNECTION Greater pressure is most often combined with an increase in the concentration of oxygen that a patient receives. The phrase *hyperbaric oxygen therapy* (HBOT) refers to treatment with 100% oxygen. A chamber that might be used for HBOT is shown below. Inside the hyperbaric chamber, pressures can reach five to six times normal atmospheric pressure. At hyperbaric therapy centres across the country, HBOT is used to treat a wide range of conditions, including burns, decompression sickness, slow-healing wounds, anemia, and some infections.

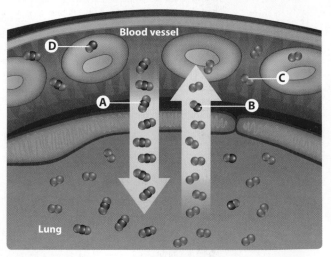

Gases are exchanged between the lungs and the circulatory system.

CARBON MONOXIDE POISONING Use the diagram above to help you understand how HBOT aids in the treatment of carbon monoxide poisoning.

NORMAL GAS EXCHANGE Oxygen, $O_2(g)$, moves from the lungs to the blood and binds to the hemoglobin in red blood cells. Carbon dioxide, $CO_2(g)$, is released, as shown by **(A)**.

ABNORMAL GAS EXCHANGE If carbon monoxide, $CO(g)$, enters the blood, as shown by **(B)**, it, instead of oxygen, binds to the hemoglobin and enters blood cells **(D)**. Cells in the body begin to die from oxygen deprivation.

OXYGEN IN BLOOD PLASMA In addition to the oxygen carried by hemoglobin, oxygen is dissolved in the blood plasma, as shown by **(C)**. HBOT increases the concentration of dissolved oxygen to an amount that can sustain the body.

ELIMINATING CARBON MONOXIDE Pressurized oxygen also helps to remove any carbon monoxide

Connect to the Environment

The engines of idling cars emit carbon monoxide. Write an email to a friend or relative who likes to let his or her car idle for many minutes in the winter to warm it up. Explain why carbon monoxide is a dangerous pollutant. Do research to strengthen your argument. Find out the connection between ground-level ozone and carbon monoxide and why that connection might harm the environment.

Section Summary

- The Kelvin temperature scale is based on a temperature of 0 K at absolute zero. The size of a kelvin on the Kelvin scale is the same as the size of a degree on the Celsius scale.

- Charles's law states that the volume of a fixed amount of gas at a constant pressure is directly proportional to the Kelvin temperature of the gas: $V \propto T$.

- The mathematical equation for Charles's law is $\frac{V_1}{T_1} = \frac{V_2}{T_2}$, where T is temperature in kelvin.

- Gay-Lussac's law states that the pressure of a fixed amount of gas at a constant volume is directly proportional to the Kelvin temperature of the gas: $P \propto T$

- The mathematical equation for Gay-Lussac's law is $\frac{P_1}{T_1} = \frac{P_2}{T_2}$, where T is temperature in kelvin.

Review Questions

1. **K/U** Explain why it is theoretically possible for an ideal gas to have a volume of zero but it is not possible for a real gas to have a volume of zero.

2. **C** Using a table or a graphic organizer such as a Venn diagram, summarize the similarities and differences between the Celsius and Kelvin scales.

3. **K/U** Convert each of the following to K or °C.
   a. 37.8°C
   b. 122.4°C
   c. −40.0°C
   d. 275 K
   e. 173.6 K
   f. 873 K

4. **K/U** If the amount and pressure of a gas remain constant, what happens to the volume of a gas as the temperature decreases? Name the law that describes this relationship.

5. **C** Draw a graph that represents the behaviour of a gas according to Charles's law. Provide an explanation for why your graph represents that particular law.

6. **A** Describe an example from everyday life that illustrates the principle of Charles's law.

7. **C** Imagine that a classmate was absent when you studied Charles's law in class. Explain to this classmate why it is necessary to use Kelvin temperatures and not Celsius temperatures when doing problems involving this law.

8. **T/I** A sample of gas has a volume of 0.4 L at 293 K. What is the volume of the gas at 523 K, assuming that the pressure and amount of gas remain constant?

9. **T/I** A sample of gas is originally at 30°C. What temperature increase in degrees Celsius will produce a 10 percent increase in the volume of the gas? Assume the pressure and amount of gas are constant.

10. **C** An experiment was conducted to investigate Gay-Lussac's law. The pressure was measured with an increase in temperature. The volume and amount of gas were kept constant. The data from the experiment are listed in the table below. Construct a graph using the data in the table to determine if Gay-Lussac's law is validated. Based on the graph, explain why the law is or is not supported by the data.

Temperature and Pressure of a Gas

Temperature Reading (°C)	Pressure Reading (kPa)
10.0	101
20.0	105
30.0	109
50.0	116
60.0	119
75.0	125
100.0	134

11. **T/I** At 277.0 K, a gas has a pressure of 99.5 kPa. What is the pressure of the gas at 210.0 K, if the volume and amount of gas are constant?

12. **A** In a fire, gas cylinders containing combustible gases are at risk of exploding, because the amount of gas and volume are fixed. If a tank of propane gas has an internal pressure of 1500 kPa at 20.0°C, what will the internal tank pressure become at 1000.0°C?

13. **T/I** A squash ball has an internal air pressure of 14.8 psi at a temperature of 18.0°C. During play, the internal temperature of the air inside the ball rises to 70.0°C. Assuming that the volume and amount of air inside the ball do not change, what is the new air pressure inside the ball?

14. **A** Assume you are in charge of designing a container that would hold and deliver a combustible gas at high pressure. Given your knowledge of Gay-Lussac's and Charles's laws, what are some of the features you would incorporate in your design?

Skill Check

Safety Precautions

- Wear a lab apron and safety eyewear during this investigation.

- Be careful when placing the materials that are used for weights. Make sure to centre them on the apparatus so they do not fall.

Materials

- glue (strong)
- 60 mL syringe
- square piece of plexiglass (about 15 to 20 cm on a side)
- scale (with range up to 10 kg)
- rubber stopper with hole
- retort stand
- 3 clamps
- barometer
- weights (such as heavy books) totalling a mass of at least 6 kg

*Go to **Constructing a Graph** in Appendix A for help with drawing graphs.*

Studying Boyle's Law

In this investigation, you will observe the change in the volume of air trapped inside a syringe when you apply pressure to the plunger of the syringe and thus to the trapped air. You will use your data to show the relationship between the pressure and volume of a gas and verify Boyle's law.

Pre-Lab Questions

1. Write a definition of Boyle's law.

2. What is the mathematical equation that represents the relationship between the pressure and volume of a gas, according to Boyle's law?

3. Explain what is meant by an inversely proportional relationship between two variables.

Question

How does the volume of a gas vary with change in pressure, when temperature remains constant?

Prediction

Predict what will happen to the air inside the syringe when you place weights on top of the plunger of the syringe.

Procedure

1. Make a data table similar to the one below to record your data.

Data Table

Atmospheric pressure (kPa): _____ Cross-sectional area of syringe (m²): _____					
Number of Objects	Weight of Added Object (N)	Total Weight on Platform (N)	Total Pressure (kPa) (atmospheric pressure + pressure due to object)	Inverse of Pressure, $\frac{1}{P}$ (1/kPa)	Volume (mL)
1 (plunger and platform)	0				

2. Obtain a 60 mL syringe and measure its internal diameter. Ensure that the plunger is airtight but slides freely. Calculate the radius of the syringe. From the radius, calculate the cross-sectional area of the syringe using the formula $A = \pi r^2$.

3. Glue the plexiglass platform onto the top of the plunger. Be sure to centre the platform.

4. Determine the weight of the plunger-platform assembly. To do this, first determine the mass of the plunger using the scale. Then convert to weight in newtons (N) by using the following equation: weight (N) = mass (kg) × gravity (m/s²). For this calculation, use 9.81 m/s² for gravity.

5. Insert the tip of the syringe into a small hole that has been drilled into a rubber stopper. The hole should be just deep enough to fit the tip of the syringe, but it should *not* penetrate the whole depth of the stopper. The fit must be airtight.

6. Assemble the apparatus as shown in the diagram. Notice that the rubber stopper is placed firmly against the base of the retort stand. The clamps on the stopper and the syringe are tight. However, the uppermost clamp is not touching the plunger. It is in place as a precaution to prevent the plunger from falling. When you insert the plunger into the syringe, trap as much air as possible.

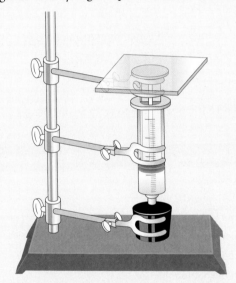

7. Read the current atmospheric pressure from the barometer and record the value. (Atmospheric pressure should not be corrected for altitude.)

8. Record the weight of the plunger-platform apparatus and the volume of air in the syringe with the plunger in place.

9. Calculate the pressure exerted by the plunger-platform apparatus by dividing the weight (force) by the cross-sectional area of the plunger.

10. Calculate the total pressure by adding the atmospheric pressure to the pressure due to the weight of the plunger-platform apparatus. Record the total pressure.

11. Determine the weight of the object (such as a heavy book) that you will be adding to the platform. Record

the weight in your table. Remember that you calculate this using the formula, weight (N) = mass (kg) × 9.81 m/s².

12. Very carefully place the object on the platform. Be sure to centre the object on the platform. Observe and record the volume of the air in the syringe.

13. Calculate the total weight of the platform plus object, the pressure due to the weight of the platform plus object (similar to step 9), and total pressure on the air in the syringe (similar to step 10). Record the data.

14. Repeat steps 11 to 13 with more objects until you have added at least 60 N of weight to the platform.

Analyze and Interpret

1. What is the independent variable in this investigation? What is the dependent variable?

2. Plot a graph of total volume, V (y-axis), versus pressure, P (x-axis).

3. Plot a graph of total volume, V (y-axis), versus the inverse of the pressure, 1/P (x-axis).

4. Which graph, V versus P or V versus 1/P, appears to give the straightest line?

Conclude and Communicate

5. Describe how your data illustrate the relationship between volume and pressure.

6. Did your data support Boyle's law? Explain why or why not.

Extend Further

7. **INQUIRY** Describe the possible sources of error in this investigation. In your description, address the following questions: If the plunger and the rubber stopper had not provided airtight seals, how would this have affected your data? How might a change in temperature of the air in the room you were in have affected your results?

8. **RESEARCH** Torricelli's apparatus represents the first barometer invented and it formed the basis for today's models. Research current models of barometers and some of their applications.

Safety Precautions

- Wear safety goggles and a lab apron during this investigation.

- When unplugging the hot plate, make sure to pull on the plug, not the cord.

- Do not touch potentially hot objects with your bare hands. Use thermal gloves to handle objects that are hot.

- Be very careful when working around the hot water.

Materials

- tap water
- ice
- 1000 mL beaker
- 50 mL graduated cylinder
- two 125 mL dropping bottles with hinged caps
- retort stand
- clamp for dropping bottle
- 600 mL beaker
- hot plate
- 2 liquid thermometers
- 2 thermometer clamps
- ring clamp
- stirring rod

Go to **Constructing a Graph** in **Appendix A** for help with drawing graphs.

Studying Charles's Law

In this investigation, you will collect and use data to show the relationship between the temperature of a gas and its volume and verify Charles's law.

Pre-Lab Questions

1. Write a verbal and mathematical definition of Charles's law.

2. Explain what is meant by a directly proportional relationship between two variables.

Question

How does the volume of a gas vary with change in temperature, when pressure and amount of gas are kept constant?

Prediction

Predict how the volume will change when the temperature of the gas changes.

Procedure

Part A

1. Make a data table similar to the one below to record your data.

Data Table

Values to Measure	Recorded Values
A: room temperature (°C)	
B: temperature of the hot water (°C)	
C: temperature of the hot water (K)	
D: temperature of the cooling water (°C)	
E: temperature of the cooling water (K)	
F: total volume of air in the bottle at higher temperature (mL)	
G: change in volume of air in the bottle (mL)	
H: volume of air at lower temperature (mL)	

2. Measure and record the temperature of the air in the room in line A.

3. Obtain a 1000 mL beaker, and add about 400 mL of tap water. Set the beaker aside for use in step 7.

4. Obtain a 125 mL dropping bottle with a hinged cap. Ensure that the lid is secured tightly to the bottle, with the hinged cap open.

5. Use a retort stand and clamp to suspend the assembled dropping bottle in a 600 mL beaker that is placed on a hot plate, as shown in the photo.

6. Pour water into the beaker to cover about three-quarters of the bottle. Use a thermometer clamp to attach a thermometer to the retort stand and insert the thermometer into the water. The bulb should not touch the beaker bottom.

7. Heat the water to boiling. Then reduce the heat and continue boiling for about 5 min. While waiting, place the beaker with water that you prepared in step 3 beside the ring stand assembly, as shown in the photo. Then clamp a second thermometer to the retort stand, and insert the thermometer into the tap water. Make sure the thermometer bulb does not touch the bottom of the beaker.

8. After the water in the 600 mL beaker has boiled for about 5 min, turn off the hot plate and record the temperature of the boiling water in both °C and K (lines B and C in the data table).

9. Close the hinged cap on the dropping bottle. Then, leaving the bottle attached to its clamp, remove the bottle from the hot water by removing the clamp from the retort stand.

10. With the bottle still attached to the clamp, immerse the bottle in the water in the 1000 mL beaker. Be careful to avoid the thermometer as you immerse the bottle.

11. Using a stirring rod, stir the water in the 1000 mL beaker until the temperature no longer changes. Then record the temperature of the water in both °C and K (lines D and E in data table).

12. Leave the bottle immersed in the tap water for about 3 min. Then detach the bottle from the clamp and completely submerge the bottle in the water with your hand. When the bottle is completely submerged, open the hinged cap and allow water to enter the bottle.

13. Hold the bottle in an inverted position, with the cap still open. Elevate or lower the bottle until the water level in the bottle is even with the water level in the beaker. Close the cap. The air in the bottle is now at atmospheric pressure.

14. Remove the bottle from the water and place it rightside-up on the work surface.

15. The volume of water in the bottle is equal to the change in volume of the air as it cooled from the temperature of boiling water to the temperature of tap water. Use a graduated cylinder to accurately measure the volume of the water in the bottle. Record this value in line G.

16. To determine the starting volume of air in the bottle at the higher temperature, fill the bottle to the top with water. Use the graduated cylinder to measure the volume of the water in the bottle. Record this in line F.

17. Calculate the volume of air in the bottle at the lower temperature by subtracting the change in volume of air in the bottle from the total volume of air in the bottle at the higher temperature. Record this value in line H.

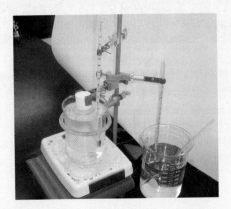

Part B

1. Repeat Part A with a clean, dry dropping bottle and an ice-water bath for step 7. Complete a new data table.

Analyze and Interpret

1. Using the equation for Charles's law, calculate the expected volume of air when cooled in tap water. Compare with the calculated final volume. Provide reasons for any difference between these two values.

2. Construct a graph of volume versus temperature. Use the volume of the gas and temperature in hot water, in tap water, and in ice water. Plot the three data points on the graph and construct a line of best fit.

3. Perform a graphical extrapolation and identify the value of absolute zero, at the x-intercept. What is this temperature in degrees Celsius?

4. Calculate the percentage error in the absolute value of zero, where the accepted value of zero is $-273°C$.

$$\% \text{ error} = \frac{\left|\text{experimental value} - (-273°C)\right|}{\left|-273°C\right|} \times 100\%$$

Conclude and Communicate

5. Based on the data, what is the relationship between temperature and volume?

6. Was Charles's law verified? Explain.

Extend Further

7. **INQUIRY** How would the results be affected if
   a. the bottle was not left in the cooling water bath long enough to reach room temperature?
   b. the bottle was removed from the boiling water bath before the temperature of the gas reached the temperature of the water bath?

8. **RESEARCH** Research how temperatures at or near absolute zero are achieved in the laboratory, and describe the practical applications of examining matter at absolute or near absolute zero.

Chapter 11 | SUMMARY

Section 11.1 | The Properties of Gases

Gases have characteristic properties, which can be explained at the molecular level by factors such as the independent movement of gas particles and attractive forces between gas particles.

KEY TERMS
ideal gas
kinetic molecular theory of gases

KEY CONCEPTS
- Particles of matter behave in characteristic ways in each of the three states of matter. Attractive forces between particles and the kinetic energy of the particles influence how particles behave in each state.

- Attractive forces between particles that affect the states of matter are attractions between oppositely charged particles, attractions between polar molecules, and attractions between non-polar molecules.

- The properties that distinguish gases from liquids and solids include compressibility, low viscosity, even and complete mixing, low density, and expansion as a result of an increase in temperature.

- The kinetic molecular theory of gases is a model that explains the visible properties of gases based on the behaviour of individual atoms or molecules of an ideal gas.

Section 11.2 | Gases and Pressure Changes

For a fixed amount of gas at a constant temperature, the volume of the gas will increase as the pressure decreases.

KEY TERMS
atmospheric pressure
Boyle's law
standard atmospheric pressure (SAP)

KEY CONCEPTS
- Atmospheric pressure is the force that a column of air exerts on Earth's surface divided by the area of Earth's surface at the base of that column.

- Pascal's investigations demonstrated the decrease in atmospheric pressure as a result of an increase in altitude.

- Boyle's law states that the volume of a given amount of gas is inversely proportional to the external pressure exerted on the gas when the temperature is constant. The equation for Boyle's law is $P_1V_1 = P_2V_2$.

- The relationship between the pressure and volume of a gas can be explained using the kinetic molecular theory. As the external pressure on a gas increases, the volume of the gas decreases. As the volume decreases, the gas molecules become closer together, causing the frequency of collision of the molecules to increase, thus increasing the gas pressure.

Section 11.3 | Gases and Temperature Changes

Both the pressure and volume of a gas are affected by temperature. At a constant pressure, the volume of a gas will increase as the temperature increases. At a constant volume, the pressure of a gas will increase with an increase in temperature.

KEY TERMS
absolute zero
Charles's law
Gay-Lussac's law

KEY CONCEPTS
- The Kelvin temperature scale is based on a temperature of 0 K at absolute zero. The size of a kelvin on the Kelvin scale is the same as the size of a degree on the Celsius scale.

- Charles's law states that the volume of a fixed amount of gas at a constant pressure is directly proportional to the Kelvin temperature of the gas: $V \propto T$.

- The mathematical equation for Charles's law is $\frac{V_1}{T_1} = \frac{V_2}{T_2}$, where T is the temperature in Kelvin.

- Gay-Lussac's law states that the pressure of a fixed amount of gas at a constant volume is directly proportional to the Kelvin temperature of the gas: $P \propto T$

- The mathematical equation for Gay-Lussac's law is $\frac{P_1}{T_1} = \frac{P_2}{T_2}$, where T is the temperature in Kelvin.

Knowledge and Understanding

Select the letter of the best answer below.

1. Which of the following are characteristics of atoms and molecules in the gas state?

   I. They display rotational motion.

   II. They have high kinetic energy.

   III. They are held together by weak ionic bonds.

   a. I only
   b. II only
   c. I and II only
   d. II and III only
   e. I, II, and III

2. When a sealed 1 L flask of gas is cooled, what happens to the gas molecules?

   a. They move farther apart.
   b. They collide more often with the walls of the flask.
   c. They increase their vibrational motion.
   d. They move more slowly.
   e. Their intermolecular forces decrease.

3. Select all the statements that are correct, according to the kinetic molecular theory of gases.

   I. Molecules are closer together in gases than in liquids.

   II. Molecules are in random motion.

   III. Molecules lose energy when they collide with other.

   IV. The average kinetic energy of molecules is proportional to the temperature.

   V. Gas molecules move in straight lines between collisions.

   a. I, II, and III only
   b. I, III, and V only
   c. II, III, and IV only
   d. II, IV, and V only
   e. I, II, III, IV, and V

4. Which of the following is described by the statement, "The pressure of a fixed amount of gas is inversely proportional to its volume at constant temperature."

   a. Charles's law
   b. Boyle's law
   c. kinetic molecular theory of gases
   d. Montgolfier's law
   e. Gay-Lussac's law

5. Which of the lines on the graph below is the best representation of the relationship between the volume of a gas and its temperature, if other factors remain constant?

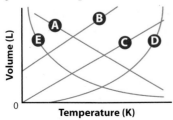

   a. A
   b. B
   c. C
   d. D
   e. E

6. A gas occupies a volume of 2.00 L at 300 mmHg and 100°C. Which mathematical expression will give the correct volume at 760 mmHg and 300°C?

   a. $2.00 \times \frac{300}{760} \times \frac{373}{573}$
   b. $2.00 \times \frac{300}{760} \times \frac{573}{373}$
   c. $2.00 \times \frac{760}{300} \times \frac{373}{573}$
   d. $2.00 \times \frac{760}{300} \times \frac{573}{373}$
   e. $2.00 \times \frac{300}{760} \times \frac{300}{100}$

7. Which of the following is described by the statement, "The pressure of a fixed amount of gas is directly proportional to its Kelvin temperature at constant volume"?

   a. Charles's law
   b. Boyle's law
   c. kinetic molecular theory
   d. Avogadro's law
   e. Gay-Lussac's law

8. A given mass of a gas in a rigid container is heated from 100°C to 300°C. Which of the following responses best describes what will happen to the pressure of the gas?

   a. decrease by a factor of three
   b. increase by a factor of three
   c. decrease by a factor of about two thirds
   d. increase by a factor of about four
   e. increase by a factor of about one and a half

Answer the questions below.

9. Copy and complete the following table.

Particles in States of Matter

Property	Solid	Liquid	Gas
Position of particles in relation to one another			
Strength of attraction between particles			
Type of motion of particles			

10. Explain why gases are easier to compress than liquids or solids.

11. A homogeneous mixture is formed by water vapour, $H_2O(g)$, and carbon tetrachloride, $CCl_4(g)$. Describe the attractive forces that affect the molecules in each gas.

12. What is an elastic collision and how does it relate to the kinetic molecular theory of gases?

Thinking and Investigation

13. Complete the following pressure unit conversions.

   a. The reading on a compressed air tank is 80.0 psi. What is the pressure in kilopascals?

   b. Normal blood pressure is 120 mmHg over 80 mmHg (pumping/resting). What are these pressures in atmospheres? Report your answers to two significant digits.

   c. Average atmospheric pressure in a tornado is 800 millibars. What is the pressure in kilopascals?

   d. The pressure underwater at 10.0 m is 2.00 atm. What is the pressure in millimetres of mercury?

14. Liquid nitrogen has a temperature below −196°C. When an inflated balloon is placed in the liquid nitrogen, it rapidly shrinks. Explain the change according to the kinetic molecular theory of gases.

15. A 1.00 L sample of a gas is at standard pressure. If it is compressed to 473 mL, what is the pressure in kPa, assuming the temperature remains constant?

16. Underwater divers must rise very slowly, or they experience "the bends." A diver has 0.04 L of gas in her blood under a pressure of 400.0 kPa and rises quickly to the surface, which is at 101.3 kPa. What is the volume of gas in her blood? Assume the temperature and amount of gas are constant.

17. Describe the effect of increasing the temperature (in K) by a factor of four on the volume of a gas, assuming that other variables are held constant.

18. Perform the following temperature conversions.

   a. 25.0°C to K   **c.** 277 K to °C

   b. −10.00°C to K   **d.** 165 K to °C

19. A 3.0 L balloon is completely filled with air at 25.0°C. The balloon is taken outdoors. If the final volume of the balloon is 2.5 L, what is the temperature of air outdoors in °C? Assume the pressure and amount of air are constant.

20. A 7.50×10^2 mL empty water bottle is capped at 20°C and 101 kPa. The water bottle is then crushed, causing the bottle lid to be projected across a room.

   a. Describe the changes that occur inside the bottle as it is crushed.

   b. What is the pressure inside the empty container when the volume is reduced to 2.50×10^2 mL, assuming the temperature and amount of air are constant.

21. A student performed an investigation involving one of the gas laws and generated the graph below.

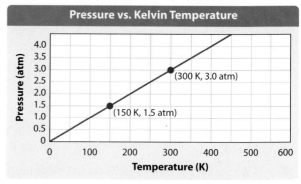

Pressure vs. Kelvin Temperature

(300 K, 3.0 atm)

(150 K, 1.5 atm)

   a. What gas law was being investigated? Explain.

   b. Is the gas law verified by the data? Explain.

22. A flask that can withstand a maximum internal pressure of 4.00 atm is filled with a gas at 20.0°C and 0.98 atm. At what temperature, in degrees Celsius, will the flask explode? Assume the amount and volume of air are constant.

Communication

23. **BIG IDEAS** Properties of gases can be described qualitatively and quantitatively and can be predicted. Research one of the gas laws that you studied in this chapter and develop a presentation in the format of your choice that includes the following.

   a. a discussion of the gas law using the kinetic molecular theory and visualizations of molecular activity

   b. the mathematical equation that expresses the law, with an example

   c. two practical applications of the gas law, with explanations of how the gas law is used

   d. some historical background information about the discovery or derivation of the gas law

24. The data below are volumes of hydrogen gas collected at several different temperatures. Assume the amount and pressure of the gas are the same for all measurements.

Volume of Hydrogen Gas Collected

Trial	Temperature (°C)	Volume (mL)
1	300.0	48
2	175.0	37
3	110.0	
4	0.0	22
5		15
6	−150.0	11

a. Identify the gas law that is being studied.

b. Based on the gas law identified in part (a), use the appropriate mathematical equation for the law to complete the missing values in the table.

c. Construct a graph of the data that demonstrates the relationship between volume and temperature of a gas, when the amount and pressure of the gas are constant.

d. Using the graph you made in part (c), determine the temperature at which the volume of the gas will reach a value of 0 mL. What is this temperature called?

25. BIG IDEAS People have a responsibility to protect the integrity of Earth's atmosphere. The propellant in aerosol cans is pressurized gas. Many aerosol cans have the warnings "Keep away from heat sources" and "Do not incinerate" on their labels.

a. Explain what happens to aerosol cans once the pressure of the gas in the can reaches atmospheric pressure.

b. Explain the changes that would occur in the pressure, temperature, and volume of the contents of an aerosol can if it were absent-mindedly thrown into a fire to dispose of it.

c. Develop an illustrated web page (or a mockup of a web page) that explains to the general public why it is important to handle, store, and dispose of aerosol cans properly.

26. Summarize your learning in this chapter using a graphic organizer. To help you, the Chapter 11 Summary lists the Key Terms and Key Concepts. Refer to Using Graphic Organizers in Appendix A to help you decide which graphic organizer to use.

Application

27. A hydrogen gas thermometer works based on changes in volume of hydrogen gas in a sealed vessel. The hydrogen gas was found to have a volume of 200.0 cm³ when placed in a water bath at 0.0°C.

a. When the same thermometer was placed in boiling liquid bromine, the volume of hydrogen at the same pressure is found to be 243.1 cm³. If there are no changes in pressure and amount of gas, what is the boiling temperature of liquid bromine in degrees Celsius?

b. The boiling temperature of liquid chlorine is known to be −34.4°C. If the hydrogen thermometer was used to determine this temperature, to what volume would the hydrogen gas in the thermometer be reduced?

28. Red blood cells carry the oxygen that is necessary for animals to sustain life. Using your knowledge of atmospheric pressure, infer why people living in Calgary (altitude 1048 m) have significantly more red blood cells per volume of blood than people living in Vancouver (altitude 0 m).

29. A mountain bike has a tire pressure range of 250.0 kPa to 380.0 kPa. The typical tire volume is 1.20×10^3 cm³ at the higher pressure.

a. Why would a range of pressures be given for the tire? Under what conditions would a person want to use a lower pressure in the tire?

b. Explain how the tire pressure is affected by the body weight of an individual riding the bike.

c. Calculate the volume of the tire at 250.0 kPa assuming constant amount and temperature of the air in the tire. By what factor did the volume change?

d. Would the pressure and the volume of the tire really be independent of temperature? Explain.

30. A pressure cooker is a sealed cooking pot that does not allow air or liquids to escape until a certain pressure is reached inside the pot. As heat is added to the pressure cooker, the pressure inside it increases. How does the increased pressure affect the temperature of the water vapour inside the pressure cooker?

31. BIG IDEAS Air quality can be affected by human activities and technology. Methane, $CH_4(g)$, undergoes a combustion reaction to produce carbon dioxide and water. In the atmosphere, methane is a greenhouse gas, helping to trap infrared radiation and warm Earth. Methane concentrations in the atmosphere have increased significantly. The primary sources for the methane added to the atmosphere are due to human activities.

a. Describe the physical properties of methane, as an illustrative example of a gas. What attractive forces at the molecular level influence its properties?

b. Compare and contrast the motions of methane particles at room temperature with water and sodium chloride particles at the same temperature.

c. Describe how the volume of methane gas can be affected by changes in temperature and pressure.

d. Vast quantities of methane are believed to be stored in Earth's crust. Large quantities are trapped in methane clathrates. What are methane clathrates, and how do rising global temperatures affect them?

Select the letter of the best answer below.

1. **K/U** Which of the following statements best explains why gases can be easily compressed?

   a. Molecules of a gas exhibit random translational motion.

   b. Molecules of a gas have negligible intermolecular forces.

   c. Molecules of a gas have small amounts of space between them.

   d. Molecules of a gas are in constant motion.

   e. Molecules of a gas have little volume.

2. **K/U** Which of the following best describes a gas?

   a. It assumes the volume and shape of the container, and it has weak intermolecular attractions.

   b. It assumes the volume and shape of the container, and it has strong intermolecular attractions.

   c. It has a distinct shape and volume, and it has strong intermolecular attractions.

   d. It has a distinct volume and assumes the shape of the container, and it has moderate intermolecular attractions.

   e. It has a distinct volume and assumes the shape of the container, but it lacks intermolecular attractions.

3. **K/U** Which of the following assumptions are made by the kinetic molecular theory of gases?

   I. Gas molecules move randomly in all directions.

   II. Gas molecules exhibit negligible intermolecular forces.

   III. Gas molecules have negligible volume.

   a. I and II only   **d.** II only

   b. I and III only   **e.** I, II, and III

   c. I only

4. **K/U** Which of the following represents the greatest pressure?

   a. 2.5 atm   **d.** 21 psi

   b. 200 kPa   **e.** 790 torr

   c. 960 mmHg

5. **K/U** Which description best describes the situation shown in the diagram below?

   a. a gas expanding as temperature increases and pressure remains constant

   b. a gas expanding as temperature and pressure remain constant

   c. a gas contracting as temperature increases and pressure remains constant

   d. a gas expanding as pressure remains constant

   e. a gas expanding as temperature and pressure change

6. **T/I** A sample of gas is in a sealed flexible container at a fixed temperature. If the pressure on the container is reduced by half, the volume will

   a. increase by a factor of 2

   b. increase by a factor of 4

   c. increase by a factor of 1

   d. decrease by a factor of 2

   e. decrease by a factor of 4

7. **K/U** A sample of nitrogen gas is placed in a sealed 2 L flexible container. Which of the following will occur if the temperature of the gas is increased?

   I. The pressure of the gas will increase.

   II. The volume of the gas will decrease.

   III. The speed of the gas molecules will increase.

   a. I and II   **d.** II only

   b. I and III   **e.** I, II, and III

   c. I only

8. **T/I** What temperature on the Kelvin scale corresponds to −35°C?

   a. 238   **d.** 333

   b. 293   **e.** 35

   c. 308

9. **K/U** Identify the choice that best describes this statement: "The volume of a fixed amount of gas is directly proportional to its temperature at a constant pressure."

   a. Boyle's law

   b. Charles's law

   c. Gay-Lussac's law

   d. kinetic molecular theory of gases

   e. Avogadro's law

10. **T/I** A sample of argon gas is stored in a container with a fixed volume at 1.00 atm of pressure. The temperature, in K, of the gas is doubled. What is the new pressure of the gas assuming the amount of gas is constant?

    a. 0.5 atm   **d.** 3.00 atm

    b. 1.00 atm   **e.** 4.00 atm

    c. 2.00 atm

Use sentences and diagrams, as appropriate, to answer the questions below.

11. **C** Construct a graphic organizer to compare the properties of gases with those of liquids. Include major similarities and differences.

12. **A** Each of the following observations relate to properties of a gas. Name the property observed, and explain the observation.

a. Gaseous oxygen and carbon dioxide are placed in a sealed flask. After several minutes, the gases are evenly distributed through the flask.

b. Air bubbled through water in a fish tank rises to the surface and is released above the water.

13. **K/U** In your own words, describe the kinetic molecular theory of gases and its assumptions.

14. **A** When food is being preserved by canning, a jar is filled with very hot food, leaving a space at the top of the jar. A rubber seal is placed on top of the jar, and the lid is screwed shut. After several minutes, a 'pop" is heard, and the metal lid is observed to be dented inward. Explain these observations using the kinetic molecular theory and the properties of gases.

15. **K/U** Torricelli and Pascal performed many important studies of atmospheric pressure.

a. What is meant by the terms *atmospheric pressure* and *standard atmospheric pressure*?

b. Explain what the term "millimetres of mercury" refers to and how it relates to the discovery of atmospheric pressure.

16. **T/I** Determine the following conversions.

a. 551 kPa to psi

b. 6.0 psi to mmHg

c. 0.52 atm to kPa

d. 902 mbar to mmHg and kPa

17. **C** An investigation to verify Boyle's law was conducted. The data for the investigation are shown in the table at the top of the next column.

Pressure and Volume Measurements of a Gas

Pressure (kPa)	Volume (ml)
100	2.5×10^2
75	3.3×10^2
50	5.0×10^2
25	1.0×10^3
15	x

a. Calculate the missing value for volume, indicated as x in the table.

b. Use the data to construct a graph that shows the relationship between pressure and volume of a gas. Explain how it demonstrates Boyle's law.

18. **T/I** A sample of neon has a volume of 239 mL at 202.7 kPa of pressure. What is the pressure when the volume is 500 mL? Assume the amount and temperature of the gas are constant.

19. **T/I** What is absolute zero? Describe a series of experiments that could be performed that would permit you to be able to determine its value.

20. **C** Use a graphic organizer to compare and contrast the Kelvin and Celsius scales.

21. **T/I** A sample of gas is heated from 273 K to 290 K. If its original volume was 2.0×10^2 mL, what is the volume after being heated? Assume amount and pressure of the gas are constant.

22. **T/I** A ball filled with air has a volume of 3.4 L at 25°C. What is its volume at 3.0°C, assuming constant amount and pressure of air?

23. **A** Xenon gas is placed in a light bulb at 300 atm and 20°C. When the bulb is in use, the bulb temperature rises to 85°C. What does the pressure in kPa become when the bulb is in use, assuming the amount and volume of gas are constant?

24. **T/I** Air in a ball has a pressure of 11.0 psi and a temperature of 25.0°C. The temperature of the air rises to 45.0°C. Calculate the new pressure of the air, assuming a constant amount and volume of air.

25. **A** Describe one common occurrence or technology that illustrates each of the gas laws that you learned about in this chapter.

Self-Check

If you missed question …	1	2	3	4	5	6	7	8	9	10	11	12	13	14	15	16	17	18	19	20	21	22	23	24	25
Review section(s)…	11.1	11.1	11.1	11.2	11.2	11.2	11.2, 11.3	11.3	11.3	11.1	11.1	11.1	11.1, 11.2	11.2	11.2	11.2	11.2	11.3	11.3	11.3	11.3	11.3	11.3	11.3	11.2, 11.3

CHAPTER 12

Exploring the Gas Laws

Specific Expectations

In this chapter, you will learn how to…

- F1.1 **analyze** the effects on air quality of technologies and human activities (12.3)

- F1.2 **assess** air quality for a Canadian location, using Environment Canada's Air Quality Health Index, and **report** on Canadian initiatives to improve air quality and reduce greenhouse gases (12.3)

- F2.3 **solve** quantitative problems by performing calculations based on the combined gas law, Dalton's law of partial pressures, and the ideal gas law (12.1, 12.2)

- F2.4 **use** stoichiometry to solve problems related to chemical reactions involving gases (12.2)

- F2.5 **determine**, through inquiry, the molar volume or molar mass of a gas produced by a chemical reaction (12.1)

- F3.1 **identify** the major and minor chemical components of Earth's atmosphere (12.3)

- F3.4 **describe**, for an ideal gas, the quantitative relationships between pressure, volume, temperature, and amount of substance (12.2)

- F3.5 **explain** Dalton's law of partial pressures, the combined gas law, and the ideal gas law (12.1, 12.2)

- F3.6 **explain** Avogadro's hypothesis and how his contribution to the gas laws has increased our understanding of the chemical reactions of gases (12.1)

The internal combustion engine provides a common application of the gas laws. Gasoline-fueled engines provide power through a cycle of events that involve changes to the temperature, pressure, and volume of gases. Initially, an air-and-gasoline mixture enters a cylinder. The mixture is compressed by pistons and then ignited by a spark, producing an explosive force to drive the piston, as shown here. This combustion of fuel in the cylinder causes a sudden change in the temperature and pressure of the gases. The pressure pushes the piston down, resulting in a change in the volume of the gases. In the final step, exhaust gases are pushed out.

In this combustion reaction, the desired outcome is energy, but of course there are other products resulting from the reaction. These products—including carbon monoxide, carbon dioxide, nitrogen oxides, and soot—affect the quality of the air we breathe.

"Can" You Explain the Effects?

Note: Your teacher may choose to carry out this activity as a demonstration.

An empty container, open to the atmosphere, is bombarded by air molecules from the outside and the inside at the same time. The volume of the container remains constant, because the pressure on the two sides is the same. As you carry out this activity, keep in mind what you have learned about the relationships among changes in gas volume, pressure, and temperature.

Safety Precautions

- Wear safety eyewear and a lab apron during this activity.
- Always use beaker tongs to handle the heated can.
- Use EXTREME CAUTION when you are near a hot plate and boiling water.

Materials

- 5 mL of water
- 10 mL graduated cylinder
- large beaker of ice water
- empty, rinsed soft-drink can
- hot plate
- beaker tongs

Procedure

1. Use the graduated cylinder to add 5 mL of water to the soft-drink can.

2. Heat the can on the hot plate until steam begins to rise from the opening of the can.

3. Using the beaker tongs, and when it is above the large beaker, quickly invert the opening of the can into the large beaker of ice water so that the opening of the can is just under the surface of the water. Observe the effects on the can.

Questions

1. Describe what happened to the water and the water molecules as you heated the can.

2. What evidence is there that air was present in the can before you started to heat it?

3. Explain what happened to the air in the can as the water began to boil.

4. Describe what happened to the temperature, volume, and pressure of the gas in the can when you inverted it suddenly in ice water.

5. Use your answer to question 4 to explain the effects on the can. (If you need help, re-read and think about the introduction to this activity at the top of the page.)

The Combined Gas Law

Key Terms

combined gas law

law of combining volumes

Avogadro's law (hypothesis)

molar volume

standard temperature and
    pressure (STP)

standard ambient
    temperature and
    pressure (SATP)

combined gas law
a gas law stating that
the pressure and volume
of a given amount
of gas are inversely
proportional to each
other and directly
proportional to the
Kelvin temperature of
the gas: $V \alpha \frac{T}{P}$

In a roadway accident, the pressure in an air bag, such as the one shown in **Figure 12.1**, can save lives. How can enough pressure be generated within milliseconds after a collision to prevent a person from injury? When a collision occurs, a sensor activates an electrical circuit that triggers a chemical reaction inside the air bag. The reaction produces about 65 L of nitrogen gas that inflates the air bag almost instantaneously. As soon as a person's body hits the air bag, the bag starts to deflate, cushioning the impact. The nitrogen gas escapes through vents in the bag, and within two seconds, the pressure inside the bag returns to atmospheric pressure.

If you analyze the process that occurs in an air bag, you discover that the temperature, pressure, and volume of the system all change significantly and at the same time. In Chapter 11, you studied the gas laws by considering only two of these variables at a time. However, in most situations that involve gases—in nature, as well as in industrial processes—all three variables change together. Therefore, developing a method to monitor simultaneous changes in the temperature, pressure, and volume of a gas is essential for studying and making accurate predictions of gas behaviour.

The Combined Gas Law

The combined gas law represents a combination of the relationships that express Boyle's law and Charles's law. The **combined gas law** states that the pressure and volume of a given amount of gas are inversely proportional to each other, and directly proportional to the Kelvin temperature of the gas.

The combined gas law is represented by the following equation, where T is the Kelvin temperature and assuming that the amount of gas remains constant.

$$\frac{P_1 V_1}{T_1} = \frac{P_2 V_2}{T_2}$$

Figure 12.1 The usefulness of air bags depends on the simultaneous changes in the temperature, pressure, and volume of the gas.

Calculations Using the Combined Gas Law

The combined gas law is a useful tool for making predictions that involve a constant amount of gas and any of the three variables—temperature, pressure, or volume. For example, imagine that you prepare a gas in a laboratory apparatus that allows you to alter the pressure and temperature of the gas. You could use the combined gas law to predict the change in volume of a fixed amount of gas that results from changes in temperature and pressure.

The following Sample Problems and Practice Problems will reinforce your understanding of the combined gas law.

Sample Problem

Determining Volume Using the Combined Gas Law

Problem
A small balloon contains 275 mL of helium gas at a temperature of 25.0°C and a pressure of 350 kPa. What volume would this gas occupy at 10.0°C and 101 kPa?

What Is Required?
You need to find the volume, V_2, of the balloon under the new conditions of temperature and pressure.

What Is Given?
You know the initial pressure, volume, and temperature:

$P_1 = 350$ kPa
$V_1 = 275$ mL
$T_1 = 25.0°C$

You also know the final temperature and pressure:

$T_2 = 10.0°C$
$P_2 = 101$ kPa

Plan Your Strategy	Act on Your Strategy
According to the combined gas law relationship, changes in the temperature and pressure of a gas cause a change in its volume.	$\dfrac{P_1 V_1}{T_1} = \dfrac{P_2 V_2}{T_2}$
Convert temperatures from the Celsius scale to the Kelvin scale.	$T_1 = 25.0°C + 273.15$ $= 298.15$ K $T_2 = 10.0°C + 273.15$ $= 283.15$ K
Isolate the variable V_2.	$\dfrac{P_1 V_1}{T_1}\left(\dfrac{T_2}{P_2}\right) = \dfrac{\cancel{P_2} V_2}{\cancel{T_2}}\left(\dfrac{\cancel{T_2}}{\cancel{P_2}}\right)$ $\dfrac{P_1 V_1 T_2}{T_1 P_2} = V_2$
Substitute numbers and units for the known variables in the formula and solve.	$V_2 = \dfrac{P_1 V_1\, T_2}{T_1\, P_2}$ $= \dfrac{(350\ \cancel{kPa})(275\ mL)(283.15\ \cancel{K})}{(298.15\ \cancel{K})(101\ \cancel{kPa})}$ $= 910$ mL

Continued on next page ❯

Alternative Solution

Plan Your Strategy	Act on Your Strategy
You know that the volume increases when the pressure decreases. Determine the ratio of the initial pressure and the final pressure that is greater than 1.	$P_1 = 350$ kPa $P_2 = 101$ kPa pressure ratio > 1 is $\dfrac{350 \text{ kPa}}{101 \text{ kPa}}$
Convert temperatures from the Celsius scale to the Kelvin scale. You know that the volume decreases when temperature decreases. Determine the ratio of the initial temperature and the final temperature that is less than 1.	$T_1 = 25.0°C + 273.15$ $\quad = 298.15$ K $T_2 = 10.0°C + 273.15$ $\quad = 283.15$ K temperature ratio < 1 is $\dfrac{283.15 \text{ K}}{298.15 \text{ K}}$
Multiply the initial volume by the pressure and temperature ratios to obtain the final volume.	$V_2 = V_1 \times$ pressure ratio $\times$ temperature ratio $= 275 \text{ mL} \times \dfrac{350 \text{ kPa}}{101 \text{ kPa}} \times \dfrac{283.15 \text{ K}}{298.15 \text{ K}}$ $= 910$ mL

Check Your Solution

The large decrease in pressure would cause a large increase in volume. Since the temperature decreased slightly, it should cause only a slight decrease in volume. Thus, you would expect a significant overall increase in volume.

Practice Problems

1. A sample of argon gas, Ar(g), occupies a volume of 2.0 L at −35°C and 0.5 atm. What would its Celsius temperature be at 2.5 atm if its volume was decreased to 1.5 L?

2. A chemical researcher produces 15.0 mL of a new gaseous substance in a laboratory at a temperature of 298 K and pressure of 100.0 kPa. Calculate the volume of this gas if the temperature was changed to 273 K and the final pressure was 101.325 kPa.

3. A sample of air in a syringe exerts a pressure of 1.02 atm at 295 K. The syringe is placed in a boiling water bath at 373 K. The pressure is increased to 1.23 atm and the volume becomes 0.224 mL. What was the initial volume?

4. Helium gas, He(g), in a 1.0×10^2 L weather balloon is under a pressure of 25 atm at 20.0°C. If the helium balloon expands to 2400 L at 1.05 atm of pressure, what would the temperature of the helium gas be?

5. A 30.00 mL gas syringe was at a pressure of 100.0 kPa at 30.0°C. On the following day, the temperature dropped to 25.0°C and the new volume was 28.5 mL. What was the atmospheric pressure on this day?

6. A 2.7 L sample of nitrogen gas, N_2(g), is collected at a temperature of 45.0°C and a pressure of 0.92 atm. What pressure would have to be applied to the gas to reduce its volume to 2.0 L at a temperature of 25.0°C?

7. A scuba diver is swimming 30.0 m below the ocean surface where the pressure is 4.0 atm and the temperature is 8.0°C. A bubble of air with a volume of 5.0 mL is emitted from the breathing apparatus. What will the volume of the air bubble be when it is just below the surface of the water, where the pressure is 101.3 kPa and the water temperature is 24.0°C?

8. A 5.0×10^2 mL sample of oxygen, O_2(g), is kept at 950 mmHg and 21.5°C. The oxygen expands to a volume of 700 mL and the temperature is adjusted until the pressure is 101.325 kPa. Calculate the final temperature of the oxygen gas.

9. A sample of Freon-12, CF_2Cl_2(g), formerly used in refrigerators, is circulated through a series of pipes for refrigeration. If the gas occupies 350 cm^3 at a pressure of 150 psi and a temperature of 15°C, what volume of gas will be released if there is a break in the line where the external temperature is 25°C and the external pressure is 102 kPa?

10. A crack in the floor of the ocean at a depth where the pressure is 16 atm releases 350 m^3 of methane gas. The temperature of the water at this depth is 8°C. If the surface temperature is 40°C and the pressure is 758 mmHg, what volume of methane is released at the surface?

Combining Volumes of Gases

Many of the first studies of gases focussed on the relationships among the volume, pressure, and temperature for a particular mass of the gas sample. By the early 1800s, scientists were exploring chemical reactions between gases. Gay-Lussac and a colleague, Alexander von Humboldt (1769–1859), performed experiments to determine the number of volumes of hydrogen and of oxygen that would react to form a volume of water. They needed this information to determine the percentage of oxygen in the atmosphere. By making precise measurements, Gay-Lussac determined that two volumes of hydrogen combined with one volume of oxygen to form two volumes of water.

From studying the volumes of other gases in other chemical reactions, Gay-Lussac reached the general conclusion that, in chemical reactions, all gases combine in simple proportions. This conclusion is now known as the **law of combining volumes**, which states that when gases react, the volumes of the gaseous reactants and products, measured under the same conditions of temperature and pressure, are always in whole-number ratios. Shown below are examples of gas volumes before and after chemical reactions to illustrate the law of combining volumes. All gas volumes are measured at the same temperature and pressure.

law of combining volumes a gas law stating that when gases react, the volumes of the gaseous reactants and products, measured under the same conditions of temperature and pressure, are always in whole-number ratios

> **Example 1: The reaction between hydrogen gas and oxygen gas**
> hydrogen gas + oxygen gas → water vapour
> 100 mL + 50 mL → 100 mL
>
> **Example 2: The decomposition of ammonia gas, $NH_3(g)$**
> ammonia gas → hydrogen gas + nitrogen gas
> 8 mL → 12 mL + 4 mL

Avogadro's Law

A few years before Gay-Lussac's description of combining gas volumes, English scientist John Dalton had published his ideas about atoms combining in simple, whole-number ratios to form compounds. Empirical work on the mass percent of elements present in compounds suggested that chemical combinations of the same elements always occurred in fixed mass ratios or multiples of those ratios. For example, water is always 88.89 percent oxygen and 11.11 percent hydrogen by mass. Thus, the mass of oxygen in a water molecule is eight times the mass of hydrogen in a water molecule. However, chemists at that time could not understand how one volume of oxygen could combine with two volumes of hydrogen to form two volumes of water vapour if the mass of oxygen in water was eight times the mass of hydrogen in water.

Italian scientist Amadeo Avogadro (1776–1856) resolved the apparent conflict between Gay-Lussac's and Dalton's observations. In 1811, he proposed that the law of combining volumes could be explained if equal volumes of gases contained the same number of particles, regardless of their mass. Avogadro's proposal was ignored for 50 years, because scientists could not accept such a revolutionary idea. Eventually, however, scientists realized the significance of the proposal and more observations supported it. Avogadro's proposal is now known as **Avogadro's law**. You also may see it called *Avogadro's hypothesis*. Either way, chemists now understand that the volume of a gas is directly proportional to the number of molecules of the gas, when the pressure and temperature are constant.

Avogadro's law (hypothesis) a gas law stating that equal volumes of all ideal gases at the same temperature and pressure contain the same number of molecules

The Mole in Relation to Gas Volume

You know that a large number of molecules are required to produce measurable volumes and masses of substances in the laboratory. Recall that Avogadro's number, 6.02×10^{23} particles, represents 1 mol of those particles. When the particles are molecules, the ratio of the amounts of the substances present, expressed in moles, is equal to the ratio of the coefficients of the substances in a balanced chemical equation. **Figure 12.2** shows how amounts of gases, expressed in moles, are related to volumes of gases in chemical reactions.

hydrogen gas	+	oxygen gas	→	water vapour
$2 H_2(g)$	+	$1 O_2(g)$	→	$2 H_2O(g)$
2 mol		1 mol		2 mol
2 volumes		1 volume		2 volumes

Figure 12.2 The coefficients in a balanced equation show the relationships between the amounts (in moles) of all reactants and products, and the relationships between the volumes of any gaseous reactant or product.

Using the Law of Combining Volumes to Determine the Stoichiometry of a Chemical Reaction

According to the law of combining volumes, when gases react the volumes of the gaseous reactants and products, measured at the same conditions of temperature and pressure, are always in whole-number ratios. Also recall that, when a chemical reaction involves only gaseous reactants and products, the mole ratios are the same as the ratios of the volumes of the gases. Therefore, the principles of the law of combining volumes can be used to determine the volume of a gaseous reactant or product based on the volume of another gaseous reactant or product.

To determine the volume of a gaseous reactant or product in a reaction, you must know the balanced chemical equation for the reaction and the volume of at least one of the gases involved in the reaction. Consider, for example, the combustion of methane gas, shown in **Figure 12.3**. This reaction takes place every time you light a Bunsen burner, since natural gas is composed of mostly methane. Because the coefficients represent volume ratios for gases taking part in the reaction, you can determine that it takes 2 L of oxygen to react completely with 1 L of methane. The complete combustion of 1 L of methane will produce 1 L of carbon dioxide and 2 L of water vapour. Note that the conditions of temperature and pressure are not mentioned. When using the law of combining volumes to solve gas stoichiometry problems, the temperature and pressure must be the same for the reactants and products.

| $CH_4(g)$ | + | $2O_2(g)$ | → | $CO_2(g)$ | + | $2H_2O(g)$ |
| 1 L | | 2 L | | 1 L | | 2 L |

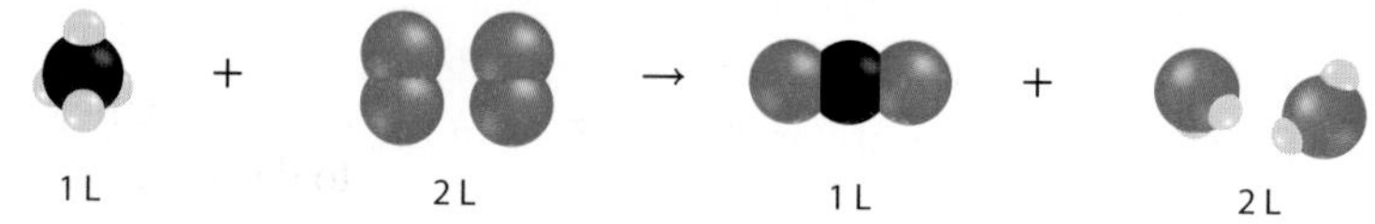

Figure 12.3 From the coefficients of the balanced chemical equation, volume ratios can be set up for any pair of gases in the reaction.

A Mathematical Statement of Avogadro's Law

The amount (in moles) of a substance consists of a specific number of molecules. Thus, Avogadro's law can be expressed in moles. These steps develop a mathematical expression for Avogadro's law, in which n represents the amount (in moles) of the gas.

When the temperature and pressure of a gas are constant, Avogadro's law can be expressed mathematically as a proportionality.	$n \propto V$
By using a proportionality constant, k, Avogadro's law can also be expressed as an equality.	$n = kV$
As long as the temperature and pressure of a gas remain constant, any combination of the gas amount (in moles) divided by the volume of the gas is equal to the same constant.	$\dfrac{n_1}{V_1} = k$ and $\dfrac{n_2}{V_2} = k$
Combining the expressions gives a mathematical statement of Avogadro's law.	$\dfrac{n_1}{V_1} = \dfrac{n_2}{V_2}$

Figure 12.4 depicts Avogadro's law by showing how an increase in the amount (n) of a gas at constant temperature and pressure causes the volume to increase.

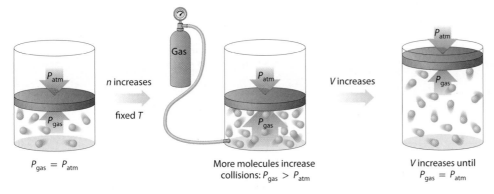

$P_{gas} = P_{atm}$

n increases

fixed T

More molecules increase collisions: $P_{gas} > P_{atm}$

V increases

V increases until $P_{gas} = P_{atm}$

Figure 12.4 When more gas enters a container, the increase in the number of molecules causes the pressure to increase. Because the pressure inside the container is greater than the external pressure at constant temperature, the volume will increase. The volume will continue to increase until the internal pressure caused by the gas becomes equal to the external pressure.

Learning Check

1. What gas laws comprise the combined gas law?

2. In your own words, describe the law of combining volumes.

3. A sample of gas has a volume of 525 mL at 300.0 K and 746 mmHg. What is the volume of the gas if the temperature increases to 350.0 K and the pressure increases to 780 mmHg?

4. When predicting volumes of gaseous reactants and products, why must a balanced equation be used?

5. Butene gas, $C_4H_8(g)$, burns in the presence of oxygen to produce gaseous carbon dioxide and water vapour.

    a. Using this reaction, define and explain the law of combining volumes of gases.

    b. Calculate the volume of butene required to produce 250 mL of carbon dioxide. Assume temperature and pressure remain constant.

6. Use the mathematical expression for Avogadro's law to determine the volume of 1.0 mol of a gas if a 2.0 mol sample of the gas has a volume of 30.0 L. Assume temperature and pressure remain constant.

The Molar Volume of Gases

molar volume amount of space occupied by 1 mol of a gaseous substance; 22.4 L/mol for an ideal gas at standard temperature and pressure and 24.8 L/mol for an ideal gas at standard ambient temperature and pressure

standard temperature and pressure (STP) conditions defined as a temperature of 0°C (273.15 K) and a pressure of 101.325 kPa

standard ambient temperature and pressure (SATP) conditions defined as a temperature of 25°C (298 K) and a pressure of 100 kPa

According to Avogadro's law, the volume of 1 mol of any gas should be the same as the volume of 1 mol of any other gas, if the conditions of temperature and pressure are the same. Thus, it is possible to calculate the **molar volume (v)** of a gas, or the volume of 1 mol of a gas. Molar volume is expressed in units of L/mol, which can be found by dividing the volume, V, by the amount in moles, n, present:

$$v = \frac{V}{n}$$

Molar volume will vary with different temperatures and pressures. Atmospheric pressure and ambient temperature (the temperature of the surrounding air) can vary from one place to another. They also can vary in the same location at different times of the day. Therefore, chemists have agreed on specific sets of conditions under which to report gas volumes. One of these is called **standard temperature and pressure (STP)**, defined as a temperature of 0°C (273.15 K) and a pressure of 1 atm (101.325 kPa).

STP values approximate the freezing point of water and atmospheric pressure at sea level. However, these are not the most comfortable conditions in which to make measurements. Thus, a second specific set of conditions that resembles the conditions in a more comfortable environment, such as a laboratory, has been devised. This second set of conditions is called **standard ambient temperature and pressure (SATP)**, defined as a temperature of 25°C (298.15 K) and a pressure of 100 kPa. Table 12.1 summarizes the standard conditions of temperature and pressure, as well as the calculated molar volumes of an ideal gas at these conditions.

Table 12.1 Standard Conditions of Temperature and Pressure

Conditions	Pressure	Celsius Temperature	Kelvin Temperature	Molar Volume of an Ideal Gas
STP	101.325 kPa	0°C	273.15 K	22.4 L/mol
SATP	100.0 kPa	25°C	298.15 K	24.8 L/mol

Molar volumes at STP have been experimentally determined for many gases. Examples are provided in Table 12.2. Notice how the values in Table 12.2 are approximately 22.4 L/mol, which is the molar volume of an ideal gas at STP. Therefore, 22.4 L/mol is often used in calculations involving the molar volume of a gas. When you are doing calculations for SATP conditions, the standard molar volume to use is 24.8 L/mol.

Table 12.2 Experimentally Determined Molar Volumes of Gases at STP

Gas	Molar Volume (L/mol)
helium	22.398
neon	22.401
argon	22.410
hydrogen	22.430
nitrogen	22.413
oxygen	22.414
carbon dioxide	22.414
ammonia	22.350

In the Sample Problems and Practice Problems that appear on the following pages, you will learn how to use and calculate the molar volume of a gas under STP conditions and SATP conditions.

Students performed a series of measurements to determine the molar volumes of three different gases. Their results are in the table below. Note: the temperature in the room was 23.0°C (296.15 K), and the pressure was 98.7 kPa.

Measurements of Three Gases

Measurement	Carbon Dioxide	Oxygen	Methane
Volume of gas (V)	150 mL	150 mL	150 mL
Mass of empty syringe	25.08 g	25.08 g	25.08 g
Mass of gas + syringe	25.34 g	25.27 g	25.18 g
Mass of gas (m)			
Molar mass of gas (M)			
Amount (in moles) of gas $\left(n = \dfrac{m}{M}\right)$			

Procedure

1. Copy and complete the table. Calculate the molar volume of each gas at the experimental conditions.

2. Using the combined gas law, determine the volume of each gas at STP. Using those values, calculate the molar volume of each gas at STP.

Questions

1. Compare the molar volumes at STP with those for the experimental conditions. What do you observe for each gas?

2. Compare the three molar volumes at STP. What do you notice?

Sample Problem

Using Avogadro's Law to Find the Quantity of a Gas

Problem

At STP, 1 mol of oxygen gas has a volume of 22.4 L. Determine the mass and number of molecules in a 44.8 L sample of the gas.

What Is Required?

You need to find the mass and number of molecules of a sample of oxygen gas.

What Is Given?

You know the initial amount of the gas, as well as the initial and final volumes:

$V_1 = 22.4$ L

$n_1 = 1$ mol

$V_2 = 44.8$ L

You also know the molar mass of $O_2(g)$: 2×16.00 g/mol $= 32.00$ g/mol

Algebraic Method

Plan Your Strategy	Act on Your Strategy
Use Avogadro's law and solve for the amount (in moles) of oxygen.	$\dfrac{n_1}{V_1} = \dfrac{n_2}{V_2}$ $n_2 = \dfrac{n_1 V_2}{V_1} = \dfrac{1.00 \text{ mol} \times 44.8 \text{ L}}{22.4 \text{ L}} = 2.00$ mol
Calculate the mass of oxygen by using the amount (in moles) and the molar mass of oxygen.	$m = n \times M$ $= 2.00 \text{ mol} \times 32.00 \text{ g/mol} = 64.0$ g
Calculate the number of molecules by multiplying the amount (in moles) by the Avogadro constant.	number of molecules $= 2.00 \text{ mol} \times 6.02 \times 10^{23}$ molecules/mol $= 1.20 \times 10^{24}$ molecules

Check Your Solution

The volume of oxygen is double the volume of 1 mol of gas, so it makes sense that 2 mol of oxygen and twice the number of molecules are present.

Calculating the Molar Volume of Nitrogen

Problem

An empty, sealed container has a volume of 0.652 L and a mass of 2.50 g. When filled with nitrogen gas, the container has a mass of 3.23 g. The pressure of the nitrogen in the container is 97.5 kPa when the temperature is 21.0°C. Calculate the molar volume of nitrogen gas at STP.

What Is Required?

You need to find the molar volume, v, of 1 mol of nitrogen gas at STP.

What Is Given?

You know the initial temperature, pressure, and the volume of the gas, as well as the mass of the container when empty and when filled with nitrogen:

$T_1 = 21.0°C$, or $T_1 = 294.15$ K
$P_1 = 97.5$ kPa
$V_1 = 0.652$ L

Mass of the container $= 2.50$ g
Mass of the container $+$ nitrogen $= 3.23$ g

You also know the conditions of STP and the molar mass of nitrogen:

$T_2 = 0°C$, or 273.15 K

$P_2 = 101.325$ kPa

Molar mass of $N_2(g) = 2 \times 14.01$ g/mol $= 28.02$ g/mol

Plan Your Strategy	Act on Your Strategy
Find the mass of the nitrogen in the container by subtracting the mass of the empty container from the mass of the container filled with nitrogen gas.	$m_{nitrogen} = m_{container} - m_{vacuum}$ $= 3.23$ g $- 2.50$ g $= 0.73$ g
Calculate the amount (in moles) of nitrogen by using the formula $n = \frac{m}{M}$, where $m =$ mass of nitrogen and $M =$ molar mass of nitrogen.	$n = \frac{m}{M}$ $= \frac{0.73 \text{ g}}{28.02 \text{ g/mol}} = 0.026053$ mol
Use the combined gas law to find the volume, V_2, that the nitrogen would occupy at STP.	$\frac{P_1 V_1}{T_1} = \frac{P_2 V_2}{T_2}$ $\frac{P_1 V_1}{T_1}\left(\frac{T_2}{P_2}\right) = \frac{P_2 V_2}{T_2}\left(\frac{T_2}{P_2}\right)$ $V_2 = \frac{P_1 V_1 T_2}{T_1 P_2}$ $= \frac{(97.5 \text{ kPa})(0.652 \text{ L})(273.15 \text{ K})}{(294.15 \text{ K})(101.325 \text{ kPa})} = 0.582597$ L
Find the volume of 1 mol of nitrogen at STP by dividing the volume of nitrogen at STP by the amount (in moles) of nitrogen.	$v = \frac{V}{n}$ $= \frac{V_2}{n}$ $= \frac{0.582597 \text{ L}}{0.026053 \text{ mol}}$ $= 22.4$ L/mol

Alternative Solution

Plan Your Strategy	Act on Your Strategy
Calculate the amount (in moles) of nitrogen by using the molar mass and the mass of nitrogen (0.73 g).	$n = \dfrac{m}{M}$ $= \dfrac{0.73\ \text{g}}{28.02\ \text{g/mol}} = 0.026053\ \text{mol}$
You know that the volume decreases when the pressure increases. Determine the ratio of the initial pressure and the final pressure that is less than 1. You know that a decrease in temperature also causes a decrease in volume. Determine the ratio of the initial temperature and the final temperature that is less than 1.	$P_1 = 97.5\ \text{kPa}$ $P_2 = 101.325\ \text{kPa}$ pressure ratio < 1 is $\dfrac{97.5\ \text{kPa}}{101.325\ \text{kPa}}$ $T_1 = 294.15\ \text{K}$ $T_2 = 273.15\ \text{K}$ temperature ratio < 1 is $\dfrac{273.15\ \text{K}}{294.15\ \text{K}}$
Multiply the initial volume by the pressure and temperature ratios to obtain the final volume of nitrogen gas.	$V_2 = V_1 \times$ pressure ratio $\times$ temperature ratio $= 0.652\ \text{L} \times \dfrac{97.5\ \text{kPa}}{101.325\ \text{kPa}} \times \dfrac{273.15\ \text{K}}{294.15\ \text{K}}$ $= 0.582597\ \text{L}$
There is less than 1 mol of nitrogen gas present, so the molar volume (the volume of 1 mol of nitrogen gas) will be greater than the volume that you calculated above. To find the molar volume, multiply by a mole ratio that is greater than 1. The molar volume, v, is equal to the volume calculated divided by 1.00 mol.	mole ratio > 1 is $\dfrac{1\ \text{mol}}{0.026053\ \text{mol}}$ volume of 1 mol $= V_2 \times$ mol ratio $= (0.582597\ \text{L}) \times \dfrac{1\ \text{mol}}{0.026053\ \text{mol}}$ $= 22.4\ \text{L}$ Therefore, $v = 22.4\ \text{L/mol}$

Check Your Solution

The answer is expressed in the correct units and it agrees with the accepted molar volume value.

Practice Problems

11. At STP, 1.0 mol of carbon dioxide gas has a volume of 22.41 L. What mass of carbon dioxide is present in 3.0 L?

12. At STP, 1.0 mol of nitrogen gas occupies a volume of 22.41 L. Find the volume that 15.50 g of nitrogen gas occupies at STP.

13. Find the volume that 20.0 g of carbon monoxide gas, $CO(g)$, occupies at SATP.

14. An experiment generates 0.152 g of hydrogen gas. What volume of gas was generated at STP?

15. A solid block of carbon dioxide has a mass of 2.50×10^2 g. Once the block has totally sublimated, what volume would it occupy at SATP?

16. A commercial refrigeration unit accidentally releases 12.5 L of ammonia gas at SATP. Determine the mass and number of molecules of ammonia gas released.

17. A 6.98 g sample of chlorine gas has a volume of 2.27 L at 0.0°C and 1.0 atm. Find the molar volume of the chlorine gas at 25°C and 100.0 kPa.

18. A sample of helium gas has a mass of 11.28 g. At STP, the sample has a volume of 63.2 L. What is the molar volume of this gas at 32.2°C and 98.1 kPa?

19. Magnesium, $Mg(s)$, was reacted with hydrochloric acid, $HCl(aq)$, in excess according to the following equation:

$$Mg(s) + 2HCl(aq) \rightarrow MgCl_2(aq) + H_2(g)$$

When 0.0354 g of magnesium was reacted, 35.63 mL of hydrogen gas was collected.

a. How many moles of magnesium were reacted?

b. How many moles of hydrogen were collected?

c. If the hydrogen gas was collected at 20.0°C and 99.5 kPa, determine the molar volume of hydrogen gas from the experimental data.

20. Helium has a density of 0.179 g/L at STP. Calculate its molar volume at these conditions.

Section Summary

- Gases can undergo changes in temperature, pressure, and volume simultaneously. The combined gas law addresses these changes and is represented mathematically as

$$\frac{P_1 V_1}{T_1} = \frac{P_2 V_2}{T_2}$$

- The law of combining volumes states that, when gases react, the volumes of the gaseous reactants and products, measured under the same conditions of temperature and pressure, are always in whole-number ratios.

- If the temperature and pressure are constant and the same for the reactants and products of a chemical reaction, the law of combining volumes can be used to determine the amounts of reactant and/or products.

- Avogadro proposed that, under the same conditions of temperature and pressure, equal volumes of all ideal gases contain the same number of molecules, even though they do not have the same mass. A mathematical statement of Avogadro's law is

$$\frac{n_1}{V_1} = \frac{n_2}{V_2}$$

Review Questions

1. **T/I** A weather balloon is released. If you know the initial volume, temperature, and air pressure, what information will you need to predict the balloon's volume when it reaches its final altitude?

2. **K/U** Based on the combined gas law, will the volume of a fixed amount of gas increase, decrease, or remain the same for each of the following changes. Explain each answer.
   a. The pressure is decreased from 2 atm to 1 atm; the temperature is decreased from 200°C to 100°C.
   b. The pressure is increased from 1 atm to 4 atm, while the temperature is increased from 100°C to 200°C.
   c. The pressure is increased from 0.5 atm to 1.0 atm; the temperature is decreased from 250°C to 100°C.

3. **A** Explain how the combined gas law could be applied to the use of weather balloons.

4. **T/I** A sample of argon gas occupies a volume of 2.0 L at −35°C and standard atmospheric pressure. What would its Celsius temperature be at 2.0 atm if its volume decreased to 1.5 L? Assume the amount of gas remains constant.

5. **T/I** Consider the diagram below. What is the pressure of the nitrogen gas in the flask on the right, assuming the amount of gas remains constant?

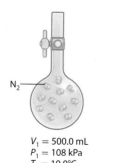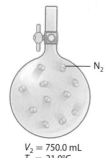

$V_1 = 500.0$ mL
$P_1 = 108$ kPa
$T_1 = 10.0°C$

$V_2 = 750.0$ mL
$T_2 = 21.0°C$

6. **T/I** A propane tank has a volume of 76 L and is at a temperature of 25°C and a pressure of 1500 kPa. Gas is discharged at a rate of 5 mL/min at a pressure of 100 kPa and temperature of 30°C. How long will it take for the tank to empty?

7. **K/U** If the volume of 1.0 mol of helium at 37°C and 90.0 kPa is 29 L, what will be the volume of 1.0 mol of nitrogen at the same temperature and pressure? Explain your reasoning.

8. **T/I** If all reactants are at the same temperature and pressure, what volume of oxygen gas is required for 15 L of methane to undergo complete combustion?

9. **T/I** Nitrogen gas reacts with oxygen gas to produce nitrogen dioxide gas. What volume of nitrogen dioxide gas is formed when 350 mL of nitrogen gas reacts with an excess of oxygen gas?

10. **K/U** In your own words, describe Avogadro's law.

11. **T/I** A balloon filled with helium gas, He(g), develops a leak. If it originally contained 0.20 mol of gas and had a volume of 3.5 L, how many moles of gas are remaining when the balloon has a volume of 2.5 L?

12. **T/I** Determine the following for a 10.0 L sample of argon gas at STP.
   a. the amount (in moles) of argon present
   b. the mass (in grams) of argon present
   c. the number of molecules of argon
   d. the molar volume

13. **C** Use a graphic organizer to distinguish between the terms STP and SATP, and explain the value of defining and using these standards.

14. **T/I** What is the molar volume of hydrogen gas at 255°C and 102 kPa, if a 1.09 L volume of the gas has a mass of 0.0513 g?

The Ideal Gas Law

You have related the combined gas law to Avogadro's volume-mole gas relationship using two sets of conditions. This enabled you to make calculations of pressure, temperature, volume, and amount of gas by holding two of the variables constant while you manipulated a third and calculated a fourth. The ideal gas law combines all four variables into a single relationship that lets you summarize mathematically the relationships expressed by the combined gas law and Avogadro's law for ideal gases. This relationship is expressed as

$$PV = nRT$$

The relationship $PV = nRT$ is the **ideal gas law**: the pressure of a gas multiplied by its volume is equal to its amount in moles multiplied by a proportionality constant, R, and the temperature. The proportionality constant, called the **universal gas constant**, is $8.314 \, \dfrac{kPa \cdot L}{mol \cdot K}$, and applies to all gases. This value for R is based on the molar volume of a gas at STP being 22.4 L/mol. Thus, using $n = 1.00$ mol, $T = 273.15$ K, $P = 101.325$ kPa, and $V = 22.4$ L, R can be calculated by using the ideal gas law equation.

This equation lets you calculate one of the four gas variables if you have data for the other three. Importantly, the equation does not require that you compare two sets of conditions for the same sample of gas, which you must do when using the other gas laws.

Guidelines for Using the Ideal Gas Law

To use the ideal gas law equation properly, remember the following guidelines:

- Always convert the temperature to kelvin units (K).

- Always convert the mass to moles (mol).

- Always convert the volume to litres (L).

- Calculations are easier if you always convert the pressure to kilopascals (kPa).

  Then you can remember just one value of $R \left(8.314 \, \dfrac{kPa \cdot L}{mol \cdot K} \right)$ for each calculation.

  If you forget the value of R, you can calculate it by finding R for 1 mol of gas at STP.

Density and Molar Mass of Gases

Since the ideal gas law considers the amount of gas present, this allows you to determine other properties of the gas. As you learned in the last section, the molar volume of a gas is defined as the space that is occupied by one mole of the gas. It is always given in units of L/mol. The density of a gas is similar to the density of a solid or a liquid. Density is found by dividing mass by volume. The density of a gas is usually reported in units of g/L. The molar mass of a gas refers to the mass (in grams) of one mole of the gas. As you know, you can calculate the molar mass of a substance by adding the molar masses of its atoms from the periodic table of the elements. You can also calculate molar mass by dividing the mass of a sample by the number of moles that are present. Molar mass is always expressed in the units g/mol. **Table 12.3** at the top of the next page summarizes molar volume, density, and molar mass. These three properties are closely related, and any one of the properties can be calculated using the other two properties.

The Sample Problems and Practice Problems that follow on the next two pages will reinforce your understanding of how the ideal gas law can be used to determine various properties of gases.

Key Terms

ideal gas law
universal gas constant (*R*)
partial pressure
Dalton's law of partial pressures

ideal gas law a gas law that describes the relationship among volume, pressure, temperature, and amount (in moles) of an ideal gas: $PV = nRT$

universal gas constant (*R*) a proportionality constant that relates the pressure on and the volume of an ideal gas to its amount and temperature:
$R = 8.314 \, \dfrac{kPa \cdot L}{mol \cdot K}$

Go to **scienceontario** to find out more

Table 12.3 Units of Molar Volume, Density, and Molar Mass

	Unit	Meaning	Calculations
Molar volume	L/mol	volume/amount	$\text{molar volume} = \dfrac{\text{volume}}{\text{amount (in moles)}}$ $v = \dfrac{V}{n}$
Density	g/L	mass/volume	$\text{density} = \dfrac{\text{mass}}{\text{volume}}$ $D = \dfrac{m}{V}$
Molar mass	g/mol	mass/amount	$\text{molar mass} = \text{sum of the molar masses of}$ $\text{the atoms in the substance, or}$ $\text{molar mass} = \dfrac{\text{mass}}{\text{amount (in moles)}}$ $M = \dfrac{m}{n}$

Inquiry Investigation 12-A, Using the Ideal Gas Law to Determine the Pressure to Make Popcorn

Sample Problem

Finding the Volume of a Gas Using the Ideal Gas Law

Problem

Find the volume of 100.0 g of oxygen gas at SATP.

What Is Required?

You need to find the volume, V, that 100.0 g of oxygen occupies at standard ambient temperature and pressure.

What Is Given?

You know the conditions of SATP:

$T = 298.15$ K

$P = 100.0$ kPa

You know the molar mass of $O_2(g)$: 2×16.00 g/mol $= 32.00$ g/mol

Plan Your Strategy	Act on Your Strategy
Find the amount (in moles) in 100.0 g of oxygen by dividing the mass by the molar mass of oxygen.	$n = \dfrac{m}{M}$ $= \dfrac{100.0 \text{ g}}{32.00 \text{ g/mol}} = 3.125$ mol
Use the ideal gas law.	$PV = nRT$
Isolate the variable V by dividing each side of the equation by P.	$PV = nRT$ $PV\left(\dfrac{1}{P}\right) = nRT\left(\dfrac{1}{P}\right)$ $V = \dfrac{nRT}{P}$
Substitute numbers and units for the known variables into the formula and solve for V.	$V = \dfrac{(3.125 \text{ mol})\left(8.314 \frac{\text{kPa} \cdot \text{L}}{\text{mol} \cdot \text{K}}\right)(298.15 \text{ K})}{100.0 \text{ kPa}}$ $= 77.46$ L

Check Your Solution

One mole of an ideal gas at SATP occupies 24.8 L. Therefore, it makes sense that the volume of slightly more than 3 mol of gas should be in the range of 77 L.

Finding the Temperature of a Gas Using the Ideal Gas Law

Problem
Find the temperature, in °C, of 2.50 mol of gas that occupies a volume of 56.5 L at a pressure of 1.20 atm.

What Is Required?
You need to find the Celsius temperature of 2.50 mol of gas, given its volume and pressure.

What Is Given?
You know the amount, pressure, and volume of the gas:

$n = 2.50$ mol

$P = 1.20$ atm

$V = 56.5$ L

Plan Your Strategy	Act on Your Strategy
Convert the units of pressure from atm to kPa.	$P = 1.20 \text{ atm} \left(\dfrac{101.325 \text{ kPa}}{\text{atm}} \right)$ $= 121.59 \text{ kPa}$
Use the ideal gas law.	$PV = nRT$
Isolate the variable T by dividing each side of the equation by nR.	$PV = nRT$ $\dfrac{PV}{nR} = \dfrac{nRT}{nR}$ $\dfrac{PV}{nR} = T$
Substitute numbers and units for the known variables in the formula and solve for T.	$T = \dfrac{(121.59 \text{ kPa})(56.5 \text{ L})}{2.50 \text{ mol} \left(8.314 \frac{\text{kPa} \cdot \text{L}}{\text{mol} \cdot \text{K}} \right)}$ $= 330.519 \text{ K}$
Convert the temperature to degrees Celsius.	$T = 330.519 \text{ K} - 273.15$ $= 57.4°C$

Check Your Solution
The pressure has been converted to kPa and the correct value of R is used, given the units of the variables. The temperature has been converted from the Kelvin scale to the Celsius scale.

Determining the Molecular Formula of a Gas Using Percentage Composition and the Ideal Gas Law

Problem
What is the molecular formula of an unknown gas that is composed of 80.0% carbon and 20.0% hydrogen if a 4.60 g sample occupies a 2.50 L volume at 25.00°C and 152 kPa?

What Is Required?
You need to determine the molecular formula of an unknown gas.

Continued on next page ⟩

What Is Given?

You know the following:

$T = 25.00°C$

$P = 152 \text{ kPa}$

$V = 2.50 \text{ L}$

$m = 4.60 \text{ g}$

$R = 8.314 \dfrac{\text{kPa} \cdot \text{L}}{\text{mol} \cdot \text{K}}$

The percentage composition of the gas is 80.0% carbon and 20.0% hydrogen.

Plan Your Strategy	Act on Your Strategy
Find the empirical formula using the molar masses of carbon and hydrogen and the percentage compositions.	For a 100 g sample: carbon: $80.0\% \times 100 \text{ g} = 80.0 \text{ g}$ hydrogen: $20.0\% \times 100 \text{ g} = 20.0 \text{ g}$ Determine the moles of each element using the formula $n = \dfrac{m}{M}$ For carbon: $n = \dfrac{80.0 \text{ g}}{12.01 \text{ g/mol}} = 6.661116 \text{ mol}$ For hydrogen: $n = \dfrac{20.0 \text{ g}}{1.01 \text{ g/mol}} = 19.80198 \text{ mol}$ The simplest ratio of the two elements provides the empirical formula. $\dfrac{6.661116}{6.661116}$ mol of C $: \dfrac{19.80198}{6.661116}$ mol of H The mole ratio is 1 mol of C : 3 mol of H. The empirical formula is CH_3.
Convert the temperature to kelvin units.	$T = 25.00°C + 273.15$ $\quad = 298.15 \text{ K}$
Use the ideal gas law.	$PV = nRT$
Isolate the variable n by dividing both sides of the equation by RT and rearranging the equation. Substitute numbers and units for the known variables into the formula to solve for n.	$PV = nRT$ $n = \dfrac{PV}{RT}$ $\quad = \dfrac{152 \text{ kPa} \times 2.50 \text{ L}}{8.314 \dfrac{\text{kPa} \cdot \text{L}}{\text{mol} \cdot \text{K}} \times 298.15 \text{ K}}$ $\quad = 0.153299 \text{ mol}$
Find the molar mass (M) of the gas by dividing the mass (m) of the gas by the amount (n) of the gas.	$M = \dfrac{m}{n}$ $\quad = \dfrac{4.60 \text{ g}}{0.153299 \text{ mol}}$ $\quad = 30.007 \text{ g/mol}$
Compare the molar mass of the unknown gas with the molar mass of the empirical formula. Multiplying the empirical formula by the ratio of the two molar masses provides the molecular formula.	Molar mass of the unknown gas = 30.007 g/mol Molar mass of the empirical formula = 12.01 g/mol + 3(1.01 g/mol) = 15.04 g/mol ratio of molar masses = $\dfrac{30.007 \text{ g/mol}}{15.04 \text{ g/mol}}$ Since the ratio of the molar masses is 2:1, the molecular formula is $CH_3 \times 2 = C_2H_6$

Check Your Solution

The simple integer ratio of the two molar masses makes the molecular formula a reasonable answer. When the ideal gas law is used, all units cancel out except for mol, which is appropriate since the amount of gas was being calculated.

Finding the Density of a Gas Using the Ideal Gas Law

Problem

What is the density of nitrogen gas in grams per litre, at 25.00°C and 126.63 kPa?

What Is Required?

You need to determine the density of nitrogen gas at 25.00°C and 126.63 kPa.

What Is Given?

You know the temperature of the gas and its pressure:

$T = 25.00°C$

$P = 126.63$ kPa

You also know the molar mass of $N_2(g)$: 2×14.01 g/mol $= 28.02$ g/mol

Plan Your Strategy	Act on Your Strategy
Convert the temperature to kelvin units.	$T = 25.00°C + 273.15$ $= 298.15$ K
Isolate the variable n and rearrange the equation. Substitute the numbers and units for P, T, R, and V into the equation. Since the volume is not given, set it as 1.00 L and solve for n.	$n = \dfrac{PV}{RT}$ $= \dfrac{126.63 \text{ kPa} \times 1.00 \text{ L}}{8.314 \frac{\text{kPa} \cdot \text{L}}{\text{mol} \cdot \text{K}} \times 298.15 \text{ K}} = 5.1085 \times 10^{-2}$ mol
Convert n to the mass of nitrogen (m) in 1.00 L by multiplying the amount by the molar mass (M) of nitrogen.	$m = n \times M$ $= 5.1085 \times 10^{-2}$ mol $\times 28.02$ g/mol $= 1.4314$ g
Determine the density by dividing the mass by the 1.00 L that was used in the ideal gas law equation.	$D = \dfrac{m}{V}$ $= \dfrac{1.4314 \text{ g}}{1.00 \text{ L}}$ $= 1.431$ g/L

Check Your Solution

When the units cancel out in the ideal gas equation, the unit "mol" remains. When units cancel out in the density equation, the unit "g/L" remains.

Finding the Molar Mass of a Gas Using the Ideal Gas Law

Problem

A 1.58 g sample of gas occupies a volume of 500.0 mL at STP. Calculate the molar mass of the gas.

What Is Required?

You need to find the molar mass, M, of a gas, based on a 1.58 g sample occupying a volume of 500 mL at standard temperature and pressure.

What Is Given?

You know the volume of the gas, the temperature and pressure at STP, as well as the mass of the sample:

$V = 0.5000$ L

$T = 273.15$ K

$P = 101.325$ kPa

$m = 1.58$ g

Continued on next page ❯

Plan Your Strategy	Act on Your Strategy
Use the ideal gas law.	$PV = nRT$
Isolate the variable n by dividing each side of the equation by RT and then rearranging the equation. Substitute numbers and units for the known variables into the formula and solve for n.	$n = \dfrac{PV}{RT}$ $= \dfrac{(101.325 \text{ kPa})(0.500 \text{ L})}{\left(8.314 \dfrac{\text{kPa} \cdot \text{L}}{\text{mol} \cdot \text{K}}\right)((273.15 \text{ K}))}$ $= 0.02231 \text{ mol}$
Calculate the molar mass (M) by dividing the mass (m) by the amount (in mol).	$M = \dfrac{m}{n}$ $= \dfrac{1.58 \text{ g}}{0.02231 \text{ mol}}$ $= 70.8 \text{ g/mol}$

Check Your Solution

The molar volume of an ideal gas at STP is approximately 22.4 L/mol, which is about 45 times the volume occupied by 1.58 g of the gas. That mass, multiplied by 45, is consistent with the answer.

Practice Problems

21. What is the volume of 5.65 mol of helium gas at a pressure of 98 kPa and a temperature of 18.0°C?

22. Propane, C_3H_8, is a common gas used to supply energy for barbecue cookers as well as energy-requiring appliances in cabins and cottages, and heavy equipment such as the forklift shown in the photograph below. If a tank contains 20.00 kg of propane, what volume of propane gas could be supplied at 22°C and 100.5 kPa?

Forklift trucks that run on propane are alternatives to those that run on gasoline or diesel fuel.

23. Find the Celsius temperature of nitrogen gas if a 5.60 g sample occupies 2.40×10^3 mL at 3.00 atm of pressure.

24. What is the pressure of 3.25 mol of hydrogen gas that occupies a volume of 67.5 L at a temperature of 295 K?

25. A weather balloon filled with helium gas has a volume of 960 L at 101 kPa and 25°C. What mass of helium was required to fill the balloon?

26. Find the molar mass of 6.24 g of an unknown gas that occupies 2.5 L at 18.3°C and 100.5 kPa.

27. A scientist isolates 2.366 g of a gas. The sample occupies a volume of 8.00×10^2 mL at 78.0°C and 103 kPa. Calculate the molar mass of the gas. Is the gas most likely to be bromine, krypton, neon, or fluorine?

28. What is the density of carbon dioxide gas, in grams per litre, at SATP?

29. A hydrocarbon gas used for fuel contains the elements carbon and hydrogen in percentages of 82.66 percent and 17.34 percent. Some of the gas, 1.77 g, was trapped in a 750 mL round-bottom flask. The gas was collected at a temperature of 22.1°C and a pressure of 99.7 kPa.

　a. Determine the empirical formula for this gas.

　b. Calculate the molar mass of the gas.

　c. Determine the molecular formula for this gas.

30. A 10.0 g sample of an unknown liquid is vaporized at 120.0°C and 5.0 atm. The volume of the vapour is found to be 568.0 mL. The liquid is determined to be made up of 84.2% carbon and 15.8% hydrogen. What is the molecular formula of the liquid?

Dalton's Law of Partial Pressures

For the calculations you have done so far, 1 mol of any gas will have the same volume when the temperature and pressure are constant. Because gases are completely miscible, any mixture of non-reacting gases will have the same molar volume as any pure gas. However, scientists often want to consider just one particular gas in a mixture of gases. Is there a way to describe the pressure of one particular type of gas when it is mixed with other gases?

This question was investigated by John Dalton when he was studying atmospheric humidity. It is common for water vapour from the surroundings to mix with gases that are being studied in the laboratory. Upon adding water vapour to dry gases, Dalton observed that the total gas pressure was equal to the sum of the pressure of the dry gas and the pressure of the water vapour on the walls of the container. Mathematically, this is presented as

$$P_{total} = P_{dry\ air} + P_{water\ vapour}$$

Further studies showed that this phenomenon holds true for the addition of any type of gas to a mixture of gases. When gases are mixed, the pressure that any one gas exerts on the walls of the container is called the **partial pressure** of that gas. **Dalton's law of partial pressures** states that, in a mixture of gases that do not react chemically, the total pressure is the sum of the partial pressures of each individual gas.

Figure 12.5 illustrates Dalton's law of partial pressures. When two separate gases originally at the same temperature are mixed, the temperature—and thus the average speed of the gases—does not change. Only the total number of molecules in the container increases. As a result, there are more molecules colliding with the walls of the container and, thus, the pressure is higher.

partial pressure the portion of the total pressure of a mixture of gases contributed by a single gas component

Dalton's law of partial pressures a gas law stating that the total pressure of a mixture of gases is the sum of the individual pressures of each gas

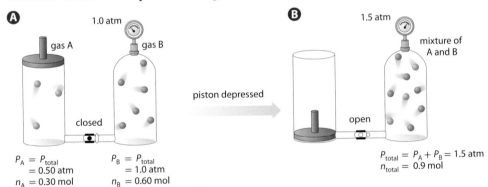

Figure 12.5 When two gases are in the same container, the molecules of each gas collide with the walls of the container as many times and with the same force as the molecules would do if the gases were in separate containers. Therefore, the pressure on the wall of the container with mixed gases is the sum of the pressures of the gases in separate containers.

Identify How would the total pressure in the container with the mixture of A and B change if twice the amount (in moles) of gas A were added?

Learning Check

7. Identify the two laws that are used to derive the ideal gas law.

8. Explain why the use of specific units for temperature, pressure, and volume is important when using the ideal gas law. What units should be used for each variable?

9. A gas mixture is composed of 40.0 percent xenon gas that has a partial pressure of 110 kPa. What is the total pressure of the gas mixture?

10. A gas mixture contains 12 percent helium, 25 percent argon, and 63 percent neon. If the total pressure is 2.0 atm, what is the partial pressure of each gas?

11. Why might the humidity of the air influence the total pressure of a gas in an open container?

12. Our atmosphere is approximately 80.0 percent nitrogen gas. At standard pressure, how much pressure is contributed by the nitrogen?

Collecting a Gas in the Laboratory

One of the safest ways to collect a gas in the laboratory is by downward displacement of water. A container, such as a graduated cylinder, is filled with water and inverted into a beaker of water. Care must be taken to avoid letting any air into the inverted cylinder. Tubing from the source of the gas is placed in the water and directed up into the inverted cylinder, as shown in **Figure 12.6**. The gas bubbles up through the water and collects in the closed end of the cylinder. The gas pushes the water down and out of the cylinder, leaving the gas sample trapped above the water. The cylinder can be adjusted so that the water level inside the cylinder is even with the water level in the beaker. The volume of the gas is read from the scale on the cylinder, and the pressure on the gas is the same as the atmospheric pressure. Many laboratories have a barometer in the room showing the barometric (atmospheric) pressure. The ambient temperature can be measured with a laboratory thermometer.

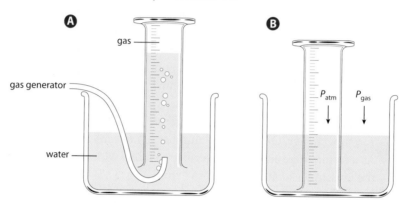

Figure 12.6 **(A)** Because gases have a much lower density than water does, gases float to the top of the container. **(B)** When the position of the container is adjusted so that the water levels inside and outside the container are the same, the gas pressures above the water are the same inside and outside the container.

Explain *Why must care be taken to avoid letting any air into the cylinder when it is inverted?*

When you collect gases over water, there is one factor that you must take into account, even if the gas is not soluble in water. The molecules of liquid water vaporize and mix with the molecules of the gas. Therefore, according to Dalton's law of partial pressures, the total pressure of the gases trapped in the cylinder is equal to the sum of the pressure exerted by each component of the gas mixture. Mathematically, this is represented by

$$P_{total} = P_{gas} + P_{water\ vapour}$$

Thus, to determine the pressure of the collected gas, often referred to as the pressure of the dry gas, the pressure of the water vapour must be subtracted from the total pressure. **Table 12.4** provides the partial pressures of water vapour at different temperatures.

Table 12.4 Partial Pressures of Water Vapour at Different Temperatures

Temperature (°C)	Pressure (kPa)	Temperature (°C)	Pressure (kPa)
15	1.71	22	2.65
16	1.81	23	2.81
17	1.93	24	2.99
18	2.07	25	3.17
19	2.20	26	3.36
20	2.33	27	3.56
21	2.49	28	3.78

Gas Stoichiometry

As you know, stoichiometry refers to the relationship between the amounts (in moles) of reactants and products in a chemical reaction. The stoichiometry of chemical reactions allows you to determine the quantity of one reactant or product if you know the quantity of another reactant or product.

When the volumes of gaseous reactants or products are not under the same conditions of temperature and pressure, you must use the ideal gas law to determine the quantities of reactant or products in a chemical reaction. A general set of steps to follow when using the ideal gas law to solve gas stoichiometry problems is shown below.

Suggested Investigation

Inquiry Investigation 12-B, Measuring the Molar Volume of a Gaseous Product in a Chemical Reaction

Steps for Solving Gas Stoichiometry Problems

1. Write a balanced equation for the reaction.

2. Convert all amounts to moles.

3. Compare molar amounts using stoichiometry ratios from the balanced equation. Solve for the unknown molar amount.

4. Convert the new molar amount into the units required, using a set of conditions with the ideal gas law, $PV = nRT$.

Also keep in mind that gaseous products of a chemical reaction are often collected over water in the laboratory. When using the ideal gas law for a gas collected over water, you must use Dalton's law of partial pressures to correct the pressure before substituting it into the gas law. To find the partial pressure of the dry gas, subtract the pressure of the water vapour from the total pressure: $P_{dry\ gas} = P_{total} - P_{water\ vapour}$

The following Sample Problem and Practice Problems will reinforce your understanding of the use of the ideal gas law in gas stoichiometry.

Sample Problem

Gas Stoichiometry Using the Ideal Gas Law

Problem

What volume of hydrogen gas is produced when excess sulfuric acid reacts with 40.0 g of iron at 18.0°C and 100.3 kPa?

What Is Required?

You need the volume of gas that is produced when sulfuric acid reacts with iron under specific temperature and pressure conditions.

What Is Given?

You know each of the following:

Reactants: sulfuric acid and iron

Products: an iron(II) compound and a gas

$m_{iron} = 40.0$ g

$T = 18.0°C$

$P = 100.3$ kPa

You also know the molar mass of iron: 55.85 g/mol

Continued on next page 〉

Plan Your Strategy	Act on Your Strategy
Write a balanced equation for the chemical reaction.	$Fe(s) + H_2SO_4(aq) \rightarrow H_2(g) + FeSO_4(aq)$
Calculate the amount (in moles) of iron present by dividing the mass, m, by the molar mass, M. Use this value, along with the mole ratios from the balanced equation, to calculate the amount of gas produced.	$n = \dfrac{m}{M}$ $= \dfrac{40.0 \text{ g}}{55.85 \text{ g/mol}}$ $= 0.71620 \text{ mol}$ The mole ratio is 1 mol of H_2 : 1 mol of Fe Therefore, the amount of hydrogen gas formed is $\dfrac{n \text{ mol } H_2}{0.71620 \text{ mol Fe}} = \dfrac{1 \text{ mol } H_2}{1 \text{ mol Fe}}$ $0.71620 \text{ mol Fe} \times \dfrac{n \text{ mol } H_2}{0.71620 \text{ mol Fe}} = \dfrac{1 \text{ mol } H_2}{1 \text{ mol Fe}} \times 0.71620 \text{ mol Fe}$ $n = 0.71620 \text{ mol } H_2$
Isolate the variable V by dividing both sides of the ideal gas law equation by P. Rearrange the equation, substitute the values for the amount of gas, the temperature in kelvin units, and the pressure into the ideal gas law, and solve for the volume of the gas.	$PV = nRT$ $V = \dfrac{nRT}{P}$ $= \dfrac{0.71620 \text{ mol} \times 8.314 \frac{\text{kPa} \cdot \text{L}}{\text{mol} \cdot \text{K}} \times 291.15 \text{ K}}{100.3 \text{ kPa}}$ $= 17.3 \text{ L}$

Check Your Solution

The answer is less than the molar volume of hydrogen gas at STP. Since less than 1 mol of hydrogen gas was formed, this seems reasonable.

Practice Problems

31. What volume of hydrogen gas will be produced at 93.0 kPa and 23°C from the reaction of 33 mg of magnesium with hydrochloric acid?

32. At STP, 0.72 g of hydrogen gas reacts with 8.0 L of chlorine gas. How many litres of hydrogen chloride gas are produced?

33. Determine the volume of nitrogen gas produced when 120 g of sodium azide, $NaN_3(s)$, decomposes at 27°C and 100.5 kPa. Sodium metal is the other product.

34. When calcium carbonate, $CaCO_3(s)$, is heated, it decomposes to form calcium oxide, $CaO(s)$, and carbon dioxide gas. How many liters of carbon dioxide will be produced at STP if 2.38 kg of calcium carbonate reacts completely?

35. When iron rusts, it undergoes a reaction with oxygen to form solid iron(III) oxide. Calculate the volume of oxygen gas at STP that is required to completely react with 52.0 g of iron.

36. Oxygen gas and magnesium react to form 2.43 g of magnesium oxide, $MgO(s)$. What volume of oxygen gas at 94.9 kPa and 25.0°C would be consumed to produce this mass of magnesium oxide?

37. In the semiconductor industry, hexafluoroethane, $C_2F_6(g)$, is used to remove silicon dioxide, $SiO_2(s)$, according to the following chemical equation:

$2SiO_2(s) + 2C_2F_6(g) + O_2(g) \rightarrow$
$\qquad\qquad 2SiF_4(g) + 2COF_2(g) + 2CO_2(g)$

What mass of silicon dioxide reacts with 1.270 L of hexafluoroethane at 0.200 kPa and 400.0°C?

38. What mass of oxygen gas reacts with hydrogen gas to produce 0.62 L of water vapour at 100.0°C and 101.3 KPa?

39. One method of producing ammonia gas involves the reaction of ammonium chloride, $NH_4Cl(aq)$, with sodium hydroxide, $NaOH(aq)$; water and aqueous sodium chloride are also products of the reaction. During an experiment, 98 mL of ammonia gas was collected using water displacement. If the gas was collected at 20.0°C and 780 mmHg, determine the amount of sodium hydroxide that must have reacted.

40. A student reacts 0.15 g of magnesium metal with excess dilute hydrochloric acid to produce hydrogen gas, which she collects over water. What volume of dry hydrogen gas does she collect over water at 28°C and 101.8 kPa?

Understanding Non-Ideal Gas Behaviour

The gas laws were developed to describe the behaviour and properties of ideal gases. How well does the behaviour of a *real* gas follow the laws for ideal gases? Look again at **Table 12.2** in the previous section. If you compare the measured molar volumes of gases in this table with the molar volume of an ideal gas (22.4 L/mol), a quick calculation shows that the maximum deviation of the molar volume of these gases is less than two percent from the ideal molar volume. In fact, real gases begin to deviate from ideal behaviour only when the temperature and pressure of their molecules diverge significantly from standard conditions. To understand why, recall what the kinetic molecular theory assumes about the of molecules of ideal gases:

- Particles move in straight lines at speeds determined by their temperature.
- Collisions are elastic and therefore do not use energy.
- Molecules are point masses, which have no volume.
- Molecules have no forces of attraction between each other and no forces of attraction with their container.

At STP, molecules of gases are moving very rapidly and are very far apart, which makes their interactions and volumes insignificant; thus, the assumptions for the ideal behaviour of gas molecules are valid. However, real gases begin to deviate from ideal behaviour when their molecules move slowly and are relatively close together. Low temperatures and high pressures are responsible for these conditions.

The Effects of Low Temperature

Intermolecular forces of attraction exist among real molecules of the same kind. These attractive forces do not significantly affect the behaviour of real gas molecules at standard temperatures, because the molecules are moving at high speeds and have a large amount of kinetic energy. When the molecules collide, their kinetic energy allows them to easily break their attractive interactions. The ability to break these short-range attractive forces decreases as the temperature of molecules decreases. At lower speeds and with reduced kinetic energy, molecules cannot easily break their attractive interactions. Eventually, these attractive interactions cause the gas to condense into a liquid.

The Effects of High Pressure: Reduced Collisions with the Container

Under standard atmospheric pressure, gas molecules are so far apart that interactions among them are very infrequent, so gases tend to behave ideally. When the external pressure on a gas increases substantially, however, the molecules are pushed closer together, and interactions are more frequent. As shown in **Figure 12.7**, when a gas molecule is about to collide with the walls of its container, nearby molecules exert attractive forces on the molecule, pulling it away. Therefore, the force of the collision with the wall is reduced, and the gas exerts less pressure than expected on the walls of the container. If you measured the pressure, you would find that it is lower than it would be for an ideal gas under the same conditions. Thus, calculations of other variables would be incorrect.

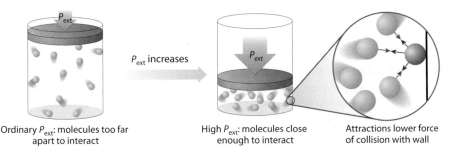

Ordinary P_{ext}: molecules too far apart to interact

High P_{ext}: molecules close enough to interact

Attractions lower force of collision with wall

Figure 12.7 Under high pressures, gas molecules are close enough together to interact with one another when they are about to collide with the wall of the container. These interactions reduce the force of the collisions with the wall.

The Effects of High Pressure: Gas Molecule Volume Is Significant

Another effect of high pressure involves volume. Under standard atmospheric pressure, the total volume percent of a container taken up by gas molecules is insignificant, so gas molecules behave as if they were point masses. When high pressures reduce the volume of the container, the total volume percent taken up by the gas molecules becomes significant. The V in the ideal gas law represents the total empty space between molecules, as per the kinetic molecular theory. However, when you measure the volume of a gas at high pressures, the value of V is larger than the actual volume of empty space, as shown in **Figure 12.8**. If you used this measured volume to calculate other variables, the calculations of other variables would be incorrect.

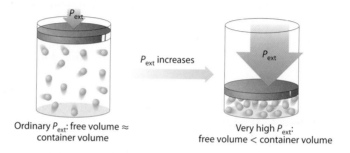

Figure 12.8 At standard atmospheric pressure, gas molecules are so far apart that they take up a very small percentage of the volume of the container. At high pressures, the actual volume of the gas molecules is a significant percentage of the volume of the container.

Summarizing Gas Laws and Properties

The graphic organizer in **Figure 12.9** summarizes the ideal gas law and concepts related to it.

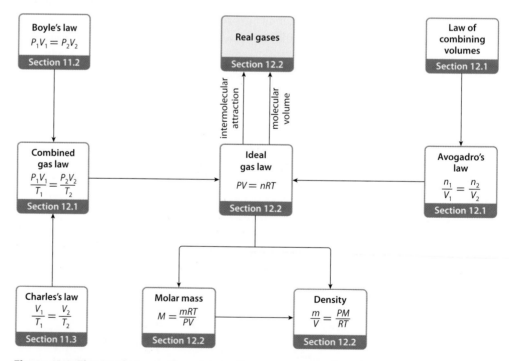

Figure 12.9 This graphic organizer relates various concepts you have learned in this unit to the ideal gas law.

Analyze *Which gas law is not included, and where would you place it in this organizer?*

Section Summary

- The ideal gas law is derived from the combined gas law and Avogadro's law. The mathematical equation for it is $PV = nRT$. R is the universal gas constant with a value of $8.314 \frac{kPa \cdot L}{mol \cdot K}$.
- For gas stoichiometry, conditions of pressure, volume, and temperature must be considered. When the reactants and products are not under the same conditions the ideal gas law is used to calculate the amounts of gaseous reactants or products.

- Real gases behave similarly to ideal gases under standard conditions of temperature and pressure. However, this behaviour deviates under conditions of low temperature and high pressure.

Review Questions

1. **C** Using a graphic organizer, summarize how the ideal gas law is derived from existing gas law equations. Why is the ideal gas law a useful equation for a chemist?

2. **K/U** Rearrange the ideal gas law equation to solve for each of the following variables.
   a. volume of a gas **c.** temperature of a gas
   b. pressure of a gas **d.** amount of a gas

3. **T/I** A tank of chlorine gas contains 25.00 kg of pressurized liquid chlorine. If all the chlorine is released in its gaseous form at 17°C and 0.98 atm, what volume does it occupy?

4. **T/I** A 5.0×10^2 kg sample of sulfur dioxide, $SO_2(g)$, has a volume of 2.37×10^6 L. If the temperature of the gas is 30.0°C, determine the pressure of the gas in units of kPa.

5. **T/I** An 11.91 g sample of a gas occupies 3.20 L at 24.2°C and 102 kPa. Find the molar mass of the gas.

6. **T/I** A syringe was filled with halogen gas to the 60.0 mL mark at the pressure of the surrounding room. The mass of the empty syringe was 65.780 g and the mass of the syringe and gas was 65.875 g. The room had an atmospheric pressure of 101.8 kPa, and the temperature was 21.7°C. Calculate the molar mass and identify the gas.

7. **T/I** The molar mass of dry air is 28.57 g/mol. Determine the density of air at 25°C and 102 kPa.

8. **T/I** A gaseous compound contains 92.31% carbon and 7.69% hydrogen by mass. If 4.35 g of the gas occupies 4.16 L at 22.0°C and 738 torr, determine the molecular formula of the gas.

9. **T/I** A sample of hydrogen gas is collected by water displacement at 20.0°C when the atmospheric pressure is 99.8 kPa. What is the pressure of the "dry" hydrogen, if the partial pressure of water vapour is 2.33 kPa at that temperature?

10. **K/U** Use the following diagram to explain Dalton's law of partial pressures.

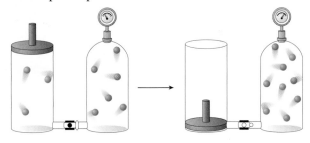

piston depressed and valve opened

11. **T/I** Zinc metal reacts with nitric acid, $HNO_3(aq)$, to produce 34.0 L of dry hydrogen gas at an atmospheric pressure of 900.0 torr and 20.0°C. How many grams of zinc are consumed?

12. **A** The process of cellular respiration involves a chemical reaction in which glucose (a fuel) reacts with oxygen, forming carbon dioxide and water. The general equation for cellular respiration is
    $$C_6H_{12}O_6(s) + 6O_2(g) \rightarrow 6CO_2(g) + 6H_2O(\ell)$$
    If 10.0 g of glucose is reacted, calculate the volume of carbon dioxide that is formed at 100.0 kPa and 37.0°C.

13. **T/I** When solid calcium is placed in water, it undergoes a vigorous reaction that produces solid calcium oxide, $CaO(s)$, and hydrogen gas. Determine the volume of dry hydrogen gas that is produced when 5.00 g of calcium reacts and the gas is collected at 28°C and 770 mmHg.

14. **C** Under what conditions does a real gas deviate from ideal gas behaviour? Use diagrams and kinetic molecular theory to illustrate your answer.

Atmospheric Gases and Air Quality

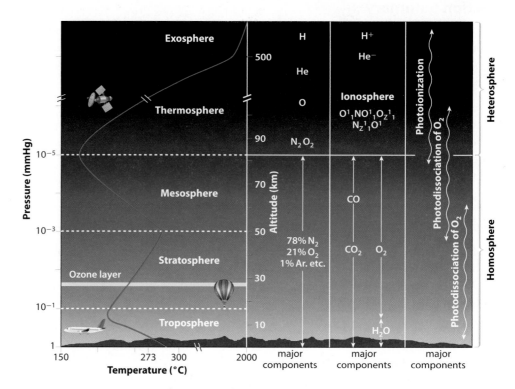

Figure 12.10 Variations in pressure, temperature, and the components that make up Earth's atmosphere are summarized here.

Infer *How can you explain the changes in temperature with altitude?*

Figure 12.10 summarizes information about the structure and composition of Earth's atmosphere. Much of this information is familiar to you from earlier in this unit or from your study of science or geography in earlier grades.

As you know from Boyle's law, gases are compressible. Thus, pressure in the atmosphere decreases with altitude, and this decrease is more rapid at lower altitudes than at higher altitudes. In fact, the vast majority of the mass of the atmosphere—about 99 percent—lies within 30 km of Earth's surface. About 90 percent of the mass of the atmosphere lies within 15 km of the surface, and about 75 percent lies within 11 km.

The atmosphere is divided into five distinct regions, based on temperature changes. You may recognize the names of some or perhaps all of these regions: the troposphere, stratosphere, mesosphere, thermosphere, and exosphere. In terms of composition, Earth's atmosphere is typically divided into two regions: the homosphere and the heterosphere.

The Homosphere Region of the Atmosphere

In the homosphere, gases are fairly evenly blended, giving the region a relatively uniform composition. This uniformity is due to mixing of the gases through a process called *convection*, which is a cyclical, up-and-down movement of air molecules caused by differences in the air density. Air near the land is warmer and less dense than the air above it. The cool, dense air above sinks, displacing the warmer, less-dense air below, causing the warm air to rise. Once away from the surface, the warm air cools and its density increases. This cooled air then sinks down to the surface and the convection process repeats itself and continues.

Gases in the homosphere behave like ideal gases. Therefore, in general, they obey the gas laws that you have studied in this unit. **Table 12.5** lists the major (green) and minor (yellow) components of the homosphere. Water vapour (not shown in the table) is also a major, variable component, making up 1 to 4 percent of the atmosphere.

Table 12.5 Percentage Composition of Clean, Dry Air in the Homosphere

Atmospheric Gas	Volume Percent
nitrogen, N_2	78.084
oxygen, O_2	20.946
argon, Ar	0.0340
carbon dioxide, CO_2	0.03845
neon, Ne	0.00182
helium, He	0.00052
methane, CH_4	0.00015
krypton, Kr	0.00011
hydrogen, H_2	0.00005
dinitrogen oxide, N_2O	0.00005
xenon, Xe	0.000009
ozone, O_3	0.000007

The Heterosphere Region of the Atmosphere

The mixing of atmospheric gases by convection does not occur in the heterosphere. Therefore, the gas composition in the heterosphere varies and is limited to a few gas types. Gas particles are layered by molecular mass. Nitrogen and oxygen molecules are in the lowest layers, along with ions such as O^+, NO^+, O_2^+, and N_2^+ that form by interactions of atoms and molecules and solar energy. Oxygen atoms are in the next highest layer; helium and free hydrogen atoms occur in the layers farther above.

Pollutants and Air Quality

The federal government, through Environment Canada, has identified several pollutants that are referred to as **criteria air contaminants**. These pollutants—carbon monoxide, nitrogen oxides, particulate materials, sulfur dioxide, and volatile organic compounds (VOCs)—are considered to be those that have the greatest impact on air quality and human health. **Figure 12.11** shows the anthropogenic (human-generated) sources of the criteria air contaminants.

criteria air contaminants air pollutants that are federally identified as those having the greatest impact on air quality and human health

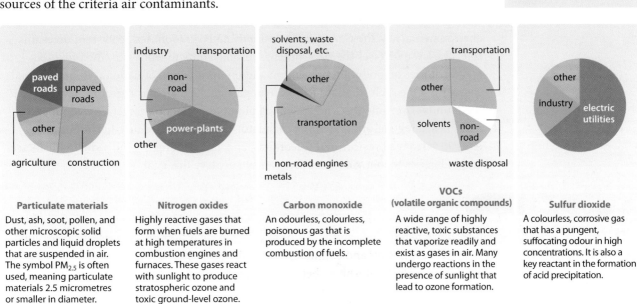

Particulate materials
Dust, ash, soot, pollen, and other microscopic solid particles and liquid droplets that are suspended in air. The symbol $PM_{2.5}$ is often used, meaning particulate materials 2.5 micrometres or smaller in diameter.

Nitrogen oxides
Highly reactive gases that form when fuels are burned at high temperatures in combustion engines and furnaces. These gases react with sunlight to produce stratospheric ozone and toxic ground-level ozone.

Carbon monoxide
An odourless, colourless, poisonous gas that is produced by the incomplete combustion of fuels.

VOCs (volatile organic compounds)
A wide range of highly reactive, toxic substances that vaporize readily and exist as gases in air. Many undergo reactions in the presence of sunlight that lead to ozone formation.

Sulfur dioxide
A colourless, corrosive gas that has a pungent, suffocating odour in high concentrations. It is also a key reactant in the formation of acid precipitation.

Figure 12.11 These pie graphs show the anthropogenic sources of the main criteria air contaminants.

Air Quality Effects on the Environment and Human Health

The environmental and human health effects of poor air quality are numerous and widespread. Ground-level ozone retards plant growth, and thus reduces the productivity of crops and causes damage to forests. In addition, ground-level ozone damages plastics, breaks down rubber, and corrodes metals. In terms of the human health effects, breathing ozone can trigger a variety of human health problems, including chest pain, coughing, throat irritation, and congestion. In addition, it can worsen existing chronic lung conditions such as asthma and emphysema.

Humans have some ability to resist the effects of poor air quality. However, Canada's national health agency, Health Canada, cautions that such resistance can be overwhelmed with higher levels of and prolonged exposure to air pollution. Children, people who are sick, and seniors are the most susceptible to pollution-related illness and disease. Many human health problems are linked to poor air quality, including asthma, allergies, lung cancer, and cardiovascular disease.

Go to **scienceontario** to find out more

To help Canadians better understand and assess the potential effects of air quality on human health, the federal departments of health and environment, with the participation of the provinces and territories, have created the Air Quality Health Index (AQHI). As you can see in **Figure 12.12**, the AQHI uses numbers from 1 to 10+, and a rating system of "low" to "very high," to indicate the air quality and its associated health risks. Each report for a selected location indicates who may be at risk given the current air quality conditions, provides health messages, and includes a short-term forecast to help in planning outdoor activities.

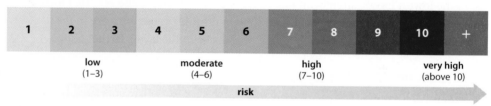

| 1 | 2 | 3 | 4 | 5 | 6 | 7 | 8 | 9 | 10 | + |

low (1–3) moderate (4–6) high (7–10) very high (above 10)

risk

Figure 12.12 The AQHI is based on the known health risks associated with a combination of common air pollutants that include ground-level ozone, particulate matter, and nitrogen dioxide.

Pollution Prevention, Emissions Limits, and Air Quality Standards

In Canada, air quality management is the responsibility of mainly the provincial and territorial governments and is overseen by the federal government. The main federal environmental law, the *Canadian Environmental Protection Act, 1999*, reinforces this role. The law addresses the government's responsibility to work with the provinces and territories on pollution prevention and on the creation of national environmental standards. One such standard focusses on particulate matter and ground-level ozone. It commits governments to meet defined pollution reduction targets. Governments have passed regulations and set up mandatory programs to reduce harmful gas emissions from all sectors of the economy, and are encouraging individuals, businesses, and industries to join voluntary programs to achieve this goal.

Learning Check

13. Describe several key features of the homosphere and the heterosphere.

14. Which gases are the main components, by volume, of Earth's atmosphere?

15. What are criteria air contaminants?

16. Why does mixing of atmospheric gases by convection occur in the homosphere?

17. What does AQHI stand for, and what is its purpose?

18. In Ontario, under what conditions would you expect AQHI values of 7 to 10?

A biofuel is any fuel made from sustainable, biological materials such as plant matter. Burning gasoline releases large amounts of the greenhouse gas, carbon dioxide, into the atmosphere. Burning biofuel is an alternative to burning gasoline. In this activity, you will compare the CO_2 emission factors of several components of biofuels and gasoline, and assess the effects of biofuel, including its impact on air quality.

Procedure

1. Read the following text and sample calculation. A carbon footprint measures the total amount of greenhouse gas emitted by an activity or process over time. Typically, carbon dioxide is the source of carbon for calculating a carbon footprint. One method of calculation involves calculating a CO_2 emission factor for a given activity:

$$CO_2 \text{ emission factor} = \frac{\text{mass of } CO_2 \text{ emitted}}{\text{quantity of activity}}$$

The "quantity of activity" can be expressed in different ways. For example, calculating the mass of CO_2 emitted per litre of a fuel being burned yields the emission factor. In this case, the litre of fuel is the quantity of activity. The example below shows how to calculate the CO_2 emission factor for the combustion of a liquid fuel, butane, $C_4H_{10}(\ell)$.

a. Liquid butane burns according to the following balanced chemical equation:

$$2C_4H_{10}(\ell) + 13O_2(g) \rightarrow 8CO_2(g) + 10H_2O(g)$$

b. Calculate the molar mass of liquid butane.

$$M_{C_4H_{10}} = 4M_C + 10M_H$$
$$= 4(12.01 \text{ g/mol}) + 10(1.01 \text{ g/mol})$$
$$= 58.14 \text{ g/mol}$$

c. Determine the amount (in moles) in 1.000 L of liquid butane. Liquid butane has a density of 573.0 g/L at 25°C.

$$D = \frac{m}{V}$$
$$m = DV = 573.0 \text{g/L} \times 1.00 \text{L} = 573.0 \text{ g}$$
$$n = \frac{m}{M} = \frac{573.0 \text{ g}}{58.14 \text{ g/mol}} = 9.85552 \text{ mol}$$

d. From the chemical equation, the mole ratio is 2 mol $C_4H_{10}(\ell)$: 8 mol $CO_2(g)$. Determine the amount of carbon dioxide in moles.

$$n_{CO_2} = 9.85552 \text{ mol } C_4H_{10} \times \frac{8 \text{ mol } CO_2}{2 \text{ mol } C_4H_{10}}$$
$$= 39.4221 \text{ mol } CO_2$$

e. Calculate the molar mass of carbon dioxide.

$$M_{CO_2} = M_C + 2M_O$$
$$= 12.01 \text{ g/mol} + 2(16.00 \text{ g/mol})$$
$$= 44.01 \text{ g/mol}$$

f. Determine the mass of carbon dioxide.

$$m = n \times M$$
$$= 39.4221 \text{ mol} \times 44.01 \text{ g/mol}$$
$$= 1734.9666 \text{ g}$$

g. Determine the carbon dioxide emission factor for burning liquid butane.

$$CO_2 \text{ emission factor} = \frac{\text{mass of } CO_2 \text{ emitted}}{\text{quantity of activity}}$$
$$= \frac{1734.9666 \text{ g } CO_2}{1.000 \text{ L } C_4H_{10}}$$

Thus, using correct significant figures, the CO_2 emission factor is 1735 g of carbon dioxide per 1.000 L of combusted liquid butane.

2. Use the data below to calculate the carbon footprints of the two biofuel and two gasoline components in terms of the CO_2 emission factor for each compound. Begin each calculation by writing a balanced combustion reaction.

Components of Biofuel

Compound	Density (g/L) at 20°C
Methanol, $CH_4O(\ell)$	791.4
Ethanol, $C_2H_6O(\ell)$	789.3

Components of Gasoline

Compound	Density (g/L) at 20°C
Pentane, $C_5H_{12}(\ell)$	626.2
Nonane, $C_9H_{20}(\ell)$	719.2

3. Biofuel combustion has negative effects on air quality. For example, burning a biofuel generates up to 15 percent more nitrogen dioxide, $NO_2(g)$, than burning gasoline does. Conduct research to

a. learn the role of nitrogen dioxide in producing ground-level ozone, $O_3(g)$, which is a chief component of smog.

b. investigate biofuel production and use in terms of the negative effects on human health and the environment. Assess the use of biofuel as a suitable replacement for gasoline.

Questions

1. a. Some people argue that biofuel has a larger carbon footprint than gasoline. Based on your calculations for methanol and ethanol, what would be true about the relationship between the CO_2 emission factor and the carbon footprint of biofuel, if this argument were correct?

   b. Identify at least one other factor that might affect the carbon footprint of biofuel. Explain your reasoning.

2. When determining the impact of a certain fuel upon the environment, what information, other than its carbon footprint, should to be taken into consideration?

3. You are trying to reduce your carbon footprint.

   a. Is driving a car that runs on biofuel an effective way to reduce your carbon footprint? Explain your reasoning.

   b. List three other transportation choices to reduce your carbon footprint. Rank them by effectiveness. Provide your reasoning for each ranking.

QUIRKS & QUARKS

with BOB MCDONALD

CBC ⊛

THIS WEEK ON QUIRKS & QUARKS

PFOA—Here Today, Here Forever?

Even though we cannot see them, chemical compounds such as perfluorooctanic acid (PFOA) are in our homes, in our bodies, and even in the blood of polar bears in the Arctic. Dr. Scott Mabury, professor and chair of the Chemistry Department at the University of Toronto, believes he has found an explanation for how persistent, bioaccumulative, and potentially toxic fluorochemicals have gotten there. Bob McDonald interviewed Dr. Mabury to find out the source of PFOA and how emissions can be reduced.

Stain Repellant and Water Resistant

One of the main sources of PFOA comes from the manufacturing of stain-resistant carpets, water-resistant clothing, and even microwave popcorn bags. Due to inefficient production methods, unreacted starting materials end up on the final product. Dr. Mabury and other scientists hypothesize that these materials, referred to as precursor alcohols, off-gas into the atmosphere. Once there, the gases travel thousands of kilometres and eventually degrade. When the precursor alcohols degrade, PFOA and other compounds like it form and enter the food chain.

PFOA is both persistent (does not break down easily so it stays for a long time in the environment) and bioaccumulative (builds up in the tissues of organisms in the food chain). Dr. Mabury notes that PFOA has been identified as a likely human carcinogen. Because of these properties, scientists at Environment Canada and the United States Environmental Protection Agency (USEPA) are concerned about the amount of PFOA being released into the environment.

Reducing Emissions

Environment Canada banned the import of four new fluorinated polymers because they are recognized as a source of fluorochemicals in the Arctic. After reviewing Dr. Mabury's research, the USEPA has asked manufacturers of PFOA to voluntarily reduce or eliminate emissions during the manufacturing process. This would mean producing a final product that did not have residual precursor alcohols on it. Dr. Mabury feels that if the manufacturing process can be cleaned up, it could significantly reduce the release of the precursors of PFOA. Scientists are also developing chemical substitutes for PFOA that, due to their chemical structure, are no longer bioaccumulative.

▶ Related Career

Dr. Scott Mabury is an environmental chemist. As a university professor, he teaches and conducts research. He asks questions about and studies the fate of fluorochemicals in the environment, such as how they degrade after being released into the atmosphere. Dr. Mabury communicates the results of his research by publishing articles in scientific journals.

Go to **scienceontario** to find out more

The PFOA coating on the inside of microwave popcorn bags keeps oil from soaking through the bag.

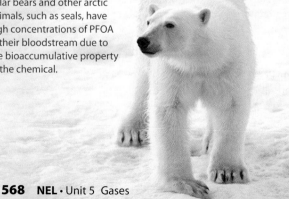

Polar bears and other arctic animals, such as seals, have high concentrations of PFOA in their bloodstream due to the bioaccumulative property of the chemical.

QUESTIONS

1. Summarize the key issues associated with the use of perfluorooctanic acid (PFOA).

2. Research what position Health Canada and Environment Canada are taking regarding the PFOA issue.

3. Research more information about being an environmental chemist. What are some of the different job positions an environmental chemist would be qualified for?

Section Summary

- The major components of Earth's atmosphere are nitrogen, oxygen, argon, and carbon dioxide gases, as well as water vapour. Minor components of Earth's atmosphere include very small amounts of other gases. Both major and minor components are concentrated mostly in the lowest region of the atmosphere, within 11 km of Earth's surface.

- Criteria air contaminants are carbon monoxide, nitrogen oxides, particulate materials, sulfur dioxide, and volatile organic compounds (VOCs).

- The Air Quality Health Index (AQHI) is a rating and advisory scale developed by the federal government, in consultation with provincial and territorial governments, to help Canadians understand the possible effects of air quality on any given day, and to make personal choices based on that information.

Review Questions

1. **C** Use a graphic organizer such as a Venn diagram to compare the homosphere and the heterosphere of Earth's atmosphere.

2. **K/U** List and describe the criteria air contaminants.

3. **K/U** Describe the role of the federal and provincial government in determining and implementing environmental policy.

4. **A** When making government policy on atmospheric pollution, target goals are often set over long periods of time. Why is it not practical for governments to set hard and fast standards in the short-term?

   Use the graph below to answer questions 5–9. (Note: NO_X refers to nitrogen dioxide and nitrogen dioxide.)

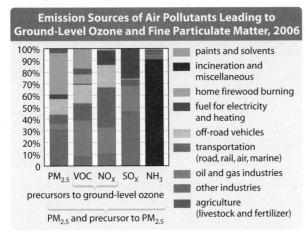

Emission Sources of Air Pollutants Leading to Ground-Level Ozone and Fine Particulate Matter, 2006

paints and solvents
incineration and miscellaneous
home firewood burning
fuel for electricity and heating
off-road vehicles
transportation (road, rail, air, marine)
oil and gas industries
other industries
agriculture (livestock and fertilizer)

$PM_{2.5}$ VOC NO_X SO_X NH_3

precursors to ground-level ozone

$PM_{2.5}$ and precursor to $PM_{2.5}$

5. **K/U** Which industry contributes the most pollution in terms of:
   a. nitrogen oxides
   b. VOCs
   c. particulate material

6. **T/I** By what percentage do paints and solvents contribute to VOC release?

7. **T/I** Does home firewood burning contribute significantly to VOC's or NO_X? Explain.

8. **A** Given the industries that contribute NO_X and VOC's to the atmosphere, make a list of concrete recommendations that could be implemented that would reduce these emissions.

9. **A** Examine the contributors to particulate matter. Make a list of recommendations that could be implemented to reduce these emissions

10. **T/I** Some provinces use an Air Quality Index (AQI) that is different from the federal Air Quality Health Index (AQHI). Conduct research to find out about the AQI system currently in use in Ontario, explain how the AQI differs from the AQHI, and use a PMI chart to assess the "pros and cons" of each system.

11. **K/U** Earth's atmosphere is classified into distinct regions based on different criteria. Identify these criteria and use each criterion to name the atmospheric regions based on it.

12. **A** The Environmental Protection Agency (EPA) of the United States has a listing of criteria air contaminants, just as Canada does. Do research to compare the EPA and Environment Canada sets of criteria air contaminants. Which pollutants are common to both, which are not, and why is there a difference?

13. **C** Carbon dioxide and methane are classified as greenhouse gases and are considered key contributors to global warming. And yet neither of these gases is identified as a criteria air contaminant. Does this surprise you? Write a short supported opinion piece to present your ideas.

14. **C** Conduct research to present, in the form of a poster or computer slide show, the link between air pollution and acid precipitation.

Skill Check

Initiating and Planning

✓ Performing and Recording

✓ Analyzing and Interpreting

✓ Communicating

Safety Precautions

- Wear safety eyewear, gloves, and lab apron.
- Vegetable oil is flammable.
- Heated oil causes serious, painful burns to the skin. Be extremely careful when heating the oil in the beaker.
- Ensure that the clamp is securely attached to the retort stand.
- Do not eat the popcorn.

Materials

- distilled water
- 18 to 20 popcorn kernels
- 1.5 mL of vegetable oil
- 10 mL graduated cylinder
- paper towels
- 250 mL beaker
- 3 wire gauze squares
- beaker tongs
- hot plate
- balance

Go to *Organizing Data in a Table* in *Appendix A* for information about designing data tables.

Using the Ideal Gas Law to Determine the Pressure To Make Popcorn

A corn kernel consists of oil, water, and a large amount of starch surrounded by a water-resistant shell. The kernel of corn will pop when the water in the kernel is heated to a pressure great enough to cause the shell to burst. This turns the kernel inside out, releasing the heated water vapour. In this investigation, you will use the ideal gas law to determine the pressure of the water vapour at the time of the kernel popping.

Pre-Lab Questions

1. What is the ideal gas law equation?
2. When you heat popcorn and cause it to pop, what happens to the water inside the popcorn kernels?
3. To predict the pressure of a gas using the ideal gas law, what other variables must be measured or determined?
4. Calculate the water vapour pressure of 0.015 mol of water that occupies a volume of 0.0025 L at 180°C.

Question

What is the pressure required to burst a popcorn kernel?

Procedure

1. Design a data table in your notebook to record your measured data.
2. Place approximately 5 mL of distilled water in the graduated cylinder, and record the volume.
3. Place 18–20 popcorn kernels into the water in the graduated cylinder. Tap the cylinder to force any air bubbles off the kernels. Record the new volume in your notebook.
4. Carefully drain the cylinder and retrieve the corn kernels. Using a piece of paper towel, dry the corn kernels in preparation for the next step.
5. Measure out 1.0 to 1.5 mL of vegetable oil in the graduated cylinder and then pour it into the beaker. Add the kernels to the beaker, and measure the total mass of the beaker, oil, and kernels. Record the mass in your data table.
6. Place the wire gauze square over the top of the beaker, place another wire gauze square on the hot plate, and set the beaker on the hot plate.
7. Gently heat the beaker. Use the beaker tongs to occasionally swirl the beaker gently on the hot plate to prevent scorching and to distribute the heat evenly.
8. Observe changes in the kernels and oil during the heating process. Turn off the hot plate when the popcorn has popped and before any burning occurs.
9. Use the beaker tongs to remove the beaker from the hot plate, and set the beaker on a second wire gauze square to allow it to cool completely.

10. Measure the final mass of the beaker, oil, and popcorn once cooling is complete. Record the mass in your data table.

11. Dispose of the popcorn and oil as directed by your teacher. Wash and return all materials and laboratory equipment.

Analyze and Interpret

1. Calculate the volume of the popcorn kernels, in litres, by using the difference in the volumes of water before and after addition of the kernels to the water in the graduated cylinder.

2. Calculate the total mass of water vapour released by using the mass measurements of the beaker, oil, and popcorn before and after popping.

3. Use the molar mass of water to determine the amount (in moles) of water vapour released.

4. Assume that the temperature of the heated oil is 220°C and that it represents the water vapour temperature. Use the calculated volume of the popcorn kernels to determine the pressure of the water vapour at the time of popping, by applying the ideal gas law.

5. Calculate the percent by number of popcorn kernels that popped. Report this as percent of successful kernel popping.

Conclude and Communicate

6. Based on your calculations, what was the water vapour pressure in the corn kernels at the time of popping?

7. Some reports suggest that the internal pressure of water vapour just prior to popping is 135 psi. What is this pressure in kPa, and how does this compare with your determined value?

8. If, during the experiment, some water vapour had become trapped in the beaker and did not evaporate (condensed on the side of the beaker), how could this have affected the experimental results? Use a sample calculation to support your answer.

Extend Further

9. **INQUIRY** What potential experimental errors could have occurred during this experiment? How could the experiment be modified to improve experimental accuracy?

10. **RESEARCH** Research information that explains the difference in the shapes of popped corn, and find out how food chemists control the shape and texture of popped corn.

Safety Precautions

- Hydrogen gas is highly flammable. Before beginning this investigation, check that there are no open flames (such as lit Bunsen burners) in the laboratory.

- The acid that you are using in this investigation is strong enough to cause burns to skin. Wear your safety eyewear and lab apron at all times. Handle the acid carefully. Wipe up any spills of water or acid immediately. If you accidentally spill any acid on your skin, wash it off immediately with large amounts of cool water.

- When you have finished the investigation, dispose of materials according to your teacher's instructions.

Materials

- 6 to 7 cm piece of magnesium ribbon
- steel wool
- water at room temperature
- 1.0 mol/L hydrochloric acid, HCl(aq)
- 10 to 15 cm piece of copper wire
- scale or balance
- 1 L beaker
- barometer
- thermometer
- 100 mL graduated cylinder
- stopper with two holes to fit graduated cylinder
- clamp and ring stand (optional)

Measuring the Molar Volume of a Gaseous Product in a Chemical Reaction

In this investigation, you will produce hydrogen gas by reacting a strong acid with magnesium metal as represented by this chemical reaction:

$$Mg(s) + 2HCl(aq) \rightarrow MgCl_2(aq) + H_2(g)$$

You will collect the hydrogen gas over water in a graduated cylinder. Using the volume of hydrogen that is collected and knowing the mass of one of the reactants, you will determine the molar volume of the hydrogen gas. **Note:** Your teacher may perform parts of this investigation for safety reasons.

Pre-Lab Questions

1. What is the mathematical expression that can be used to determine the molar volume of a gas? Make sure to define each variable in the equation.
2. What are two important considerations when collecting a gas over water?
3. What conditions do STP and SATP represent?

Question

What is the molar volume of hydrogen gas at STP?

Prediction

Predict the volume of hydrogen gas that will be produced. Use the mass of the magnesium in the Materials list and the balanced chemical equation given above. Assume 100 percent yield. Organize your calculations clearly.

Procedure

1. Prepare a table, like the one shown below, in your notebook. Record your observation and calculations in the table. Show your work.

Observations and Results

Observations	Trial 1	Trial 2 (time permitting)
Mass of magnesium ribbon (g)		
Temperature of water (°C)		
Barometric pressure (kPa)		
Volume of hydrogen collected (mL converted to L)		
Vapour pressure of water (kPa) at the temperature of water. **Note:** See **Table 12.4** for this value.		

Results	Trial 1	Trial 2 (time permitting)
Amount of magnesium (mol)		
Volume of collected dry hydrogen gas at STP (L)		
Molar volume of hydrogen gas at STP (L/mol)		

2. Obtain a piece of magnesium ribbon that is about 6 to 7 cm long. Use steel wool to clean the surface of the ribbon. Measure and record the ribbon's mass.

3. If you have not done so yet, use the mass of the magnesium and the balanced equation to predict the volume of hydrogen gas that will be produced. Show your calculations in your notebook.

4. Fill the beaker about half-full of water at room temperature. Measure the temperature of the water. (You will use this temperature to approximate the temperature of the gas produced.)

5. Measure the atmospheric pressure from the barometer. Record this value in your data table.

6. Add 15 mL of water to the graduated cylinder. Then, *very carefully*, pour 60 to 70 mL of 1 mol/L HCl into the graduated cylinder. *Very slowly and carefully*, pour water at room temperature down the inner *sides* of the cylinder until the cylinder is completely filled. By pouring the water this way, you avoid directly mixing the water with the acid at the bottom of the cylinder. **Warning! Normally, you should avoid adding water to an acid. Be especially careful during this step.**

7. Attach the magnesium ribbon to the copper wire. Dangle the magnesium in the graduated cylinder. The magnesium should hang 1 to 2 cm below the stopper. Close the cylinder with the stopper. Do not worry if a small amount of water overflows out of the cylinder.

8. Hold your gloved finger over the holes in the stopper. Tip the cylinder upside down into the 1 L beaker. Be careful to not allow any air bubbles to get into the cylinder. Hold or clamp the cylinder into place. Your set-up should appear as shown in the diagram.

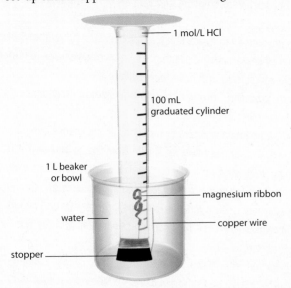

1 mol/L HCl

100 mL graduated cylinder

1 L beaker or bowl

magnesium ribbon

water

copper wire

stopper

9. Watch the reaction proceed. Record your observations in your notebook.

10. Add water, at room temperature, to the beaker until the level of the water inside the cylinder is exactly the same as the level of the water in the beaker. This equalizes the pressure of the hydrogen gas with the air pressure outside the cylinder. **Note:** Another way to equalize the pressure is to raise the graduated cylinder slightly to align the water levels.

11. Record the volume of the trapped gas in your data table.

12. All the magnesium should be used up by the reaction, since it is the limiting reactant. If any magnesium ribbon does remain after the reaction, rinse it with water, dry it with a paper towel, and measure its mass. To find the mass of the magnesium used up by the reaction, subtract the final mass from the initial mass.

13. Empty and clean all your apparatus. Clean your work space. Wash your hands.

Analyze and Interpret

1. Calculate the molar volume of the hydrogen. Use the volume of the hydrogen gas and the water temperature. You also need to use the atmospheric pressure minus the pressure of the water vapour.

2. Use the combined gas law or ideal gas law to convert to the conditions of STP. Redo the calculations.

Conclude and Communicate

3. What was the class average result for the molar volume of dry hydrogen gas at STP? How close was your molar volume to the class average result?

4. How close was your molar volume to the accepted molar volume of a gas at STP? Calculate the percent error for your molar volume.

5. What were some possible sources of error in your investigation?

Extend Further

6. **INQUIRY** How would the measured volume of hydrogen gas be different if you had performed this experiment at a much higher altitude or in a classroom in a very hot climate?

7. **RESEARCH** Research the use of hydrogen as a fuel for transportation. Prepare a summary that includes information about how the hydrogen is produced, stored, and distributed. Also include any safety concerns regarding these practices and the use of hydrogen by the general public.

Case Study

Ontario's Drive Clean Program
Less Smog, Less Greenhouse Gases?

Scenario

You are an intern at Ontario's Ministry of the Environment. The Ministry wants to create a press release for Earth Day, providing updates about Ontario's Drive Clean Program. The director's assistant is off sick and you have been asked to write the press release. The director hands you the assistant's research notes and asks you to prepare something that briefly explains the program and contains updated information and statistics about the program for the public. The assistant's notes are shown below. Included in the assistant's notes are some Internet printouts about Ontario's Drive Clean Program.

On average, Canadians produce half of their annual greenhouse gas emissions—including those from carbon dioxide (CO_2) and nitrogen oxides (NO_x)—from driving.

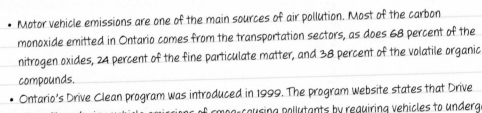

- Motor vehicle emissions are one of the main sources of air pollution. Most of the carbon monoxide emitted in Ontario comes from the transportation sectors, as does 68 percent of the nitrogen oxides, 24 percent of the fine particulate matter, and 38 percent of the volatile organic compounds.
- Ontario's Drive Clean program was introduced in 1999. The program website states that Drive Clean "is reducing vehicle emissions of smog-causing pollutants by requiring vehicles to undergo an emissions test to identify emissions problems and have them repaired."
  The program "is also reducing vehicle emissions of other chemicals, including greenhouse gases that damage our health and our environment."
- Under Drive Clean, vehicles must have an emissions test before the vehicle licence can be renewed or the ownership transferred. If a vehicle releases smoke from the tailpipe or if it is missing a catalytic converter, it will not be tested until it is repaired. Hybrid vehicles are exempt from the tests. Drive Clean has a program for light-duty vehicles, such as passenger cars and sport utility vehicles, and a program for heavy-duty vehicles, such as large trucks and buses. Drive Clean is self-funding; vehicle owners pay a fee for each emissions test, and the fee revenue exceeds the costs of the program.
- According to the Drive Clean Program Web site: "From 1999 to 2008, Drive Clean reduced smog-causing emissions of hydrocarbons (HC) and nitrogen oxide(s) (NO_x), from light duty vehicles (LDVs) by an estimated 266 000 tonnes. During the same period, the program also reduced emissions of carbon monoxide (CO), a poisonous gas, by over 2.48 million tonnes; and carbon dioxide (CO_2), a greenhouse gas, by about 256 000 tonnes."

Ontario

MINISTRY OF T[

HOME | ABOUT | NEWS | PUBLICATIONS | CONTACT

AIR

DRIVE CLEAN

The solid blue line on the graph below shows the estimated reduction in emissions in the Greater Toronto Area (GTA) and Hamilton since the start of the Drive Clean Program in 1999. The dotted red line shows the percent of reduction that scientists think would have occurred without the Drive Clean Program. Some reductions in emissions would have occurred without the Drive Clean Program due to improvements in vehicle technology and maintenance, and the use of cleaner fuels. Scientists estimate that the Drive Clean Program has resulted in an extra 20 percent reduction in emissions over the 9-year period shown in the graph.

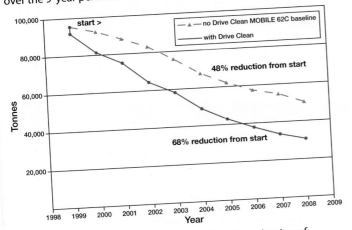

The table below contains information about the reduction of emissions in 2008 due to the Drive Clean Program. The data are shown in both tonnes and percent of reduction.

Carbon monoxide	Hydrocarbons	Nitrogen oxides
359 171 (37.1 %)	19 480 (38.2 %)	18 578 (34.6 %)

Research and Analyze

1. Research more information and statistics about Ontario's Drive Clean Program. How does it work? What is required of owners of light-duty and heavy-duty vehicles? How many vehicles per year are identified as releasing excess pollutants? Are there any drawbacks to the program?

2. Describe the importance of the program. How does it help to protect human health and the health of the environment?

3. Research other government programs that are designed to reduce greenhouse gas emissions from the transportation sector.

4. Analyze the information presented in the case study and that you researched. Use a graphic organizer, such as a main idea web or a fishbone diagram, to organize your information and ideas. Go to Appendix A, Using Graphic Organizers, for more help on how to construct a graphic organizer.

Take Action

5. **Plan** Assess the effectiveness of the Drive Clean Program in reducing greenhouse gas emissions in the province. Describe some examples of other programs that reduce vehicle emissions of greenhouse gases. How does the impact of these programs on emissions of greenhouse gases compare with the impact of the Drive Clean Program? Write point-form notes outlining your reasoning.

6. **Act** Write a 300-word press release about Ontario's Drive Clean Program, including the most up-to-date information about the success of the program and any relevant statistics.

Section 12.1 | The Combined Gas Law

For a given amount of gas, a change in one of temperature, pressure, or volume affects the other two variables.

KEY TERMS
Avogadro's law
combined gas law
law of combining volumes
molar volume
standard ambient temperature and pressure (SATP)
standard temperature and pressure (STP)

KEY CONCEPTS
- Gases can undergo changes in temperature, pressure, and volume simultaneously. The combined gas law addresses these changes and is represented mathematically as

$$\frac{P_1 V_1}{T_1} = \frac{P_2 V_2}{T_2}$$

- The law of combining volumes states that, when gases react, the volumes of the gaseous reactants and products, measured at the same conditions of temperature and pressure, are always in whole-number ratios.

- If the temperature and pressure are constant and the same for the reactants and products of a chemical reaction, the law of combining volumes can be used to determine the amounts of reactant and/or products.

- Avogadro proposed that at the same conditions of temperature and pressure, equal volumes of all ideal gases contain the same number of molecules, even though they do not have the same mass. A mathematical statement of Avogadro's law is $\frac{n_1}{V_1} = \frac{n_2}{V_2}$.

Section 12.2 | The Ideal Gas Law

The ideal gas law relates four different variables of a gas: amount, pressure, temperature, and volume.

KEY TERMS
Dalton's law of partial pressures
ideal gas law
partial pressure
universal gas constant (R)

KEY CONCEPTS
- The ideal gas law is derived from the combined gas law and Avogadro's law. The mathematical equation for it is $PV = nRT$. R is the universal gas constant with a value of $8.314 \frac{\text{kPa} \cdot \text{L}}{\text{mol} \cdot \text{K}}$.

- For gas stoichiometry, conditions of pressure, volume, and temperature must be considered. When the reactants and products are not under the same conditions or the reaction occurs under non-standard conditions, the ideal gas law can be used to calculate the amounts of gaseous reactants or products.

- Real gases behave similarly to ideal gases under standard conditions of temperature and pressure. However, this behaviour deviates under conditions of low temperature and high pressure.

Section 12.3 | Atmospheric Gases and Air Quality

The composition of the gases in Earth's atmosphere is affected by the addition of polluting substances that result from natural events as well as human activities.

KEY TERMS
criteria air contaminants

KEY CONCEPTS
- The major components of Earth's atmosphere are nitrogen, oxygen, argon, and carbon dioxide gases, as well as water vapour. Minor components of Earth's atmosphere include very small amounts of other gases. Both major and minor components are concentrated mostly in the lowest region of the atmosphere, within 11 km of Earth's surface.

- Criteria air contaminants are carbon monoxide, nitrogen oxides, particulate materials, sulfur dioxide, and volatile organic compounds (VOCs).

- The Air Quality Health Index (AQHI) is a rating and advisory scale developed by the federal government, in consultation with provincial and territorial governments, to help Canadians understand the possible effects of air quality on any given day, and to make personal choices based on that information.

Chapter 12 | REVIEW

Knowledge and Understanding

Select the letter of the best answer below.

1. Which of the following correctly represents the combined gas law?

a. $\dfrac{P_1 V_1 T_2}{T_1 P_2} = V_2$

d. $\dfrac{P_1 V_2 T_2}{V_1 T_1} = P_2$

b. $\dfrac{T_2 P_2 V_2}{P_1 V_1} = T_1$

e. $\dfrac{T_1 P_1 V_2}{P_1 V_1} = T_2$

c. $\dfrac{P_1 V_2 T_2}{T_1 P_2} = V_1$

2. What law is represented by the image below?

 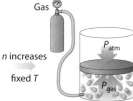 

a. Dalton's law of partial pressures

b. law of combing volumes

c. Avogadro's law

d. combined gas law

e. universal gas law

3. Which of the following applies to the law of combining volumes?

a. The reactants and products can be in any state.

b. Both temperature and pressure must be the same for the gaseous reactants and products.

c. All reactants and products are considered, regardless of state.

d. Only the pressure must be the same for all reactants and products.

e. The volumes of gaseous and liquid reactants and products react in whole-number ratios.

4. Which of the following are conditions at SATP?

a. 98 kPa, 273.15 K

d. 101.325 kPa, 273.15 K

b. 100 kPa, 298.15 K

e. 100 kPa, 273.15 K

c. 101.325 kPa, 298.15 K

5. When using the value of 8.314 for the universal gas constant in the ideal gas law equation, what units must be used for temperature, volume, and pressure?

a. temperature: °C; pressure: Pa; volume: L

b. temperature: K; pressure: Pa; volume: L

c. temperature: °C; pressure: kPa; volume: L

d. temperature: K; pressure: Pa; volume: mL

e. temperature: K; pressure: kPa; volume: L

6. Which of the following are the conditions under which a gas will behave non-ideally?

I high temperature

II low temperature

III high pressure

IV low pressure

a. I and III

d. II and IV

b. I and IV

e. I, II, II, and IV

c. II and III

7. About how much of Earth's atmosphere is made up of carbon dioxide gas?

a. 0.04 percent

d. 20.9 percent

b. 0.9 percent

e. 78.1 percent

c. 10.4 percent

8. Which of the following gases is considered a criteria air contaminant by Environment Canada?

a. methane

d. carbon monoxide

b. ozone

e. PCBs

c. carbon dioxide

Answer the questions below.

9. What is meant by the term "molar volume of a gas"?

10. What experimental pieces of information are needed to calculate the molar mass of a gas? Without doing the exact calculation, show how this information can be used to determine the molar mass.

Thinking and Investigation

11. Methane, CH_4(g), can be compressed by cooling and increasing the pressure. A 6.00×10^2 L sample of methane gas at 25.0°C and 100.0 kPa is cooled to −20.0°C at a constant pressure. In a second step, the gas is compressed until the pressure is quadrupled. What will the final volume be?

12. A sample of gas has a volume of 475 mL, a temperature of 25°C and a pressure of 95.0 kPa. If the volume has changed to 50.1 mL at a temperature of 1350°C, what is the pressure of the gas?

13. Sulfur hexafluoride, SF_6(g), is an inert gas used in chemical demonstrations and as an insulator. A 4 L sample of this gas was collected at 250°C and 400 kPa. What pressure must be applied to this gas sample to reduce its volume to 1 L at 25°C?

14. A 2.00 L flask is filled with ethane, $C_2H_6(g)$, from a small cylinder, as shown in the diagram below. What is the mass of the ethane in the flask?

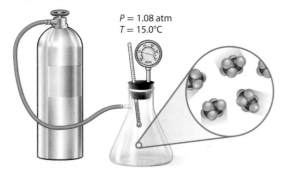

$P = 1.08$ atm
$T = 15.0°C$

15. A sample of helium gas with a mass of 1.1 g is at a temperature of 23°C and has a volume of 6.5 L. What is the pressure of the gas?

16. A steel gas cylinder contains 30 L of acetylene, $C_2H_2(g)$, under pressure when it is full. Over a period of time, the gas is used up until the cylinder is declared "empty." If the temperature of the container when empty is 25°C and the pressure is that of the atmosphere at 102 kPa, answer the following questions:
a. Explain why the tanks is not really empty.
b. How many moles of gas remain in the cylinder?
c. What is the mass of the remaining gas in the cylinder?

17. For each of the following conditions, calculate the molar mass of the gaseous compound.
a. 8.97 g of a gas occupies 1.5 L at 20.5°C and 102 kPa
b. 4.23 g of a gas occupies 858 mL at 15°C and 740 mmHg

18. A noble gas has a mass of 1.355 g when sealed in a container of 5.00×10^2 mL at a temperature of 0°C and 3.00 atm. Determine the molar mass and identity of the gas.

19. A 13.4 g sample of an unknown liquid is vaporized at 85.0°C and 100.0 kPa. The vapour has a volume of 4.32 L. The percentage composition of the liquid is found to be 52.1 percent carbon, 13.2 percent hydrogen, and 34.7 percent oxygen. What is the molecular formula of the liquid?

20. Determine the density of oxygen gas at −30.0°C and 87 kPa.

21. A sample of hydrogen gas was collected by water displacement at 20.0°C when the pressure of the atmosphere was measured to be 99.8 kPa. What was the pressure of the hydrogen?

22. Ammonia and oxygen gases react according to the unbalanced chemical equation below. In a reaction where the temperature and pressure are the same for all reactants and products, 2.0×10^3 L of nitrogen monoxide gas was produced.
$$NH_3(g) + O_2(g) \rightarrow NO(g) + H_2O(g)$$
a. Balance the chemical equation.
b. Calculate the volume of oxygen gas that would be required to produce this amount of nitrogen monoxide.
c. Calculate the volume of ammonia gas that would be required to produce this amount of nitrogen monoxide.
d. Calculate the volume of water vapour that would be produced in the reaction.
e. What would be the total volume of gaseous products produced?

23. During complete combustion of propane, $C_3H_8(g)$, carbon dioxide gas and water vapour are produced.
a. Write the complete balanced chemical equation for the burning of propane in the presence of oxygen.
b. Determine the volume of propane that will react with 5 L of oxygen gas.
c. Determine the volume of carbon dioxide gas that would be produced if the 5 L of oxygen gas were used up.
d. Determine the volume of gaseous water that would be produced if the 5 L of oxygen were used up.

24. Ammonium nitrite, $NH_4NO_2(s)$, when heated, decomposes to nitrogen gas and water.
a. Write a balanced equation for this reaction.
b. Calculate the quantity of ammonium nitrite required to inflate a ball to a volume of 90.1 mL at 130 kPa and 25°C.

25. The principal chemical reaction involved in the inflation of air bags involves this decomposition reaction.
$$2NaN_3(s) \rightarrow 2Na(s) + 3N_2(g)$$
Determine the volume of nitrogen gas produced when 120 g of sodium azide, $NaN_3(s)$, is decomposed in the airbag at 27°C and 100.5 kPa.

26. Hydrogen gas can easily be generated using the reaction of magnesium with hydrochloric acid. For a reaction, 250 mL of hydrogen gas was collected by water displacement. Calculate the mass of magnesium. What volume of 3.0 M hydrochloric acid would be required to produce "dry" hydrogen gas at 102 kPa and 22°C?

Communication

27. **BIG IDEAS** Properties of gases can be described qualitatively and quantitatively and can be predicted. The diagram below models the reaction of methane gas with oxygen gas to produce carbon dioxide gas and water vapour. Draw similar diagrams for each of the reactions below, clearly showing how the law of combining volumes can be applied to indicate the volumes of gases that react and are produced.

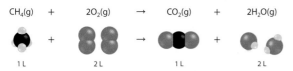

$$CH_4(g) \quad + \quad 2O_2(g) \quad \rightarrow \quad CO_2(g) \quad + \quad 2H_2O(g)$$

1 L   2 L   1 L   2 L

   a. Hydrogen and nitrogen gas combine to produce ammonia, $NH_3(g)$.

   b. Ammonia gas combines with oxygen gas to produce nitrogen monoxide, $NO(g)$, and water vapour.

28. Develop an illustration that demonstrates Dalton's law of partial pressures.

29. The graph below shows changes in Canada's greenhouse gas emissions between 1990 and 2008. The greenhouse gases comprising the data in this graph are carbon dioxide, methane, $CH_4(g)$, nitrous oxide, $N_2O(g)$, sulfur hexafluoride, $SF_6(g)$, perfluorocarbons, and hydrofluorocarbons. Carbon dioxide makes up a little over 78 percent of all greenhouse emissions in Canada during this time period.

   a. During what time period(s) did emissions increase?

   b. During what time period(s) did emissions decrease?

   c. What factors might account for the period(s) of decrease?

   d. Emissions in 2008 were 24 percent greater than emissions in 1990. During this period of time, there was a strategy at national, provincial, and local levels to counsel Canadians on the environmental dangers of greenhouse gases and to encourage changes—industrially as well as personally—to reduce emissions. Study the growth trend in the graph, and give your opinion of the success of this strategy.

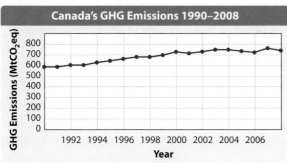

Canada's GHG Emissions 1990–2008

GHG Emissions (MtCO$_2$eq) vs. Year (1992, 1994, 1996, 1998, 2000, 2002, 2004, 2006)

30. Summarize your learning in this chapter using a graphic organizer. To help you, the Chapter 12 Summary lists the Key Terms and Key Concepts. Refer to Using Graphic Organizers in Appendix A to help you decide which graphic organizer to use.

Application

31. **BIG IDEAS** People have a responsibility to protect the integrity of Earth's atmosphere. There are many ways that Canadians can reduce the amount of carbon dioxide and other greenhouse gases they contribute to the atmosphere. These actions include buying more locally grown foods, lowering the use of electronics and appliances that require a lot of electric energy, and ensuring regular vehicle maintenance.

   a. Explain how each of these recommendations could achieve the goal of reducing greenhouse gas emissions.

   b. Identify at least four other actions that you, personally, could take to achieve this same goal.

32. **BIG IDEAS** Air quality can be affected by human activities and technology. Air quality differs from place to place, not only within a large city, but also between different locations in a province, a country, and even a continent. Identify at least two industrial factors, two geographic factors, and two meteorological (weather-related) factors that influence differences in air quality between one community and another.

33. Choose two different locations in Canada for which AQHI information is available. Use the AQHI to assess the air quality conditions for each of these locations, and infer reasons for any similarities and differences.

34. Sulfur dioxide, $SO_2(g)$, is a toxic waste gas that is produced during the smelting of some metal ores. Sulfur dioxide gas is a leading cause of acid precipitation. One method of controlling sulfur dioxide gas involves "scrubbing," or treating, the effluent gas with magnesium hydroxide according to the following equation:

$$Mg(OH)_2(aq) + 5H_2O(\ell) + SO_2(g) \rightarrow$$
$$MgSO_3 \cdot 6H_2O(\ell)$$

The hydrated magnesium sulfite can further react with sulfur dioxide gas to produce magnesium hydrogen sulfite, a water-soluble compound.

$$SO_2(g) + MgSO_3 \cdot 6H_2O(s) \rightarrow$$
$$Mg(HSO_3)_2(aq) + 5H_2O(\ell)$$

   a. What mass of magnesium hydroxide would be required to react with 1.0×10^6 L of sulfur dioxide gas, produced at 50.0°C and 103 kPa?

   b. Calculate the final mass of magnesium hydrogen sulfite that would be produced.

Select the letter of the best answer below.

1. **K/U** What gas laws are combined to generate the combined gas law?
   a. Gay-Lussac's law and Boyle's law
   b. Avogadro's law and the law of combining volumes
   c. Charles's law and Boyle's law
   d. ideal gas law and Avogadro's law
   e. Gay-Lussac's law and Dalton's law of partial pressures

2. **K/U** Which of the following represents Avogadro's law?

   a. $\dfrac{n_1}{V_1} = \dfrac{n_2}{V_2}$

   b. $\dfrac{n_1}{V_2} = \dfrac{n_2}{V_1}$

   c. $\dfrac{m_1}{V_1} = \dfrac{m_2}{V_2}$

   d. $\dfrac{T_1}{V_1} = \dfrac{T_2}{V_2}$

   e. $\dfrac{V_1}{n_1} = \dfrac{n_2}{V_2}$

3. **K/U** Which of the following conditions will cause an increase in pressure of a gas?
   I. increasing temperature
   II. increasing volume
   III. adding additional moles of gas

   a. I only
   b. II only
   c. III only
   d. I and II only
   e. I and III only

4. **K/U** Which of the following conditions represent STP?
   a. 98 kPa, 273.15 K
   b. 100 kPa, 298.15 K
   c. 101.325 kPa, 298.15 K
   d. 101.325 kPa, 273.15 K
   e. 100 kPa, 273.15 K

5. **T/I** What volume will be occupied by 8.00 g of helium gas at STP?
   a. 11.2 L
   b. 22.4 L
   c. 33.6 L
   d. 44.8 L
   e. 56.0 L

6. **T/I** A container holds a mixture of the gases listed in the table below. If the total pressure is 65.3 kPa, what is the missing value in the table? Assume that the gases do not react with each other.

Gas	Partial Pressure (kPa)
water vapour	6.00
neon	7.60
argon	
helium	28.7

   a. 0.0310 kPa
   b. 36.6 kPa
   c. 108 kPa
   d. 42.3 kPa
   e. 23.0 kPa

7. **T/I** The synthesis of ammonia is represented by the chemical equation
   $$N_2(g) + 3H_2(g) \rightarrow 2NH_3(g)$$
   What volume of hydrogen gas will produce 1.2×10^3 L of ammonia?
   a. 3.6×10^3 L
   b. 1.2×10^3 L
   c. 1.8×10^3 L
   d. 8.0×10^2 L
   e. 2.4×10^3 L

8. **T/I** Methane, $CH_4(g)$, reacts with steam to produce *synthesis gas*, which is a mixture of carbon monoxide and hydrogen. The unbalanced chemical equation for the reaction is
   $$CH_4(g) + H_2O(g) \rightarrow CO(g) + H_2(g)$$
   What mass of hydrogen is produced if 275 L of methane is used at STP?
   a. 12.3 g
   b. 24.7 g
   c. 37.1 g
   d. 49.4 g
   e. 74.2 g

9. **K/U** Which of these conditions would likely cause the greatest deviation from the ideal gas law?
   I. high pressure
   II. low temperature
   III. large volume

   a. I only
   b. II only
   c. III only
   d. I and II only
   e. II and III only

10. **K/U** Which of the following statements applies to the composition of Earth's atmosphere?
    a. Most components, by volume percent, are noble gases.
    b. There are large amounts of a small number of major components, and very small amounts of a large number of minor components.
    c. There are large amounts of a large number of major components, and very small amounts of a large number of minor components.
    d. Carbon dioxide makes up a significant portion.
    e. The overall amount of oxygen in the atmosphere has not changed for billions of years.

Use sentences and diagrams as appropriate to answer the questions below.

11. **K/U** Describe the effect on 1 mol of gas when the pressure is reduced by a factor of 2 while the temperature, in kelvins, is increased by a factor of 4.

12. **A** Large quantities of methane gas are being released from Arctic waters as water temperatures rise due to global warming. A methane bubble has a volume of 2 m³ at 10 atm of pressure and 4.0°C. What will the volume of the bubble be at 10°C and 102 kPa?

13. **K/U** Consider the diagram below. Each gas is in a cylinder of the same volume and at the same temperature. Use the diagram to help explain Avogadro's law.

14. **T/I** A syringe is filled with nitrogen gas to the 60.00 mL mark. The mass of the empty syringe is 25.468 g and the filled syringe is 25.537 g. The nitrogen gas is at a temperature of 21.0°C and a pressure of 100.0 kPa. Calculate its molar volume under these conditions.

15. **A** A gas cylinder with a volume of 5.5×10^5 L contains oxygen at a pressure of 150 kPa and a temperature of 4.0°C.
   a. What mass of oxygen is present at this temperature and pressure?
   b. If all the oxygen gas were to be released at conditions of SATP, what volume would it occupy?

16. **C** Using the density (in g/L) of radon gas at 102 kPa and 15°C, write instructions for determining the density of a gas. As part of the instructions, show your work for solving this problem.

17. **T/I** A gas has a mass of 0.340 g and occupies a volume of 4.00 L at a temperature of 15°C and a pressure of 101.8 kPa. What is the molar mass of the gas?

18. **T/I** A mixture of helium and nitrogen gases has a total pressure of 122 kPa. If the volume ratio of helium to nitrogen in the container is 3:1, what is the partial pressure of nitrogen in the container?

19. **K/U** Why is it necessary to correct the pressure of a gas that is collected over water? Describe how you would do this.

20. **T/I** Nitrogen and oxygen gases react to form dinitrogen oxide, $N_2O(g)$. What volumes of $O_2(g)$ and $N_2(g)$ are needed to produce 34 L of $N_2O(g)$? Explain your answer.

21. **A** Carbon dioxide in high levels can cause death. One method of removing carbon dioxide from the air of enclosed compartments involves this reaction:
$$2NaOH(s) + CO_2(g) \rightarrow Na_2CO_3(s) + H_2O(\ell)$$
What mass of sodium hydroxide is required to remove all the carbon dioxide in the air over a 24 h period, if the volume of CO_2 produced is 1.25 L/min at a temperature of 22°C and a pressure of 101.8 kPa?

22. **C** What are the gaseous components of Earth's atmosphere and what percentage of the atmosphere do they each account for? Draw a pie graph that represents this information.

23. **C** Write a short radio ad (about 30 s) that explains the value of the AQHI to elementary school students.

24. **C** Use a graphic organizer such as a spider map or a main idea web to identify the criteria air contaminants and at least three of the principal sources for each.

25. **A** Ontario's Drive Clean Program inspection measures emissions of nitric oxide, NO(g), carbon monoxide, CO(g), and hydrocarbons under two different conditions: at 40 km/h and at curbside idling.
   a. Why would measurements be taken under two different conditions such as these?
   b. Two helpful hints on driving clean, as suggested by the Ontario government, include those listed below. Explain the value of these hints, and then—without consulting Drive Clean literature—suggest three more hints for driving clean.
   - Have the vehicle maintained regularly, following the manufacturer's suggested service schedule.
   - Avoid adding unnecessary weight to the vehicle.

Self-Check

If you missed question …	1	2	3	4	5	6	7	8	9	10	11	12	13	14	15	16	17	18	19	20	21	22	23	24	25
Review section(s)…	12.1	12.1	12.1	12.1	12.2	12.2	12.2	12.2	12.2	12.3	12.1	12.1	12.1	12.1	12.2	12.2	12.2	12.2	12.2	12.2	12.2	12.3	12.3	12.3	12.3

Evaluating Emission Control Technologies

The governments of many countries, including Canada, have air quality regulations in place. These regulations aim to limit the emission of pollutants from activities such as fossil fuel-based electricity production, smelting of ores, and transportation. In many cases, technologies that reduce or remove pollutants from emissions—referred to as emission control technologies—are already in use. For example, sulfur emissions can be reduced by up to 90 percent by mixing limestone with coal before the coal is burned to generate electricity. However, government agencies and private companies continue to research and design new emission control technologies that are more efficient and less expensive and that meet stricter air quality requirements.

In this project, you will research a new emission control technology for a particular pollutant. You will evaluate the technology and act as an advisor to a potential investor. You will write a report with your recommendation as to whether or not a monetary investment should be made in the new technology.

How can you evaluate whether a new emission control technology is a worthwhile investment?

SaskPower chose one of the coal-fired units at the Boundary Dam Power Station in Estevan, Saskatchewan, to be a demonstration project for the long-term capture and storage of carbon.

Since 1970, plant upgrades and the use of new emission control technologies have resulted in a 90 percent reduction of SO_2 emissions from this nickel and copper smelter in Sudbury, Ontario.

Initiate and Plan

1. Study the table below. Decide which pollutant and its corresponding emission control technology you would like to research.

Pollutants and Corresponding Emission Control Technologies

Pollutant	Source	New Emission Control Technology
Carbon dioxide (CO_2)	fossil fuel-based power plants	clay-liquid removal of carbon dioxide from coal-fired power plants or any new pre-combustion or post-combustion carbon capture and storage (CCS) process
Sulfur dioxide (SO_2)	fossil fuel-based power plants	electron-beam flue gas treatment or any other new flue gas desulfurization (FGD) technology
Sulfur dioxide (SO_2)	smelting	direct-to-blister flash smelting or any other new emission control technology
Mercury (Hg)	coal-fired power plants or smelting	thief process or any other new emission control technology
Nitrogen oxides (NOx)	combustion of fossil fuels	oxygen-enhanced combustion or any other new emission control technology
Other pollutants: volatile organic compounds (VOCs) or methane (CH_4)	power plants and chemical plants; landfills	any corresponding new emission control technology

Perform and Record

2. Research your chosen pollutant and its corresponding emission control technology. Begin by familiarizing yourself with background information about the pollutant and the existing technologies. Consider the following questions to guide your research:

   - What is the main source or sources of the pollutant?
   - Why is it important to reduce or eliminate the pollutant?
   - What existing technologies are used to help control the emission of the pollutant?
   - What are the sources of the information you have gathered? How trustworthy and credible are your sources? How can you know or decide?

3. Research information about a new emission control technology for your chosen pollutant. Consider the following questions to guide your research:

   - How does the technology work? How do the chemical composition, properties, and behaviour of the gas that is the pollutant affect the design of the technology?
   - What is the cost of materials or labour to install or maintain the new technology? How does the cost compare to the cost of existing technology?
   - How does the technology affect the efficiency of the system?
   - How effective is the technology itself? What percent of the pollutant is removed from emissions?
   - What are other potential advantages or disadvantages of the technology? For example:
     - How much energy does the technology use?
     - Does the use of the technology produce other wastes?
     - Are there any marketable by-products that can be made from the waste materials?
     - Does the technology have widespread commercial application?
     - How much research and/or testing have been done on the technology?
   - What are the sources of the information you have gathered? How trustworthy and credible are your sources? How can you know or decide?

Analyze and Interpret

1. Identify the pros and cons of the technology you researched. You may use a T-chart or other graphic organizer to help you organize your information. (Refer to Using Graphic Organizers in Appendix A.)

2. Evaluate the pros and cons and determine whether you would advise someone to invest money in a company that is designing and producing the new emission control technology. Explain your reasoning.

Communicate Your Findings

3. Prepare a written report about the new emission control technology. Include a description of the technology, any appropriate diagrams or illustrations (either print or digital), the pros and cons of the technology, and your recommendation to an investor who is considering supporting the technology by making a monetary investment.

Assessment Criteria

Once you complete your project, ask yourself these questions. Did you …

- ☑ **K/U** research background information about the pollutant and its sources?
- ☑ **K/U** research information about existing technologies used to help control emissions of the pollutant?
- ☑ **K/U** research information about a new emission control technology?
- ☑ **T/I** evaluate the sources of the information you gathered?
- ☑ **T/I** identify and evaluate the pros and cons of the new technology?
- ☑ **A** determine whether you would advise someone to make a monetary investment in the new technology?
- ☑ **C** explain your reasoning as to whether or not you would advise making an investment?
- ☑ **C** communicate your findings in the form of a report that is appropriate for the audience and purpose?
- ☑ **C** use scientific vocabulary appropriately?

BIG IDEAS

- Properties of gases can be described qualitatively and quantitatively, and can be predicted.
- Air quality can be affected by human activities and technology.
- People have a responsibility to protect the integrity of Earth's atmosphere.

Overall Expectations

In this unit, you learned how to…

- **analyze** the cumulative effects of human activities and technologies on air quality and describe some Canadian initiatives to reduce air pollution, including ways to reduce your own carbon footprint
- **investigate** gas laws that explain the behaviour of gases, and solve related problems
- **demonstrate** an understanding of the laws that explain the behaviour of gases

Chapter 11 | Properties of Gases

- The properties of gases that distinguish them from liquids and solids include compressibility, low viscosity, even and complete mixing, low density, and expansion with an increase in temperature.
- The kinetic molecular theory of gases is a model that explains the macroscopic properties of gases based on the behaviour of individual atoms or molecules of an ideal gas.
- Torricelli and Pascal demonstrated the existence of atmospheric pressure and how it varies with altitude.
- Boyle's law states that the volume of a given amount of gas at constant temperature is inversely proportional to the external pressure exerted on it. The equation for Boyle's law is $P_1V_1 = P_2V_2$.
- As the external pressure on a gas increases, the volume begins to decrease. As the volume decreases, the molecules become closer together causing the collision frequency to increase, increasing the pressure.

- The Kelvin temperature scale is based on a temperature of 0 K at absolute zero. The size of a kelvin on the Kelvin scale is the same as the size of a degree on the Celsius scale.
- Charles's law states that the volume of a fixed amount of gas at a constant pressure is proportional to the temperature of the gas: $V \propto T$. When temperature is in the Kelvin scale, there is a directly proportional relationship between temperature and volume. The equation for Charles's law is $\frac{V_1}{T_1} = \frac{V_2}{T_2}$, where T is temperature in Kelvin.
- Gay-Lussac's law states that the pressure of a fixed amount of gas at a constant volume is directly proportional to the Kelvin temperature of the gas: $P \propto T$. The equation for Gay-Lussac's law is $\frac{P_1}{T_1} = \frac{P_2}{T_2}$, where T is temperature in Kelvin.

Chapter 12 | Exploring the Gas Laws

- Gases undergo changes in temperature, pressure, and volume simultaneously. The combined gas law can be used to predict these changes and is represented by the equation $\frac{P_1V_1}{T_1} = \frac{P_2V_2}{T_2}$.
- According to the law of combining volumes, when gases react, the volumes of the gaseous reactants and products, measured at the same conditions of temperature and pressure, are always in whole-number ratios.
- Avogadro's law states that, at the same conditions of temperature and pressure, equal volumes of all ideal gases contain the same number of molecules, even though they do not have the same mass. It is represented by the equation $\frac{n_1}{V_1} = \frac{n_2}{V_2}$.
- The ideal gas law is derived from the combined gas law and Avogadro's law. The mathematical equation for it is $PV = nRT$. R is the universal gas constant with a value of $8.314\frac{kPa \cdot L}{mol \cdot K}$.
- For gas stoichiometry, when the reactants and products are not under the same conditions or the reaction occurs

under non-standard conditions, the ideal gas law can be used to calculate the amounts of gaseous reactants or products.

- The major components of Earth atmosphere are nitrogen, oxygen, argon, and carbon dioxide gas, as well as water vapour. The components of the atmosphere are concentrated mostly within 11 km of Earth's surface.
- Criteria air contaminants are carbon monoxide, nitrogen oxides, particulate materials, sulfur dioxide, and volatile organic compounds (VOCs).
- The Air Quality Health Index (AQHI) is a rating and advisory scale developed by the Canadian government, in consultation with provincial and territorial governments, to help Canadians understand the possible effects of air quality on any given day, and to make personal choices based on that information.

Knowledge and Understanding

Select the letter of the best answer below.

1. Which of the following statements describes an ideal gas?
   a. There are attractive forces between the gas molecules.
   b. The molecules are in constant random motion.
   c. The volume occupied by the molecules is significant compared to the container volume.
   d. The gas behaves according to the atomic theory of matter.
   e. There are repulsive forces between the gas molecules.

2. Which statement best describes the behaviour of the mass of a fixed quantity of gas in a sealed syringe?
   a. It increases when the volume in the syringe decreases as the plunger in the syringe is pulled back.
   b. It decreases when the volume in the syringe increases as the plunger in the syringe is pushed forward.
   c. It decreases when the temperature of the syringe decreases.
   d. It increases when the temperature of the syringe increases.
   e. It does not change when the temperature or pressure in the syringe changes.

3. Which of the lines on the graph below best represents the relationship between the volume of a gas and its pressure, if other factors remain constant?

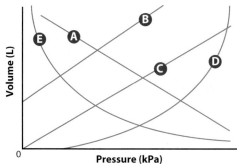

   a. A
   b. B
   c. C
   d. D
   e. E

4. Increasing the temperature of an ideal gas from 25°C to 50°C at a constant volume and mass will cause which of the following to also increase?
   I. the molar mass of the gas
   II. the average kinetic energy of the gas molecules
   III. the pressure of the gas
   IV. the density of the gas

   a. I and IV
   b. II and III
   c. I, II and III
   d. II only
   e. III only

5. Which of the following properties of a gas are directly proportional to one another?
   a. pressure and temperature
   b. pressure and volume
   c. pressure and moles
   d. volume and temperature
   e. a, c, and d

6. Which the following statements about equal volumes of two different gases at the same temperature and pressure is generally true?
   a. they have different number of molecules
   b. they have the same mass
   c. they have the same number of molecules
   d. they have different P and V values
   e. they have different number of moles

7. Under what conditions will nitrogen gas behave most like an ideal gas?
   a. high temperature and high pressure
   b. high temperature and low pressure
   c. low temperature and high pressure
   d. low temperature and low pressure
   e. at moderate temperature and pressure

8. The atmospheric pressure on a certain day is 99.8 kPa. What is the partial pressure of oxygen, given that its percent abundance in the atmosphere is 21%?
   a. 78.8 kPa
   b. 21.0 kPa
   c. 120.8 kPa
   d. 57.8 kPa
   e. 99.8 kPa

9. One step in the industrial production of nitric acid is the reaction of ammonia with oxygen gas:
   $$4NH_3(g) + 5O_2(g) \rightarrow 4NO(g) + 6H_2O(g)$$
   What is the volume of nitrogen monoxide gas that can be produced from 8 L of oxygen gas?
   a. 4.0 L
   b. 6.4 L
   c. 10 L
   d. 8.0 L
   e. 7.5 L

10. The correct sequence of the layers of Earth's atmosphere, from lowest to highest, with respect to composition is
   a. stratosphere, exosphere
   b. mesosphere, exosphere, thermosphere, heterosphere
   c. homosphere, heterosphere, thermosphere
   d. homosphere, heterosphere
   e. homosphere, stratosphere, heterosphere

Answer the questions below.

11. Explain how a mercury barometer measures atmospheric pressure. How would replacing mercury with water affect the operation of the barometer?

12. A gas is placed in a cylinder with a piston, and the piston is pulled upward, as shown below.

   a. What happens to the pressure of the gas in the cylinder? Explain why.

   b. What happens to the density of the gas? Explain why.

13. For each of the following descriptions, explain the changes that take place at a molecular level and name the gas law that applies.

   a. For a container with a fixed mass of gas at a constant pressure there is an increase in volume when the temperature is raised.

   b. For a container with a fixed mass of gas at a constant volume there is a decrease in pressure when the temperature is lowered.

   c. For a container at a constant temperature and volume there is an increase in pressure when more molecules of a gas are added.

   d. For a container with a fixed mass of gas at a constant temperature there is a reduction in volume when the pressure is increased.

14. When using the law of combining volumes, why is it not necessary to specify the conditions of temperature and pressure if the same conditions apply for all substances?

15. What is meant by the term "molar volume of a gas?" How is this numerical value used in the derivation of the ideal gas law?

16. What are the two main components of Earth's atmosphere? What percentages of the composition of the atmosphere do they each account for? List four examples of the minor components.

17. Describe the AQHI, and explain the rationale behind its use for Canadian cities.

Thinking and Investigation

18. Complete the following table.

Pressure Unit Conversions

Pressure (kPa)	Pressure (mmHg)	Pressure (atm)	Pressure (psi)
102			
		3.0	
			40.0

19. An Erlenmeyer flask is tightly closed with a one-hole rubber stopper, and a funnel is inserted into the flask through the hole. Water is poured into the funnel and begins to fill the flask. As the water in the flask reaches the bottom of the funnel spout, it becomes harder to add any more water. Use your understanding of the properties of gases to explain why this occurs.

20. A tire has a volume of 535 cm³ at a temperature of 20.0°C. While in use, the temperature of the tire rises to 65.0°C. Calculate the new volume of the tire, assuming constant pressure and amount of air.

21. The inside of a tennis ball contains air at a pressure of 11.0 psi and a temperature of 25°C. During tennis play, the temperature of the ball and the air inside it rise to 45°C. Assume a constant amount and volume of air.

   a. Calculate the new pressure of the gas inside the ball.

   b. How would an increase in the temperature of the ball affect the properties of the ball?

22. If a halogen bulb is filled with iodine vapour at a pressure of 5.00 atm when the temperature is 22°C, what will be the pressure in the bulb when it reaches its operating temperature of 1.0×10^{3}°C? Assume a constant amount and volume of gas.

23. Chlorofluorocarbons (CFCs) are gases once commonly used in refrigeration. Before they were banned for their role in destroying stratospheric ozone, a sample of a CFC compound was placed in the tubing of a refrigerator. The volume of the gas was 125 mL at a temperature of 22.0°C and a pressure of 550 kPa. The liquid was then forced into a different tube and allowed to expand to a new volume of 540 mL at a new pressure of 110 kPa. What was the resulting new temperature of this compound in degrees Celsius?

24. Calculate the density of each of the following. Assume the amount of gas is constant.

   a. krypton at 99 kPa and 88°C

   b. oxygen at 35 kPa and −5.0°C

   c. methane at 40.0°C and 103 kPa

25. The atmosphere of the planet Venus is mostly carbon dioxide gas. At the surface of the planet, the temperature is 8.0×10^2°C with a pressure of 7.5×10^3 kPa. If these are STP conditions on Venus, what would be the molar volume of an ideal gas?

26. Nitrogen and oxygen gas can be extracted from the air using compression and very low temperatures. What mass (kg) of nitrogen would occupy a 500.0 L cylinder if the pressure was 1300 MPa and the temperature was −150°C?

27. A small carbon dioxide gas cylinder used to propel paint balls has a mass of 56.3 grams. When all the gas is expelled from the cylinder and the cartridge is empty, the cylinder has a mass of 50.4 g. If the volume of the cylinder is 25 cm^3 and the temperature of the cylinder initially was 23°C, determine the initial pressure of gas in the cylinder in units of kPa.

28. For a particular experiment, an Erlenmeyer flask was completely filled with carbon dioxide gas. The temperature of the gas was 15°C and the volume of the gas was 265 mL. If the room pressure was 102.2 kPa, calculate the following.
   a. the moles of carbon dioxide gas
   b. the mass of carbon dioxide gas
   c. the number of molecules of carbon dioxide gas

29. A diatomic halogen gas has a density at 4.14 g/L at a temperature of 3.0°C and pressure of 250 kPa. Calculate the molar mass of the gas. Identify the gas, and provide the chemical formula for it.

30. A 0.366 g sample of a metal carbonate was reacted with hydrochloric acid. A chemical equation for this reaction can be represented as shown below, where X is the unknown metal.
$$XCO_3(s) + 2HCl(aq) \rightarrow XCl_2(aq) + H_2O(\ell) + CO_2(g)$$
The carbon dioxide gas generated was collected by water displacement at a temperature of 25°C and pressure of 101 kPa. The final volume of carbon dioxide gas collected was 110 mL. Determine each of the following.
   a. the pressure of the dry gas at this temperature
   b. the moles of carbon dioxide produced
   c. the moles of the carbonate reacted
   d. the molar mass of the carbonate
   e. the identity of element X and the chemical formula for the carbonate

31. Benzene is a hydrocarbon obtained from crude oil. Benzene has a mass percent composition of 92.24 percent carbon and 7.76 percent hydrogen. When a sample of benzene with a mass of 14.78 g is vaporized in a sealed container with a volume of 4.00 L and heated to 120°C, the pressure in the flask is 154.4 kPa.
   a. Determine the empirical formula of the compound.
   b. Determine the molar mass of the compound.
   c. Determine the molecular formula of the compound.

32. Carbon monoxide gas can react with hydrogen gas under certain conditions to produce propane, $C_3H_8(g)$, and water vapour.
   a. Write out the balanced chemical equation for this reaction.
   b. If 150.0 L of propane gas are required, what volume of hydrogen gas would be necessary to produce this amount?
   c. How much hydrogen gas would be required to react with 200.0 L of carbon monoxide gas?
   d. What amount of water vapour would be produced if 90.0 L of hydrogen gas were consumed in a reaction?

33. Antacids often contain carbonates that are used to reduce excess stomach acid. The preferred carbonate to use is calcium carbonate. The chemical equation for the neutralization reaction is
$$CaCO_3(s) + 2HCl(aq) \rightarrow CaCl_2(aq) + CO_2(g) + H_2O(g)$$
The amount of calcium carbonate in a tablet can be determined by collecting and measuring the carbon dioxide gas present when a crushed tablet is reacted with an excess amount of hydrochloric acid. A 500.0 mg tablet is crushed and reacted with excess 2.0 mol/L hydrochloric acid and the gas collected by water displacement. If the air pressure is 100.0 kPa, the temperature is 24°C, and the volume of gas generated is 103 mL, determine the mass of calcium carbonate present in the tablet and the mass percent composition of calcium carbonate in the tablet.

34. Nitric acid, a key component of fertilizer production, is manufactured using the Ostwald process where ammonia gas is reacted with air in the presence of a Pt catalyst. The chemical equation for this reaction is
$$NH_3(g) + 2O_2(g) \rightarrow HNO_3(aq) + H_2O(\ell)$$
Calculate the volume of $O_2(g)$, in litres, that is needed to completely react 1500 g of $NH_3(g)$ at 30.0°C and 2500 kPa.

35. Nitrous oxide, or laughing gas, has been used for decades as an anesthetic. It can be prepared by carefully heating ammonium nitrate according to the chemical equation
$$NH_4NO_3(s) \rightarrow 2H_2O(g) + N_2O(g)$$

However, recently, a new method for producing nitrous oxide gas has been developed, which involves the reaction of ammonia with oxygen using a special catalyst. The chemical equation for this reaction is $2NH_3(g) + 2O_2(g) \rightarrow N_2O(g) + 3H_2O(g)$

a. Compare the volumes of nitrous oxide gas that can be produced from 1.0 kg of ammonia or 1.0 kg of ammonium nitrate. Which method produces the largest volume of gas? Assume the temperature of the gas produced is 40.0°C and the pressure is 105 kPa.

b. What volume of nitrous oxide gas can be produced from 5.0×10^3 L of ammonia vapour at STP? What volume of water vapour would also be produced through this reaction?

c. If the actual desired volume of nitrous oxide gas was 1.0×10^4 L at a pressure of 1500 kPa and 25°C, calculate the mass of ammonium nitrate that would have to be decomposed to produce this amount.

36. A common rocket engine fuel relies on the reaction between monomethyl hydrazine, $N_2H_3CH_3(\ell)$, and dinitrogen tetroxide, $N_2O_4(\ell)$. The reaction is represented by the chemical equation
$$4N_2H_3CH_3(\ell) + 5N_2O_4(\ell) \rightarrow$$
$$4CO_2(g) + 12H_2O(g) + 9N_2(g)$$

a. What volume of nitrogen gas would be produced from 25 kg of $N_2H_3CH_3$ at STP?

b. What volume of carbon dioxide gas would be produced under the same conditions?

c. What volume of water vapour would be produced under the same conditions?

d. What is the total volume of gases produced?

e. Why is this chemical reaction a good choice for launching a rocket?

37. One method of inflating devices involves using metal hydrides. When metal hydrides are allowed to react with water, they generate hydrogen gas and a metal hydroxide. Two potential compounds that can be used for this purpose are lithium hydride and magnesium hydride.

a. Write the balanced chemical equations for the reactions of each of these hydrides with water.

b. If 100.0 L of gas are to be generated at 102 kPa and 25°C, determine the mass of the hydride required for each reaction.

Communication

38. Consider the molecules C_2H_6 and N_2H_4.

a. Draw a Lewis structure for each molecule.

b. Describe the intermolecular forces that would be present in each molecule.

c. Which compound is most likely a gas at room temperature? Explain.

39. Solid carbon dioxide (Dry Ice™) is made by allowing liquid carbon dioxide stored in highly pressured tanks to be released from the tank to evaporate at room temperature. During this change of state, some of the liquid drops to a temperature of −78°C and freezes.

a. Explain using diagrams how pressure can be used to make a compound that is commonly found in the gaseous state into a liquid.

b. Why does the temperature change when liquid carbon dioxide evaporates?

40. A gas is placed in a container with a partition that separates the gas from a vacuum. A hole is made in the partition to produce the container shown below.

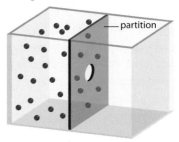

a. Use a series of diagrams to predict what will occur over time.

b. What effect would changes in temperature have on this process? Explain.

c. Explain how the density of the gas on both sides of the partition will be affected by this process.

41. Use a series of sketches to explain why a tire deflates when it is punctured.

42. Properties of molar volume, density, and molar mass are closely related, and any one property can be calculated using the other two. Use a table or graphic organizer such as a concept map to summarize this relationship. Include an example to support your answer.

43. Describe how the diagram below represents Dalton's law of partial pressures.

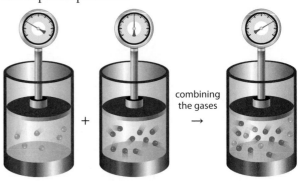

44. Using a graphic organizer or a set of diagrams, summarize each of the gas laws that you learned about in this unit. Design this in a manner that would allow you to use it as a quick reference guide when studying or doing problems. Make sure it highlights important points about each gas law, including what variables are held constant, what a graph showing the relationship between the variables looks like, and the associated mathematical equation. Include how the gas laws differ, and any other information that would help you to remember and organize the information for future use.

45. Using a comparative table or a graphic organizer, compare the properties of an ideal gas to the properties of a real gas. Include an indication of the conditions when the behaviours of ideal gases and real gases are the most similar and the most different.

Application

46. Magdeburg hemispheres, such as those shown below, were created in 1650 to demonstrate the principles of a vacuum. One of the hemispheres has a tube connected to a vacuum pump. When the two hemispheres are placed together to form a sphere, the pump draws all the air from inside the apparatus, creating a vacuum. Separating the two hemispheres is extremely difficult. The first experimental observations reported that "two teams of 15 horses could not pull the two hemispheres apart." Explain why it is so difficult to separate the two hemispheres when a vacuum exists inside the apparatus. What would be one way of easily separating the hemispheres, which does not involve a great deal of physical strength?

47. Explain why a hard-boiled egg would take longer to cook on top of Whistler Mountain, in British Columbia (elevation 2182 m), than in Windsor, Ontario (elevation 190 m).

48. Commercial airplanes are equipped with air masks that supply air for passengers and crew members in the event of a loss of cabin pressure during flight. If there is a loss in cabin pressure, the pilots also intentionally lower the airplane's altitude and then level off until a suitable landing site can be reached. Explain the requirement for the air masks and flight maneuvering.

49. Infer why gases such as the oxygen used in hospitals are compressed. What must happen to compressed oxygen before it can be inhaled?

50. **BIG IDEAS** Properties of gases can be described qualitatively and quantitatively and can be predicted. Sperm whales are known to dive to depths of over 1800 m and remain submerged in excess of an hour. The whales do most of their hunting at depths of 1000 m.
   a. Sperm whales are air-breathing mammals, but they do not experience "the bends" when they surface from great depths. Research and explain why this is the case.
   b. Whale lung volumes at STP are approximated to be near 5.0×10^3 L. What would lung volume be at ocean depth of 500.0 m where the temperature is 12°C and the pressure is 48 atm? What would the lung volume become at a depth of 1.0×10^3 m, with a pressure of 100.0 atm and temperature of 5.0°C?

51. **BIG IDEAS** Air quality can be affected by human activities and technology. Do research about the importance of stratospheric ozone in the atmosphere.
   a. Describe examples of human activities and technological devices that are associated with the depletion of ozone in the atmosphere.
   b. Give examples of chemical compounds that are used in place of the compounds that are associated with ozone depletion. Explain whether these substitutes themselves have negative consequences for the environment.
   c. Canadian scientists developed the Brewer ozone spectrophotometer, which is the most accurate ozone-measuring device in the world. Research how this device works, and prepare an oral or written report on your findings.

52. **BIG IDEAS** People have a responsibility to protect the integrity of Earth's atmosphere. Earth Day and Earth Hour are annual events that enable all citizens of the planet to do their own part to protect the integrity of the atmosphere. Identify at least five other choices and practices that you, personally, can make and do that contribute further to this protection, and explain how they are effective.

Select the letter of the best answer below.

1. **K/U** According to Charles's law, which of the following will occur if the temperature of a gas is increased?
   a. volume increases
   b. pressure increases
   c. moles increase
   d. decreased volume decreases
   e. pressure decreases

2. **T/I** What Kelvin temperature is equal to 35°C?
   a. 308 K   d. 135 K
   b. 298 K   e. −35 K
   c. 318 K

3. **T/I** A sample of hydrogen gas is collected by water displacement. The atmospheric pressure in the room is 100.8 kPa and the vapour pressure of water is 2.8 kPa. What is the partial pressure of hydrogen under these conditions?
   a. 2.8 kPa   d. 106.4 kPa
   b. 103.6 kPa   e. 100.8 kPa
   c. 98.0 kPa

4. **K/U** Which property of gases accounts for pressure?
   a. the space between molecules
   b. the density of the gas
   c. the motion of the gas molecules
   d. the identity of the gas
   e. the intermolecular forces present in the gas

5. **K/U** Which two of the following statements are correct?
   I The volume of a given mass of gas varies as the temperature is changed.
   II When a gas is cooled the molecules move faster and closer together.
   III At constant volume, a decrease in pressure may be due to an increase in the number of molecules.
   IV The molecules of a gas are in constant motion.
   a. I and III   d. II and IV
   b. II and III   e. I, II and III
   c. I and IV

6. **T/I** The atmospheric pressure on the summit of Mt. Everest is 0.329 atm. This corresponds to a pressure in kPa of
   a. 329 kPa   d. 100.9 kPa
   b. 33.3 kPa   e. 104.39 kPa
   c. 134.3 kPa

7. **T/I** Zinc metal (0.200 mol) and a volume of aqueous hydrochloric acid that contains 0.100 mol of HCl are combined and react to completion. What volume of hydrogen gas, measured at STP, is produced?
   $Zn(s) + 2HCl(aq) \rightarrow ZnCl_2(aq) + H_2(g)$
   a. 2.24 L of H_2 d. 11.2 L of H_2
   b. 4.48 L of H_2 e. 22.4 L of H_2
   c. 5.60 L of H_2

8. **K/U** Particles in ideal gases are assumed to have which of the following?
   I. no mass
   II. no volume
   III. no attractive forces between them
   a. I only   d. I and II
   b. II only   e. II and III
   c. III only

9. **K/U** The ideal gas law can be used to derive
   a. Boyle's law   d. Avogadro's law
   b. Charles's law   e. all of the above
   c. Gay-Lussac law

10. **T/I** Acrylonitrile, $C_3H_3N(g)$ is an industrial chemical used to make synthetic fibres. It can be produced through the following reaction:
    $2C_3H_6(g) + 2NH_3(g) + 3O_2(g) \rightarrow$
    $$2C_3H_3N(g) + 6H_2O(g)$$
    If a volume of 5000 L of acrylonitrile gas is produced, what volume of oxygen gas would be required for the reaction?
    a. 5000 L   d. 7500 L
    b. 10 000 L   e. 3333 L
    c. 2500 L

11. **K** Explain why different gases can have different densities if they all have the same molar volume at the same temperature and pressure.

12. **C** Use a bar graph to compare the percent composition of the major and minor components that make up Earth's atmosphere. Include six components.

13. **A** A chemist is interested in creating a barometer from the metal gallium. Gallium has a density of 6.095 g/cm³ and a melting point of 29.8°C. Mercury, the early standard for barometers, has a melting point of −38.8 °C and a density of 13.5 g/cm³.
    a. If the temperature of the air was above 30.0°C, what height would the liquid gallium have to be to equalize against atmospheric pressure?
    b. Why was mercury the liquid of choice when barometers were first developed?

14. **T/I** An air tank has a volume of 5.0 L at 550 kPa. If the pressure of the tank decreases to 102 kPa, what is the volume of gas released if the temperature is not changed?

15. **C** An empty 2 L soft-drink bottle, with its cap tightly fastened, was taken to a recycling bin. The outside temperature was −20°C. Later, the bottle appeared distorted and shrunken. Use labelled sketches to explain these observations.

16. **T/I** An inflatable buoy was anchored in a lake to warn boaters of shallow rocks. The buoy was left out over the winter and the volume of the balloon shrank to 81 L at −22°C. If the volume of the buoy in the summer is 95 L, what is the temperature?

17. **A** Two gases are placed in three different flasks that are connected by two valves. Molecules of the two gases are represented by the red and blue spheres in the diagram. Flask A and Flask C have the same volume, while Flask B is half the volume of the other two flasks.

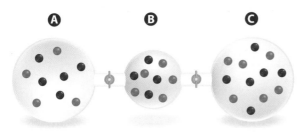

a. Rank the flasks from greatest to least in terms of total pressure exerted by the gase, and explain your answer.

b. What would have to occur for Flask A to have the same pressure as Flask C?

c. Describe what will happen when the valve between Flasks A and B is opened.

d. Describe what will happen when the valve between Flasks B and C is opened.

18. **T/I** In a fire, gas cylinders containing combustible gases are at risk of exploding. If a tank of propane has an internal pressure of 1500 kPa at 20.0°C, what will be the internal tank pressure at a temperature of 1100°C?

19. **T/I** A steel gas cylinder contains 30.0 L of acetylene, $C_2H_2(g)$, when it is full. Over a period of time, the gas is used up until it is declared empty. If the temperature of the container when empty is 25°C and if the atmospheric pressure is 102 kPa:

a. Is the container really empty? Explain your answer.

b. How many moles of gas remain in the cylinder?

c. What is the mass of the remaining gas in the cylinder?

d. How many molecules of acetylene remain in the cylinder?

e. Is it possible to remove all the acetylene gas from the container? Explain.

20. **T/I** At liftoff, a weather balloon held 700.0 m³ of gas at a temperature of 30.0°C and a pressure of 101 kPa. At higher elevations of almost 9 km, the balloon experienced temperatures of −6.0°C and pressures of 23 kPa. What volume would the balloon have at this altitude?

21. **A** Many municipalities in Ontario, as well as throughout the country, have initiatives that encourage the use of public transit systems as a means for helping to improve air quality.

a. In what ways can air quality be improved through the use of public transit?

b. What are the limitations of this approach? (Hint: Are public transit systems available in all locations throughout in a given province?)

22. **T/I** An unknown metal carbonate is reacted with excess hydrochloric acid
$X_2CO_3(s) + 2HCl(aq) \rightarrow 2XCl(aq) + H_2O(l) + CO_2(g)$
A 1.42 g sample of the carbonate was reacted and the carbon dioxide gas generated was collected by water displacement. At a temperature of 22.0°C and a pressure of 100.5 kPa, 258 mL of gas were collected.

a. Determine the pressure of the dry gas at this temperature.

b. Determine the moles of carbon dioxide produced.

c. Determine the moles of the carbonate reacted.

d. Determine the molar mass of the carbonate.

e. What is the identity of the element represented by X?

Self-Check

If you missed question …	1	2	3	4	5	6	7	8	9	10	11	12	13	14	15	16	17	18	19	20	21	22
Review section(s)…	11.3	11.3	12.2	11.1	11.1	11.2	12.2	12.1	12.1	12.2	12.1	12.3	12.1	11.2	11.2	11.3	11.2	12.2	12.2	12.2	12.3	12.2

Guide to the Appendices

Guide to the Appendices

Analyzing STSE Issues

STSE is an abbreviation for **s**cience, **t**echnology, **s**ociety, and the **e**nvironment. An issue is a topic that can be seen from more than one point of view. In *Chemistry 11,* you are frequently asked to make connections between scientific, technological, social, and environmental issues. Making such connections could involve, for example, assessing the impact of science on developments in consumer goods, medical devices, or industrial processes; on people, social policy, or the economy; or on air, soil, and water quality, the welfare of organisms, or overall ecosystem health. Analyzing STSE issues involves researching background information about a problem related to science, technology, society, and the environment; evaluating differing points of view concerning the problem; deciding on the best response to the problem; and proposing a course of action to deal with the problem.

The following flowchart outlines one process that can help you to focus your thinking and organize your approach to analyzing STSE issues. The most effective analyses result in decision making and, ultimately, an action plan. Group discussion and collaborative analysis can also play a role in analyzing an STSE issue.

A Process for Analyzing Issues

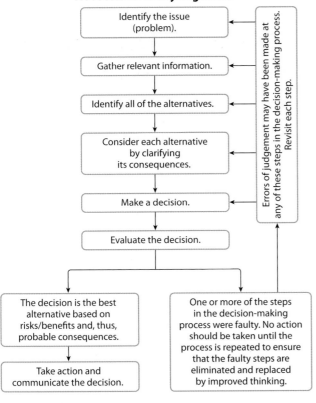

Identify the Issue (Problem)

An STSE issue is a topic that is debatable—it can be viewed from more than one perspective. When you encounter an issue related to STSE, such as a medical breakthrough, a health-care policy, or an environmental regulation, you need to try to understand it from multiple points of view.

> Suppose you have learned that phosphates from household products, industrial waste, sewage, and agricultural run-off result in algal blooms that reduce biodiversity in rivers, lakes, and oceans.

Assess whether there is any controversy associated with this situation. Could there be different viewpoints concerning the cause of the situation and how to respond to it?

> You read a blog that calls for action on the part of the Canadian government to stop this situation from getting worse. You decide that this situation does represent an STSE issue, because it lends itself to multiple points of view and there is more than one course of action the government could take in response to the situation.

Try to sum up the issue in a specific question.

> For example, "What steps, if any, should the federal government take to reduce phosphates in our waterways?"

Gather Relevant Information

You will need to do some research to gain a better understanding of the issue. Go to Developing Research Skills in Appendix A for help with finding information.

> For example, what role, if any, does the federal government currently play in regulating phosphates? What are the major sources of phosphates that the government regulates? Are there sources of phosphate run-off that are not currently regulated by the government? What methods of regulation could the government impose?

Identify Possible Solutions to the Problem

In order to make an informed decision about how to respond to the issue, you will need to assess the possible solutions to the problem. Your research should reveal some alternative solutions.

For example, you see a news report in which a political candidate proposes to expand current federal regulations on phosphate levels to include dishwasher detergents rather than just laundry detergents. You read a blog that supports a tax rebate to companies that produce reduced-phosphate fertilizer. One article you read suggests that untreated sewage plays a large role in the high phosphate levels in waterways. Perhaps the government should enact stricter regulations on phosphate levels allowed in waste water from water treatment facilities.

Clarify the Consequences of Each Possible Solution

You may need to do additional research to identify potential consequences of each alternative solution and the reactions of the various stakeholders (that is, the individuals or groups affected by the issue).

For example:

- How long does it usually take for the government to approve new environmental regulations?
- Is there a cost associated with enforcing these regulations?
- Are tax cuts in one category usually replaced with tax increases elsewhere?

You can sort the potential consequences of an alternative into benefits (positive outcomes) and risks (negative outcomes). Use a risk-benefit analysis table like the one below to help you analyze the alternative solutions. For each possible solution, assess the impact on various stakeholders. The potential consequences of each solution could be different for each stakeholder. For some issues, you might choose to assess differing perspectives rather than differing effects on stakeholders. For example, you could assess benefits and risks from economic, environmental, social, scientific, and ethical perspectives. Each perspective could reveal different consequences.

Risk-Benefit Analysis

Issue: What steps, if any, should the federal government take to reduce phosphates in our waterways?

Possible Solutions	Stakeholders	Potential Benefits (positive outcomes)	Potential Risks (negative outcomes)
1. The Canadian government should offer a tax rebate to companies that produce reduced-phosphate fertilizer.	Government	• Reduction in amount of phosphates entering the water system • Less money spent on environmental clean-up	• Reduction in tax revenue • Money would be needed to implement and promote the new rebate
	Fertilizer manufacturer	• Financial incentive to develop a new product • Possible increase in sales due to product being labelled "environmentally friendly" • No job losses, since there is no additional cost to the manufacturer	• Cost of product development may be greater than the rebate offered • Possible loss of revenue due to reduced effectiveness of new product
	Citizen	• Cleaner water leads to healthier, more abundant fish • Healthier ecosystem increases quality of life	• Tax rebate for manufacturers may come at the cost of higher taxes in other categories
	Farmer	• Reduced impact of farming operations on environment and water quality • Effective, environmentally-friendly fertilizer option available	• Possible decrease in crop yield due to reduced effectiveness of new product
2. The Canadian government should develop stricter regulations for sewage treatment.			

Make a Decision

Once you have identified potential outcomes for each possible solution, you are faced with the task of making a decision. Which alternative promises the greatest benefits and the least risks or lowest costs? Your personal values will influence your assessment. You will need to decide whether the benefits of a particular alternative are major or minor. You will also need to decide what an acceptable level of risk is. You might find it helpful to write down a list of questions to help you evaluate the alternative solutions. Some factors to consider are listed here:

- How likely is it that a potential outcome will occur?
- Is there evidence to support the likelihood of a potential outcome?
- How many people (or other organisms) will the proposed course of action affect?
- Is there an estimated sum of money associated with the benefits or costs of each solution?
- Is the outcome of a proposed solution short-term (a one-time benefit/risk) or long-term (ongoing)?
- According to your analysis, how important are the risks of a possible solution compared to its potential benefits?
- How do the benefits and risks of one possible solution compare with the risks and benefits of other possible solutions?

After considering all the alternatives, you might decide that offering a tax rebate to companies that produce phosphate-free or reduced-phosphate fertilizers will prompt industry change without causing job losses. This solution will have the desired effect of reducing the amount of phosphates entering waterways, while still providing farmers and others with effective fertilizers. Your research also suggests that the cost of providing the rebate is lower than the cost of an environmental clean-up.

Evaluate the Decision

Once you have made a decision, evaluate whether you can justify it with logic and verifiable information. If you discover that some of the information you used to make the decision was incorrect, you should reconsider the alternatives. If new information becomes available, that could also affect your decision.

Suppose a new study reveals that phosphate levels in ground water continue to rise in a community that banned phosphate fertilizers two years ago. How might this new information affect your decision?

Also, assess whether you have taken all perspectives into account in your analysis. Is there another stakeholder that is strongly affected by a particular alternative? If you decide that you are not confident in the decision you have made, you will need to revisit each step in your analysis.

Act on Your Decision

If you are confident in your decision, the next step is to propose and implement a course of action.

For example, you could start a community e-mail campaign urging your Member of Parliament to propose that tax rebates be offered to companies that reduce phosphate levels in their products.

Instant Practice

1. Consider the second possible solution listed in the risk-benefit analysis table on the previous page. Create a table in your notes to analyze the benefits and risks of this possible solution. Fill in the "Stakeholders," "Potential Benefits," and "Potential Risks" columns.

2. Look for a chemistry-related STSE issue in the news. Apply the analysis method outlined in this appendix to determine your response to the issue. Write a brief paragraph to explain your viewpoint and a proposed course of action.

Scientific Inquiry

Scientific inquiry is a process that involves making observations, asking questions, performing investigations, and drawing conclusions.

A Process for Scientific Inquiry

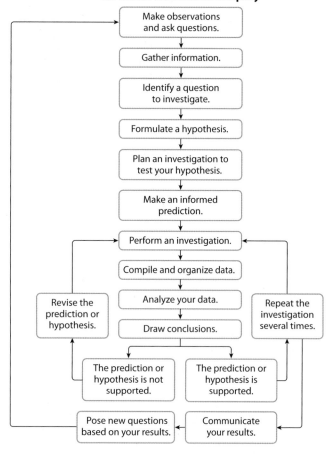

Make Observations and Ask Questions

Scientific inquiry usually starts with observations. You notice something that sparks your curiosity and prompts you to ask questions. You try to make sense of your observations by connecting them to your existing knowledge. When your existing knowledge cannot explain your observations, you ask more questions.

For example, suppose a train derailment has resulted in an acid spill near the shore of a local lake. You wonder what effect the spill will have on the fish and other organisms that live there. Has the spill killed all of the organisms in the lake? How have other organisms, such as aquatic plants, been affected by the spill? How can you find answers to your questions without endangering your safety?

Gather Information

Background research may help you to understand your observations and answer some of your questions. Go to Developing Research Skills in Appendix A for guidance on conducting research. You may also be able to gather information by making additional observations.

For example, you read a news report about an environmental assessment of the spill site. You discover that the pH of the lake water before the accident was 6.7. Measurements taken after the accident indicate that the pH dropped to 4.1. You do additional research to find out what kinds of organisms inhabit the lake and the optimal pH for their survival and growth.

Identify a Question to Investigate

You need to have a clear purpose and decide on a specific question that you are able to investigate with the resources available. If a question is provided for you, make sure you understand the science behind the question.

You decide to investigate the effect of acidity on living organisms. You do not wish to risk harming fish or other animals, so you decide to use aquatic plants as your test organism. You pose the scientific question, "What effect will increasing acidity have on aquatic plants grown in an aquarium?"

Formulate a Hypothesis

A hypothesis attempts to answer the question being investigated. It often proposes a relationship that is based on background information or an observed pattern of events.

You hypothesize that because plants can remove some impurities from polluted water, aquatic plants will be able to reduce the effect of small amounts of acid. However, because highly acidic water will damage or kill most organisms, you hypothesize that the aquatic plants will not be able to counteract the addition of large quantities of acid.

Plan an Investigation

Some investigations lay out steps for you to follow in order to answer a question, analyze a set of data, explore an issue, or solve a problem. In planning your own investigation, however, *you* must decide how to approach a scientific question. Taking time to plan your approach thoroughly will ensure that you address the question appropriately.

Design a Procedure Write out step-by-step instructions for performing the investigation. Include instructions for repeat trials, if appropriate. Ensure that the procedure is written in a logical sequence, and that it is complete and clear enough that someone else could carry it out. Create diagrams, if necessary. Ask someone else to read through the procedure and explain it back to you, to ensure you have not omitted any important details.

> You decide to investigate the change in the pH of water when you add acid to a large glass bottle containing water and aquatic plants. You will measure the pH of the water and observe the physical appearance of the plants twice a day for three days.

Identify Variables Many investigations study relationships between variables (quantities or factors that can change). An *independent variable* is changed by the person conducting the investigation. A *dependent variable* is affected by changes in the independent variable. *Controlled variables* are kept the same throughout an experiment.

A simple, controlled experiment shows relationships especially clearly because it has a single independent variable and a single dependent variable. All other variables are controlled. Changes in the dependent variable occur only in response to changes in the independent variable. When you are planning your investigation, you will need to identify the variables and decide which ones to control.

If possible, investigations include a *control*: a situation identical to the one being tested, except that the independent variable is not changed in any way. There is no reason, therefore, for the dependent variable to change. If it does, the reasoning behind your hypothesis, prediction, and variable analysis may be faulty. Look at the illustration at the top of the next column to see some examples of independent and dependent variables, as well as two examples of a control (no independent variable).

a. A test to find the best filter for muddy water

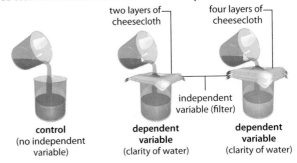

b. A test to find the best plant food for plant growth

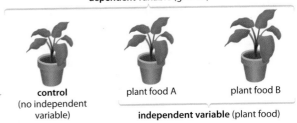

In planning your investigation, you decide to manipulate (change) the quantity of acid added to the aquarium you have made. Therefore, the quantity of acid is the independent variable. The pH of the water will be the dependent variable. Water temperature, lighting conditions, the particular species of aquatic plant used, and nutrients will be the controlled variables. In addition, you decide to set up an identical aquarium as a control. No acid will be added to the water in this aquarium, so you expect the pH of the water to stay constant.

You decide to set up three different experimental aquariums. You plan to add a different amount of acid to the water in each of these aquariums. You will add no acid to the fourth aquarium (the control). Your teacher suggests using 1.0 mol/L hydrochloric acid. You will add five drops of hydrochloric acid to Aquarium 1, 10 drops to Aquarium 2, and 15 drops to Aquarium 3. You will measure the pH of the water in each aquarium at 9:00 A.M. on the first day, immediately before adding the acid, and then at 2:00 P.M. You will measure the pH again at 9:00 A.M. and 2:00 P.M. on Day 2 and Day 3.

List Materials and Safety Precautions Develop a list of materials and apparatus you will need. Include measuring and recording instruments. Examine your procedure for safety hazards and plan any necessary precautions. (Go to Safety in Your Chemistry Lab and Classroom at the front of this book for information about safety hazards and precautions.) **Note:** Before doing any experimental work, ask your teacher to examine and approve your plan.

Your materials list will include safety goggles, a lab coat or apron, protective gloves, four glass jars, a dropper, aquatic plants, hydrochloric acid, water, and a pH meter. Safety precautions include handling glassware carefully to avoid breakage; wearing safety eyewear, gloves, and protective clothing to protect yourself from any acid spillage; storing the acid safely after use; and disposing of the aquarium water at the end of the investigation according to your teacher's instructions.

Make an Informed Prediction

A clear hypothesis often leads to a specific, testable prediction about what the investigation will reveal. You need to determine how to test your question before you can predict what will happen.

You predict that an aquarium full of a certain species of aquatic plants will maintain a stable pH of about 7 when a small quantity of acid is added. When greater quantities of acid are added, however, you predict that the plants will be damaged. The pH of the water will decrease rapidly and the plants will eventually die.

Perform an Investigation

Be responsible whenever you conduct an investigation. Think before acting, and follow all safety precautions. Carry out your procedure carefully. Ask for assistance if you are unsure how to proceed or if you encounter an unexpected difficulty. Report any accidents to your teacher immediately. Keep your workspace neat and clean it up when you have finished your investigation.

Compile and Organize Data

Record your results carefully and organize them in a logical way. Go to Organizing Data in a Table in Appendix A for help with recording and organizing the results of an investigation. As part of your observations, keep careful notes of any unexpected occurrences, problems with equipment, or unusual circumstances that might affect your

results. If you are working with a partner, ensure that both of you have a copy of all observations and results.

Your results may include either qualitative or quantitative observations, or both. *Quantitative observations* are measurable and involve numbers. *Qualitative observations* involve descriptions rather than numbers or measurements. When making qualitative observations, try to record specific characteristics so that you can make comparisons between different trials.

In your investigation, you will record both qualitative and quantitative results. The pH values that you record are quantitative observations. Your descriptions of the physical appearance of the aquatic plants are qualitative observations. Looking at specific plant characteristics such as colour (green or brown) and vigour (robust or spindly) will help you to compare the physical appearance of the plants in each aquarium.

You might use a table like the one below to record and organize the data from your investigation.

Effect of Increasing Acidity on the Physical Appearance of Aquatic Plants in an Aquarium

	Physical Appearance of Plants (colour and vigour)			
	Control	+ 5 drops of acid	+ 10 drops of acid	+ 15 drops of acid
Day 1, 9:00 A.M. (before addition of acid)	green, robust	green, robust	green, robust	green, robust
Day 1, 2:00 P.M.	green, robust	green, robust	green, robust	brownish spots, spindly
Day 2, 9:00 A.M.	green, robust	green, robust	brownish spots, less robust	brown, spindly (looks dead)
Day 2, 2:00 P.M.	green, robust	green, robust	mostly brown, less robust	brown, spindly (looks dead)
Day 3, 9:00 A.M.	green, robust	green, robust	brown, spindly (looks dead)	brown, spindly (looks dead)
Day 3, 2:00 P.M.	green, robust	green, robust	brown, spindly (looks dead)	brown, spindly (looks dead)

Analyze Your Data

Perform any necessary graph work or calculations. Go to Constructing Graphs in Appendix A for help with graphing. Then consider and interpret your results. Do your data and observations support or refute your hypothesis and prediction? Are additional data needed before you can draw definite conclusions? Identify any possible sources of error or bias in your investigation. Does the procedure or apparatus need to be modified to obtain better data?

Using the pH data from your investigation, you construct the graph shown below.

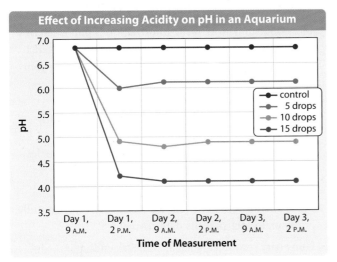

Effect of Increasing Acidity on pH in an Aquarium

Draw Conclusions

Conclusions usually answer several questions:

- What has the investigation revealed about the answer to the question?
- How well does your prediction agree with the data?
- How well is your hypothesis supported by the data? Are the observations explained by the hypothesis?
- How precise were the measuring instruments and resulting observations?
- What improvements could be made to the investigation?

Relate your conclusions to your background knowledge of the scientific principles involved.

You conclude that your hypothesis and prediction were supported by the data. The aquatic plants were able to survive the addition of small amounts of hydrochloric acid (5 drops), but when more acid was added to the aquarium, the pH decreased rapidly and the plants soon looked brown and spindly.

However, you are unsure whether the plants in the highly acidic water were actually dead. You think perhaps you could place them in fresh water to see whether they could recover.

Now you can relate your results to the original problem of the acid spill in the lake. Are the aquatic plants in the lake likely to survive? Do your results reflect the conditions in the lake? What other factors might you need to consider?

Communicate Your Results

Communicate the results of your investigation. Always include a summary of your findings and an evaluation of the investigation. Be sure to round answers to the proper number of significant digits. Go to Significant Digits in Appendix A for help with reporting numerical answers. Demonstrate your results clearly using graphs, tables, or diagrams, as appropriate. Go to Constructing Graphs or Organizing Data in a Table in Appendix A for help with communicating your results. Be sure to include units when expressing measurements. Go to Measurement in Appendix A for information on units and measurements.

Pose New Questions Based on Your Results

The conclusion of an investigation is not the end of scientific inquiry. Scientific inquiry is a continuous process in which results and conclusions lead to new questions. What new research questions might arise from your investigation? How might you find an answer to one of these questions?

After performing this investigation, you wonder how much the aquatic plants themselves affected the pH of the water. Would the decrease in pH have been more noticeable if there were no plants present? How would you test this?

Instant Practice

You are asked to plan an investigation to determine whether calcium chloride or sodium chloride is a better de-icing agent. Think about how you could test a hypothesis related to this question.

1. Will your results include qualitative or quantitative observations, or both? Explain.
2. State a hypothesis for this investigation.
3. What will your independent variable be? What will your dependent variable be? What control will you set up?

Developing Research Skills

In this course, you will need to conduct research to answer specific questions and to explore broad research topics. The following skills will take you through the research process from start to finish:

- focussing your research
- searching for resources that contain information related to your topic
- evaluating the reliability of your information sources
- gathering, recording, and organizing information in an appropriate format
- presenting your work

Focussing Your Research

- Start by carefully reading your assignment. Pick out key words and phrases, such as *apply, analyze, argue, compare and contrast, describe, discuss, evaluate, explain, identify, infer, interpret,* and *predict.* These key words and phrases will guide you on what kind of information you need to collect, and what you need to do with the information.
- Jot down ideas on your own, and then get additional input from others, including your teacher.
- Once you have done some general research, narrow down your topic until you can express it in one specific question. This will help you focus your research.
- Ensure that the question you are researching fulfills the guidelines of the assignment provided by your teacher.

Searching for Resources

- It is important to find reliable resources to help you answer your question. Potential sources of information include print and on-line resources such as encyclopedias, textbooks, non-fiction books, journals, websites, and newsgroups.

- The library and the Internet can both provide information for your search. Whether you are looking at print or digital resources, you need to evaluate the accuracy and objectivity of the information.

Evaluating the Reliability of Your Information Sources

Assess the reliability of your information sources to help you decide whether the information you find is likely to be accurate. To determine the validity of a source, check that the author is identified, a recent publication date is given, and the source of facts or quotations is identified. An author's credentials are important. Look for an indication of educational background, work experience, or professional affiliation. If the information is published by a group, try to find out what interests the group represents. The following guidelines may be helpful in assessing your information sources:

- On-line and print scientific journals provide data that have been reviewed by experts in a field of study (peer-reviewed), so they are usually a reliable source. Be aware, however, that the conclusions in journal articles may contain opinions as well as facts.

- Data on the websites of government statistical departments tend to be reliable. Be sure to read carefully, however, to interpret the data correctly.

- University resources, such as websites ending in ".edu" are generally reliable.

- Reliable experts in a field of study often have a PhD or MSc degree, and their work is regularly cited in other publications.

- Consumer and corporate sources may present a biased view. That is, they may only present data that support their side of an issue. Look for sources that treat all sides of an issue equally and fairly, or that clearly specify which perspective(s) they are presenting.

- Some sources, such as blogs and editorials, provide information that represents an individual's point of view or opinion. Therefore, the information is not objective. However, opinion pieces can alert you to controversy about an issue and help you consider various perspectives. The opinion of an expert in a field of study should carry more weight than that of an unidentified source.

- On-line videos and podcasts can be dynamic and valuable sources of information. However, their accuracy and objectivity must be evaluated just as thoroughly as all other sources.

- A piece of information is generally reliable if you can find it in two other sources. However, be aware that several on-line resources might use the same incorrect source of information. If you see identical wording on multiple sites, try to find a different source to verify the information.

Gathering, Recording, and Organizing Information

- As you locate information, you may find it useful to jot it down on large sticky notes or make colour-coded entries in a digital file so you can group similar ideas together. Remember to document the source of your information for each note or data entry.

Avoid Plagiarism Copying information word-for-word and then presenting it as your own work is called *plagiarism*. Instead, you must cite every source you use for a research assignment. This includes all ideas, information, data, and opinions, other than your own, that appear in your work. If you include a quotation, be sure to indicate it as such, and supply all source information. Avoid direct quotations whenever possible—put information in your own words. Remember, though, that even when you paraphrase, you need to cite your sources.

Record Source Information A research paper should always include a bibliography—a list of relevant information sources you have consulted while writing the paper. Bibliographic entries include information such as the author, title, publication year, name of the publisher, and city in which the publisher is located. For magazine or journal articles, the name of the magazine or journal, the name of the article, the issue number, and the page numbers should be recorded. For on-line resources, you should record the site URL, the name of the site, the author or publishing organization, and the date on which you retrieved the information. Remember to record source information while you are taking notes to avoid having to search it out again later! Ask your teacher about the preferred style for your references.

- You might find it helpful to create a chart to keep track of detailed source information. For on-line searches, a tracking chart is useful to record the key words you searched, the information you found, and the URL of the website where you found the information.

- Write down any additional questions that you think of as you are researching. You may need to refine your topic if it is too broad, or take a different approach if there is not enough information available to answer your research question.

Presenting Your Work

- Once you have organized all of your information, you should be able to summarize your research so that it provides a concise answer to your original research question. If you cannot answer this question, you may need to refine the question or do a bit more research.

- Check the assignment guidelines for instructions on how to format your work.

- Be sure that you fulfill all of the criteria of the assignment when you communicate your findings.

Instant Practice

1. Your assignment asks you to research "green" cleaners and present your opinion on which are the least harmful to the environment and why.
   a. What search terms might you use for your initial research on the Internet or at the library?
   b. How might you narrow down this assignment into a research question?
2. How could you evaluate the reliability of an on-line video about air pollution?
3. Wiki sites allow users to contribute and edit content.
   a. How could this affect the reliability of the information they present?
   b. What steps would you take to verify a piece of information you found in a wiki entry?

Writing a Lab Report

Use the following headings and guidelines to create a neat and legible lab report.

Title

- Choose a title that clearly states the independent variable and the dependent variable, but not the outcome of the investigation. For example, "A Comparison of the Neutralizing Ability of Different Antacid Ingredients."
- Under the title, write the names of all participants and the date(s) of the investigation.

Introduction

- Summarize the background of the problem.
- Cite any relevant scientific principles or literature related to the question being investigated.

Question/Problem

- Clearly state the question being investigated or the problem for which you are seeking a solution. For example, "Which ingredient in antacids is most effective at neutralizing acid?"

Hypothesis

- State, in general terms, the relationship that you believe exists between the independent variable and the dependent variable. For example, "Calcium carbonate, an ingredient in many antacids, is more effective at neutralizing acid than sodium hydrogen carbonate, another ingredient found in antacids."

Prediction

- State, in detailed terms, the specific results you expect to observe. For example, "Calcium carbonate will neutralize more hydrochloric acid than the same mass of sodium hydrogen carbonate will."

Materials

- List all of the materials and equipment you used, or refer to the appropriate page number in your textbook, and note any additions, deletions, or substitutions you have made.

Procedure

- Write your procedure in the form of precise, numbered steps, or refer to the appropriate page number in your textbook, and note any changes to the procedure. Include any safety precautions.

Results

- Set out the observations and/or data in a clearly organized table(s). Give your table(s) a title.
- If appropriate, construct a graph that shows the data accurately. Label the x-axis and the y-axis of the graph clearly and accurately, and use the correct scale and units. Give your graph a title.

Data Analysis

- Analyze all the results you have gathered and recorded, and ensure that you can defend your analysis. For example, "As shown in the following calculations, the volume of hydrochloric acid neutralized by calcium carbonate was 1.5 mL more on average than that neutralized by sodium hydrogen carbonate."
- Show sample calculations for any mathematical data analysis.

Conclusion

- State a conclusion based on your data analysis. Relate your conclusion to your hypothesis. For example, "Based on the results of this investigation, calcium carbonate has a greater neutralizing ability than sodium hydrogen carbonate."
- Compare the results you obtained with those you expected, or those obtained by other researchers.
- Examine and comment on experimental error.
- Assess the effectiveness of the experimental design.
- Indicate how the data support your conclusion.
- Make recommendations for how your conclusion could be applied, or for further study of the question you investigated.

References

- Cite your information sources according to the reference style your teacher suggests.
- Sources that need to be cited include background information for your introduction, a materials list or procedure from a textbook, any specialized methods of data analysis, results from other studies that you used for comparison with your own results, and any other sources used in your conclusion.

Organizing Data in a Table

Scientific investigation is about collecting information to help you answer a question. In many cases, you will develop a hypothesis and collect data to see if your hypothesis is supported. An important part of any successful investigation is recording and organizing your data. Often, scientists create tables in which to record data.

Planning to Record Your Data Suppose you are doing an investigation on the water quality of a stream that runs near your school. You will take water samples at three different locations along the stream. You need to decide how to record and organize your data. Begin by making a list of what you need to record. For this experiment, you will need to record the sample site, the pH of the water at each sample site, the chemicals found in the water at each sample site, and the concentration of these chemicals.

Creating Your Data Table Your data table must allow you to record your data neatly. To do this you need to create

- headings to show what you are recording
- columns and rows that you will fill with data
- enough cells to record all the data
- a title for the table

In this investigation, you will find several chemicals in the water at each site, so you must make space for multiple recordings at each site. This means every row representing a sample site will have at least four rows associated with it for the different chemicals.

If you think you might need extra space, create a special section. In this investigation, leave space at the bottom of your table, in case you find more than four chemicals in the water at a sample site. Remember, if you use the extra rows, make sure you identify which sample site the extra data are from. Finally, give your table an appropriate title. Your data table might look like the one in the next column.

Reading a Table A table can be used to organize observations and measurements so that data are represented neatly and clearly. However, a table can also show relationships among the data presented. When you are reading a table, be sure to start by reading the column and row headings carefully. If the table contains measurements, look for the units in which they are reported. Follow vertically down a column or horizontally across a row to look for trends in the data. If the table contains numbers, do the numbers increase or decrease as you look down the column or across the row?

Water Quality Observations Made at Three Sample Stream Sites

headings show what is being recorded

columns and rows contain data

Sample Site	pH	Type of Chemical	Concentration of Chemical (mg/L)
1	6.9	sulfate	30.7
		nitrate	0.11
		phosphate	0.001
2	7.2	sulfate	31.2
		nitrate	0.35
		phosphate	0.002
3	7.1	sulfate	30.9
		nitrate	0.07
		phosphate	0.001
		iron	0.1

extra rows to collect data in case you need to add observations

Also look for relationships between columns or rows. Do the numbers in one column increase as the numbers in another column decrease? Is there one piece of data that does not fit the pattern in the rest of the table? Think about why this might be the case.

Instant Practice

1. You want to compare the antibiotic effects of silver (found in wound dressings) and penicillin. Construct a table to record the number of Bacteria A, Bacteria B, and Bacteria C growing on three different media:
   - a standard culture medium
   - a standard medium with penicillin added
   - a standard medium with silver added

2. Now you wish to refine your investigation to record the number of bacteria of each type growing on each medium after 12 hours, 24 hours, and 36 hours. Draw a new table to record these data.

3. Examine the table at the top of the page. What does it tell you about the three sample sites?

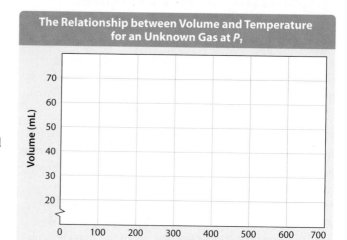# Constructing Graphs

A graph is a diagram that shows relationships among variables. Graphs help you to interpret and analyze data. The three basic types of graphs used in science are the line graph, the bar graph, and the circle graph.

The instructions given here describe how to construct graphs using paper and pencil. You can also use computer software to generate graphs. Whichever method you use, the graphs you construct should have the features described in the following pages.

Line Graphs

A line graph is used to show the relationship between two variables. The independent variable is plotted on the horizontal axis, called the *x*-axis. The dependent variable is plotted on the vertical axis, called the *y*-axis. The dependent variable (y) changes as a result of a change in the independent variable (x).

Suppose a chemist carried out an investigation to determine the relationship between the temperature and volume of an unknown gas at a specific pressure (P_1). She measured the volume (in mL) of the gas upon heating it to various temperatures (in K), as shown in the table below.

Volume and Temperature for an Unknown Gas at P_1

Volume (mL)	Temperature (K)
38	300
49	400
62	500
75	600

To make a graph of volume versus temperature measurements for this gas, start by determining the dependent and independent variables. Th e volume of the gas is the dependent variable and is plotted on the *y*-axis. Th e independent variable, or the temperature to which the the gas was heated, is plotted on the *x*-axis.

Give your graph a title and label each axis, indicating the units if appropriate. In this example, label the temperature on the *x*-axis. Your *x*-axis will need to be numbered to at least 600 K. Because the lowest volume of gas measured was 38 mL and the highest was 75 mL, you know that you will have to start numbers on the *y*-axis from at least 38 and number to at least 75 mL. For instance, you could decide to number 20 to 80 by intervals of 10, spaced at equal distances. Look at the example at the top of the page to see how you could label your axes.

Begin plotting points by locating 300 on the *x*-axis and 38 on the *y*-axis. Where an imaginary vertical line from the x-axis and an imaginary horizontal line from the *y*-axis meet, place the first data point. Place other data points using the same process. After all the points are plotted, draw a "best fit" straight line through the points.

A best fit line should be drawn to represent the general trend of the data. Try to draw the line so that there are as many points above it as there are below. Do not change the position or slope of the line dramatically just to include an outlier—a single data point that does not seem to be in line with all the others.

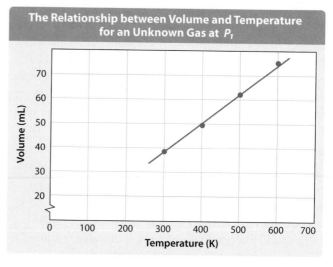

The chemist then repeated the investigation at a different pressure (P_2), using the same amount of gas. Her observations are shown in the table at the top of the next page.

Volume and Temperature for an Unknown Gas at P_2

Volume (mL)	Temperature (K)
21	300
28	400
36	500
43	600

What if you want to compare the relationship between volume and temperature of the gas at these two different pressures? The P_2 data can be plotted on the same graph as the data for P_1. Label the different lines indicating different sets of data as P_1 and P_2.

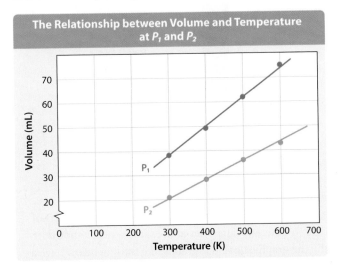

The Relationship between Volume and Temperature at P_1 and P_2

Instant Practice

1. Between 400 K and 500 K, what change in volume occurred at P_1?

2. As temperature increased, how did the changes in volume at P_1 compare to the changes in volume at P?

3. Construct a line graph for the following data:

Partial Pressures of Water Vapour at Different Temperatures

Temperature (°C)	Pressure (kPa)
15	1.71
16	1.81
17	1.93
18	2.07
19	2.20

Slope of a Linear Graph The slope of a line is a number determined by any two points on the line. This number describes how steep the line is. The greater the absolute value of the slope, the steeper the line. Slope is the ratio of the change in the y-coordinates (rise) to the change in the x-coordinates (run) as you move from one point to the other.

The graph below shows a line that passes through points (5, 4) and (9, 6).

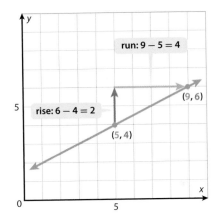

$$\text{Slope} = \frac{\text{rise}}{\text{run}}$$

$$= \frac{\text{change in } y\text{-coordinates}}{\text{change in } x\text{-coordinates}}$$

$$= \frac{6 - 4}{9 - 5}$$

$$= \frac{2}{4} \quad \text{or} \quad \frac{1}{2}$$

So, the slope of the line is $\frac{1}{2}$.

A positive slope indicates that the line climbs from left to right. A negative slope indicates that the line descends from left to right. A slope of zero indicates that there is no change in the dependent variable as the independent variable increases. A horizontal line has a slope of zero.

Linear and Non-Linear Trends Two types of trends you are likely to see when you graph data in chemistry are linear trends and non-linear trends. A linear trend has a constant increase or decrease in data values. For a non-linear trend, the degree to which the data values are increasing or decreasing is not constant. The graphs shown on the next page are examples of these two common trends.

In the graph below, there are two lines describing the solubility of salts at various temperatures. Both lines show an increasing, linear trend. As the temperature increases, so does the solubility of each salt. The rate of increase is constant.

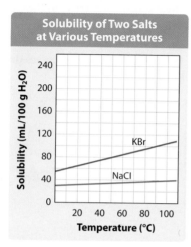

The graph below also shows two lines describing the solubility of salts at various temperatures. Both lines show an increasing, non-linear trend. As in the graph above, the solubility of each salt increases as the temperature increases. However, for the graph below, the rate of increase is not constant. For instance, for potassium nitrate, you will see that the compound's solubility increases more as the temperature increases 20°C from 60°C to 80°C than it does as it increases 20°C from 30°C to 50°C.

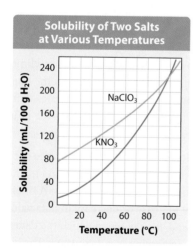

When you are drawing a curve to represent a non-linear trend, you should not connect the data points. Instead, draw a smooth best-fit curve that shows the general trend of the data. Try to draw the curve so there are as many points above it as there are below. The curve should change smoothly. It should not have a dramatic change in direction just to include a single data point that does not fit with the others.

Bar Graphs

A bar graph displays a comparison of different categories of data by representing each category with a bar. The length of the bar is related to the category's frequency. To make a bar graph, set up the x-axis and y-axis as you did for the line graph. Plot the data by drawing thick bars from the x-axis up to an imaginary line representing the y-axis point.

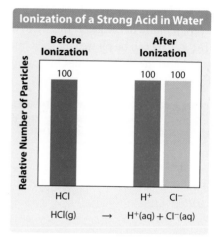

Look at the graph above. The independent variable is the type of particle. The dependent variable is the relative number of particles.

Bar graphs can also be used to display multiple sets of data in different categories at the same time, as shown in the the graph below. Bar graphs like the one below have a legend to denote which bars represent each set of data.

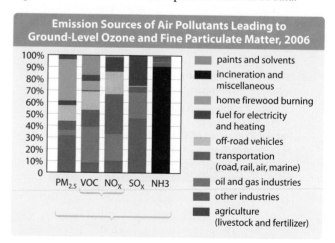

Instant Practice

In the graph at the top of this column, what is the relative number of chlorine ions after ionization?

Circle Graphs

A circle graph is a circle divided into sections that represent parts of a whole. When all the sections are placed together, they equal 100 percent of the whole.

Consider the circle graph shown below. This graph shows the anthropogenic sources of the air contaminant sulfur dioxide. Each component of the graph, electrical utilities, industry, and other, add up to 100 percent of all sources of sulfur dioxide air pollution.

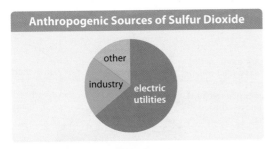

Anthropogenic Sources of Sulfur Dioxide

Suppose you wanted to make a circle graph to represent data you observed or calculated, such as the percentage composition of a compound. For instance, if you determine the percentage composition of copper(I) sulfide, $Cu_2S(s)$, to be 79.9% copper and 20.1% sulfur by mass, you can represent this graphically with a circle graph.

To begin, you know that the percent of the different elements in the compound must add up to 100. This 100 percent is represented by the 360° (the number of degrees in a circle) that make up the circle graph.

To find out how much of the circle each element should cover in the graph, first multiply the percent of copper by 360. Then, round your answer to the nearest whole number.

$$79.9\% \times 360° = 0.799 \times 360°$$
$$= 287.64°$$
$$= 287°$$

The sum of all the segments of the circle graph should add up to 360°. Therefore, you can calculate the segment of the circle that represents the percent of sulfur by subtracting the degrees representing copper from 360°.

$$360° - 287° = 73°$$

To draw your circle graph, you will need a compass and a protractor. First, use the compass to draw a circle. Then, draw a straight line from the centre to the edge of the circle. Place your protractor on this line, and mark the point on the circle where an angle of 73° will intersect the circle. Draw a straight line from the centre of the circle to the intersection point. This is the section representing the percent of sulfur in the compound. The remaining section represents the percent of copper.

Complete the graph by labelling the sections of the graph with percentages and giving the graph a title. Your completed graph should look similar to the one below.

If your circle graph has more than two sections, you will need to construct a segment for each entry. Place your protractor on the last line segment that you have drawn and mark off the appropriate angle. Draw a line segment from the centre of the circle to the new mark on the circle. Continue this process until all of the segments have been drawn.

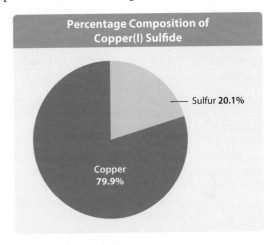

Percentage Composition of Copper(I) Sulfide

Sulfur **20.1%**

Copper **79.9%**

Instant Practice

Create a circle graph to illustrate the percentage composition of the mineral chalcopyrite, $CuFeS_2(s)$. Based on your calculations, you know that the percentage composition of chalcopyrite is 34.62% copper, 30.43% iron, and 34.95% sulfur, by mass.

Using Graphic Organizers

When deciding which type of graphic organizer to use, consider your purpose. It may be to brainstorm, to show relationships among ideas, to summarize a section of text, to record research notes, or to review what you have learned before writing a test. Several different graphic organizers are shown here. The descriptions indicate the function or purpose of each organizer.

PMI Chart

PMI stands for Plus, Minus, and Interesting. A PMI chart is a simple three-column table that can be used to state the positive and negative aspects of an issue, or to describe advantages and disadvantages related to the issue. The third column in the chart is used to list interesting information related to the issue. PMI charts help you organize your thinking after reading about a topic that is up for debate or that can have positive or negative effects. They are useful when analyzing an issue.

Base metal smelting is an important industry in Canada. However, chemicals released during smelting can endanger people's health.

P	M	I
Base metal smelting produces useful metals, making it an important industry in Canada.	Harmful chemicals can be released into the environment during smelting, endangering the health and safety of local populations.	Many of the substances emitted by base metal smelters are listed as toxic by the Canadian Environmental Protection Act.
The smelting industry employs thousands of Canadians and contributes billions of dollars to the Canadian economy.	Smelting companies argue that setting strict limits on smelting emissions would place them at a competitive disadvantage, which would affect jobs.	Many smelters are located in remote areas. Finding other types of employment in these areas can be challenging.

Main Idea Web

A main idea web shows a main idea and several supporting details. The main idea is written in the centre of the web, and each detail is written at the end of a line extending from the centre. This organizer is useful for brainstorming or for summarizing text.

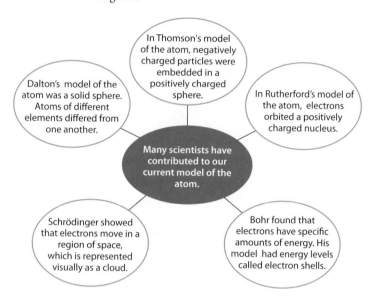

Spider Map

A spider map shows a main idea and several ideas associated with the main idea. It does not show the relationships among the ideas. A spider map is useful when you are brainstorming or taking notes.

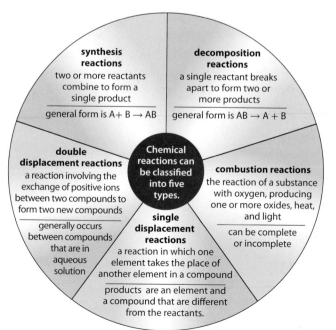

Fishbone Diagram

A fishbone diagram looks similar to a spider map, but it organizes information differently. A main topic, situation, or idea is placed in the middle of the diagram. This is the "backbone" of the "fish." The "bones" (lines) that shoot out from the backbone can be used to list reasons why the situation exists, factors that affect the main idea, or arguments that support the main idea. Finally, supporting details shoot outward from these issues. Fishbone diagrams are useful for planning and organizing a research project. You can clearly see when you do not have enough details to support an issue, which indicates that you need to do additional research.

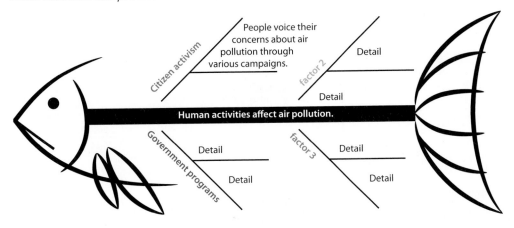

Concept Map

A concept map uses shapes and lines to show how ideas, concepts, or formulas are related. Each idea, concept, or formula is written inside a circle, a square, a rectangle, or another shape. Lines and arrows that connect the shapes indicate the relationships between them. In some cases, words that explain how the concepts are related are written on the lines that connect the shapes.

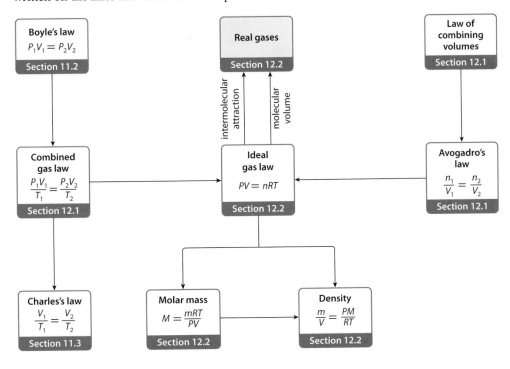

Flowchart

A flowchart shows a sequence of events or the steps in a process. An arrow leads from an initial event or step to the next event or step, and so on, until the final outcome is reached. Side arrows may also be added to provide further explanation. All the events or steps are shown in the order in which they occur.

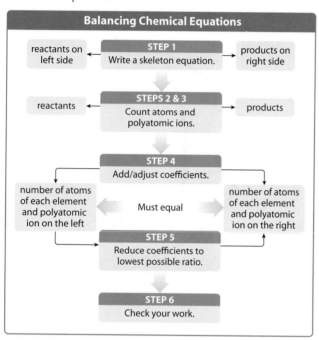

Cycle Chart

A cycle chart is a flowchart that has no distinct beginning or end. All the events are shown in the order in which they occur, as indicated by arrows, but there are no first and last events. Instead, the events occur again and again in a continuous cycle. In the photosynthesis/cellular respiration cycle, shown below, arrows branch off to show energy entering and leaving the cycle.

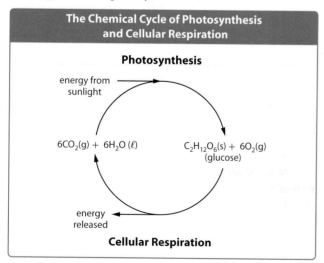

Venn Diagram

A Venn diagram uses overlapping shapes to show similarities and differences among concepts.

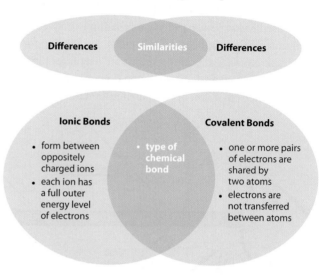

Cause-and-Effect Map

The first cause-and-effect map below shows one cause that results in several effects. The second map shows one effect that has several causes.

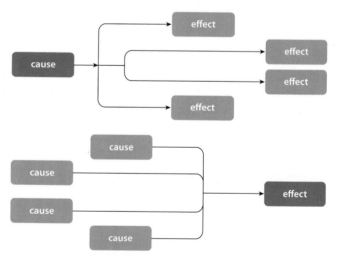

Instant Practice

1. Find an example of a flowchart in this textbook and discuss with a partner its effectiveness in communicating information.

2. Create a Venn diagram to compare and contrast polar and non-polar molecules.

Appendix A

Measurement

Scientists have developed globally agreed-upon standards for measurement, and for recording and calculating data. These are the standards that you will use throughout this science program.

Units of Measurement

When you take measurements for scientific purposes, you use the International System of Measurement (commonly know as SI, from the French *Système international d'unités*). SI includes the metric system and other standard units, symbols, and prefixes, which are reviewed in the tables on this page.

In SI, the base units include the metre, the kilogram, and the second. The size of any particular unit can be determined by the prefix used with the base unit. Larger and smaller units of measurement can be obtained by either multiplying or dividing the base unit by a multiple of 10.

For example, the prefix *kilo-* means multiplied by 1000. So, one kilogram is equivalent to 1000 grams:

$$1 \text{ kg} = 1000 \text{ g}$$

The prefix *milli-* means divided by 1000. So, one milligram is equivalent to one thousandth of a gram:

$$1 \text{ mg} = \frac{1}{1000 \text{ g}}$$

The following tables show the most commonly used metric prefixes, as well as some common metric quantities, units, and symbols.

Commonly Used Metric Prefixes

Prefix	Symbol	Relationship to the Base Unit
tera-	T	$10^{12} = 1\ 000\ 000\ 000\ 000$
giga-	G	$10^9 = 1\ 000\ 000\ 000$
mega-	M	$10^6 = 1\ 000\ 000$
kilo-	k	$10^3 = 1\ 000$
hecto-	h	$10^2 = 100$
deca-	da	$10^1 = 10$
—	—	$10^0 = 1$
deci-	d	$10^{-1} = 0.1$
centi-	c	$10^{-2} = 0.01$
milli-	m	$10^{-3} = 0.001$
micro-	μ	$10^{-6} = 0.000\ 001$
nano-	n	$10^{-9} = 0.000\ 000\ 001$
pico-	p	$10^{-12} = 0.000\ 000\ 000\ 001$

Commonly Used Metric Quantities, Units, and Symbols

Quantity	Unit	Symbol
Length	nanometre	nm
	micrometre	μm
	millimetre	mm
	centimetre	cm
	metre	m
	kilometre	km
Mass	gram	g
	kilogram	kg
	tonne	t
Area	square metre	m^2
	square centimetre	cm^2
	hectare	ha ($10\ 000$ m^2)
Volume	cubic centimetre	cm^3
	cubic metre	m^3
	millilitre	mL
	litre	L
Time	second	s
Temperature	degree Celsius	°C
Force	N	newton
Energy	joule	J
	kilojoule*	kJ
Pressure	pascal	Pa
	kilopascal**	kPa
Electric current	ampere	A
Quantity of electric charge	coulomb	C
Frequency	hertz	Hz
Power	watt	W

* Many dieticians in North America continue to measure nutritional energy in Calories, also known as kilocalories or dietetic Calories. In SI units, 1 Calorie = 4.186 kJ.
** In current North American medical practice, blood pressure is measured in millimetres of mercury, symbolized as mmHg. In SI units, 1 mmHg = 0.133 kPa.

Accuracy and Precision

In science, the terms accuracy and precision have specific definitions that differ from their everyday meanings.

Scientific *accuracy* refers to how close a given quantity is to an accepted or expected value. For example, under standard (defined) conditions of temperature and pressure, 5 mL of water has a mass of 5 g. When you measure the mass of 5 mL of water under the same conditions, you should, if you are accurate, find the mass is 5 g.

Scientific *precision* refers to the exactness of your measurements. The precision of your measurements is directly related to the instruments you use to make the measurements. While faulty instruments (for example, a balance that is not working properly) will likely affect both the accuracy and the precision of your measurements, the calibration of the instruments you use is the factor that most affects precision. For example, a ruler calibrated in millimetres will allow you to make more precise measurements than one that shows only centimetres.

Precision also describes the repeatability of measurements. The closeness of a series of data points on a graph is an indicator of repeatability. Data that are close to one another, as in graph A, below, are said to be precise.

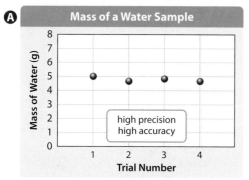

Graph A shows a group of data with high accuracy, since the data points are all grouped around 5 g.

There is no guarantee, however, that the data are accurate until a comparison with an accepted value is made. For example, graph B shows a group of measurements that are precise, but not accurate, since they report the mass of a 5 g sample of water as approximately 7 g.

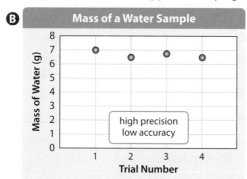

Graph B shows data with low accuracy, since the data points are grouped around 7 g.

In graph C, the data points give an accurate value for average mass, but they are not precise.

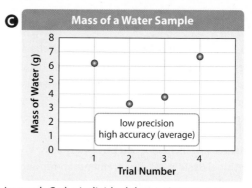

In graph C, the individual data points are not very accurate, since they are all more than 1 g away from the expected value of 5 g. However, taken as a group, the data set in graph C has high accuracy, since the average mass from the four trials is 5 g.

Error

Error exists in every measured or experimentally obtained value. Even the most careful scientist cannot avoid having error in a measurement. *Random error* results from uncontrollable variation in how we obtain a measurement. For example, human reflexes vary, so it is not possible to push the stem of a stopwatch exactly the same way every time. No measurement is perfect. Repeating trials will reduce but never eliminate the effects of random error. Random error affects precision and, usually, accuracy.

Systematic error results from consistent bias in observation. For example, a scale might consistently give a reading that is 0.5 g heavier than the actual mass of a sample, or a person might consistently read the scale of a measuring instrument incorrectly. Repeating trials will not reduce systematic error. Systematic error affects accuracy.

Percent Error

The amount of error associated with a measurement can be expressed as a percentage, which can help you to evaluate the accuracy of your measurement. The higher the *percent error* is, the less accurate the measurement. Percent error is calculated using the following equation:

$$\text{Percent error} = \left| \frac{\text{measured} - \text{expected value}}{\text{expected value}} \right| \times 100\%$$

(Note that the vertical lines surrounding the fraction mean *the absolute value of* the expression within the lines. That is, the expression's numerical value should be reported without a positive or negative sign.) As an example, a student measures a 5 mL sample of water and finds the mass to be 4.6 mL.

$$\text{percent error} = \left| \frac{4.6 \text{ mL} - 5 \text{ mL}}{5 \text{ mL}} \right| \times 100\%$$

$$= \left| \frac{-0.4 \text{ mL}}{5 \text{ mL}} \right| \times 100\%$$

$$= 8\%$$

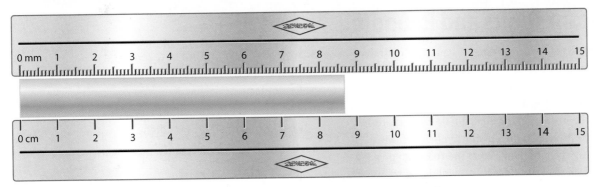

Estimated uncertainty is half of the smallest visible division. In this case, the estimated uncertainty is ±0.5 mm for the top ruler and ±0.5 cm for the bottom ruler.

Uncertainty

Estimated uncertainty describes the limitations of a measuring device. It is defined as half of the smallest division of the measuring device. For example, a metre stick with only centimetres marked on it would have an error of ±0.5 cm. A ruler that includes millimetre divisions would have a smaller error of ±0.5 mm (0.05 cm, or a 10-fold decrease in error). A measurement can be recorded with its estimated uncertainty. In the diagram at the top of the page, for example, the top ruler gives a measurement of 8.69 ±0.05 cm, while the bottom ruler gives a measurement of 8.7 ±0.5 cm.

You can convert the estimated uncertainty into a percentage of the actual measured value using the following equation:

$$\text{Relative uncertainty} = \frac{\text{estimated uncertainty}}{\text{actual measurement}} \times 100\%$$

Example

Convert the error represented by 22.0 ±0.5 cm to a percentage.

$$\text{Relative uncertainty} = \frac{0.5 \text{ cm}}{22.0 \text{ cm}} \times 100\%$$
$$= 2\%$$

Estimating

Sometimes it is not practical or possible to make an accurate measurement of a quantity. You must instead make an *estimate*—an informed judgement that approximates a quantity. For example, if you were conducting an experiment to compare the number of weeds in a field treated with herbicide with the number of weeds in an untreated field, counting the weeds would be impractical, if not impossible. Instead, you could count the number of weeds in a typical square metre of each field. You could then estimate the number of weeds in the entire field by multiplying the number of weeds in a typical square metre by the number of square metres in the field. To make a reasonable estimate of the number of weeds in the field, though, you would need to sample many areas, each 1 m², and then calculate an average to determine the number of weeds in a typical square metre for each field.

Estimating can be a valuable tool in science. It is important to keep in mind, however, that the number of samples you take can greatly influence the reliability of your estimate. To make a good estimate, include as many samples as is practical.

Instant Practice

1. Your teacher gives you a 500 g sample of sugar and asks you to measure its mass on a balance. You measure the sample three times, producing the measurements 492.8 g, 503.1 g, and 505.4 g. Analyze these results in terms of accuracy and precision.

2. Calculate the percent error for each of the measurements in question 1, and for the average of the three measurements. How does the accuracy of the individual measurements differ from the accuracy of the group of measurements?

3. The estimated uncertainty of the measurements in question 1 is ±0.5 cm. Calculate the relative uncertainty of the average you determined in question 2.

Significant Digits and Rounding

You might think that a measurement is an exact quantity. In fact, all measurements involve uncertainty. The measuring device is one source of uncertainty, and you, as the reader of the device, are another. Every time you take a measurement, you are making an estimate by interpreting the reading. For example, the illustration below shows a ruler measuring the length of a rod. The ruler can give quite an accurate reading, since it is divided into millimetre marks. But the end of the rod falls between two marks. There is still uncertainty in the measurement. You can be certain that if the ruler is accurate, the length of the rod is between 5.2 cm and 5.3 cm. However, you must estimate the distance between the 2 mm and 3 mm marks.

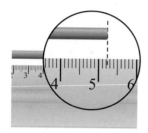

Significant Digits

Significant digits are the digits you record when you take a measurement. The significant digits in a measured quantity include all the certain digits plus the first uncertain digit. In the example above, suppose you estimate the length of the rod to be 5.23 cm. The first two digits (5 and 2) are certain (those marks are visible), but the last digit (0.03) is estimated. The measurement 5.23 cm has three significant digits.

Determining the Number of Significant Digits

The following rules will help you determine the number of significant digits in a given measurement.

1. All non-zero digits (1–9) are significant.
   Examples:
   • 123 m—three significant digits
   • 23.56 km—four significant digits

2. Zeros between non-zero digits are also significant.
   Examples:
   • 1207 m—four significant digits
   • 120.5 km/h—four significant digits

3. Any zero that follows a non-zero digit *and* is to the right of the decimal point is significant.
   Examples:
   • 12.50 m/s^2—four significant digits
   • 6.0 km—two significant digits

4. Zeros that are to the left of a measurement are not significant.
   Examples:
   • 0.056—two significant digits
   • 0.007 60—three significant digits

5. Zeros used to indicate the position of the decimal are not significant. These zeros are sometimes called spacers.
   Examples:
   • 500 km—one significant digit (the decimal point is assumed to be after the final zero)
   • 0.325 m—three significant digits
   • 0.000 34 km—two significant digits

6. In some cases, a zero that appears to be a spacer is actually a significant digit. All counting numbers have an infinite number of significant digits.
   Examples:
   • 6 apples—infinite number of significant digits
   • 125 people—infinite number of significant digits
   • 450 deer—infinite number of significant digits

Instant Practice

Determine the number of significant digits in each measurement.

a. 32 individuals

b. 891 m

c. 15.764 g

d. 0.0280 L

e. 3690 km

f. 0.742 kg

g. 50.8 cm

Using Significant Digits in Mathematical Operations

When you use measured values in mathematical operations, the calculated answer cannot be more certain than the measurements on which it is based. Often the answer on your calculator will have to be rounded to the correct number of significant digits.

Rules for Rounding

1. When the first digit to be dropped is less than 5, the preceding digit is not changed.

   Example:
   - 6.723 m rounded to two significant digits is 6.7 m. The digit after the 7 is less than 5, so the 7 does not change.

2. When the first digit to be dropped is 5 or greater, the preceding digit is increased by one.

   Example:
   - 7.237 m rounded to three significant digits is 7.24 m. The digit after the 3 is greater than 5, so the 3 is increased by one.

3. When the first digit to be dropped is 5, and there are no following digits, increase the preceding number by 1 if it is odd, but leave the preceding number unchanged if it is even.

   Examples:
   - 8.345 L rounded to three significant digits is 8.34 L, because the digit before the 5 is even.
   - 8.375 L rounded to three significant digits is 8.38 L, because the digit before the 5 is odd.

Adding or Subtracting Measurements

Perform the mathematical operation, and then round off the answer so it has the same number of decimal places as the value that has the fewest decimal places.

Example:
Add the following measured lengths and express the answer to the correct number of significant digits.

$x = 2.3 \text{ cm} + 6.47 \text{ cm} + 13.689 \text{ cm}$
$= 22.459 \text{ cm}$
$= 22.5 \text{ cm}$

Since 2.3 cm has only one decimal place, the answer can have only one decimal place.

Multiplying or Dividing Measurements

Perform the mathematical operation, and then round off the answer so it has the same number of significant digits as the value that has the least number of significant digits.

Example:
Multiply the following measured lengths and express the answer to the correct number of significant digits.

$x = (2.342 \text{ m})(0.063 \text{ m})(306 \text{ m})$
$= 45.149\ 076 \text{ m}^3$
$= 45 \text{ m}^3$

Since 0.063 m has only two significant digits, the final answer must also have two significant digits.

Instant Practice

Perform the following calculations, rounding off your answer to the correct number of significant digits.

a. 9.745 km − 4.2 km

b. 8.33 L + 0.4 L + 56.358 L

c. 16.9 g × 0.007 56 g

d. 463.8 mL/0.660 mL

e. 580.62 mm × 1.02 mm/0.7 mm

Scientific Notation

An exponent is the symbol or number denoting the power to which another number or symbol is to be raised. The exponent shows the number of repeated multiplications of the base. In 10^2, the exponent is 2 and the base is 10. The expression 10^2 means 10×10.

Powers of 10

Digits	Standard Form	Exponential Form
Ten thousands	10 000	10^4
Thousands	1 000	10^3
Hundreds	100	10^2
Tens	10	10^1
Ones	1	10^0
Tenths	0.1	10^{-1}
Hundredths	0.01	10^{-2}
Thousandths	0.001	10^{-3}
Ten thousandths	0.0001	10^{-4}

Why use exponents? Consider this: One molecule of water has a mass of 0.000 000 000 000 000 000 000 029 9 g. Using such a number for calculations would be quite awkward. The mistaken addition or omission of a single zero would make the number either 10 times larger or 10 times smaller than it actually is. Scientific notation allows scientists to express very large and very small numbers more easily, to avoid mistakes, and to clarify the number of significant digits.

Expressing Numbers in Scientific Notation

In scientific notation, a number has the form $x \times 10^n$, where x is greater than or equal to 1 but less than 10, and 10^n is a power of 10. To express a number in scientific notation, use the following steps:

1. To determine the value of x, move the decimal point in the number so that only one non-zero digit is to the left of the decimal point.

2. To determine the value of the exponent n, count the number of places the decimal point moves to the left or right. If the decimal point moves to the right, express n as a positive exponent. If the decimal point moves to the left, express n as a negative exponent.

3. Use the values you have determined for x and n to express the number in the form $x \times 10^n$.

Examples

Express 0.000 000 000 000 000 000 000 029 9 g in scientific notation.

1. To determine x, move the decimal point so that only one non-zero number is to the left of the decimal point:

   2.99

2. To determine n, count the number of places the decimal moved:

   Since the decimal point moved to the right, the exponent will be negative.

3. Express the number in the form $x \times 10^n$:

   2.99×10^{-23} g

Express 602 000 000 000 000 000 000 000 in scientific notation.

1. To determine x, move the decimal point so that only one non-zero number is to the left of the decimal point:

   6.02

2. To determine n, count the number of places the decimal moved:

   Since the decimal point moved to the left, the exponent will be positive.

3. Express the number in the form $x \times 10^n$:

   6.02×10^{23}

Logarithms and Calculating pH

An understanding of logarithms is essential for calculating the pH of a solution.

Logarithms

The logarithm of a number is the power to which you must raise a base to equal that number. By convention, we usually use 10 as the base. Every positive number has a logarithm. For example, the logarithm of 10 is 1, because $10^1 = 10$. The logarithm of 100 is 2, because $10^2 = 100$. This can be understood by examining the following equation:

$$\log_a x = y; \text{ where } a^y = x$$

Therefore, since $10^1 = 10$, we know $\log_{10} 10 = 1$. This can also be written as $\log 10 = 1$, since it is understood that 10 is used as the base by convention unless otherwise indicated. Similarly, since $10^2 = 100$, we know $\log_{10} 100 = 2$ or $\log 100 = 2$.

All numbers that are greater than 1 have a positive logarithm. Numbers that are between 0 and 1 have a negative logarithm. For instance, since $10^{-3} = 0.001$, we know $\log 0.001 = -3$. Note also that the number 1 has a logarithm of 0. The table below shows several examples of numbers and their logarithms.

Some Numbers and Their Logarithms

Number	Scientific Notation	As a Power of 10	Logarithm
1 000 000	1×10^6	10^6	6
7 895 900	7.8959×10^6	$10^{6.897\,40}$	6.897 40
1	1×10^0	10^0	1
0.000 001	1×10^{-5}	10^{-5}	-5
0.004 276	4.276×10^{-3}	$10^{-2.3690}$	-2.3690

Logarithms are a convenient method for communicating large and small numbers, and are especially useful for expressing values that span a range of powers of 10. For instance, the Richter scale for earthquakes, the decibel scale for sound, and the pH scale for acids and bases all use logarithmic scales.

Calculating pH

The pH of an acid solution is defined as $-\log[H^+]$, where the square brackets mean *concentration*. In the figure below, the "p" in pH represents "the negative logarithm of..." As the logarithm of a number refers to an exponent or "power," the "p" can be thought of as "power." The power referred to is exponential power: the power of 10. Similarly, the "H" stands for the concentration of hydrogen ions, measured in mol/L.

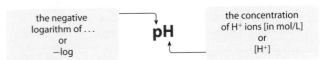

The concept of pH allows hydrogen ion concentrations to be expressed as positive numbers, rather than negative exponents. For example, the $[H^+]$ of neutral water at 25°C is 1.0×10^{-7}. You can find the pH of water (and of any solution) by taking the negative log of the concentration of hydrogen ions.

$$\therefore \text{pH} = -\log [H^+]$$
$$= -\log (1.0 \times 10^{-7})$$
$$= -(-7.00)$$
$$= 7.00$$

For a logarithm, only the digits to the right of the decimal place are significant. The numbers to the left of the decimal place reflect the power of base 10, and are, therefore, not significant.

Example:

1. Find the pH of a solution with a hydrogen ion concentration of 0.004 76 mol/L.

$$\text{pH} = -\log [H^+]$$
$$= -\log (0.004\,76 \text{ mol/L})$$
$$= 2.322$$

Note that the pH scale is a negative log scale. Thus, a decrease from pH 7 to pH 4 is actually an increase of 10^3, or 1000 times, in the acidity of a solution. An increase from pH 3 to pH 6 is a decrease in acidity of 10^3 times.

Instant Practice

1. Find the pH of a solution with the following hydrogen ion concentrations.
   a. 0.000 000 01 mol/L
   b. 8.7×10^{-3} mol/L

Preparing Solutions

Using a Volumetric Flask to Prepare a Standard Aqueous Solution

1. Place the known mass of solute in a clean beaker. Use distilled water to dissolve the solute completely.	
2. Rinse a clean volumetric flask of the required volume with a small quantity of distilled water. Discard the rinse water. Repeat the rinsing several times.	
3. Transfer the solution from the beaker to the volumetric flask using a funnel.	
4. Using a wash bottle, rinse the beaker with distilled water, and pour the rinse water into the volumetric flask. Repeat this rinsing several times.	
5. Using a wash bottle or a beaker, add distilled water to the volumetric flask until the level is just below the graduation mark. Then remove the funnel from the volumetric flask.	
6. View the neck of the volumetric flask straight on from the side, so that the graduation mark looks like a line, not an ellipse. Add distilled water, drop by drop, until the bottom of the *meniscus* (the curved surface of the solution) appears to touch the graduation mark.	

Using a Volumetric Pipette to Measure the Volume of a Stock Solution for Dilution

1. Make sure that the outside of the pipette, especially the tip, is dry. If not, wipe it with a paper towel.	
2. Squeeze the pipette bulb, and then place it over the top of the pipette. If using a pipette pump, place it over the top of the pipette.	
3. Rinse the pipette as follows. Place the tip of the pipette below the surface of the stock solution. Release the bulb carefully to draw up some liquid until the pipette is about half full. Remove the pipette bulb, invert the pipette, and drain the liquid into a beaker for waste. Repeat this rinsing two or three times.	
4. Fill the pipette with stock solution so that the level is past the graduation mark, but do not allow stock solution into the pipette bulb.	
5. Remove the pipette bulb, and quickly seal the top of the pipette with your finger or thumb.	
6. Remove the tip of the pipette from the stock solution. Lift your finger slightly, and let stock solution drain out slowly until the meniscus reaches the graduation mark. Wipe the tip of the pipette with a piece of paper towel.	
7. Move the pipette to the container into which you want to transfer the stock solution. Touch the tip of the pipette against the inside of the container, and release your finger to allow the liquid to drain. A small volume of liquid will remain inside the pipette. The pipette has been calibrated to allow for this volume of liquid. *Do not* force this liquid from the pipette.	

Performing an Acid-base Titration

The following steps describe how to prepare for and perform a titration.

Figure A Squeeze the pipette bulb as you put it on the stem of the pipette.

Rinsing the Volumetric or Graduated Pipette

Rinse a pipette with the solution whose volume you are measuring. This will ensure that any drops remaining inside the pipette will form part of the measured volume.

1. Put the pipette bulb on the pipette, as shown in **Figure A**. Place the tip of the pipette into a beaker of distilled water.
2. Relax your grip on the bulb to draw up a small volume of distilled water.
3. Remove the bulb, and discard the water by letting it drain out.
4. Pour a sample of the solution with the unknown concentration into a clean, dry beaker.
5. Rinse the pipette by drawing several millilitres of the solution with the unknown concentration from the beaker into the pipette. Coat the inner surface with the solution, as shown in **Figure B**. Discard the rinse. Rinse the pipette twice in this way. The pipette is now ready to be filled with the solution that has the unknown concentration.

Filling the Pipette

6. Place the tip of the pipette below the surface of the solution with the unknown concentration.
7. Hold the suction bulb loosely on the end of the glass stem. Use the suction bulb to draw the solution up to the point shown in **Figure C**.
8. As quickly and smoothly as you can, slide the bulb off the glass stem and place your index finger over the end.
9. Roll your finger slightly away from end of the stem to let the solution slowly drain out.
10. When the bottom of the meniscus aligns with the etched mark, as in **Figure D**, press your finger back over the end of the stem. This will prevent more solution from draining out.
11. Touch the tip of the pipette to the side of the beaker to remove any clinging drops. The measured volume inside the pipette is now ready to transfer to an Erlenmeyer flask.

Figure B Cover the ends of the pipette so that none of the solution spills out as you rock the pipette back and forth to coat its inner surface with solution.

Transferring the Solution

12. Place the tip of the pipette against the inside glass wall of the flask, as shown in **Figure E.** Let the solution drain slowly, by removing your finger from the stem.
13. After the solution drains, wait several seconds and then touch the tip to the inside wall of the flask to remove any drops on the end. Note: Do not remove the small amount of solution shown in **Figure F**.

Figure C Start with more of the unknown solution than you need. You will drain out the excess solution in the next two steps.

Figure D Always read the volume of the solution at the bottom of the meniscus.

Figure E Draining the pipette with the tip against the wall of the flask will prevent splashing.

Adding the Indicator

14. Add two or three drops of the indicator to the flask and its contents. Do not add too much indicator. Using more indicator does not make the colour change easier to see. Also, most indicators are weak acids. Too much indicator can change the amount of base needed for the neutralization. You are now ready to prepare the apparatus for the titration.

Rinsing the Burette

15. To rinse the burette, close the tap and add about 10 mL of distilled water from a wash bottle.

16. Tip the burette to one side, and roll it gently back and forth so that the water comes in contact with all the inner surfaces.

17. Hold the burette over a sink. Let the water drain out, as shown in **Figure G**. While you do this, check that the tap does not leak. Make sure that the tap turns smoothly and easily.

18. Rinse the burette twice, with 5 to 10 mL of the titrant. Remember to open the tap to rinse the lower portion of the burette. Discard the rinse solution each time.

Filling the Burette

19. Assemble a retort stand and burette clamp to hold the burette. Place a funnel in the top of the burette, and put a beaker under the burette.

20. With the tap closed, add the solution until it is above the zero mark. Remove the funnel. Carefully open the tap. Drain the solution into the beaker until the bottom of the meniscus is at or below the zero mark.

21. Touch the tip of the burette against the beaker to remove any clinging drops. Check that the part of the burette below the tap is filled with solution and contains no air bubbles. **Figure H** shows the air bubbles that you should avoid.

22. Find the initial burette reading using a meniscus reader, as shown in **Figure I**. Record the initial volume to the nearest 0.05 mL.

Titrating the Unknown Solution

23. Replace the beaker with the Erlenmeyer flask that contains the solution you want to titrate. Place a sheet of white paper under the flask to help you see the colour change.

24. Add titrant from the burette to the Erlenmeyer flask by opening the tap, as shown in **Figure J**. You may start by adding the titrant quickly, but slow down when you start to see a colour change in the solution in the flask.

25. At first, the colour change will disappear as you mix the solution in the flask. Add a small amount of titrant, and swirl thoroughly before adding any more. Stop adding titrant when the solution in the Erlenmeyer flask has a persistent colour change. If you are using phenolphthalein as an indicator, stop when the solution is a faint pink colour.

26. Use the meniscus reader to read the final volume. Record this volume, and subtract the initial volume from it to find the volume of the titrant needed to reach the end point.

Figure J Always swirl the flask as you add the titrant. If you have trouble swirling and adding titrant at the same time, use a magnetic stirrer or have your laboratory partner swirl the flask as you add the titrant.

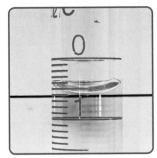

Figure I Hold the meniscus reader so that the line is under the meniscus.

Figure F A small amount of solution will always remain in the tip of the pipette. Do not remove this.

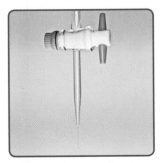

Figure G The tap is fully open when the handle on the tap is parallel to the burette and the solution inside the burette comes out quickly.

Figure H Do NOT start a titration if you have air bubbles like these in the tip of the burette. They will cause errors in your measurements.

Chemistry Data Tables

Electron Configurations

Bohr-Rutherford Diagrams for Elements 1-18

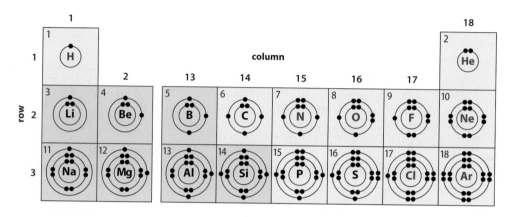

Names and Formulas of Ions

Ionic Charges of Representative Elements

IA 1	IIA 2	IIIA 13	IVA 14	VA 15	VIA 16	VIIA 17	VIIIA 18
H^+						H^-	Noble gases do not ionize in nature.
Li	Be^{2+}			N^{3-}	O^{2-}	F^-	
Na^+	Mg^{2+}	Al^{3+}		P^{3-}	S^{2-}	Cl^-	
K^+	Ca^{2+}				Se^{2-}	Br^-	
Rb^+	Sr^{2+}					I^-	
Cs^+	Ba^{2+}						

Charges of Some Transition Metal Ions

1+	2+	3+
silver, Ag^+	cadmium, Cd^{2+} nickel, Ni^{2+} zinc, Zn^{2+}	scandium, Sc^{3+}

Common Metal Ions with More Than One Ionic Charge

Formula	Stock Name	Classical Name
Cu^+	copper(I) ion	cuprous ion
Cu^{2+}	copper(II) ion	cupric ion
Fe^{2+}	iron(II) ion	ferrous ion
Fe^{3+}	iron(III) ion	ferric ion
Hg_2^{2+} (Hg^+)	mercury(I) ion	mercurous ion
Hg^{2+}	mercury(II) ion	mercuric ion
Pb^{2+}	lead(II) ion	plumbous ion
Pb^{4+}	lead(IV) ion	plumbic ion
Sn^{2+}	tin(II) ion	stannous ion
Sn^{4+}	tin(IV) ion	stannic ion
Cr^{2+}	chromium(II) ion	chromous ion
Cr^{3+}	chromium(III) ion	chromic ion
Mn^{2+}	manganese(II) ion	
Mn^{3+}	manganese(III) ion	
Mn^{4+}	manganese(IV) ion	
Co^{2+}	cobalt(II) ion	cobaltous ion
Co^{3+}	cobalt(III) ion	cobaltic ion

Some Common Polyatomic Ions

Name	Formula
ammonium	NH_4^+
acetate or ethanoate	CH_3COO^-
benzoate	$C_6H_5COO^-$
borate	BO_3^{3-}
carbonate	CO_3^{2-}
hydrogen carbonate	HCO_3^-
perchlorate	ClO_4^-
chlorate	ClO_3^-
chlorite	ClO_2^-
hypochlorite	ClO^-
chromate	CrO_4^{2-}
dichromate	$Cr_2O_7^{2-}$
cyanide	CN^-
hydroxide	OH^-
iodate	IO_3^-

Name	Formula
nitrate	NO_3^-
nitrite	NO_2^-
oxalate	$OOCCOO^{2-}$
hydrogen oxalate	$HOOCCOO^-$
permanganate	MnO_4^-
phosphate	PO_4^{3-}
hydrogen phosphate	HPO_4^{2-}
dihydrogen phosphate	$H_2PO_4^-$
sulfate	SO_4^{2-}
hydrogen sulfate	HSO_4^-
sulfite	SO_3^{2-}
hydrogen sulfite	HSO_3^-
cyanate	CNO^-
thiocyanate	SCN^-
thiosulfate	$S_2O_3^{2-}$

Prefixes and Suffixes for Families of Polyatomic Ions

Relative Number of Oxygen Atoms	Prefix	Suffix	Example	
Family of Four				
most	per-	-ate	ClO_4^-	perchlorate
second most	(none)	-ate	ClO_3^-	chlorate
second fewest	(none)	-ite	ClO_2^-	chlorite
fewest	hypo-	-ite	ClO^-	hypochlorite
Family of Two				
most	(none)	-ate	NO_3^-	nitrate
fewest	(none)	-ite	NO_2^-	nitrite

Names and Formulas for Compounds

Rules for Naming Binary Ionic Compounds

1. The name of the metal ion is first, followed by the name of the non-metal ion.

2. The name of the metal ion is the same as the name of the metal atom.

3. If the metal is a transition metal, it might have more than one possible charge. In these cases, a roman numeral is written in brackets after the name of the metal to indicate the magnitude of the charge.

4. The name of the non-metal ion has the same root as the name of the atom, but the suffix is changed to -ide.

Rules for Writing Chemical Formulas for Ionic Compounds

1. Identify the positive ion and the negative ion.

2. Find the chemical symbols for the ions, either in the periodic table or in the table of polyatomic ions. Write the symbol for the positive ion first and the symbol for the negative ion second.

3. Determine the charges of the ions. If you do not know the charges, you can find them in the periodic table.

4. Check to see if the charges differ. If the magnitudes of the charges are the same, the formula is complete. If they differ, determine the number of each ion that is needed to create a zero net charge. Write the numbers of ions needed as subscripts beside the chemical symbols, with one exception. When only one ion is needed, leave the subscript blank. A blank means one. If a polyatomic ion needs a subscript, the formula for the ion must be in brackets and the subscript must be outside the brackets.

Names of Some Common Acids without Oxygen

Pure Substance (name)	Formula H(negative ion)(aq)	Classical Name hydro (root) ic acid	IUPAC Name aqueous hydrogen (negative ion)
hydrogen fluoride	$HF(aq)$	hydrofluoric acid	aqueous hydrogen fluoride
hydrogen cyanide	$HCN(aq)$	hydrocyanic acid	aqueous hydrogen cyanide
hydrogen sulfide	$H_2S(aq)$	hydrosulfuric acid	aqueous hydrogen sulfide

Classical Naming System for Families of Oxoacids

Name of Ion	Name of Acid (dissolved in water)	Examples	
		Name of Ion	Name of Acid (dissolved in water)
hypo(root)ite	hypo(root)ous acid	hypochlorite, ClO^-	hypochlorous acid, $HClO$
(root)ite	(root)ous acid	chlorite, ClO_2^-	chlorous acid, $HClO_2$
(root)ate	(root)ic acid	chlorate, ClO_3^-	chloric acid, $HClO_3$
per(root)ate	per(root)ic acid	perchlorate, ClO_4^-	perchloric acid, $HClO_4$

Rules for Naming Binary Molecular Compounds

1. Name the element with the lower group number first. Name the element with the higher group number second.

2. The one exception to the first rule occurs when oxygen is combined with a halogen. In this situation, the halogen is named first.

3. If both elements are in the same group, name the element with the higher period number first.

4. The name of the first element is unchanged.

5. To name the second element, use the root name of the element and add the suffix *-ide*.

6. If there are two or more atoms of the first element, add a prefix to indicate the number of atoms.

7. Always add a prefix to the name of the second element to indicate the number of atoms of this element in the compound. (If the second element is oxygen, an "o" or "a" at the end of the prefix is usually omitted.)

Prefixes for Binary Molecular Compounds

Number	Prefix
1	mono-
2	di-
3	tri-
4	tetra-
5	penta-
6	hexa-
7	hepta-
8	octa-
9	nona-
10	deca-

Names and Formulas for Some Common Hydrocarbons

Name	Formula
methane	$CH_4(g)$
ethane	$C_2H_6(g)$
propane	$C_3H_8(g)$
butane	$C_4H_{10}(g)$
acetylene (ethyne)	$C_2H_2(g)$
benzene	$C_6H_6(\ell)$

Ion Properties

Colours of Some Common Ions in Aqueous Solution

Ionic Species	Solution Concentration	
	1.0 mol/L	0.010 mol/L
chromate	yellow	pale yellow
chromium(III)	blue-green	green
chromium(II)	dark blue	pale blue
cobalt(II)	red	pink
copper(I)	blue-green	pale blue-green
copper(II)	blue	pale blue
dichromate	orange	pale orange
iron(II)	lime green	colourless
iron(III)	orange-yellow	pale yellow
manganese(II)	page pink	colourless
nickel(II)	blue-green	pale blue-green
permanganate	deep purple	purple-pink

The Flame Colour of Selected Metal Ions

Ion	Symbol	Colour
lithium	Li^+	red
sodium	Na^+	yellow
potassium	K^+	violet
cesium	Cs^+	violet
calcium	Ca^{2+}	red
strontium	Sr^{2+}	red
barium	Ba^{2+}	yellowish-green
copper	Cu^{2+}	bluish-green
baron	B^{2+}	green
lead	Pb^{2+}	bluish-white

Bond Character

Predicting Bond Character from Electronegativity Difference Values

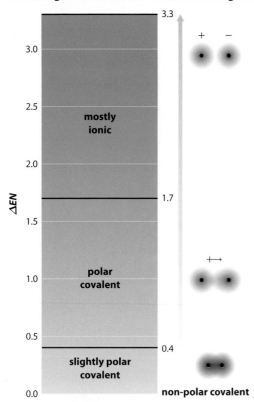

Character of Bonds

Electronegativity Difference	0.00	0.65	0.94	1.19	1.43	1.67	1.91	2.19	2.54	3.03
Percent Ionic Character	0	10	20	30	40	50	60	70	80	90
Percent Covalent Character	100	90	80	70	60	50	40	30	20	10

Types of Chemical Formulas

Empirical, Molecular, and Structural Formulas

	Empirical Formula	Molecular Formula	Structural Formula
Description	• shows the smallest whole-number molar ratio, or proportional relationship, of the elements in a compound	• shows the actual number of atoms of each element in one molecule of the compound	• shows how the atoms are connected in a chemical compound • can indicate the presence of double and triple bonds • in diagram form, bonds are represented as single, double, or triple lines joining atoms • when it is not practical to produce a diagram, a structural formula can be expressed in text only, as shown below, by listing the cluster of atoms around each carbon atom
Example:			
acetic acid or ethanoic acid	CH_2O	$C_2H_4O_2$	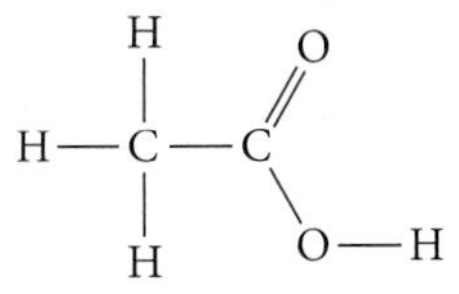 or CH_3COOH

Note: The molecular formula for acetic acid is sometimes expressed as $HC_2H_3O_2$, and the structural formula is sometimes expressed as HCH_3COO, because acetic acid ionizes in aqueous solution into $H^+(aq)$ and $C_2H_3O_2^-(aq)$ or $CH_3COO^-(aq)$. This notation is also used for other acids, such as citric acid, $C_6H_8O_7(aq)$. Since citric acid is a triprotic acid, the formula can be expressed as $H_3C_6H_5O_7(aq)$.

Chemical Information

Chemicals in Everyday Life

Common Name	Chemical formula and name (other names)	Physical properties	Safety concerns	Comments
acetone	$CH_3COCH_3(\ell)$ 2-propanone	clear; evaporates quickly	flammable; toxic by ingestion and inhalation	solvent; contained in some nail polish removers
acetylene	$C_2H_2(g)$ ethyne	smells sweet	highly explosive	burns very hot, with oxygen, in oxyacetylene welding torches; used to produce a wide range of synthetic products
ASA	$C_9H_8O_4(s)$ 2-acetyloxybenzoic acid acetylsalicylic acid	white crystals with a slightly bitter taste	excessive use may cause hearing loss or Reye's syndrom, especially in young people	used in Aspirin™ and related medicines for pain, fever, and inflammation
baking soda	$NaHCO_3(s)$ sodium hydrogen carbonate sodium bicarbonate	tiny white crystals	none	used for baking and cleaning, as an antacid and mouthwash, and in fire extinguishers
battery acid	$H_2SO_4(aq)$ sulfuric acid	clear and odourless	corrosive	used in lead-acid storage batteries (automobile batteries)
bleach	$NaOCl(aq)$ sodium hypochlorite solution	yellowish solution with a chlorine smell	toxic, strong; oxidizing agent	household chlorine bleach; used for bleaching clothes and for cleaning
bluestone	$CuSO_4 \cdot 5H_2O(s)$ copper(III) sulfate pentahydrate cupric sulfate pentahydrate	blue crystals or blue crystalline granules	toxic by ingestion; strong irritant	used in agriculture and industry, as a germicide, and for wood preservation
borax	$Na_2B_4O_7 \cdot 10H_2O(s)$ sodium borate decahydrate	white crystals	none	main source is mining; used in the glass and ceramics industries; used for making Silly Putty® and for washing clothes
carborundum	$SiC(s)$ silicon carbide	hard, black solid	none	used as an abrasive
citric acid	$C_6H_8O_7(aq)$ or $H_3C_6H_5O_7(aq)$ 3-carboxy-3-hydroxy pentanedioic acid 2-hydroxy-1,2,3-propane tricarboxylic acid	translucent crystals with a strongly acidic taste	none	used in foods and soft drinks as an acidifying agent and an antioxidant
CFCs	$CCl_2F_2(g)$, $CCl_3F(g)$, $CClF_3(g)$ chlorofluorocarbons (freons, Freon-12)	colourless, odourless gases	CFCs are now banned by the Montreal Protocol	in the past, were used as refrigerants and aerosols
charcoal/graphite	$C(s)$ pure carbon, in a less structured form than diamond	soft grey or black solid that rubs easily onto other substances	none	used as pencil "lead" and artists' charcoal, as a de-colorizing and filtering agent, in gunpowder, and for barbeque briquettes
cream of tartar	$KHC_4H_4O_6(s)$ potassium hydrogen tartrate	white, crystalline solid	none	used as a leavening agent in baking powder
dry ice	$CO_2(s)$ solid carbon dioxide	cold white solid that sublimates	damaging to the skin and tissue after prolonged exposure	used as a refrigerant in laboratories when cold temperatures (as low as $-79°C$) are required

Common Name	Chemical formula and name (other names)	Physical properties	Safety concerns	Comments
Epsom salts	$MgSO_4 \cdot 7H_2O(s)$ magnesium sulfate heptahydrate	colourless cyrstals	can cause abdominal cramps and diarrhea	used as a bath salt and in cosmetics and dietary supplements; has industrial uses
ethylene	$C_2H_4(g)$ ethene	colourless gas with sweet odour and taste	flammable	used to accelerate fruit ripening and to synthesize polymers such as polystyrene; occurs naturally in plants
ethylene glycol	$C_2H_6O_2(\ell)$ or $CH_2OHCH_2OH(\ell)$ ethane-1,2-diol glycol	clear, colourless, syrupy liquid	toxic by ingestion and inhalation	used in antifreeze and cosmetics, and as a de-icing fluid for airport runways
Glauber's salt	$Na_2SO_4 \cdot 10H_2O(s)$ sodium sulfate decahydrate	large, transparent crystals, needles, or granular powder	none	a laxative; used for paper and glass making, and in solar heat storage and air conditioning; energy storage capacity more than seven times that of water
glucose	$C_6H_{12}O_6(s)$ dextrose graph sugar corn sugar	white crystals with a sweet taste	none	source of energy for most organisms
grain alcohol	$C_2H_6O(\ell)$ or $C_2H_5OH(\ell)$ ethanol ethyl alcohol	clear, volatile liquid with distinctive odour	flammable	beverage alcohol, antiseptic, laboratory/industrial solvent; produced by the fermentation of grains or fruits
gyp rock	$CaSO_4 \cdot 2H_2O(s)$ gypsum	hard, beige mineral	none	used in plaster of Paris and as a core for drywall
hydrogen peroxide	$H_2O_2(\ell)$ dihydrogen dioxide	clear, colourless liquid	damaging to skin in high concentrations	sold as 3% solution in drugstores; non-chlorine bleach often 6% H_2O_2
ibuprofen	$C_{13}H_{18}O_2(s)$ 2-[4-(2-methylpropyl)phenyl] propanoic acid p-isobutyl-hydratropic acid	white crystals	can conflict with other medications	ingredient in over-the-counter pain relievers
laughing gas	$N_2O(g)$ nitrous oxide dinitrogen oxide	colourless, mainly odourless, soluble gas	prolonged exposure causes brain damage and infertility	used as a dental anesthetic, an aerosol propellant, and to increase fuel performance in racing cars
lime	$CaO(s)$ calcium oxide quicklime hydrated lime hydraulic lime	white powder	reacts with water to produce caustic calcium hydroxide, or slaked lime, with liberation of heat	used to make cement and to clean and nullify odours in stables
limestone	$CaCO_3(s)$ calcium carbonate	soft white mineral	none	used for making lime and for building; has industrial uses
lye	$NaOH(s)$ sodium hydroxide caustic soda	white solid, found mainly in form of beads or pellets; quickly absorbs water and carbon dioxide from the air	corrosive, strong irritant	produced by the electrolysis of brine or the reaction of calcium hydroxide and sodium carbonate; has many laboratory and industrial uses; used to manufacture chemicals and make soap
malachite	$CuCO_3 \cdot Cu(OH)_2(s)$ basic copper(II) carbonate	clear, hard, bright green mineral	none	ornamental and gem stone; copper found in the ore

Common Name	Chemical formula and name (other names)	Physical properties	Safety concerns	Comments
milk of magnesia	$Mg(OH)_2(aq)$ magnesium hydroxide magnesia magma	aqueous solution of magnesium hydroxide; $Mg(OH)_2(s)$ is a white powder	harmless if used in small amounts	antacid, laxative
moth balls	$C_{10}H_8(s)$ naphthalene	white, volatile solid with an unpleasant odour	toxic by ingestion and inhalation	used to repel insects in homes and gardens, and to make synthetic resins; obtained from crude oil
MSG	$C_5H_8NNaO_4(s)$ monosodium glutamate	white, crystalline powder	may cause headaches in some people	flavour enhancer for foods in concentrations of about 0.3%
muriatic acid	$HCl(aq)$ hydrochloric acid	colourless or slightly yellow aqueous solution	toxic by ingestion and inhalation; strong irritant	has many industrial and laboratory uses; used for processing food, cleaning, and pickling
natural gas	about 85% methane, $CH_4(g)$, 10% ethane, $C_2H_6(g)$, and some propane, $C_3H_8(g)$, butane, $C_4H_{10}(g)$, and pentane, $C_5H_{12}(g)$	odourless, colourless gas	flammable and explosive; a warning odour is added to household gas as a safety precaution	used for heating, energy, and cooking; about 3% is used as a feedstock for the chemical industry
oxalic acid	$HOOCCOOH(s)$ ethanedioic acid	strongly flavoured acid; white crystals	toxic by inhalation and ingestion; strong irritant in high concentrations	occurs naturally in rhubarb, wood sorrel, and spinach; used as wood and textile bleach, rust remover, and deck cleaner; has many industrial and laboratory uses
Pepto-Bismol™	$C_7H_5BiO_4(s)$ (active ingredient) 2-hydroxy-2H, 4H-benzo[d]1, 3-dioxa-2-bismacyclohexan-4-one bismuth subsalicylate	pink solid or solution	may cause stomach upset if taken in excess of recommended dose	relieves digestive difficulties by coating the digestive tract and reducing acidity
PCBs	$C_{12}H_{10-x}Cl_x(\ell)$ polychlorinated biphenyls: class of compounds with two benzene rings and two or more substituted chlorine atoms	colourless liquids	highly toxic, unreactive, and persistent; cause ecological damage	used as coolants in electrical transformers
potash	$K_2CO_3(s)$ potassium carbonate Traditionally, "potash" referred to potassium carbonate, but the name is now commonly used to refer to a whole family of potassium compounds, including potassium chloride and others.	white, granular, translucent powder	solutions irritating to tissue	laboratory and industrial uses; used in special glasses, in soaps, and as a dehydrating agent
PVC	$(C_2H_3Cl)_n(s)$ polyvinyl chloride	tough, white, unreactive solid	none	used extensively as a building material

Common Name	Chemical formula and name (other names)	Physical properties	Safety concerns	Comments
road salt	$CaCl_2(s)$ calcium chloride Other salts are also commonly used as de-icing agents on roads. For example, magnesium chloride, $MgCl_2(s)$ is sometimes used in combination with calcium chloride; sodium chloride, $NaCl(s)$, is also commonly used.	white crystalline compound	none	by-product of an industrial process that produces sodium carbonate, $Na_2CO_3(s)$, from salt brine, $NaCl(aq)$, and limestone, $CaCO_3(s)$
rotten-egg gas	$H_2S(g)$ hydrogen sulfide	colourless gas with an offensive odour	highly flammable therefore high fire risk; explosive; toxic by inhalation; strong irritant to eyes and mucous membranes	obtained from sour gas (natural gas with higher than average levels of hydrogen sulfide) during natural gas production
rubbing alcohol	$(CH_3)_2CHOH(\ell)$ propan-2-ol isopropanol isopropyl alcohol	colourless liquid with a pleasant odour	flammable, therefore high fire risk; explosive; toxic by inhalation and ingestion	has industrial and medical uses
salicylic acid	$C_7H_6O_3(s)$ or $HOC_6H_4COOH(s)$ 2-hydroxybenzoic acid	white crystalline solid	damages skin in high concentrations	can be used in different amounts in foods and dyes, and in wart treatment
sand	$SiO_2(s)$ silica	large, glassy cubic crystals	toxic by inhalation; chronic exposure to dust may cause silicosis	occurs widely in nature as sand, quartz, flint, and diatomite
slaked lime	$Ca(OH)_2(s)$ calcium hydroxide	white powder that is insoluble in water	none	used to neutralize acidity in soils to make whitewash, bleaching powder, and glass
soda ash	$Na_2CO_3(s)$ sodium carbonate	white powdery crystals	none	used to manufacture glass, soaps, and detergents
sugar	$C_{12}H_{22}O_{11}(s)$ sucrose cane sugar beet sugar	cubic white crystals	none	used in foods as a sweetener; source of metabolic energy
table salt	$NaCl(s)$ sodium chloride rock salt halite	cubic white crystals	none	produced by the evaporation of natural brines and by the solar evaporation of sea water; also mined from underground sources; used in foods and for de-icing roads
Tylenol™	$C_8H_9NO_2(s)$ N-(4-hydroxyphenyl)acetamide N-acetyl-p-aminophenol acetaminophen paracetamol	colourless, slightly bitter crystals	can be toxic if an overdose is taken	pain reliever (analgesic)
TSP	$Na_3PO_4(s)$ trisodium phosphate sodium phosphate sodium orthophosphate	white crystals	toxic by ingestion; irritant to tissue; pH of 1% solution is 11.8 to 12	used as a water softener and cleaner (for example, to clean metals and to clean walls before painting); has many industrial uses

Common Name	Chemical formula and name (other names)	Physical properties	Safety concerns	Comments
vinegar	5% solution of acetic acid, CH_3COOH or $C_2H_4O_2(aq)$	clear solution with a distinctive smell	none	used for cooking and household cleaning
vitamin C	$C_6H_8O_6(s)$ L-threo-hex-2-enomo 1, 4-lactone ascorbic acid	white crystals or powder with a tart, acidic taste	none	required in diet to prevent scurvy; found in citrus fruits, tomatoes, potatoes, and green leafy vegetables
washing soda	$Na_2CO_3 \cdot H_2O(s)$ sodium carbonate monohydrate soda ash	white powdery crystals	may be irritating to skin	used for cleaning and photography, and as a food additive; has many industrial and laboratory uses
wood alcohol	$CH_3OH(\ell)$ methanol methyl alcohol	clear, colourless liquid with faint alcoholic odour	flammable; toxic by ingestion, skin absorption, and inhalation; causes blindness and death	has many industrial and household uses; used in gasoline antifreeze and as a thinner for shellac and paint; can be mixed with vegetable oil and lye to make diesel fuel

Reactivity and Solubility

Activity Series of Metals

Metal	Displaces Hydrogen...	Reactivity
lithium		most reactive
potassium		
barium		
calcium		
sodium	from cold water	
magnesium		
aluminum		
zinc		
chromium		
iron		
cadmium		
cobalt		
nickel		
tin		
lead	from acids	
hydrogen		
copper		
mercury		
silver		
platinum		
gold		least reactive

Activity Series of Halogens

Halogen	Reactivity
fluorine	most reactive
chlorine	
bromine	
iodine	least reactive

Solubility of Common Ionic Compounds in Water

	Anion	+	Cation	→	Solubility of Compound*
	Most		Alkali metal ions: Li^+, K^+, Rb^+, Cs^+, Fr^+		Soluble
1.	Most		hydrogen ion, H^+		Soluble
	Most		ammonium ion, NH_4^+		Soluble
	nitrate, NO_3^-		Most		Soluble
2.	acetate (ethanoate), CH_3COO^-		Ag^+		Low solubility
			Most others		Soluble
3.	chloride, Cl^- bromide, Br^- iodide, I^-		Ag^+, Pb^{2+}, Hg_2^{2+}, Cu^+, Tl^+		Low solubility
			All others		Soluble
4.	fluoride, F^-		Mg^{2+}, Ca^{2+}, Ba^{2+}, Pb^{2+}		Low solubility
			Most others		Soluble
5.	sulfate, SO_4^{2-}		Ca^{2+}, Sr^{2+}, Ba^{2+}, Pb^{2+}		Low solubility
			All others		Soluble
6.	sulfide, S^{2-}		Alkali ions and H^+, NH_4^+, Be^{2+}, Mg^{2+}, Ca^{2+}, Sr^{2+}, Ba^{2+}		Soluble
			All others		Low solubility
7.	hydroxide, OH^-		Alkali ions and H^+, NH_4^+, Sr^{2+}, Ba^{2+}, Tl^+		Soluble
			All others		Low solubility
8.	phosphate, PO_4^{3-} carbonate, CO_3^{2-} sulfite, SO_3^{2-}		Alkali ions and H^+, NH_4^+		Soluble
			All others		Low solubility

*Compounds listed as soluble have solubilities of at least 1 g/100 mL of water at 25°C and 100 kPa.

Concentration Calculations

Measures of Concentration

Type of Concentration	Formula	Common Application
Concentration as a Percent • mass/volume percent	$\text{percent (m/v)} = \dfrac{\text{mass of solute [in grams]}}{\text{volume of solution [in millilitres]}} \times 100\%$	• intravenous solutions, such as a saline drip
• mass percent	$\text{percent (m/m)} = \dfrac{\text{mass of solute}}{\text{mass of solution}} \times 100\%$	• concentration of metals in an alloy
• volume percent	$\text{percent (v/v)} = \dfrac{\text{volume of solute}}{\text{volume of solution}} \times 100\%$	• solutions prepared by mixing liquids
Very Small Concentrations • parts per million	$\text{ppm} = \dfrac{\text{mass of solute}}{\text{mass of solution}} \times 10^6$	• safety limits for contaminants, such as mercury or lead in food or water
• parts per billion	$\text{ppb} = \dfrac{\text{mass of solute}}{\text{mass of solution}} \times 10^9$	
Molar Concentration	$\text{molar concentration} = \dfrac{\text{amount of solute [in moles]}}{\text{volume of solution [in litres]}}$ $c = \dfrac{n}{V}$	• solutions used as reactants

Acids, Bases, and Indicators

The Most Common Strong Acids

Name	Formula
hydrochloric acid	HCl(aq)
hydrobromic acid	HBr(aq)
hydroiodic acid	HI(aq)
perchloric acid	HClO$_4$(aq)
nitric acid	HNO$_3$(aq)
sulfuric acid	H$_2$SO$_4$(aq)

Some Common Strong Bases

Name	Formula
lithium hydroxide	LiOH(aq)
sodium hydroxide	NaOH(aq)
potassium hydroxide	KOH(aq)
calcium hydroxide	Ca(OH)$_2$(aq)
barium hydroxide	Ba(OH)$_2$(aq)

Range of Some Common pH Indicators

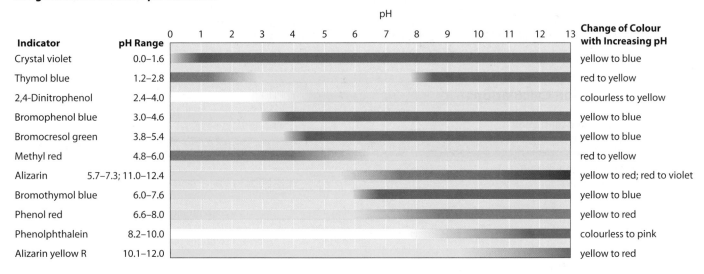

Indicator	pH Range	Change of Colour with Increasing pH
Crystal violet	0.0–1.6	yellow to blue
Thymol blue	1.2–2.8	red to yellow
2,4-Dinitrophenol	2.4–4.0	colourless to yellow
Bromophenol blue	3.0–4.6	yellow to blue
Bromocresol green	3.8–5.4	yellow to blue
Methyl red	4.8–6.0	red to yellow
Alizarin	5.7–7.3; 11.0–12.4	yellow to red; red to violet
Bromothymol blue	6.0–7.6	yellow to blue
Phenol red	6.6–8.0	yellow to red
Phenolphthalein	8.2–10.0	colourless to pink
Alizarin yellow R	10.1–12.0	yellow to red

Pressure and Temperature Units

Units of Pressure Used for Various Instruments

Unit of Pressure	Symbol
standard atmosphere	atm
millimetres of mercury	mmHg
torr	torr
pascal	Pa
kilopascal	kPa
bar	bar
millibar	mb
pounds per square inch	psi

$$1 \text{ atm} = 760 \text{ mmHg} = 760 \text{ torr} = 101\,325 \text{ Pa} = 101.325 \text{ kPa} = 1.01325 \text{ bar} = 14.7 \text{ psi}$$

The Relationship between the Celsius and Kelvin Temperature Scales

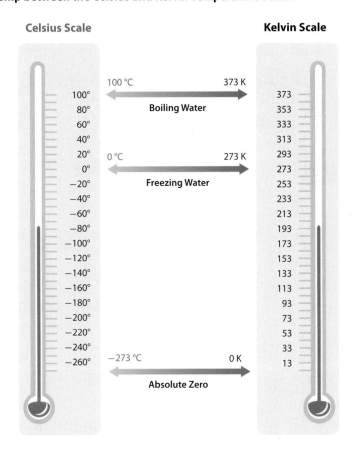

For converting Celsius to kelvin: $K = \,^{\circ}C + 273.15$
For converting kelvin to Celsius: $^{\circ}C = K - 273.15$

Gases: Pressure, Volume, Temperature, and Stoichiometry Calculations

Standard Conditions of Temperature and Pressure

Conditions	Pressure	Celsius Temperature	Kelvin Temperature	Molar Volume of an Ideal Gas
STP (standard temperature and pressure)	101.325 kPa	0°C	273.15 K	22.4 L/mol
SATP (standard ambient temperature and pressure)	100.0 kPa	25°C	298.15 K	24.8 L/mol

Partial Pressures of Water Vapour at Different Temperatures

Temperature (°C)	Pressure (kPa)	Temperature (°C)	Pressure (kPa)
15	1.71	22	2.81
16	1.81	23	2.99
17	1.93	24	3.17
18	2.07	25	3.36
19	2.20	26	3.36
20	2.33	27	3.56
21	2.49	28	3.37

Universal Gas Constant

$$R = 8.314 \ \frac{\text{kPa} \cdot \text{L}}{\text{mol} \cdot \text{K}}$$

Formulas for Calculations Involving Gases

	Formula
Molar mass	$M = \frac{m}{n}$
Density	$D = \frac{m}{V}$
Pressure	$P = \frac{F}{A}$
Boyle's Law	$P_1 V_1 = P_2 V_2$
Charles's Law	$\frac{V_1}{T_1} = \frac{V_2}{T_2}$
Gay-Lussac's Law	$\frac{P_1}{T_1} = \frac{P_2}{T_2}$
Combined Gas Law	$\frac{P_1 V_1}{T_1} = \frac{P_2 V_2}{T_2}$
Avogadro's Law	$\frac{n_1}{V_1} = \frac{n_2}{V_2}$
Molar volume	$v = \frac{V}{n}$
Ideal Gas Law	$PV = nRT$
Dalton's Law of Partial Pressures	$P_{\text{total}} = P_{\text{dry air}} + P_{\text{water vapour}}$

Appendix B

Alphabetical List of Elements

Element	Symbol	Atomic Number
Actinium	Ac	89
Aluminum	Al	13
Americium	Am	95
Antimony	Sb	51
Argon	Ar	18
Arsenic	As	33
Astatine	At	85
Barium	Ba	56
Berkelium	Bk	97
Beryllium	Be	4
Bismuth	Bi	83
Bohrium	Bh	107
Boron	B	5
Bromine	Br	35
Cadmium	Cd	48
Calcium	Ca	20
Californium	Cf	98
Carbon	C	6
Cerium	Ce	58
Cesium	Cs	55
Chlorine	Cl	17
Chromium	Cr	24
Cobalt	Co	27
Copernicium	Cn	112
Copper	Cu	29
Curium	Cm	96
Darmstadtium	Ds	110
Dubnium	Db	105
Dysprosium	Dy	66
Einsteinium	Es	99
Erbium	Er	68
Europium	Eu	63
Fermium	Fm	100
Fluorine	F	9
Francium	Fr	87
Gadolinium	Gd	64
Gallium	Ga	31
Germanium	Ge	32
Gold	Au	79
Hafnium	Hf	72
Hassium	Hs	108
Helium	He	2
Holmium	Ho	67
Hydrogen	H	1
Indium	In	49
Iodine	I	53
Iridium	Ir	77
Iron	Fe	26
Krypton	Kr	36
Lanthanum	La	57
Lawrencium	Lr	103
Lead	Pb	82
Lithium	Li	3
Lutetium	Lu	71
Magnesium	Mg	12
Manganese	Mn	25
Meitnerium	Mt	109
Mendelevium	Md	101
Mercury	Hg	80
Molybdenum	Mo	42

Element	Symbol	Atomic Number
Neodymium	Nd	60
Neon	Ne	10
Neptunium	Np	93
Nickel	Ni	28
Niobium	Nb	41
Nitrogen	N	7
Nobelium	No	102
Osmium	Os	76
Oxygen	O	8
Palladium	Pd	46
Phosphorus	P	15
Platinum	Pt	78
Plutonium	Pu	94
Polonium	Po	84
Potassium	K	19
Praseodymium	Pr	59
Promethium	Pm	61
Protactinium	Pa	91
Radium	Ra	88
Radon	Rn	86
Rhenium	Re	75
Rhodium	Rh	45
Roentgenium	Rg	111
Rubidium	Rb	37
Ruthenium	Ru	44
Rutherfordium	Rf	104
Samarium	Sm	62
Scandium	Sc	21
Seaborgium	Sg	106
Selenium	Se	34
Silicon	Si	14
Silver	Ag	47
Sodium	Na	11
Strontium	Sr	38
Sulfur	S	16
Tantalum	Ta	73
Technetium	Tc	43
Tellurium	Te	52
Terbium	Tb	65
Thallium	Tl	81
Thorium	Th	90
Thulium	Tm	69
Tin	Sn	50
Titanium	Ti	22
Tungsten	W	74
Ununhexium	Uuh	116
Ununoctium	Uuo	118
Ununpentium	Uup	115
Ununquadium	Uuq	114
Ununtrium	Uut	113
Uranium	U	92
Vanadium	V	23
Xenon	Xe	54
Ytterbium	Yb	70
Yttrium	Y	39
Zinc	Zn	30
Zirconium	Zr	40

Periodic Table of the Elements

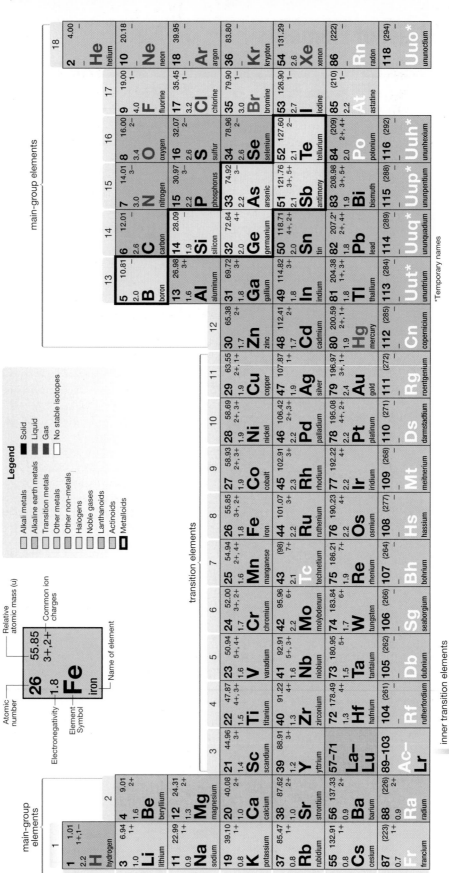

main-group elements

Legend

Alkali metals
Alkaline earth metals
Transition metals
Other metals
Other non-metals
Halogens
Noble gases
Lanthanoids
Actinoids
Metalloids

Solid
Liquid
Gas
No stable isotopes

transition elements

inner transition elements

Lanthanoids

Actinoids

Although Group 12 elements are often included in the transition elements, these elements are chemically more similar to the main-group elements.
Any value in parentheses is the mass of the least unstable or best known isotope for elements that do not occur naturally.

*Temporary names

Answers are provided for all Learning Check questions, caption questions, Practice Problems, and multiple choice questions. Answers are provided for selected Section Review, Chapter Review, Chapter Self-Assessment, Unit Review, and Unit Self-Assessment questions.

UNIT 1

Chapter 1

Answers to Learning Check Questions

1. (1) In the Thomson model of the atom, the positive charge is spread over the entire atom, whereas in the Rutherford model, the positive charge is contained in a very small volume at the centre of the atom. (2) In the Thomson model, the negative charges are embedded in the large positively charged region. In the Rutherford model, the negative charges orbit the tiny positive charge.

2. Li⋅   ⋅C⋅   ⋅F⋅⋅

3. The radius that Bohr calculated for the orbit of the electron in the hydrogen atom is the same as the average distance that Schrödinger calculated for the electron from the nucleus of the hydrogen atom.

4. Models represent the information that chemists have obtained about an object or concept. Chemists can use models to predict the behaviour of the object and design further experiments to test the model and, if necessary, modify the model. Also, models help chemists communicate about an object or concept.

5. Dalton: All matter consists of tiny particles (called atoms). Atoms of each element are unique. Thomson: Atoms contain negatively charged particles that can be ejected from the atom. Bohr: Electrons exist only in certain allowed energy levels in an atom.

6. $2n^2$: $2 \times 8^2 = 2 \times 64 = 128$

7. Mendeleev listed the elements vertically, in order of atomic mass (then called atomic weight). When he came to an element with properties similar to one higher in the list, he started a new column by putting the next element beside the one which had similar properties.

8. When elements are arranged by atomic number, their chemical and physical properties recur periodically. Many elements have similar properties and these properties follow a pattern that repeats itself regularly.

9. Each column in the periodic table constitutes a group. Groups contain elements with similar chemical and physical properties. Each row in the periodic table constitutes a period. The atomic number of the elements increases sequentially across a period. The outermost electron shell that is occupied is the same for each element in a period.

10. A specific electron shell is filled as you go across a period. When the shell is filled, the period ends. Elements with filled outer electron shells are noble gases.

11. The elements in the periodic table are categorized in several different ways. In one case, elements are categorized by whether they are metals, metalloids, or non-metals. In another case, the elements are categorized by very specific chemical and physical properties. Elements are also categorized by dividing the periodic table into blocks.

12.

Specific Properties
- alkali metals
- alkaline earth metals
- transition metals
- other metals
- lanthanoids
- actinoids
- noble gases
- halogens
- other non-metals

↑

Categories in the Periodic Table

General Properties
- metals
- metalloids
- non-metals

Blocks in the Periodic Table
- main group elements
- transition metals
- inner transition metals

13. The radius of an atom is the radius of a sphere within which electrons spend 90 percent of the time.

14. Electrons exist in a region that is best described as a cloud so atoms do not have defined boundaries. There is no way to directly measure the radius within which electrons spend 90 percent of their time.

15. As the charge of a nucleus increases, it exerts a greater force on the electrons. Thus, for electrons in a given energy level, the electrons are drawn closer to the nucleus. As a result, the size of the atom decreases across a period from left to right in the periodic table.

16. Electrons in filled shells reduce the effect of positive charge on the outer electrons. Thus, outer electrons are not as strongly attracted to the nucleus as they would be if the electrons in the lower energy levels were absent. As a result, the size of an atom increases down a group in the periodic table.

17. Increasing atomic number: oxygen (8), potassium (19), krypton (36), tin (50)

Increasing size (atomic radius): oxygen (73 pm), krypton (112 pm), tin (140 pm), potassium (227 pm).

As the atomic number increases going across a period from left to right, the nuclear charge increases, which means there is more pull on the electrons and therefore the atomic radius decreases. Thus, within a period the progression of the atomic number and size are opposite. Going down a group, however, even though the atomic number increases, the effective nuclear charge is reduced due to shielding; the atomic radius therefore increases. Also, the number of occupied electron shells increases, making the atoms larger.

18. The nuclear charge; the number of occupied electron shells; shielding; the number of valence electrons

Answers to Practice Problems

1. 35.45 u
2. 10.8 u
3. 6.94 u
4. 24.31 u
5. 39.9%; 69.7 u
6. 49.31%; 79.91 u
7. 85.47 u
8. The isotopic abundance of nitrogen-14 is very high with just a trace of nitrogen-15.
9. 186 u
10. 193 u

Answers to Caption Questions

Figure 1.10 Columns represent periods. The length of the period depends on the number of electrons allowed in the highest energy electron shell. That number increases with period number.

Figure 1.15

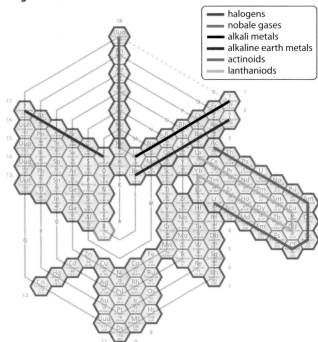

halogens
nobale gases
alkali metals
alkaline earth metals
actinoids
lanthaniods

Figure 1.18 Th e atomic radii get smaller as you go from left to right across any period. Atomic radii get larger as you go down any group. Th e last three elements of each period have very similar atomic radii. Th e change in atomic radii across a period shows the most dramatic change between groups 2 and 13 and between groups 15 and 16.

Figure 1.22 Fluorine has the greatest electronegativity and francium has the smallest electronegativity. Th ey are on corners of the periodic table diagonal to each other.

Figure 1.23 Nuclei that can get closer to the outer electrons of another atom will attract those electrons with a greater force. Th erefore, atoms with smaller radii will have a higher electronegativity.

Figure 1.25 Ionization energy, electron affi nity, and electronegativity follow the same trends. Atomic radius follows opposite trends.

Answers to Section 1.1 Review Questions

9. Isotope Data

	Name of Isotope	Notation for Isotope	Atomic Number	Mass Number	Number of Protons	Number of Electrons	Number of Neutrons
a.	bromine-81	$^{81}_{35}\text{Br}$	35	81	35	35	46
b.	neon-22	$^{22}_{10}\text{Ne}$	10	22	10	10	12
c.	calcium-44	$^{44}_{20}\text{Ca}$	20	44	20	20	24
d.	silver-107	$^{107}_{47}\text{Br}$	47	107	47	47	60

10. 28.1 u

Answers to Section 1.2 Review Questions

13 a. Na   **b.** Br   **c.** H

Answers to Section 1.3 Review Questions

7. $A^{2+} + \text{energy} \rightarrow A^{3+} + e^-$ (third ionization) The third ionization energy will be larger than the first or second because the electron is essentially being pulled away from a 3+ charge as opposed to 2+ and 1+ for the second and first ionization energies respectively.

14. Aluminum: 26.98 u, 1.43×10^{-10} m
    Lead: 207.2 u, 1.46×10^{-10} m

15. **a.** 2.45×10^{-19} J

Answers to Chapter 1 Review Questions

1. c   3. d   5. b   7. c
2. a   4. b   6. e   8. b

15. $A^+ + \text{energy} \rightarrow A^{2+} + e^-$

19. 121.76 u

Answers to Chapter 1 Self Assessment Questions

1. b   3. d   5. a   7. e   9. a
2. d   4. b   6. d   8. b   10. e

13. 152.0 μ

Chapter 2

Answers to Learning Check Questions

1. When bonds form between atoms, the atoms gain, lose, or share electrons in such a way that they create a filled outer shell containing eight electrons. For example, a fluorine atom can gain one electron and become a fluoride ion that has a completed octet of electrons in its outer shell.

2. One calcium atom can donate an electron to each of two bromine atoms. The combination of one calcium ion and two bromide ions results in a neutral compound.

3. Determine the total number of valence electrons that each of the atoms in the compound should have and add them together. Count the number of electrons shown in the Lewis structure. If the numbers are equal, the compound is neutral and is a molecular compound. If the numbers are not equal, the compound carries a charge and is thus a polyatomic ion.

4.
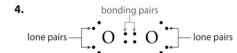
bonding pairs
lone pairs — O :: O — lone pairs

5. Double bonds form when four electrons are shared by two atoms. Triple bonds form when six electrons are shared by two atoms. In some cases, when two atoms share two electrons, neither atom has an octet of electrons in its outer shell. When both atoms contribute another electron, they share four electrons in total to form a double bond. They might then both have an octet of electrons in their outer shells. If not, the atoms may share another electron each to form a triple bond. Double and triple bonds are used to ensure that all the elements in the compound have an octet of electrons in their outer shell.

6. A group of two or more atoms of non-metal elements can share electrons and form covalent bonds, but, as a group, must either lose or gain electrons so that they can all form a stable octet of electrons. In this case, the atoms have formed a polyatomic ion. This ion can then form an ionic bond with other ions.

7. A binary ionic compound is an ionic compound that contains atoms of two and only two different elements.

8.
  a. K_2S, potassium sulfide
  b. MgO, magnesium oxide
  c. $FeCl_2$, iron(II) chloride, or $FeCl_3$, iron(III) chloride
  d. Mg_3N_2, magnesium nitride
  e. HI, hydrogen iodide
  f. $Ca(OH)_2$, calcium hydroxide

9.
  a. chromium(II) bromide
  b. sodium sulfide
  c. mercury(I) chloride
  d. lead(II) iodide
  e. aqueous hydrogen nitrate, or nitric acid
  f. potassium hydroxide

10.
  a. $ZnBr_2$
  b. Al_2S_3
  c. Cu_3N_2
  d. $MgCl_2$
  e. H_3N
  f. $Cu(OH)_2$

11.
  a. hypofluorite
  b. fluorite
  c. fluorate
  d. perfluorate

  When there is a family of compounds that can have 1, 2, 3, or 4 oxygen atoms, the combination of prefixes and suffixes are hypo-...-ite for one oxygen atom, (none)...-ite for two oxygen atoms, (none)-...-ate for three oxygen atoms, and per-...-ate for four oxygen atoms.

12.
  a. $FeSO_4$
  b. $NaNO_3$
  c. $CuCrO_4$
  d. $Mg_3(PO_4)_2$
  e. H_2CO_3
  f. $Al(OH)_3$

13. When a substance is melting, the particles (ions or molecules) have gained enough energy to break the bonds or attractive forces between the particles.

14. The compound with a melting point of 714°C is probably an ionic compound and the one with a melting point of 146°C is probably a molecular compound.

15. Potassium iodide is an ionic compound and, in ionic compounds, each ion is attracted to every oppositely charged ion adjacent to it. There are no combinations of ions that are unique and therefore cannot be called molecules.

16. A dipole-dipole force is an attractive force between the slightly positive end of a polar molecule and the slightly negative end of another polar molecule.

17. The forces of attraction among non-polar molecules are very weak. It takes only a small amount of energy for these molecules to pull apart. This means that a relatively low temperature is capable of supplying the small amount of needed energy.

18. Intermolecular forces include dipole-dipole forces and the weak forces among non-polar molecules.

Answers to Practice Problems

1. tetraphosphorus heptasulfide
2. lead(II) nitrate
3. $MnCl_4$
4. NI_3
5. copper(I) bromide
6. Fe_2O_3
7. SiO_2
8. selenium hexafluoride
9. calcium oxide
10. $Co(NO_3)_3$

Answers to Caption Questions

Figure 2.9 When you are confident that the Lewis structure is correct and all atoms have an octet of electrons, the number of shared electrons shows the number of bonds. One pair of electrons between two atoms represents one bond.

Figure 2.17 The electronegativity of chlorine is larger than the electronegativity of carbon, indicating that the chlorine attracts the shared electrons with a greater force than does the carbon.

Figure 2.22 The common name of dihydrogen monoxide is water.

Answers to Section 2.1 Review Questions

6. 6 electrons (3 pairs of electrons)
10.
  a. $\Delta EN = 1.4$; polar covalent
  b. $\Delta EN = 0.4$; polar covalent (on the line)
  c. $\Delta EN = 0.0$; non-polar covalent
  d. $\Delta EN = 1.5$; polar covalent
  e. $\Delta EN = 0.3$; slightly polar covalent
  f. $\Delta EN = 3.1$; mostly ionic
  g. $\Delta EN = 1.6$; polar covalent
  h. $\Delta EN = 1.8$; mostly ionic

Answers to Section 2.2 Review Questions

5.
  a. aluminum oxide
  b. mercury(II) iodide
  c. sodium phosphide
  d. potassium phosphate
  e. ammonium chloride

f. lithium perchlorate

g. aqueous hydrogen nitrate or nitric acid

h. lithium hydroxide

6. a. ZnO

b. FeS

c. KClO

d. MgI_2

e. $CoCl_3$

f. NaCN

8. nitrogen monoxide, nitrogen dioxide, dinitrogen monoxide, dinitrogen trioxide, dinitrogen tetroxide, dinitrogen pentoxide

9. a. PCl_5

b. F_2O

c. SO_3

d. $SiBr_4$

e. $Co(OH)_2$

f. SF_6

10. a. carbon monoxide

b. boron trichloride

c. carbon disulfide

d. carbon tetrachloride

e. silicon dioxide

f. phosphorus triiodide

g. barium hydroxide

h. trihydrogen borate

Answers to Chapter 2 Review Questions

1. c	**3.** a	**5.** b	**7.** b
2. c	**4.** d	**6.** e	**8.** d

17. a. 1:2

b. 1:1

c. 1:1

d. 1:1

19. a. magnesium chloride

b. sodium oxide

c. iron(III) chloride

d. copper(II) oxide

e. barium hypochlorite

f. ammonium nitrate

g. aqueous hydrogen chromate, or chromic acid

h. hydrogen phosphate

i. potassium hydroxide

j. cadmium hydroxide

20. a. $AuCl_3$

b. MgO

c. $LiNO_2$

d. Ca_3P_2

e. MnS

f. $Ca(ClO)_2$

g. HCl(aq)

h. H_2SO_4(aq)

i. $Co(OH)_2$

j. LiOH

22. a. sulfur dioxide

b. dinitrogen tetroxide

c. carbon monoxide

d. dichlorine oxide

23. a. H_2O

b. SO_3

c. $SiCl_4$

Answers to Chapter 2 Self Assessment Questions

1. b	**3.** b	**5.** e	**7.** b	**9.** b
2. e	**4.** a	**6.** d	**8.** c	**10.** c

15. a. $\Delta EN = 1.4$; polar covalent

b. $\Delta EN = 0.4$; slightly polar covalent

c. $\Delta EN = 0.0$; non-polar covalent

d. $\Delta EN = 1.8$; mostly ionic

16. a. magnesium phosphate

b. sodium iodate

c. aluminum phosphate

d. sodium hydrogen carbonate

17. a. KSCN

b. YCl_3

c. Fe_2S_3

d. SnF_2

18. a. trisilicon tetranitride

b. phosphorus pentachloride

c. sulfur hexafluoride

d. chlorine trifluoride

19. a. SO_3

b. CO

c. Se_2Br_2

d. NI

Answers to Unit 1 Review Questions

1. d	**3.** d	**5.** c	**7.** e	**9.** b
2. c	**4.** d	**6.** a	**8.** d	**10.** c

15. O, C, Ge, Ca, Ba

16. Cl, P, Mg, Ca, K

20 a. $\Delta EN = 2.1$; mostly ionic

b. $\Delta EN = 0.8$; polar covalent

c. $\Delta EN = 0$; non-polar covalent

28 a. Ca_3N_2

b. SI_2

c. $PbBr_2$

d. $AlPO_3$

29 a. nitrogen trichloride

b. potassium carbonate

c. iron(II) oxide

d. dinitrogen tetroxide

30 a. silicon

b. sulfur

c. phosphorus

31 a. nitric acid, aqueous hydrogen nitrate

b. aqueous hydrogen iodide, hydroiodic acid

c. aqueous hydrogen oxalate, oxalic acid

d. cobalt(III) hydroxide

32 a. HClO(aq)

b. NH_4OH

c. HNO_2 (aq)

d. $Mg(OH)_2$

34 a. ionic

  b. 2, 7, 7

  c. 0, 8, 8

35 a. 2+

  b. 2–; 0

  c. 1–; 0

57. 50.94 u

58. a. 3– charge

  b. 2– charge

  c. 2+ charge

  d. 1+ charge

59. a. one aluminum to three chlorine

  b. one aluminum to one nitrogen

  c. two aluminum to three oxygen

61. NH_3

Answers to Unit 1 Self-Assessment Questions

1. a	**3.** a	**5.** d	**7.** d	**9.** b	**11.** 3; 15
2. b	**4.** c	**6.** e	**8.** d	**10.** a	**13.** 107.9
u					

18. a. lose 2 electrons

  b. gain 2 electrons

  c. lose 1 electron

  d. gain 3 electrons

19 a. ionic; Mg_3N_2

  b. covalent; OF_2

  c. ionic; $SnBr_2$

  d. ionic; $AlPO_4$

  e. ionic; $Co_2(SO_3)_3$

20 a. covalent; phosphorus pentachloride

  b. ionic; lithium carbonate

  c. ionic; copper(II) oxide

  d. covalent; dinitrogen trioxide

  e. ionic; ammonium nitrite

UNIT 2

Chapter 3

Answers to Learning Check Questions

1. A skeleton equation is a representation of a chemical reaction that does not include the relative quantities of the substances involved.

2. (aq)

3. a. The skeleton equation would show that carbon dioxide and water are the reactants and sugar and oxygen are the products; the equation would show the chemical formula and state for each substance.

  b. The skeleton equation would not show the relative number of atoms, molecules, or ions involved in the reaction.

4. The two-way arrow indicates that the reaction is reversible.

5. a. The state of water should be (ℓ) but is shown as (g). The second plus sign (between water and sodium hydroxide) should be an arrow. The formula for sodium hydroxide should be NaOH. The state of sodium hydroxide should be (aq). The formula for hydrogen should have a subscript 2.

  b. $Na(s) + H_2O(\ell) \rightarrow NaOH(aq) + H_2(g)$

6. a. reactant: liquid water; products: hydrogen gas and oxygen gas

    $H_2O(\ell) \rightarrow H_2(g) + O_2(g)$

  b. reactants: chlorine gas and aqueous zinc bromide; products: liquid bromine and zinc chloride

    $Cl_2(g) + ZnBr_2(aq) \rightarrow Br_2(\ell) + ZnCl_2(s)$

7. The general form of a synthesis reaction is $A + B \rightarrow AB$

8. A binary ionic compound is formed.

9. a. A single product is formed from two or more reactants.

  b. $2Na(s) + Cl_2(g) \rightarrow 2NaCl(s)$

10. a. $CaCl_2(s)$

  b. $Ca(s) + Cl_2(g) \rightarrow CaCl_2(s)$

11. Your graphic organizer should show that the first reaction, $S(s) + O_2(g) \rightarrow SO_2(g)$, involves two elements; the second reaction, $2SO_2(g) + O_2(g) \rightarrow 2SO_3(g)$, involves a compound and an element; and the third reaction, $SO_3(g) + H_2O(\ell) \rightarrow H_2SO_4(aq)$, involves two compounds.

12. No, there must be another product that contains hydrogen. Because there is more than one product, the reaction cannot be a synthesis reaction.

13. A metal hydroxide, or base, forms when a metal oxide reacts with water.

14. The graphic organizer should show that the solution formed from a metal oxide is basic and that the solution formed from a non-metal oxide is acidic.

15. $AB \rightarrow A + B$; a single reactant breaks down into two or more products.

16. No, the reactant must be a compound because chemical reactions can break a compound into simpler substances but cannot change elements into simpler substances.

17. Electrolysis is possible in the aqueous state. An aqueous solution of an ionic compound can conduct an electric current because when the compound dissolves in water, the ions separate and are free to move.

18. Thermal decomposition can be used to isolate elemental mercury by heating solid mercury(II) oxide, according to the following equation: $2HgO(s) \rightarrow 2Hg(\ell) + O_2(g)$ Uses of mercury include (any two of the following) thermometers, barometers, and dental fillings.

19. Combustion reactions release light and heat.

20. A hydrocarbon is a compound composed of only carbon and hydrogen.

  a. Hydrocarbons can take part in complete or incomplete combustion reactions.

  b. The products of complete combustion are carbon dioxide and water.

21. a. The reaction of nitrogen gas and oxygen gas is not a combustion reaction because energy in the form of heat is absorbed, not released.

  b. It is a synthesis reaction.

22. a. A blue flame indicates complete combustion is occurring, and a yellow flame indicates incomplete combustion.
  b. The blue flame indicates complete combustion, so the flame is generating much more heat than light.
  c. In **Figure 3.23**, carbon dioxide and water are formed. In **Figure 3.24**, carbon dioxide, water, elemental carbon, and possibly carbon monoxide are produced.

23. A gas stove is designed to be hot, so it should allow complete combustion to occur. In addition, because gas stoves are used indoors, it is important for the combustion to be complete to avoid the production of toxic carbon monoxide.

24. A combustion reaction requires oxygen, which is not present in space. The antenna array could not burn.

25. Incomplete combustion is a chemical reaction in which a substance reacts with oxygen but there is too little oxygen for complete combustion to occur. In addition to the carbon dioxide and water that are produced during complete combustion reactions, elemental carbon and carbon monoxide are produced during incomplete combustion reactions.

26. Carbon monoxide, a poisonous gas, is formed during incomplete combustion.

27. Carbon monoxide production can occur in pulp and paper production, petroleum refineries, and steel production.

28. The amount of oxygen present determines whether complete or incomplete combustion will occur.

29. If the synthesis reaction involves an element or compound reacting with oxygen, it can also be classified as a combustion reaction.

30. This process is cellular respiration, because it occurs at fairly low temperatures and is mediated by living organisms.

Answers to Practice Problems

  1. $H_2(g) + O_2(g) \rightarrow H_2O(g)$
  2. $Na(s) + H_2O(\ell) \rightarrow NaOH(aq) + H_2(g)$
  3. $KClO_3(s) \rightarrow KCl(s) + O_2(g)$
  4. $Cu(s) + O_2(g) \rightarrow CuO(s)$
  5. $AgNO_3(aq) + NaCl(aq) \rightarrow NaNO_3(aq) + AgCl(s)$
  6. $C_3H_8(g) + O_2(g) \rightarrow H_2O(g) + CO_2(g)$
  7. $SO_3(g) + H_2O(\ell) \rightarrow H_2SO_4(aq)$
  8. $HCl(g) + NH_3(g) \rightarrow NH_4Cl(s)$
  9. $AlF_3(s) \rightarrow Al(s) + F_2(g)$
  10. $Hg(\ell) + O_2(g) \rightarrow HgO(s)$
  11. $2NO(g) + O_2(g) \rightarrow 2NO_2(g)$
  12. $3Mg(s) + 2AlCl_3(aq) \rightarrow 2Al(s) + 3MgCl_2(aq)$
  13. $2NaOH(aq) + CuCl_2(aq) \rightarrow 2NaCl(aq) + Cu(OH)_2(s)$
  14. $C_2H_4(g) + 3O_2(g) \rightarrow 2CO_2(g) + 2H_2O(g)$
  15. $Cu(s) + 2AgNO_3(aq) \rightarrow Cu(NO_3)_2(aq) + 2Ag(s)$
  16. $4Al(s) + 3MnO_2(s) \rightarrow 2Al_2O_3(s) + 3Mn(s)$
  17. $C_3H_8(g) + 5O_2(g) \rightarrow 3CO_2(g) + 4H_2O(g)$
  18. $4NH_3(g) + 7O_2(g) \rightarrow 4NO_2(g) + 6H_2O(\ell)$
  19. $K_2S(aq) + CoCl_2(aq) \rightarrow 2KCl(aq) + CoS(s)$
  20. $2HCl(g) + Na_2CO_3(aq) \rightarrow CO_2(g) + H_2O(\ell) + 2NaCl(aq)$
  21. lithium oxide; $4Li(s) + O_2(g) \rightarrow 2Li_2O(s)$

22. strontium fluoride; $Sr(s) + F_2(g) \rightarrow SrF_2(s)$
23. iron(II) bromide or iron(III) bromide; $2Fe(s) + 3Br_2(\ell) \rightarrow 2FeBr_3(s)$ or $Fe(s) + Br_2(\ell) \rightarrow FeBr_2(s)$
24. phosphorus trihydride; $2P(s) + 3H_2(g) \rightarrow 2PH_3(g)$
25. calcium iodide; $Ca(s) + I_2(g) \rightarrow CaI_2(s)$
26. tin(IV) oxide or tin(II) oxide; $Sn(s) + O_2(g) \rightarrow SnO_2(s)$ or $2Sn(s) + O_2(g) \rightarrow 2SnO(s)$
27. bismuth(III) sulfide or bismuth(V) sulfide; $2Bi(s) + 3S(s) \rightarrow Bi_2S_3(s)$ or $2Bi(s) + 5S(s) \rightarrow Bi_2S_5(s)$
28. aluminum iodide; $2Al(s) + 3I_2(s) \rightarrow 2AlI_3(s)$
29. silver oxide; $4Ag(s) + O_2(g) \rightarrow 2Ag_2O(s)$
30. nitrogen dioxide; $N_2(g) + 2O_2(g) \rightarrow 2NO_2(g)$
31. potassium and bromine; $2KBr(\ell) \rightarrow 2K(\ell) + Br_2(\ell)$
32. aluminum and oxygen; $2Al_2O_3(\ell) \rightarrow 4Al(\ell) + 3O_2(g)$
33. magnesium oxide and water; $Mg(OH)_2(s) \rightarrow MgO(s) + H_2O(g)$
34. calcium nitrite and oxygen; $Ca(NO_3)_2(s) \rightarrow Ca(NO_2)_2(s) + O_2(g)$
35. copper(II) oxide and carbon dioxide; $CuCO_3(s) \rightarrow CuO(s) + CO_2(g)$
36. chromium and chlorine; $2CrCl_3(\ell) \rightarrow 2Cr(\ell) + 3Cl_2(g)$
37. barium oxide and carbon dioxide; $BaCO_3(s) \rightarrow BaO(s) + CO_2(g)$
38. rubidium nitrite and oxygen; $2RbNO_3(s) \rightarrow 2RbNO_2(s) + O_2(g)$
39. lithium oxide and water; $2LiOH(s) \rightarrow Li_2O(s) + H_2O(g)$
40. magnesium and chlorine; $MgCl_2(s) \rightarrow Mg(\ell) + Cl_2(g)$
41. $C_7H_{16}(\ell) + 11O_2(g) \rightarrow 7CO_2(g) + 8H_2O(g)$
42. $C_9H_{20}(\ell) + 14O_2(g) \rightarrow 9CO_2(g) + 10H_2O(g)$
43. $2C_2H_2(\ell) + 5O_2(g) \rightarrow 4CO_2(g) + 2H_2O(g)$
44. $2C_6H_6(\ell) + 15O_2(g) \rightarrow 12CO_2(g) + 6H_2O(g)$
45. $2C_8H_{18}(\ell) + 25O_2(g) \rightarrow 16CO_2(g) + 18H_2O(g)$
46. $2C_8H_{18}(\ell) + 17O_2(g) \rightarrow 16CO(g) + 18H_2O(g)$
47. $2C_5H_{12}(\ell) + 11O_2(g) \rightarrow 10CO(g) + 12H_2O(g)$
48. $C_3H_8(g) + 2O_2(g) \rightarrow 3C(s) + 4H_2O(g)$
49. $4C_7H_{16}(\ell) + 37O_2(g) \rightarrow 14CO_2(g) + 14CO(g) + 32H_2O(g)$
50. a. $C_6H_{12}(\ell) + 6O_2(g) \rightarrow 6CO(g) + 6H_2O(g)$
  b. $C_6H_{12}(\ell) + 3O_2(g) \rightarrow 6C(s) + 6H_2O(g)$

Answers to Caption Questions

Figure 3.4: There is one aluminum atom in both the reactants and the product; however, there are two bromine atoms in the reactants and three in the product.

Figure 3.5: In the reactants, there are three molecules of bromine, each containing two atoms for a total of six bromine atoms. In the products, there are two formula units of aluminum bromide, each containing three bromine atoms, for a total of six. The number of bromine atoms in the reactants and products is balanced, with six on each side.

Figure 3.8: Both the general form of a synthesis reaction and the reaction shown illustrate two separate substances joining together to form one substance.

Figure 3.10: Manganese and copper are both multivalent metals. Like copper, manganese can form different binary compounds depending on the reaction that occurs.

Figure 3.14: The nitrogen gas causes the air bag to inflate.

Figure 3.24: Elemental carbon forms dark soot deposits on surfaces.

Figure 3.25: This reaction is a combustion reaction because hydrogen reacts with oxygen to form an oxide, and noticeable heat and light are produced.

Figure 3.26: Each product is an oxide of the element.

Answers to Section 3.1 Review Questions

7. (g)

11. a. $2K(s) + Cl_2(g) \rightarrow 2KCl(s)$
 b. $2Al(s) + 3CuSO_4(aq) \rightarrow 3Cu(s) + Al_2(SO_4)_3(aq)$
 c. $N_2(g) + 3H_2(g) \rightarrow 2NH_3(g)$
 d. $CaCl_2(aq) + F_2(g) \rightarrow CaF_2(aq) + Cl_2(g)$

12. a. hydroxide, OH^-, and phosphate, $PO_4{}^{3-}$
 c. $Ba(OH)_2(s) + H_3PO_4(aq) \rightarrow BaHPO_4(s) + 2H_2O(\ell)$

13. The coefficients are not in the lowest possible ratio. The correct equation is: $2NaOH(aq) + H_2SO_4(aq) \rightarrow Na_2SO_4(aq) + 2H_2O(\ell)$.

14. The formulas are incorrectly written. Therefore, the equation is not balanced properly. $2Al(s) + 3Cl_2(g) \rightarrow 2AlCl_3(s)$

Answers to Section 3.2 Review Questions

5. $2Al(s) + 3S(s) \rightarrow Al_2S_3(s)$

15. lithium carbonate → lithium oxide + carbon dioxide; lithium hydroxide → lithium oxide + water

Answers to Section 3.3 Review Questions

5. $5CO_2(g)$

6. a. $C_2H_4(g) + 3O_2(g) \rightarrow 2CO_2(g) + 2H_2O(g)$
 b. $2C_{10}H_{22}(\ell) + 31O_2(g) \rightarrow 20CO_2(g) + 22H_2O(g)$
 c. $C_4H_8(g) + 6O_2(g) \rightarrow 4CO_2(g) + 4H_2O(g)$
 d. $2C_6H_{14}(\ell) + 19O_2(g) \rightarrow 12CO_2(g) + 14H_2O(g)$

Answers to Chapter 3 Review Questions

1. c   **3.** e   **5.** c   **7.** d
2. a   **4.** c   **6.** a   **8.** e
10. 1

18. a. $Mg_3N_2(s) \rightarrow 3Mg(s) + N_2(g)$
 b. $4Mn(s) + 3O_2(g) \rightarrow 2Mn_2O_3(s)$
 c. $CO_2(g) + 4H_2(g) \rightarrow CH_4(g) + 2H_2O(g)$
 d. $2PbO(s) \rightarrow 2Pb(s) + O_2(g)$
 e. $2C_2H_6(g) + 7O_2(g) \rightarrow 4CO_2(g) + 6H_2O(g)$
 f. $Cu(s) + 2AgNO_3(aq) \rightarrow 2Ag(s) + Cu(NO_3)_2(aq)$
 g. $C_3H_8(g) + 5O_2(g) \rightarrow 3CO_2(g) + 4H_2O(g)$
 h. $3PbCl_4(aq) + 4K_3PO_4(aq) \rightarrow 12KCl(aq) + Pb_3(PO_4)_4(s)$

20. a. potassium sulfide; $2K(s) + S(s) \rightarrow K_2S(s)$
 b. chromium(III) chloride; $2Cr(s) + 3Cl_2(g) \rightarrow 2CrCl_3(s)$; chromium(II) chloride; $Cr(s) + Cl_2(g) \rightarrow CrCl_2(s)$
 c. silver oxide; $4Ag(s) + O_2(g) \rightarrow 2Ag_2O(s)$
 d. sulfur hexachloride; $S(s) + 3Cl_2(g) \rightarrow SCl_6(g)$

21. a. magnesium and iodine; $MgI_2(\ell) \rightarrow Mg(s) + I_2(g)$
 b. copper(II) nitrite and oxygen; $Cu(NO_3)_2(s) \rightarrow Cu(NO_2)_2(s) + O_2(g)$
 c. barium oxide and carbon dioxide; $BaCO_3(s) \rightarrow BaO(s) + CO_2(g)$

23. a. $2C_2H_6(g) + 7O_2(g) \rightarrow 4CO_2(g) + 6H_2O(g)$
 b. $C_5H_{12}(\ell) + 8O_2(g) \rightarrow 5CO_2(g) + 6H_2O(g)$
 c. $2C_8H_{18}(\ell) + 25O_2(g) \rightarrow 16CO_2(g) + 18H_2O(g)$

27. Your diagram should include the following chemical equations: $Ca(s) + Br_2(\ell) \rightarrow CaBr_2(s)$; $Mg(s) + Br_2(\ell) \rightarrow MgBr_2(s)$; $Sr(s) + Br_2(\ell) \rightarrow SrBr_2(s)$

31. The chemical formulas for nitrogen gas and hydrogen gas are incorrect, so the chemical equation is not correctly balanced; in addition, the arrow should show that the chemical reaction is reversible. The correct chemical equation is $N_2(g) + 3H_2(g) \rightleftharpoons 2NH_3(g)$.

34. b. $N_2(g) + 2O_2(g) \rightarrow 2NO_2(g)$

35. a. $2H_2O_2(\ell) \rightarrow 2H_2O(\ell) + O_2(g)$

37. a. decomposition; $CaCO_3(s) \rightarrow CaO(s) + CO_2(g)$
 b. synthesis; $CaO(s) + SO_2(g) \rightarrow CaSO_3(s)$
 c. synthesis; $2CaSO_3(s) + O_2(g) \rightarrow 2CaSO_4(s)$

Answers to Chapter 3 Self-Assessment Questions

1. e   **3.** c   **5.** d   **7.** a   **9.** b
2. b   **4.** e   **6.** a   **8.** d   **10.** b

11. $CO_2(g)$

12. a. $Na_2CO_3(s) \rightarrow Na_2O(s) + CO_2(g)$

15. a. $Cr(ClO_3)_2(s) \rightarrow CrCl_2(s) + 3O_2(g)$
 b. $4Rb(s) + O_2(g) \rightarrow 2Rb_2O(s)$
 c. $C_2H_4(g) + 3O_2(g) \rightarrow 2CO_2(g) + 2H_2O(g)$
 d. $2KOH(s) \rightarrow K_2O(s) + H_2O(g)$

16. a. combustion; $C_3H_8(g) + 5O_2(g) \rightarrow 3CO_2(g) + 4H_2O(g)$
 b. decomposition; $2KBrO_3(s) \rightarrow 2KBr(s) + 3O_2(g)$
 c. synthesis; $CaO(s) + SO_2(g) \rightarrow CaSO_3(s)$
 d. decomposition; $Ca(NO_3)_2(s) \rightarrow Ca(NO_2)_2(s) + O_2(g)$
 e. decomposition; $C_{12}H_{22}O_{11}(s) \rightarrow 12C(s) + 11H_2O(\ell)$
 f. combustion; $2C_2H_6(g) + 7O_2(g) \rightarrow 4CO_2(g) + 6H_2O(g)$

17. a. aluminum chloride; $2Al(s) + 3Cl_2(g) \rightarrow 2AlCl_3(s)$
 b. barium hydroxide; $BaO(s) + H_2O(\ell) \rightarrow Ba(OH)_2(s)$

18. a. calcium and nitrogen; $Ca_3N_2(s) \rightarrow 3Ca(s) + N_2(g)$
 b. sulfur dioxide and water; $H_2SO_3(aq) \rightarrow SO_2(g) + H_2O(\ell)$

19. a. $C_5H_{12}(\ell) + 8O_2(g) \rightarrow 5CO_2(g) + 6H_2O(g)$
 b. $2C_3H_6(g) + 9O_2(g) \rightarrow 6CO_2(g) + 6H_2O(g)$

Chapter 4
Answers to Learning Check Questions

1. $A + BX \rightarrow AX + B$

2. An element displaces a different element from a compound, forming a new compound and the replaced element as products.

3. Scientists perform experiments to determine the relative reactivity of a series of elements.

4. Platinum and gold, which are at the bottom of the reactivity series of metals, are the least reactive metals. A platinum or gold coating can prevent another, more reactive, metal underneath it from reacting with any substances the metal object might come into contact with.

5. **a.** No reaction.
   b. A reaction would not occur because copper is not reactive enough to displace lead.

6. **a.** Titanium would appear closer to the bottom of the series.
   b. Its use in medical implants indicates that titanium is not very reactive.

7. AX + BY → AY + BX

8. They are in aqueous solution.

9. Cations are positive ions, so describing a double displacement reaction as the exchange of cations is correct.

10. The ions in each reactant switch partners, so knowing which ions are involved makes it possible to correctly pair them up and determine the reaction products.

11. No, the products of a double displacement reaction are generally two compounds, not elements.

12. **a.** potassium nitrate and silver bromide; KNO_3(aq) and AgBr(s)
   b. KBr(aq) + $AgNO_3$(aq) → KNO_3(aq) + AgBr(s)

13. A precipitate is an insoluble solid that forms as a result of a chemical reaction between two soluble compounds.

14. The other compounds are in aqueous solution, but the silver chloride is a solid precipitate.

15. No, the solubility rules in the table are for the solubility of compounds in water only.

16. A double displacement reaction that produces a gas can produce either carbon dioxide or ammonia. The general forms of these reactions are as follows: acid + compound containing carbonate ion → ionic compound + water + carbon dioxide; compound containing ammonium ions + compound containing hydroxide ions → ionic compound + water + ammonia

17. **a.** $Ba(OH)_2$(s)
   b. MgS(s)
   c. H_3PO_4(s)
   d. Na_2SO_3(s)

18. $CaCO_3$(s) + 2HCl(aq) → $CaCl_2$(aq) + H_2O(ℓ) + CO_2(g)

19. Single displacement; aluminum displaces iron from iron oxide.

20. The liquid metal product of a thermite reaction is useful for welding.

21. Solid aluminum reacts with solid copper(II) oxide to produce pure liquid copper and aluminum oxide.

22. Seashells are a source of calcium carbonate, which is decomposed to produce calcium oxide. Calcium oxide is the reactant in the next step in the process of magnesium extraction.

23. The precipitation allows for the magnesium to be easily separated from the other ions in seawater.

24. Chlorine is re-used to produce hydrochloric acid, which is needed for the neutralization reaction.

Answers to Practice Problem Questions

1. Mg(s) + $CrSO_4$(aq) → $MgSO_4$(aq) + Cr(s)

2. NR

3. Zn(s) + H_2SO_4(aq) → $ZnSO_4$(aq) + H_2(g)

4. F_2(g) + MgI_2(aq) → MgF_2(aq) + I_2(s)

5. Cl_2(g) + 2NaI(aq) → 2NaCl(aq) + I_2(s)

6. NR

7. NR

8. 2K(s) + $2H_2O$(ℓ) → 2KOH(aq) + H_2(g)

9. 2HCl(aq) + Cd(s) → H_2(g) + $CdCl_2$(aq)

10. $3Pb(ClO_3)_4$(aq) + 4Al(s) → $4Al(ClO_3)_3$(aq) + 3Pb(s)

11. potassium chloride and calcium sulfate; K_2SO_4(aq) + $CaCl_2$(aq) → 2KCl(aq) + $CaSO_4$(s)

12. barium carbonate and sodium nitrate; $Ba(NO_3)_2$(aq) + Na_2CO_3(aq) → $BaCO_3$(s) + $2NaNO_3$(aq)

13. iron(III) hydroxide and sodium chloride; $FeCl_3$(aq) + 3NaOH(aq) → $Fe(OH)_3$(s) + 3NaCl(aq)

14. rubidium iodide and copper(II) sulfide; Rb_2S(aq) + CuI_2(aq) → 2RbI(aq) + CuS(s)

15. zinc acetate and copper(I) bromide; $ZnBr_2$(aq) + $2CuCH_3COO$(aq) → $Zn(CH_3COO)_2$(aq) + 2CuBr(s)

16. lithium chloride and magnesium hydroxide; 2LiOH(aq) + $MgCl_2$(aq) → 2LiCl(aq) + $Mg(OH)_2$(s)

17. aluminum nitrate and lead(II) sulfate; $Al_2(SO_4)_3$(aq) + $3Pb(NO_3)_2$(aq) → $2Al(NO_3)_3$(aq) + $3PbSO_4$(s)

18. lithium chloride and magnesium phosphate; $2Li_3PO_4$(aq) + $3MgCl_2$(aq) → 6LiCl(aq) + $Mg_3(PO_4)_2$(s)

19. calcium sulfate and magnesium nitrate; $Ca(NO_3)_2$(aq) + $MgSO_4$(aq) → $CaSO_4$(s) + $Mg(NO_3)_2$(aq)

20. silver chloride and magnesium nitrate; $2AgNO_3$(aq) + $MgCl_2$(aq) → 2AgCl(s) + $Mg(NO_3)_2$(aq)

21. potassium chloride, water, and carbon dioxide; K_2CO_3(aq) + 2HCl(aq) → 2KCl(aq) + H_2O(ℓ) + CO_2(g)

22. sodium sulfate, water, and carbon dioxide; H_2SO_4(aq) + Na_2CO_3(aq) → Na_2SO_4(aq) + H_2O(ℓ) + CO_2(g)

23. sodium chloride, water, and ammonia; NH_4Cl(aq) + NaOH(aq) → NaCl(aq) + H_2O(ℓ) + NH_3(g)

24. rubidium chloride and water; RbOH(aq) + HCl(aq) → RbCl(aq) + H_2O(ℓ)

25. calcium acetate, water, and carbon dioxide; $CaCO_3$(s) + $2HCH_3COO$(aq) → $Ca(CH_3COO)_2$(aq) + H_2O(ℓ) + CO_2(g)

26. lithium bromide, water, and ammonia; LiOH(aq) + NH_4Br(aq) → LiBr(aq) + H_2O(ℓ) + NH_3(g)

27. lithium sulfate and water; H_2SO_4(aq) + 2LiOH(aq) → Li_2SO_4(aq) + $2H_2O$(ℓ)

28. lithium acetate, water, and carbon dioxide; $LiHCO_3$(aq) + HCH_3COO(aq) → $LiCH_3COO$(aq) + H_2O(ℓ) + CO_2(g)

29. calcium nitrate and water; $Ca(OH)_2$(aq) + $2HNO_3$(aq) → $Ca(NO_3)_2$(aq) + $2H_2O$(ℓ)

30. magnesium chloride, water, and ammonia; $2NH_4Cl$(aq) + $Mg(OH)_2$(aq) → $MgCl_2$(aq) + $2H_2O$(ℓ) + $2NH_3$(g)

Answers to Caption Questions

Figure 4.2: Nothing. The nitrate ions do not change during the reaction.

Figure 4.5: Metals that can displace hydrogen from acids are tin, nickel, cobalt, cadmium, iron, chromium, zinc, aluminum, sodium, calcium, barium, potassium, and lithium; metals that cannot displace hydrogen from acids are copper, silver, mercury, platinum, and gold.

Figure 4.9: The positive ions, Ag^+ and Na^+, change places.

Figure 4.13: The thermite reaction occurs in the solid state, but most double replacement reactions occur in an aqueous solution.

Figure 4.16: Aluminum-magnesium tubing is strong, light, and more resistant to corrosion than pure aluminum—all of which are valuable properties for a kayak.

Figure 4.17: The furnaces are built at different elevations, so the material can flow downhill, moving by gravity from one furnace to the next.

Figure 4.19: A precipitate holds the cyanide in place as a solid, increasing the chance that it can be cleaned up before it is washed into groundwater or a river.

Answers to Section 4.1 Review Questions

5. a. NR
  b. A reaction occurs. Zinc displaces iron. $Zn(s) + FeCl_2(aq) \rightarrow Fe(s) + ZnCl_2(aq)$
  c. A reaction occurs. Magnesium displaces aluminum. $3Mg(s) + Al_2(SO_4)_3(aq) \rightarrow 2Al(s) + 3MgSO_4(aq)$
  d. A reaction occurs. Zinc displaces hydrogen. $Zn(s) + 2HCl(aq) \rightarrow H_2(g) + ZnCl_2(aq)$
  e. NR
  f. A reaction occurs. Magnesium displaces hydrogen. $Mg(s) + H_2SO_4(aq) \rightarrow H_2(g) + MgSO_4(aq)$

12. If the liquid were water, the metal shown could not be zinc, because zinc does not react with water to form hydrogen gas. If the liquid were an acid, then the metal could be zinc, because zinc can displace hydrogen from an acid.

15. a. A reaction occurs. Iron displaces hydrogen. $Fe(s) + 2HBr(aq) \rightarrow H_2(g) + FeBr_2(s)$ or $2Fe(s) + 6HBr(aq) \rightarrow 3H_2(g) + 2FeBr_3(s)$
  b. A reaction occurs. Bromine displaces iodine. $Br_2(\ell) + MgI_2(aq) \rightarrow MgBr_2(aq) + I_2(aq)$
  c. A reaction occurs. Potassium displaces aluminum. $6K(s) + Al_2(SO_4)_3(aq) \rightarrow 2Al(s) + 3K_2SO_4(aq)$
  d. A reaction occurs. Lithium displaces hydrogen. $2Li(s) + 2H_2O(\ell) \rightarrow 2LiOH(aq) + H_2(g)$
  e. NR
  f. NR

Answers to Section 4.2 Review Questions

1. b. $CW + DZ \rightarrow CZ + DW$

9. b. $CaCO_3(s) + 2HCl(aq) \rightarrow CaCl_2(aq) + H_2O(\ell) + CO_2(g)$

11. d. $2NaOH(aq) + CuCl_2(aq) \rightarrow 2NaCl(aq) + Cu(OH)_2(s)$

12. ammonia, $NH_3(g)$; $NH_4Br(aq) + NaOH(aq) \rightarrow NaBr(aq) + H_2O(\ell) + NH_3(g)$

15. calcium chloride and water; $2HCl(aq) + Ca(OH)_2(aq) \rightarrow 2H_2O(\ell) + CaCl_2(aq)$

16. Although the chemical formulas are correct, the equation is not correctly balanced because water needs a coefficient. The states of the products are incorrect. Water should be liquid, and sodium phosphate should be in aqueous solution. Correct equation: $3NaOH(aq) + H_3PO_4(aq) \rightarrow 3H_2O(\ell) + Na_3PO_4(aq)$

Answers to Section 4.3 Review Questions

8. b. $MgCl_2(\ell) \rightarrow Mg(\ell) + Cl_2(g)$

13. c. $2CuFeS_2(s) + 4O_2(g) \rightarrow Cu_2S(\ell) + 2FeO(\ell) + 3SO_2(g)$
  $2Cu_2S(\ell) + 3O_2(g) \rightarrow 2Cu_2O(\ell) + 2SO_2(g)$
  $Cu_2S(\ell) + 2Cu_2O(\ell) \rightarrow 6Cu(\ell) + SO_2(g)$

Answers to Chapter 4 Review Questions

1. c	**3.** b	**5.** c	**7.** d
2. a	**4.** d	**6.** d	**8.** b

15. $A + BX \rightarrow BA + X$

20. a. $3Mg(s) + 2Co(NO_3)_3(aq) \rightarrow 3Mg(NO_3)_2(aq) + 2Co(s)$
  b. $Cl_2(g) + 2LiBr(aq) \rightarrow Br_2(g) + 2LiCl(aq)$
  c. $Zn(s) + 2HClO_4(aq) \rightarrow Zn(ClO_4)_2(aq) + H_2(g)$
  d. NR
  e. $2Al(s) + 3NiCl_2(aq) \rightarrow 2AlCl_3(aq) + 3Ni(s)$
  f. $2K(s) + 2H_2O(\ell) \rightarrow 2KOH(aq) + H_2(g)$
  g. NR

23. a. potassium bromide and barium sulfate; $K_2SO_4(aq) + BaBr_2(aq) \rightarrow 2KBr(aq) + BaSO_4(s)$
  b. lithium nitrate, water, and carbon dioxide; $2HNO_3(aq) + Li_2CO_3(aq) \rightarrow 2LiNO_3(aq) + H_2O(\ell) + CO_2(g)$
  c. copper(II) hydroxide and sodium bromide; $CuBr_2(aq) + 2NaOH(aq) \rightarrow Cu(OH)_2(s) + 2NaBr(aq)$
  d. rubidium nitrate and lead(II) sulfide; $Rb_2S(aq) + Pb(NO_3)_2(aq) \rightarrow 2RbNO_3(aq) + PbS(s)$
  e. potassium sulfate, water, and ammonia; $(NH_4)_2SO_4(aq) + 2KOH(aq) \rightarrow K_2SO_4(aq) + 2H_2O(\ell) + 2NH_3(g)$
  f. iron(II) nitrate and silver bromide; $FeBr_2(aq) + 2AgNO_3(aq) \rightarrow Fe(NO_3)_2(aq) + 2AgBr(s)$
  g. lithium sulfate and water; $2LiOH(aq) + H_2SO_4(aq) \rightarrow Li_2SO_4(aq) + 2H_2O(\ell)$

25. $2Al(s) + 3H_2SO_4(aq) \rightarrow 3H_2(g) + Al_2(SO_4)_3(aq)$; $2Al(OH)_3(aq) + 3H_2SO_4(aq) \rightarrow 6H_2O(\ell) + Al_2(SO_4)_3(aq)$

26. Ammonia gas can form from a double displacement reaction if an ammonium compound and a hydroxide react together. The ammonia is produced when ammonium hydroxide formed in the double displacement decomposes. $NH_4Cl(aq) + NaOH(aq) \rightarrow NaCl(aq) + NH_4OH(aq)$ followed by $NH_4OH(aq) \rightarrow H_2O(\ell) + NH_3(g)$

38. b. React magnesium metal with hydrochloric acid: $Mg(s) + 2HCl(aq) \rightarrow MgCl_2(aq) + H_2(g)$. Add sodium phosphate solution to the magnesium chloride solution formed: $3MgCl_2(aq) + 2Na_3PO_4(aq) \rightarrow Mg_3(PO_4)_2(s) + 6NaCl(aq)$. Filter the resulting products to collect the magnesium phosphate precipitate.

40. $HCl(aq) + NaHCO_3(aq) \rightarrow NaCl(aq) + H_2O(\ell) + CO_2(g)$

Answers to Chapter 4 Self-Assessment Questions

1. e	**3.** e	**5.** b	**7.** b	**9.** c
2. b	**4.** d	**6.** d	**8.** a	**10.** d

12. a. silver chloride and lithium nitrate; $LiCl(aq) + AgNO_3(aq) \rightarrow AgCl(s) + LiNO_3(aq)$

    b. NR

    c. potassium chloride and iodine; $Cl_2(g) + 2KI(aq) \rightarrow 2KCl(aq) + I_2(s)$

    d. lead(II) sulfate and sodium nitrate; $Pb(NO_3)_2(aq) + Na_2SO_4(aq) \rightarrow PbSO_4(s) + 2NaNO_3(aq)$

14. a. A precipitate will form if calcium chloride or lead(II) acetate is present.

    b. $CaCl_2(aq) + Na_2SO_4(aq) \rightarrow CaSO_4(s) + 2NaCl(aq)$; $Pb(CH_3COO)_2(aq) + Na_2SO_4(aq) \rightarrow PbSO_4(s) + 2NaCH_3COO(aq)$

15. a. Water should be a liquid. The product should be potassium hydroxide rather than potassium oxide. $2K(s) + 2H_2O(\ell) \rightarrow 2KOH(aq) + H_2(g)$

    b. Lithium nitrate is soluble, so it should be marked aqueous rather than solid, and lead(II) chloride is not soluble, so it should be marked solid rather than aqueous. A coefficient is needed in front of lithium nitrate to balance the equation. $2LiCl(aq) + Pb(NO_3)_2(aq) \rightarrow 2LiNO_3(aq) + PbCl_2(s)$

23. a. $H_2CO_3(aq) + 2NaOH(aq) \rightarrow Na_2CO_3(aq) + 2H_2O(\ell)$

    c. $Na_2CO_3(aq) + Ca(OH)_2(aq) \rightarrow CaCO_3(s) + 2NaOH(aq)$

Answers to Unit 2 Review Questions

1. d	**3.** c	**5.** b	**7.** b	**9.** e
2. d	**4.** c	**6.** e	**8.** c	**10.** b

11. $3NaOH(aq) + AlCl_3(aq) \rightarrow 3NaCl(aq) + Al(OH)_3(aq)$

16. a. $Ca_3N_2(s) \rightarrow 3Ca(s) + N_2(g)$

    b. $4Cr(s) + 3O_2(g) \rightarrow 2Cr_2O_3(s)$

    c. $CH_4(g) + 2O_2(g) \rightarrow CO_2(g) + 2H_2O(g)$

    d. $2BaO(s) \rightarrow 2Ba(s) + O_2(g)$

26. b. The solution is acidic and will turn blue litmus paper red. $CO_2(g) + H_2O(\ell) \rightarrow H_2CO_3(aq)$

27. c

28. a. $2C_6H_6(\ell) + 15O_2(g) \rightarrow 12CO_2(g) + 6H_2O(\ell)$

    b. $2CO(g) + O_2(g) \rightarrow 2CO_2(g)$

    c. $Cl_2(g) + 2NaBr(aq) \rightarrow 2NaCl(s) + Br_2(g)$

    d. $CaCO_3(s) \rightarrow CaO(s) + CO_2(g)$

30. a. $Cu(NO_3)_2(aq)$; copper(II) nitrate

    b. $Cu(s) + 2AgNO_3(aq) \rightarrow 2Ag(s) + Cu(NO_3)_2(aq)$

    c. 188 g

31. d. $HCl(aq) + AgNO_3(aq) \rightarrow HNO_3(aq) + AgCl(s)$

32. a. NR. Neither of the possible products will form a precipitate.

    b. $Nb_2(SO_4)_5(aq) + 5Ba(NO_3)_2(aq) \rightarrow 2Nb(NO_3)_5(aq) + 5BaSO_4(s)$

    c. $SrBr_2(aq) + 2AgNO_3(aq) \rightarrow Sr(NO_3)_2(aq) + 2AgBr(s)$

45. Sample answer: The reactivity of sodium is too great for this process. You are counting on the reaction $Na(s) + AgNO_3(aq) \rightarrow NaNO_3(aq) + Ag(s)$. However, sodium is reactive enough to displace hydrogen from water by the reaction $2Na(s) + 2H_2O(\ell) \rightarrow 2NaOH(aq) + H_2(g)$. Besides the dangerous nature of the sodium itself, the sodium hydroxide is caustic, and the hydrogen is flammable.

48. b. $MgCO_3(s) \rightarrow MgO(s) + CO_2(g)$; $Mg(OH)_2(s) \rightarrow MgO(s) + H_2O(g)$

50. a. $CaCO_3(s) + H_2SO_4(aq) \rightarrow CaSO_4(aq) + H_2O(\ell) + CO_2(g)$

52. a. $4Fe(s) + 3O_2(g) \rightarrow 2Fe_2O_3(s)$

53. formula for DTBP: $C_8H_{18}O_2(g)$ or $(CH_3)_3COOC(CH_3)_3(g)$; complete combustion: $2C_8H_{18}O_2(g) + 23O_2(g) \rightarrow 16CO_2(g) + 18H_2O(g)$ A fuel that could power an engine in the absence of oxygen could be used in low-oxygen environments, such as to power a chainsaw for use by a firefighter within a burning building.

54. a. $Ni(s) + 4CO(g) \rightarrow Ni(CO)_4(g)$

55. b. $CuCO_3(s) + H_2SO_4(aq) \rightarrow CuSO_4(aq) + H_2O(\ell + CO_2(g)$; $CuSO_4(aq) + Fe(s) \rightarrow FeSO_4(aq) + Cu(s)$

Answers to Unit 2 Self-Assessment Questions

1. d	**3.** e	**6.** c	**8.** c	**10.** a
2. b	**4.** b	**7.** b	**9.** a	

12. a. $Br_2(\ell) + 2NaI(aq) \rightarrow 2NaBr(aq) + I_2(s)$

    b. $2Al(s) + 3Cu(NO_3)_2(aq) \rightarrow 3Cu(s) + 2Al(NO_3)_3(aq)$

    c. $2Fe_2O_3(s) \rightarrow 4Fe(s) + 3O_2(g)$

    d. $Cl_2(g) + 2NaBr(aq) \rightarrow 2NaCl(aq) + Br_2(\ell)$

    e. $6Li(s) + N_2(g) \rightarrow 2Li_3N(s)$

    f. $2AgNO_3(aq) + CaCl_2(aq) \rightarrow 2AgCl(s) + Ca(NO_3)_2(aq)$

15. a. $C_4H_8(g) + 6O_2(g) \rightarrow 4CO_2(g) + 4H_2O(g)$

    b. $2Al(s) + 3Br_2(\ell) \rightarrow 2AlBr_3(s)$

    c. $2RbNO_3(s) \rightarrow 2RbNO_2(s) + O_2(g)$

16. a. $2LiCl(\ell) \rightarrow 2Li(\ell) + Cl_2(g)$

    b. If water enters the reaction chamber, the lithium that forms might react with the water and produce flammable hydrogen gas, which could ignite and cause injury to the scientist. $2Li(\ell) + 2H_2O(\ell) \rightarrow 2LiOH(aq) + H_2(g)$

18. a. $SiO_2(s) + C(s) \rightarrow CO_2(g) + Si(\ell)$

20. b. $H_2SO_4(aq) + CaCO_3(s) \rightarrow H_2O(\ell) + CO_2(g) + CaSO_4(aq)$

23. a. $MgS(aq) + Cu(NO_3)_2(aq) \rightarrow CuS(s) + Mg(NO_3)_2(aq)$

    b. No reaction

    c. $Br_2(g) + 2KI(aq) \rightarrow 2KBr(aq) + I_2(s)$

    d. No reaction

Unit 3

Chapter 5

Answers to Caption Questions

Figure 5.1: A score is 20, a gross is 12×12 or 144, a great gross is twelve gross or 1728, a baker's dozen is 13, a paper bale is 5000 sheets of paper.

Figure 5.5: 6.02×10^{22} molecules

Figure 5.10: 0.496 mol

Answers to Learning Check Questions

1. Avogadro's constant is defined as exactly equal to the number of atoms of carbon-12 in 1 g of carbon-12. However, the numerical value must be determined experimentally. Scientists are constantly updating the value as they improve the methods used to determine the value.

2. A mole is the mass of particles (atoms, ions, molecules, or formula units) that contain the same number of particles as 12 g of the isotope carbon-12. The numerical value of the number of particles that constitute a mole is the same as the Avogadro constant.

3. You would have two times the Avogadro constant of hydrogen atoms. Rounded off, the number would be $2(6.02 \times 10^{23}) = 1.20 \times 10^{24}$ hydrogen atoms.

4. You would not be able to see one person, but a mole of people is so many that they would be visible, as a group, from space. In fact, a mole of people would have a mass about the same size as the mass of Earth.

5. four

6. Paper is often purchased in large quantities because people use paper in large quantities. If it was purchased by the dozen, people would have to calculate how many dozen sheets of paper would be enough to use, so measuring paper by the dozen is an inefficient measurement.

7. Oxygen was originally chosen by chemists as the reference for measuring atomic masses. However, with the development of the mass spectrometer, the reference was changed to carbon. Mass spectrometers accelerate particles in a vacuum, through a magnetic field, which causes a deflection in the path of the particles. A reference mass is needed but oxygen not practical. To create a vacuum, strong pumps must be used. A tiny amount of the carbon in the lubricants for the pumps always got into the vacuum. Most naturally occurring carbon is carbon-12. So, since it was already there, chemists decided to use carbon-12 as the universal reference for atomic masses.

8. A mole is an exceptionally huge number. It was designed specifically to apply to atoms and molecules. To try to use a mole for measuring objects larger than atoms and molecules is exceedingly impractical. All values would be expressed in minute fractions of a mole. For example, a dozen would be about 2×10^{-23} mol. A great gross would be 2.87×10^{-21} mol. It defies common sense.

9. Atomic molar mass refers to the mass of one mole of atoms, whereas molar mass is more general referring to the mass of one mole of any entity, such as atoms, molecules, formula units, etc.

10. The numerical value for the atomic mass in both units is the same.

11. 63.55 g

12. This value can also be described as the atomic molar mass, or just the molar mass. The units are g/mol, meaning the mass of one mole of atoms of a certain element.

13. The reported values of atomic mass are the weighted averages of the naturally occurring isotopes. In addition, the actual value of the masses of individual isotopes measured by a mass spectrometer are not whole numbers.

Answers to Practice Problem Questions

1. 9×10^{13} refrigerators

2. 1.6×10^{27} km

3. 1.91×10^{16} y

4. 5.8×10^{16} km high

5. 2.1×10^{11} Rogers Centres

6. 2.48×10^{15} rows

7. One mole of tablespoons has a volume of 9.03×10^{9} km^3. Because this volume is greater than the total volume of the oceans, you would drain the oceans. In fact, you could drain the equivalent of over six times the world's oceans.

8. 1.91×10^{12} dollars/s

9. The Earth is 4.1×10^{3} times heavier

10. 1.2×10^{16} cm

11. 6.38×10^{22} atoms

12. 5×10^{21} atoms

13. 5.1×10^{27} molecules

14. 5.11×10^{26} formula units

15. 3.15×10^{22} formula units

16. 2.32×10^{24} molecules

17. 1.3×10^{27} atoms

18. a. 2.90×10^{24} molecules
b. 1.45×10^{25} atoms

19. a. 4.36×10^{25} atoms
b. 2.18×10^{26} molecules

20. a. 0.015 mol
b. 1.05×10^{23} atoms of C

21. 0.158 mol

22. 0.277 mol

23. 2.0×10^{2} mol

24. 1.4×10^{-4} mol

25. 5.1×10^{4} mol

26. 27.5 mol

27. 2.0 mol

28. 0.0346 mol

29. 1.3×10^{3} mol

30. 0.106 mol

31. a. 22.99 g/mol
b. 183.84 g/mol
c. 131.29 g/mol
d. 58.69 g/mol

32. 123.88 g/mol

33. 310.2 g/mol

34. 331.2 g/mol

35. 113.94 g/mol

36. 306.52 g/mol

37. 315.51 g/mol

38. 283.88 g/mol

39. 392.21 g/mol

40. 132.91 g/mol

41. 182 g

42. 11 g

43. 0.231 g or 231 mg

44. 5.3×10^{2} mg

45. **a.** cobalt(II) nitrate 8.2×10^{-1} g
    b. lead(IV) thiosulfate 1.28×10^4 g
46. **a.** NH_4NO_3 3.9×10^2 g
    b. Fe_2O_3 2.59×10^3 g
47. 2.4×10^2 mg
48. 1.001 kg
49. **a.** Br_2
    b. $Sr(IO_3)_2$
50. aluminum iodate
51. 1.73 mol
52. 139 mol
53. 8.75×10^{-4} mol
54. 1.1×10^{-4} mol
55. **a.** SiO_2, 6.2×10^{-5} mol
    b. $Ti(NO_3)_4$, 0.08577 mol
    c. $In_2(CO_3)_3$, 4.70×10^{-5} mol
    d. 313 mol $CuSO_4 \cdot 5H_2O$,
56. 1.47 mol
57. 1.80×10^2 mol
58. 1.52×10^{-5} mol
59. $Al(OH)_3(s)$, $AgCl(s)$, $Ni(NO_3)_2(s)$
60. barium perchlorate, glucose, tin(IV) oxide
61. **a.** 3.52×10^3 g
    b. 616 g
    c. 7.00 g
    d. 1.8×10^6 g
62. **a.** 2.40×10^{22} formula units
    b. 4.80×10^{19} molecules
    c. 2.0×10^{26} formula units
    d. 1.75×10^{24} formula units
63. **a.** 9.52 g
    b. 2.51×10^{24} formula units
64. **a.** 2.1×10^{18} molecules
    b. 1.1 mg/day
65. 3.24×10^{-22} g
66. **a.** 2.58×10^{24} atoms
    b. 6.88×10^{24} atoms
    c. 8.60×10^{23} atoms
67. **a.** 1.26×10^{22} formula units
    b. 6.29×10^{22} ions
68. gallium arsenide
69. 0.37 g
70. $HCN(\ell)$, $CH_3COOH(\ell)$, $C_{12}H_{22}O_{11}(s)$

Answers to Section 5.1 Review Questions

4. 2.992×10^{-26} L
6. 3.48×10^{22} formula units of sodium chloride
7. **a.** 9.39×10^{22} atoms of gold
   b. 4.7×10^{24} formula units of magnesium chloride
   c. 9.15×10^{24} molecules of hydrogen peroxide

8. **a.** 4.5×10^{-4} mol
   b. 0.0117 mol
9. the sample of carbon
10. 2.37×10^{-3} mol of water
11. **a.** 1.6×10^{24} formula units
    b. 7.8×10^{24} atoms
    c. 3.1×10^{24} atoms
12. **a.** 2.8×10^{23} formula units
    b. 5.5×10^{23} atoms of chlorine
14. octane, then sodium hydrogen carbonate, then copper

Answers to Section 5.2 Review Questions

4. 0.241 mol
5. 1.0×10^5 g
6. 1.80×10^2 mol
7. 2.7×10^2 g
8. 5.2×10^4 g
9. 2.42×10^{23} atoms
10. aluminum sulfate
11. carbon dioxide
12. **a.** 2.1×10^{17}
    b. 6.3×10^{17} atoms
13. 3.1×10^{17} lead atoms translates into 0.11 mg/L, which is about ten times greater than the allowable lead limit. Therefore, the water is not safe to drink.
14.

Substance	Number of Particles	Amount (mol)	Mass (g)
$P_4(s)$	7.95×10^{24}	13.2	1.64×10^3
$Ba(MnO_4)_2(s)$	6.7×10^{20}	1.1×10^{-3}	0.42
$C_5H_9NO_4(s)$	8.027×10^{22}	0.1333	19.62

15. 3.49×10^{-4} g
16. $CH_3COOH(\ell)$, $HOOCCOOH(s)$, $C_6H_5COOH(s)$

Answers to Chapter 5 Review Questions

1. c   3. b   5. d   7. b
2. a   4. a   6. c   8. c
10. 65.41; 6.02×10^{23}
14. Avogadro constant, 6.02×10^{23}
17. **a.** 0.36 mol
    b. 15.8 mol
    c. 5.91×10^{-22} mol
18. **a.** 1.2 g
    b. 119 g
    c. 1.4×10^5 g
19. 1.94×10^{-3} mol
20. 1.2×10^4 g
21. 20.0 g glucose = 0.111 mol; 20.0 g propane = 0.453 mol
22. NCl_3

23. a. 0.06802 mol
   b. 0.08752 mol of octane
   c. 0.09331 mol of cysteine

25.

Substance	Total number of atoms	Number of molecules or formula units	Molar mass (g/mol)	Amount of substance (mol)	Mass (g)
$C_3H_6O_2(\ell)$	4.47×10^{23}	4.06×10^{22}	74.09	0.0675	5.00
$NaC_6H_5COO(s)$	3.56×10^{19}	2.37×10^{18}	144.11	3.94×10^{-6}	5.68×10^{-4}
$Al(H_2PO_4)_3(s)$	1.363×10^{24}	6.193×10^{22}	317.95	0.1029	32.71
$CCl_2F_2(g)$	2.38×10^{26}	4.75×10^{25}	120.9	78.9	9.54×10^3
$C_4H_{10}O_2(\ell)$	5.53×10^{23}	3.46×10^{22}	90.14	0.0574	5.17
$NaHCO_3(s)$	1.778×10^{24}	2.963×10^{23}	84.01	0.4921	41.34

35. Ti_2S_3, titanium(III) sulfide
36. a. 1.2×10^{-5} mol
37. a. 2×10^{-6} mol
38. a. 0.42 g
   b. 1.8×10^{-3} mol
39. a. 13 g of potassium nitrate, 0.28 g sodium fluoride
   b. 0.12 mol potassium nitrate, $6.5 \times 10-3$ mol sodium fluoride

Answers to Chapter 5 Self-Assessment Questions

1. e	**3.** b	**5.** c	**7.** a	**9.** d
2. d	**4.** c	**6.** d	**8.** c	**10.** b

12. 0.52 g
13. 5.27×10^{-4} mol of Zn
14. 3.80×10^2 g
15. b. 192.1 g/mol
16. 1.09×10^{21} molecules
17. 1.69×10^{22} carbon atoms; 9.87×10^{21} hydrogen atoms; 4.23×10^{21} chlorine atoms; 2.82×10^{21} oxygen atoms
18. a. 111.11 g/mol
   b. 1.20×10^{25} carbon atoms; 1.20×10^{25} hydrogen atoms; 4.82×10^{24} oxygen atoms; 2.41×10^{24} nitrogen atoms
21. a. 4.25 kg
   b. 14.5 kg
23. 57.51 g ethanol, 22.49 g water
25. 119 pg/day

Chapter 6
Answers to Caption Questions:

Figure 6.2: H_2O; two hydrogen atoms and one oxygen atom
Figure 6.4: The percent composition provides the basic information for determining the molecular formula of citric acid. From the formula a plan for synthesizing the compound can be created.
Figure 6.10: The numbers of atoms in the molecular formula are x times as great as the corresponding numbers in the empirical formula.

Answers to Learning Check Questions

1. Yes, water is the same regardless of the source, based on the law of definite proportions.

2. Sample answer: The elements can combine in different proportions to create different compounds with very different properties. For example: N_2O, NO_2, N_2O_5

3. The law is also called the law of definite composition or the law of constant composition because the chemical composition, including the ratio of elements in a compound, remains the same (that is, it is constant and definite).

4. The mass percent of carbon in carbon dioxide cannot change, based on the law of definite composition.

5. Carbon can be found in many different compounds with different formulas, so the mass percent can be different from one compound to the next.

6. a. 57.1 %.
   b. Carbon and oxygen have different mass percents because their molar masses are different.

7. There is one atom of carbon for every four atoms of hydrogen; that is, the ratio is 1:4.

8. Both provide proportions of elements found in the compound; however, the proportions in percentage composition are based on the overall mass of each element found in the compound, and the proportions in a molecular formula are based on the number of atoms of each element in the compound.

9. The molecular formula shows the actual number of atoms of each element in a molecule of the substance, whereas the empirical formula only tells you the ratios of the atoms in a molecule.

10. The molecular and empirical formulas are the same when the actual amounts of each element inside a compound are already in the lowest whole number ratio.

11. Every compound has its own set of unique properties that is a direct result of its structure and composition; a molecular formula reflects the specific composition of a compound, whereas an empirical formula can represent two or more compounds with the same lowest whole number ratio. For example, NO_2 and N_2O_4, have the same empirical formula (NO_2) but have different properties.

12. NO_2 and N_2O_4; N_2O_4 is a whole-number multiple of NO_2.

13. A molecular formula includes the actual numbers of each atom in one molecule of a substance; in an empirical formula, although the numbers of each atom are in the correct ratios, they may not be the actual numbers that occur in one molecule of substance.

14. Choose a standard mass for the substance, such as 100 g. Then use the given mass percents to calculate the amount of mass of each atom in the 100 g sample. Then use the molar mass of each atom to determine the number of moles of each atom. Finally, divide or multiply each number of moles by the correct factor to convert each number to the smallest whole numbers possible.

15. The molar mass is determined experimentally, usually using a mass spectrometer.

16. A hydrate has a number of H_2O units attached to its molecule; an anhydrate does not.

17. Water molecules add mass to a solid and this extra mass can affect measurements and calculations.

18. Heat the substance to drive off all water molecules in the hydrate. Once all the water has been driven off, calculate the difference between the initial and final mass.

Answers to Practice Problem Questions

1. 22.27%
2. 69.55%
3. 7.8%
4. 53.28%
5. 25.6%
6. 27%
7. $H_2Cr_2O_7(aq)$
8. $H_2SO_3(aq)$
9. 63.89% Cl
10. $ZnS(s)$, $Cu_2S(s)$, $PbS(s)$
11. 82% N; 18% H
12. 68.4% Cr; 31.6% O
13. 40.0% C; 6.7% H; 53.3% O
14. 48% Ni; 17% P; 35% O
15. 37.0% C; 2.20% H; 18.5% N; 42.3% O
16. 67.10% Zn; 32.90% S
17. 128 g Cu; 32.2 g S
18. 24.74% K; 34.76% Mn; 40.50% O
19. 10.1% C; 0.80% H; 89.1% Cl
20. No, the percentage composition of carbon in the sample is 64.8%. If the sample was were ethanol, the percentage composition of carbon would be 52.1%.
21. 63.14% Mn, 36.86% S
22. 93.10% Ag, 6.90% O
23. 2.06% H, 32.69% S, 65.25% O
24. 34.59% Al, 61.53% O, 3.88% H
25. 41.40% Sr, 13.24% N, 45.36% O
26. 73.27% C, 3.85% H, 10.68% N, 12.20% O
27. 205 kg
28. 127 kg
29. 17.1 g
30. 248 kg
31. CH_3
32. $MgCl_2$
33. $CuSO_4$
34. $K_2Cr_2O_7$
35. NH_3
36. Li_2O
37. BF_3
38. Cl_3S_5
39. Na_2CO_3
40. P_2O_5

41. $C_6H_{12}O_6(s)$
42. $C_8H_{10}(\ell)$
43. $C_4O_2H_{10}(\ell)$
44. $C_8H_8(s)$
45. $HgCl(s)$
46. $C_8H_{10}N_4O_2(s)$
47. $C_2H_5NO_2(s)$
48. $B_2H_6(g)$
49. C_3ONH_8; $C_6O_2N_2H_{16}$
50. Its empirical formula is $C_9H_{12}O$ and its molecular formula is $C_{18}H_{24}O_2$
51. 50.88%
52. 62.97%
53. 13.43%
54. 62.97%
55. $MgSO_4 \cdot 7H_2O(s)$, $Ba(OH)_2 \cdot 8H_2O(s)$, $CaCl_2 \cdot 2H_2O(s)$, $Mn(SO_4)_2 \cdot 2H_2O(s)$
56. 5
57. 4
58. $Cr(NO_3)_3 \cdot 9H_2O$
59. $MgI_2 \cdot 8H_2O$
60. 2.83 g

Answers to Section 6.1 Review Questions

4. 42.10% C, 6.49% H, 51.41% O
6. 36% Ca, 64% Cl
7. 0.32% H, 57.95% Au, 41.73% Cl
10. 27.74% Mg, 23.56% P, 48.69% O
11. 7.52 g
12. **a.** 40.04% Ca, 12.00% C, 47.96% O
13. C
14. Na^+

Answers to Section 6.2 Review Questions

3. SnO_2
4. $AlCl_3$
5. $KMnO_4$
6. As_2O_3
7. $PbCl_2$
9. 89.16%
10. 7
11. 6
13. **a.** CF_2
    b. twice as much
14. $C_{10}H_{16}N_2O_8$
15. C_3H_6O
16. $C_{24}H_{24}O_6$

Answers to Chapter 6 Review Questions

| 1. b | 3. a | 5. c | 7. a | 9. b |
| 2. d | 4. d | 6. e | 8. b | |

10. **a.** 1:1; 2 H : 2 O
    b. 1:2:1; 2 C : 4 H : 2 O
    c. 3:1:4; 3 Na : 1 P : 4 O
    d. 1:1:3; 1 Ag : 1 N : 3 O
19. Mg_2Cl
20. $HBrO_3$
21. $C_6H_{12}O_6$
22. $C_4H_6O_6$
23. $C_{20}H_{40}O_2$
24. $C_{12}H_{12}Cl_9F_6$
25. $C_6H_6O_2$
26. CH_2
27. 43.09%
28. 80.48 g/mol
29. barium chloride dihydrate $BaCl_2.2H_2O$
30. hematite, Fe_2O_3 (s)
31. 49.47% C, 5.20% H, 28.85% N, 16.48% O
32. **a.** $Ba(OH)_2.8H_2O$; 45.6%
    b. $Na_2CO_3.10H_2O$; 62.9%
    c. $CoCl_2.6H_2O$; 45.4%
    d. $FePO_4.4H_2O$; 32.3%
    e. $CaCl_2.2H_2O$; 24.5%
33. $C_4H_4O_2$
39. Fe_3O_4
40. **a.** trial 1: 80.4% Zn(s), 19.6% O(g) ;
    trial 2: 80.4% Zn(s), 19.6% O(g) ;
    trial 3: 86.2% Zn(s), 13.8% O(g) ;
    trial 4: 80.4% Zn(s), 19.6% O(g)
    b. discard trial 3
    d. $x = 1, y = 1$
41. **a.** 64.67% NiS, 43.93% NiAs, 30.47% $(Ni,Fe)_9S_8$
42. **a.** 10.35%, 89.65%
42. **a.** mass percent of methane = 10.4%; mass percent of water = 89.6%
42. **a.** 10.4% methane; 89.6% water
43. 11.50 g, 12.99 g, 15.99 g, 17.48 g, 18.98 g, 20.48 g, 26.46 g

Answers to Chapter 6 Self-Assessment Questions

1. a	**3.** e	**5.** d	**7.** e	**9.** d
2. d	**4.** d	**6.** b	**8.** b	**10.** d

14. $BaCO_3$
15. H_2SO_3
16. N_2O_3
17. $NaNO_2$
18. 7
19. C_8H_{20}
20. N_2O_2
21. 68.9%
22. **a.** $SiCl_3$
    b. Si_2Cl_6

23. **a.** NO_2
    b. N_2O_4
24. **a.** Sodium carbonate heptahydrate
    b. 54.34%
    c. 52 kg
25. Cu_2S

Chapter 7
Answers to Caption Questions
Figure 7.1 8 slices of toast, 8 turkey slices, 4 lettuce leaves, and 4 tomato slices
Figure 7.3 20 atoms of H, 10 atoms of O

Answers to Learning Check Questions

1. 10 slices of toast, 10 turkey slices, 5 lettuce leaves, and 5 tomato slices
2. The exact proportion of moles of each reactant and product is needed to determine the relative amounts of reactants and products in a complete reaction. The coefficients show the reacting molecular and mole ratios.
3. One mole of methane reacts completely with two moles of oxygen to produce one mole of carbon dioxide and two moles of water.
4. the relative number of moles of each reactant and product in a complete reaction
5. Because we are interested in the mole ratio for each reactant and product, not for each individual atom. Coefficients refer to relative amounts of each entire molecule, whereas subscripts refer to the relative amounts of each atom within a molecule.
6. **a.** 2 mol C_2H_6(g) : 4 mol CO_2(g) or 1:2
    b. 2 mol C_2H_6(g) : 7 mol O_2(g) or 2:7
    c. 4 mol CO_2(g) : 6 mol H_2O(g) or 2:3
7. The exact molar amount of a reactant or product, as predicted by a balanced chemical equation.
8. **a.** limiting reagent, gas; excess, oxygen in the air
    b. limiting reagent, deposits ($CaCO_3$(s)); excess, vinegar
    c. limiting reagent potato excess oxygen in the air
9. tomato
10. not necessarily; the limiting reactant is the one that is less than the stoichiometric amount (i.e., the reactant that would be used up while the other reactants are still available)
11. The amount in excess is not used in the reaction.
12. oxygen, because one expects there will be plenty of oxygen remaining after all the phosphorus has reacted with oxygen
13. The theoretical yield is the mass or amount of product calculated using the chemical equation and the associated reacting mole ratios. The actual yield is the mass or amount of product measured experimentally.
14. The theoretical yield is usually higher since some product is usually lost during the experiment no matter how carefully the experiment is done.

15. Improper lab techniques may reduce reaction yields in a number of ways. Some product may cling to various lab equipment and not be properly rinsed and collected. Spillage might occur. Measurements might not be made correctly.

17. If a reactant is not pure, then the actual mass of reactant that reacted is less than it should be; this will result in less product.

18. For example, a leak in a tank will cause the amount of fluid in the tank to decrease; a competing reaction drains away some of the desired product just like the leak in the tank.

If some of the turkey sandwiches are eaten there will be fewer on the plate to serve for lunch.

Possible answer: Children are eating cookies as they come out of the oven, reducing the actual yield.

Answers to Practice Problem Questions

1. 2 mol $Mg(s)$: 1 mol $O_2(g)$: 2 mol $MgO(s)$

2. 2 mol $NO(g)$: 1 mol $O_2(g)$: 2mol $NO_2(g)$

3. 1 mol $Ca(s)$ atom : 2mol $H_2O(\ell)$: 1 mol $Ca(OH)_2(s)$: 1 mol $H_2(g)$

4. 2 mol $C_2H_6(g)$: 7 mol $O_2(g)$: 4 mol $CO_2(g)$: 6 mol $H_2O(g)$

5. 5 molecules

6. 155 molecules of $AlCl_3(s)$

7. 3.4×10^{23}

8. 3.8×10^{24}

9. 2.72×10^{24}

10. 2

11. 0.25 mol

12. 6.00 mol

13. 4.50×10^4 mol

14. 3.6 mol $O_2(g)$

15. 4.70 mol

16. a. 46.8 mol
　　b. 187 mol

17. 56.5 mol $O_2(g)$

18. 6.45 mol of $P_4(s)$

19. 5.1×10^{23}

20. 7.24×10^5 mol of $H_2O(g)$

21. 1.3×10^5 g

22. 8.60×10^2

23. 726 g

24. 0.123 mol

25. 0.421 g

26. 4.35 g

27. 10.3 g

28. 0.963 g

29. 5.16×10^{-2} g

30. 21.0 g

31. $CaF_2(s)$

32. $C_7H_6O_3(aq)$

33. $H_2O(\ell)$

34. $NiCl_2(aq)$

35. $HNO_3(aq)$

36. $Li(s)$ is limiting, $O_2(g)$ is in excess

37. $Na_2S_2O_3(aq)$

38. $C_3H_6(g)$

39. 61.5 g

40. a. oxygen
　　b. 8

41. a. $AgNO_3(aq)$
　　b. 0.03532 mol

42. 66.98 g

43. 11.4 g

44. 172 g

45. 9.8 g

46. 57.3 g

47. 11.8 g

48. 69.94 g

49. a. 0.446 g
　　b. $F_2(g)$, 24.0 g

50. a. 388.4 g of $CaCl_2(aq)$ is excess (the total initial mass of $CaCl_2(aq)$ is 776.9 g)
　　b. 1189 g of $AgNO_3(aq)$
　　c. 1003 g of $AgCl(s)$ and 574.3 g of $Ca(NO_3)_2(aq)$

51. 0.97 g, 0.77 g

52. a. 37.3 g
　　b. 75.6 %

53. 79.3%

54. a. 8.32 g
　　b. 5.99 g

55. 44.6 g

56. 11.9 g

57. 61.8 g

58. 0.46 g

59. a. 1.51 kg
　　b. 1.18 kg

60. 35 g

Answers to Section 7.1 Review Questions

4. 6.0×10^2 g

5. 28 g

7. a. $Fe_2O_3(s) + 3CO(g) \rightarrow 2Fe(s) + 3CO_2(g)$
　　b. 526.2 kg or about ½ tonne

9. 2, 1, 1, 2; 1.20×10^{24}, 6.02×10^{23}, 6.02×10^{23}, 1.20×10^{24}; 162 g, 74.0 g, 200 g, 36.0 g

10. a. 1.7×10^{22}
　　b. 2.30×10^{22} formula units

Answers to Section 7.2 Review Questions

1. water

2. $O_2(g)$

3. $O_2(g)$

4. a. $FeCl_3(aq)$
　　b. 2.83 g $NaOH(aq)$ would remain
　　c. 37.8 g of $Fe(OH)_3(s)$, 62.0 g of $NaCl(s)$

5. 3.47 g

9. 15.7 g

10. b. 0.15 mol

Answers to Section 7.3 Review Questions

4. 94.60%

5. 91.9%

11. a. 3.4 g of I_2, 1.6 g of NaCl

    b. 1.0 g of NaCl

12. Theoretical Yield: 2.638 g $CuSO_4(s)$, 1.4890 g H_2O; Actual Yield: 2.913 g $CuSO_4(s)$, 1.214 g H_2O; Percentage Yield: 110.4% $CuSO_4(s)$, 81.5% H_2O.

Answers to Chapter 7 Review Questions

1. b	**3.** c	**5.** d	**7.** a
2. d	**4.** b	**6.** a	**8.** e

17. a. $2Al(s) + 3Br_2(g) \rightarrow 2AlBr_3(s)$

    b. 7.5 mol

    c. 5.0 mol

18. theoretical yield: 14.77 g; percent yield: 98.75%

19. 2.3 mol

20. 0.488 g, 93.2%

22. b. 3.86 g

24. 87.3%

Answers to Chapter 7 Self-Assessment Questions

1. b	**3.** a	**5.** c	**7.** e	**9.** a
2. b	**4.** e	**6.** a	**8.** b	

12. 17.81 g

13. a. 4.68×10^{-2} mol

    b. 4.5 mol

15. 93.2 kg

16. a. $3Zn(s) + 2FeCl_3(aq) \rightarrow 2Fe(s) + 3ZnCl_2(aq)$

    b. 272.73 g

    c. 479.12 g

19. b. 55.78 %

20. 0.31 g

22. 0.185 mol

23. O_2 limiting reactant

25. 12.5 mol

Answers to Unit 3 Review Questions

1. e	**3.** c	**5.** b	**7.** b
2. e	**4.** a	**6.** e	**8.** b

12. a. $N_2(g) + 3H_2(g) \rightarrow 2NH_3(g)$

    b. 1.07 mol; 2.16 g

14. 19 kg

15. a. $O_2(g)$

    b. 8.96 g

16. 17.2 g

17. 1.97×10^{23} g

    b. 1.66×10^{-22} mol per meal, 32.8 g per meal

    c. 2×10^{-11} mol

    d. 4.13×10^{12} g

18. 27 % K, 35 % Cr, and 38% O

19. 32.37% Na, 22.58% S, 45.05% O

20. a. $Pb_1S_2O_8$; $Pb(SO_4)_2$

    b. lead(IV) sulfate

21. Sb_2S_3

22. $Na_2C_8H_4O_4$

23. $C_3H_5O_2I$; $C_6H_{10}O_4I_2$

24. a. 40% C, 6.7% H, 53.3% O; CH_2O

    b. $C_5H_{10}O_5$

27. 0.3 g Al

32. a. $2C_8H_{18}(\ell) + 25O_2(g) \rightarrow 16CO_2(g) + 18H_2O(g)$

    b. 0.700 mol; 30.8 g

35. HCO_2

37. C_3H_6O

38. a. 115 kg

    b. 144 kg

    c. 228 kg

    d. 81.6 %

39. $Na_2SO_4 \cdot 10H_2O$, sodium sulfate decahydrate

40. 11.99 g Ag/1.0 g Al

41. $C_3H_8(g) + 5O_2(g) \rightarrow 3CO_2(g) + 4H_2O(g)$; $2C_8H_{18}(\ell) + 25O_2(g) \rightarrow 16CO_2(g) + 18H_2O(g)$

42. a. Two cartridges are needed; 15.1 kg H_2O

    b. 51.7 kg $CO_2(g)$

Answers to Unit 3 Self-Assessment Questions

1. d	**3.** b	**5.** b	**7.** c	**9.** b
2. c	**4.** a	**6.** d	**8.** e	**10.** c

11. 11.2 g

13. CH_2; C_6H_{12}

15. Cu_2S

16. $C_{20}H_{12}O_4$

17. 16.0 g

21. $C_{16}H_{18}N_2O_5$

23. 26.7%

Unit 4

Chapter 8

Answers to Learning Check Questions

1. The mixture is not homogeneous on the microscopic scale.

2. Nitrogen is the solvent, because it is present in the largest proportion.

3. Sample answers: solid solute: sugar in water; liquid solute: alcohol in water; gas solute: carbon dioxide in water.

4. Gasoline and water do not dissolve in one another (they are immiscible).

5. Yes. A solute is classified as insoluble if less than 0.1 g will dissolve in 100 mL of solvent. Therefore, if as much solute is dissolved in the solvent as possible, the solution will be dilute but saturated.

6. Yes. When the pressure is released by opening the can, the excess solute (carbon dioxide gas) leaves the solution.

7. No. Like dissolves like, so if benzene is a solvent for fats and oils, which are non-polar, it should not dissolve in a polar solvent such as water.

8. A fluoride ion has a very small radius. This results in most fluorides being less soluble than the other halides.

9. Methanol. The solubility of liquids is changed very little by temperature.

10. The taste is affected by how much carbon dioxide is dissolved in the soft drink. The solubility of carbon dioxide in water decreases at higher temperature.

11. Unsaturated. At 40°C, the solubility of $KNO_3(s)$ is about 53 g/100 g.

12. Fish avoid extremes of hot and cold, just as we do. In particular, warm water contains less dissolved oxygen, so fish are less likely to be found in these areas.

13. The concentration is likely to be more precise.

14. A volumetric flask and a volumetric pipette, or graduated pipette, can be used to prepare a solution with a known concentration.

15. 1. Measure the mass of pure solute and dissolve in water. Transfer to a 1 L volumetric flask and make up the solution to the calibration mark. 2. Dilute a more concentrated standard solution using a graduated pipette and a 1 L volumetric flask.

16. $n = c \times V$; the product of concentration and volume is constant. Thus, as the volume of solution increases, its concentration decreases.

17. Drying the beaker is not necessary in this case. The water in the beaker will dilute the solution slightly, but it will not change the amount of $CuSO_4(aq)$, which is determined by the volume of the pipette and the concentration of the solution in it.

18. First prepare a more concentrated solution and dilute it to the necessary concentration.

Answers to Practice Problems

1. 1.2×10^2% (m/v)
2. 8.6% (m/v)
3. 31.5 g
4. 8.0×10^1 g
5. 6.67% (m/v)
6. 1.75% (m/v)
7. 131 g
8. 3 g NaCl, 0.09 g KCl, 0.1 g $CaCl_2$
9. 800 mL
10. Measure 14 g of the solute and dissolve in water. Add water to bring the total volume of the solution to 400 mL.
11. 15% (m/m)
12. 3.8% (m/m)
13. 90 kg chromium, 40 kg nickel, 370 kg iron
14. 7.91% (m/m)
15. 15.1% (m/m)
16. 11 g
17. 5 g
18. 1.7% (m/m)

19. 7.23×10^{-4}% (m/m)
20. 0.243% (m/m) nickel, 2.3% (m/m) copper, 2.3×10^{-4}% (m/m) platinum
21. 400 mL
22. 20% (v/v)
23. 2 L
24. about 2.2 L
25. Add enough water to increase the volume to 1.00 L
26. 2 L
27. 9% (v/v)
28. 6 L
29. 4.0×10^1% (v/v), assuming that the mixed solution has a volume of 20.0 L
30. 74 mL
31. 7.2 ppm
32. 35 mg
33. 0.7 ppm
34. 2×10^{-6} g
35. 11.5 ppm
36. The concentration is 3.0 mg/L, within limits.
37. 0.1 g or 100 mg
38. 4 g
39. 0.52 μg or 5.2×10^{-7} g
40. 1×10^2 ppm or 100 ppm
41. **a.** 1.5 mol/L
    b. 0.2 mol/L
42. 1.4 L
43. **a.** 1 mol/L
    b. 0.628 mol/L
    c. 0.1 mol/L
44. 0.14 g
45. **a.** 0.093 g
    b. 10 g
    c. 534 g
46. **a.** $Na^+(aq)$ = 1.2 mol/L; $SO_4^{2-}(aq)$ = 0.60 mol/L
    b. $NH_4^+(aq)$ = 3.1 mol/L; $PO_4^{3-}(aq)$ = 1.0 mol/L
    c. $Ca^{2+}(aq)$ = 1×10^{-4} mol/L; $PO_4^{3-}(aq)$ = 8×10^{-5} mol/L
47. 0.29 mol/L
48. 12 mol/L
49. 218 mL
50. 2.0×10^{-5} mol/L
51. **a.** 33.3 mL
    b. 107 mL
    c. 25 mL
52. **a.** 0.99 mol/L
    b. 0.19 mol/L
    c. 0.0116 mol/L
53. 0.0938 L
54. 0.02 mol/L
55. **a.** 0.08 mol/L
    b. 0.2 mol/L

56. 15% (m/v)

57. 3.00×10^2 mL

58. 0.5 L; about 0.5 L

59. All parts: Your procedure should be similar to the procedure outlined in **Table 8.7**.

59. a. Mass 2.1 g $AgNO_3(s)$
 b. Mass 6.05 g $K_2CO_3(s)$
 c. Mass 12.6 g $KMnO_4(s)$

60. All parts: Your procedure should be similar to the procedure outlined in **Table 8.8**.
 a. Dilute 29 mL
 b. Dilute 7.5 mL
 c. Dilute 945 mL

Answers to Section 8.3 Review Questions

3. a. 0.04% (m/m)
 b. 4×10^2 ppm

4. 0.5 g

6. 23.9% (v/v)

7. 8.7 mol/L

8. K_3PO_4 or $K_3C_6H_5O_7$

9. 5.3×10^{14} kg

10. 0.7 mol/L

11. 0.098 mg

12. 45 ppb

13. No. The ground water concentration is over 4000 ppb.

14. 0.4 mol/L

Answers to Caption Questions

Figure 8.8 Sucrose is not an ionic compound.

Figure 8.11 The solubility is about 45 g KCl(aq) per 100 g water.

Figure 8.17 Iron is present in the greatest amount in the solution.

Figure 8.19 Sample answer: A pair of glasses has a frame and two lenses.

Figure 8.20 All of the flasks have long, thin necks, stoppers, and a graduation mark to indicate volume.

Answers to Section 8.4 Review Questions

3. 0.159 mol/L

4. 33 mL of stock solution is required

5. 3.00 L

6. 250 mL; 500 mL

7. 4.10 g

8. 7.5×10^9 L

11. 0.50 L of stock solution is required

12. 567 mL

14. 5.105 g of KHP is required

Answers to Chapter 8 Review Questions

1. c	**3.** d	**5.** e	**7.** d
2. b	**4.** d	**6.** e	**8.** c

18. a. 45 mL
 b. 64 mL

19. 3.9% (m/m) zinc, 1% (m/m) tin, 95% (m/m) copper

20. 4×10^{-5} mol

21. 55.49 mol/L

22. a. 0.204% (m/v)
 b. 8.60×10^2 ppm
 c. 0.14 mol/L
 d. 1.0×10^{-2} mol/L
 e. 4.5×10^{-2} mol/L

23. 2.2 L

24. b. approximately 90 g
 c. approximately 76°C

25. a. approximately 380 g
 b. 40°C

35. a. 0.1 ppb
 b. 5×10^{-10} mol/L

Answers to Chapter 8 Self-Assessment Questions

1. c	**3.** e	**5.** b	**7.** c	**9.** c
2. b	**4.** c	**6.** e	**8.** b	**10.** b

17. 85 g/100 mL

18. 50% (m/m)

19. 29 mL

20. 0.4 ppm

21. 200 L

22. 943 mL; about 2.06 L of distilled water

24. use 4.0 g of $KMnO_4(s)$

25. a. $AgNO_3(aq)$

26. 205 mL

Chapter 9
Answers to Learning Check Questions

1. Any insoluble substance, such as AgCl(s), $CaCO_3(s)$, or a molecular compound such as $H_2O(\ell)$ or $CO_2(g)$ is never shown as an ion in a net ionic equation.

2. A chemical equation shows reacting substances; an ionic equation shows soluble ionic substances in their dissociated form.

3. Atoms and electrons cannot be gained or lost during a chemical change.

4. There is no reaction because all the ions are soluble.

5. Flame test: the colour of the flame when a substance is burned can indicate what elements are present; solution colour: the colour of a solution can indicate what ions are present in the solution; precipitate: if a precipitate forms under specific reaction conditions, it is possible to deduce the ions that were present in the solution.

6. For most substances, there is no relationship between the colour of a solution, which is usually (but not always) due to certain anions, and the colour in a flame test, which is due to certain metal cations.

7. n is amount of a given substance (in moles), and V is volume of solution (in litres)

8. You need to know the molar ratios of the reactants and products.

9. The molar mass of the solute is needed.

10. The limiting reactant is the reactant that is completely consumed during a chemical reaction, limiting the amount of product that is produced

11. The cations and anions have equal but opposite charges.

12. The calculations are based on molar ratios and amounts in moles.

13. If your water is hard, you will have difficulty in producing a lather using soap, or you might observe lime build-up on a heating element. If your water is soft, you will find that soap lathers easily, and you should not have notable build-up on heating elements, dishes, or shower tiles.

14. $Ca^{2+}(aq)$ and $Mg^{2+}(aq)$ are present in greater concentrations in hard water.

15. The amount of calcium carbonate and magnesium carbonate in rocks varies in different areas; these ions dissolve in rainwater as it flows through the rocks.

16. Arsenic-containing soil was deposited as silt in a river delta. Deep wells penetrate below the layer of silt.

17. The ions $AsO_4^{3-}(aq)$ and $PO_4^{3-}(aq)$ have the same charge and are similar in radius. Thus, they should have similar solubility when combined with a cation.

18. The signs of dental fluorosis are streaking in the teeth, and especially brown mottling. You should suggest that the patient get his/her water tested for $F^-(aq)$, if using well water. You could also suggest using a fluoride-free toothpaste and mouth wash.

19. Temporary hardness is the result of dissolved $CaCO_3(aq)$ and $MgCO_3(aq)$ and can be removed by boiling water. Permanent hardness results from dissolved $CaSO_4(aq)$ and $MgSO_4(aq)$ and can only be removed by chemical means.

20. The salt is used to regenerate sodium ions on the resin beads inside the unit, after the beads have become coated with calcium or magnesium ions.

21. Oil can be supplied to both plants through the same pipeline. Also, waste heat from the electrical generating plant can supply some of the heat needed by the desalination plant.

22. Canada has sufficient supplies of fresh water.

23. There has been a steady increase in life expectancy since 1800, as water treatment techniques have improved.

24. Filtering, removing suspended particles, killing harmful bacteria.

Answers to Practice Problem Questions

1. **a.** $3Ba^{2+}(aq) + 2PO_4^{3-} \rightarrow Ba_3(PO_4)_2(s)$
2. $Sr^{2+}(aq) + SO_4^{2-}(aq) \rightarrow SrSO_4(s)$
3. $Mg^{2+}(aq) + 2OH^-(aq) \rightarrow Mg(OH)_2(s)$
4. $BaCl_2(aq) + Na_2SO_4(aq) \rightarrow BaSO_4(s) + 2NaCl(aq)$; $Ba^{2+}(aq) + SO_4^{2-}(aq) \rightarrow BaSO_4(s)$
5. The precipitate is $FeS(s)$. The spectator ions are $Na^+(aq)$ and $SO_4^{2-}(aq)$. The net ionic equation is: $Fe_2^+(aq) + S_2^-(aq) \rightarrow FeS(s)$
6. **a.** Spectator ions: $NH_4^+(aq)$ and $SO_4^{2-}(aq)$; net ionic equation: $3Zn_2^+(aq) + 2PO_4^{3-}(aq) \rightarrow Zn_3(PO_4)_2(s)$
   b. Spectator ions: $Li^+(aq)$ and $NO_3^-(aq)$; net ionic equation: $CO_3^{2-}(aq) + 2H^+(aq) \rightarrow CO_2(g) + H_2O(\ell)$

c. No spectator ions; $2H^+(aq) + SO_4^{2-}(aq) + Ba^{2-}(aq) + 2OH^-(aq) \rightarrow BaSO_4(s) + 2H_2O(\ell)$

7. $Pb_2^+(aq) + 2I^-(aq) \rightarrow PbI_2(s)$
8. Examples: $Al(NO_3)_3(aq)$ or $Al(CH_3COOH)_3(aq)$ and $Na_2Cr_2O_7(aq)$ or $(NH_4)_2Cr_2O_7(aq)$
9. $Fe_3^+(aq) + 3OH^-(aq) \rightarrow Fe(OH)_3(s)$; The spectator ions are $NO_3^-(aq)$ and $K^+(aq)$
10. **a.** $Pb(NO_3)_2(aq) + Na_2CO_3(aq) \rightarrow PbCO_3(s) + 2NaNO_3(aq)$; $Pb^{2+}(aq) + CO_3^{2-}(aq) \rightarrow PbCO_3(s)$
    b. $Co(CH_3COO)_2(aq) + (NH_4)_2S(aq) \rightarrow CoS(s) + 2NH_4CH_3COO(aq)$; $Co^{2+}(aq) + S^{2-}(aq) \rightarrow CoS(s)$
11. $NH_4^+(aq) = 0.4$ mol/L; $PO_4^{3-}(aq) = 0.1$ mol/L
12. 0.11 mol/L
13. $2Ag^+(aq) + CO_3^{2-}(aq) \rightarrow Ag_2CO_3(s)$; 0.239 mol/L
14. 1.10 g $FeS(s)$
15. 2.7 g
16. 24 g
17. 1.22 g
18. 0.0370 mol/L
19. **a.** 0.214 mol/L
    b. 12.5 g
20. **a.** 0.40 mol/L
    b. 16 g
21. 12 g
22. 0.518 g
23. 96.4 mL
24. 0.662 g
25. 4.11 g; The calculation assumes $PbI_2(s)$ is completely insoluble, whereas some lead(II) iodide will remain in solution.
26. 33.5 mL
27. 0.826 mol/L
28. 19.5 mL
29. 0.500 g
30. 0.0150 mol/L

Answers to Caption Questions

Figure 9.4: Sample answer: red: Li^+ and/or Sr_2^+, green: Cu_2^+ and/or Ba_2^+, orange: Ca_2^+ and/or Na^+

Figure 9.6: $CuCl_2$ is soluble, while $AgCl$ is not.

Figure 9.8: Over 99% of earth's water is unavailable for human consumption.

Figure 9.18: Benefits of a home water softener: reduced deposits on heating elements, better soap lather; Disadvantages: cost, maintenance, people on low-sodium diets cannot use a water softener

Answers to Section 9.1 Review Questions

2. **a.** $Cl^-(aq)$ and $NH_4^+(aq)$
   b. $NO_3^-(aq)$ and $Ba^{2+}(aq)$
   c. $Na^+(aq)$ and $Cl^-(aq)$
3. **a.** $3Cu^{2+}(aq) + 2PO_4^{3-}(aq) \rightarrow Cu_3(PO_4)_2(s)$
   b. $Al^{3+}(aq) + 3OH^-(aq) \rightarrow Al(OH)_3(s)$
   c. $Mg^{2+}(aq) + 2OH^-(aq) \rightarrow Mg(OH)_2(s)$

4. **a.** copper(II) carbonate, $CuCO_3(s)$
   b. $Cu^{2+}(aq) + CO_3^{2-}(aq) \rightarrow CuCO_3(s)$
   c. $Na^+(aq)$ and $SO_4^{2-}(aq)$

5. **a.** Sample answers: $BaCl_2(aq)$, $BaBr_2(aq)$, $BaI_2(aq)$, or $Ba(OH)_2(aq)$ and $Na_3PO_4(aq)$, $K_3PO_4(aq)$ or $(NH_4)_3PO_4(aq)$
   b. Sample answers: $MgCl_2(aq)$, $MgBr_2(aq)$, $MgI_2(aq)$, $MgSO_4(aq)$ and $NaOH(aq)$, $KOH(aq)$
   c. Sample answers: $Al(NO_3)_3(aq)$ or $Al(CH_3COO)_3(aq)$ and $Na_2Cr_2O_7(aq)$ or $K_2Cr_2O_7(aq)$

11. **a.** $Ca(OH)_2(aq) + CO_2(g) \rightarrow CaCO_3(s) + H_2O(\ell)$; $Ca^{2+}(aq) + 2OH^-(aq) + CO_2(g) \rightarrow CaCO_3(s) + H_2O(\ell)$

12. **a.** $Pb^{2+}(aq)$, $PbI_2(s)$
    b. $Pb^{2+}(aq) + 2I^-(aq) \rightarrow PbI_2(s)$

14. **a.** $H_2SO_4(aq)$
    b. $HCl(aq)$

Answers to Section 9.2 Review Questions

1. 0.10 mol/L $MgCl_2(aq)$ There are 2 mol of chloride ions for every 1 mol of magnesium chloride.

2. **a.** 0.5 mol/L
   b. 0.45 mol/L

3. 42 mL

4. 4.00×10^2 mL

5. 135 mL

6. **a.** 0.66 g $H_2(g)$
   b. 1.5 mol/L

7. **a.** $H_3C_6H_5O_7(aq)$ (citric acid)
   b. 0.238 g

8. $PbSO_4$; 2 g

9. 2×10^{-3} mol/L

10. 2.00×10^2 mL $CaCl_2(aq)$ and 99.9 mL $K_2CO_3(aq)$

12. 0.64 mL

Answers to Section 9.4 Review Questions

16. **a.** Sample answers: $NaCl(aq)$; $Na_2SO_4(aq)$
    b. Sample answers: $Pb^{2+}(aq) + 2Cl^-(aq) \rightarrow PbCl_2(s)$; $Pb^{2+}(aq) + SO_4^{2-}(aq) \rightarrow PbSO_4(s)$

Answers to Chapter 9 Review Questions

1. b	3. c	5. a	7. c
2. b	4. d	6. b	8. e

10. $Na^+(aq)$, $SO_4^{2-}(aq)$

11. **a.** $Cu(s) + Fe^{2+}(aq) \rightarrow Cu^{2+}(aq) + Fe(s)$
    b. $Sr^{2+}(aq) + SO_4^{2-}(aq) \rightarrow SrSO_4(s)$
    c. $H^+(aq) + OH^-(aq) \rightarrow H_2O(\ell)$

12. **a.** $Cu^{2+}(aq)$
    b. $Ca^{2+}(aq)$
    c. $Na^+(aq)$ and $MnO_4^-(aq)$

17. 0.010 mol

18. 1.5 mol/L Mg^{2+}; 3.0 mol/L NO_3^-; One formula unit of magnesium nitrate dissociates into one magnesium ion and two nitrate ions.

19. **b.** $Mg(OH)_2(s) + 2HCl(aq) \rightarrow MgCl_2(aq) + 2H_2O(\ell)$; $Mg(OH)_2(s) + 2H^+(aq) \rightarrow Mg^{2+}(aq) + 2H_2O(\ell)$

21. 0.015 mol/L

22. **a.** $Na^+(aq) = 5.91$ mol/L; $NO_3^-(aq) = 5.91$ mol/L
    b. $Ca^{2+}(aq) = 0.88$ mol/L; $CH_3COO^-(aq) = 1.8$ mol/L
    c. $NH_4^+(aq) = 1$ mol/L; $PO_4^{3-}(aq) = 0.4$ mol/L

23. 0.22 g $CaCO_3(s)$

24. $Zn(s) + Pb^{2+}(aq) \rightarrow Pb(s) + Zn^{2+}(aq)$; 0.096 mol/L

25. $Cu^{2+}(aq) + 2F^-(aq) \rightarrow CuF_2(s)$; 1.46 mol/L

26. **a.** $MgCl_2(aq) + 2NaOH(aq) \rightarrow Mg(OH)_2(s) + 2NaCl(aq)$
    b. $Mg^{2+}(aq) + 2OH^-(aq) \rightarrow Mg(OH)_2(s)$
    c. 0.9 g

35. 0.3 mol/L

36. **b.** $(NH_2)_2CO(aq) + H_2O(\ell) \rightarrow 2NH_3(aq) + CO_2(g)$; $2NH_3(aq) + 3O_2(g) \rightarrow 2NO_2^-(aq) + 2H^+(aq) + 2H_2O(\ell)$; $2NO_2^-(aq) + O_2(g) \rightarrow 2NO_3^-(aq)$

39. **a.** 4 L

Answers to Chapter 9 Self-Assessment Questions

1. c	3. a	5. b	8. b	10. d
2. e	4. d	7. a	9. e	

12. **a.** $2Na_3PO_4(aq) + 3Ca(OH)_2(aq) \rightarrow 6NaOH(aq) + Ca_3(PO_4)_2(s)$
    b. $3Ca^{2+}(aq) + 2PO_4^{3-}(aq) \rightarrow Ca_3(PO_4)_2(s)$

13. **a.** tin(II) phosphate, $Sn_3(PO_4)_2(s)$
    b. $K^+(aq)$ and $Cl^-(aq)$
    c. $3Sn^{2+}(aq) + 2PO_4^{3-}(aq) \rightarrow Sn_3(PO_4)_2(s)$

14. $Cr^{2+}(aq)$, $Cu^{2+}(aq)$; A bluish flame in a flame test would show that the solution contains copper(II) instead of chromium(II). Also, hydrogen sulfide could be added: copper(II) will form a precipitate with the sulfide ions.

15. Sample answers: Use $NaCl(aq)$: $Pb(NO_3)_2(aq) + 2NaCl(aq) \rightarrow PbCl_2(s) + 2NaNO_3(aq)$; Use $Na_2SO_4(aq)$: $Pb(NO_3)_2(aq) + Na_2SO_4(aq) \rightarrow PbSO_4(s) + 2NaNO_3(aq)$

16. 5.6 g

17. **a.** 2.0 mol
    b. 3×10^3 kg

18. 25 g

Chapter 10

Answers to Learning Check Questions

1. An Arrhenius acid is a substance that ionizes in water to produce one or more hydrogen ions, $H^+(aq)$. An Arrhenius base is a substance that dissociates in water to form one or more hydroxide ions, $OH^-(aq)$.

2. **a.** Sample answers: citrus fruit, tomatoes, vinegar, carbonated drink
   b. Sample answers: soap, detergent, ammonia solution (window cleaner), oven cleaner

3. Sample answer: Venn diagram: two circles that partially overlap, one labelled "Acids", one labelled "Bases". Within the overlapping section: conduct electricity and corrode tissues. In the "Acids" circle: sour taste, pH less than 7, turn litmus red, produce no colour change in phenolphthalein, corrode metals, react with active metals to produce hydrogen gas, react with carbonates to produce carbon dioxide gas. In the "Bases" circle: bitter taste, slippery texture, pH greater than 7, turn litmus blue, turn phenolphthalein pink, do not corrode metals, no reaction with active metals, no reaction with carbonates.

4. **a.** HBr(aq): acid
   b. KOH(aq): base
   c. H_3PO_4(aq): acid
   d. $HClO_4$(aq): acid
   e. $Ca(OH)_2$(aq): base
   f. HNO_3(aq): acid
   g. $Sr(OH)_2$(aq): base
   h. CsOH(aq): base

5. The paper will remain blue, because the solution is basic.

6. **a.** The H^+(aq) concentration is higher than the OH^-(aq) concentration.
   b. The H^+(aq) concentration is higher than the OH^-(aq) concentration.
   c. The OH^-(aq) concentration is higher than the H^+(aq) concentration.

7. A strong acid is one that ionizes completely into ions; for example, HCl(aq). A concentrated acid is one with a relatively large amount of acid dissolved in the solution; for example, 6 mol/L HCl(aq).

8. **a.** 0.01 mol/L NaOH(aq)
   b. 4 mol/L HF(aq)

9. Both acids are weak. Reasoning: Soft drinks are consumed by humans and it is dangerous and deadly to consume strong acids. Also, neither is on the list of strong acids.

10. The acids are listed in order of increasing strength. The addition of more oxygen atoms increases the polarity of the bond between the ionizable hydrogen atom and the oxygen atom it is attached to.

11. Your diagrams should be similar to **Figure 10.9** on page 461. They should show a high degree of dissociation in a strong base, compared with a high concentration of solute in a concentrated base, and a low degree of dissociation in a weak base, compared with a low concentration of solute in a dilute base.

12. The safety hazards associated with strong acids and bases are far greater than those associated with weak acids and bases. Strong acids and bases are highly corrosive and should never be consumed, while weak acids and bases are actually ingredients in some common foods and beverages.

13. A salt and water are produced by a neutralization reaction.

14. The pH at the equivalence point is usually not 7.0 for titrations involving either a weak base or a weak acid.

15. If both concentrations were unknown, you would have no way to calculate the concentration of either the base or the acid. The volume and concentration of the known solution and the volume of the unknown solution at the equivalence point are all needed to calculate the unknown concentration.

16. The term "neutralization reaction" refers to a reaction between an acid and a base. It does not mean that all of the hydrogen ions and all of the hydroxide ions have been neutralized.

17. The end point occurs when the indicator changes colour. The equivalence point occurs when stoichiometric quantities of acid and base have been mixed together.

18. Phenolphthalein changes colour in basic solution between pH 8.2 and pH 10.0. Thus, more of the weak base will be added than required for equivalence.

Answers to Practice Problem Questions

1. 2.12 mol/L
2. 67.5 mL
3. 0.1298 mol/L
4. 107 mL
5. 0.32 mol/L
6. 0.128 mol/L
7. 2×10^{-4} mol/L
8. 87.3 mL
9. 118 mL
10. 2×10^1 mL (Rounded to appropriate number of significant digits)

Answers to Caption Question

Figure 10.1 Both lemons and grapes are acidic.

Answers to Section 10.1 Review Questions

8. **a.** pH < 7
   b. pH > 7
10. $H_2O(\ell)$, H^+(aq), and Cl^-(aq) are present in an aqueous solution of hydrochloric acid.
13. NH_3(aq), CH_3COOH(aq), HCl(aq) Ammonia is a weak base, acetic acid is a weak acid, and hydrochloric acid is a strong acid.

Answers to Section 10.2 Review Questions

1. acid + base → salt + water
2. **a.** H_2SO_4(aq) + 2KOH(aq) → K_2SO_4(aq) + $2H_2O(\ell)$
   b. 2HI(aq) + $Mg(OH)_2$(aq) → MgI_2(aq) + $2H_2O(\ell)$
8. 0.5 mol
10. 21.3 mL
13. 6.554×10^{-2} mol/L

Answers to Chapter 10 Review Questions

| 1. d | 3. c | 5. a | 7. d |
| 2. a | 4. c | 6. b | 8. e |

9. Sample answers: sodium hydroxide, NaOH(aq), potassium hydroxide, KOH(aq), ammonia, NH_3(aq), magnesium hydroxide, $Mg(OH)_2$(aq)

10. a. Sample answers: hydrochloric acid, HCl(aq), hydroiodic acid, HI(aq)
   b. Sample answers: sulfuric acid, H_2SO_4(aq), carbonic acid, H_2CO_3(aq)
   c. Sample answers: phosphoric acid, H_3PO_4(aq), citric acid, $H_3C_6H_5O_7$(aq)

12. a. $Ca(OH)_2(aq) + 2HCl(aq) \rightarrow 2H_2O(\ell) + CaCl_2(s)$
   b. $2H_3PO_4(aq) + 3Sr(OH)_2(aq) \rightarrow 6H_2O(\ell) + Sr_3(PO_4)_2(s)$
   c. $2NaOH(aq) + H_2SO_4(aq) \rightarrow 2H_2O(\ell) + Na_2SO_4(s)$

18. $HOCl(aq) \rightarrow H^+(aq) + OCl^-(aq)$

20. 0.101 mol/L

23. 0.5 mol

24. a. 10.90 mL, 10.05 mL, 9.95 mL
   c. 0.21 mol/L

35. The pH of the solution is between 2.8 and 6.0.

37. a. 8×10^1 kg (Rounded to appropriate number of significant figures)
   b. 3.68 mol/L

39. 0.429 g

40. Volume of MOH neutralized should be 100.0 mL. Based on this revised information, the molar mass of MOH is 24.0 g/mol.

41. 67.6% (m/m)

Answers to Chapter 10 Self-Assessment Questions

1. b	**3.** b	**5.** e	**7.** c	**9.** b
2. a	**4.** e	**6.** b	**8.** b	**10.** d

22. The pH is less than 7.

24. 0.2479 mol/L

25. 0.585 mol/L NaOH(aq)

Answers to Unit 4 Review Questions

1. e	**3.** c	**5.** a	**7.** c
2. a	**4.** d	**6.** a	**8.** c

12. K^+(aq) and Cl^-(aq)

19. You could add sulfuric acid, which would precipitate Ca^{2+}, but not Mg^{2+}. Equation: $Ca(NO_3)_2(aq) + H_2SO_4(aq) \rightarrow CaSO_4(s) + 2HNO_3(aq)$

20. 3.0 mol/L

21. Assuming the density of solution is 1.0 g/mL, the atomic mass of X is 19 g/mol and the anion is F^-.

22. Precipitate: iron(III) hydroxide, $Fe(OH)_3$(s) Net ionic equation: $Fe^{3+}(aq) + 3OH^-(aq) \rightarrow Fe(OH)_3(s)$

23. 0.137 g

24. a. 0.3 mol/L
   b. 0.7 mol/L

26. 0.716 mol/L

29. The pH is less than 3.2.

30. a. 3.67 g

31. a. 4 L

32. 6×10^2 mL

33. 0.886 mol/L

49. 210 mL

50. 0.05 ppm

55. a. 3×107 L

57. a. 0.7660 mol/L
   b. 4.02% (m/v)

Answers to Unit 4 Self-Assessment Questions

1. b	**3.** d	**5.** d	**7.** c	**9.** c
2. b	**4.** e	**6.** c	**8.** b	**10.** c

14. 20 mL

15. 4.5×10^{19} t You need the density of sea water; assuming it is 1 g/mL, then the answer is as stated.

18. a. $Ba(NO_3)_2(aq) + Na_2S(aq) \rightarrow 2NaNO_3(aq) + BaS(s)$
   b. 0.2 mol/L

25. 76.9 mL

Unit 5

Chapter 11

Answers to Learning Check Questions

1. The solid state of water is commonly called ice. The liquid state is always called water and the gaseous state is called water vapour or steam.

2. Stronger the attractive force, more likely to be a solid or liquid at room temperature

3. The molecules in gases are more random than in a liquid, and in solid the molecules are more ordered than in a liquid.

4. Intermolecular forces hold the molecules together in an orderly fasion; without them the molecules would be free to move randomly about. Without intermolecular forces all substances would be gases.

5. The average kinetic energy of the particles increases as you go from solid to liquid to gas.

6. Graphic organizers should distinguish between electrostatic attraction, interactions between polar molecules, and interactions between non-polar molecules.

7. It is the force that a column of air exerts on the Earth's surface divided by the area of Earth's surface at the base of the column.

8. The height of mercury in a tube immersed in a shallow dish of mercury was dependent on atmospheric pressure. The higher the atmospheric pressure, the higher the level of mercury in the tube.

9. **a.** 363 kPa
   b. 1.39 atm
   c. 1.0×10^2 kPa
   d. 6.3×10^4 Pa

10. number of bars = number of psi $\times$ (1.01325/14.7)

11. As altitude increases, atmospheric pressure decreases. This means there is less oxygen available at a higher altitude and breathing may need to be assisted.

12. The force on the surface of the water in the pan due to atmospheric pressure is balanced by the pressure from the water in the bottle, which tends to push the water in the pan upward. When the water level in the pan falls below the water level in the bottle, the force from the water in the bottle is greater than the force due to atmospheric pressure, pushing the water level in the pan back up until it is level with the water in the bottle.

13. For a given amount of gas at constant pressure the volume of a gas varies directly with the temperature expressed in K.

14. Absolute zero is the extrapolated value for the volume-temperature graph when the volume of the gas is zero—an impossible situation for real gases. Kelvin suggested that this temperature, $-273.15°C$, is the lowest possible temperature value. A new temperature scale was created with 0 K as the starting temperature.

15. Linear lines with common x-intercept at $-273.15°C$

16. **a.** 300 K
    b. 2.48×10^2 K
    c. 100.1°C
    d. −249.6°C

17. Volume of a gas is inversely related to pressure of gas (Boyle's Law) and it, therefore, represents another variable that will affect the study of the gas.

18. The balloon quickly shrivels upon contact with the liquid nitrogen as molecular motion slows down with temperature decrease; the volume of the gas decreases.

Answers to Practice Problem Questions

1. 2.29 atm
2. 398 kPa
3. 24 atm
4. 0.27 L
5. 1.3×10^2 kPa
6. 440 L
7. 14.3 mL
8. 1.1×10^2 L
9. 1.73×10^3 L
10. **a.** 3.6×10^3 L
    b. 5.6×10^2 min
11. 11 L
12. 35.5 mL
13. 2.14 L
14. 95 L
15. 561 cm^3
16. 539 K
17. 308 K
18. 27°C
19. 1.25 times room temperature
20. −214°C
21. 42.0 kPa
22. 327 kPa
23. 2800 kPa

24. 430 kPa
25. **a.** 720 kPa
    b. about 7 times higher
26. 1.1 atm
27. 136 K
28. −150°C
29. 273°C
30. 273°C

Answers to Caption Questions

Figure 11.1 Atoms of hydrogen and atoms of chlorine have different electronegativities, causing a permanent dipole effect.

Figure 11.5 Increasing temperature causes the average kinetic energy of molecules to increase. They strike the surface of a container more often and harder, causing the volume of a flexible container to increase.

Figure 11.7 As atmospheric pressure changes, the level of mercury in the tube changes. Greater pressure causes the mercury column to rise and lower atmospheric pressure causes the column to lower.

Figure 11.14 If the Celsius temperature is doubled, the Kelvin temperature is not doubled. Instead it goes up by the amount of the difference between the two Celsius temperatures.

Figure 11.15 The x-intercept for the Celsius graph is -273 °C, and the y-intercept varies depending on initial pressure and volume. The x-and y-intercept for the Kelvin graph are both 0.

Figure 11.17 A relief valve is a safety device to prevent the cylinder from exploding. At a critical pressure the relief valve opens and permits the escape of gas, so that the pressure inside the tank does not increase to dangerous levels.

Answers to Section 11.2 Review Questions

5. **a.** 1.25 atm
   b. 1.5 bar
   c. 105 kPa
   d. 1.25 atm
11. 3.0 L
12. 417 kPa
13. 1000

Answers to Section 11.3 Review Questions

3. **a.** 311 K
   b. 395.6 K
   c. 233 K
   d. 1.85°C
   e. −99.55°C
   f. 6.00×10^2°C
8. 0.7 L
9. doubling of the temperature, to 60°C
11. 75.4 kPa
12. 6500 kPa
13. 17.4 psi

Answers to Chapter 11 Review Questions

1. c   **3.** d   **5.** c   **7.** e

2. d   **4.** b   **6.** b   **8.** e

13. a. 551 kPa
   b. pumping: 0.16 atm; resting: 0.11 atm
   c. 80 kPa
   d. 1520 mmHg
   e. 1.2 atm
   f. 101 kPa

15. 214 kPa

16. 0.16 L

18. a. 298 K
   b. 263.2 K
   c. 4°C
   d. −108°C

19. −25°C

20. b. 303 kPa

22. 923°C

24. 32 mL; −93°C

27. a. 59°C
   b. 170 cm^3

29. c. 1.82×10^3 cm^3; the volume changes by a factor of 1.52

Answers to Chapter 11 Self-Assessment Questions

1. b   **3.** e   **5.** a   **7.** b   **9.** b

2. a   **4.** a   **6.** a   **8.** a   **10.** c

16. a. 79.9 psi
   b. 310 mmHg
   c. 53 kPa
   d. 677 mmHg; 90.2 kPa

17. a. $x = 1.7 \times 10^3$ mL

18. 100 kPa

21. 2.1×10^2 mL

22. 3.1 L

23. 40 000 kPa

24. 11.7 psi

Chapter 12

Answers to Learning Check Questions

1. Boyle's and Charles's laws

2. Volumes of gaseous reactants and products are always in whole-number ratios, when measured at the same temperature and pressure.

3. 586 mL

4. The ratio of the volumes is the same as the ratios of the coefficients of the balanced equation.

5. b. 62 mL

6. 15 L

7. combined gas law and Avogadro's law

8. For using the universal gas constant value of 8.31 kPa·L/mol·K, the units to use are: pressure in KPa, temperature in K, and volume in L.

9. 280 kPa

10. helium: 0.24 atm; argon: 0.50 atm; neon: 1.3 atm

11. Humidity represents water vapour in the air, which contributes its own partial pressure to the total pressure.

12. 81.1 kPa

13. In the homosphere layer the gases are evenly mixed and behave like an ideal gas. In the heterosphere the gases are layered and more limited and include charged ionic gases.

14. The main gases present are nitrogen, oxygen, argon and carbon dioxide.

15. Criteria air contaminants are those pollutant gases that are considered to have the greatest impact on air quality and human health.

16. Convection occurs in the homosphere due to heating of the land by radiant energy. As gases are heated at ground level, they become less dense and will rise upwards while the colder, denser gases sink downwards.

17. AQHI stands for Air Quality Health Index. It is a scale used to indicate air quality and associated health risks.

18. Air quality in Ontario would reach levels over 7 when temperatures are high and air is not mixing. This would occur mostly in summer months and during periods of heavy traffic and industrial emissions.

Answers to Practice Problem Questions

1. 620°C

2. 13.6 ml

3. 0.214 mL

4. 22°C

5. 104 kPa

6. 1.2 atm

7. 21 mL

8. 57°C

9. 3.7×10^3 cm^3

10. 6×10^3 m^3

11. 5.9 g

12. 12 L

13. 17.7 L

14. 1.69 L

15. 141 L

16. 8.59 g; 3.03×10^{23} molecules

17. 26 L/mol

18. 25.9 L/mol

19. a. 1.46×10^{-3} mol
   b. 1.46×10^{-3} mol
   c. 24.5 L/mol

20. 22.3 L/mol

21. 140 L

22. 1.1×10^3 L

23. 166°C

24. 118 kPa

25. 160 g

26. 60 g/mol

27. 83.8 g/mol; element is most likely krypton

28. 1.775 g/L

29. a. C_2H_5
　　b. 58 g/mol
　　c. C_4H_{10}

30. C_8H_{18}

31. 36 mL

32. 16 L

33. 69 L

34. 533 L

35. 15.6 L

36. 0.787 L

37. 2.73×10^{-3} g

38. 0.32 g

39. 0.16 g

40. 0.16 L

Answers to Caption Questions

Figure 12.5 The pressure would be 2.0 atm instead of 1.5 atm.

Figure 12.6 If any air were already in the cylinder before the gas was collected, the measured volume of the collected gas would be incorrect.

Figure 12.9 Gay-Lussac's law; could be placed connecting to the combined gas law

Answers to Section 12.1 Review Questions

4. 84°C

5. 74.8 kPa

6. 4×10^3 hours

7. 29 L

8. 30 L

9. 7.0×10^2 mL

11. 0.14 mol

12. a. 0.446 moles
　　b. 17.8 g
　　c. 2.69×10^{23} atoms
　　d. 22.4 L/mol

14. 42.9 L/mol

Answers to Section 12.2 Review Questions

2. a. $V = \dfrac{nRT}{P}$

　　b. $P = \dfrac{nRT}{V}$

　　c. $T = \dfrac{PV}{nR}$

　　d. $n = \dfrac{PV}{RT}$

3. 8.6×10^3 L

4. 8.3 kPa

5. 90.2 g/mol

6. 38.1 g/mol, F_2

7. 1.2 g/L

8. C_2H_2

9. 97.5 kPa

11. 109 g

12. 8.59 L

13. 3.2 L

Answers to Chapter 12 Review Questions

1. a	**3.** b	**5.** e	**7.** a
2. c	**4.** b	**6.** c	**8.** d

11. 127 L

12. 4900 kPa

13. 900 kPa

14. 2.75 g

15. 1.0×10^2 kPa

16. b. 1 mol
　　c. 30 g

17. a. 140 g/mol
　　b. 120 g/mol

18. 20.2 g/mol; neon

19. $C_4H_{12}O_2$

20. 1.4 g/L

21. 97.5 kPa

22. a. $4NH_3(g) + 5O_2(g) \rightarrow 4NO(g) + 6H_2O(g)$
　　b. 2.5×10^3 L
　　c. 2.0×10^3 L
　　d. 3.0×10^3 L
　　e. 5.0×10^3 L

23. a. $C_3H_8(g) + 5O_2(g) \rightarrow 3CO_2(g) + 4H_2O(g)$
　　b. 1 L
　　c. 3 L
　　d. 4 L

24. a. $NH_4NO_2(s) \rightarrow N_2(g) + 2H_2O(g)$
　　b. 4.7×10^{-3} mol

25. 69 L

26. 0.25 g Mg; 6.7 mL HCl

34. a. 2.2×10^6 g
　　b. 7.1×10^6 g

Answers to Chapter 12 Self-Assessment Questions

1. c	**3.** e	**5.** d	**7.** c	**9.** d
2. a	**4.** d	**6.** e	**8.** e	**10.** b

12. 20 m³

14. 24.4 L/mol

15. a. 1.1×10^6 g
　　b. 8.9×10^5 L

17. molar mass 2.0 g/mol; H_2

18. 30.5 kPa

20. $2N_2(g) + O_2(g) \rightarrow 2N_2O(g)$; volume of $N_2(g)$: 34 L; volume of $O_2(g)$ 17 L

21. 6.0×10^3 g

Answers to Unit 5 Review Questions

1. b	**3.** e	**5.** e	**7.** e	**9.** b
2. e	**4.** b	**6.** c	**8.** b	**10.** d

18.

Pressure Unit Conversions

Pressure (kPa)	Pressure (mmHg) or Torr	Pressure (atm)	Pressure (PSI) lbs/inch
102	765	1.01	14.8
3.0×10^2	2300	3.0	44
2.76×10^2	2.07×10^3	2.72	40.0

20. 617 cm^3

21. a. 12 psi

22. 22 atm

23. $-18°C$

24. a. 2.8 g/L
  b. 0.50 g/L
  c. 0.635 g/L

25. 1.2 L/mol

26. $1.8 \times 10^4 \text{ kg}$

27. $1.3 \times 10^4 \text{ kPa}$

28. a. 0.011 mol
  b. 0.50 g
  c. 6.8×10^{21}

29. Molar mass is 38 g/mol and identity is F_2

30. a. 98 kPa
  b. 4.3×10^{-3} mol
  c. 4.3×10^{-3} mol
  d. 84 g/mol
  e. X is magnesium. The carbonate formula is $MgCO_3$.

31. a. empirical formula: C_1H_1
  b. molar mass: 78 g/mol
  c. molecular formula: C_6H_6

32. a. $3CO(g) + 7H_2(g) \rightarrow C_3H_8(g) + 3H_2O(g)$
  b. $1.050 \times 10^3 \text{ L}$
  c. 466.7 L
  d. 38.6 L

33. mass of $CaCO_3(s)$ in tablet: 0. 40 g; 81% (m/m)

34. 180 L

35. a. Ammonium nitrate produces 310 L of nitrous oxide gas, while ammonia produces 730 L. The volume of nitrous oxide produced from ammonia is over twice that produced by the same mass of ammonium nitrate.
  b. 2500 L $N_2O(g)$, 7500 L $H_2O(s)$
  c. $4.8 \times 10^5 \text{ g}$

36. a. $2.7 \times 10^4 \text{ L}$
  b. $1.2 \times 10^4 \text{ L}$
  c. $3.6 \times 10^4 \text{ L}$
  d. $7.5 \times 10^4 \text{ L}$

37. a. lithium hydride reaction: $LiH(s) + H_2O(\ell) \rightarrow LiOH(aq) + H_2(g)$; magnesium hydride reaction: $MgH_2(s) + 2H_2O(\ell) \rightarrow Mg(OH)_2(aq) + 2H_2(g)$
  b. 33 g $LiH(s)$, 54 g $MgH_2(s)$

50. b. lung volume at a depth of 500.0 m: $1.1 \times 10^2 \text{ L}$; lung volume at a depth of $1.0 \times 10^3 \text{ m}$: 51 L

Answers to Unit 5 Self-Assessment Questions

1. a   **3.** c   **5.** c   **7.** d   **9.** e
2. a   **4.** c   **6.** b   **8.** e   **10.** d

13. a. height of liquid gallium: 1.68×10^3 mmHg

14. 27 L

16. 21°C

18. $7.0 \times 10^3 \text{ kPa}$

19. b. 1.2 mol
  c. 32 g
  d. 7.5×10^{23} molecules

20. 2700 m^3

22. a. 97.8 kPa
  b. 0.0103 mol $CO_2(g)$
  c. 0.0103 mol $X_2CO_3(s)$
  d. molar mass of carbonate: 138 g/mol
  e. The element represented by X is potassium.

Glossary

How to Use This Glossary

This Glossary provides the definitions of the key terms that are shown in **boldface** type in the text. Definitions for terms that are *italicized* within the text are included as well. Each glossary entry also shows the number(s) of the sections where you can find the term in its original context.

1.1 = Chapter 1, Section 1   C3R = Chapter 3 Review   App. B = Appendix B
U2P = Unit 2 Project   App. A = Appendix A

A pronunciation guide, using the key below, appears in square brackets after selected words.

a = mask, back	i = simple, this	uhr = insert, turn
ae = same, day	ih = idea, life	s = sit
ah = car, farther	oh = home, loan	z = zoo
aw = dawn, hot	oo = food, boot	zh = equation
e = met, less	u = wonder, Sun	
ee = leaf, clean	uh = taken, travel	

Emphasis is placed on the syllable(s) in CAPITAL letters.

A

absolute zero the lowest theoretical temperature, equivalent to −273.15°C; the temperature at which the volume of a gas approaches zero (11.3)

acid-base indicator a substance that changes colour beyond a threshold pH level (10.1)

actual yield the actual amount of product that is recovered aft er a reaction is complete (7.3)

activity series a ranking of the relative reactivity of metals or halogens in aqueous reactions (4.1)

alkali a base that is soluble in water (2.2)

algal bloom rapid growth of large quantities of algae (9.3)

alloy a solid solution of two or more metals (8.3)

aqueous solution a solution that contains water (8.1)

Arrhenius theory of acids and bases an acid-base theory stating that acids ionize to produce hydrogen ions in solution and bases dissociate to produce hydroxide ions in solution (10.1)

atmospheric pressure the force exerted on Earth's surface by a column of air of a given area (11.2)

atomic mass unit one twelft h o f t he m ass of a carbon-12 atom; unit u (1.1)

atomic molar mass the mass of one mole of atoms of any element in the periodic table; unit g/mol (5.2)

atomic radius the distance from the centre of an atom to the boundary within which the electrons spend 90% of their time (1.3)

average atomic mass the weighted average of the masses of all the isotopes of an element based on their isotopic abundances; given in atomic mass units (u) (1.1)

Avogadro constant the number of particles in one mole of a substance; a value that is equal to 6.02×10^{23} particles (5.1)

Avogadro's hypothesis see *Avogadro's law* (12.1)

Avogadro's law (hypothesis) a gas law stating that equal volumes of all ideal gases at the same temperature and pressure contain the same number of molecules (12.1)

B

balanced chemical equation a statement that uses chemical formulas and coefficients to show the identities and ratios of the reactants and products in a chemical reaction (3.1)

beta particle a negatively charged particle with a mass and charge of an electron, that is emitted from the nucleus of some radioisotopes (1.1)

binary ionic compound an ionic compound that consists of atoms of only two (bi-) different elements (2.2)

blue-baby syndrome a condition in babies less than 3 months old, in which tissues become starved for oxygen due to excess nitrate ions in drinking water (9.3)

boiling point the temperature at which a compound changes from a liquid to a gas (2.3)

bond dipole polar covalent bonds that have a positive "pole" and a negative "pole" (2.1)

bonding pair a pair of electrons that is shared by two atoms, thus forming a covalent bond (2.1)

Boyle's law a gas law stating that the volume of a fixed amount of gas at a constant temperature is inversely proportional to the applied (external) pressure on the gas: $V \propto \frac{1}{P}$ (11.2)

burette a clear tube with volume marking along its length and a tap at the bottom (10.2)

C

Charles's law a gas law stating that the volume of a fixed amount of gas at a constant pressure is directly proportional to the Kelvin temperature of the gas: $V \propto T$ (11.3)

chemical equation a condensed statement that expresses chemical change using symbols and chemical names or formulas (3.1)

chemical reaction a process in which substances interact, causing different substances with different properties to form (3.1)

coefficient in a balanced chemical equation, a positive number that is placed in front of a formula to show the relative number of particles of the substance that are involved in the reaction (3.1)

combined gas law a gas law stating that the pressure and volume of a given amount of gas are inversely proportional to each other and directly proportional to the kelvin temperature of the gas: $V \propto \frac{T}{P}$

combustion reaction the reaction of a substance with oxygen, producing one or more oxides, heat, and light (3.3)

competing reaction a reaction that occurs along with the principal reaction and that involves the reactants and/or products of the principal reaction (7.3)

concentrated having a high ratio of solute to solution (8.3)

concentration the ratio of the quantity of solute to the quantity of solvent or the quantity of solution (8.3)

convection is a cyclical, up-and-down movement of air caused by differences in density; a method of heat transfer (12.3)

covalent bond the attraction between atoms that results from the sharing of electrons (2.1)

criteria air contaminants air pollutants that are federally identified as those having the greatest impact on air quality and human health (12.3)

Dalton's law of partial pressures a gas law stating that the total pressure of a mixture of gases is the sum of the individual pressures of each gas (12.2)

decomposition reaction a chemical reaction in which a compound breaks down into elements or simpler compounds (3.2)

desalination the process of obtaining fresh water from salt water (9.4)

dissociation the process in which ions break apart when dissolved in solution (10.1)

dilute having a low ratio of solute to solution (8.3)

dipole a molecule with a slightly positively charged end (positive pole) and a slightly negatively charged end (negative pole) (2.3)

dipole-dipole force the attractive force between the positive end of one molecule and the negative end of another molecule (2.3)

diprotic acid an acid having two hydrogen atoms that can ionize in solution (10.1)

double bond a covalent bond that results from atoms sharing two pairs of electrons (2.1)

double displacement reaction a chemical reaction in which the positive ions of two ionic compounds exchange places, resulting in the formation of two new ionic compounds (4.2)

effective nuclear charge the apparent nuclear charge, as experienced by the outermost electrons of an atom, as a result of the shielding by the inner-shell electrons (1.3)

electrical conductivity the ability of a substance or an object to allow an electric current to exist within it (2.3)

electrolysis a process that uses electrical energy to cause a chemical reaction (3.2)

electron affinity the energy absorbed or released when an electron is added to a neutral atom (1.3)

electron pairs two electrons that are interacting in a unique way, allowing them to be situated close to each other (1.1)

electronegativity an indicator of the relative ability of an atom to attract shared electrons (1.3)

electronegativity difference the difference between the electronegativities of two atoms (2.1)

empirical formula a formula that shows the smallest whole-number ratio of the elements in a compound (6.2)

end point the point at which the indicator in a titration changes colour (10.2)

equivalence point the point at which the amount of titrant is just enough to react with all the reactant in the sample (10.2)

excess reactant a reactant that remains after a reaction is over (7.2)

exothermic a chemical process during which heat is released (8.4)

filtrate a filtered solution (9.1)

formula unit the representative particle of a pure ionic compound (5.1)

formula unit molar mass the mass of one mole of formula units of a substance; unit g/mol (5.2)

flame test qualitative analysis that uses the colour that a sample produces in a flame to identify the metal ion(s) in the sample (9.1)

fresh water water that is not salty (9.3)

Gay-Lussac's law a gas law stating that the pressure of a fixed amount of gas at a constant volume is directly proportional to its Kelvin temperature: $P \propto T$ (11.2)

graduation mark a mark on a volumetric flask and other lab ware that indicates the level of volume (8.4)

ground water water that seeps through the ground below the surface (9.3)

group a column in the periodic table (1.2)

hard water water that contains relatively large concentrations of ions that form insoluble compounds with soap (9.3)

heat pollution the release of waste heat into the environment (8.2)

hydrate a compound that has a specific number of water molecules bound to each formula unit (6.2)

hydration the process in which water molecules surround the molecules or ions of a solute (8.2)

hydrocarbon a compound that is composed only of the elements carbon and hydrogen (3.3)

hydrogen bonding the attraction between a hydrogen atom on one molecule and a small, very electronegative atom on another molecule (8.2)

ideal gas a hypothetical gas made up of particles that have mass but no volume and no attractive forces between them (11.1)

ideal gas law a gas law that describes the relationship among volume, pressure, temperature, and amount (in moles) of an ideal gas: $PV = nRT$ (12.2)

immiscible describes substances that do not mix completely with each other (11.1)

inner transition elements elements 57 through 71 and 89 through 103, or Group 3 of Periods 6 and 7; usually written in two rows at the bottom of a periodic table (1.2)

insoluble the attribute given to a solute if less than 0.1 g of the solute will dissolve in 100 mL of solvent (8.1)

intermolecular forces attractive forces that act among molecules including the dipole-dipole force and a weak attractive force between non-polar molecules (2.3)

ionic bond the attractive electrostatic force between a negative ion and a positive ion (2.1)

ionic compound a chemical compound composed of ions that are held together by ionic bonds (2.1)

ionic equation a chemical equation in which soluble ionic substances are written in dissociated form (9.1)

ionization the process of forming ions (10.1)

ionization energy the amount of energy required to remove the outermost electron from an atom or ion in the gaseous state (1.3)

isotopes atoms that have the same number of protons but different numbers of neutrons (1.1)

isotopic abundance the amount of a given isotope of an element that exists in nature, expressed as a percentage of the total amount of this element (1.1)

kinetic molecular theory of gases the theory that explains gas behaviour in terms of the random motion of particles with negligible volume and negligible attractive forces (11.1)

law of combining volumes a gas law stating that when gases react, the volumes of the gaseous reactants and products, measured under the same conditions of temperature and pressure, are always in whole-number ratios (12.1)

law of definite proportions a law stating that a chemical compound always contains the same proportions of elements by mass (6.1)

leaching a process that is used to extract a metal by dissolving the metal in an aqueous solution (4.3)

Lewis diagram a model of an atom that has the chemical symbol for the element surrounded by dots to represent the valence electrons of the element (1.1)

Lewis structure a Lewis diagram that portrays a complete molecular compound (2.1)

limiting reactant a reactant that is completely consumed during a chemical reaction, limiting the amount of product that is produced (7.2)

lone pair a pair of electrons that is not part of a covalent bond (2.1)

main-group elements elements in Groups 1, 2, and 12 through 18 in the periodic table (1.2)

mass percent the mass of an element in a compound, expressed as a percentage of the total mass of the compound (6.1)

matte an impure metallic sulfide mixture that is formed by smelting the sulfide ores of metals (4.3)

maximum allowable concentration (MAC) a drinking water standard for a substance that is known or suspected to affect health when above a certain concentration (9.3)

melting point the temperature at which a compound changes from a solid to a liquid (2.3)

meniscus the curved surface of a solution in a lab ware (8.4)

miscible describes substances that are able to mix completely with each other (11.1)

molar concentration the amount (in moles) of solute dissolved in 1 L of solution; also called *molarity* (8.3)

molar mass the mass of one mole of a substance; the symbol for molar mass is M, and the unit is g/mol (5.2)

molar volume amount of space occupied by 1 mol of a gaseous substance; 22.4 L/mol for an ideal gas at standard temperature and pressure and 24.8 L/mol for an ideal gas at standard ambient temperature and pressure (12.1)

molarity *see molar concentration* (8.3)

mole the SI base unit that is used to measure the amount of a substance; it contains as many particles (atoms, molecules, ions, or formula units) as exactly 12 g of the isotope carbon-12 (5.1)

mole ratio the ratio of the amounts (in moles) of any two substances in a balanced chemical equation (7.1)

molecular compound a chemical compound that is held together by covalent bonds (2.1)

molecular formula the formula for a compound that shows the number of atoms of each element that make up a molecule of that compound (6.2)

molecular molar mass the mass of one mole of molecules of a substance; unit g/mol (5.2)

mostly ionic bond a bond in which the electronegativity difference of the atoms is greater than 1.7 (2.1)

multivalent metal a metal that can form ions with more than one charge (3.2)

net ionic equation an ionic equation that does not include spectator ions (9.1)

neutralization reaction a reaction between an acid and a base (10.2)

neutralization the process of making a solution neutral (pH = 7) by adding a base to an acidic solution or by adding an acid to an alkaline (basic) solution (4.2)

non-polar covalent bond a bond in which the electronegativity difference of the atoms is zero (2.1)

octet rule a "rule of thumb" that allows you to predict the way in which bonds will form between atoms (2.1)

oxoacid an acid composed of hydrogen, oxygen, and atoms of at least one other element (2.2)

partial pressure the portion of the total pressure of a mixture of gases contributed by a single gas component (12.2)

parts per billion (ppb) a ratio of solute to solution $\times\ 10^9$ (8.3)

parts per million (ppm) a ratio of solute to solution $\times\ 10^6$ (8.3)

percent (m/m) a ratio of mass of solute to mass of solution, expressed as a percent; also referred to as *weight/weight,* or *(w/w) percent* (8.3)

percent (m/v) a ratio of the mass of solute to the volume of solution, expressed as a percent (8.3)

percent (v/v) a ratio of the volume of solute to the volume of solution, expressed as a percent (8.3)

percent concentration a ratio of solute to solution; commonly expressed as a percentage; see *percent (m/v), percent (m/m),* and *percent (v/v)* (8.3)

percent covalent character a measure of the degree to which electrons in a bond are shared equally, related to the electronegativity difference; given as a percentage (2.1)

percent ionic character a measure of the degree to which electrons in a bond are not shared equally, related to the electronegativity difference; given as a percentage (2.1)

percentage composition the percent by mass of each element in a compound (6.1)

percentage yield the actual yield of a reaction, expressed as a percentage of the theoretical yield (7.3)

period a row in the periodic table (1.2)

periodic law a statement that describes the repeating nature of the properties of the elements (1.2)

permanent hardness hardness that cannot be removed from water by boiling (9.4)

pH scale a scale used to describe the acidity or basicity of a solution (10.1)

polar covalent bond a covalent bond around which there is an uneven distribution of electrons, making one end slightly positively charged and the other end slightly negatively charged; also, bond in which the electronegativity difference is between 0.4 and 1.7 (2.1)

pollution non-point source a source of pollution that does not have a single, easily defined location (9.3)

pollution point source a single source of pollution with a specific location (9.3)

polyatomic ion a molecular compound that has an excess or a deficit of electrons, and thus has a charge (2.1)

potable water water that is safe to drink (9.4)

precipitate an insoluble solid that is formed by a chemical reaction between two soluble compounds (4.2)

pressure the force per unit area (8.2)

primary water treatment a water-treatment process in which solid materials in water in a holding tank are removed by sedimentation and by skimming scum from the surface of the water (9.4)

product a substance that is formed in a chemical reaction (3.1)

qualitative analysis analysis that identifies elements, ions, or compounds in a sample (9.1)

quantitative analysis an analysis that determines how much of a substance is in a sample (9.2)

radioisotope an isotope with an unstable nucleus that decays into a different, often stable, isotope (1.1)

rate of dissolving a measure of how quickly a solute dissolves in a solvent (8.2)

ratio relative amount; proportional relationship (6.2)

reactant a starting substance in a chemical reaction (3.1)

relative atomic mass the atomic mass of an element in relation to that of another element; typically expressed in relation to the mass of carbon-12 (1.1)

reverse osmosis the process of using high pressure to force water from a concentrated solution through a semi-permeable membrane to get a less concentrated solution (9.4)

 S

salt a compound composed of a metal cation from a base and an anion from an acid (10.2)

saturated solution a solution that cannot dissolve more solute (8.1)

secondary water treatment a water-treatment process that uses natural micro-organisms that feed on organic matter in the sewage to convert the organic matter into carbon dioxide, water, and nitrogen compounds (9.4)

single bond a covalent bond that results from atoms sharing one pair of electrons (2.1)

single displacement reaction a chemical reaction in which one element in a compound is replaced (displaced) by another element (4.1)

skeleton equation a chemical equation in which the reactants and products in a chemical reaction are represented by their chemical formulas; their relative quantities are not included (3.1)

soft water water that contains relatively small concentrations of ions that form insoluble compounds with soap (9.3)

solubility the maximum amount of solute that will dissolve in a given quantity of solvent at a specific temperature (8.1)

solubility curve a graph that shows the relationship between the solubility of a solute and the temperature of the solvent (8.2)

soluble the attribute given to a solute if more than 1 g will dissolve in 100 mL of solvent (8.1)

solute a substance that is dissolved in a solvent (8.1)

solution a homogeneous mixture of two or more substances (8.1)

solution stoichiometry stoichiometry that is applied to substances in solution (9.2)

solvent the component of a solution that is present in the greatest amount (8.1)

soot fine particles consisting mostly of carbon formed during the incomplete combustion of a hydrocarbon (3.3)

sparingly (or slightly) soluble the attribute given to a solute if it has a solubility between 0.1 g and 1 g per 100 mL of solvent (8.1)

spectator ion an ion that is present in a solution but not involved in a chemical reaction (9.1)

standard ambient temperature and pressure (SATP) conditions defined as a temperature of 25°C (298 K) and a pressure of 100 kPa (12.1)

standard atmospheric pressure (SAP) atmospheric pressure in dry air at a temperature of 0°C at sea level (11.2)

standard solution a solution with a known concentration (8.4)

standard temperature and pressure (STP) conditions defined as a temperature of 0°C (273.15 K) and a pressure of 101.325 kPa (12.1)

stock solution usually a concentrated solution that is diluted with water before use (8.4)

stoichiometric amount the exact molar amount of a reactant or product, as predicted by a balanced chemical equation (7.2)

stoichiometry the study of the quantitative relationships among the amounts of reactants used and the amounts of products formed in a chemical reaction (7.1)

strong acid an acid that ionizes completely in water (10.1)

strong base a base that dissociates completely in water (10.1)

strong nuclear force the attractive force among protons and neutrons (1.1)

structural formula a diagram that has the chemical symbols connected by lines to show the connections among atoms in a chemical compound (2.2)

supersaturated solution a solution that contains more dissolved solute than a saturated solution at the same temperature (8.1)

surface water water on the surface of the land (9.3)

synthesis reaction a chemical reaction in which two or more reactants combine to produce a single compound (3.2)

 T

temporary hardness hardness that can be removed from water by boiling (9.4)

tertiary water treatment a water-treatment process that involves chemical precipitation of nitrogen, phosphorus, and organic compounds (9.4)

theoretical yield the amount of product that is predicted by stoichiometric calculations (7.3)

thermal decomposition heat breaks chemical bonds in a compound in this process (3.2)

thermal pollution see *heat pollution* (8.2)

titrant in titration, the solution with a known concentration (10.2)

titration a procedure that is used to determine the concentration of a solution by reacting a known volume of that solution with a measured volume of a solution that has a known concentration (10.2)

transition elements elements in Groups 3 through 11 in the periodic table with the exception of Group 3 in Periods 6 and 7; also called *transition metals* (1.2)

triple bond a covalent bond that results from atoms sharing three pairs of electrons (2.1)

triprotic acid an acid having three hydrogen atoms that can ionize in solution (10.1)

univalent metal a metal that can form ions with only one charge (3.2)

universal gas constant (*R*) a proportionality constant that relates the pressure on and the volume of an ideal gas to its amount and temperature:
$$R = 8.314 \frac{\text{kPa} \cdot \text{L}}{\text{mol} \cdot \text{K}} \ (12.2)$$

universal indicator a chemical mixture that changes colour throughout the range of pH values from 0 to 14 (10.1)

unpaired electrons electrons in an unfilled outer shell that are not part of a pair and are, therefore, more likely to participate in bonds with other atoms (1.1)

unsaturated solution a solution that could dissolve more solute (8.1)

valence electrons electrons in the outermost shell of an atom (1.1)

volumetric flask glassware that is used to make a liquid solution with an accurate volume (8.4)

weak acid an acid that ionizes very slightly in a water solution (10.1)

weak base a base that produces relatively few hydroxide ions in water (10.1)

X-ray crystallography a technique for determining the relative positions of atoms and distances between them in a crystal (1.3)

Index

Note: "*f*" in a page reference indicates material found in a figure; "*t*" in a page reference indicates material found in a table.

hoverfly, 122
Hund, Friedrich, 11*t*
hydrangeas, 456*f*
hydrated sodium carbonate, 431
hydrates, 277, 278, 286–287
hydration, 360
hydrazine, 110, 129
hydrobromic acid, 459
hydrocarbons, 137–139, 148–149
hydrochloric acid, 113, 166, 166*f*, 301, 458, 459, 461, 464, 464*t*, 467
hydrocyanic acid, 459
hydrofluoric acid, 459
hydrogen
    in the atmosphere, 53
    atomic radius, 31*f*
    in the body, 39
    and chlorine, bond, 62*f*
    combustion of, 142
    melting and boiling points, 77*t*
    and naming, 67
    neutrons, number of, 15
    single displacement reactions, 166–167
    smallest atom, 14
    in water, 57*f*
hydrogen atoms, 359
hydrogen bonding, 359, 503
hydrogen chloride, 69, 77*t*, 364, 454
hydrogen cyanide, 69*t*
hydrogen fluoride, 69*t*
hydrogen gas, 41, 308, 572
hydrogen ion, 59
hydrogen peroxide, 150, 260, 268
hydrogen sulfide, 69*t*
hydroiodic acid, 459
hydroxide ion, 68, 177, 460, 461
hyperbaric medicine, 526
hyperbaric oxygen therapy (HBOT), 526
hypochlorous acid, 113

ideal gas, 505
ideal gas law, 551–562, 570–571
impaired electrons, 13
Incan Empire, 20
incomplete combustion, 139, 141, 148–149
industry

percentage composition, 264
    reactions in, 181–186
    safety, 328–329
inner transition elements, 26
insoluble, 356
interim maximum acceptable concentration (IMAC), 428, 428*t*
intermediate melting and boiling points, 78
intermolecular forces, 79
International Space Station, 194–195
International Union for Pure and Applied Chemistry (IUPAC), 64, 69, 70, 378
iodine-131, 15, 15*f*
ion-exchange water softeners, 431
ionic bonds, 54–55
ionic compound
    chemical formulas, 67–68, 68*t*
    conductivity, 361
    defined, 54
    formation of, 124
    ionic charge and radius, 362*t*
    modelling, 83
    molar concentrations of ions, 379
    naming, 66–67
    properties, 76–81
    representative particles, 225
    solubility, 174*t*, 360, 362, 363*t*
    solubility curve, 365, 365*f*
    solubility guidelines, 174, 363
    transition metals, 55
ionic equation, 407
ionization energy, 34, 35*f*, 42–43, 455
iron, 225, 324
iron oxide, 181
iron(II) chloride, 324
iron(II) sulfate, 186
iron(II) sulfide, 184
iron(III) chloride, 186*f*
iron(III) oxide, 65*t*, 114, 114*f*, 184, 310, 320
isotope notation, 15
isotopes, 15
isotopic abundance, 18

Kelvin (Lord), 517

Kelvin temperature scale, 517–518
kelvin units, 518
kidney stone, 419
kilopascal (kPa), 509
kinetic energy, 76, 77, 504
kinetic molecular theory of gases, 505, 514, 522

lactic acid, 269*t*, 464
Lake Erie, 426
law of combining volumes, 543, 544
law of definite proportions, 258, 259
leaching, 185
lead nitrate, 416
lead pollution, 425
lead(II) chloride, 173
lead(II) nitrate, 172
leavening agent, 471
Lewis diagrams, 13, 13*f*, 23, 23*f*, 54
Lewis structure, 57, 57*f*, 58, 74, 74*f*
life processes, 142–143
"like dissolves like," 364
lime scale, 424
limestone, 430
limiting reactants, 306–312, 322–323, 418–420
liquids
    properties, 502*t*
    solubility, 366
lithium chloride, 172
lithium hydroxide monohydrate, 194–195
lithium recycling, 152–153
lone pair, 57
low melting and boiling points, 79
lunar water, 258

M (molar mass), 233
Mabury, Scott, 568
magnesium, 166, 166*f*, 182–183, 362, 362*t*, 424
magnesium carbonate, 406
magnesium oxide, 77*t*, 282–283
magnesium sulfate, 431
magnesium sulfate heptahydrate, 277
main-group elements, 26
mass

conversion to and from mole, 236–237
conversion to and from particles, 239–241, 244–245
molar mass, 233–236, 274, 275, 551–556
and moles and particles, 235–236
percent (m/m) concentrations, 374
percent (m/v) concentration, 373
point mass, 505
precipitate, 416, 418–419
relationships in chemical equations, 301
stoichiometric mass calculations, 302–303
mass number, 15
mass percent, 259–260, 374, 439–440
mass ratios, 374
mass spectrometer, 281
mass/volume percent, 372–374
material safety data sheet (MSDS), 388
matte, 184
matter, 502–503
maximum allowable concentration (MAC), 428, 428*t*
McDonald, Bob, 20, 144, 280, 389, 568
measures of concentration, 381, 381*t*
melting point, 76–79, 77*t*
Mendeleev, Dmitri, 22
mercury, 16, 20, 26, 27*f*, 131, 377
mercury barometer, 508
mercury poisoning, 430
mercury pollution, 428
mercury(II) sulfide, 20
metal carbonate, 132
metal hydroxide, 134
metal nitrate, 132
metal nitrite, 132
metal oxide, 128, 134
metalloids, 26, 27*f*
metals
    activity series, 188–189
    alkali metals, 167, 460
    alkaline earth metal hydroxides, 460
    multivalent metal, 125
    ores, 52, 52*f*, 160, 182, 184–185, 265
    precious metals, 52

Credits

Photo Credits

v top IBM Almaden Group, David Tanaka; vi-vii Prill Mediendesign and Fotografie/iStock; vi top NASA, Image copyright Edward Burtynsky, courtesy Nicholas Metivier Gallery, Toronto; vii top Bill Ivy, Norman Pogson/iStock, Michael Rosenfeld/Science Faction/Corbis; viii-ix Robert McGouey/All Canada Photos; viii top Dave Chidley/The Canadian Press, Dr. Ken Macdonald/Science Photo Library, Bill Ivy; ix The Granger Collection, NY, Roger Harris/Photo Researchers, Inc.; xii-xiii Sebastian Duda/iStock; xiv Peter Blottman/iStock; p2-3 Eye of Science/Photo Researchers, Inc., inset Dr P. Marazzi/Photo Researchers, Inc.; p4 left Doug Martin/Photo Researchers, Inc., Laurence Gough/iStock; p6 left Charles D. Winters/Photo Researchers, Inc., Bill Ivy, Andrew Lambert/LGPL/Alamy/Get Stock; p8-9 IBM Almaden Group, p9 left Joe Belanger/ iStock, jerryhat/iStock, ILLYCH/iStock; p10 The Alchemist by William Fettes Douglas, The Gallery Collection/Corbis; p20 top CBC logo and Bob McDonald, courtesy of CBC, pixhook/iStock, Andrey Volodin/iStock, NASA, bottom CCL/wiki; p22 The Granger Collection, NY; p27 top left clockwise Feng Yu/iStock, E.R. Degginger/Photo Researchers, Inc., Charles D. Winters/Photo Researchers, Inc., Charles D. Winters/Photo Researchers, Inc., Theodore Gray/Visuals Unlimited, Inc., Andraž Cerar/iStock; p28 right Theodor Benfey, Tetrahedral Periodic Table by Valery Tsimmerman, 2006, also known as ADOMAH periodic table; p29 top Photo courtesy of the Chemical Heritage Foundation, Jeff Moran/Electric Prism Inc.; p30 left Simon Curtis/Alamy, Future of Technology/Garden Memories Ltd.; p32 "Atomic Scale Imaging: A Hands-On Scanning Probe Microscopy Laboratory for Undergraduates" by Chuan-Jian Zhong, Li Han, Mathew M. Maye, Jin Luo, Nancy N. Kariuki, and Wayne E.Jones Jr., which appeared in the February 2003 issue Journal of Chemical Education 80: 194–197. Used with permission from the Journal of Chemical Education, Vol. 80, No. 2, 2003, ©2003, Division of Chemical Education, Inc.; p50-51 David Tanaka; p52 left Photo courtesy of John Ninomiya, Gunter Marx Photography/CORBIS; p64 Radius/MaXxImages; p74 Dr. Mark J. Winter/Photo Researchers, Inc.; p80 David Tanaka; p84 A. Barrington Brown/Photo Researchers, Inc.; p85 David Tanaka; p86-87 Peter Blottman/iStock; p94-95 Diane Diederich/iStock; p96 top IBM Almaden Group, David Tanaka; p101 David Tanaka; p104-105 Duane S. Radford/Lone Pine Photo, inset Rene Baumgarter/iStock; p106 Martyn F. Chillmaid/Photo Researchers, Inc; p107 Arnold Fisher/Photo Researchers, Inc.; p108 Josh Westrich/Corbis, CORBIS; p109 jacus/iStock; p110-111 NASA; p112 Tomasz Pietryszek/iStock; p113 Lorie Slater/iStock; p114 Philip Evans/Visuals Unlimited, Inc.; p116 David Tanaka; p121 Medicimage/Visuals Unlimited, Inc.; p122 Shattil & Rozinski/NPL/Minden Pictures; p124-125 Richard Megna/Fundamental Photographs, NY; p127 Richard Baker/Corbis; p128 J. Whyte/IVY IMAGES; p129 NASA, JPL/NASA; p130 Chuck Carlton/Monsoon/Photolibrary/Corbis; p131 top Charles D. Winters/Photo Researchers, Inc., A.T.Willett/GetStock; p132 Bill Ivy; p135 top National Archives of Canada/The Canadian Press, Paul Silverman/Fundamental Photographs, NY; p136 Richard Megna/Fundamental Photographs, NY; p137 Photo by Sergey Zimov. Used by permission of Katey Walter, University of Alaska, Fairbanks; p138 left Charles D. Winters/Photo Researchers, Inc., Matthew Adams/iStock; p139 Sami Suni/iStock; p142 top Photo courtesy of T.G. Harrison, School Teacher Fellow, School of Chemistry University of Bristol, bottom left Charles D. Winters/Photo Researchers, Inc., Richard Megna/Fundamental Photographs, NY; p144 top CBC logo and Bob McDonald, courtesy of CBC, pixhook/iStock, Andrey Volodin/iStock, NASA, bottom left Claudio Baldini/iStock, Photo courtesy of Shelley D. Minteer, Ph.D. Graduate Program Director College of Arts and Sciences Endowed Professor of Chemistry Saint Louis University; p145 Bill Ivy; p146 David Tanaka; p152 Trail Times; p153 left Masterfile, David Tanaka; p157 Bill Ivy; p159 left Tommounsey/iStock, Charles D. Winters/Photo Researchers, Inc.; p160-161 Image copyright Edward Burtynsky, courtesy Nicholas Metivier Gallery, Toronto; p162 Sol Neelman/Corbis; p163 Charles D. Winters/Photo Researchers, Inc.; p164 Richard Megna,

Fundamental Photographs, NY; p165 Comstock/Corbis p166 Larry Stepanowicz/Visuals Unlimited, Inc.; p167 Charles D. Winters/Photo Researchers, Inc.; p170 Charles D. Winters/Photo Researchers, Inc.; p171 demet celikkaya/iStock; p172 Bob Gibbons/Photo Researchers, Inc.; p173 Richard Megna/Fundamental Photographs, NY; p176 Christina Hammett of The Advocate, Mt. Hood Community College, Gresham, OR, USA; p178 Geoff Kidd/SPL/Photo Researchers, Inc.; p180 Andrew Lambert Photography/Photo Researchers, Inc.; p181 Crown Copyright/Health & Safety Laboratory/Photo Researchers, Inc.; p182 Lofman/Pix Inc./Time Life Pictures/Getty Images; p183 Photo courtesy of Feathercraft Products Ltd. Vancouver BC; p185 Sheila Terry/Photo Researchers, Inc.; p186 Andrew Lambert Photography/SPL/Photo Researchers, Inc.; pp187–192 David Tanaka; p194 The Canadian Press/Victoria Times Colonist/Deddeda Stemler; p196 Richard Megna, Fundamental Photographs, NY; p197 Andrew Lambert Photography/Photo Researchers, Inc.; p198 Paul A. Souders/Corbis; p202 left Richard Megna, Fundamental Photographs, NY, Ewen Cameron/iStock; p203 CCL/wiki; p205 Norbert Wu/Science Faction/Corbis; p206 background Bill Ivy, p208 top NASA, Image copyright Edward Burtynsky, courtesy Nicholas Metivier Gallery, Toronto; p209 David Wrobel/Visuals Unlimited, Inc.; p210 left Tommounsey/iStock, Charles D. Winters/Photo Researchers, Inc.; p211 Richard Megna/Fundamental Photographs, NY; p212 CORBIS, Ted Spiegel/CORBIS p213 Bill Ivy, Herman Eisenbeiss/Photo Researchers, Inc.; p215 Oleksiy Maksymenko/All Canada Photos; p216–217 Gerry Ellis/Minden Pictures; p222–223 Bill Ivy; p224 David Tanaka; p226 left Chris Cheadle/All Canada Photos, Matt Meadows; p229 NASA; p230 Andrzej Drożdża/iStock; p231 David Tanaka; p232 Stephen Strathdee/iStock; p233 Colin Cuthbert/Photo Researchers, Inc.; p235 David Tanaka; p236 Charles Amundson/iStock; p237 Charles D. Winters/Photo Researchers, Inc.; p238 Thomas Kitchin & Victoria Hurst/All Canada Photos/Corbis; pp243–247 David Tanaka; p248 Michael Newman/PhotoEdit, Inc.; p249 D. Hurst/Alamy/Get Stock; p250 Bill Ivy; p253 iStock; p256–257 Norman Pogson/iStock; p257 podfoto/iStock; p258 top NASA, lower left DK Stock/David Deas/Getty Images, top right Jasper Juinen/Getty Images Sport, Quinn Rooney Getty Images Sport; p260 Philip Evans/Visuals Unlimited, Inc.; p261 top Stefan Redel/iStock, Yuri Bathan-iStock; p262 doga yusuf dokdok/iStock; p264 Mark Schneider/Visuals Unlimited, Inc.; p267 Martin Strmko/iStock; p269 Oscar Gutierrez/iStock; p270 Daniel Stein/iStock, NASA; p277 Charles D. Winters/Photo Researchers, Inc.; p280 top CBC logo and Bob McDonald, courtesy of CBC, pixhook/iStock, Andrey Volodin/iStock, NASA, bottom left Science Source/Photo Researchers, Inc., Photo courtesy of archive, IFM-GEOMAR; p281 Andrew Brookes/National Physical Laboratory/Photo Researchers, Inc.; p284 David Tanaka; p286 Ian Chrysler; p288 Norman Pogson/iStock; p294–295 Michael Rosenfeld/Science Faction/Corbis; p296 Lowell Georgia/Photo Researchers, Inc.; p300 Gilles Glod/iStock; p303 NASA; p304 Masterfile; p307–309 Bill Ivy; p310 E.R. Degginger/Alamy/Get Stock; p311 CC Studio/Photo Researchers, Inc.; p312 left Tom Thulen/Alamy, Imagemore Co. Ltd./Corbis; p314 Masterfile; p315 Charles D. Winters/Photo Researchers, Inc.; p317 Bill Ivy; p318 Nigel Cattlin/Alamy/Get Stock; p320 Bill Ivy; p322 David Tanaka; p328 Photo courtesy of the U.S. Chemical Safety and Hazard Investigation Board; p329 Photo courtesy of AristaTek, Laramie, WY; p332 Andrew Syred/Photo Researchers, Inc.; p335 Lawrence Migdale/Photo Researchers, Inc.; p336 left Leslie Garland Picture Library/Alamy, John Hartman/The Stock Connect/Science Faction/Corbis, David Tanaka; p338 top Bill Ivy, Norman Pogson/iStock; p339 top Michael Rosenfeld/Science Faction/Corbis, lunanaranja/iStock; p340 Chris Cheadle/All Canada Photos; p341 left Theodore Gray/Visuals Unlimited, Inc., Mark Schneider/Visuals Unlimited, Inc.; p342 E.R. Degginger/Alamy/Get Stock; p343 left Robert Mathena/ Fundamental Photographs, NY, NASA; p346 Chris Harris/All Canada Photos; p349 Charles D. Winters/Photo Researchers, Inc.; p350 Richard Megna/Fundamental Photographs, NY; p351 David Tanaka; p352–353 Dave Chidley/The Canadian Press; p353–354 David Tanaka; p355 left to right in rows Patrick

Illustration Credits

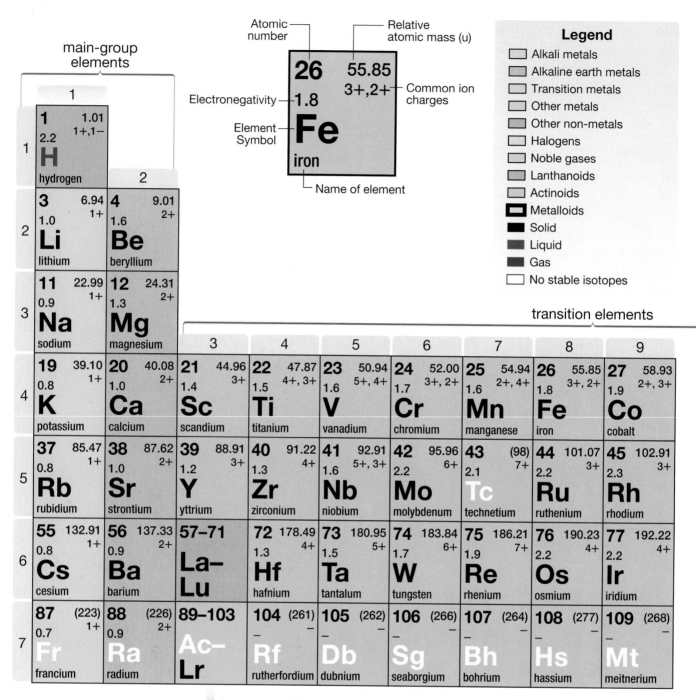

main-group elements

Atomic number — **26**
Relative atomic mass (u) — **55.85**
Electronegativity — **1.8**
Common ion charges — **3+,2+**
Element Symbol — **Fe**
iron
Name of element

Legend
- Alkali metals
- Alkaline earth metals
- Transition metals
- Other metals
- Other non-metals
- Halogens
- Noble gases
- Lanthanoids
- Actinoids
- Metalloids
- Solid
- Liquid
- Gas
- No stable isotopes

transition elements

	1	2	3	4	5	6	7	8	9
1	**1** 1.01 1+,1− 2.2 **H** hydrogen								
2	**3** 6.94 1+ 1.0 **Li** lithium	**4** 9.01 2+ 1.6 **Be** beryllium							
3	**11** 22.99 1+ 0.9 **Na** sodium	**12** 24.31 2+ 1.3 **Mg** magnesium							
4	**19** 39.10 1+ 0.8 **K** potassium	**20** 40.08 2+ 1.0 **Ca** calcium	**21** 44.96 3+ 1.4 **Sc** scandium	**22** 47.87 4+, 3+ 1.5 **Ti** titanium	**23** 50.94 5+, 4+ 1.6 **V** vanadium	**24** 52.00 3+, 2+ 1.6 **Cr** chromium	**25** 54.94 2+, 4+ 1.6 **Mn** manganese	**26** 55.85 3+, 2+ 1.8 **Fe** iron	**27** 58.93 2+, 3+ 1.9 **Co** cobalt
5	**37** 85.47 1+ 0.8 **Rb** rubidium	**38** 87.62 2+ 1.0 **Sr** strontium	**39** 88.91 3+ 1.2 **Y** yttrium	**40** 91.22 4+ 1.3 **Zr** zirconium	**41** 92.91 5+, 3+ 1.6 **Nb** niobium	**42** 95.96 6+ 2.2 **Mo** molybdenum	**43** (98) 7+ 2.1 **Tc** technetium	**44** 101.07 3+ 2.2 **Ru** ruthenium	**45** 102.91 3+ 2.3 **Rh** rhodium
6	**55** 132.91 1+ 0.8 **Cs** cesium	**56** 137.33 2+ 0.9 **Ba** barium	**57–71** **La– Lu**	**72** 178.49 4+ 1.3 **Hf** hafnium	**73** 180.95 5+ 1.5 **Ta** tantalum	**74** 183.84 6+ 1.7 **W** tungsten	**75** 186.21 7+ 1.9 **Re** rhenium	**76** 190.23 4+ 2.2 **Os** osmium	**77** 192.22 4+ 2.2 **Ir** iridium
7	**87** (223) 1+ 0.7 **Fr** francium	**88** (226) 2+ 0.9 **Ra** radium	**89–103** **Ac– Lr**	**104** (261) – **Rf** rutherfordium	**105** (262) – **Db** dubnium	**106** (266) – **Sg** seaborgium	**107** (264) – **Bh** bohrium	**108** (277) – **Hs** hassium	**109** (268) – **Mt** meitnerium

inner transition elements

6 Lanthanoids	**57** 138.91 3+ 1.1 **La** lanthanum	**58** 140.12 3+ 1.1 **Ce** cerium	**59** (140.91) 3+ 1.1 **Pr** praseodymium	**60** 144.24 3+ 1.1 **Nd** neodymium	**61** (145) 3+ – **Pm** promethium	**62** 150.36 3+, 2+ 1.2 **Sm** samarium	**63** 151.96 3+, 2+ – **Eu** europium
7 Actinoids	**89** (227) 3+ 1.1 **Ac** actinium	**90** 232.04 4+ 1.3 **Th** thorium	**91** 231.04 5+, 4+ 1.5 **Pa** protactinium	**92** 238.03 6+, 4+ 1.7 **U** uranium	**93** (237) 5+ 1.3 **Np** neptunium	**94** (244) 4+, 6+ 1.3 **Pu** plutonium	**95** (243) 3+, 4+ – **Am** americium

Although Group 12 elements are often included in the transition elements, these elements are chemically more similar to the main-group elements. Any value in parentheses is the mass of the least unstable or best known isotope for elements that do not occur naturally.

main-group elements

					18

2 4.00 — He helium

13	14	15	16	17	

| **5** 10.81 — 2.0 B boron | **6** 12.01 — 2.6 C carbon | **7** 14.01 3− 3.0 N nitrogen | **8** 16.00 2− 3.4 O oxygen | **9** 19.00 1− 4.0 F fluorine | **10** 20.18 — Ne neon |

| **13** 26.98 3+ 1.6 Al aluminum | **14** 28.09 — 1.9 Si silicon | **15** 30.97 3− 2.2 P phosphorus | **16** 32.07 2− 2.6 S sulfur | **17** 35.45 1− 3.2 Cl chlorine | **18** 39.95 — Ar argon |

10	11	12					

| **28** 58.69 2+, 3+ 1.9 Ni nickel | **29** 63.55 2+, 1+ 1.9 Cu copper | **30** 65.38 2+ 1.7 Zn zinc | **31** 69.72 3+ 1.8 Ga gallium | **32** 72.64 4+ 2.0 Ge germanium | **33** 74.92 3− 2.2 As arsenic | **34** 78.96 2− 2.6 Se selenium | **35** 79.90 1− 3.0 Br bromine | **36** 83.80 — Kr krypton |

| **46** 106.42 2+,3+ 2.2 Pd palladium | **47** 107.87 1+ 1.9 Ag silver | **48** 112.41 2+ 1.7 Cd cadmium | **49** 114.82 3+ 1.8 In indium | **50** 118.71 4+, 2+ 2.0 Sn tin | **51** 121.76 3+, 5+ 2.1 Sb antimony | **52** 127.60 2− 2.1 Te tellurium | **53** 126.90 1− 2.7 I iodine | **54** 131.29 — 2.6 Xe xenon |

| **78** 195.08 4+, 2+ 2.2 Pt platinum | **79** 196.97 3+, 1+ 2.4 Au gold | **80** 200.59 2+, 1+ 1.9 Hg mercury | **81** 204.38 1+, 3+ 1.8 Tl thallium | **82** 207.2* 2+, 4+ 1.8 Pb lead | **83** 208.98 3+, 5+ 1.9 Bi bismuth | **84** (209) 2+, 4+ 2.0 Po polonium | **85** (210) 1− 2.2 At astatine | **86** (222) — Rn radon |

| **110** (271) — Ds darmstadtium | **111** (272) — Rg roentgenium | **112** (285) — Cn copernicium | **113** (284) — Uut* ununtrium | **114** (289) — Uuq* ununquadium | **115** (288) — Uup* ununpentium | **116** (292) — Uuh* ununhexium | | **118** (294) — Uuo* ununoctium |

*Temporary names

| **64** 157.25 3+ 1.2 Gd gadolinium | **65** 158.93 3+ — Tb terbium | **66** 162.50 3+ 1.2 Dy dysprosium | **67** 164.93 3+ 1.2 Ho holmium | **68** 167.26 3+ 1.2 Er erbium | **69** 168.93 3+ 1.3 Tm thulium | **70** 173.05 3+, 2+ — Yb ytterbium | **71** 174.97 3+ 1.0 Lu lutetium |

| **96** (247) 3+ — Cm curium | **97** (247) 3+, 4+ — Bk berkelium | **98** (251) 3+ — Cf californium | **99** (252) 3+ — Es einsteinium | **100** (257) 3+ — Fm fermium | **101** (258) 2+, 3+ — Md mendelevium | **102** (259) 2+, 3+ — No nobelium | **103** (262) 3+ — Lr lawrencium |